AF658024

Groundwater Systems Analysis of Indian Coastal Deltas

Resource Potentials, Quality Trends, Saline-Fresh Water Interrelationships and Management Strategies

Groundwater Systems Analysis of Indian Coastal Deltas

Resource Potentials, Quality Trends, Saline-Fresh Water Interrelationships and Management Strategies

Indugula Radhakrishna
Formerly Scientist G (Director grade)
CSIR-National Geophysical Research Institute
Hyderbad - 500 007.

BS Publications
A unit of **BSP Books Pvt. Ltd.**
4-4-309/316, Giriraj Lane, Sultan Bazar,
Hyderabad - 500 095

Third Reprint 2018

Published by

BS Publications

A unit of **BSP Books Pvt. Ltd.**

4-4-309/316, Giriraj Lane, Sultan Bazar,
Hyderabad - 500 095
Phone : 040 - 23445600, 23445688
e-mail : info@bspbooks.net
www.bspbooks.net

ISBN: 978-93-85433-71-9

Dedication

This book is dedicated to
My Father
Late Shri. Indugula Subbarao
An Eminent Teacher

Foreword

Delta systems along the coastal regions of India are the prominent landforms and vibrant ecosystems rich in biodiversities and water resources (both surface and groundwater) and play a very important role in the economic development and agriculture production. They are often traditionally called as the '*Rice bowls*' of India. These systems often are interfaced with oceanic basins on one side and upland regions with major river valley projects on the other end and become targets for all anthropogenic(human induced) and environmental impacts (cyclones, storm surges, tidal impacts, global warming and sea level rise events). These systems also act as filters, repositories and reactors for different continental materials draining into the oceanic basins like fresh water, sediment, carbon, minerals and pollutants affecting the environments at the land ocean boundaries and biogeochemical cycles. The east coast of India exhibits a big network of delta systems like Ganges, Mahanadi-Brahmini-Baitarni, Krishna-Godavari-Pennar and Cauvery, Vaigai, Ponnaiyar-Vellar and Palar, while the west coast is characterized by several moderately built up estuaries excepting the north western stretch where small estuary deltas emerge. The disposition and evolution of all these delta systems are manifestations of structural, tectonic and eustatic changes of sea levels during different stages of sedimentalogical processes and evolutionary history.

The book on *'Groundwater systems analysis of Indian coastal deltas'* is a comprehensive multifaceted manual devoted to the understanding, quantification of fresh groundwater resources, quality trends and analysis, saline-water intrusion processes/interface structural maps and the required optimization/management schemes in different delta systems. The manual contains six chapters: 1. Introduction to water resources (Global and National scenarios); 2. Geomorphology, evolution and structure of coastal delta basins; 3. Hydrological characteristics and soil water balances; 4. Assessment of fresh groundwater resources; 5. Groundwater quality trend analysis and 6. Saline water intrusion processes/interface structural maps and management strategies. Each of these chapters is normally considered as an independent scientific topic by itself but the author has successfully integrated all these specialized topics into a multidisciplinary text aimed at the optimization and management of coastal groundwater resources. Every chapter contains both methodologies and applications with case studies wherever required. A manual of this scope giving a '*Holistic Approach*' towards coastal delta water resources management is a rare compendium not only in India but also globally.

The author of this monumental work, Dr. I. Radhakrishna, devoted his life time to the study of issues adumbrated in this work. His command of the subject shows up in the high quality of the compendium. Though the book ostensibly refers to Indian case histories, yet the methodologies are globally applicable, and herein lies the great merit of the book.

The book will be useful as a reference volume to post-graduate students, research scholars, resource managers and policy makers, climate change scientists etc., concerned with the Coastal Groundwater Management (CGM).

Hyderabad
October, 2013

Prof. U. Aswathanarayana D.Sc ; F.N.A ; F.A.Sc ; F.T.W.A.S.
Honorary Director
Mahadevan International Centre for Water Resources Management,
Hyderabad

Preface

Groundwater systems in the coastal delta regions of India and around the world constitute important resource belts for agriculture, irrigation and drinking water needs for several millions of people living in these tracks and play a vital role for the growth and development of rural base. These regions in general are densely populated than inland areas because of several economic reasons such as development of agriculture, ports and shipping, fisheries, tourism and coal and oil mining etc. It has been estimated that by the year 2025 about 60% of the world population will be living within 60 kms from the coastline. This trend appears to be more pronounced amongst the developing world including India resulting in several anthropogenic impacts, imbalances in the ecological dynamics and reduction in freshwater resources. Adding fire to the fuel the global warming impacts are causing great concerns to the coastal water resource systems and ecological units. India with a vast coastline of the order of about 7500 kms has a network of major and minor delta systems and estuaries all along the coastal regions. These delta systems with high density canal networks and rich fresh groundwater potentials underneath the surface are the epicenters for agriculture production, growth and development. Of late, over the years with the design and construction of major river valley projects in the upstream areas of the deltas, increasing population growth levels, industrialization and expansion of agricultural practices, there has been substantial reduction in surface water resources particularly in the middle and tail end regions of deltas, resulting in large scale withdrawal of groundwater resources and at the same time causing saline water migration and mix up of fresh and saline water resources. A quantitative understanding of these fresh water resources, their movement patterns, the geometrical variations of fresh-saline water interface boundaries and the strategies required to optimize and manage fresh groundwater resources therefore are of paramount importance to protect these valuable fresh water resources.

It is therefore felt necessary that a book dealing with different elements of coastal deltas such as morphology, evolution and structure; soil water balances; assessment of fresh water resources; water quality trends; saline-fresh water interrelationships/estimations of interface structures/fresh water discharges into oceanic basins and management/mitigation strategies with an *Integrated/holistic approach* would be very useful and necessary in the present context of increasing levels of water utilization, falling water levels, decreasing levels of recharge potentials and impending impacts of global warming and sea level rise events. Though several exploratory endeavours have been carried out over the last two or three decades in different deltaic regions there still exists large data gaps in relation to the delineation of fresh aquifer bodies and their lateral and vertical extensions, hydraulic properties, water quality trends, saline-fresh water boundaries/interface depth structures and their lateral variations and implementable water management practices. It has been a heavy task to cover up these gaps and to retrieve large amount of data related to various aspects from several published and unpublished reports/ papers and put together the data network, analyze and interpret in a presentable format useful to the students, researchers and policy makers. The database and the systems analysis presented in the book are at a regional level, but a more accurate database at micro level needs to be evolved through systematic, integrated and exploratory efforts in future.

A large number of scientists working in different institutions have helped me in this endeavour. In this connection I would like to thank Shri. B. M. Jha, Former Chairman, CGWB, New Delhi; Shri. M.B. Raju, Former Regional Director, CGWB, Chennai; and Dr. N. Varadaraj, Member (SAM) and Director, RGI, CGWB, New Delhi for their help and assistance in the retrieval of data from different published and unpublished reports. I would also like to express my thanks to Dr. V. P. Dimri, Distinguished Scientist of CSIR and Former Director of National Geophysical Research Institute, Hyderabad for providing all necessary infrastructure facilities as and when required. Thanks are also due to Dr. S. N. Rai, Chief Scientist, NGRI; Dr. B. M. Khanna, Technical Officer and incharge library, NGRI;

and Mr. S. D. Deshmukh, Former Technical Officer, NGRI for their co-operation and useful discussions. I would like to express my gratitude to Prof. V. V. Haragopal, Head, Department of Statistics, Osmania University, Hyderabad for his help rendered in relation to factor analysis in water quality systems. I am also grateful to Dr. G.G.S.N. Rao, Scientist, CRIDA, Hyderabad; Dr. P.V.L.N. Babu, Former Dy. General Manager (Exploration), ONGC; Dr. S.V.N. Rao, Chief Project Manager, WAPCOS, Hyderabad; and Dr. M. Srimannarayana, Technical Expert, Groundwater Department, Chennai for their critical reviews and suggestions.Thanks are also due to UNESCO, New Delhi for providing me the required material related to the coastal groundwater systems analysis. I would appreciate the hard work rendered by Shri. Mr. Leela Krishna, S., Software analyst, Mr. Shivaram, G., GIS analyst and Mr. G. Rajashekar Reddy, Software analyst in preparation of this document. I gratefully acknowledge the support given by my family members: Satyarani (Wife); Sai (Son); Nagini (Daughter); Prashant (Son-in-Law) and Gayathri (Daughter-in-Law).

I am grateful to the Secretary, Department of Science and Technology/Head, SERC Division, Ministry of Science and Technology, New Delhi for their financial assistance under the USERS programme with number HR/UR/01/2006. The entire project was carried out under the auspicious of *'Society for Integrated Technologies for Groundwater Resource Systems Development', Hyderabad.*

I am glad to put on record that my book entitled *'Groundwater systems analysis of Coastal Deltas'* had the good fortune of receiving a perceptive 'FOREWORD' by Prof. U. Aswathanarayana. He is the doyen of geosciences in India and has published over hundred papers and several books. These include principles of *Nuclear Geology* (1985), *Geoenvironment: An Introduction* (1995); *Soil Resources and the Environment* (1999); *Water Resource Management and the Environment* (2001), *Mineral Resources Management and the Environment* (2003); *Natural Resources Development Methodology* (2006); *Advances in Water Science Methodologies* (2005); *The Indian Ocean Tsunami* (2006), *Food and Water Security*(2007), *Energy Portfolios* (2009) and *Green Energy: Technology, Economics and Policy* (2010). He held the Fellowships of UNIDO (1989), UN Outer Space Affairs (1989), and UNESCO Expert affiliated to the Institute of Trace Element, Research, Lyon, France (1991). He has been the Chairman of the Working Group on "Geochemical Training in Developing Countries" of the Int. Assn of Geochemistry & Cosmochemistry (1996-2008). He is the receipient of the highest awards of American Geophysical Union, the largest Geoscience Society in the world namely "*Excellence in Geophysical Education*" (2005), the International Award for *advancing science and benefiting society by laboring under adverse circumstances with limited resources* in 2007; *Certificate of Recongnition of the International Association of Geo Chemistry, and Eminent Citizen Award in the area of Water Sciences of the Sivananda Trust, India* (2007). He is also the Founder & Honorary Director of Mahadevan International Centre for Water Resources Management, Hyderabad, India.

Finally I would like to thank Dr. V.P. Dimri, President Indian Geophysical Union (IGU) and its Executive Council for their approval to release this publication as a part of 50th Annual Convention (Golden Jubliee Celebrations) of Indian Geophysical Union during Jan 2014.

Indugula Radhakrishna

About the Author

Dr. Indugula Radhakrishna, obtained his M.Sc (1963) and Ph.D (1978) from Andhra University, Visakhapatnam. He worked initially as a Research Fellow of CSIR for a period of four years (1963-67) and later joined Central Groundwater Board as a Hydrogeologist (1968-75). During this period he worked on several exploratory/UNDP programmes covering parts of (a) Alluvial tracks of Uttar Pradesh and Bihar (b) Desert regions of Gujarat and Rajasthan (c) Coastal deltaic regions of Andhra Pradesh, Tamilnadu, Orissa and West Bengal and (d) Some isolated regions of North Eastern states of India. He made concerted efforts in all the above regions and evolved reliable data bases for development of Groundwater resources. His work in coastal regions of Orissa, related to the delineation of saline-fresh water interface boundaries and identification of deep fresh water artesian fresh water reservoirs underneath the saline water systems in Mahanadi delta region using deep geophysical logs stands as monumental work which led the way for scientific development of fresh water resources for the first time in the country.

Dr. Radhakrishna subsequently joined National Geophysical Research Institute (1975) as a senior level scientist and worked as project leader in different capacities and superannuated as Scientist 'G' (Director Grade) in the year 2001. During all these years as a project leader, organized/involved in a number of R&D programmes such as (a) Integrated hydrogeological, geophysical and hydrogeochemical studies and parametric evaluations of different aquifer systems (Indo-German Project); (b) Hydrogeological studies and analysis in the drought prone regions of Anantapur district, A.P.; (c) Assessment of fresh water resources in Pennar delta region; saline-fresh water interface dynamics and saline water movement with time; (d) Technology mission programmes in the Barmer district of Rajasthan, a hard core desert track. Identification of fresh water resources through appropriate geohydrological and geophyscial tools; (e) Saline fresh water interface dynamics in Krishna delta region through intensive hydrogeochemical studies; Delineation of fresh water resources, development and optimization techniques and adoption of artificial recharge methods; (f) Pore pressure studies in Koyna-Warna seismic areas through deep drilling and hydrological operations and analysis of precursors to assess earthquake mechanisms.

Dr. Radhkrishna is a fellow of Indian Geophysical Union, and former member of International Association of Hydrogeologists and Indian Association of Geohydrologists. He had officiated as Member in a number of technical committees: Technical Advisory Committee – DANIDA project for coastal Groundwater systems of Orissa 1988-90; Committee on source finding in Rajasthan under fifth submission in Technology Mission 1987-90; Working group for Shore Development Authority (Govt. of A.P.); 1987-89 and Technical Advisory Panel on Environment, Science and Technology (Govt. of A.P.) 1986-89. He was also resource person for DANIDA and UNESCO Training Courses, Member and adviser for selection of Geologists and Hydrogeologists, Union Public Service Commission (1999-2000). He was former Editor of Indian Geohydrology Journal.

Dr. Radhakrishna visited Germany (Hannover) in 1979 as a Senior DAAD fellow. He also attended UNESCO Conferences in Nepal, Thailand and Germany and presented papers related to coastal Groundwater systems of India.

He had published over fifty papers in various National and International Journals related to groundwater resource assessments/potentials, saline-fresh water interface dynamics and management aspects and about twenty technical reports dealing with water resources and pore pressure studies (1975-2004). Dr. Radhakrishna is also the Founder President for "Integrated Technologies for Groundwater Resource Systems Development" (ITGRSD).

Contents

CHAPTER 3: HYDROLOGICAL CHARACTERISTICS AND SOIL WATER BALANCES OF COASTAL DELTAS

CHAPTER 4: ASSESSMENT OF FRESH GROUNDWATER RESOURCES IN THE COASTAL DELTA SYSTEMS

CHAPTER 5: GROUNDWATER QUALITY CHARACTERISTICS AND TREND ANALYSIS IN COASTAL DELTA COMPLEXES

CHAPTER 6: SALINE FRESH WATER INTERRELATIONSHIPS, ESTIMATIONS OF INTERFACE STRUCTURES AND MANAGEMENT MITIGATION STRATEGIES

List of Tables

List of Figures

CHAPTER 1

INTRODUCTION TO WATER RESOURCES

Water is a unique and widely distributed resource for sustenance of life and humanity on earth and plays a very vital role in the economic development of any country or a region related to irrigation, agriculture, industrial development and drinking water needs. It is a finite resource in the global scenario and undergoes a cyclic phenomenon characterizing continuous exchange of processes from the surface of the earth (evaporation and evapotranspiration) and back to the surface in the form of precipitation under the influence of solar radiation and seasonal climatic changes. The water cycle provides us with a magnificent phenomenon of transforming and converting all types of waters on the earth's surface into fresh waters and brings them back in the form of precipitation. The precipitation though finite in its totality has wide space-time variations in different continents of the world. The global climatic phenomenon popularly known as 'EL Nino'-a quasi periodic climatic pattern which is largely a function of heat energies/solar radiations and trade winds across Pacific could create abnormal climatic variations in the southern continents of the world such as droughts and floods which are common and frequent in many parts of the world. The water utilization from precipitation levels in the early decades of human population has been insignificant, but with increasing population growth, mounting agricultural practices and large scale industrial activities over the last few decades, the demands for water has become multifold bringing the water potentials in several regions of the world to critical or near critical situations, and at the same time increasing the levels of contamination/pollution to alarming situation. In addition, the increased *'green house gases'* and the related anthropogenic impacts are set to cause *'global warming'* and *'sea level rise'* events and could result in disastrous situations in future in relation to precipitation levels, floods, and droughts in several parts of the world including India. Coastal systems with diversified ecosystems and environments, wetlands and deltas with potential surface and subsurface fresh water systems bear the brunt of these global processes and will have significant impacts on the production levels related to water resources and agricultural practices. Assessment of freshwater resources, development of appropriate methodologies and sustainable solutions, land-water interactions, and environmental protection strategies are of paramount importance apart from other resource potentials like minerals, soil nutrients, timber, fisheries and fauna & flora. Over exploitation, misuse of water resources, rapid industrialization and urbanization processes are posing serious threats for sustainable development, environmental protection and food security problems in most coastal regions of the world including India. An overall global and national perspectives related to water resources are briefly described below, before targeting in detail the Indian coastal delta ground water systems, fresh water resource potentials, saline-fresh water interrelationships, Quality trends and management strategies.

Several countries in the world and the United Nations agencies like UNDP, WHO, UNICEF, WMO, UNESCO & UNEP have taken several initiatives and coordinated action plans/ programs related to the availability of fresh water resources in different continents and the constraints involved in the development of this scarce resource. The International Conference on Water and Environment at Dublin (ICWE, 1992) and the United Nations Conference on Environment and Development at Rio de Janeiro (UNCED, 1992) stressed the need for world wide reforms towards sustainable food production and efficient management of water resources and water conservation methods/practices. The salient and guiding principles of Dublin and that of Rio de Janeiro (Agenda 21) conferences are briefed below:

Guiding principles of Dublin conference (ICWE, 1992)

- "Fresh water is a finite and vulnerable resource essential to sustain life, development and environment". Since water sustains life, effective management of water resources and holistic approaches linking social and economic development with natural ecosystems are of high priority.
- "Water development and Management should be based on a participatory approach involving users, planners and policy makers at all levels". The participatory approach involves raising awareness of the importance of water among policy makers and the general public. It means that the decisions are taken at the lowest appropriate levels with full public consultation and involvement of users in the planning and implementation of water projects.
- "Women play a central point in the provisions, management and safeguarding water" This pivotal role of women as providers and users of water and guardians of the living environment has seldom been reflected in the institutional arrangements for the development and management of water resources. Acceptance and implementation of this principle requires positive polices to address women and their specific needs and equip and empower women to participate at all levels in water resources programs including decision making processes.
- "Water has an economic value in all its competing uses and should be recognised as an economic good". It is vital to recognise first the basic right of all human beings to have access to clean water and sanitation at an affordable price. Earlier failures to recognise the economic value of water has led to wasteful and environmentally damaging uses of the resource. Managing water as an economic good is an important way of achieving efficient and equitable use and encourage conservation and protection practices of water resources.

Guiding principles: The Rio de Janeiro, Agenda 21 conference (UNCED, 1992)

- Establishing a new and equitable global partnership through creation of new levels of cooperation among states, key sector societies and people.
- Holistic management of fresh water and integration of several water plans and programs, with in the frame work of national, economic and social policies.

The above conferences have arrived at a number of recommendations based on the above guiding principles and discussions such as (i) Alleviation and poverty and disease (ii) Water conservation and reuse (iii) Sustainable urban development (iv) Agriculture production and rural water supply (v) Protection of aquatic ecosystems (vi) Resolution of water conflicts (vii) Protection of environments and the development of knowledge base and capacity building. Subsequently there are several forums on International needs to focus attention on food security, water resources and environmental programmes such as: (i) First world forum at Marrakech, 1997 (ii) Second world forum at Hague, 2000 (iii) Third world forum at Kyoto, 2003 (iv) Fourth world forum at Mexico, 2006 (v) Fifth world forum at Istanbul, 2009 and several world forums planned for future in the years to come. All these International gatherings and deliberations will go a long way to put the water resources programmes in proper perspective to derive the expected goals and fruits to all levels of society.

1.1 GLOBAL WATER RESOURCES

Water occurs widely throughout the globe and exists in three states: liquid, solid and vapour. A major chunk of water in liquid state is found in oceans, seas, lakes, rivers and groundwater and in the crustal layers of the earth down to 2000 meters. The solid form appears as ice and snow mostly in the Polar and Alpine Himalayan regions. A certain amount of water is found in the air as water vapour, water droplets and ice crystals and also in the biosphere. The different attributes of water, the hydrological and nutrient cycles, and the constraints involved in water budgeting etc are critically reviewed and discussed by Narasimhan, T. (2005).

Assessment of total water resources is rather a very complicated problem in view of its dynamic nature and the changes involved from liquid stage to gaseous stage and back again to liquid stage. Pioneering attempts to estimate the world water resources are made by Lvovitch, M.I. (1973); Korzun, V.I. (1978); Baumgartner, F.; Reichel, E. (1975) and more recently by Shiklomanov, I.A. (1997 & 98). The estimated global water resources are of the order of 1386 Million km³ of which 97.50% occurs as oceanic saline waters and the rest 2.5% as fresh water. A major portion of this fresh water (68.90%) is locked up in the form of ice caps, glaciers, permafrost and snow in the Arctic, Antarctic and Mountainous regions of the world. Then about 29.90% of the fresh water component occurs as groundwater. About 0.30% exists in lakes, river systems and about 0.90% in the soil moisture and swamps. The distributional pattern of different components of water is shown in Fig.1.1.

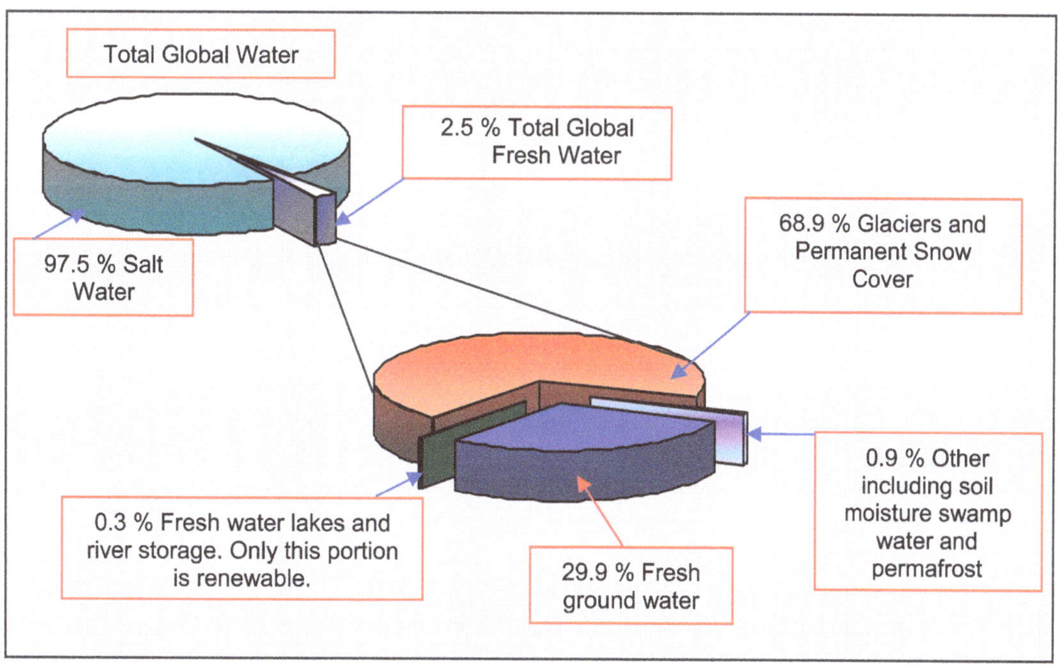

Fig. 1.1 Global water resources

(Source: Shiklomanov, I.A., 1998)

The above system approximately characterises the global storage potentials of fresh and saline waters and are subjected to cyclic changes over short and long periods of time. The volume of water stored in the hydrosphere vary depending upon the cyclic changes in the oceans, land and atmosphere. This exchange is called turnover of the water in terms of volume in the global hydrological cycle. The global water cycle involves basically evaporation of water from earth's surface into the atmosphere and returning back to earth in the form of precipitation which is the major source of rivers, lakes, glaciers and groundwaters as schematized in Fig. 1.2.

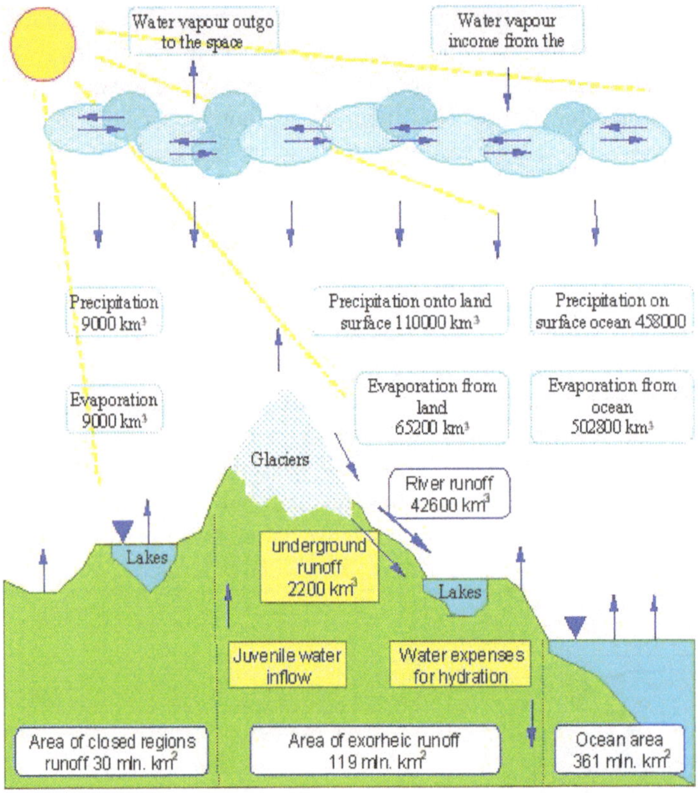

Fig. 1.2 Global water cycle

(Source: Shiklomanov, I.A; 1998)

A portion of the atmospheric precipitation evaporates, some part goes into river systems flow and some flow infiltrates into the soil cover and recharges the groundwater system. The quantitative estimates of different components of global hydrological cycle are reflected in fig. 1.2. The turn over of water on the earth over one hydrological cycle is estimated as 577000 km³ of water. The water that evaporates from oceanic surface is of the order of 502, 800 km³ and from the land surface is of the order of 74,200 km³. The amount of water that falls as atmospheric precipitation is of the order of 458,000 km³ on the ocean and about 119,000 km³ on the land. The difference between precipitation and evaporation from the land surface is 44,800 km³/year (119,000-74,200) which represents the total runoff component through the earth's river systems including groundwater runoff (42,700 km³/year, river runoff component + 2,100 km³/year groundwater runoff component) Shiklomanov, I.A., (1998).

Assessment of water resources in general involves two components: static storage and dynamic storage (renewal resources): Static storage by convention means renewal turnover time of over several decades and centuries like ocean bodies, glaciers, large lakes and deep groundwater bodies, where as the dynamic storage is one where the renewal turnover time is only one year or one hydrological cycle like major and minor river systems and near surface realms of groundwater. The river runoff of renewable water resources is the most important component of the hydrological cycle and has pronounced impact on the economy and ecology of any region. Assessments of precipitation levels, river runoff therefore are very important to understand the resource potentials in a region or a basin. A wide network of observational stations throughout the globe are in operation and several numerical model studies are in progress to assess these components for long term and short term perspectives. The world meteorological organization operates nearly 64,000 stations for river runoff measurements throughout the world. The renewable water resources and percapita availability of water have been estimated continental wise using about 2500 stations and are presented below in the Table 1.1.

Table 1.1 Distribution of renewable water resources in different continents and per capita availability

Continent	Area (Million km^2)	Population (millions)	Water resources, (km^3/year)				Per Capita water Availability 1000 m^3/year	
			Average	Max	Min	Cv	per 1 km^2	Per capita
Europe	10.46	685	2900	3410	2254	0.08	277	4.23
North America	24.3	453	7890	8917	6895	0.06	324	17.4
Africa	30.1	708	4050	5082	3073	0.10	134	5.72
Asia	43.5	3445	13510	15008	11800	0.06	311	3.92
South America	17.9	315	12030	14350	10320	0.07	672	38.2
Australia and Oceania	8.95	28.7	2404	2880	1891	0.10	269	83.7
The World	135	5633	42785	44751	39775	0.02	317	7.60

(Source: Shiklomanov, I.A., 1998)

The above table demonstrates that the maximum volume of water is available in Asia and South America and the minimum resources are in Europe and Australia. The mean value of renewable global water resources is estimated as 42,700 km^3 per year with large space-time variations. The largest volume of water resources are those of Asia and South America (13,000 and 12000 km^3 per year respectively) and the smallest resources are found in Europe and Australia + Oceania (2900 and 2400 km^3 per year respectively).

The rapid population growth rate between 1970 and 1994 has resulted in the per capita reduction from 12.9 to 7.0 thousand m^3/year/person. The greatest reduction appears to have taken place in Africa (2.8 times) followed by Asia (2.0 times) and South America (1.7 times). The studies on river run off over different continents have shown wide variations from the mean values and are cyclic in nature. The current estimations show that the major part of runoff in Europe takes place between April to July (46%); in Asia between June to September (54%), in Africa between September to December (44%), in South America between April to July (45%) and in Australia/Oceana between January to April (40%). The average runoff during the wet seasons for different continents is estimated to be of the order of 45% of the total renewable water resources. It is further estimated that a major chunk of river runoff (52%) takes place through major rivers like Amazon (Brazil), Ganges-Brahmaputra (India), Congo (Central Africa), Yangtze (China) and Orinoco (Venezuela). A graphical sketch presented in Fig. 1.3 shows long term run off patterns where annual volume exceeds 100 km^3 per year. The river runoff also includes the groundwater runoff component, whose quantified estimations over all the continents is not exactly available, but approximately pegged at 5% of the total river runoff (Shiklomanov, I.A., 1998).

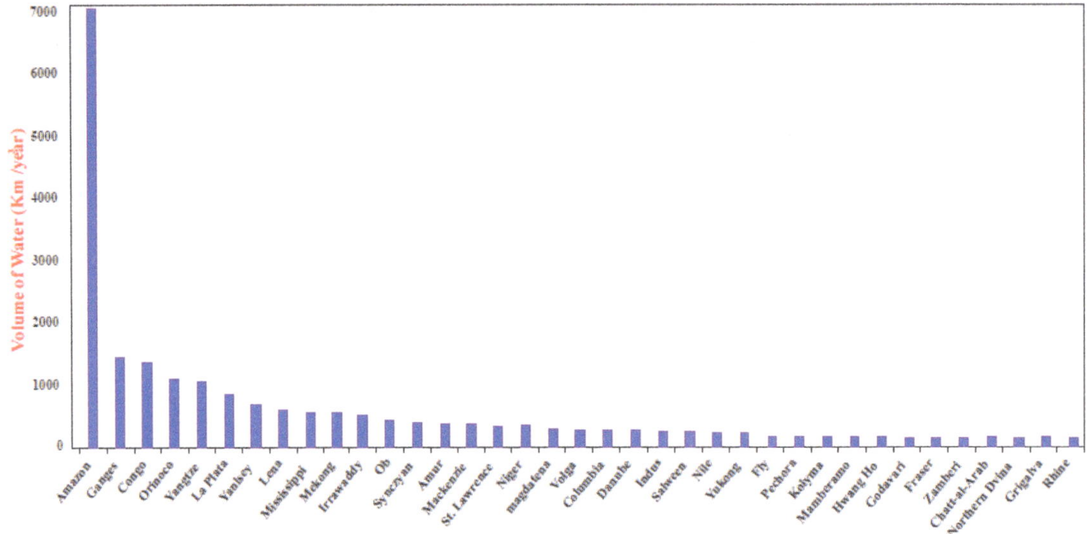

Fig. 1.3 Annual runoff patterns for different major river basins in the world

(Source: Shiklomanov I.A., 1998)

1.2 WATER RESOURCES (NATIONAL)

India is bestowed with complex physiographic and climatic distribution system with a number of major, medium and minor river systems. The average precipitation level is of the order of 1200 mm. The total precipitation in terms of volume over India is estimated to be of the order of 4000 BM^3. Against this total, the average annual runoff is estimated to be of the order of 1869 BM^3 and rest 2131 BM^3 returns back to atmosphere through evaporation (C.W.C., 2005). A brief description of different physiographic and climatic regions, precipitation levels and water resource potentials are described below, in the following paragraphs.

1.2.1 Physiographic regions

India with a geographical area of 3.29×10^6 km² has a wide range of physiographic and climatic situations. Physiographically India could be divided into seven broad well defined regions:

(i) The high altitude Himalayan Mountain ranges in the north where Indus, Ganges and Brahmaputra take their origin.

(ii) The Indo-Gangetic plains traversed by Ganges and Brahmaputra

(iii) The Central high lands starting from Aravallis in the west to steep escarpments and high altitude hill regions in the east.

(iv) The peninsular plateaus with Western Ghats on the west, the Eastern Ghats on the east and the Deccan plateau in the central region.

(v) The east coastal plains with a number of major, medium and minor east flowing rivers discharging into Bay of Bengal. By and large all the major delta systems fall in this region and join Bay of Bengal with good alluvial stretches and with widths varying from 100-140 kms, but sometimes gets narrowed down to 5-10 kms along the rocky shorelines.

(vi) The west coastal plains with a number of medium and minor westerly flowing rivers, discharging their sediments and waters into Arabian sea with relatively narrow widths of alluvial sediments, and with widths varying from 10 kms to 30 kms and sometimes shorelines abutting against elevated rocks and cliffs.

(vii) The Andaman-Nicobar islands in the Bay of Bengal and Lakshadweep islands in the Arabian Sea.

1.2.2 Climatic zones

The Physiography of India plays a very important role to characterize the climate variations. The Himalayas in the north, Indian ocean in the south, the Arabian sea and the Bay of Bengal on either sides of the peninsular coastal regions and the plateau regions play an important role on the climatic variations of India. India could be broadly classified into six climatic zones as shown in Fig.1.4 and as described briefly below.

Fig. 1.4 Climatic zones of India

(a) ***Alpine Himalayas Zone:*** The High altitude Himalayan region is characterized by high diurnal temperatures, sharp temperature inversions and high rain fall variabilities. The states of Jammu & Kashmir, Himachal Pradesh, Uttarkhand experiences this type of climate.

(b) ***Humid Subtropical Zone:*** The Himalayan foot hill region and a major portion of Gangetic alluvial plains come under this category. The east Rajasthan region, Uttar Pradesh, Bihar and Northern part of West Bengal and Assam experiences this type of climate. The summers in general are hot and winters are moderately cold.

(c) ***Arid regions:*** The Western Rajasthan with scanty rainfall conditions comes into this category. This region is characteristic of very hot summers sometimes attaining as much as 50 °C and cold winters sometimes almost touching freezing point.

(d) ***Semi arid region:*** This is transitional climate falling between tropical desert (arid) climate and humid sub-tropical climate with annual rainfall slightly more than the arid desert region of Rajasthan. Parts of Punjab, Haryana and Gujarat experiences this type of climate.

(e) ***Tropical wet and dry climate:*** Most plateau regions of Peninsular India enjoys this climate with summer temperatures touching 45 °C and experiences good rainfall during June to September with an annual rainfall ranging between 750 mm to 1500 mm or even more. The coastal regions in general are wet and central high land regions are dry.

(f) ***Tropical wet:*** Part of west coastal region of Peninsular India reaching summer temperatures of the order varying between 35-40 °C. Experiences high rainfalls ranging between 1500 mm to 2500 mm and sometimes even more.

1.2.3 Precipitation levels

The precipitation including snowfall is the source of all water potentials on the earth's surface. The average annual precipitation over India is estimated to be of the order of 4000 BCM of which a part goes as evapo-transpiration and a part goes as recharge to groundwater and the rest appears as surface runoff. Most of the rainfall over India is contributed by south west monsoon (June to September) and a limited contribution by north east monsoon (October to November). The long term average annual precipitation for the country is about 1869 mm with high levels of fluctuations and variabilities. The highest rainfall with about 11690 mm is recorded near Chirapunji in Meghalaya in the north east and the lowest of the order of 150 mm is recorded near Jaisalmer in Western Rajasthan. Rainfall varies between 2000 mm to 4000 mm along the north eastern and along the western ghat regions of India. The annual average rainfall over the foot hill regions of Himalayas and along the Indo-Gangetic plains varies from 1500mm in the east to 500 mm in the west. The northern half of peninsular India and the central high land region registers an average annual rainfall varying between 1500 mm on the east to about 500 mm on the west. The southern Indian peninsula registers an average annual rainfall varying from 100 mm to 750 mm from east to west. The western regions of Rajasthan receives vary scarce rainfall averaging about 150 mm or sometimes less. The distributional trends of normal annual rainfall over India are shown in the Fig 1.5 and the longtime rainfall trends for the period 1870 to 2010 are shown in Fig.1.6 based on all India weighted mean values over a homogeneous rainfall data set of about 306 rainguage stations covering India.

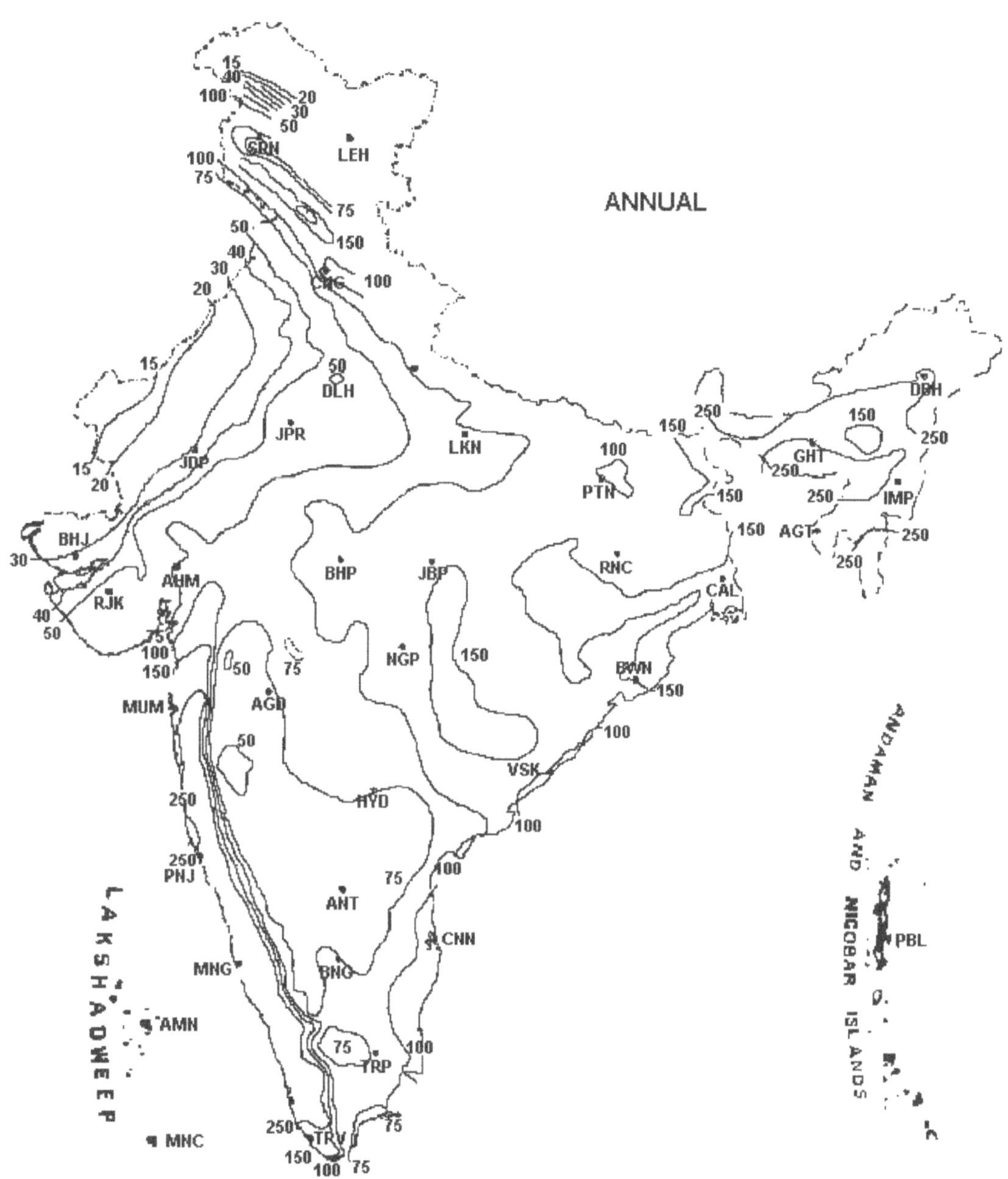

Fig. 1.5 Distribution of normal and annual precipitation levels of India (IMD, 2004)

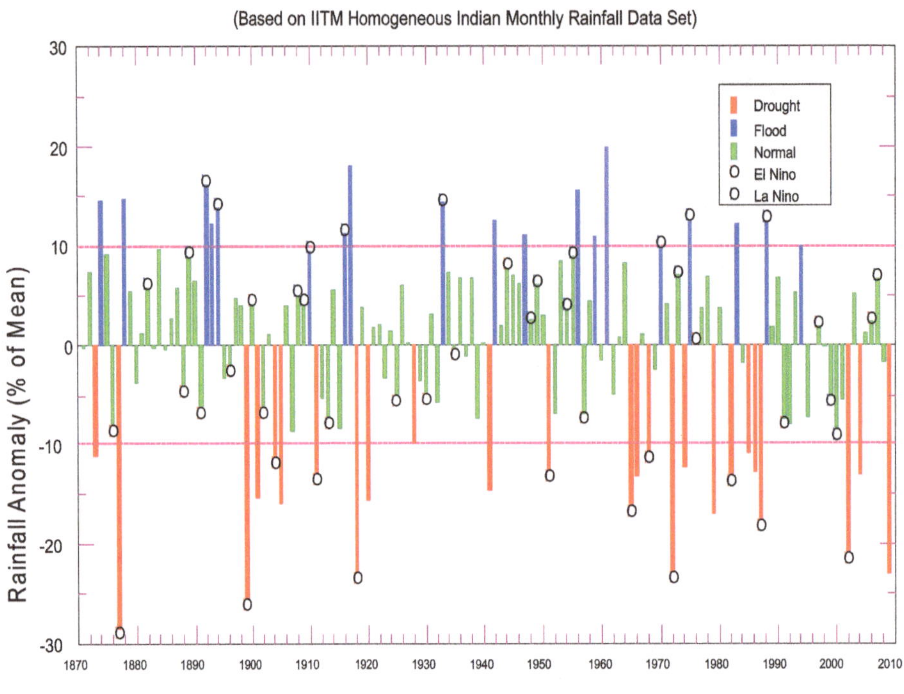

Fig. 1.6 All-India summer monsoon rainfall anomalies (1870-2010)

(Source: IMD, 2011)

A long time rainfall anomalies expressed in terms of percentage mean over the period 1870 to 2010 as described by IMD are presented in fig. 1.6. The series reveals several interesting aspects of decadal variations in terms of floods and droughts. There are about 19 flood years and 23 drought years identified over the period.

Flood years 1874, 1878, 1892, 1893, 1894, 1910, 1916, 1917, 1933, 1942, 1947, 1956, 1959, 1961, 1970, 1975 ,1983, 1988, 1994.

Drought years 1873, 1877, 1899, 1901, 1904, 1905, 1911, 1918, 1920, 1941, 1951, 1966, 1968, 1972, 1974, 1979, 1982, 1985, 1986, 1987, 2002, 2004 and 2009.

The Indian monsoon has a direct link with the Southern Oscillation Index (SOI). Weak Indian monsoons in the country are associated with a large negative SOI and occurrence of *EL Nino*, whereas, strong monsoons have been linked to large positive SOIs and absence of *EL Nino* events. In addition several global and regional parameters have also been found to contribute to the Interannual variabilities of monsoon rainfall.

1.2.4 Water resources potentials

Water resources of a country or a region play a major and crucial role in the overall economic development including agriculture production, irrigation potentials, drinking water needs and industrial productivity. India with a geographical area of 3.29 million km^2 contains 17% of the population, 2.45% of land resources and 4% of fresh water resources of the world. The southwest monsoon contributes to over 80% of the total rainfall input and the rest through northeast monsoon and other sporadic rains. The water resource potentials of India which occurs as natural run off in the rivers are estimated around 1869 BCM, which includes 432 BCM replenishable groundwater resources. It is also further estimated out of the above availabilities only about 1086 BCM could actually be put to use with 690 BCM from surface water and 396 BCM from groundwater (CWC, 2005 & CGWB, 2005).

India is a vast country with excellent net work of river systems and with potential subsurface groundwater systems. There are about 14 major river basins (catchment areas greater than 20,000 sq.kms) with total catchment area of 2,579,108 sq.kms and about 44 medium river basins with a catchment area of 239,302 sq.kms (catchment areas between 20,000 sq.kms and 2000 sq.kms). The major river basins form 85% of the total drainage area and together with medium river basins form about 91% of the total drainage area. In addition there are several minor rivers in the catchment areas less than 2000 sq.kms and whose total length stretches are mostly restricted to the coastal regions and desert regions. The Ganga-Brahmaputra has the largest catchment area with 110 Million hectares accounting to 43% of the catchment areas of all major rivers of the country, followed by Indus (32.10 M.Ha), Godavari (31.3 M.Ha), Krishna (25.9 M.Ha) and Mahandi (14.2 M.Ha) etc. and several others. The total annual flow of all the rivers in the country is estimated as 1869 BCM (Central Water Commission, 2005) and the replenishable groundwater recharge of the order of 432 BCM (Central Groundwater Board, 2005). The basin wise details for all the river systems with details on catchment areas, average annual potential, utilizable surface water resources and replenishable groundwater flows are presented in Table 1.2.

Table 1.2 Water resource potentials of the river basins of India (surface and groundwater)

S.No.	Name of the River Basin	Catchment Area(km^2)	Average annual potential in river (BCM)	Utilisable Surface flow (BCM)	Repleni-shable Ground flow (BCM)
1.	Indus (up to border)	321289	73.31	46.0	26.49
2.	(a) Ganga	861404	525.02	250.0	170.99
	(b) Brahmaputra Barak & Others	194413	585.60	24.0	26.55
3.	Godavari	41723	110.54	76.3	40.65
4.	Krishna	258948	78.12	58.0	26.41
5.	Cauvery	81155	21.36	6.8	12.30
6.	Pennar	55213	6.32	6.9	4.93
7.	East Flowing Rivers Between Mahanadi & Pennar	866643	22.52	13.1	18.84
8.	East Flowing Rivers Between Pennar & Kanyakumari	1100139	16.46	16.7	18.22
9.	Mahanadi	141589	66.88	50.0	16.46
10.	Brahmani & Baitarni	51822	28.48	18.3	4.05
11.	Subernarekha	29196	12.37	6.87	1.82
12.	Sabarmati	21674	3.81	1.9	
13.	Mahi	34842	11.02	3.1	8.52
14.	West Flowing Rivers of Kutch, Sabarmati including Luni	321851	15.10	15.0	11.23
15.	Narmada	98796	45.64	34.5	10.83
16.	Tapi	65145	14.88	14.5	8.27
17.	West Flowing Rivers from Tapi to Tadri	44940	87.41	11.9	

Table 1.2 ***Contd...***

S.No.	Name of the River Basin	Catchment Area(km²)	Average annual potential in river (BCM)	Utilisable Surface flow (BCM)	Repleni-shable Ground flow (BCM)
18.	West Flowing Rivers from Tadri to Kanyakumari	56177	113.53	24.3	17.69
19.	Minor River Basin Drainage into Bangladesh & Burma	36302	31.00		
	TOTAL:		**1869.37**	**690.3**	**431.42**

(Source: Ministry of Water Resources, 1999)

The utilizable water resource is the quantum of water that could be withdrawn in a given basin/area over a period of one hydrological cycle. The total utilizable water resource of the country is estimated as 1086 km^3 with surface water component as 690 km^3 and the groundwater component as 396 km^3. The present utilization trends of both surface and groundwater are of the order of 690 km^3. This is likely to go up to 1073 km^3 by the year 2050. This means that the water resource potentials of the country would reach a critical position by around 2050 and water conservation methods will play a very dominant role to preserve and protect the available water resources.

1.2.5 Water requirements, population rise and water scarcity trends

The water requirements in India are closely linked with the population growth rate, increasing tends of irrigation and agricultural outputs, industrial and energy outputs and drinking water needs. The population of India as on March, 2001 stood at 1.027 billion and is likely to shoot up to 1.308 billion by 2025 and 1.463 billion by 2050 (Fig. 1.7). The population growth rates since 1951 and the per capita availability of water are shown in Table 1.3 (Ramesh Kumar Gupta, 2008).

Table 1.3 Population and per capita availability of water resources

Year	Population (million)	Per capita
1951	361	5177
1961	439	4257
1971	548	3410
1981	683	2736
1991	846	2209
2001	1027	1819
2050	1463	1277

(Source: CWC, 2005)

The per capita availability of water at the beginning of the last century was of the order of 8192 M^3/year and was reduced to 5177 M^3/year by the year 1951. This exponential population growth then after was further brought down to 2209 M^3/year by the year 1991 and to 1819 M^3/year by the year 2001 and is expected to go down to 1277 M^3/year by 2050. These trends are shown graphically in Fig 1.8 with indications of stressed and scarcity boundary lines.

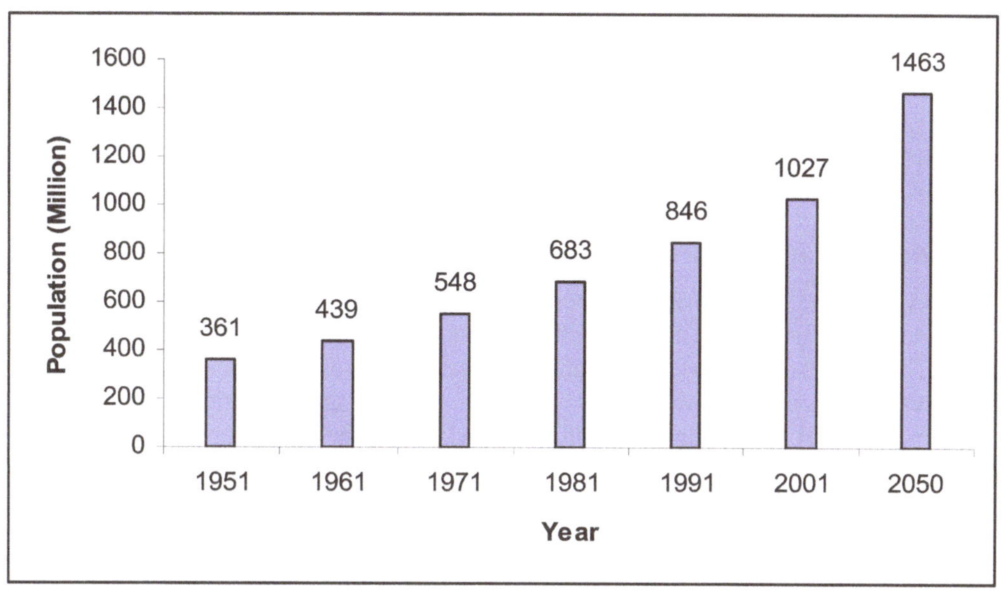

Fig. 1.7 Population growth levels

(Source: Ramesh Kumar Gupta, 2008)

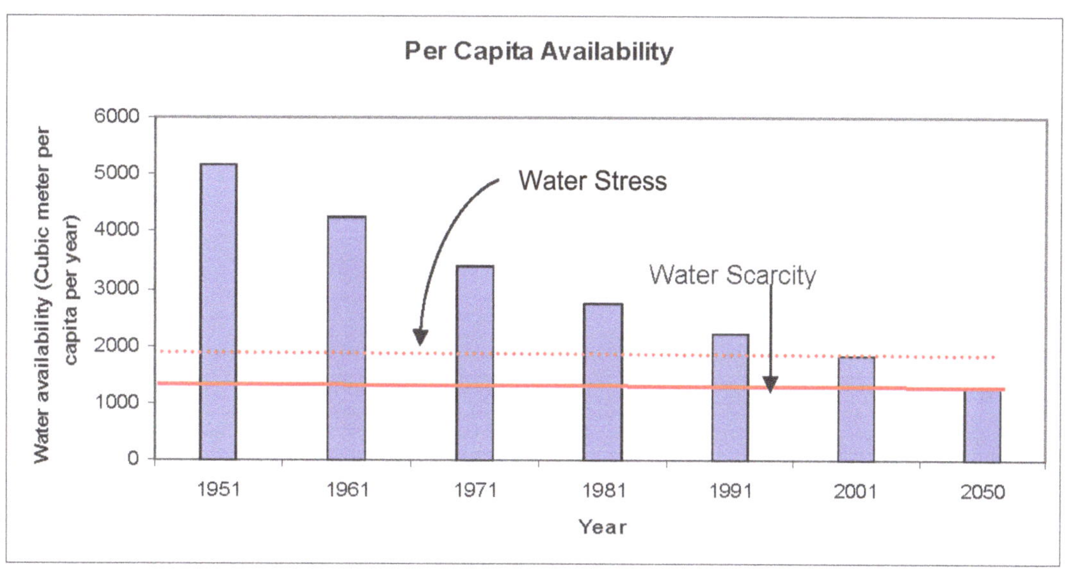

Fig. 1.8 Per capita availability/water stress/water scarcity trends

(Source: Ramesh Kumar Gupta, 2008)

The trends indicate by and large the country is already in a water stressed situation in respect of water resources consumption and would reach near to the scarcity line by 2025 and reach critical situation by 2050. These trends reflect on the need to plan carefully to optimize the utilization of consumption rates and also adopt suitable methodologies to create additional water resources.

The total requirement of water to produce required food grains based on the projected population is of the order of 557 BCM for 2010, 611 BCM for 2025 and 807 BCM for 2050 through surface water. Similarly the projected contribution from groundwater resources is 218 BCM for 2010, 245BCM for

2025 and 344 BCM for 2050. The requirements in the domestic sector would be 43 BCM, 62 BCM and 111 BCM for the year 2010, 2025 and 2050 respectively and from groundwater 19, 26 and 45 BCMs respectively for the above years. In addition there will be demands from industrial, power and environmental maintenance sectors also. All these statistics indicate both surface and groundwater utilizable resources would reach a critical stage by 2050 and demands a very careful management and conservation practices which are required to be initiated and adopted from now on.

1.2.6 Irrigation and agriculture growth rates

Irrigation and Agriculture is the prime concern of the nation and about 85% of the available water potential is diverted to irrigation and agriculture sector and the rest towards domestic, industry and energy needs. There has been a very systematic and sustained growth of irrigation potentials since the beginning of five year plan period from 1951 onwards. The ultimate irrigation potential of the country was estimated as 140 M.Ha from Major, Medium and Minor irrigation systems. The irrigation potential created at the end of 1951 is 16% of the ultimate irrigation of 140 M.Ha, and by the end of 2005 the created irrigated potential has reached upto 71%. The irrigation potential created had increased from 22.6 M.Ha in 1951 to about 93.90 M.Ha at the end of the 9th Plan period 97-2002. The Plan wise position of irrigation potential created and utilized is given in the Table.1.4.

Table 1.4 Plan-wise position of irrigation potential created and utilized (In million hectares.)

Plan	Potential created			Potential utilized		
	Major & Medium	Minor S.W & G.W Total	Total	Major & Medium	Minor S.W & G.W Total	Total
1951-56	12.20	14.06	26.26	10.98	14.06	25.04
1956-61	14.33	14.75	29.08	13.05	14.75	27.80
1961-66	16.57	17.00	33.57	15.17	17.00	32.17
1966-69	18.10	19.00	37.10	16.75	19.00	35.75
1969-74	20.70	23.50	44.20	18.69	23.50	42.19
1974-78	24.72	27.30	52.02	21.16	27.30	48.46
178-80	26.61	30.00	56.61	22.64	30.00	52.64
1980-85	27.70	37.52	65.22	23.57	35.25	58.82
1985-90	29.92	46.61	76.53	25.47	43.12	68.59
1990-92	30.74	50.35	81.09	26.32	46.54	72.86
1992-97	32.96	53.31	86.26	28.44	47.83	76.27
1997-2002	37.05	56.90	93.95	31.01	49.05	80.06

(Source: CWC, 2005)

The basic responsibilities for the creation of irrigation & potentials and agriculture products etc. exists with the respective state governments. A sustained and coordinated action programs are therefore required to be launched to meet the emerging future demands. The National Water Policy launched in 1987 and subsequently modified in the year 2002 provides a direction for optimum development of water resources while protecting the environment etc. Some of the salient points included in the policy are briefed below:

- Water is a prime natural resource, a basic human need, a prime natural asset and is a part of the global ecological system. It is to be planned, developed, conserved and managed on an integrated way with sound environmental principles.

- A well developed information system with a net work of data banks and data bases for all water related systems at state/central level needs to be introduced for appropriate resource planning systems analyses and augmentation of water resources.
- All water resources development and management programmes should be planned taking sub-basin/basin as a hydrological unit and evolve integrated action plans. Watershed management programme should consider soil conservation, catchment area treatment, afforestation and check dams construction.
- All existing institutions at various levels under the water resources sector will have to be re-oriented and strengthened wherever necessary taking into consideration the emerging technologies and practices.
- Planning and operation of systems should broadly consider the following aspects: (a) Drinking water; (b) Irrigation; (c) Hydropower; (d) Ecology; (e) Agro industries and non-agricultural industries and f) Navigation and other uses.
- All water resources development projects should be planned and developed as mutlipurpose projects involving drinking water needs, ecological and environmental balances, economics and sustainabilities.
- Groundwater resource assessments should be organized with sound scientific basis involving quantity and quality of water. All exploitation endeavours should be based on input/output analyses. Over exploitation of groundwater resources should as far as possible be avoided and conjunctive utilization of surface and groundwater resources should be planned.
- All irrigation projects/development practices should take into account the cost-benefit analyses and adopt methods for optimization of resources, reclamation of water logged and saline affected areas. Rehabilitation practices should be adopted wherever required particularly in major river valley projects.
- Adequate emphasis needs to be given towards physical and financial sustainability while creating additional water resource potentials by way of fixing water cess so as to cover operational maintenance and capital costs.
- Water utilization and management programmes, should incorporate participatory approach of various governmental institutions, users and stake holders. Efforts should be made to involve women water users associations and local bodies like panchayats and municipalities etc in different stages and project formulation, development and operational maintenance etc.
- Private sector participation should be encouraged wherever feasible and encourage innovative ideas, generate financial resources and help introducing corporate management practices.
- Water qualities of both surface and groundwaters should be regularly maintained and introduce appropriate neutralization methods wherever water qualities deteriorate beyond their threshold limits and introduce necessary legislation practices wherever required.
- The entire water resources domain of the country should be classified into different zones in terms of quantity and quality and update the data base and measures from time to time.
- Water conservation methods should be adopted invariably at all levels of water utilization practices, introduce techniques related to desilting of canals and tanks, recycling of waste water, adaptation of drip and sprinkler irrigation systems wherever feasible.
- Adequate cushioning for floods should be provided in the water storage areas, while physical flood projection works like embankments and dykes are required to be continued. Flood forecasting and warning measures should be modernized and minimize the loss of life and roperty as and when floods recur.
- Erosion impacts both in the coastal and river shore areas should be minimized by appropriate cost-effective measures. Indiscriminate exploitation of coastal resources should be regulated.

All coastal states should prepare comprehensive coastal zone management plans for sustainable development.

- Soil moisture conservation measures should be practiced in the vulnerable drought prone areas. Development of groundwater through appropriate water harvesting, recharge systematic monitoring, identification of bottlenecks and adopting remedial measures should form important activity of both surface and groundwater projects.
- Water sharing/distribution in different states should be guided by national perspectives and availabilities of water resources in a basin/river system. The inter-state water disputes act of 1956 should be renewed periodically and emerging disputes be referred to tribunals.
- There should be a paradigm shift from the emphasis on creation and expansion of water resources infrastructure facilities to the improvement of performance and maintenance of the existing water resources facilities and accordingly reprioritise the activities and financial budgets. Safety measures and modernisation practices for all existing structures should be undertaken and keep them in good health, through appropriate organization practices involving experts in different fields.
- All the latest available techniques and methods in the water related fields should be adopted for an effective management and utilization of available water resources. Training practices should form an integral part of all water resources developmental programs.

Agriculture

Agriculture is of fundamental importance to the Indian economy contributing to about 20-21% of gross domestic product (GDP) and generates two thirds of the employment potential. Assured irrigation water supply is most important for food security. The total food grain production which was only 51 million tons at the time of commencement of the first five year plan has gone upto 213.45 million tons by 2005 and projected to go up to 250 million tons by 2010. Thanks to the green revolution inducted in the country coupled with induction of high yielding varieties. Uttar Pradesh, Haryana, Madhya Pradesh, Rajasthan and Bihar account for 95% of wheat production in the country while West Bengal, Uttar Pradesh, Andhra Pradesh, Tamilnadu, Bihar, Punjab, Orissa, Madhya Pradesh, Assam and Karnataka account for more than 90% of rice production by 2010. The coarse cereals and pulses production are mostly from states like Maharashtra, Karnataka, U.P, Rajasthan, Madhya Pradesh, Andhra Pradesh, Gujarat and Bihar contributing to about 90% of production in the country. Similarly these states have contributed abundantly for all agro-based products like sugarcane, cotton, oil seeds, and horticulture based products like vegetables, fruits and animal based production like milk, fish and eggs etc. The production rates of all these items are incorporated in table 1.5 and also are shown graphically in Fig. 1.9.

Table 1.5 Agriculture growth development

S.No.	Agricultural Products	1950-51 (M.Tons)	1964-65 (M.Tons)	1980-81 (M.Tons)	1995-96 (M.Tons)	2010-11 (M.Tons)
1.	Food Grains (Wheat, Rice, Coarse Cereals & Pulses)	51	72	130	180	270
2.	Agro based products (Sugarcane, Cotton, Oil Seeds)	12	19	25	52	74
3.	Horticulture based products (Vegetables, Fruits)	8	14	40	120	230
4.	Animal based products (Milk, Fish, Eggs)	19	22	38	80	149
	Total (Agri-Products)	90	127	233	432	703

(Source: Goel, A.K., 2008)

Population growth rate will be far the most important parameter for the demands of irrigation and agricultural potentials. If the country's population has to be stabilized around 1.55 billion by around 2050, we have to plan to increase the overall food grains and related products from the present level of 430 million tons to more than double the amount by 2050. To achieve these objectives the current irrigation potentials of the order of 97 million tons will have to be substantially increased.

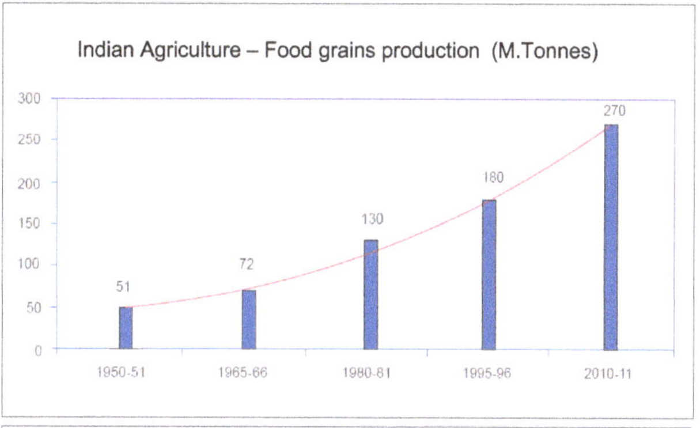

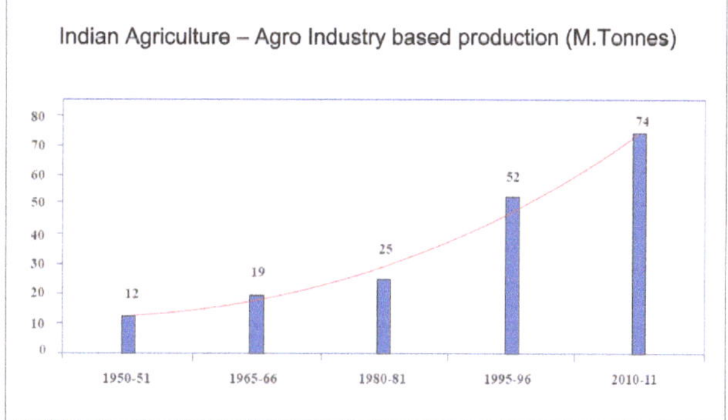

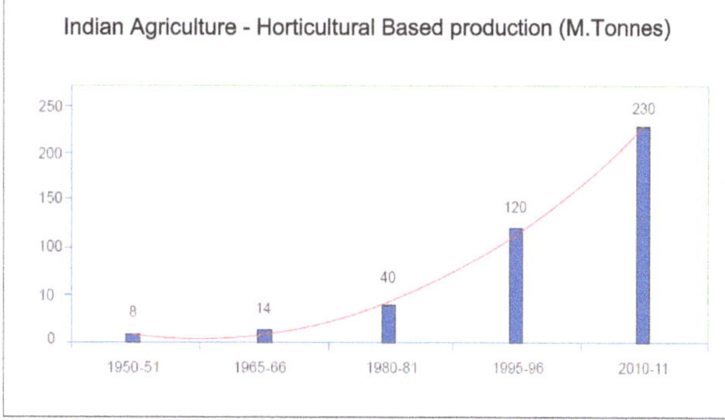

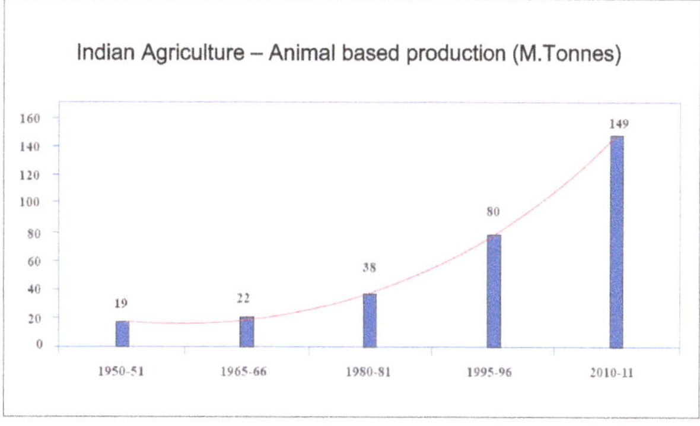

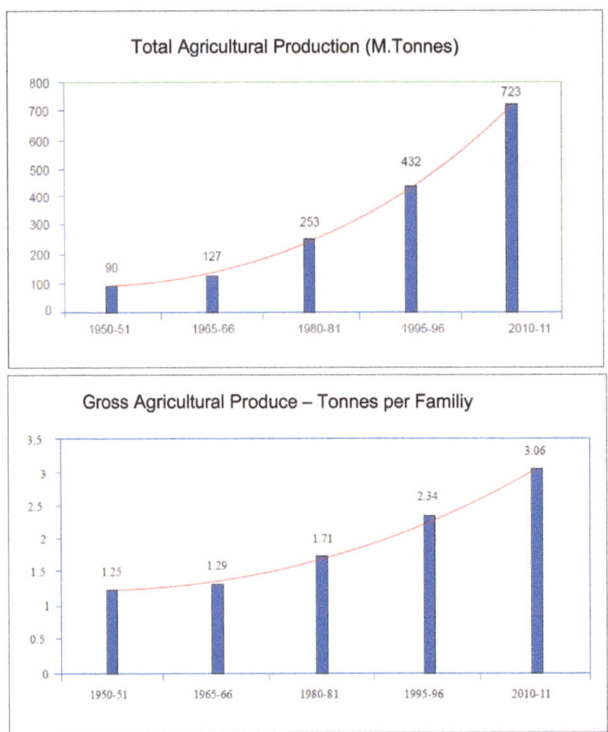

Fig. 1.9 Agriculture, Agro based products, Horticulture and Animal based products growth levels

(Source: Goel A.K, 2008)

The basic responsibilities for the creation of irrigation and agricultural potentials exist with the respective state governments. A sustained and coordinated action plans are required to be launched. A national water policy was launched in the year 1987 and was reviewed and updated in the year 2002, which provides directions for optimum development water resources and adopt necessary strategies/methodologies etc. Some of the salient points were included in the earlier paragraphs.

1.3 COASTAL SYSTEMS (WORLD SCENARIO)

Coastal Systems/Areas are the most dynamic systems interfacing the continents and oceanic basins and support some of the important, diverse productive habitats in the world. These are the areas where fresh water from the continents mix with the saline water from the oceans and make the environment very fragile and sensitive covering approximately 15% of the land surface of the earth and with over 1.6 million kms length (Burke, L., et al 2001). There are no rigid definitions and spatial boundaries to the coastal systems/areas. A distinction perhaps could be made between the terms '*coastal zone*' and '*coastal systems/area*'. The term '*coastal zone*' could be referred to a relatively narrow zone immediately adjacent to the coast line on either sides covering shallow parts of the oceanic basins and the low lying terrestrial regions of the continents where intense land-sea interactions take place and the term *coastal system/area* could be applied more broadly to the inland areas where oceanic and continental influences take place. These include beaches and dune systems, backwaters and lagoons, saline marshes, tidal flats, estuaries, deltas and the adjoining sedimentary environments.

Small, C. and Nicholls, R.J. (2003) characterizes the *coastal zone* or a *costal system* as an area running up to 100 kms from the coast line or 50 meters elevation whichever is closer to the coastal line with an outward extension of 50 meters depth level into the sea. LOICZ (Land-Ocean Interaction with Coastal Zone) a core project of the International Geosphere – Biosphere program defines the coastal zone as an area extending from the coastal plains to the outer edge of the continental shelf approximately matching the region that has been alternately flooded and exposed during the sea level fluctuations of geological past. The coastal regions are very crucial for social needs and economic development. They tend to be more densely populated than inland areas due to several economic reasons such as development of ports, shipping, fisheries, tourism, urbanization processes, coal, oil and gas field exploration and development, mining and refineries etc. It has been estimated that by the year 2025 about 75% of the world population would be living within 50 kms stretch of the coast line. This trend is more pronounced amongst the developing countries and as a consequence coastal systems/regions with degradation levels will be more conspicuous and could result in significant risks of life, property and severe health hazards. Some of the generalized characteristics of the coastal zone are mentioned below (HartWig Kremer and Criss Crossland, 2002).

- Comprises <20% of the Earths surface.
- Contains >50% of human population.
- Locating 70% of urban cities with >1.6 million inhabitants
- Yields 90% of the global fisheries.
- Produces about 25% of global biological productivity.
- Act as major sinks for sediments.
- Act as major sites for nutrients-sediment bio-geochemical processes.
- Act as a heterogeneous domain, dynamic process in space and time
- High gradients, high variabilities and high diversities.

1.3.1 Coastal features and ecosystems

Though our awareness in respect of coastal morphology, ecological dynamics of wet lands such as mangroves, tidal flats, coral reefs, sea grass systems, beaches and dunes, estuaries and deltas have increased over the past several years, the impact processes both natural and anthropogenic are still at the beginning of our understanding. Coastal systems, both terrestrial and offshore are not only the most productive but are also the most threatened systems of the world. They serve as huge productive centers and account for 35% of the global primary productivity, 90-95% of the world's marine fish, 80% of the global carbonate production, 50% of the global denitrification and 90% of the global sedimentary mineralization (UNEP, 1992). These systems are undergoing rapid physical, chemical and biological transformations over the last several years resulting in coastal degradations including severe erosion, losses in fishery resources, timber, fuel wood, oil and natural gas and beach sand minerals, agriculture and forestry, increasing pollution loads, tidal impacts and saline water intrusion processes in the near surface and deep groundwater systems. Effective measures to address these deteriorating levels world over have been very poor and ineffective. It would be the endeavour of this book to review all aspects of the subsurface groundwater systems, their resource potentials, environment and required management strategies, while taking into account the surficial impacts and their manifestations of the inland coastal ecosystems. Some of the major ecosystems confronting these coast lines are (i) Sand and gravel beaches, (ii) Estuaries, marshes, lagoons and backwaters (iii) Mangroves, (iv) Coral reefs and atolls and (v) Coastal deltas. A brief description of each of these ecosystems are briefly mentioned below.

(i) ***Sandy beaches, barriers and dunes:*** Sand beaches function as wave energy sinks and barrier beaches act as natural break waters. Coastal dunes form natural buffers and sand repositories. Majority of the coastal beaches/dunes are shaped primarily by wave and tidal processes and riverine and long shore current impacts and wind energies. Coarse sand, gravel and cobble beaches occur in tectonically active and high wave dynamic areas and fine sand and muddy beaches occur in low energy wave dynamic areas. It is estimated that over the last 100 years that about 70% of the world shorelines are retreating, 20-30% are stable and about 10% of them are advancing (Bird, E.C.F., 1993). These situations probably will alter with rising temperatures and related sea level changes in the years to come. Beaches, barriers and dunes serve as ecological protectors, but are subjected to quick dynamic erosional and depositional events depending upon the wave dynamics, currents and winds.

(ii) ***Estuaries, marshes, lagoons and backwaters:*** These ecosystems lie very close to the coast and face from time to time the ocean, atmospheric and continental dynamics, and at the same time highly productive and critical for the sustenance of coastal economy. Estuaries, marshy lands and lagoons play important role in maintaining hydrological balances, filtering water from pollutants entering into the environmental realms and provide good habitat for birds, fish, molluscs, crustaceans and other kinds of ecological sensitive fauna and flora (Beck, M.W., et al, 2001). It is estimated that world wide there are about 1200 estuaries and lagoons accounting to 80% of the worlds fresh water discharge (Alder, J., 2003). These systems often are subjected to anthropogenic impacts from humanity, industries and several pollution loads from urban and industrial centers which generally grow around these ecosystems. Efficient management and protective measures therefore are very important and necessary.

(iii) ***Mangroves:*** Mangroves are the shrubs and trees, grown very luxuriantly in the tidal flat areas and around the estuarine shorelines of the tropical and subtropical regions. These are considered as most productive bio-diverse wetland ecosystems of the earth and are estimated to cover about 16 to 18 million hectares in the Globe (Valiela, I., et al 2001; Spalding, M.D., et al 1997). These plants can survive in varying fresh water to saline conditions with different strengths and act as important guards and protectors of fauna and flora in the regions. Mangroves in general are classified into three major zones based on the physical and morphological characters of the regions (Ewel, K.C., et al 1998). (i) Tide dominated fringing mangroves (ii) River dominated mangroves and (iii) Interior basin mangroves. The fringe mangroves provide protection from cyclones and storm surges, floods and soil erosion and allow sustenance of animal habitat and nurseries. The riverine mangroves also by and large provide the same type of protective features. The interior basin mangroves in addition to the above act as nutrient sinks improve water qualities and allow growth of good plant products. These ecosystems are highly valued commodities for several coastal population groups for their living and growth. Mangroves have an important ecological and socioeconomic function in relation to sea food production, source of wood products and also act as nutrient sinks and shore line protection, (Ronnback, P., 1999), but then mangroves have become targets for degradation for several decades in terms of deforestation processes, agriculture conversions and climate change.

(iv) ***Coral reefs and Atolls:*** Coral reefs are shallow water tropical marine ecosystems characterized by a remarkably high biomass production and a rich faunal and floral diversity probably unequalled by any other habitat and harbour more than 25% of the known marine fish in the world. Corals can flourish only in relatively shallow marine waters, exposed to sunlight with optimum temperatures varying between 23 °C - 25 °C and free from suspended sediments. The reefs are formed by calcareous skeletons, a type of soft bodied radially symmetrical marine invertebrate. In general reefs are divided into three major types (i) Fringing reefs: reefs that grow close to the shore and extend into the sea like a submerged platform (ii) Barrier reefs: reefs separated from the land by wide expanses of water, but follow the coastline (iii) Atolls: reefs built up in circular fashions surrounding lagoons. The total extant of living coral reefs in the world has been estimated at about 284,300 km^2 (Spalding, M.D., et al, 2001). A major population of these are located in South East Asia. They serve important functions such as atoll island foundations, coastal protection structures and sources for beach sand. These systems are very valuable to continental pollution, coastal construction and industrial activities and climate change impacts. It is estimated that over the last few decades, about 20% of the coral reefs were destroyed and another 20% were degraded world over (Wilkinson, C., 2004).

(v) ***Coastal Deltas:*** Coastal river deltas are among the most valuable, heavily populated and vulnerable coastal systems in the world. These are also the systems where intense agricultural practices are taking place through good network of canal irrigation systems and several thousands of filter points tapping the fresh groundwater resources. The term delta was first applied by Greek historian Herodotus approximately 450 BC to the triangular alluvial deposits at the mouth of Nile river. Deltas develop where rivers deposit more sediment at the shore than can be carried away by waves. Delta plain land forms span most areas of the coastal regions and include distributory channels, river mouth bars, bays, tidal flats and ridges, beaches and dunes, swamps, marshes and estuaries. Coleman, J.M. (1976) while reviewing the processes responsible for most world delta regions indicated a number of parameters that go into the delta formation and processes such as climate, water and sediment discharges and their variabilities, river mouth processes, near shore wave power, tides, near shore currents, shelf slopes and tectonics and basin geometry. The delta basins could be classified basically into three categories, namely, river dominated deltas, wave dominated deltas and tide dominated deltas depending upon the energy levels. The deltas, estuaries and offshore island ecosystems are most vulnerable to climate change and sea level rise (IPCC, 2001). Some of the major delta systems of the world which come under the influence of climate change and sea level rise are Mississippi (USA); Amazon (South America); Rhine delta (Germany); Niger delta (Africa); Nile delta (Egypt), Indus (Pakistan); Ganges (India – Bangladesh), Mekong (Vietnam); Yangtze (China) and also several medium level delta systems in several coastal countries of the world.

1.3.2 Values and benefits of coastal zone

Some of the important values and benefits in the coastal zone are mentioned below as described by Hartwig Kremar and Chris Crossland, 2002.

Resources-Natural	Products (Natural and Human derived)	Amenities and services: (Natural & Human derived)
Water resources (Surface & Groundwater)	Agriculture & Food	Transport
Forests & Timber	Woolen industry and	Tourism
Arabia land and Food	Fisheries	Recreation & culture
Ores and Minerals	Oil, Gas & Minerals	Biodiversities
Ecosystems & Biodiversities	Habitation	Biodiversities and ecosystems services

1.3.3 Natural and Anthropogenic pressures in coastal systems

Natural impacts

- *Climate changes/shifts:* Climate changes/shifts and their intensities influence largely the thermohaline circulation pattern, the ENSO (Elnino Southern Oscillation) in Pacific and Atlantic oceans. These changes will have their impacts on our coastal ecosystems, bio-diversities, coastal hydrological processes and the CO_2 concentrations across the globe.

- *Sea level:* The IPCC estimated that the global average surface temperature has increased by 0.6 °C in the 20th century and is projected to rise by about 2.5 °C over the next 100 years. The IPCC estimates indicate that the sea level has risen by 1.0-2.5 mm a year over the last century and projects to increase to over 10-88 cm over the next 100 years.
- Several coastal ecosystems will be totally washed away and degrade and a few may survive depending upon the inland nutrients and strength they receive.
- Several coastal regions will undergo severe erosional processes, severe storm surge events and flooding etc. The tectonic subsidence that could take place in several coastal regions could trigger the subsidence rates further and facilitate large chunks of sea inundations in the low lying coastal areas.
- The atmospheric green house gases have increased considerably over the last 100 years. These will have impacts on coastal and shallow marine biological processes and their productivities, and degradation levels of all costal ecosystems and shallow coral reefs.

Anthropogenic impacts

- The population pressures in the coastal regions are increasing at an alarming rate. Most of the major cities around the world are located in the coastal areas. The influx of population in the coastal zone is of the order of 50-70% of the world population with wide variations among different coastal countries and cities. Nutrient loading into ocean basins is a key factor related to catchment area, run off and population built up.
- Agriculture and irrigation practices, conversion of forests and grass lands into agriculture based lands, intensification of fertilizer applications are some of the anthropogenic factors involved in coastal ecosystem patterns.
- Deforestation practices mounting year after year, diminishing watershed protections, increasing erosional processes and water quality deteriorations affecting the coastal ecosystems. There has been in recent years a spurt in the construction of major and minor dams in many parts of the world, affecting the river flow and sediment loads and nutrient influxes, which could bring in considerable change in the eco balance systems and increase saline water intrusion processes underground, reduce surface and groundwater discharges and biodiversities in the coastal areas.
- Urbanisation and industrial development processes, leading to nutrient and contaminant wastes, atmospheric levels, sewages, oils and oil wastes etc. are some of the important and globally significant issues affecting the coastal systems.
- Shoreline development reclamation & transportation projects like flood mitigation, barrages and dykes, ports and harbours development, dredging processes etc have affects on water qualities, biodiversities, fisheries and coastal sediment processes etc. which are globally wide spread events.
- Sand mining for minerals, oil and gas extraction and development programs will have substantial impacts on bathymetric changes, groundwater flows, shoreline stabilities, erosional processes and land subsidences.
- Fisheries is a major industry in the global coastal environment. An estimated 90% of the world fish catch comes from the coastal zone. In addition aquaculture expansion programmes are making considerable impacts on the physio chemical changes of the coastal zone.
- Integrated multidisciplinary approaches involving scientific and economic disciplines are required to be planed at macro and micro levels in different coastal countries and in the island areas.

Systematic and Holistic approaches involving studies on catchment areas of river systems, deltas, outflows, ocean dynamics and ecosystems are the need of the hour. Improved numerical and conceptual models and approaches are required to be adopted with real time analyses.

1.3.4 Global climate change impacts

The global importance of the coastal systems/area in terms of ecological and socio-economic values is well recognized and several International organizations including IPCC (Inter governmental Panel on Climate Change) have encouraged several actions and drew up programmes for appropriate management strategies. The IPCC has so far released four assessment reports since 1990 to 2007. The fifth assessment report is to be released during 2014. All these assessment reports in general cover different aspects related to scientific, technical and socio-economic information and inputs related to adaptation and mitigation strategies of climate change. The IPCC (1990) in its first assessment report, estimated that if green house gas emissions continue to grow, the global mean temperatures will increase by 0.2 °C - 0.5 °C per decade over the next century. One of the most important impacts of this global temperature change for the coastal zone is "*Accelerated Sea Level Rise (ASLR)*". The IPCC working group projected a global sea level rise ranging between 30 cm to 110 cm by the year 2100 due to thermal expansion of the ocean and melting of glacier mounds. As a consequence of this situation the coastal regions will have to face serious impacts in terms of population and resource potential losses, degradation of coastal ecosystems and increase in the levels of erosional processes etc. The climate change could also bring in variations in the intensities and patterns of extreme weather events like tropical cyclones, intense precipitation levels, storm surges and floods. In line with these findings and emerging situations, the IPCC has operationalised *a Coastal Zone Management sub group* (CZM). The initial findings of this group are summarized in the following statement *"In many parts of the world the natural systems that provide protection against the sea are being degraded by several unsustainable developmental activities such as sand and coral mining, mangrove cuttings and deforestation processes, damming the continuing river flows and wet land fillings etc. These issues are not being discussed substantially world over and the impacts on environmental degradations are increasing year after year".* The importance of these findings are explicitly reflected in chapter-17 of agenda 21 of the UN Conference on Environmental Development (UNCED) held at RIO in the year 1992. In response to the decisions in the conference, a world coastal conference (WCC93) was held at Netherlands with the following objectives:

(i) All nations with low lying coastal areas to come forward and share expertise to understand and solve their coastal management problems.

(ii) All nations prepare 'Coastal Zone Management Plans' with available tools and techniques.

(iii) All nations to participate and exchange information on climate change, impacts and adaptabilities and related programmes.

(iv) All nations to stimulate discussions and co-ordinate activities with all national and international organizations and implement coastal zone management programmes.

The WCC workshop further recognized and identified that "*Integrated Coastal Zone Management*" (ICZM) as the most appropriate process to address all short term and long term coastal management strategies including problems related to population rise, health hazards, habitat loss, coastal ecosystem degradations, beach erosion intensities, degradation of water quality changes in hydrological cycles, depletion of water resources intensities, saline water intrusion processes and adaptation to sea level rise events and other related impacts of global climate change(WCC, 1993). In addition the conference had identified the following salient points in its statement to be implemented by all coastal states by the year 2000 and also adopted an agenda for action.

- There is an urgent need for coastal states to strengthen their capabilities for ICZM, working towards the development of appropriate strategies and programmes by the year 2000.

- ICZM is the most appropriate process to anticipate and respond to long-term concerns and needs while addressing present day challenges and opportunities.
- ICZM involves the comprehensive assessment, setting of objectives, planning and management of coastal systems and resources, taking into account traditional, cultural and historical perspectives and conflicting interests and uses; it is a continuous and evolutionary process for achieving sustainable development.
- Coastal states that are in the process of defining and implementing a national programme for ICZM have encountered obstacles that constrain the effective development of national programmes. These include: limited understanding and experience in ICZM; limited understanding of coastal and marine resources; fragmented institutional arrangements; single-sector oriented bureaucracies; competing interests and lack of priorities; inadequate legislation and/or lack of enforcement; land-tenure regimes and other social factors; and lack of information and resources.
- Coastal states are urged to identify their priorites for ICZM, to identify their most pressing needs to improve their capabilities for ICZM, to undertake national measures to increase their capabilities, and to identify their special needs for assistance.
- Effective ICZM can be achieved by coordination between national, regional and international organizations and institutions. This will help to avoid unnecessary duplication and to develop the concepts, tools and networks needed to facilitate the development and implementation of national programmes. Support for ICZM capacity building could be in the following areas: information; education and training; concepts and tools; research, monitoring and evaluation; and funding.

Agenda for Action

- ***Raise awareness:*** Educate policy makers and decision makers about the urgency and the benefits of ICZM.
- ***Define the process:*** Clarify the process of ICZM and its elements, and identify the obstacles to its progress.
- ***Strengthen the national responses:*** Identify coastal states priorities for ICZM, improve their capabilities, and start implementing ICZM programmes.
- ***Strengthen the international responses:*** Strengthen international support for ICZM capacity building, including financial support for programme development, particularly in developing countries. This can be achieved through global, regional or blateral cooperation to assist and support national initiatives. Activities should include the organization of networks and clearinghouses to exchange and share ICZM knowledge, experiences, concepts, tools, education and training, and research programmes.

1.4 COASTAL SYSTEMS (NATIONAL SCENARIO)

India has an extensive coastline of about 7500 kms including those of offshore islands Andaman and Nicobar islands in the east and Lakshadweep islands in the west. The west coast in general has a wider continental shelf region (340 kms in the north and 60 kms in the south) and narrow alluvial coastal margins. The east coast on the other hand has wider alluvial margins and narrow continental shelf region with an average width of 38 kms. The coastal features include well developed beaches, and dune systems, rocky and cliffed shore lines; offshore bars; barriers and spits; tidal mud flats; marshy lands; with potential growth for mangroves; lagoons and backwaters; coral reefs and atolls and estuaries and delta systems (Ahmad, E., 1972).

1.4.1 Coastal features and ecosystems

The west coast is predominantly a coast with several river mouth estuaries, backwaters and lagoons with associated spits and barriers, narrow beaches/shore line/cliffed shorelines and offshore islands. The east coast is essentially a deltaic coast with wide spread beaches, large river mouth areas with estuarine conditions and well developed mangrove systems, lagoons, deltas and offshore islands. The Indian main land as per the naval hydrographic charts contain about 43% sandy beaches, 11% rocky coast with cliffs and 46% mud flats and marshy areas (Sanil Kumar, et al 2006). *Mangroves and estuaries* are the dominating ecological units along the Indian coast. Mangroves covers approximately an area of 6740 km^2 and estuaries cover over an area of 2.7×10^4 km^2. The estuaries are the mouth regions of the vast network of river systems joining the seas or along the coastlines and support rich biodiversities, nutrients and huge stocks of marine brackish water fish and aquaculture. The mangroves and estuaries form very important ecological assets and economic resource systems of the coastal regions and survive several environmental threats and hostile conditions, act as protective barriers from erosional processes, provide nursery grounds for a number of important fish like prawns and crabs etc and act as storehouses for several organic and inorganic nutrients. Well developed and prominent mangroves along the east coast of India are found in the delta distributary regions of Ganges in West Bengal (Sunderbans); Mahanadi – Brahmini – Baitarni delta – estuary regions of Orissa (Bhitarnika); Godavari delta-estuary region in Andhra Pradesh (Coringa); Krishna delta estuary region in Andhra Pradesh (Gollamattapaya and Nadimeru distributaries) and Cauvery delta estuary in Tamil Nadu (Pichavaram and Muthupet mangroves). In the west coast prominent mangrove systems are seen along the Gulf of Kutch and Khambat region of Gujarat. The offshore island in the Bay of Bengal contain a very prominent and well developed dense mangrove systems near the tidal creeks, bays and lagoons of Andaman and Nicobar islands. (Estuaries of India, ENVIS publication series 1/2002 and Mangroves of India, ENVIS publication series 2/2002). As per the records available it appears that over 40% of the mangrove systems along the Indian coast over the last one century have been either degraded or lost. The trends of degradation of mangroves over the last one decade are reflected in the following Table 1.6.

Table 1.6 Area distribution of mangroves in India (thousand ha)

State/Union territory	Government of India, 1987	Government of India, 1997
West Bengal (Sunderbans)	420	212.3
Andaman and Nicobar Islands	119	96.6
Maharashtra	33	12.4
Gujarat	26	99.1
Andhra Pradesh	20	38.3
Tamil Nadu	15	2.1
Orissa	15	21.1
Karnataka	6	0.3
Goa	20	0.5
Kerala	Sparse	Nil
Total	674	482.7

Lagoons and backwaters

Lagoons and backwaters occupy good stretches along the east and west coasts of India, act as good ecological centers, rich with flora and fauna and also act as sinks for nutrients. The coastal lagoons are shallow water bodies invariably separated from the ocean through a barrier but connected to the ocean at one or two inlet points. These lagoons in general are found in the low lying coastal regions and are normally align themselves parallel to the coast. They are often highly productive aquaculture systems, but at the same time strongly influenced by anthropogenic events and human activities. Some of the important lakes/lagoons along the east coast of India are (i) Chilka lagoon, Orissa (ii) Kolleru lake, Andhra Pradesh (iii) Pulicat lagoon, Andhra Pradesh (iv) Muthupet & Gulf of Mannar lagoons, Tamil Nadu. Some of the important lakes/lagoons in the west coast are (i) Vembanad lagoon (Kerala), (ii) Astamundi Lagoon, backwaters (Kerala), (iii) ETG Kulam lagoon (Kerala), (iv) Several lagoons along Maharashtra coast including Mumabi and (v) Several small lagoons in offshore islands of Lakshadweep Island.

Coral reefs and Atolls

Coral reefs in India are extensively developed and they represent both semi and sub-tropical climate types essentially along the offshore islands of Bay of Bengal and Arabian sea. Their distribution however is restricted to six regions. viz; Gulf of Kutch, Lakshadweep islands, Gulf of Mannar, Palk bay and Andaman and Nicobar islands. These reefs appear to have been building up since Tertiary times. Majority of the reefs in Lakshadweep islands appear as '*Atolls*' and the rest of the reefs both in the west and east coasts appear as '*reef*' structures with of course certain physical and structural variations. The reefs in the Gulf of Mannar, Palk bay and Andaman and Nicobar islands appear as fringing type of reefs. Platform types of reefs are seen in the Gulf of Kutch region.

Deltas

The east coast of India is predominantly a deltaic depositional coast with major river systems building their deltas like Ganges (West Bengal); Mahanadi-Brahmini-Baitarni and Subarnarekha (Orissa), Godavari-Krishna and Pennar (Andhra Pradesh) and Cauvery, Palar, Ponnaiyar and Vaighai (Tamil Nadu). The west coast of India is totally devoid of delta systems excepting Narmada and Tapti (estuarine deltas) in the North Western part of the Gujarat Coast. In addition there are a number of medium and minor rivers both along east and west coasts, and many of them are estuary delta systems bordering the Bay of Bengal and Arabian seas (Ahmad, E., 1972, Vaidyanadan, R., 1991 and Singh, I.B. and Seeta Rama swamy, A., 2006). There are 14 major rivers, 44 medium rivers and several minor rivers along east and west coasts of India making the coastal areas rich with soil and water resources and at the same time nutrient rich. All the major rivers by and large have large delta systems and the medium rivers have relatively either small deltas or estuary mouths. The delta systems of the Indian coasts are active resource centers for agriculture, industries and fisheries etc. Invariably all the coastal delta areas have good net work of canal systems and intense agricultural and irrigation practices. Population build up is increasing year after year with increasing levels of anthropogenic and pollution impacts. These rivers carry large sediment influx into the sea and get redistributed along and their associated environments such as levees, sand bars and spits, beaches and dunes etc., with rich soil patterns. A large proportion of food production comes from these delta tracks, and are active centers for fish and agriculture forming industries. These have also become active industrial centers and also centers for oil and gas explorations and development and as a result pollution levels are increasing year after year with possible threats of land subsidence and other environmental hazards (natural and anthropogenic).

1.4.2 Environmental impacts and vulnerabilities

India has been identified as one of the important countries in the world vulnerable to the impacts of global warming and *Accelerated Sea Level Rise* events (UNEP, 1989). The high degree of vulnerability of Indian coastal regions could be mainly attributed to (i) Extensive low lying coastal areas, particularly the east coast deltaic plains which are prone for saline water inundations and cyclonic threats (ii) Increasing levels of population growth, with over 30% population living in the coastal regions (iii) Large commercial exploitation of resources more than their optimum levels. A large chunk of population are directly dependent on resources like mangrove vegetations and timber wood cutting, agriculture, fisheries and aquaculture programmes (iv) Frequent occurrences of cyclones and storms. The cyclones, storms and depressions are more common along the east coast of India than in the west coast. Storm surges are generated by the winds and the atmospheric changes associated with the cyclones resulting in severe erosion of beaches and changes in the beach cycles. The tropical cyclones and surges are more common in the low latitude land locked locations in the Bay of Bengal. (v) The increase in the sea surface temperatures due to climate change could lead to increase in the frequencies and intensities of cyclones and storm surges along the Indian coasts and could bring in greater disasters in future such as loss of life and property, loss of cultivable land due to inundation of saline water, saline water intrusion and contamination of fresh water aquifers underground in the deltaic and outfall areas (vi) Exploration and exploitation of oil and gas resources. The developmental activities in the onshore and offshore regions are multiplied several folds in view of high potential oil and gas fields viz., Bombay High off shore region, Krishna-Godavari onshore and offshore regions, Cauvery delta and Mahanadi delta offshore regions etc. These activities have resulted in the oil spill contamination and waste disposals etc. (vii) Land subsidence and sea level rise: The deep seated basin structures with faulted boundaries and their reactivations could bring in land subsidence in certain regions of the coast. These events could further trigger the sea level rise and erosional impacts along the coasts. Several parts of west and east coasts of late have become targets of heavy erosion. Heavy erosional impacts are seen along the coastal regions of Kerala (480 kms), Maharashtra (263 kms), Karnataka (250 kms), Orissa (108 kms), Tamil Nadu (36 kms), West Bengal (49 kms), Gujarat (36 kms), and Andhra Pradesh (9 kms), (Sanil Kumar, V., et al 2006).

1.4.3 Coastal Zone Management/Integrated Coastal Zone Management (CZM/ICZM), National Initiatives

CZM'S and ICZM'S are two different spheres of activities dealing with coastal management systems but at the same time they are synonymous to each other. CRZ'S deal with bureaucratic norms regulations to protect the coastal resources and ecosystems where as ICZM'S deal with scientific, socioeconomic studies, capacity building, systems analysis and monitoring. Infact an intensive ICZM activity should precede implementation of CRZ'S to have a better regulatory measures. The Indian coast with a length of about 7500 kms contains valuable natural resources in terms of minerals, fisheries, soil nutrients, diversified fauna and flora and has a unique landscape attracting tourism and harbour buildings etc. In order to protect these vast resources from the onslaughts of anthropogenic and natural hazard impacts the government of India issued a notification called *"Coastal Regulation Zone act (CRZ)"* in the year 1991 under the environmental protection act 1986. The notification declares the coastal stretches of seas, bays estuaries, creeks, rivers and backwaters which are under the influence of tidal action (landward side) up to 500 meters from the High Tide Line (HTL) and the land between Low Tide Line (LTL) as *"Coastal Protection Zone (CPZ)"* . The notification further says that the coastal zones are being classified as Category I (CRZ-I),Category II (CRZ-II), Category III (CRZ-III) and Category IV (CRZ-IV) and the norms prescribed in respect of this zones are outlined below.

CRZ-I

- Ecologically sensitive areas(national/marine parks, sanctuaries, reserve forests, wild life habitats, mangroves, coral reefs, areas close to breeding and spawning grounds of fish and other marine life, areas of outstanding natural beauty/historical/heritage, areas rich in genetic diversity.
- Those falling between HTL and LTL.
- Those areas likely to be inundated due to sea level rise due to global warming and such other areas as may be declared by the concerned authority (Central/State/UT).

CRZ-II

- The areas that have already been developed up to or close to the shoreline as *'Developed area'* which falls within the limits of municipalities or in other legally designated urban areas which are already substantially developed with provisions of drainage and approach roads and other infrastructure facilities such as water supply systems and sewerage lines.

CRZ-III

- The areas that are relatively undisturbed which do not belong to either CRZ I & II. These include coastal zone in the rural areas (developed and undeveloped) and also areas within municipal limits or in other legally designated urban areas which are not substantially developed.

CRZ-IV

- Coastal stretches in the Andaman & Nicobar Islands, Lakshadweep and small islands except those designated as CRZ-I, CRZ-II, and CRZ-III.

Norms for Development and construction activities in CRZs

- The notification specifies norms and imposes certain restrictions for construction activities in different categories of CRZ areas such as covered area limits, heights and number of floors in a building etc. Specific guidelines have also been issued for development of hotels and beach resorts etc. in CRZ-III and restrict installation of effluent treatment plants and sand mining from the dune belts of CRZ areas.

Regulatory activities

- Clearances can be given for activities within CRZ areas if it requires water front and foreshore facilities subject to the clearance from the Ministry of environment and forests, Govt. of India. These include ports, harbours, light houses, spill ways, storage places for liquefied natural gas and the re-gasification facilities for transportation of raw materials for thermal plants, intake of cooling water and discharge of waste water.
- Buildings are permitted in CRZ-II areas subject to the provisions mentioned in the notification and existing local town and planning regulations including the existing norms of Floor Space Index/Floor Area Ratio.
- The area up to 200 meters from HTL in CRZ-III area is to be treated as "*No Development Zone*". However developments related to agriculture, gardens, pastures, parks, play grounds, forestry and salt manufacturing units from sea are permitted with the approval of Ministry.
- Construction of dwelling limits between 200-500 meters of HTL is permitted as they are within the ambit of traditional and customary usages of fishing villages, with certain restrictions as

notified. The total number of dwelling limits shall not be more than twice the number of existing units, covered area and shall not exceed 33% of the plot size, height shall be restricted to 9 meters and construction shall not be more than 2 floors including ground floor.

- Special provisions have been made for offshore islands such as Andaman, Nicobar and Lakshadweep islands. No new construction of buildings shall be permitted within 200 meters of HTL in Andaman-Nicobar islands. In the case of Lakshadweep, permission to construct building shall be decided based on the size of island and the distance from HTL. However in both the cases the rise of buildings are restricted to two floors only. Corals and sand from the beaches cannot be used for constructional purposes.

Prohibited Activities

- The notification prohibits setting up of new industries, expansion of existing areas and untreated wastes and effluents from coastal cities/towns and industries.
- It prohibits manufacture, storage or disposal of hazardous substances excepting transfer from ships to ports and vice-versa.
- It prohibits dumping of fly ash or any other waste from thermal power stations, mining of sands and other geological material except 'rare earth minerals' not available out side the CRZ areas. It also restricts groundwater withdrawal within 200 meters of HTL, and attempts to alter morphological configuration of dune sands and related coastal features, and setting up fish processing limits including warehousing and excluding hatchery and natural fish drying. Special provisions have been made in the notification for protection of ecologically sensitive areas such as national parks, marine parks etc.

Coastal Zone Management plans (CZM)

- As per the provisions contained in CRZ notification 1996, 13 coastal states and union territories are required to prepare Coastal Zone Management Plans (CZMPs) for identifying and classifying the CRZ-I within their respective territories as per the guidelines/notification and submit to the Government within one year stipulated time from the date of issue of the notification.
- In addition national level and state level coastal zone authorities are to be constituted as per the directives of Apex court dated November 26th 1998, delegating powers to adopt required strategies for the protection of coastal environments. The national level authority will provide all the required guidance, review modifications if any from the respective state/union territory governments and make necessary amendments whereever necessary and formulate specific management plans. (please look for details in the notification on Coastal Regulation Zone (CRZ) 1991 and subsequent amendments; Ministry of Environment and Forests; Govt. of India).

New CRZ notification 2011

In view of the several representations, complaints, modifications etc required to be made in all CRZs the Govt. of India issued a new CRZ notifications dated 6th January 2011, the details of which are published in the Gazette of India extraordinary, Part-II, Section 3, Sub-section (ii). The CRZ notification of 2011 uses essentially the 1991 notification as its base and codifies certain modifications in all CRZs. Similar to the 1991 notification the 2011 notification also classifies the 0-500 m coastal strip into four categories: CRZ-I (Ecologically sensitive areas); CRZ-II (Built up areas); CRZ-III (Rural areas) and CRZ-IV (Territorial waters and tidal influenced water bodies). In addition a separate notification for the protection of islands has also been issued. It also codifies 25 amendments to the earlier notification to ensure livelihood, security to the fishing communities and other local communities living in the coastal areas and also to conserve and protect the coastal stretches and promote development in a sustainable manner. The CRZ also includes, for the first time, the water and bed areas between the low tide line (LTL) to the territorial water limit (12 nautical miles) in case of the sea, as well as the water and the bed area of tidal influenced water bodies such as creeks, rivers and estuaries. Some other important features of this new notification are as follows:

- It introduces a concept of a "hazard line" to be demarcated taking into account tides, wave, sea level rise and shoreline changes and can expand wherever the hazard line falls beyond the 500 m line. Demarcation of such a hazard line will help to protect life and property of coastal communities and infrastructure developed.
- The notification also introduces two categories of "areas requiring special consideration"
  - (i) Special dispensations are provided for greater Mumbai, Goa and the backwater islands of Kerala.
  - (ii) It provides for notifying critical vulnerable coastal areas (CVCA) including the sunderbans mangroves and other ecologically important areas, through a process of consultation with local communities.
- The notification also takes into cognizance, the problem of erosion due to human interventions and proposes to classify the coastline into "high erosion", "medium erosion" and "low or stable stretches". Ports and harbour projects excepting the projects related to strategic and defense purposes will not be permissible in the high erosion areas while comprehensive environmental impact assessment (EIA) reports with cumulative studies will be required for projects in stretches classified as low and medium erosions.
- As reflected in the 1991 notification, the state governments/union territories are required to develop coastal zone management authorities which are given the primary responsibility for monitoring and enforcement. Further to assist in these tasks the respective state governments/UTs are required to constitute district level committees under the chairmanship of the district magistrate with at least three representatives of local traditional coastal communities.
- Time bound procedures for obtaining clearances of projects have been specified. Project authorities are required to submit half yearly compliance reports which are to be made available to the public and also display in the websites. (For details one can go through the website relating to CRZ notification dated 6th January 2011, and also publication in the Gazette of India extraordinary, Part-II, section 3, sub section(ii)).

Integrated coastal zone management

- The Integrated coastal zone management is a dynamic process which coordinates the activities related to environmental, social, cultural and insituational resources for sustainable development and solutions. The need to develop ICZMs for different countries/states have been emphasized by IPPC, 1990 and WCC, 1993 which have been discussed in the earlier paragraphs. These ICZM programs however have not been developed in the country in a systematic way in fact the ICZMs should have been systematized in the country simultaneously when the CRZMs were intitiated, which would have facilitated development of appropriate expert systems and insituational frames to deal with the activities related to CRZs in the country.

Government of India of late have cleared a major project on the Integrated coastal zone management in an effort to study the 7500 kms coast line. The objective of the Integrated Coastal Zone Management (ICZM) project is to assist Government of India (GOI) towards development and implementation of comprehensive coastal zone management approach in the country, and to pilot an integrated coastal zone management approaches in states of Gujarat, Orissa and West Bengal. There are four components involved in the project. The first component is National ICZM capacity building. This component includes mapping, delineation and demarcation of the hazard lines and delineation of coastal sediment cells all along the mainland coast of India. The second component involves piloting an ICZM approach in Gujarat. This component will support capacity building at the state level agencies and institutions, including preparation of an ICZM plan for the coastal sediment cell that includes the Gulf of Kutch and pilot investments. The third component involves piloting ICZM approach in Orissa. This component will include capacity building at the state level agencies and institutions, including preparation of an ICZM plan for the coastal sediment cells that includes the stretches of Paradip-Dhamra and Gopalpur-Chilika. The fourth component involves development of an ICZM approach in West Bengal on similar lines as that of Gujarat and Orissa. The ICZM programmes in other states are expected to be taken up in the second phase subsequently.

1.5 COASTAL SYSTEMS (RESOURCES)

The Indian coastal regions contain productive ecosystems such as mangroves, fisheries and aquaculture centers, rich soil systems, huge net work of canal systems with good irrigation potentials and agricultural practices, the surface and groundwater resources and several mineral deposits, oil and gas fields and lignite coal fields.

1.5.1 Water resources

Coastal systems constitute important resource belts for agriculture, irrigation, fisheries and are rich with biodiversities and nutrient soil profiles. Fresh water (surface and groundwater) plays a very important role in protecting these resources. Fourteen major rivers, forty four medium rivers and about one hundred and fifty odd small and minor rivulets join the sea on either sides of the peninsular coast, discharging fresh water of the order of 15,78,244 MCM into the sea. All the major rivers and a few medium rivers have built up good delta systems at present and also in the geological past, and are responsible for building up multiple water bearing strata in different sedimentary environmental systems along the coast. The multiple aquifer systems which were deposited under fluviatile and marine environments hold rich potentials of fresh and brackish groundwater resources which act as vital resource belts for agriculture, irrigation and several other industrial and drinking water needs and also for sustenance of flora and fauna in different coastal environments and echo systems. The deltas and sometimes estuaries are the store houses for these rich water belts and serve as potential irrigation belts for agriculture with intense canal network using surface water and high density development of dug wells, dug cum bore wells and shallow and deep tube wells. All the delta systems along the east coast have relatively high thicknesses of alluvial belts with thicknesses varying from 0-60 meters and sometimes even more. These alluvial systems constitute aquifers with rich permeabilities and high yields, where thousands of filter points, shallow tubewells are in operation supporting huge irrigation potentials. Similarly the sedimentries which are deposited in Tertiary/Mesozoic periods also serve as huge repositories for fresh groundwater systems and support irrigation potentials through shallow and deep tube wells. The Tertiary aquifers with fresh water are most prevalent in the coastal regions of Tamil Nadu, Andhra Pradesh, Orissa, West Bengal and Gujarat while some of these aquifer systems have become either brackish or saline. The irrigation potentials and availabilities of fresh waters for the coastal districts of India are briefly presented in Table 1.7 and are graphically shown in Fig. 1.10.

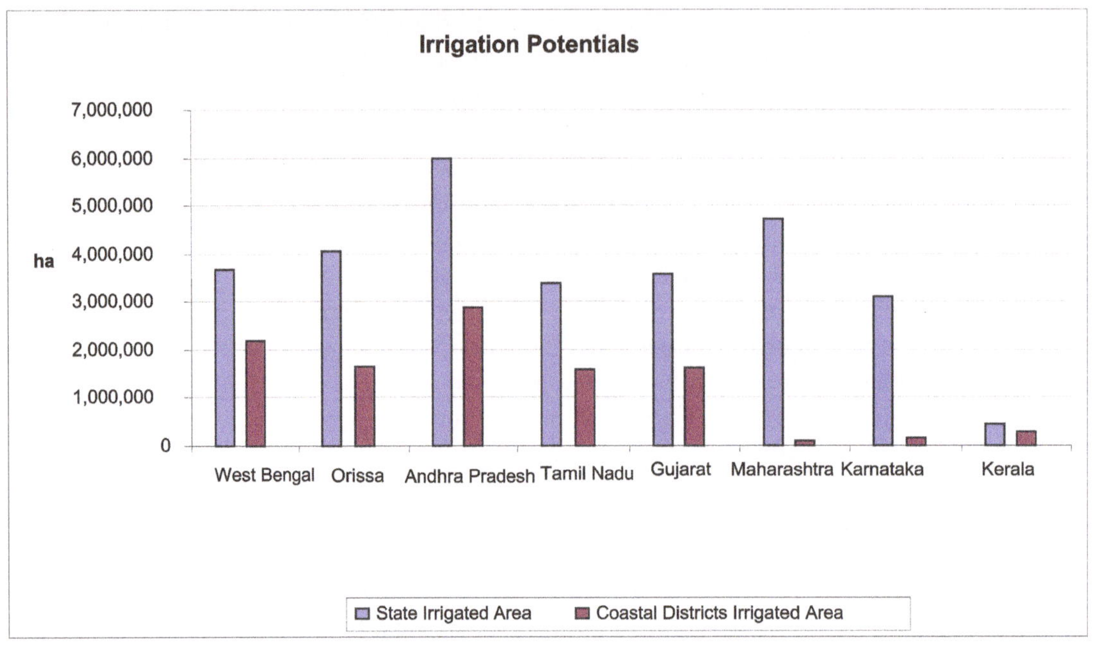

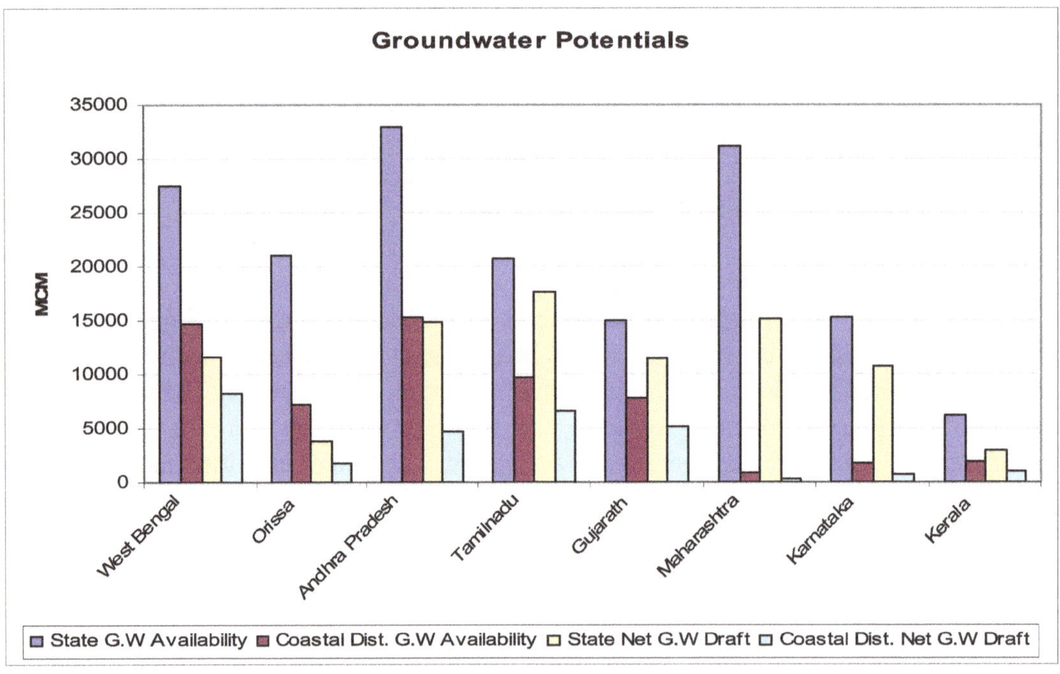

Fig. 1.10 Irrigation/groundwater potentials in coastal states/districts

Table 1.7 A Comparative Analysis of Irrigation potentials/Groundwater availabilities/draft situation in the coastal districts

S.No	State	Coastal districts	Total Irrigated Area in the state (ha)	Total Irrigated area in the coastal districts (ha)	%	Total groundwater resource availability (MCM)	Total groundwater resource availability in the coastal districts (MCM)	%	Total Draft (MCM)	Total draft in the coastal districts (MCM)	%
1	West Bengal	Midnapore, North 24 Paraganas, South 24 Paraganas, Howrah, Hugli Murshidabad, Nadia, Bhardman	3,661,000	2,188,718	59.8	27458.13	14725	53.6	11647.38	8254	70.9
2	Orissa	Bhadrak, Baleswar, Cuttuck, Ganjam, Jagatsinghpur, Khorda, Kendrapara, Puri	4040022	1644811	40.7	21009.28	7206	34.3	3847.67	1715	44.6
3	Andhra Pradesh	East Godavari, West Godavari, Krishna, Guntur, Nellore, Prakasam, Srikakulam, Vishakapatnam, Vizayanagaram	5996466	2883790	48.1	32946.36	15284.64	46.4	14901.33	4768.27	32.0
4	Tamilnadu	Cuddalore, Kanchipuram, Kanyakumari, Nagapattinam, Pudukkottai, Ramanathapuram, Sivaganga, Tirunelveli, Thanjavur, Thoothukudi, Thiruvallur, Villupuram,	3396700	1584437	46.6	20763.38	9711	46.8	17653.4	6683	37.9

Table 1.7 ***Contd...***

S.No	State	Coastal districts	Total Irrigated Area in the state (ha)	Total Irrigated area in the coastal districts (ha)	%	Total groundwater resource availability (MCM)	Total groundwater resource availability in the coastal districts (MCM)	%	Total Draft (MCM)	Total draft in the coastal districts (MCM)	%
5	Gujarat	Ahmedabad, Amreli, Bharuch, Jamnagar, Junagadh, Kutch, Rajkot, Surat, Vadodara, Valsad	3572000	1605104	44.9	15020.39	7811	52.0	11485.72	5161	44.9
6	Maharashtra	Mumbai, Rajgad, Ratnagiri, Sindhudurg, Thane	4705000	100985	2.1	31214.04	847	2.7	15085.13	273	1.8
7	Karnataka	Dakshina Kannada, Udipi, Uttara Kannada	3089000	155395	5.0	15296.6	1736	11.3	10713.12	687	6.4
8	Kerala	Alappuzha, Ernakulam, Kollam, Kannur, Kasaragod, Kozhikode, Malappuram, Thirussur, Thiruvananthapuram	432000	279866	64.8	6229.55	1871	30.0	2920.01	956	32.7

1.5.2 Soil resources

Soils are thin top layers of earth's crust in different basins/water sheds and contain mixtures of minute particles, minerals, organic matter and bacteria. These are basically generated through weathering processes of provenance rocks and could be seen as *in situ* layers and sometimes transported and deposited in the vicinities of valleys, flood plains and related depositional areas, with varied thicknesses ranging from few centimeters to several tens of meters over a period of time depending upon the weathering and sedimentation processes involved. Soil profiles and platforms are essential base strata for good growth of plantations, cultivation and agricultural practices etc. The compositional soil characters depend upon its constituents like organic matter, soil organisms & bacteria, soil porosity and moisture levels and inorganic matter and soil pH and fertility. Soils could be broadly grouped into two categories namely acidic and alkaline/saline. The Acidic soils have pH in their aqueous solutions less than 7, open textures and massive structures with low Ca and Mg constituents and insignificant soluble salts. The alkaline soils are formed under high pH with significant amounts of sodium in combination with carbonates and chlorides. There are six major groups of soils in the coastal regions of India (Pofali, R.M., et al 1991).

Red soils: Red colour is due to various oxides of iron and are poor in N, P, K and with pH values ranging between 7 to 7.5. They are light in texture with good porosity and moisture levels. They occur widely in the coastal regions of Kerala, Maharashtra and certain inter deltaic regions of Tamilnadu, Andhra Pradesh, Orissa and Bengal coasts. These soils have been classified as Alfisols.

Lateritic soils: Dark brick red in colour with pH values ranging between 5 and 6. These soils form in the hilly regions and are subjected to high rainfall and weathering conditions. Eastern ghats and Western ghats contain this type soils at many places. These are classified as Ultisols and Oxisols.

Alluvial soils: These soils are sandy loam to clay loam with light grey to dark grey colour loose in structure and are good fertile soils. The pH values vary between 7 to 8. These soils are widely distributed over coastal delta plains of India and act as rich agriculture belts. These alluvial soils are classified as 'Entisols, Inceptisols and Alfisols'.

Black cotton Soilss: These are dark grey to black in colour with high clay content with pH values between 7 to 9. They have poor drainage and high base exchange conditions and contain Ca and K dominantly. They occur widely in the deltaic tracks of east coast and river mouth areas of Maharastra and Gujarat in the west coast. These are classified as 'Entisols, Inceptisols and Vertisols.

Peaty and Marshy Soils: These are well developed under humid conditions with high accumulation of organic matter. The soils get submerged in water during monsoon times and get exposed to atmosphere during non-monsoon times. These soils are black in colour with high acidic conditions and with pH values ranging from 3-3.5. These soils are commonly seen in the river mouth fringe areas of both east and west coasts.

Saline-Sodic Soils: These soils are dark grey in colour, high pH values and contain excess natural salts dominated by chlorides and sulphates and affects plant growth. These occur widely in saline affected regions of east and west coasts of India. These are classified as Aridsols, Entisols and Vertisols.

1.5.3 Mineral resources

The Indian coastal regions are endowed with rich placer mineral deposits, oil and gas fields both in the onshore and offshore delta frames, Tertiary lignite beds, placer deposits like ilmenite, rutile, zircon, monazite, garnet, sillimanite, glass sands and patches of gold and diamonds are reported to occur along the east and west coasts of India (Victor Rajamanickam, G., 2000; Subrahmanyam, A.V., et al., 2005). These placers basically are derived from the provenance rocks like granites, gneisses and schists of east and west coasts of India, transported by rivers/streams etc. and are subjected to wave actions and currents of the sea coast and get deposited along the coastal stretches, such as continental shelf regions, beach and dune sands, and along the areas adjoining the lagoons and estuaries. Some of the prominent and abundant placer deposits along the Indian coasts are briefed here. Ilmenite ($FeTiO_3$), Rutile (TiO_2), Zircon ($Zrsio_4$), Garnet and Monazite are the common heavy minerals distributed widely along the coasts of Kerala, Tamilnadu and Andhra and Orissa. The estimated deposits of these minerals in the country as a whole are: Ilmenite (278 million tons), Rutile (18 million tons), Zircon (18 million tons), Monazite (7 million tons) and Garnet (80 million tons) and Sillimanite (84 million tons) (Singh, A.P. and Gupta, P.K., 2003). In addition large quantities of glass sands and scattered quantities of gold coastal placers are reported from the coastal regions. The coastal placers being dynamic, the exact estimates of these deposits are subjective in nature but presumably about 30 to 40% of these overall estimates could come from placers of east and west coasts of India. The importance of these minerals are on the rise world over in view of their strategic importance in various industries like refractories, nuclear power plants, and gemstones.

The coastal deltas, on shore and offshore areas are the sites for potential occurrences of petroleum and gas fields. The Bombay high offshore region and the Krishna-Godavari onshore and offshore areas contain huge deposits of oil and gas. According to the Ministry of Petroleum and Natural gas, the estimated resources of the hydrocarbon potential (crude oil and gas) in the Bombay high area of west coast of India are of the order of 731.34 million tons. Another potential area along the east coast of India is Krishna Godavari basin. The estimated hydro carbon reserves in the K-G basin are of the order of 726 million tons. This shows both Bombay high structure on the continental shelf region of west coast and the onshore and offshore areas of Krishna–Godavari basin by and large contain rich potentials of hydrocarbons. The oil and gas commission have drilled several wells in the other coastal offshore areas along the east coast of India, but are of little importance when compared to overall production of oil and gas in the country.

The Lignite (Tertiaries) coal is known to occur in several coastal regions, but the most potential lignite field is Neyveli Lignite field in South Arcot region of Tamil Nadu. The estimated resources are of the order of 3840.36 million tons. These rich lignite belts occur as in the bedded formations in Tertiaries, spread over an area of 260 km^2 with thicknesses varying between 3 to 24 meters. In addition occurrences of tertiary lignites are reported with thin layers in coastal regions of Kerala, Orissa, and Gujarat but are not of much economic significance.

CHAPTER 2

GEOMORPHOLOGY, EVOLUTION AND STRUCTURE OF COASTAL DELTA BASINS

Geomorphology and associated land forms are of fundamental importance to the water resources and economic development of any region/country in view of their importance and significance related to agriculture, irrigation, minerals, and hydrocarbons. The Indian coastal delta systems exhibit diversified land forms all along its stretch of 7500 kms including the offshore islands. They include long stretches of beach sands, dunes, river mouth bars, barriers, spits, tidal flats, marshes, mangroves, back waters, lagoons, estuaries and deltas. A preliminary description of all these coastal features is incorporated in the earlier chapter. The dominant morphological units however are deltas from the water resources and irrigation point of view. These delta systems often are described as '*rice bowls of India*' and support rich agricultural practices and irrigation potentials.

The east coast of India prominently exhibits big net work of delta systems from south to north while the west coast is characterized by moderate build up of estuaries all along its stretch excepting in the north western stretches where Narmada and Tapti make their appearances as small estuary deltas. The other geomorphic features like elongated beaches, bars, barriers, spits, lagoons and marshy areas etc., are common to both east and west coasts of India by and large with some minor variations in the shoreline characteristics (Ahmad, E., 1972). The disposition and evolution of all these geomorphic units are manifestations of structural, tectonic, eustatic changes of sea levels and sedimentalogical processes during different stages of evolutionary history of coastal margins, transgression and regression phases of oceans and on land fluviatile and shore line processes along the coasts.

2.1 EVOLUTION OF INDIAN COASTAL MARGINS

The east and west coasts of India exhibit remarkably different geomorphic and evolutionary trends. There have been several studies carried out related to tectonic history and evolution of continental margins of India. (Krishnan, M.S.,1953; Balakrishnan, T.S., 1997; Chandrasekharam, D., 1985; Mahadevan T.M., 1994; Biswas, S.K., 1987, 1988; Subrahmanya, K.R., 2001; Sastri, V.V., et al 1973; Balasubramanyam, M.N., 2006; Vaidyanathan, R. and Ghosh, R.N., 1993; Prabhakar, K.N. and Zutshi, P.L., 1993; Maurya, D.M., et al 2000). The early stages of evolutionary history could be traced back to the break up of Gondwana land when all the southern continents namely South America, Africa, Antarctica, Madagascar, Australia and India were together with a passive northern continental margin. The break up of continents and drifting started during Triassic and early Jurassic periods. The first stage of break up of Gondwana land probably took place around 150 Ma making a distinct separation of India and Medagaskar from Africa followed by a series of tectonic episodes before India attained its present position.

2.1.1 East coast of India

The east coast of India was attached to Antarctica and Australia until early Cretaceous period. A triple junction plume probably operated between the contact points of Australia, India and Antarctica resulting in the separation of Australia and Antarctica from India around 130-90 Ma. This could have caused widespread volcanism (Rajmahal Traps) in the northern parts of Indian east coast around 118-116 Ma and exposed the entire east coast to oceanic processes. The spreading trends continued beyond 100-80 Ma, enlarging the Indian Ocean area. In between this period as the coastal zone opened up from south to north, several sedimentary basins got opened up along the older lineaments with open connections to the sea. The prevailing tectonic movements and transverse faults probably had changed the paleo slope directions of Godavari and Mahanadi rift systems and aligned in the directions of coast (Casshyap, G.M., 1977) and at the same time new basins in the north eastern region were opened up. At the end of the period and during 80-64 Ma the spreading rate had increased with rotations of India in an anti clock wise direction. This trend continued until 53 Ma by which time the east coast was evolved with a fully developed passive margin oceanic basin and the Indian plate continued to move further northward leading to the subduction and collision of Indian plate with Burmese plate. This lead to the closure of north eastern sedimentary basin and opening up of Bay of Bengal. The subduction of Indian plate continued further up to 32 Ma leading to the formation of Arakan mountain chain. The present day passive margin of east coast delta basins with localized tectonic, stratigraphic and sedimentalogical frame work could be classified as under:

(a) Ganges delta – Bhagirathi-Hoogly (West Bengal)

(b) Mahanadi – Brahmini – Baitarni (Orissa)

(c) Godavari – Krishna (Andhra Pradesh)

(d) Pennar – Palar (Andhra Pradesh & Tamil Nadu)

(e) Cauvery – Ponnaiyar – Vaigai (Tamil Nadu)

All the above major river delta basins located in the east coast are involved in major active delta building processes and transportation of huge amounts of clastic sediments into the Bay of Bengal. They show close similarities in respect of structure, stratigraphy and evolutionary trends. All these basins are pericratonic and intracratonic in nature and show the signatures of basin formation and evolutionary trends since Cretaceous times.

2.1.2 West coast of India

The separation of west coast probably took place around 100 Ma (Eldholdm, O. and Todal, A., 1996). A group of volcanics lying parallel to the west coast with linear trends in NNW – SSE direction gave age results of 93 Ma (Valsangkar, A.B., et al 1981). This probably demonstrates the partial melting and doming of the lithosphere between India and Madagascar as India continued to drift northward. The eastern margins of Madagascar had a good spread of basaltic volcanism and probably could be related to '*Marian hot spot*'. The age estimations of these basaltic rocks indicated an age of the order of 88 Ma (Storey, M., et al 1995). This probably reflects that the west coast had taken its initial shape after its break up from Madagascar around 93-88 Ma (Subramanya, K.R., 1998). A subsequent magmatic episode in eruption of Deccan basalts which occupy roughly an area of 500,000 km^2 with a maximum thickness of the order of 1.5 kms (Kaila, K.L., 1988). This is related to the *'Reunion hot spot'* activity (Morgan, W.J., 1981, Cox, K.G., 1988 and Royer, J.Y., et al 1989). The average age of Deccan volcanics is placed at 67 Ma (Subrahmanya, K.R., 1998). As a result of northern movement of India, this hot spot left its trail as marginal plateau (Laccadive – Chagos ridge) endorsing the idea related to the shift of Reunion hot spot from Indian sub-continent to offshore region of African sea. This probably happened around 40 Ma (Duncan, R.A. and Pyle, D.G., 1988). As the northward movement of Indian plate continued, the southerly movement of Reunion hotspot left a trail in the form of marginal plateau of 27 Ma (67 Ma-40 Ma). Enormous amounts of flood baslts seen in the Seychelles were considered as extensions of these Deccan basalts. The sea floor spreading along Charlesberg ridge resulted in the break up of Seychelles micro region from the Konkan and Sourashtra region of India as the northward movement of Indian plate continued. The collision of India with Asian main land has resulted in the northerly tilt of Peninsular India. The continued northward movement of India has also brought in northward tilt making the Konkan coast a submergent coast and the Malabar coast as an emergent coast and had also brought in subsidence and a series of step faults along Sourashtra coast (Kaila, K.L., 1988). At the same time as a result of western ghat development and uplift processes, the peninsular region appeared to have been tilted east thereby changing the direction and orientation of the major drainage lines/systems. The breakup of Gondwana land (150 Ma) and the subsequent dismemberment from Antarctica and Australia (130 Ma) and Madagascar (63 Ma) and its continued northward movement coupled with anticlockwise rotation and under thrusting against Asian land mass had resulted in several pericretonic basins along the east and west coasts of India. The east coast is characterized by several such basins namely Ganges, Mahanadi-Brahmini-Baitarni, Godavari-Krishna and Cauvery-Vaigai from north to south where as the west coast is characterized by only three such basins in the northwest region namely Kutch, Cambay and Narmada and several small estuaries along the Konkan and Malabar coasts.

2.2 TECTONIC ELEMENTS AND SEDIMENTATION

A brief description incorporating the nature and tectonic elements involved in each of the above basins is described below in the following paragraphs.

The Kutch, Cambay and Narmada (northwestern region); Bombay-Ratnagiri (mid western region) and Konkan region (southwestern) and their offshore regions in the west coast and Cauvery – Ponnaiyar – Vaigai (southeastern region); Palar – Pennar (southeastern region); Krishna – Godavari (mid eastern region); Mahanadi – Brahmini – Baitarni – Subarnarekha and Hoogly (northeastern region) – all along the east coast are the major pericratonic rift basins evolved around the periphery of Indian peninsula during different stages of tectonic history. All these basins by and large were evolved during Mesozoic and Cenozoic periods during different stages of continental drifting processes prior to the collision of Indian plate with Eurasian plate. The structural and tectonic frame and their evolutionary trends have been extensively discussed by a number of workers. Krishnan, M.S., 1953; Biswas, S.K., 1987; 1996; 1992; 1999; Agarwal, R.P. and Mitra, D.S., 1999; Subrahmanya, K.R., 1998; Rao, G.N., 1993; Fuloria, R.C., 1993, Sastri, V.V., et al 1973, 1981; Prabhakar, K.N., and Zutshi, P.L., 1993; Ganguly, S., 1997; Maurya, D.M., et al 2000; and several others. A regional tectonic map of India is presented in Fig. 2.1.

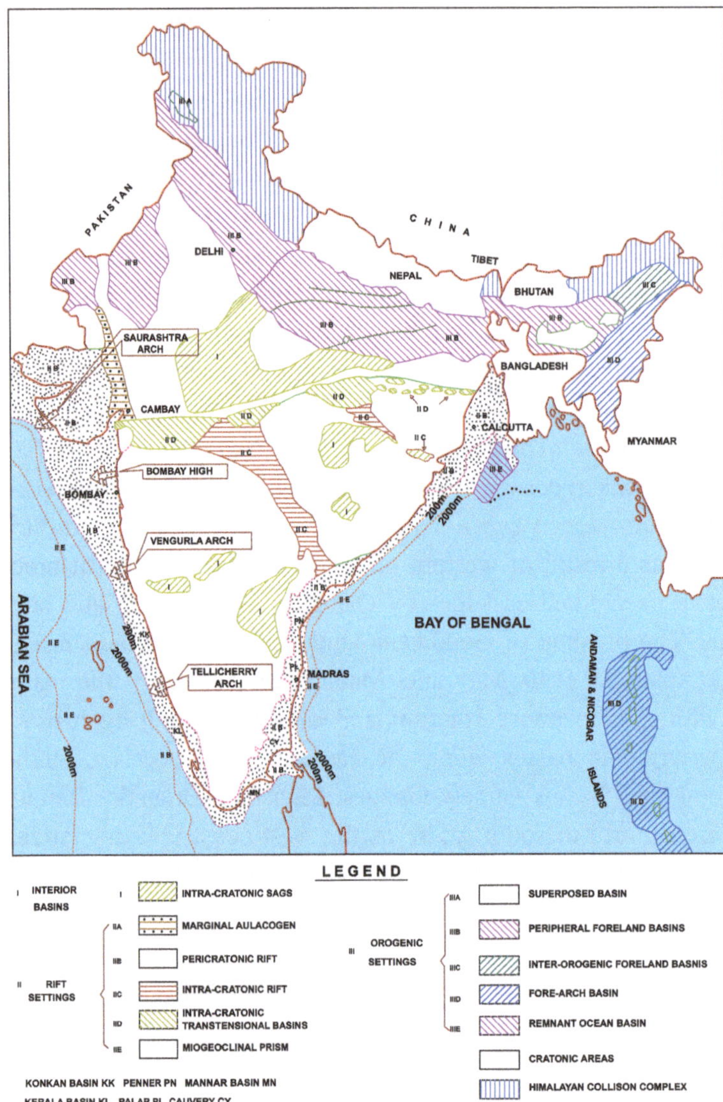

Fig. 2.1 Regional tectonic sedimentary basin map of India

(Source: Biswas, S.K. and Agarwal, R.P., 1999)

2.2.1 West coast of India

The Western continental margin which is a passive margin has three orogenic trends. (i) NNW-SSE Dharwar trend, (ii) NE-SW Aravalli trend and (iii) ENE-WSW Satpura trend. The Dharwar trend is parallel to the faulted west coast of India along with a series of extension faults parallel to the coast which probably is responsible for a wide continental shelf region. Its northward extension into the western part of the Indian shield across the Narmada rift gave rise to Cambay graben. The Narmada-Son lineament along the ENE-WSW Satpura trend is a major tectonic boundary dividing the Indian shield into southern block and northern foreland block. The Aravalli with NE-SW trend continues across Cambay graben and swings in a south westerly direction into the continental shelf region dividing it into northern Kutch-Sourashtra shelf region in the north and southern Bombay-Kerala shelf region.

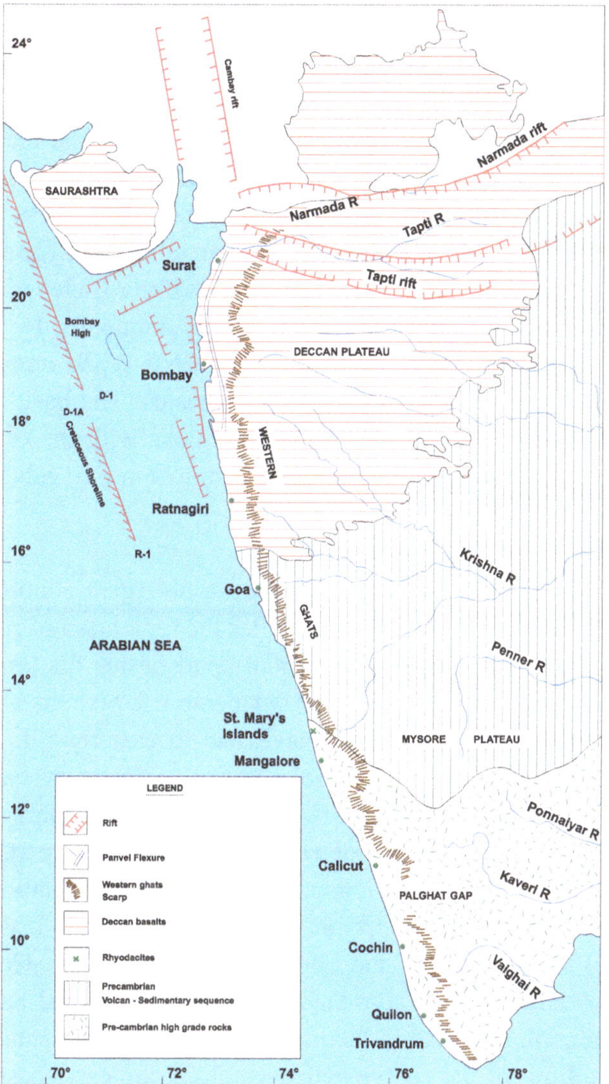

Fig. 2.2 West coast continental margin with tectonic elements

(Source: Subramanyam, K.R., 1998)

Kutch-Cambay-Narmada basin region

The Kutch-Cambay-Narmada basins are the major rift basins in the northwestern coastal margin and are bounded by intersecting precambrian trend faults. The evolution of these three basins took place during different stages of Indian plate drift northward after its break up from Gondwana land, Kutch rifting took place in the late Triassic or early Jurassic period, Cambay rifting in the early Cretaceous and Narmada rifting in late Cretaceous (Biswas, S.K., 1987). The Kutch basin essentially is a Mesozoic basin with stratigraphic sequence from mid Jurassic to Holocene with thicknesses running up to 300 m. The Cambay basin largely represents Tertiary deposits with thicknesses of the order of 5000 m and is considered as one of the largest petroliferous basins of India. Narmada, the southern most of the three basins is confined to Cretaceous-Tertiary-Quaternary sediments only.

Bombay-Ratnagiri region

The Bombay-Ratnagiri offshore sedimentary track again is an example of peri-cretonic rift basin. This basin represents the southern extension of Cambay basin into offshore Arabian sea and believed to be the largest continental terrace in the world (Ganguly, S., 1997). The entire basin is segmented into several horst-graben features closely controlled by NW-SE to N-S treding faults. The sediment deposits essentially belong to Tertiary stage running more than 5000 m thickness resting unconformably over Deccan traps. The coastal area south of Ratnagiri (Konkan coast) and Kerala coastal region are mainly evolved in the Mesozoic-Cenozoic periods and during the rifting stages of Indian plate around 82 Ma. These basins occur as pericontinental shallow embayments dominated by NW-SE trending horsts and grabens, segmented by ENE-WSW strike slip faults. The ENE-WSW transverse strike slip faults form transverse zones segmenting the basins into several sub-basins. The basin fills are mostly clastics and carbonate sediments both in the onshore and offshore regions with thicknesses varying from 60 m to 100 m. A regional tectonic map of west coast of India is shown in Fig. 2.2 (Subrahmanya, K.R., 1998).

2.2.2 East coast of India

The sedimentary basins along the east coast of India are divergent continental passive margin basins evolved during different phases of continental drift after the separation of Antarctica and Australia from Indian east coast. These basins are intra-cratonic pull apart basins till Jurassic and then after became pericratanic. The sedimentation patterns in these Paleozoic rift grabens are fluvial and continental in nature. The Indian plate during its movement gave rise to easterly tilt, resulting in south easternly drainage patterns along the east coast with several major sedimentary basins like (i) Cauvery-Palar-Ponnaiyar Vaigai (Tamil Nadu) (ii) Godavari-Krishna-Pennar (Andhra Pradesh) (iii) Mahanadi – Brahmini-Baitarni-Subarnarekha (Orissa) and (iv) Bhagiradhi-Hoogly (Ganges, West Bengal). The easternly tilt also caused marine transgressions and extensive sedimentation during late Jurassic, early Cretaceous and Eocene periods associated with wide spread volcanic activities. The oldest flows are Rajmahals (late Jurassic) in West Bengal and Eocene basalts in Krishna-Godavari basin area. A middle Eocene shelf edge is demarcated from Mymensingh in the NE Bengal basin to Cauvery basin in the south. The oceanic crust cooling probably facilitated maximum sedimentation since Miocene times in these basins. A probable limit of the continental crust along the east coast coupled with Eocene shelf edge and other tectonic elements is shown in Fig.2.3 (Rao, G.N., 1993).

The structural trends of these east coast basins from south to north are as follows: (i) NE-SW trend between Kanyakumari and Nagapattinam (ii) N-S trend between Nagapattinam and Ongole (iii) NE-SW trend between Ongole and Puri (iv) NNE-SSW trend between Puri and Hoogly and (v) E-W trend in Bhagiradhi-Hoogly delta region. The trends of these east coast basins are mainly related to the Dharwar trends in south east, the Eastern ghat trends in the middle east and the Singhbum trends in the north east. In addition the three rift valleys namely '*Godavari rift*', '*Mahanadi rift*' and '*Damodar rift*' initially are intracratonic and subsequently became pericratonic as a consequence of drifting processes as discussed earlier. The Godavari rift valley trends in NW-SE cutting across Dharwars and Eastern ghats joins the Narmada lineament in the North West region of Indian Peninsula. Similarly the Mahanadi rift also trends in NW-SE and cuts across Eastern ghats and singhbhum group of rocks and the Damodar rift trends in NNW-SSE direction cutting across the gneissic complexes. A brief tectonic, structural and stratigraphic evolution of each of the major coastal basins is described below.

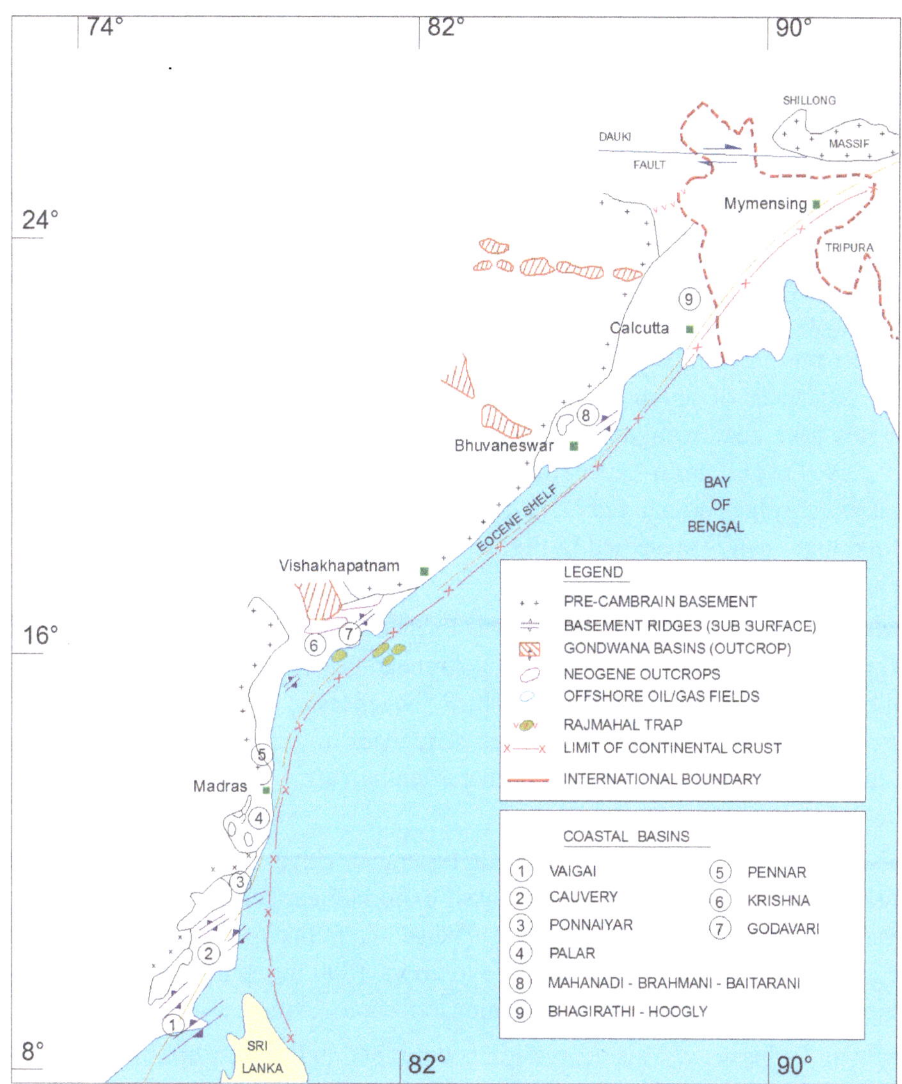

Fig. 2.3 East coast continental margin with tectonic elements

(Source: Rao, G.N., 1993)

2.3 TECTONIC, STRUCTURAL AND STRATIGRAPHIC EVOLUTION OF MAJOR COASTAL DELTA BASINS

2.3.1 Cauvery–Ponnaiyar–Vaigai coastal delta basins (Tamil Nadu)

The Cauvery and its adjoining basins with an aereal spread of about 2500 km^2 extends from Pondicherry in the north up to Gulf of Mannar in the south. The western region of these basins is characterized by Archean basement, Cretaceous and Tertiaries and Quaternary sediments in the down dip side and extend towards east into the Bay of Bengal. This is a block faulted pericratonic basin with highly differentiated basement characterized by typical horst-graben structures. The Cauvery and its adjacent basin regions appear to have been evolved in four stages namely (i) The rift graben stage, (ii) The cratonic basin stage, (iii) The shelf platform stage and (iv) The deltaic progradation stage

(Babu, P.V.L.P., 1991). The oldest statigraphic sequence of rocks (late Jurassic to early Cretaceous) is indicated by continental facies of Gondwana period. These sediments are believed to have been deposited in the structural troughs along the paleo basin margins. These are mostly represented by fluvial to deltaic coarse sands and clays. The sediments of second stage appear to have been deposited during the transgressional and regressional phases of paleo oceans. These are represented by fluvial deltaic marine muds and sands associated with rich fossiliferrous limestones/marls (late Cretaceous). The third stage appears to have been represented by carbonates, shales, sands, silts and turbidites (Paleocene to Neogene). The fourth stage appears to have been deposited under fluvial deltaic and lagoonal environments with sands and clays (Quaternary- Pleistocene to Holocene). The Cauvery delta region appears to have been evolved over the geological time scale from late Cretaceous to Recent times with thick sequences of sediments of the order of 6 to 8 kms. At least four tectonic stages and evolutionary trends have been recognized with a series of geographically changing deltaic boundaries (Babu, P.L.N.P., 1991). The basin can be divided into a number of sub-basins, separated from one another with basement ridges oriented in NNE-SSW direction (Sastry, V.V., et al 1981). The basement structural configuration map is presented in Fig. 2.4. A brief description of these structural depressions is presented below:

(i) ***Ariyalur-Pondicherry depression:*** This is located in the northern part of the basin with a NE-SW linear trend delineated in the west by Archean granites and gneisses and is bounded in the south by Kumbakonam-Shiyali ridge. Thick sequences of Cretaceous sediments out crop near the western fringe of the basin and thick sequences of Tertiary sediments are encountered in the central part of the basin with basement depths varying from 5000 to 5500 m and extend several kilometers offshore.

(ii) ***Kumbakonam-Shiyali ridge:*** This is a basement ridge and appears to be a subsurface extension of precambrian crystalline massif exposed near south west of Ariyalur-Pondicherry depression, aligned in NE-SW direction. The sediment thicknesses in south western part of the ridge are low but progressively increase to about 1800 m towards north east. Majority of the sediments belong to Tertiary age with thin Cretaceous sequences in the north eastern region.

(iii) ***Thanjavur Depression:*** This depression is delimited in the west by Precambrian shield and in the east by Devakottai-Mannargudi ridge. The maximum depth of the basement ranges from 3000 to 4000 m with thick sequences of Cretaceous sediments and with thin sequences of Tertiary sediments.

(iv) ***Tranquebar depression:*** This depression is located south of Kumbakonam-Shiyali ridge delimited in the south by Karaikal high and extends offshore in the east. The estimated maximum basement ranges from 3000 to 4000 m with sediment sequences belonging to Cretaceous and Tertiaries.

(v) ***Karaikal High:*** This is aligned in ENE-WSW direction close to the coast with basement depths ranging from 1700 m to 2400 m. Both Cretaceous and Tertiary sediments occur on this high.

(vi) ***Tiruthuvaipundi–Nagapattinam depression:*** This depression is aligned in NE-SW direction delimited in the west by Devakottai-Mannargudi ridge in the north by Karaikal high and in the south east by Thirupundi-Vedaranniyam high. The basement depths in this region vary from 2000 to 4500 m with thick Cretaceous and Tertiary sequence of sediments.

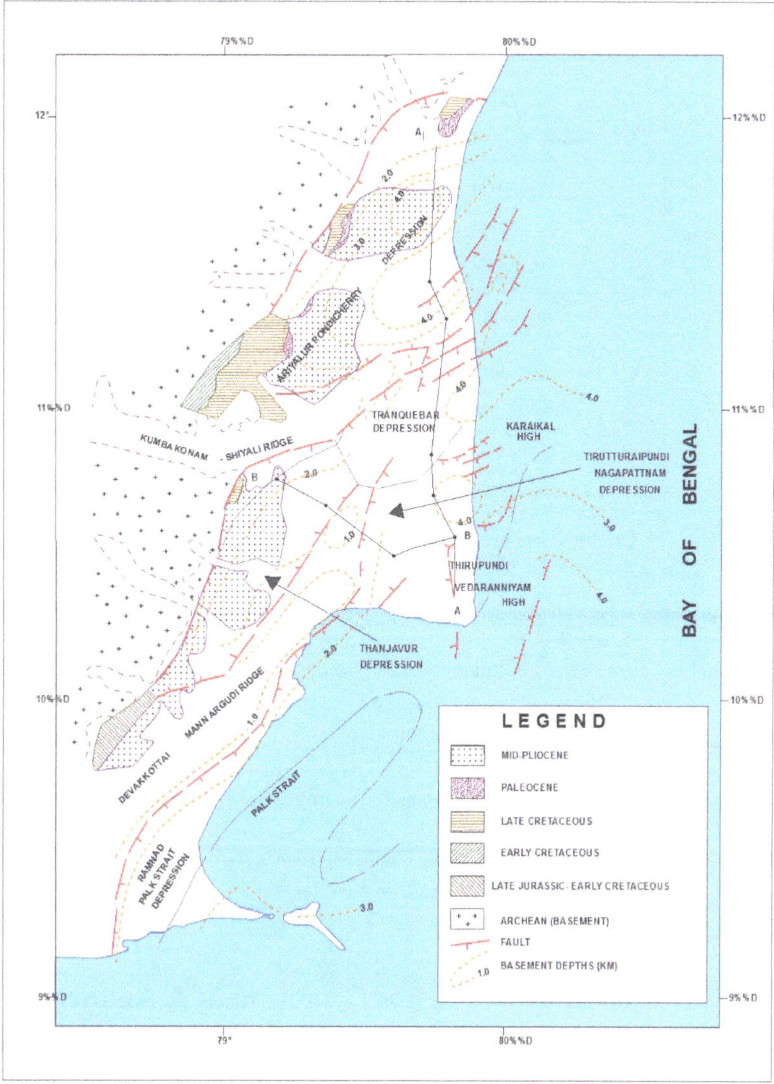

Fig. 2.4 Geological and basement map of the Cauvery and its adjoining coastal delta basins/environs

(Source: Sastri, V.V., 1981)

(vii) ***Thirupundi-Vedaranniyam high:*** This high is located east of Tirutturaipundi-Nagapattinam depression and aligned in N-S direction. The basement depth range varies from 250 m in the south east to 1000 m in the north east. Majority of the sediment sequences overlying the basement are Tertiary sediments.

(viii) ***Ramnad-Palkstraits depression:*** This depression is located south of Tirutturaipundi-Nagapattinam depression, delimited in the west by Devakottai-Mannargudi and continues east into the offshore region. The basement depths vary from 1700 m in the west and to about 4000 m in the east, filled with Cretaceous and Tertiary sediments.

A regional stratigraphic sequence in Cauvery-Ponnaiyar-Vaigai coastal delta basins and its environs is given below in Table 2.1.

Table 2.1 A regional stratigraphic sequence in Cauvery coastal delta basin and its environs

Recent-Pleistocene	Alluvium (80-100 m) -----------------Unconformity-------------	Coarse to fine grained alluvial sands and clays
Pliocene-Miocene	Tertiary Cuddalore sandstones (150 – 1200 m) ---------------Unconformity---------------	Sandstones, limestones, claystones and silt stones with intercalations of liginite coal beds.
Oligocene Eocene Paleocene	(0 – 340 m) (0 – 1150 m) Niniyur and Pondicherry formation: (70-700 m) ---------------Unconformity---------------	Sandstones,shales and limestones sandstones, shales, claystones and bioclastic lime stones sandstones, clays, marls and lime stones
Late Cretaceous	Uttatur, Tiruchinopoly, and Ariyalur formations (1190 – 2700 m) -----------------Unconformity---------------	Fossiliferrous limestones, argillaceous sandstones and conglomerates
Early Cretaceous	Dalmiapuram formations (400 m) -----------------Unconformity----------------	limestones and black shales
Early Cretaceous	Upper Gondwanas (350 – 1090 m) -----------------Unconformity-----------------	coarse gritty and pebbly sandstones, shales and argillaceous sandstones
Precambrians	Archean granitic gneisses	granitic gneisses

2.3.2 Palar-Pennar coastal delta basins (Tamil Nadu & Andhra Pradesh)

The Palar basin is an independent basin located NE of Cauvery, occupies about 4000 km^2 on land and about 2800 km^2 offshore. The basin is bordered by Precambrian igneous and metamorphic rocks in the west, north and south and are essentially filled with sediment sequences belonging to Permean, early Cretaceous, Tertiaries and Quaternaries towards east. The structural and basement configuration of the basin is shown in Fig.2.5. The basin structure shows the presence of a depression and a ridge filled with sediment thicknesses of the order varying between 500-300 m, increasing towards the north eastern side of the coast.

The Pennar basin is the southern most basin in Andhra Pradesh located north of Palar basin and has an aereal extension of 1470 sq.kms on land bordered by Precambrian complexes in the west and sedimentaries on all other sides. A subsurface ridge separates Palar and Pennar basins as shown in Fig. 2.5. The estimated sediment thickness is of the order of 4000 m. The oldest sediments in the basin are Proterozoic meta sediments unconformably overlained by Mesozoics, Tertiaries, Quaterneries and Recent. A regional stratigraphic sequence in respect of Pennar and Palar basins is presented below in Table 2.2.

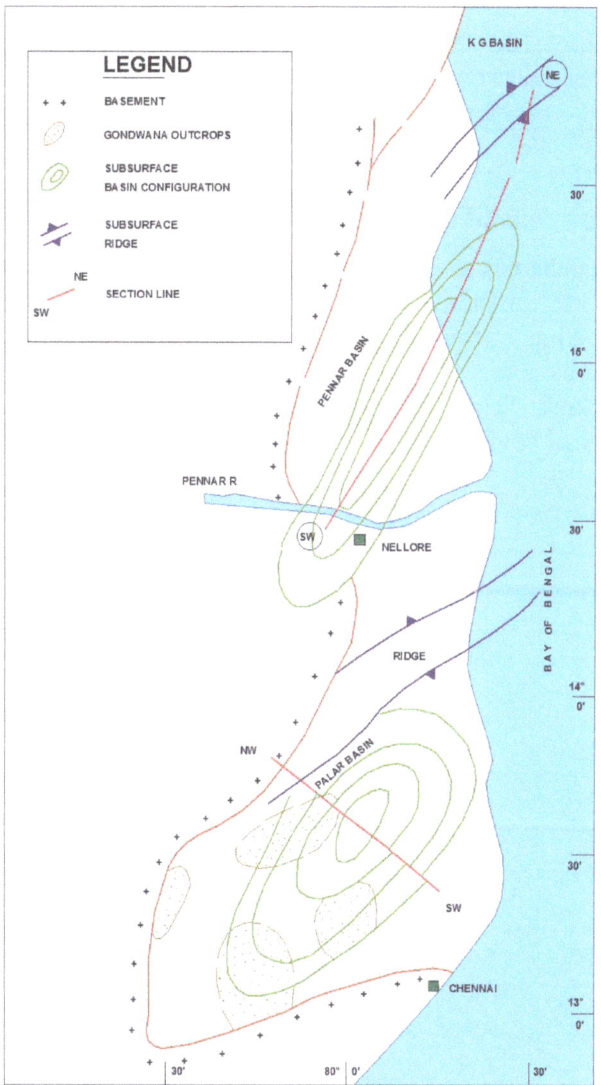

Fig. 2.5 Structural and basement configuration of the Palar and Pennar coastal delta basins

(Source: Rao, G.N., 1993)

Table 2.2 A regional stratigraphic sequence in Palar and Pennar coastal delta basins

Recent to Paleogene	**Alluvium and Conjeveram gravel (80-100 m)** **-------------Unconformity----------**	**Coarse-medium sands, alluvium and gravels**
Plio-Micoene and Paleo cene	Cuddalore sandstones (10-400 m) ------------Unconformity-----------	argillaceous sandstones and 'conglomerates with sandstones intercalations and caly stones
Early Cretaceous	Satyavedu beds (300 – 200 m) Sri perumbudur beds (600 m) ------------Unconformity-----------	coarse sandstones boulderbeds and congolomerates with sandstone inter calatious and claystones

Table 2.2 ***Contd…***

Early Permean	Talcher formation ------------Unconformity-----------	splintery grey, greenish gray shale and boulder beds
Proterozoic	Nagulapuram quartzites (Cuddapahs) -----------Unconformity------------	ortho quartzites
Archeans	Archeans	Granites and basic intrusives

2.3.3 Godavari-Krishna coastal delta basins (Andhra Pradesh)

The Godavari and Krishna delta basins along the Andhra coast are spread over an area of 20,000 km^2 on land and about 13,500 km^2 offshore. This is a pericratonic rift and a good example of divergent continental margin developed as a result of continental rifting of the Gondwana land and subsequent drift phase during Jurassic to Quaternary represented by thick sequence of sediments varying between 6 to 8 kms. These sediments are believed to have been deposited under various environments like (i) continental fluvio-deltaic, (ii) fluvio-marine - beach ridge conditions and (iii) estuary/

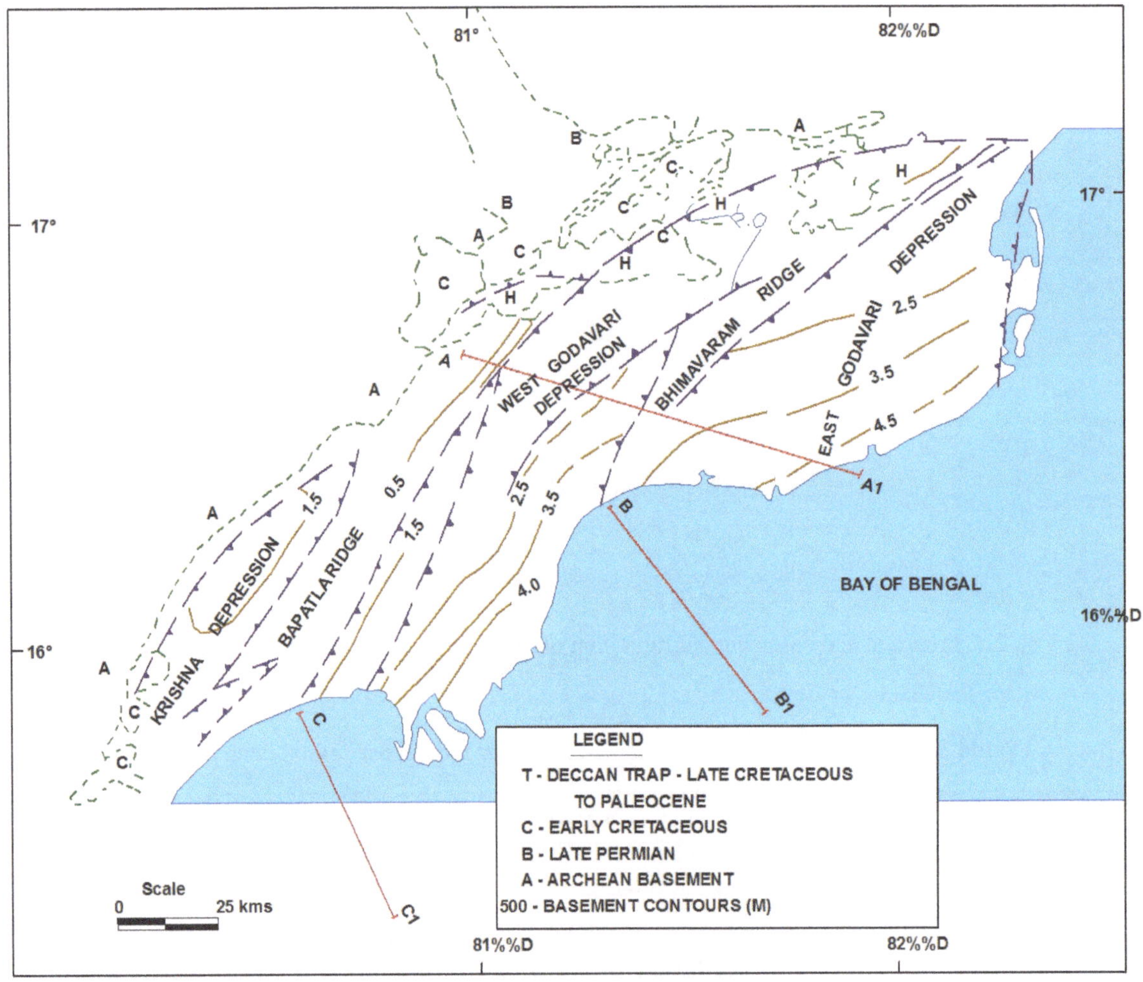

Fig. 2.6 Structural and basement configuration map of Krishna and Godavari coastal delta basin

(Source: Sastri. V.V., 1981)

lagoonal/shallow marine conditions. The Archean igneous complex delimits the basin towards the west and exhibits ENE-WSW; NW-SE and NE-SW trends. Based on geophysical and geological studies, the basin area is divided into three depressions separated from each other by sub surface ridges namely (i) East Godavari depression, (ii) West Godavari depression and iii) Krishna depression, separated from each other by Tanuku – Bhimavaram ridge and Baptla ridge respectively (Ray, D., et al 1985). The depth to basement in respect of Bhimavaram – Tanuku ridge varies from 1000 m to 2000 m and that of Bapatla ridge between 3000 to 4000 m. A basement and structural configuration map for these two basins is shown in Fig. 2.6. All these depressions and ridges by and large are aligned in NE – SW direction.

The regional stratigraphic sequences in respect of these two basins are provided below in Table 2.3.

Table 2.3 A regional Stratigraphic sequence of Godavari and Krishna delta basins and their environs

Recent	Alluvium -----------Unconformity------------	Alluvial sands and kankar
Miocene	Rajahmundry Sandstones (720 m) ------------Unconformity-----------	coarse_medium & fine sandstones with pebble beds and variegated clays.
Eocene to late Cretaceous	Deccan traps and with in the trapean beds (130 m) ------------ Unconformity----------	vesicular and columnar basalts, limestones, marls and clastic sediments.
Cretaceous	Infra-trappean beds (70 m) -------------Unconformity----------	fossliferrous limestones, greenish and yellowish sandstones with marine fossils.
Early Cretaceous to	Tirupati / Pavalur sandstones (830m) and Raghavapuram/ vemavaram shales (170 m) ------------Unconformity-----------	hard compact ferruginous and feldspathic sandstones and conglomerates, hard and compact shales, sandstones, silt stones and glauconitic sandstones and brown shales.
Late Permean	Chintalpudi sandstone (3260 m) ------------Unconformity-----------	coarse-medium feldspathic compact sandstones with streaks of shales and coal layers
Archeans	Archeans	gneisses and schists, garnetiferrous with pegmatite veins

2.3.4 Mahanadi-Brahmini-Baitarni coastal delta basin (Orissa)

Mahanadi basin is a major pericratonic basin with an aereal spread of the order of about 1800 km^2 on land and about 12,500 km^2 offshore covered by sediments ranging from late Jurassic to Recent. The basement and structural configuration of the delta basin is shown in Fig.2.7. The basin has three depressions namely, Cuttack-Chandbali depression, Paradeep depression and Puri depression. These depressions are separated by Rajnagar ridge and Bhubaneshwar ridge respectively (Murthy, K.V.S., et al., 1973).

The Cuttack-Chandbali depression is a long narrow graben like structure, extending into the offshore area, aligned in ENE-WSW direction near Cuttack and NE-SW direction near Chandbali. The sediment thicknesses in this depression range from 2000 to 2500 m with majority of the sediments belonging to Mesozoic. The Rajnagar ridge strikes in NE-SW direction at Rajnagar but swings towards N-S direction towards offshore. The Paradeep depression with little bit of stretch in the onshore, mostly extends offshore in NE-SW direction with thicknesses varying between 2000-2500 m. The Bhubaneshwar ridge trends in E-W direction. The Puri depression trends in ENE-WSW direction with sediment thicknesses varying from 1500 to 3000 m. A brief regional stratigraphic succession of the basin is described below in Table 2.4.

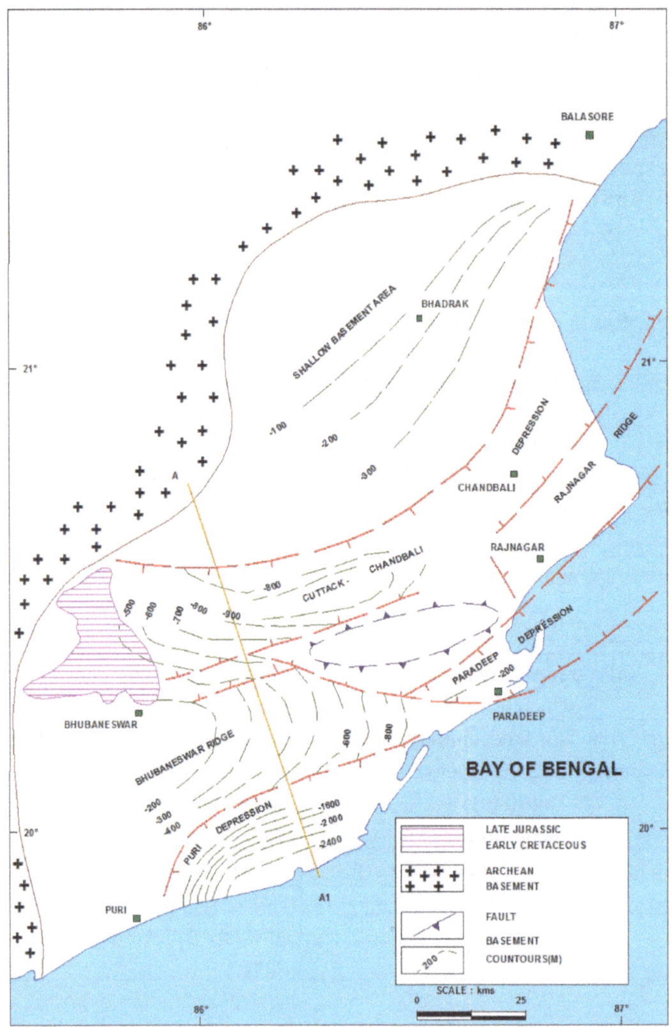

Fig. 2.7 Structural and basement configuration map of Mahanadi-Brahmini-Baitarni delta basin

(Source: Murthy, K.V.S., et al 1973)

Table 2.4 Regional stratigraphic sequence of Mahanadi-Brahmini-Baitarni delta and its environs

Recent to Pleistocene	Alluvium (100 m) -----------Unconformity--------	coarse to fine sands and clays
Plio-Miocene	Baripada Beds (1000 m) ----------Unconformity--------	sandstones and claystones with interclations of lignite
Oligocene to Paleocene	Unclassified sandstones, claystones and volcanics ---------Unconformity---------	sandstones, claystones and volcanic basalts
Cretaceous	Unclassified claystones and sandstones ---------Unconformity----------	sandstones and claystones
Archeans	Archeans	granites, gneisses, charnockites and quartzites with dolerite dykes.

2.3.5 Bhagiradhi-Hoogly coastal delta basin (West Bengal)

The Bhagirathi–Hoogly basin in West bengal represents a part of Ganges basin which is a prominent pericratonic basin in the north eastern region of east coast of India. It has an aereal extension of about 56,000 kms on land and 3400 kms offshore. Some of the major tectonic features which could be seen in and around Bhagiradhi-Hoogly basin are (i) The basin margin fault zone; (ii) Stable Shelf Zone (iii) The Hinze zone and (iv) The deep basin (Fig. 2.8).

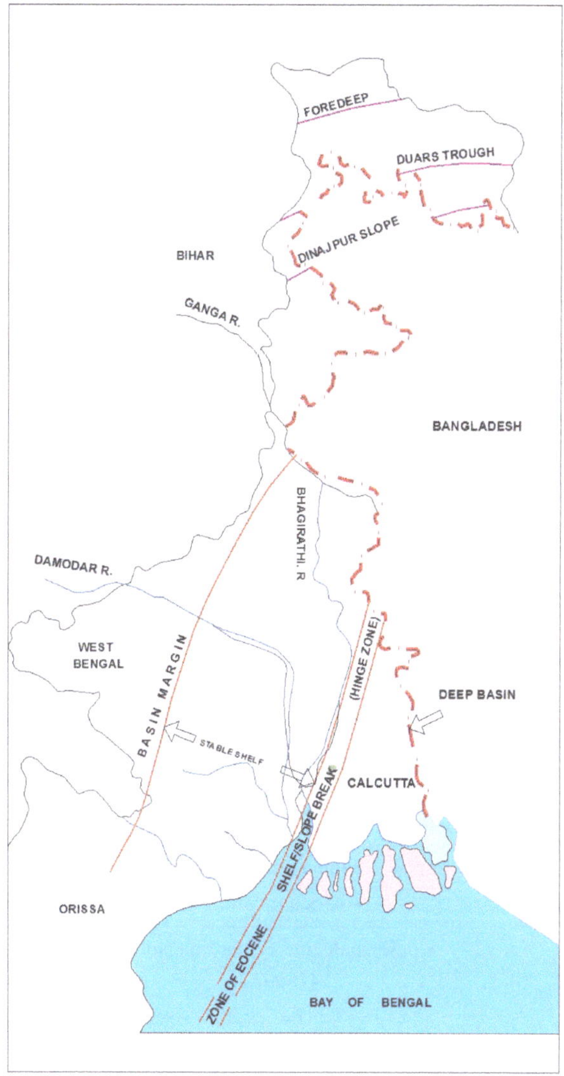

Fig. 2.8 Structural and basement configuration map of Bhagiradhi-Hoogly delta basin

(Source: Ganguly, S., 1997)

Basin margin fault zone

This is NE-SW trending fault zone in between the basement complex in the west and the western margin of stable shelf zone. This is related to the extensional tectonics of early cretaceous period. It gets truncated towards north by basement faults/ridges and extends deep into the Bangladesh boundary.

Stable shelf zone

The western part of the West bengal basin is bounded by Precambrian outcrops and the '*Eocene hinze zone*' in the west with a gentle slope towards east. This zone worked as a platform for continental and marine sediment deposition in the geological past and is also characterized by numerous basin faults trending NE-SW, covered by thick sequences of Paleozoic to Holocene sediments varying between 6-7 kms with increasing thicknesses towards east.

Hinze zone

This is a narrow transition zone separated by the shelf area from the deep basin with a width varying between 24 to 30 kms filled with sediments varying between 7-8 kms characterized by numerous enechelon faults and flextures at depth levels 2000 to 4000 m with gentle slopes at initial stages and finally gets flattend towards deeper parts of the basin.

Deep basin

The basin extends from the eastern parts of the '*Hinze zone*' in West bengal extends beyond the West bengal basin deep interior into Bangladesh and south into the Bay of Bengal almost up to the latitude 10° South. The sediment thicknesses in this basin in the West bengal region run up to 10 kms and increases further east in the Bangladesh region. The basement boundaries are not exactly known but appear to be around 10-12 kms in the eastern boundary of West bengal and deepens further east to an order of 20 kms in the Bangladesh region. The exploratory studies so far carried out in this region indicate fractured archean basement overlained by Gondwanas, Cretaceous and Paleocene sediments and thick sequences of carbonaceous materials during Tetiaries followed by thick sequence of Quaternary sediments. A regional stratigraphic sequence of West Bengal basin is given in Table 2.5.

Table 2.5 A regional stratigraphic sequence in Bhagiradhi-Hoogly (Ganges) delta basin

Recent to Pleistocene	Alluvium (500 m) --------Unconformity------	coarse to fine sands and clays
Plio-Miocene	Debagram, Ranaghat formation (400 m) --------Unconfirmity-------	sandstones, claystones and shale.
Oligocene	Burdwan Memari sandstone (700 m) --------Unconfirmity-------	sandstones and shales
Eocene	Kopili shale and sylhet limestone (100 m) -------Unconfirmity-------	sandstones, limestones and claystones
Paleocene	Jalangi sandstones (700 m) --------Unconfirmity-------	sandstones and shales.
Cretaceous to upper Jurassic	Bolpur-ghat volcanics, Rajmahals (700 m) --------Unconfirmity-------	volcanic basalts
Paleozoics	Unclassified Gondwanas (200 m) --------Unconfirmity-------	sandstones and shales.
Archeans	Archeons	granite and gneissic crystalline complex.

2.3.6 Narmada and Tapti delta basins (Gujarat)

Narmada and Tapti delta basins represent prominent pericratonic basins of northwestern region of India. They also represent the southern most of the three marginal basins namely Kutch, Cambay and Narmada in the north western platform of Indian craton formed as a result of sequential tectonic episodes during the processes of drifting and rifting after the break up of Gondwana land (Biswas, S.K., 1982, 1987). The rifting of Kutch took place first during late Triassic-early Jurassic followed by Cambay and Narmada rifting in early Cretaceous and late Cretaceous respectively. Rao, R.N. and Talukdar, S.N. (1980), Biswas, S.K. and Deshpande, S.V. (1973) have described the tectonic, structural episodes and stratigraphic features of Narmada basin. The regional structural trends of the rift basins of Kutch, Cambay and Narmada are presented in Fig. 2.9. The Cambay rift had taken place by and large parallel to the west coast (NNW and SSE direction) where as the Narmada rift took place almost perpendicular to the Cambay rift in the ENE-WSW direction. The Narmada graben is a part of the Narmada-Son mega lineament and is bounded by a system of sub parallel, dextral wrench faults. This is a major geo fracture zone extending to a depth of 35-40 kms (Kaila, K.L., et al., 1985) and is associated with several sets of block faults and divides the Indian shield into southern peninsular block and a northern forcland block (West, W.D., 1962; Mathur, L.P., et al 1968). The stratigraphic sequence of Narmada-Tapti basins are presented below in Table 2.6.

Table 2.6 A regional stratigraphic sequence in Narmada-Tapti basin and adjoining environs

Quaternary	Holocene To Pleistocene	Alluvium Narmada Fm. (80-100 m) -------Unconfirmity-----	Sandstones, silts, clays and gravels
Tertiary	Miocene	Gaj Series Jhagadia Fm. and Kand Fm. --------Unconfirmity----	conglomerates, fossil limestones, calcareous sandstones and gravelly clays
		Babaguru Fm. (900-1000 m) -------Unconfirmity------	conglomerates, sandstones, clays-cherry red and highly ferruginous.
	Oligocene	Nari Series Dinod Fm.(50 m) -------Unconfirmity-----	fossiliferous limestones and marls
	Eocene		shaly limestones and sandstones
	Paleocene	Vagadkhol Fm. (+50 m) --------Unconfirmity----	conglomerates, grits, sandstones, variegated clays and siltstones.
Mesozoic	Upper Cretaceous	Deccan Trap	basalsts, sometimes amygdaloidal and trachytic.
	Lower Cretaceous	Bagh/Lameta beds ------Unconfirmity------	limestones, marls, sandstones, etc.
Archeans		Crystaline complex	granites and gneisses

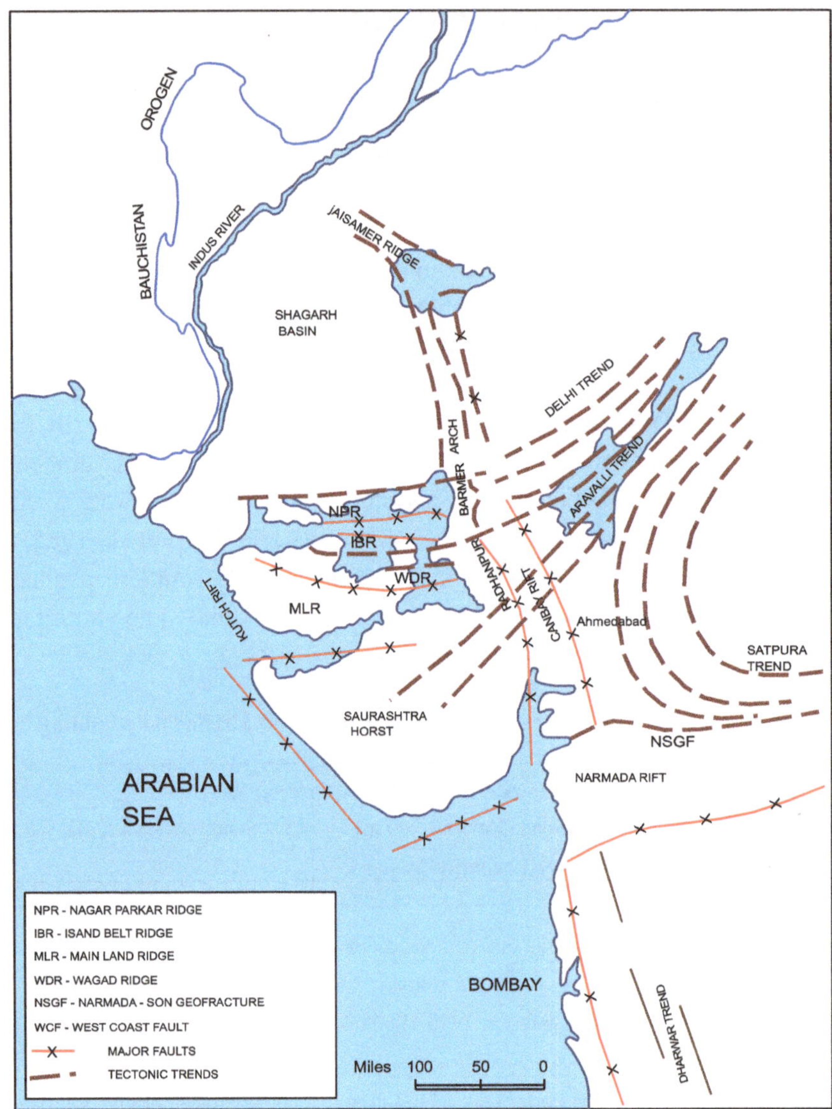

Fig. 2.9 Regional tectonic trends of Kutch, Cambay and Narmada rift basins

2.4 MORPHOLOGICAL EVOLUTION OF QUATERNARY DELTA BASINS

Coastal deltas constitute rich resource belts in terms of agriculture, irrigation, human settlements, hydrocarbons and industrial expansions. More than 30% of the population live in these deltaic and adjoining regions for their livelihood. These deltas in view of their plain topographical features, richness of the soil fertilities, high intensities of canal networks and potential subsurface groundwater resources form granaries for agriculture production and rich irrigation belts. A major chunk of agriculture and irrigation potentials of India come from these regions, next only to Indo-Gangetic plains in the north. In addition, these delta basins both onshore and offshore have become sites for huge hydrocarbon potentials like Bombay High in the west coast and offshore and onshore regions of Krishna-Godavari deltaic basins on the east coast. Probably several deltaic regions along the east coast could be opened up as hydrocarbon potential fields in the near future to come. The tectonic evolution of these basins have already been briefly mentioned and discussed in the aforesaid paragraphs. The morphological set up and the evolutionary trends for all these delta systems during Quaternary,

Holocene and Recent periods have been well documented by various authors in the volume on "Quaternary deltas of India" edited by Vaidyanadhan, R. (1991).

The east coast of India contain large net work of major and medium delta systems and several minor rivers moving their ways into Bay of Bengal, when compared to the west coast of India where literally no delta system exists except small estuary deltas in the north western region of Gujarat coast and several small estuary rivulets/mouth areas all along the west coast. The focus therefore in these chapters is made largely on the east coast delta systems namely Cauvery-Ponnaiyar, Vaigai and Palar delta systems of Tamil Nadu; Godavari, Krishna and Pennar delta systems of Andhra Pradesh; Mahanadi-Brahmini-Baitarni delta systems of Orissa; Bhagiradhi – Hoogly delta system (Ganges) of West Bengal and a few selected estuary delta systems namely Narmada-Tapti along the west coast of Gujarat. The sediments of these delta regions are mostly unconsolidated deposits of fluvial, marine, aeolian and lacustrive origin with occasional outcrops or sometimes stretches of older sedimentaries or Precambrian rocks. A majority of these unconsolidated deposits are considered to belong to 'Quaternary age', partly Pleistocene and partly Holocene. These delta deposits show wide variations in their morphology and structure basically due to geotectonics, fluvial and marine hydro dynamics and related changes in coastal processes. The east coast deltas exhibit a number of land forms such as flood plains, abandoned river channels/lobes, point bars, oxbows, natural levees, strand lines, mangrove swamps, lagoons and back waters , tidal flats, beach sand ridges and barrier spits. By and large all these deltas exhibit fan shaped, arcuate and cuspate and some times disposed linearly and ends up as estuary/river mouths with progradation into ocean side and are also closely linked with the rise and fall of sea levels (Vaidyanadhan, R. and Ghosh, R.N., 1993). These deltaic and associated morpho units are developed basically through several interactive fluvial, marine and aeleon processes over a period of time. The sedimentation processes, the deltaic environments and the lithofacies changes in each of the delta system along the east coast of India have been illustrated and described by Singh, I.B. and Seetaramaswamy, A. (2006).

2.4.1 Cauvery, Ponnaiyar, Vaigai and Palar coastal delta basins (Tamil Nadu)

The coastal Tamil Nadu region exhibits a number of fluvio-marine environments all along its coast starting from south Pulicate lake area in the north up to Kanyakumari in the south with a length of approximately 907 kms. The coast line of Tamil Nadu in general exhibits a convex-concave trend signifying the depositional and erosional trends with visible features like beach ridges, lagoons, back waters, tidal creeks, tidal flats, marshes and mangroves etc. The coastal region of Tamil Nadu has a string of major, medium delta systems like Palar, Ponnaiyar, Cauvery and Vaigai etc., along with some minor estuaries and river mouths in between the major delta systems all along its coast (Ramaswamy, S.M., 2006). These deltas though separated from each other are responsible for the built up of Quaternary alluvial sediments all along the coast by and large as a continuous geological unit with varying widths and thicknesses. All these major coastal delta basins have their initial signatures probably during Cretaceous times with marine connections which probably manifested in the form of major and medium deltas along this part of the coast. The most prominent and well developed delta systems in this part of the coast is Cauvery delta associated with medium delta systems like Palar and Ponnaiyar in north and Vaigai in south.

Cauvery delta

The Cauvery delta located between Pondicherrry and Vedaranniyam bounded by N lat 10°15' to 11°30' and E long 79°00; to 79°48' is probably the second largest of the east coast deltas with an aereal spread of the order of 11,000 sq kms. It is a large fan shaped arcuate delta basically built up with accretion of sediments brought by Cauvery river and its major distributaries like Coleroon, Cauvery, Arsalar, Vettar and Vennar and their sub distributaries with their apex near Thanjur as shown in fig 2.10. The

Quaternary/Recent delta sediments have thicknesses as vrying from few centimeters to nearly 55 meters increasing towards the coast.

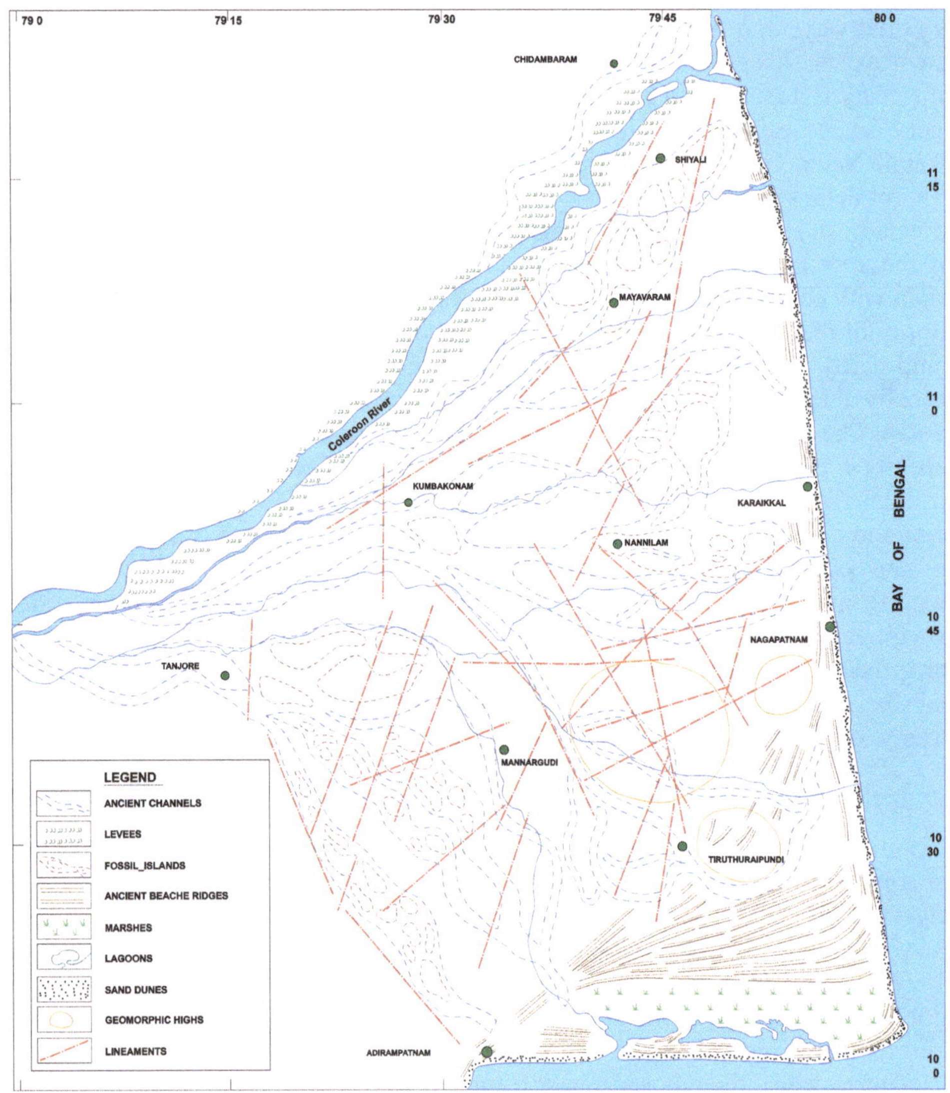

Fig. 2.10 Evolution and morphology of Cauvery delta, Tamil Nadu

(Source: Sambasiva Rao. M, 1982)

The structural morphological and evolutionary trends of Cauvery delta and adjoining river delta basins of Tamil Nadu have been worked out by several workers (Ramaswamy, S.M., 1991; 2006; Sambasiva Rao, M., 1982; Vaidynathan R., 1971; Babu, P.V.L.P., 1991 and Meijerink, A.M.J., 1971). At least four sets of lineaments could be identified in the delta mosaic namely NE-SW, NNW-SSE, N-S and NNE-SSW (Raiverman, V., et al 1966). These lineaments could have played a major role in shaping the delta system, paleo channel distribution and river migrations in this area.

The Cauvery delta morphologically could be classified into three units (i) upland plains (denudation) (ii) deltaic plains (fluvio-marine) and (iii) coastal plains (marine). The upland plains are represented by lateritic table land, red in color and are conspicuous by its high topographic elevation in the west. The drainage is conspicuously absent but is characterized by thick forest vegetation and orchards with high and rich fertile soil content. The deltaic plains are characterized by a network of abandoned fluvial channels filled with muds, silts and fine sand. The coastal plains are represented by marshes, tidal flats, beech sands and ridges. The surface flows at present are seen only along the Coleroon river in the north, but at times, during high flood seasons some of the distributary channels become active for short periods. A number of well distributed paleochannels net work could be seen from south to north. The Cauvery river probably was flowing originally in the south east direction and gradually moved from south to north, leaving a trail of abandoned channels which are productive source areas for groundwater development. At least 12 major abandoned river courses have been identified as presented in Fig. 2.11 (Sambasivaro, M., 1982).

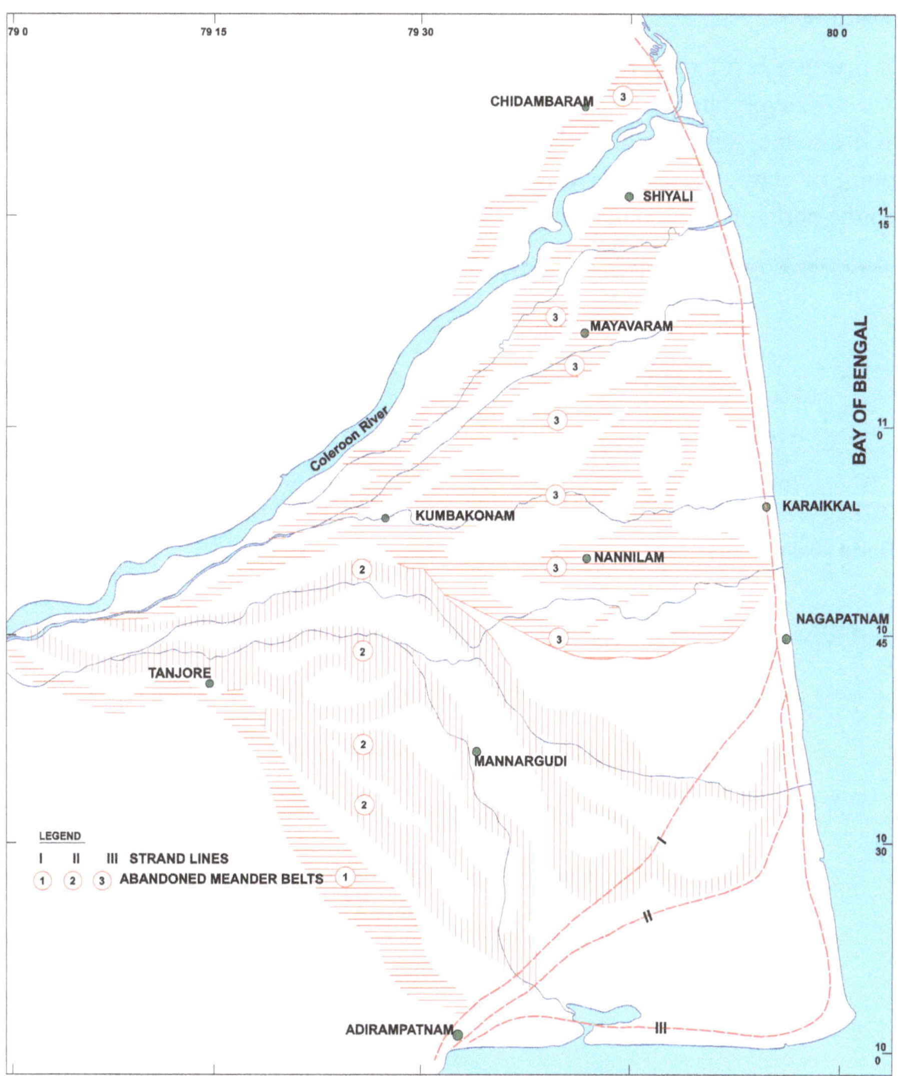

Fig. 2.11 Evolution and morphology of Cauvery delta, Tamil Nadu

(Source: Sambasiva Rao, M., 1982)

A majority of the ancient Paleo distributory channels were bifurcated from the apex portion of the delta. In addition several land forms like natural levees, delta islands, ancient beach ridges, marshes, lagoons and coastal dunes have been identified. At least three numbers of ancient strand lines have been identified as presented in Fig. 2.11 In other words, the evolution of delta has taken place in three stages. The initial stage of river flow was towards south east meeting the coastline at Adirampatnam. During its second stage of evolution, the Cauvery shifted its course towards north with four distributaries, combining into two near the coastline. The third stage signifies further northward shift with at least seven distributaries, combining into three or four near the coast line before meeting the sea. During this stage probably a major part of the delta growth has taken place. The present stage or the fourth is represented by main Coleroon river with its sub distributaries like Palavar, Cauvery, Vettar and Vennar etc.

Ponnaiyar delta basin

Ponnaiyar is a medium level river delta basin located north of the Cauvery and south of Palar river mouth. In between Cauvery and Ponnaiyar there is an intermediate minor river namely Vellar which is not significant from the delta build up point of view, but could play a potential role from the groundwater point of view. The Ponnaiyar river delta is a linearly disposed river with several burried channels laid to the north of the present flowing river and two short stretched distributaries with its apex near Thirukkovilur as shown in Fig. 2.12.

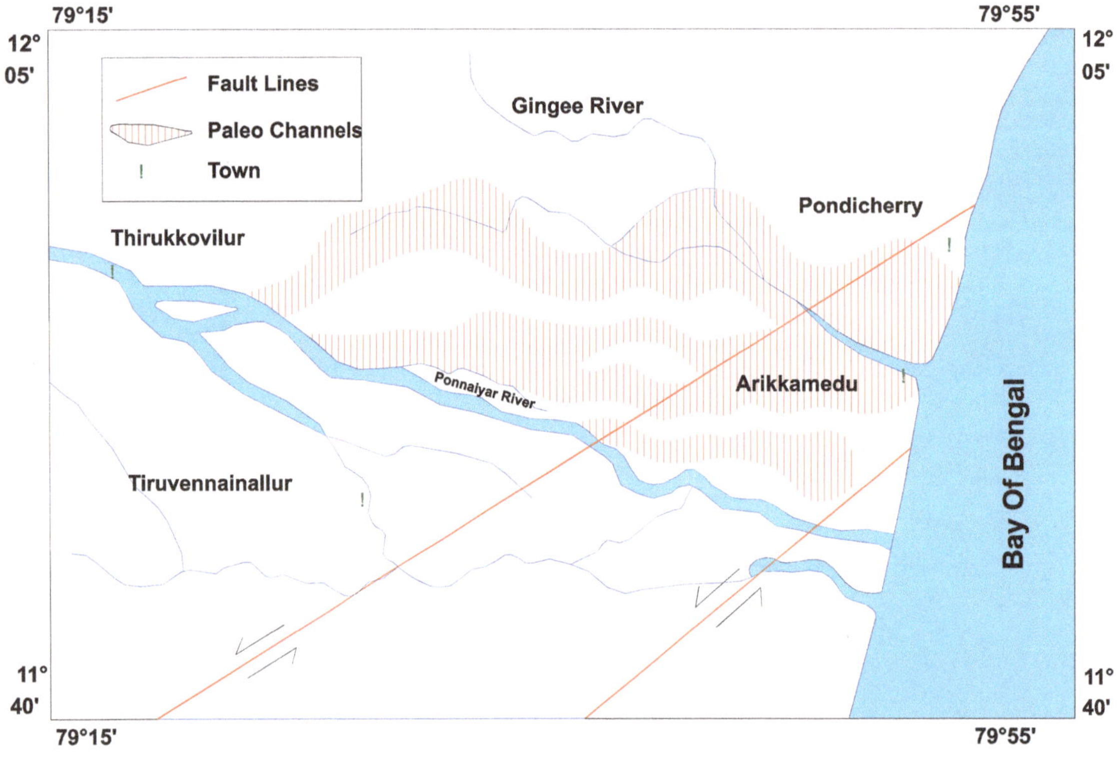

Fig. 2.12 Morphology and evolution of Ponnaiyar delta basin

(Source: S.M. Ramaswamy, 2006)

The paleochannels and the present flowing river along with its distributaries are laid out in between the Tertiary massifs and could play an important role in recharging the Cuddalore aquifer systems which are believed to be the most potential aquifer system in this part of the coast. The Ponnaiyar river, its distributaries and its paleochannels have their regional axial trends in E-W directions with minor localized variations. The basin is characterized by NE-SW lineaments/fault lines. The Neyveli liginite basin in the Cuddalore exists in the proximity of Ponnaiyar and Vellar basins. The NE-SW lineaments/fault lines cut across the Ponnaiyar and its associated distributaries and Paleochannels, facilitating a possible hydraulic linkage between the river systems/paleochannels in this region and the groundwater aquifer systems of Tertiaries. The high groundwater potentials in the area had become a major hurdle for Lignite mining in the vicinity and subsequently proposed to dewater the system to facilitate Lignite mining through a net work of heavy duty tube wells/well fields.

Vaigai delta

The Vaigai delta is located south of Cauvery delta region (E.long. 77°50' to 80°0'; N.lat. 10°10', 9°0') and covers an area of about 2500 sq. kms with its apex near Paramakkudi and with several distributaries located in the downstream side of the delta plains as shown in Fig. 2.13 (Prabhakar, K., et al 2010).

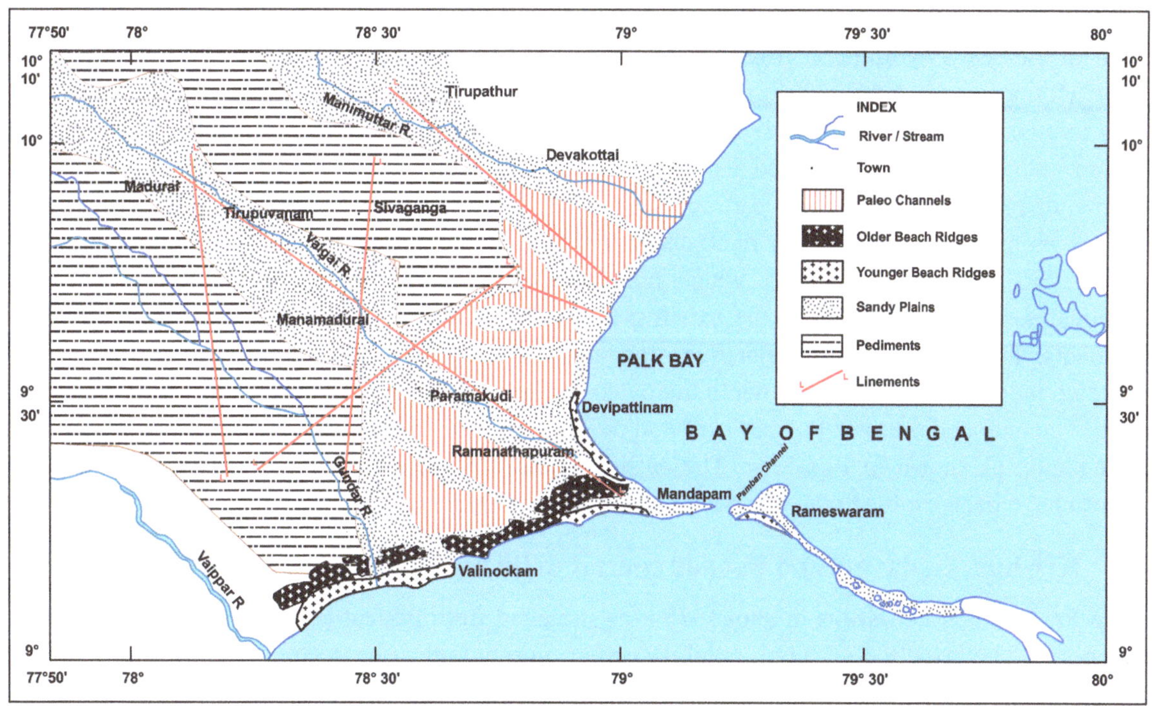

Fig. 2.13 Morphology and evolution of Vaigai delta basin

(Source: Prabhakar, K., et al 2010)

There exists several small delta systems north of Vaigai namely Manimuttar, Vellar, Ambuliar etc. interlacing with each other which play an important role from the point of view of groundwater resources. The Vaigai river appears to be flowing along a WNW-ESE lineament and has build up a medium level delta system. Three stages of evolution of Vaigai delta with probable strand lines and progradation have been identified (Ramaswamy, S.M., 1991). The first one is near Madurai, the second one at Tirupavanam and the third and last one is at Paramakudi where several paleochannels emerge south and north of the present flowing river. The pattern of occurrences of paleochannels on either side of the present river characterizes the migratory trends of the river in the recent geological past. Beach ridges, swales, spits and bars and abandoned channels, dunes and delta plains are the characteristic land forms in these delta tracks. Another typical characteristic of this delta region is the occurrence of several hundreds of lakes in a radial pattern which trap most of the suspended sediments and at the same time reflecting the flow patterns and delta formation trends. The coral terrace deposits and their data records characterise the transgressional and regressional phases of the ocean in the recent past (Loveson, V.J., et al., 1990). The Vaigai delta domain exhibits N-S, NE-SW and NW-SE trending lineaments (Loveson, V.J. and Rajamanickam, V., 1987). In general the NW-SE lineaments control the present river and the paleo channel trends in the region. The present river could have been originally aligned in south eastern direction and moved in an anti-clock wise direction in different stages leaving a series of paleo channels which are aligned along NW-SE lineaments. All the fluvial paleo channel deposits could hold good bearing on the groundwater occurrences in the near surface systems.

Palar delta basin

This is the northern most river delta systems in Tamil Nadu. Though the present river is flowing in a linear fashion in south east direction from Walajapet, a good builtup of paleo delta system appears to be present in the NE-SW direction from Walajapet towards the coast with at least two lobate like delta features (Ramaswamy, S.M., 2006) probably with its apex located near Tiruvallur. Vaidyanadhan, R. (1971) termed this as '*proto-Cauvery*' signifying the Cauvery originally had its flow path along this direction and subsequently changed its course. Ramaswamy, S.M., et al, 1992 affirms this idea of existence of proto Cauvery in this part and is responsible for building up this paleo delta system and describes its migratory trends with radiometric datings. Narasimhan, T.N., 1990 and Rao, P.S., 1989 identifies this as paleo Palar river which had built up its delta during the Quaternary times and subsequently shifted its course to the existing flow path of Palar river as a consequence of vertical movements. Though the present Palar river does not show any significant delta development except a few buried paleo channels near its mouth the paleo Palar river had built up reasonably good stable delta north of Chennai and south of Pulicat lakes and holds good promise for groundwater resource potentials in this region as shown in Fig. 2.14. The earlier paleo river/delta trends and the present Palar river appear to have been evolved over NE-SW and NW-SE lineament trends.

2.4.2 Krishna, Godavari and Pennar coastal delta basins (Andhra Pradesh)

Andhra Pradesh has a coastline of about 975 kms stretched from Pulicat lake in the south to Ichapuram in the north. The coastline all along exhibits varied morphological characters with several major and medium deltas and minor rivulets joining the sea. The northern coast is by and large rocky with medium rivers like Vamsadhara and Nagavali and several minor rivers joining the sea. The southern coastline is characterized by major and medium delta systems like Godavari, Krishna and Pennar. In addition the coast line has a prominent landform namely Pulicat lake, a prominent bird sanctuary and a number of shallow bays along the river mouth areas of Godavari, Krishna and Pennar. Spits and bars in vicinity of Godavari and Krishna mouth areas and features like flood plains, natural leaves, channel bars, mangrove swamps etc. are commonly seen along the delta fringes. A brief description of morphological and evolutionary trends of Godavari, Krishna and Pennar are presented below.

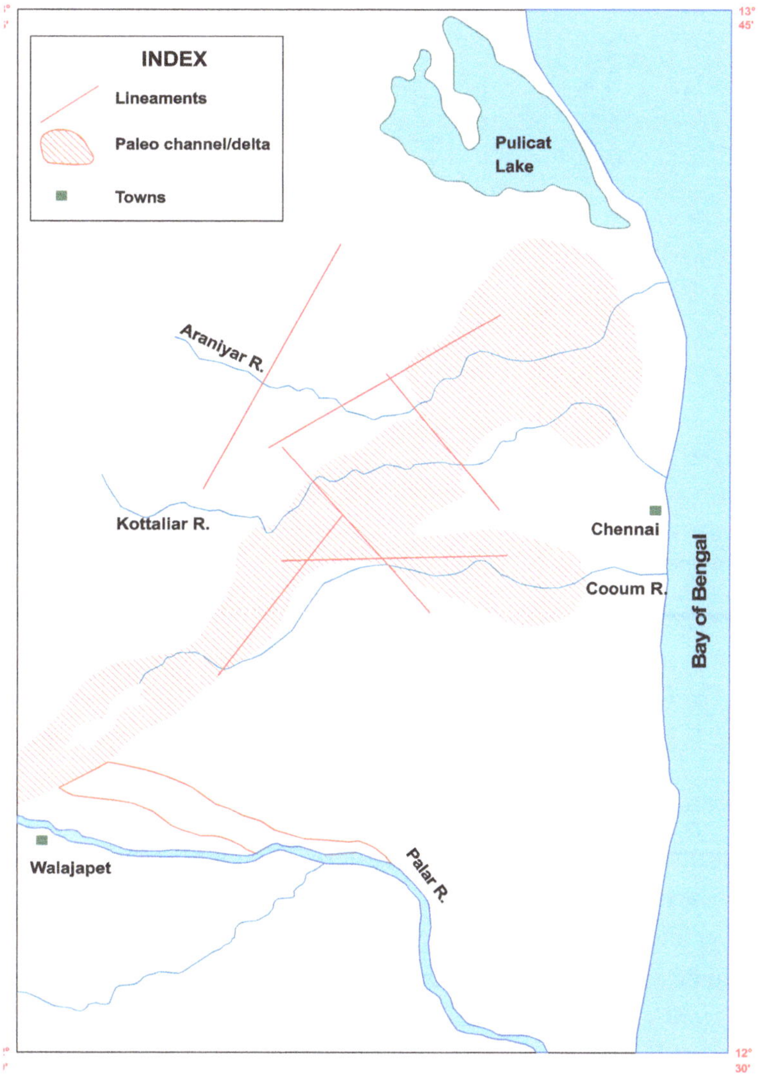

Fig. 2.14 Morphology and evolution of Palar delta system

(Source: Ramaswamy, S.M. 2006)

Krishna-Godavari basin

The Krishna and Godavari delta region has been studied by a number of workers in terms of morphological evolutions and environmental analyses (Biswas, S.K., 1993; Nageswararao, K. and Vaidyanathan, R., 1978, 1979; Krishnarao, B. and Swamy, A.S.R., 1991; Vaidyanathan, R. 1987, 1989 and 1990; Mahender Reddy, K. and Shah, B.M., 1991 and Regamannar, V. and Pradhan, P.K., 1991). The Godavari and Krishna basins by and large are characterized by similar geomorphic configurations and could be broadly divided into three groups (i) Continental fluvio-deltaic complex, (ii) Beach ridge complex or fluvio-marine environment and (iii) Bay estuary lagoon complex of shallow marine environments. Both the basins have dominant lineament trends towards ENE-WSW, NW-SE and NE-SW as reflected in Fig. 2.15 (Biswas, S.K., 1993). These lineament trends appear to have close control on the major distributaries and paleo channel trends.

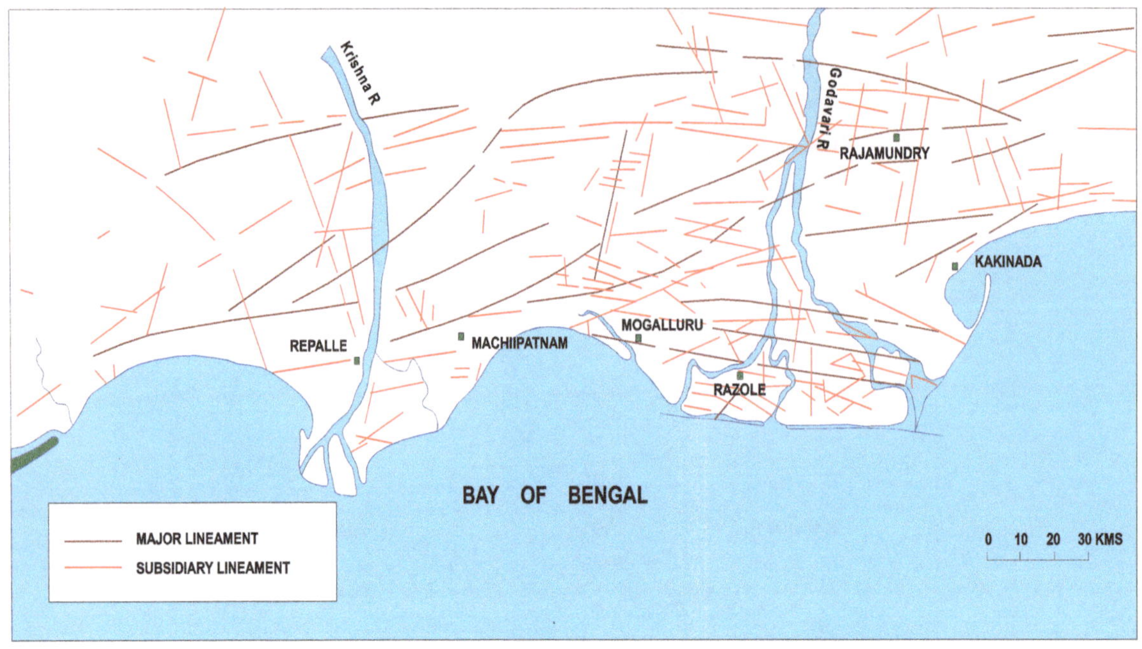

Fig. 2.15 Lineaments trend map of Krishna-Godavari delta basin

(Source: Biswas, S.K., 1993)

Godavari delta

Godavari delta ($16^0 5^|$ to $17^0 07^|$ N and $81^0 22^| 30^{||}$ E) basin is the third biggest of the delta systems along the east coast with an aereal extension of the order of 5,500 sq. kms with its apex located near Rajahmundry. The river distributes itself in a fan shaped delta with several distributaries and sub distributaries in the middle and outfall areas namely Gautami, Vasishta, Vainateyam and Nilarevu. The epic name 'Sapta Godavari' appear to contain relevance in the existence of abandoned channels namely (i) Gosthani, (ii) Atreya, (iii) Karsika, (iv) Julyabhaga, (v) Vasishta, (vi) Vainateya and (vii) Goutami. All these river channels initially were responsible for the delta buildup. A prominent bay with a long spit in the northern part of the delta is a prominent morphological feature of the delta. Sambasiva Rao, M. and Vaidyanadhan, R. (1979) have identified four stages of delta growth based on the presence of ancient beach ridges and inferred existence of fourteen meandering paleo channels in the delta Fig. 2.16. The initial main paleo river appear to have flown towards west of the present river discharging their waters and sediment near Pippara, the then ocean boundary or the first stage of Godavari delta. The second stage of delta development is represented by more elongated and abandoned channels distributed on both sides of the present river which appear to be the manifestations of subsurface basin configuration. Further downstream about five abandoned channels have been identified during the third growth stage of the delta. During the fourth stage, at least six abandoned channels appear to be present. The ENE-WSW trending lineaments appear to have made a large control on the evolutionary trends of this basin.

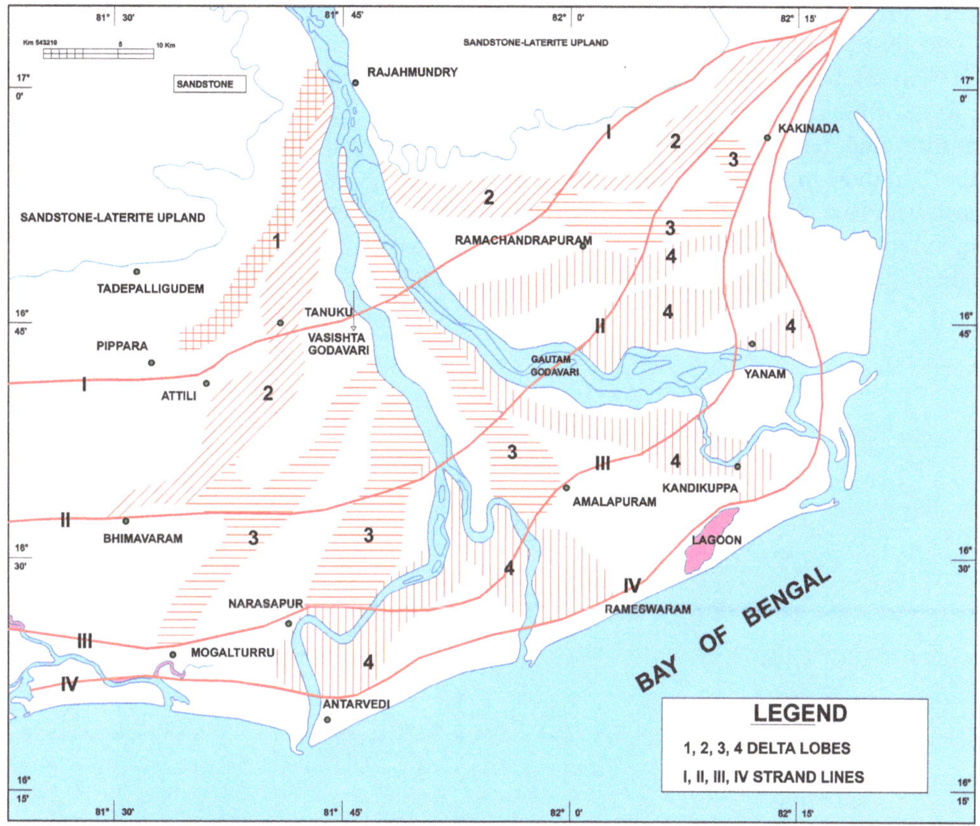

Fig. 2.16 Morphology and evolution of Godavari delta

Rengamannar, V. and Pradhan, P.K. (1991) while confirming these different stages gave a more detailed account of different stages of evolution.

***Stage* I:** Errakalava, Gostandi and old Tulayabhaga drain are the earliest abandoned channels terminating at Kaikaluru-Mahadevapatnam-Velangi in the north east. The abandoned course of Errakalava is traced in the delta along Komorru, Moyyur and Kesavapuram and discharged into Kolleru lake. Gostandi an abandoned course of Godavari is traced along Tanuku-Attili and Arravalli towns and probably flowed into the sea.

***Stage* II:** The strand line between Mahadevapatnam and Velangi was overshadowed by the sediments of advancing Godavari during its early evolution forming a second stage. Three abandoned courses represented by Palamurru, Nakkal and Athreya rivers cut the above strand line and gets terminated at the Bhimavaram – Kakinada beach ridge.

***Stage* III:** The area between Bimavaram-Kakinada and Narsapur-Amalapuram is the 3rd stage of progradation of the Godavari. The abandoned distributaries are represented by Gonteru-Bhaggeswaram, Athreya and Kaisika responsible for the third stage of delta evolution.

***Stage* IV:** The last stage of progradation of the delta is between Narasapur and the present coast.

Krishna Delta

Krishna delta (N lat 15°44' and E long: 80°20' – 81°30') another important and major delta system with an aerial spread of about 4600 sq.kms and with its apex near Vijayawada. Krishna from Vijayawada runs in a sub arcuate fashion for a length of about 60 kms up to Avanigadda where the fist distributory gets branched off towards east and joins the sea near Hamsaladivi. Further down slope the river bifurcates into three channels near Edurmodi namely Gulumuttapaya (east), Medimurru (central) and Krishna river (west), before joining the sea coast. A generalised but conspicuous feature of this delta is its arcuate form of the river with a large protrusion into the sea and with development of spits on either side of the protruded delta into the sea. Ancient beach ridges, paleo channels, flood plains, levees and spits are the common morphological features observed (Nageswara Rao, K. and Vaidyanadhan, R., 1978; Nageshwara Rao, J., 1988; Babu, P.V.L.P., 1975 and Mahender Reddy, K. and Shah, B.M., 1991) have attempted to analyze the evolutionary trends of this basin. At least five delta lobes and five stages of strand lines have been identified by Mahender Reddy, K. and Shah, B.M., 1991 as shown in Fig. 2.17.

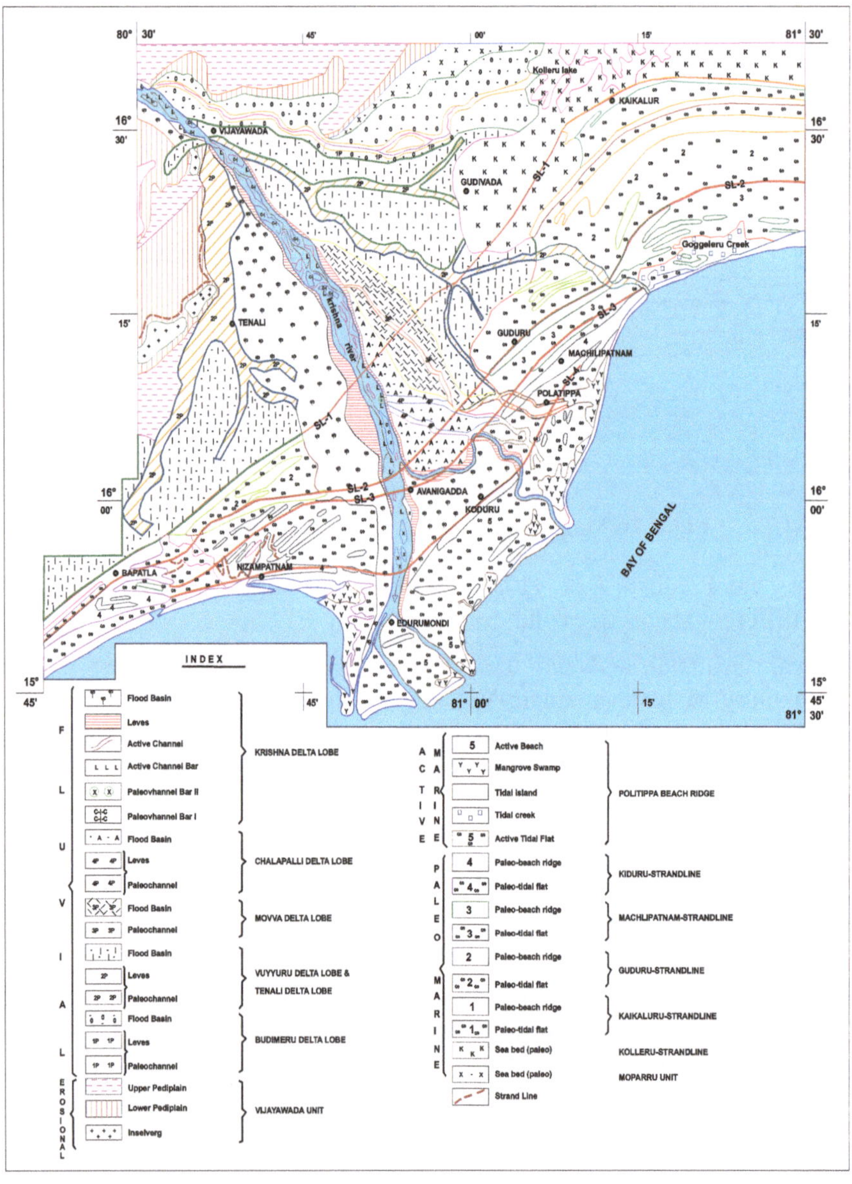

Fig. 2.17 Morphology and evolution of Krishna delta

(Source: Mahender Reddy, K and Shah, B.M, 1991)

The different fluvial delta lobes identified are (i) Budameru delta lobe: The paleo channel exposed near Ibrahimpatnam in the west to Kadasavalli in the east, (ii) Vuyyuru delta lobe: It has two sub delta lobes, namely the Manikonda and Vuyyur. The main Vuyyuru channel bifurcates at Kankipadu into two and along the down slope gets further sub divided into several channels, (iii) Tenali delta lobe: This is a prominent delta lobe on the western side of Krishna river near Tenali. The main paleo channel gets bifurcated into two – one flows south westerly course and terminates at Appikatla and the other flows in south direction and again get subdivided into three channels at Kuchipudi, (iv) Movva delta lobe: This paleo delta lobe extends from Devarapalle to Daliparru in south eastern direction along Meduru, Krishnapuram and Ghantasala, (v) Challapalli delta lobe: The lobe includes paleo channels originating near Nadakoduru, further bifurcates into several channels and ultimately joins the tidal flat regions in the down slope and (vi) Active delta lobe: The present active delta lobe gets bifurcated near Avanigadda and the main channel further down slope gets bifurcated into three channels near Edurmodi. In addition five progradational sequences of events have been identified and are described below:

Stage I: The first stage of Krishna delta formation was initiated at this stage with Krishna bifurcating into Budameru on the east and Tenali paleo channel on the west. The Budameru probably debouched into Kolleru, which was partly a part of the sea and the Tenali lobe ends up into a lagoonal area near Bhattiprolu – Inturu transit.

Stage II: The area between Kaikaluru and Bapatla and the Guduru strand line suggests second stage of delta development. A series of arcuate sand ridges south of Kaikaluru and north of Bantimilli, Inturu represent frequent oscillations of sea at this stage. The Vuyyuru and Movva delta lobes on the east and Tenali paleo channel on the west suggest second stage of delta evolution.

Stage III: The delta continued to be active, advancing and discharging its sediments into Challapalli lobe and then into the sea. At this stage the sea was along Machilipatnam – Avanigadda – Idupalle line. The supply of sediments was less near Peddapatnam and Gogileru creek while it was more towards Machilipatnam reflecting the progradation of the delta.

The ENE-WSW lineaments appear to be present in this region also which could have played some important role in the building up paleo channels and delta lobes present.

Pennar delta

Pennar delta (N.lat 14°40' – 14°20' and E.long 80°00' – 80°12') is an important basin in Andhra Pradesh with an aerial spread of the order of 1420 km^2. This is an independent structural/tectonic basin with about 4000 m sediment thicknesses ranging from Recent to Permian and beyond. The Quaternary delta of this basin mainly evolved during late Pleistocene to Holocene and contain several sets of paleo channels, strand lines and artifacts. The artifacts indicate an age of 6,500 BP (Krishnarao, B., et al 1990). The delta has its apex near Sangham at an altitude of 33 m A.M.S.L and has a gentle slope for a stretch of about 55 kms up to the coastal boundary. The river flows as a single stretch up to Duvvuru from the apex point and then bifurcates into two, one arm reflects as a small buried channel and the other as the main river which flows in easterly direction and debouches in to the sea near Utukur. At least five sets of lineaments in the delta frame could be identified (Radhakrishna. I., 1997), with each set showing a distinct trend and orientation. The lineaments are oriented either in NW-SE or NE-SW direction. The initial lineaments possibly are characterized by low angle and the subsequent lineaments are characterized by increasingly large angles with respect to east-west axis. The river initially must have drained along these low angle lineaments and gradually spread towards south in the form of a fan shaped delta. The fan shaped paleo delta has been stable for quite some time and the fluviatile sedimentation processes could have ceased subsequently in most parts of the paleo delta. The lineament trends/boundaries and paleo channel trends are disposed by and large in the same aereal proximity indicating a close structural control. Kameshwararao, K., (1991) while analyzing the evolution of Pennar delta has identified five sets of major paleo channels on both sides of the present active river, three ancient beach ridges (strandlines), six depositional geomorphic surfaces and one erosion surface reflecting the state of hydrogeomorphic dynamics in this part of the coast.

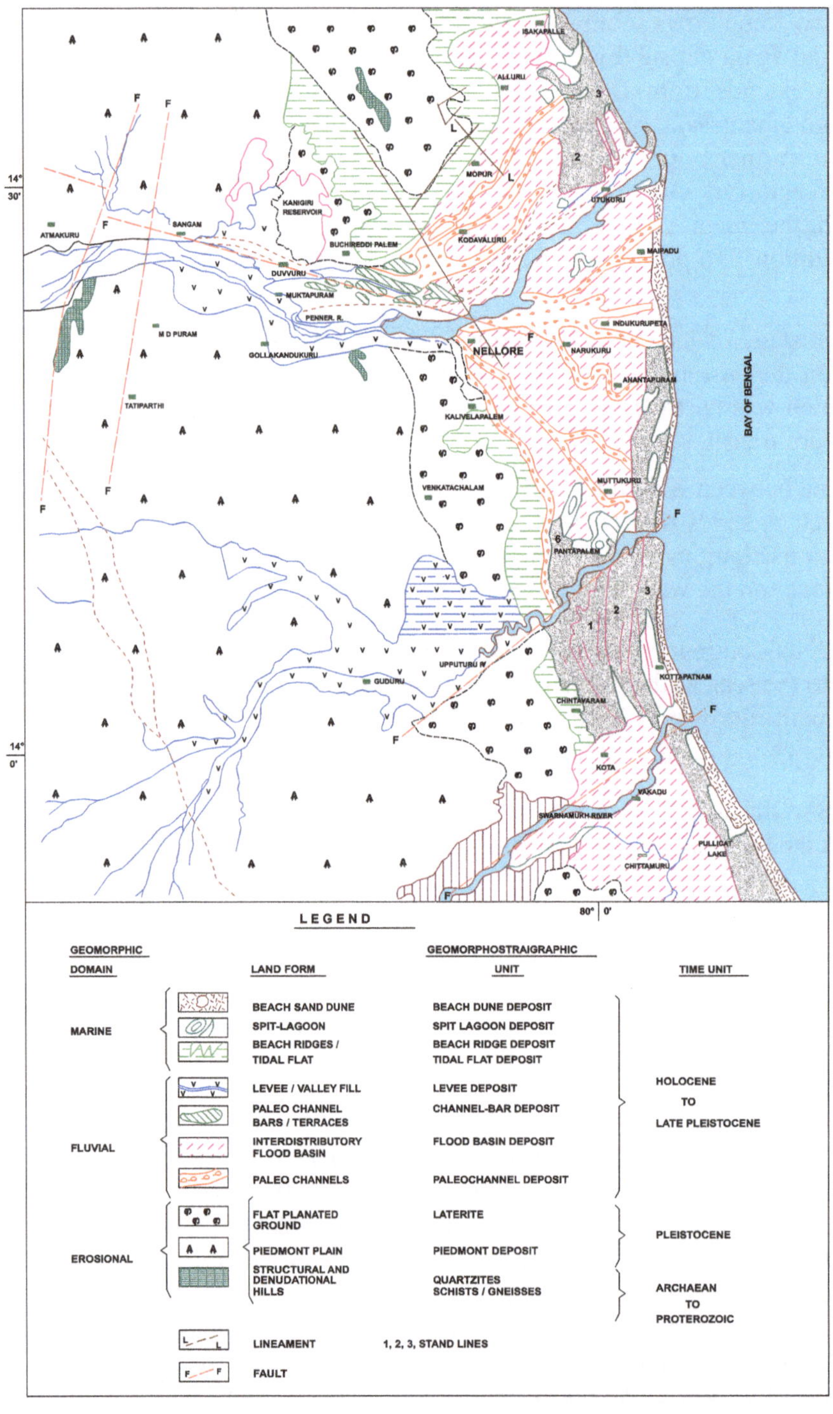

Fig. 2.18 Morphology and evolution of Pennar delta

(Source: Kameshwararao, K., 1991)

The evolutionary trends of Pennar delta basin have been studied by Kameshwara Rao, K. (1991) and Radhakrishna, I. (1997). These studies indicate existence of large net work of paleo channels both to the north and south of the present river channel. In addition a number of land forms have also been identified like flood plains, terraces, river mouth bars, lagoons, ancient beach ridges, old tidal flats, mud flats and spits. The river unlike other major river deltas in south do not exhibit any major distributaries and end up near Utukur near the shoreline in a curve linear form but significantly exhibits a number of paleo channels both towards north and south. At least three sets of paleo channels with delta lobe characteristics could be traced south of the river bifurcating near Nellore and another two or three sets could be traced north of the river bifurcating a little bit upstream from Duvvur/Bucchireddipalem area as shown in Fig.2.18. The trends of the paleo channel distribution indicate that the initial delta built up had taken place north of the present river with its apex near Duvvur. Subsequently the river shifted its course along the present flow direction and built up its delta south of the present river with its apex near Nellore and later on the river migrated north and attained its present position, building a network of paleo channels during its process of migration. The depth and stability of the paleo channels is more predominant and contain relatively large thicknesses of alluvial fills in south when compared to northerly spread out paleo channels and consequently the alluvial sands in southern region act as rich resource belts for groundwater development in this region. A number of lineament trends essentially oriented in NW-SE and NE-SW could be traced from landsat maps, which probably control the paleo channel distribution and migration in this part of the delta.

2.4.3 Mahanadi, Brahmini and Baitarni coastal delta basins (Orissa)

Mahandi-Brahmini-Baitarni delta complex (N lat $19^{\circ}30'$-$20^{\circ}40'$ and E long $85^{\circ}20'$-$87^{\circ}0'$) is the largest complex with an aerial spread of about 9500 sq. kms along the Orissa coast. This is an interlaced composite delta system, combining all the three river systems which take independent origin at the source, but joins together near the middle and outfall areas of the coast. Mahanadi is the single largest of the river delta systems and has several distributaries with its apex located near Naraj in the Cuttack district. This is classified as a fan shaped arcuate delta system with several distributaries and paleo channel networks. Meijerink, A.M.J. (1971); Sambasiva Rao, M., et al (1978); Babu, P.V.L.P, (1978); Mahalik, N.K., (2000); Bharali, B. et al (1991) and Vinod kumar, K. et al (2003); made substantial contributions towards understanding the morphology and evolution of this delta system. The major distributaries and sub distributaries of Mahanadi delta region are Birupa, Mahanadi/Chitrapada, Katjhori/Devi, Koyakhai/Daya and Kushbadra/Bhargavi. Similary rivers Brahmini and Baitarni gets branched off into two with their apexes respectively at Jeenapur and Akhudapada and join together in the outfall regions as Maipura and Dhamra rivers. All these three river systems at one point or the other get interlaced with each other and combine themselves as a major composite delta system. A major part of the delta is flat with gentle slope towards the coast. A typical characteristic geomorphic feature of Mahanadi-Brahmini-Baitarni delta system is the formation of long narrow islands formed as a result of interlacing of distributaries in the central and outfall regions of the delta. A large lagoon (Chilka lake) in the extreme southern tip of the delta with well developed spits and barriers exist. Geomorphic features like ancient channels/delta lobes, ancient beach ridges, coastal sands, swamps/marshes etc. are commonly observed. The geomorphic features of the entire Mahanadi-Brahmini-Baitarni delta complex could be classified under three categories (i) Upland region with lateretic table land, river terraces and valley fills. These features are commonly seen near apex points of the deltas, (ii) The mid deltaic region with extensive alluvial plains, paleo channels, meander scrolls and oxbow lakes, levees, tidal falts and backswamps and (iii) The coastal zone which includes lagoons, estuaries, spits and bars, tidal flats and marshes and beach sand ridges.

The lineaments identified and projected in these delta systems are of four groups: (i) NW-SE lineaments, (ii) NE-SW lineaments, (iii) EW lineaments and (iv) NS lineaments. By and large all the lineaments appear to have some control or the other on the distribution and alignment of the river distributary channels. At least 3 to 4 strand lines and about ten meandering ancient channels are observed in this deltaic region as reflected in the Fig. 2.19.

The first stage strand line runs along and about 30-50 kms from the present coast line and the successive strand lines are disposed very close to the coast. The earliest paleo channels in the first stage are seen primarily from Jajpur to Rajkanika in the mid upstream portions of Brahmini-Baitarni deltas. About five sets of paleo channels are seen prominently in the first and second stages of development of the delta. A major evolution of delta build up has taken place during the first and second stages of the delta. The third stage of development is represented by one paleo channel in Brahmini and one in Mahanadi delta region. During fourth stage of development all most all the distributaries of Mahanadi, Brahmini and Baitarni are stabilized in the present position, except in the north east portion of Brahmini outfall delta region. All these paleo river channels could act as good resource areas for near surface groundwater development.

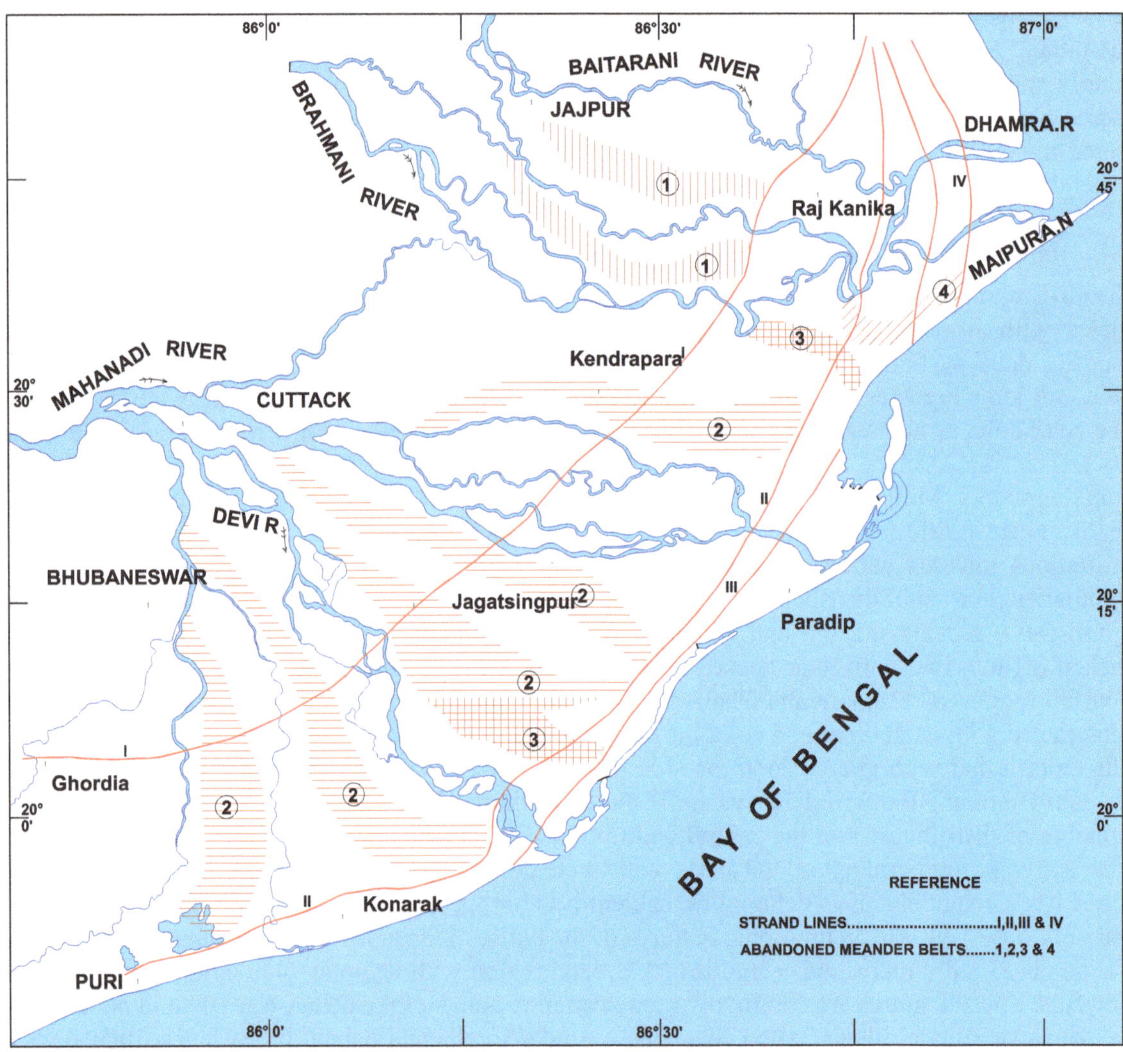

Fig. 2.19 Morphology and evolution of Mahanadi delta basin

(Source: Bharali, B. 1991 and Vinod Kumar, K. et al 2003)

2.4.4 Bhagiradhi-Hoogly coastal delta basins (West Bengal)

Bhagiradhi-Hoogly is a part of the major Gangetic delta system located in the southern region of West Bengal. Bhagirathi takes off as a distributary of Ganga near Farrakka barrage at the upstream region of West bengal and flows in a southernly direction, joins the Hoogly in the outfall region before debauching into Bay of Bengal. A number of tributaries like Mayurakshi, Ajay, Damodar, Rupnarayan and Kasai join the main Bhagiradhi-Hoogly river at different points. In view of its peculiar tectonic and structural framework and the paleo geographical boundaries of ocean-continental margins in this part of the coast, the delta build up of all these small and medium river tributaries had taken place along the western flanks of Bhagiradhi–Hoogly river as well as in the region parallel to the present coast line. A Quaternary morphological map is presented in the Fig. 2.20, which shows the nature and distribution of deltaic plains in this region. The basin essentially is divided into four units namely (a) Lateritic upland areas; (b) Older deltaic plains; (c) Younger deltaic plains and d) Recent sediments. A dominant hydrogeomorphic unit in this part of the coast is Sundarban mangrove ecoforest system spread over an area of 4260 sq. kms with several tidal units protruding into coastline. Some of the common landforms seen in this part of this delta area are paleo channels/ lobes in the vicinity of the present flowing river, beach ridges fresh water and brackish water swamps, lakes and raised terraces etc.

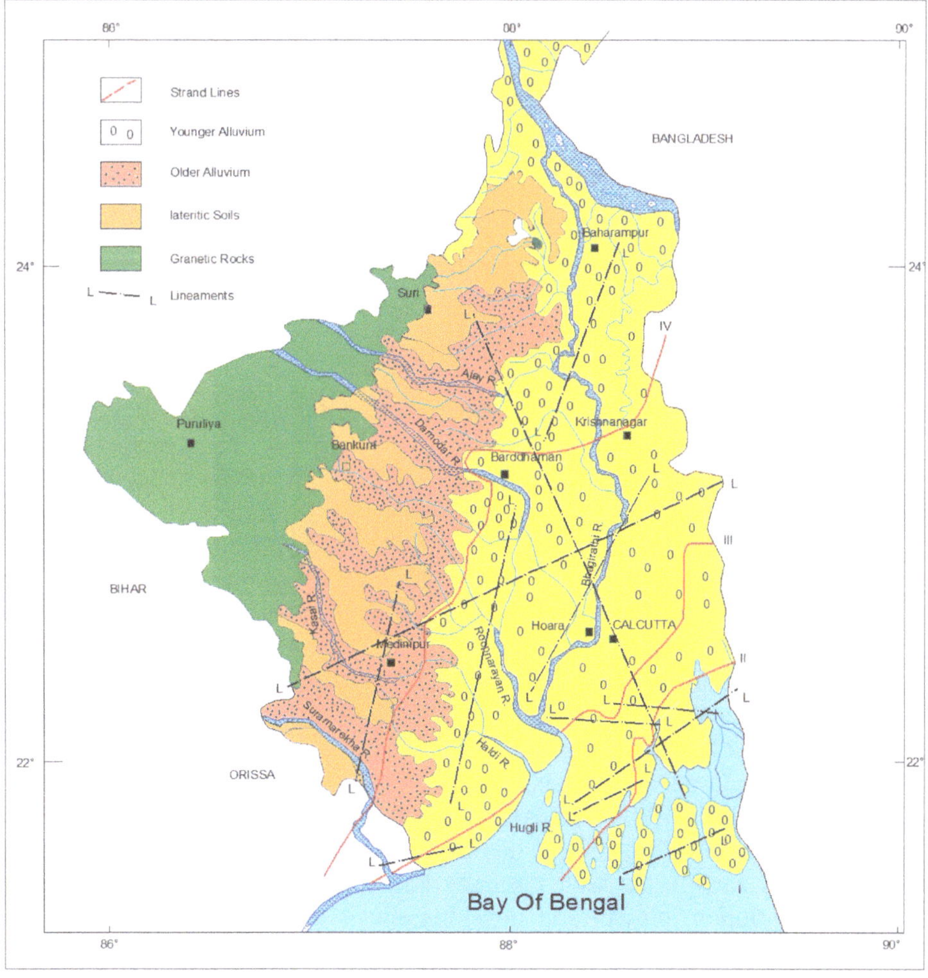

Fig. 2.20 Morphology and evolution of Bhagiradhi-Hoogly delta

At least three sets of lineaments could be traced in and around Bhagirathi-Hoogly delta region of West Bengal (Agarwal, R.P. and Mitra, D.S.,1991). (i) A set of northeast-southwest lineaments running along Medinapur-Bolapur region, Damodar region and Chinchura-Krishnagarh region, (ii) A second set of lineaments running in northwest-southeast direction in the region of Bundarbahar-Kalindi and Jalbandh-Kishoregenj regions and (iii) A number of east west lineaments in the region of tidal flats and marshy shore line region. The northwest southeast lineaments are believed to be younger than northeast southwest lineaments. A number of sympathetic lineaments in line with the major lineaments are observed in the delta region. The evolution of modern Bengal delta has been dealt by a number of workers (Morgan, J.P. and McIntyre, W.G., 1959; Curray, J.R. and Moore, D.G., 1971; Curray, J.R. 1980; Ganju, J.L. and Mitra, D.S., 1982; Badhe, Shubhada, 1983; Vaidyanadhan, R. and Ghosh, R.N., 1993). At least three stages of modern delta evolution have been recognized. The first stage of evolution has probably taken place when the sea levels were lowered during Wisconsin, the last glacian period. The second stage commenced during flandrian transgression period when the sea levels gradually rose to the present position. The third stage commenced with the progradation of the present delta. The progradation of the delta is initially commenced in the northwestern flanks and gradually migrated towards east leaving behind several abandoned deltaic plains, point bars, marshes etc. The lower deltaic plains are continuously reworked by several tidal channels. At least four stages of delta formations have been recognized (Agarwal, R.P. and Mitra, D.S., 1991). The oldest strand line (fourth) has been identified on the basis of paleo delta lobes of Damodar and Kasai rivers in the upstream western flanks of the delta. The third strand line (three) is traced based on several beach ridges present in the east and northeast portion of this region. The second strand line (second) is identified based on the traces of termination of tidal flats and paleo channels. The first strand line forms the present coastline. Most of the paleo channel networks/fills during different stages of delta evolution in this part has been substantially covered underneath the soils and Recent alluvium and are not clearly detectable, but most of them form good near surface groundwater resource systems in this part of the coast. The lineament patterns observed in NW-SE, NE-SW and E-W directions appear to control the paleo- river migrations and evolutionary trends of this delta basin.

2.4.5 Narmada-Tapti coastal delta basins (Gujarat)

Narmada and Tapti are classified as major river basins along the west coast of India with catchment areas of the order of 98,796 sq. kms and 65,144 sq. kms respectively. They rise from the high hills of central India at an elevation of 900 m and 720 m AMSL and with river course lengths of 1312 kms and 720 kms and debouch near Bharuch and Surat respectively. Significantly both the river systems do not have any major deltas but have medium level estuary deltas with aerial spreads of the order of 200 sq. kms. The structural and topographical and tectonic history of these basins could have played a major role in minimizing the delta spreads in these tracks. The erosional and depositional land forms developed within the frame work of these estuary deltas could be related to different evolutionary changes during Quaternary period reflecting the neotectonic and eustatic sea level changes. Both these delta systems have typical characteristic funnel shape and fall under macro tidal and mesotidal estuaries respectively (Hayes, H.O., 1975 and Davis, J.L., 1972). A typical geomorphic feature in the mouth area is built up of estuary islands namely Aliabet (Narmada mouth) and Kadiabet (Tapti mouth area) (Ganapathi, S. and Pandey, A.N., 1991). The Aliabet island has a stretch of about 30 kms in length and 13 kms width. Similarly the Kadiabet island has a length of about 4.75 kms and a width of about 2.7 kms. These islands essentially are seen normally above the water level line and are made up of sands and clays. The evolution, growth and development of these islands have been explained by different workers (Bedi, N. and Vaidhyanathan, R., 1982; Mukharjee, M.K, 1985 and Nayak, S.R. et al., 1987).

The other geomorphic features observed in the estuary regions are barrier ridges and marshy areas. Barrier ridges are the depositional features seen as isolated patches parallel to the coast line in the back shore areas of Narmada and Tapti coastal regions. These are seen extending to a length of about 6 kms with widths varying from 300 m to 600 m and elevations varying from 10 m to 15 m AMSL. The marshy areas are seen to occur extensively in the low lying areas of the coast line and along the fringe areas of the islands (low marshy areas) and are also seen to occur in the high land areas of the islands also (high marshy lands) as shown in Fig. 2.21. The low marshy areas probably indicate the lowered sea levels and high marshy areas represent periods of transgressions. The Quaternary sediments in these delta regions have thicknesses varying from 165 m (Tapti outfall area) to 525 m (Narmada outfall area) and in general consist of sands, gravels, sandstones and poorly compacted claystones and shales. Sands and gravels probably deposited under fluviatile to subareal delta conditions and clays and shales under mixed environments (Datta, A.K., 1974). The Quaternary land forms of Narmada and Tapti estuary deltas reveal three stages of strand lines including the present sea level, (i) A regressive phase, (ii) A transgressive phase and (iii) A regressive phase (present). These phases could be compared with wurm, flandrian and the present shore line. The Quaternary land forms in general reflect on the affects of neotectonism and eustatic sea level changes in these regions.

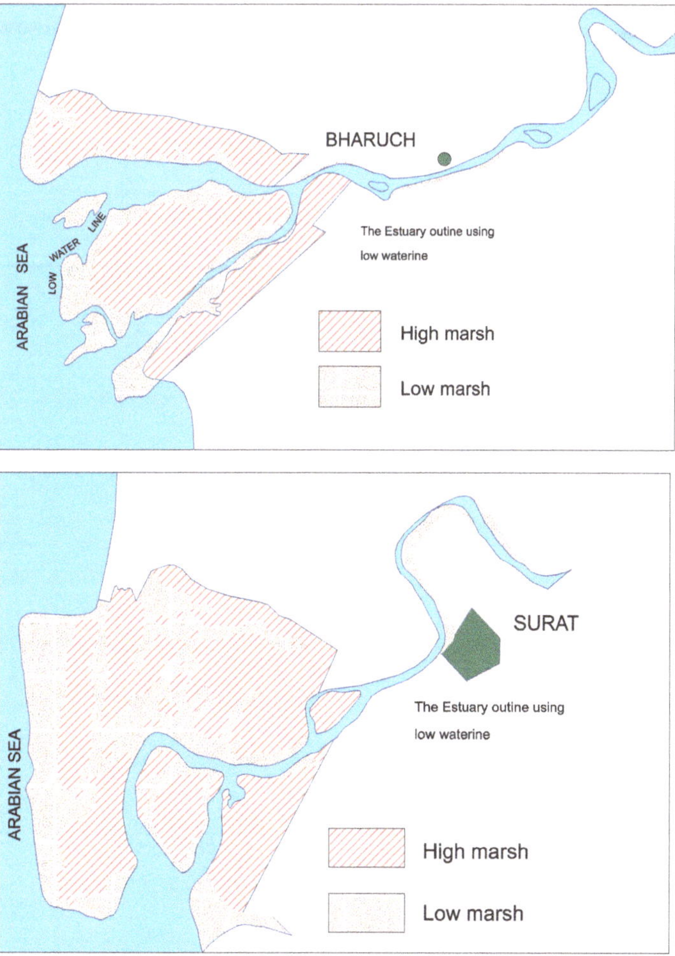

Fig. 2.21 Morphology and evolution of Narmada and Tapti estuary deltas (Gujarat)

(Source: Ganapathi, S. and Pandey, A.N., 1991)

CHAPTER 3

HYDROLOGICAL CHARACTERISTICS AND SOIL WATER BALANCES OF COASTAL DELTAS

The Indian subcontinent is endowed with large network of river systems with a good built up of delta systems near the coastal margins. All most all major and medium river systems take their origin in the far interior high lands of Himalayan and Peninsular regions and end up either into Bay of Bengal or Arabian sea. There are at least fourteen major river systems, fourty four medium river systems and several minor river systems discharging water and sediment into Bay of Bengal and Arabian Sea (Fig. 3.1). Three of the major river systems namely Indus, Ganges and Brahmaputra take their origin from the high elevated plateau regions of Himalayas, flow in different directions over long stretches and ultimately join the Arabian sea (Indus) and Bay of Bengal (Ganges and Brahmaputra) making huge delta systems before debouching into the oceans. The rest of the major river systems namely Cauvery, Pennar, Godavari, Krishna, Mahanadi, Brahmini, Subarnarekha (all east flowing rivers) and Narmada, Tapti, Mahi and Sabarmati (all west flowing rivers) rise from the Peninsular high lands, flow towards east and west respectively and join the Bay of Bengal and Arabian seas. Almost all the major east flowing rivers have built up reasonably good moderate delta systems, where as the westernly flowing rivers have relatively either small or insignificant delta systems excepting Narmada and Tapti, where the delta systems are small and are of estuary type. The physical and hydrological characters in brief of the river systems/delta areas are mentioned in the Table 3.1 (Rao, K.L., 1975; Chandramohan P. et al., 2001).

Table 3.1 Hydrological characteristics of major river systems of India

S.No.	Name of the river	Catchment Area (Sq.Kms)	Length (Kms)	Annual discharge (MCM)	Annual Sed. Load (x 10^9 kgs)	Delta area (sq.kms)	Remarks
1.	Indus	321,289	1114	207,800	450.000	41,440	
2.	Ganga	861,404	2525	493,400	329.000	1,38,000	
3.	Brahmaputra	187,113	916	510,450	597.000		
4.	Sabarmati	21,674	371	3,200	0.018	----	
5.	Mahi	34,842	583	8,500	5.879	----	
6.	Narmada	98,796	1312	40,705	44.358	250	Approximate
7.	Tapti	65,145	724	17,982	10.522	120	Approximate
8.	Subarnarekha	19,300	395	7,940	5.911	500	Approximate
9.	Brahmini	39,033	799	18,310	13.277	9500	
10.	Mahanadi	141,589	851	66,640	13.204		

Table 3.1 *Contd...*

S.No.	Name of the river	Catchment Area (Sq.Kms)	Length (Kms)	Annual discharge (MCM)	Annual Sed. Load (x 10^9 kgs)	Delta area (sq.kms)	Remarks
11.	Godavari	312,812	1465	105,000	38.839	6320	
12.	Krishna	258,948	1401	67,675	0.320	5476	
13.	Pennar	55,213	597	3,238	0.257	1775	
14.	Cauvery	81,155	800	20,950	0.471	10,350	
	Total All river basins:	**2575,949**	**13,853**	**10,92,660**	**1509.056**		
	(or Say 258 Million Sq. Kms.)						

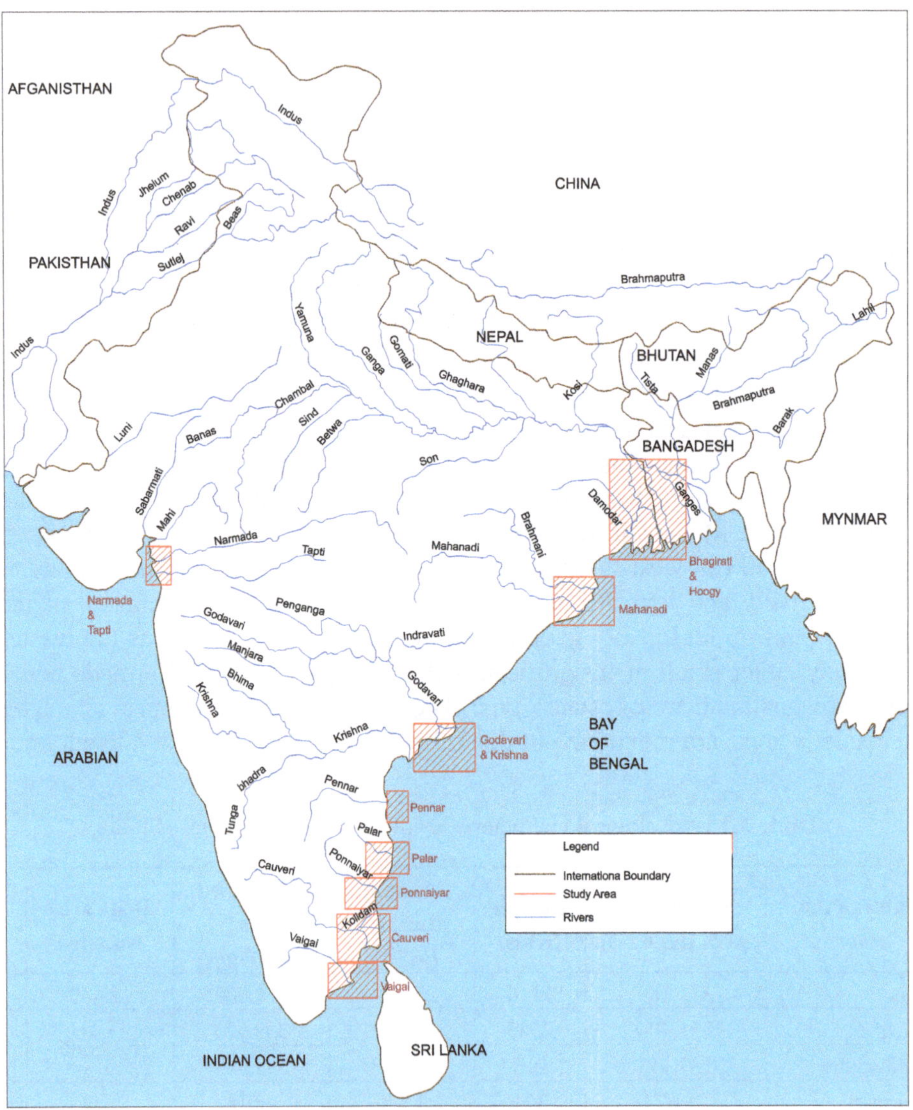

Fig. 3.1 Coastal river/delta systems of Peninsular of India

In addition to the above major river systems, there are 19 nos. of medium level west flowing rivers and 21 nos. of east flowing rivers with a total catchments area of the order of 2,13,567 sq.kms. and with a total discharge potential of the order of 95,535 MCM. (Rao, K.L., 1975).

Table 3.2 Annual run off and sediment load characteristics in major river/delta systems of east and west coast of India

Name of the Delta	Parameter	1986-87	1987-88	1988-89	1989-90	1990-91	1991-92	1992-93	1993-94	1994-95	1995-96	1996-97	1997-98	1998-99	1999-2000	2000-01	2001-02
Godavari (Pola-varam)	**Discharge**	...	...	...	...	178.3	71.58	86.937	58.075	160.54	86.552	58.56	47.08	74.82	96.823	75.28	...
	Sediment load	0	20	160	75	150	40.2	79.964	30.047	57.729	62.279	27.01	21.93	40.35	44.772	86.42	48.66
Krishna (Vijaya-wada)	**Discharge**	4.952	4.271	38.22	25.32	26.86	36.60	8.625	11.211	39.2	6.026	...	...	...	...	...	...
	Sediment load	2	2.15	3.5	9.2	2.155	2.875	0.203	1.122	3.514	0.7	0	1.011	2.355	0.591		
Cuavery (Musiri)	**Discharge**	...	...	...	4.557	3.774	7.768	10.45	7.596	12.672	6.582	6.595	9.754	8.538	7.539	...	...
	Sediment load	...	...	...	0.27	0.086	0.417	0.495	0.954	0.64	0.203	0.33	0.39	0.278	0.281	...	...
Mahanadi (Tikara-para)	**Discharge**	59.60	23.915	23.87	29.37	71.77	45.43	47.724	36.429	125.24	31.732	24.67	...	...	...	...	...
	Sediment load	20	7.2	7.8	7.2	20	20	20	8.897	56.048	10.283	6.058	11.38	6.623	3.876	2.729	16.47
Narmada (Garudesh-war)	**Discharge**	0.301	0.149	1.616	1.026	50.87	25.49	20.685	35.55	73.464	23.392	26.95	...	...	...	...	...
	Sediment load	...	...	...	...	51.53	27.44	15.404	9.946	36.534	2.253	8.325	7.398	4.384	15.325	...	...
Tapti	**Discharge** (sarang-kheda)	6.860	2.751	16.81	8.243	16.33	4.238	5.650	7.884	14.653	4.200	6.404	...	...	...	...	...
	Sediment load (Sarang-kheda)	??	??	54.78? ?	19.35? ?	39.13? ?	9.637 ??	9.186? ??	18.2 ??	23.308 ??	??	??	6.987 ??	...	...	...	...

units: *Discharge in B.C.M, Sediment load in M.M.T*

(Source: C.W.C., 2006)

3.1 HYDROLOGICAL CHARACTERS OF MAJOR RIVER DELTA SYSTEMS

All the major river systems invariably are characterized by large catchment areas with a number of tributaries joining the main stream and with large delta systems and a number of distributaries. The major delta systems under review are shown in Fig. 3.1. The catchment areas of a river with their tributaries are convergent systems where the water and sediment loads are collected from various tributaries and sub-tributaries, get transported into the main stream and then discharges into the divergent delta systems where the water and sediment loads are distributed through several distributaries and ultimately into ocean bodies. The annual discharge/runoff and sediment load parameters and trends for selected major river systems namely Godavari, Krishna, Cauvery, Mahanadi of east coast and Narmada and Tapti of west coast are presented in Table 3.2 and are graphically shown in Fig. 3.2 (a), 3.2 (b) & 3.2(c).

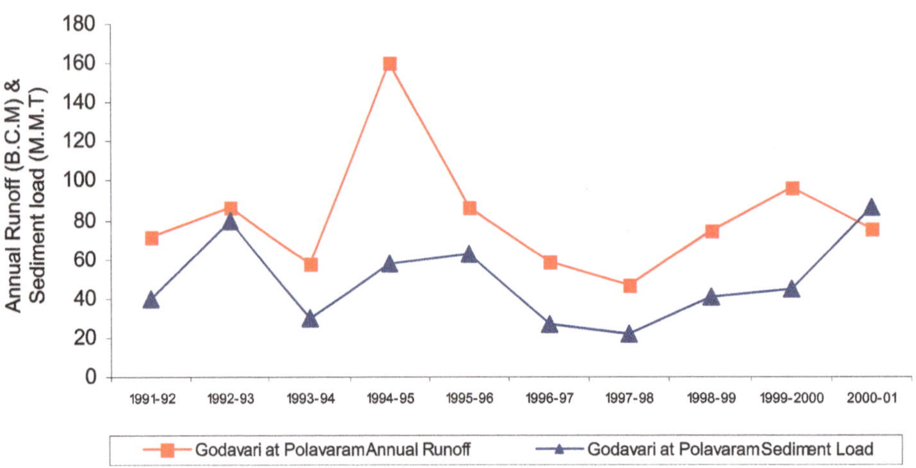

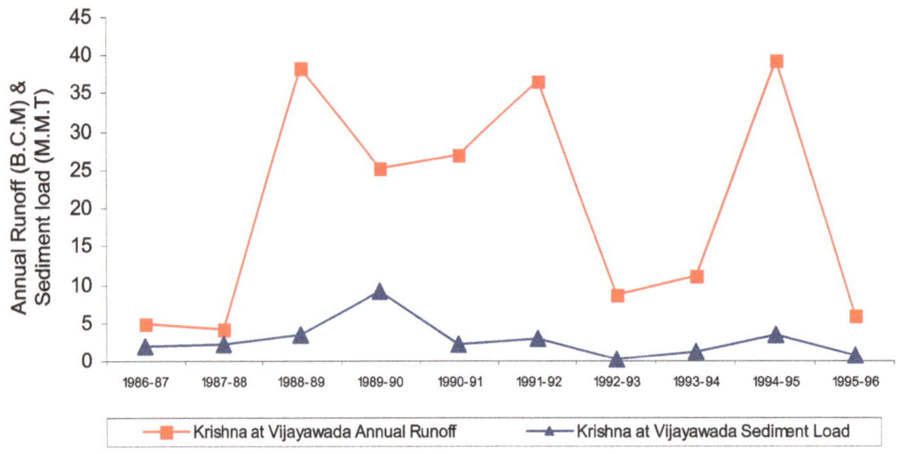

Fig. 3.2(a) Runoff and sedimentary load trends at the terminal points of Godavari and Krishna

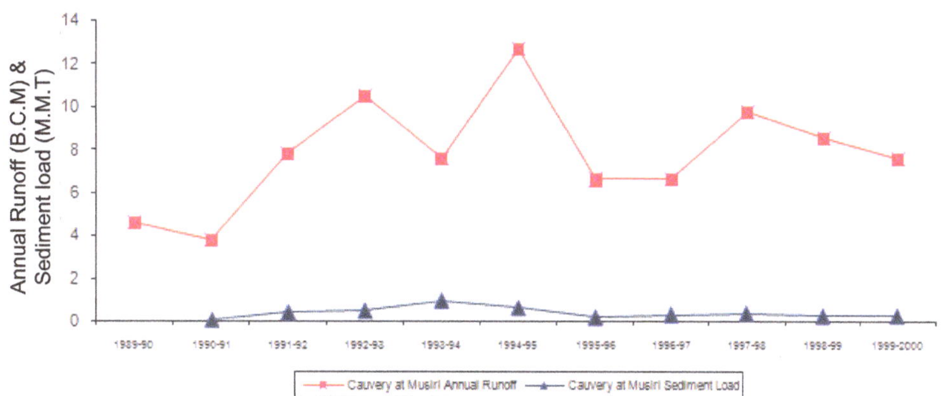

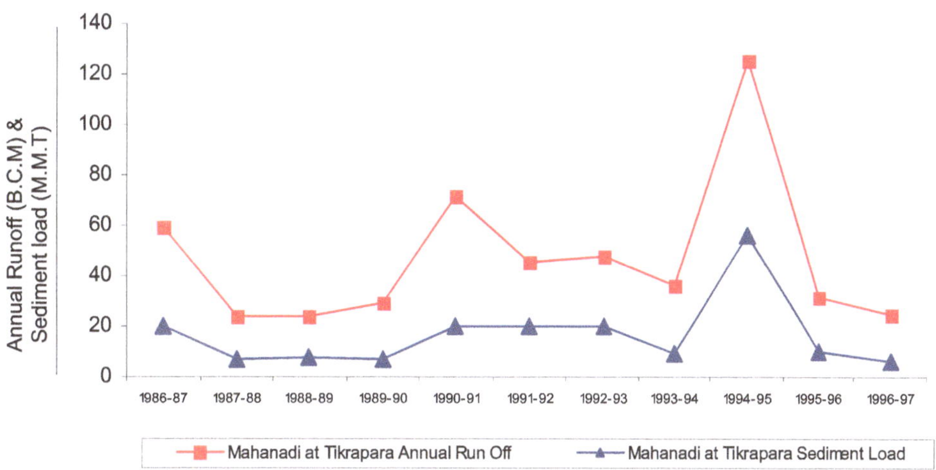

Fig. 3.2(b) Runoff and sedimentary load trends at the terminal points of Cauvery and Mahanadi

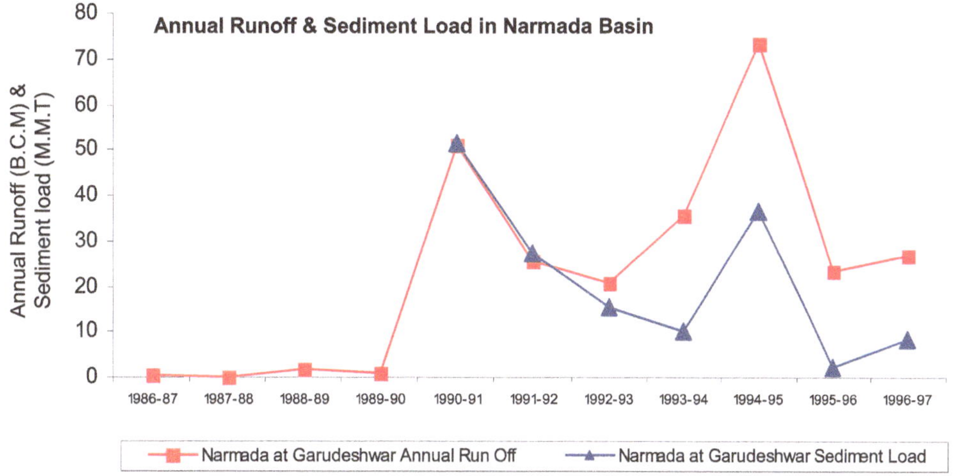

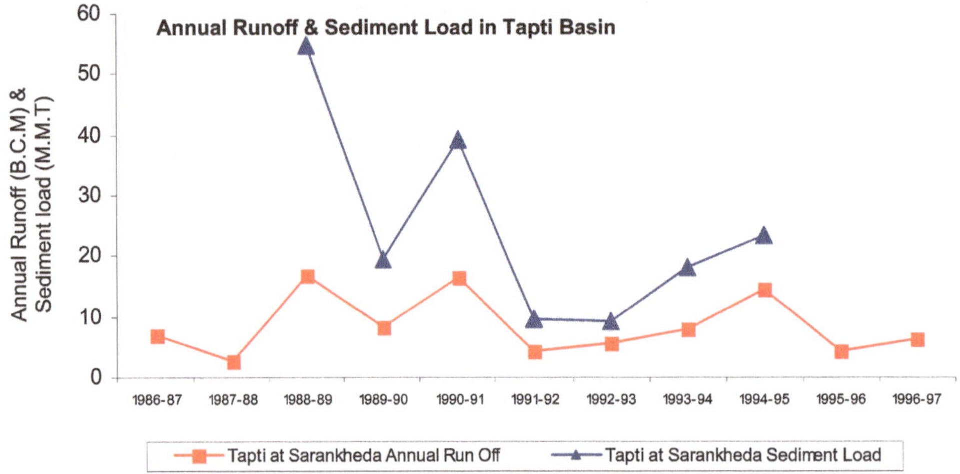

Fig. 3.2(c) Runoff and sedimentary load trends at the terminal points of Narmada and Tapti

(Source: C.W.C., 2006)

The discharge and sediment loads are reported to be maximum in the case of Ganges, Brahmaputra and minimum in the case of Sabarmati reflecting the nature, turbulence and erosive characters of the river systems. It is also to be noted that some of the major east flowing river systems like Godavari, Krishna and Cauvery have very less loads of sediment transport, when compared to some west flowing rivers like Narmada and Tapti. The sediment loads by and large reflect on the size of the catchment area, length of the river systems, morphotectonics, climate and severity of weathering processes. The present river systems occur in a wide variety of geological settings and environments and knowledge of these processes would help to understanding the delta build up processes, sediment load characteristics and the related coastal processes. Coleman, J.M. (1976) demonstrated that factors like climate, precipitation levels, sediment discharges and their variabilities, river mouth processes, shoreline wave power and currents, tidal regimes, shelf slopes and morphotectonics are responsible for variations and build up of deltas. A graphical inter relations/cross plots between selected river/delta characteristics are shown and described in the following paragraphs.

The deltas are the integral parts of river systems and several workers contributed immensely towards estimations of various riverine and delta build up processes. Coleman, J.M and Wright, L.D. (1971) and Wright, L.D., et al (1974) quantified different processes/variables involved and clustered different deltas into discrete groups. Elliot, T. (1978) proposed a classification scheme by way of plotting the variables in a Ternary diagram to define and describe the fields of wave, fluvial and tidal dominance. A detailed account of depositional environments and sedimentary facies changes in the east coast deltas are described by Singh, I.B. and Seeta Rama Swamy, A. (2006). Coleman, J.M. and Roberts H.H. (1989) had characterized different world deltas by way of several statistical diagrams and cross plots and analyzed various fluvial, wave and tidal dynamics involved in the delta build up processes. A similar criterion is adopted and graphical sketches are drawn for various major delta basins of India and are presented in Fig. 3.3 & 3.4 by way of histograms and cross plots.

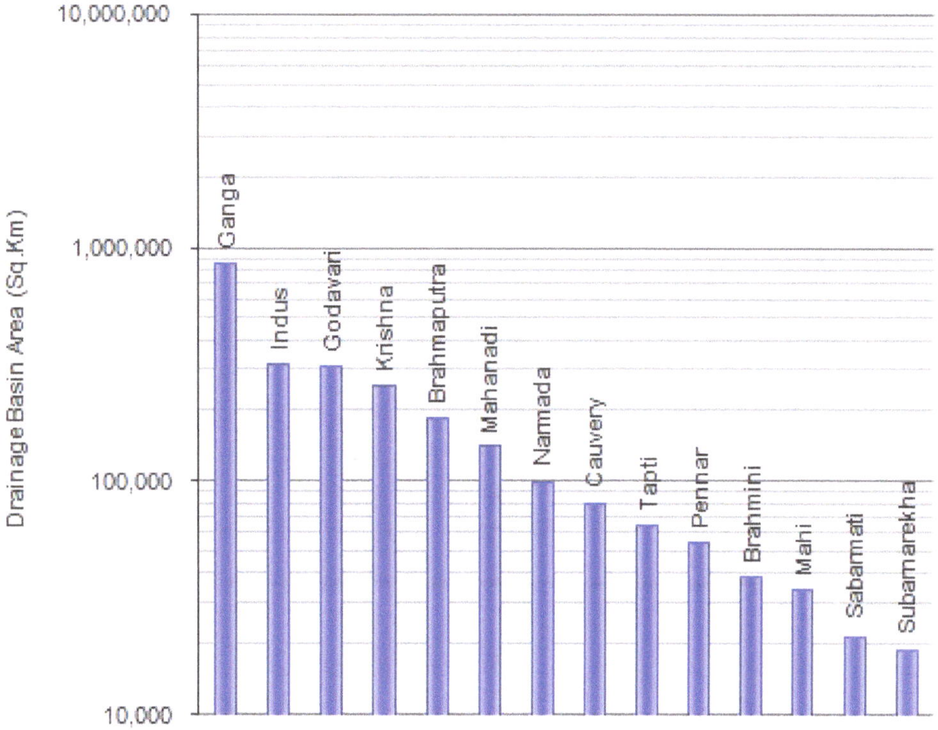

Fig. 3.3 Drainage basin area trends

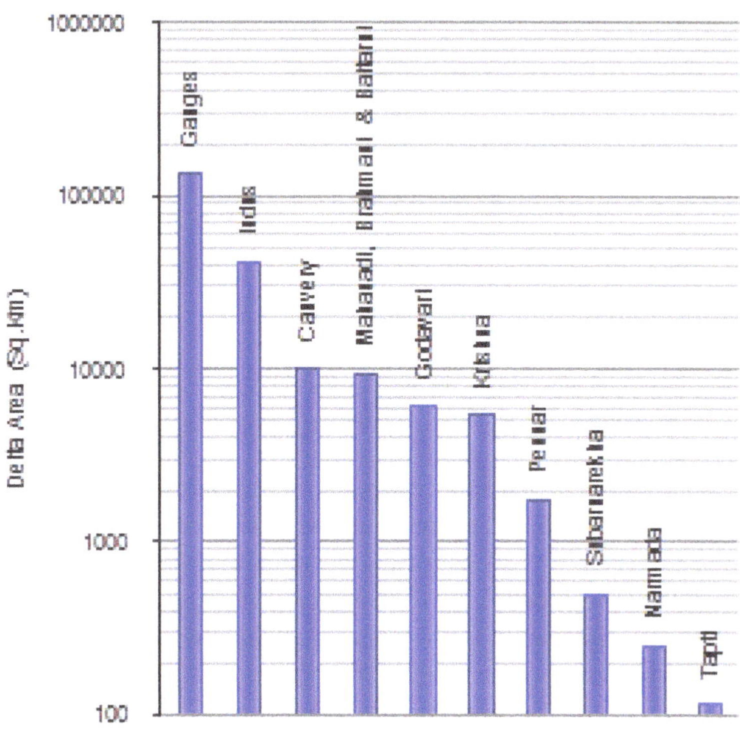

Fig. 3.4 Delta area

The drainage basin area trends for all fourteen major river systems in India are shown in Fig. 3.3. These river systems span over various climatic zones of India and represent a wide variety of depositional settings. The drainage basin area span over wide order of magnitude starting from 1.9×10^4 km^2 (Subarnarekha) to greater than 8×10^6 km^2 (Ganges). Similarly the delta plain area trends of major river systems are shown in Fig. 3.4. The delta area spans also exhibit variations of several order of magnitude starting from 1.2×10^2 km^2 (Tapti delta) to more than 1.4×10^5 km^2 (Ganges delta). There are several other small delta systems with magnitudes less than 1×10^2 km^2 which are not included in the present illustrations.

The cross plot trends between drainage and delta basin area are shown in Fig. 3.5. The trends reflect on the proportional increase of delta areas with increase of drainage basin area. However the trend analyses indicate wide variations/scattering in the delta sizes for different drainage areas. It is also to be noted that there are some basins with significant drainage areas but literally ending up without any delta build up near the mouth area viz., Narmada and Tapti in the west coast and Subarnarekha, Palar, Gundlakamma etc., in the east coast. The cross plots also indicate wide variations within the attributes characterizing wide variations in the delta facies and depositional parameters and the fluvial and marine dynamics but however show stable wave dynamics and depositional processes in almost all deltaic areas of east coast.

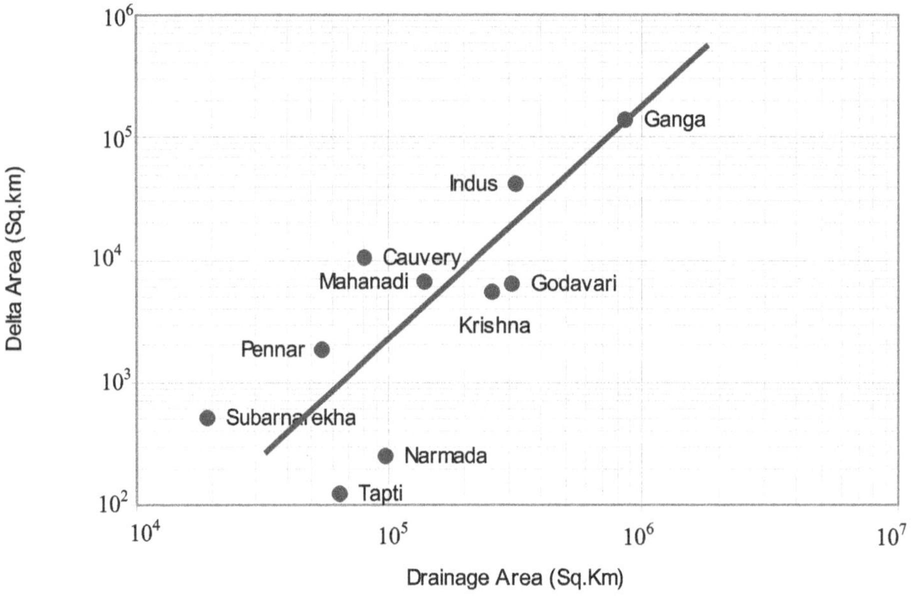

Fig. 3.5 Cross plot between drainage basin and delta basin area

The fluid discharge and sedimentalogical trends at the terminal points for selected river/delta systems are shown in Fig. 3.6 & 3.7. The trends show wide variations in the discharge levels starting from an order of 3.238×10^3 MCM (Pennar) to a magnitude of the order of more than 5×10^5 MCM (Ganges). Similarly the sediment loads vary from a maximum annual load of 926×10^9 kg (Ganges and Brahmaputra) to 0.25×10^9 Kg (Pennar). It is to be noted that the Narmada river on the west coast and Godavari river on the east coast also carry substantial annual loads along with them (44.35×10^9 kg and 38.83×10^9 kg respectively) reflecting the turbulent and erosive characters of the rivers in the catchments and transit areas but significantly Narmada and Tapti rivers along the west coast could not built up any delta systems near the mouth areas probably because of the prevailing geotectonic situations in the respective regions. The cross plot trends between river discharge and sediment loads are presented in Fig. 3.8. The trend shows by and large a linear relationship, but widely scattered reflecting the sediment load and discharge dynamics. Similarly a cross plot between discharge and delta area presented in Fig. 3.9 also shows a near linear relationship but again the points are widely scattered. As the delta area increases the discharge also increases, but with lots of variations.

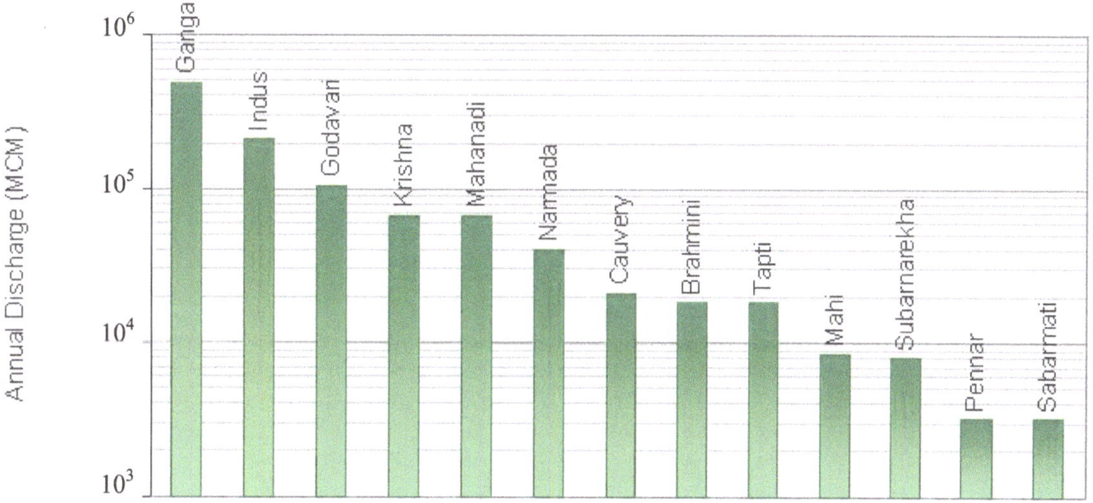

Fig. 3.6 Annual discharge trends for selected major river systems

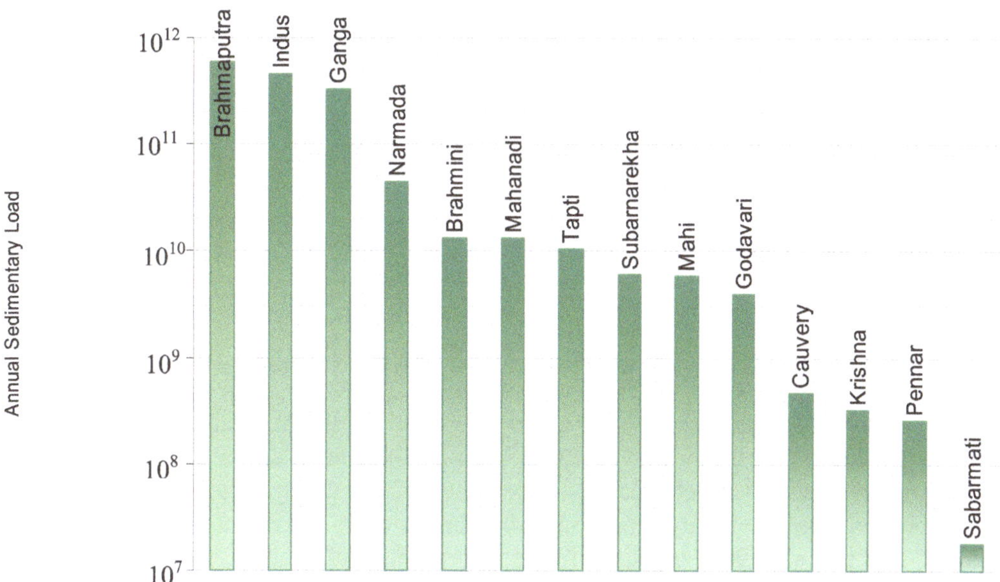

Fig. 3.7 Annual sediment load trends for selected major river systems

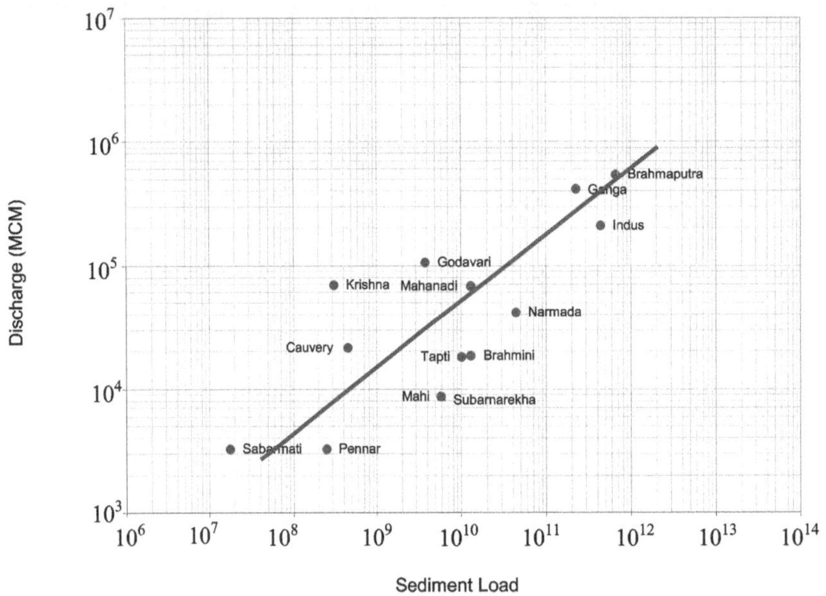

Fig. 3.8 Cross plot between sediment load and discharge for different river systems

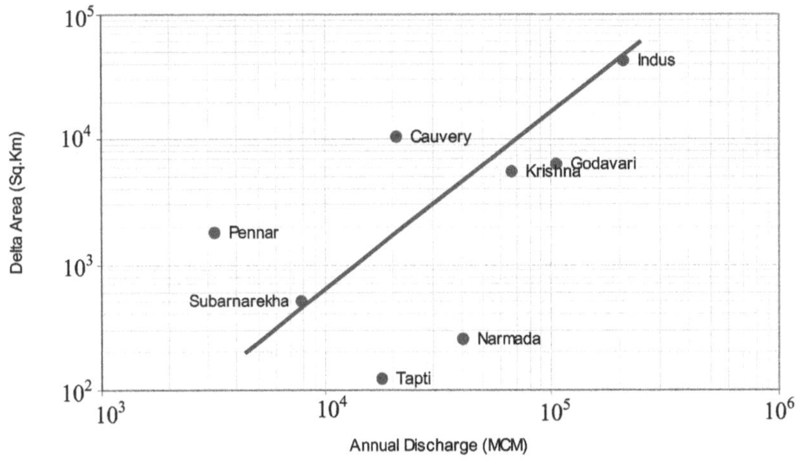

Fig. 3.9 Cross plot between annual discharge and delta areas for different river systems

Some other factors which control the shape and size of the delta areas are the continental shelves, their slopes and tidal processes, adjacent to the deltaic coastal boundaries. Deltas with low offshore slopes display low wave energy levels near the shore line and similarly higher offshore slopes produce higher wave energy events. The tidal processes also control closely the deltaic facies up to their impact areas. Coleman, J.M., et al (1989) classifies three important tidal dominated river delta systems (a) situations where water mass mixing by tidal activity destroys vertical density stratification (b) situations where tidal processes in a year account for high levels of sediment transport both landward and seaward along the river mouth (c) Areas where riverine-marine interactions greatly influence vertical and horizontal stratification of sediments. There are also other factors like land subsidence and related tectonic incidences which control the sedimentation rates and their geometries. An alarming event in the present day deltaic low lands is the loss of sub aerial wet land a phenomenon which is world wide. A more elaborate and detailed studies are required in this direction to evaluate the rapid changes occurring along the deltaic coastal boundaries and in areas where intense erosion and depositional processes are taking place.

3.2 METHODOLOGIES FOR ESTIMATION OF SOIL WATER BALANCE PARAMETERS

Soil water balance is an important pre-requisite to understand and quantify the hydrological cycle events taking place in a specified basin or a region. The important hydrological components involved in this cycle are 1. Precipitation, 2. Evaporation and Evapo-transpiration, 3. Infiltration and Recharge and 4. Runoff. Rainfall is the principle source of replenishment of moisture in the soil zone. The movement of moisture is controlled by capillary action and hydraulic conductivity. A simple soil water balance equation is given as under:

$$P = AE + I + \Delta S + R_o \qquad \text{.....(3.1)}$$

where P = Rainfall;

AE = Actual evaporation;

I = Infiltration;

R_o = Surface runoff; and

ΔS = Change in soil moisture.

A brief description of each of these components and the methodologies involved for their estimations are presented in the following paragraphs briefly.

3.2.1 Precipitation

Precipitation is the amount of water that falls on the earth's surface in the form of rain, snow or hail and is a dominant component in the hydrological cycle and water balance studies. Fairly accurate precipitation levels covering the entire country through India Meteorological data sets and also through the website on India water portal are available freely and adequately for analysis in respect of any region. The India Meteorological department has in operation more than 8000 rainfall stations covering the entire country. The precipitation levels are normally expressed either in inches or centimeters/millimeters and are measured through several non recording type rain gauges or through recording type rain gauges. The non recording type rain gauge basically involves collection of water through a calibrated funnel and measure the rain water collected through a graduated cylinder. The depth of the rainfall is expressed as under:

$$\text{Rain fall in cm} = \frac{\text{Volume collected in cm}^3}{\text{Area of the aperture of the gauge in cm}^2} \qquad \text{.....(3.2)}$$

The recording types of rain gauges give rather a permanent and automatic graph recording without involving any physical measurements. These gauges basically have a calibrated funnel through which rain water is received and enter the chamber consisting of a float and a siphon. A pen is normally mounted on the stem of the float and records fluctuations in the water levels accumulated in the chamber over a chart with a clock control to facilitate continuous recording over a period of time. The system automatically siphons the water when the recording pen reaches maximum and restarts from the zero base. This type of recording system not only helps estimating the rain fall input but also gives rainfall intensity over a period of time.

Sometimes special radar equipment can be used to measure atmospheric precipitation though not very accurate which converts the intensity of radar reflection into precipitation rates. These rates are integrated over a period of time to yield depth of precipitation over an area. These rates could be combined with conventional ground stations to get improved aerial precipitation measurements (Wilson, J.W., 1970). The following methods are in vogue to assess the precipitation depth levels in a basin.

Arithmetic mean method

The average depth of rainfall over a specific basin of a region could be assessed by the following approach.

$$P = \frac{P_1 + P_2 + P_3 + P_4 + \ldots\ldots\ldots\ldots P_n}{n} \qquad \ldots..(3.3)$$

where P = mean precipitation of the basin and P_1, P_2, P_3, P_4......, P_n are the precipitation levels at 'n' stations 1, 2, 3, 4,...n and n is the total number of stations. The above equation/approach may give good results if there is fairly good uniformity in the terrain characteristics.

Isohyetel method

Isohyets are the contours of equal rainfall levels. The point rainfalls are plotted on a map and then the levels of equal rainfalls could be drawn giving consideration to the terrain variations and intensities. The average rainfalls between successive isohyets are weighted in the area between the isohyets. A graphical representation of this method is given by Fig. 3.10.

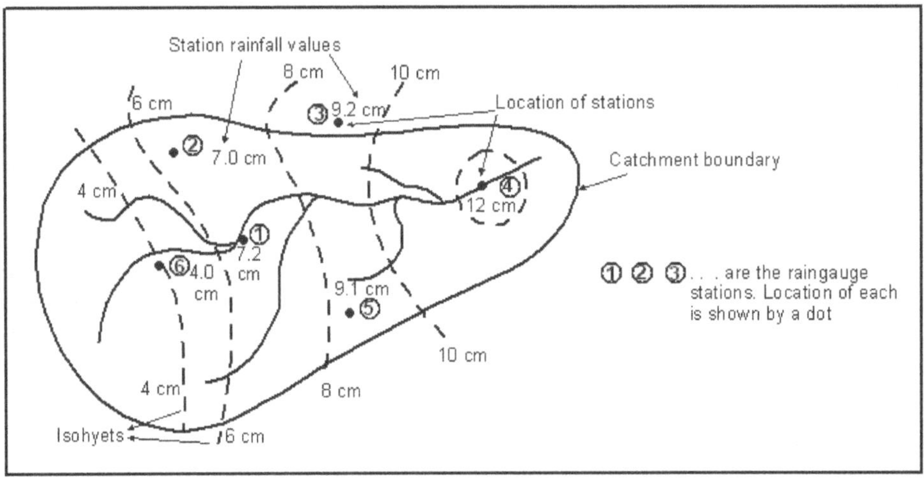

Fig. 3.10 Isohyetel method

Theisson Polygon method

The weighted mean values could be obtained by using Thiesson polygons and drawing perpendicular bisectors to the lines joining adjacent stations. Each polygon area is assumed to be close to the rain gauge station inside – viz. If P_1, P_2 and P_3 are points of rainfall stations in a basin and the areas of polygons surrounding these stations are A_1, A_2, and A_3 respectively, then the average depth of rainfall for the entire basin is given by

$$P_{mean} = \frac{\Sigma A_1 P_1}{\Sigma A_1} \qquad \ldots..(3.4)$$

The results obtained using these methods in general are more accurate than those obtained by simple arithmetic average method. A graphical representation is shown in Fig. 3.11.

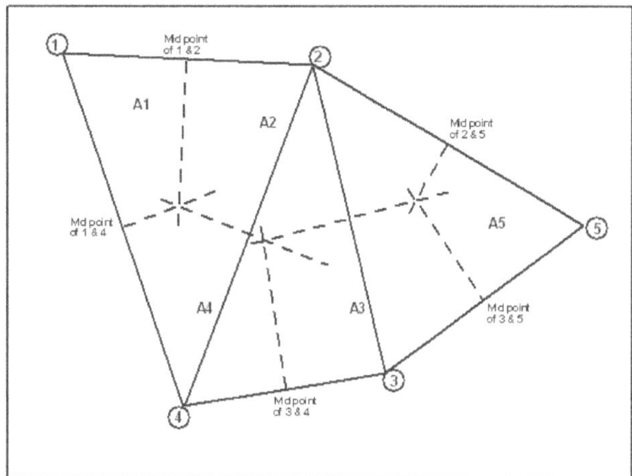

Fig. 3.11 Theisson Polygon method

3.2.2 Evaporation and evapotranspiration

Evaporation is the process where by water received through precipitation on earth's surface is converted in the form of vapor and returned back to atmosphere. Evaporation is expressed in quantitative terms as the depth of water vaporized from a unit surface in unit time: mm/day or m/year. The evaporation rate depends upon several factors like air and water temperature, sunshine, wind velocity and air dryness. Oceans, lakes, rivers and soils are the surficial regions normally subjected to the process of evaporation. Dalton's law is the most basic equation for estimating the evaporation which states that evaporation (E) is proportional to the difference between saturated vapour pressure (e_s) at the temperature of water and the vapour pressure of air (e_a) as expressed below.

$$E_a = K (e_s - e_a) \quad .....(3.5)$$

where K is a constant

where

E = Evaporation rate in mm per day.

e_s= Saturation vapour pressure in the thin air film adjacent to water surface at corresponding temperature of water, called the saturation vapour pressure of water.

Values of e_s at different temperatures by Garg, S.K., 2007 on "Hydrology & Water Resources Engineering" Refer Table 6.1.

e_a = The actual vapour pressure of the air above.

K = A constant, whose value depends upon various factors, such as, wind velocity, humidity, etc.

Pan evaporimeter method

A pan evaporimeter is the generally used equipment for evaporation measurements. This is a USWB (US Weather Bureau) class A pan with 122 cm diameter and 25.5 cm depth, made of galvanized iron with stilling well, vernier point gauge and a thermometer. This equipment is maintained over a wooden stand of 10 cm thick and is covered by a wide screen. The pan is filled with water up to a reference level. The amount of water lost by evaporation after a given time interval is measured by adding known quantities of water to the pan from a graduated cylinder till the water level reaches the reference point or sometimes could be measured directly by a point gauge. The readings are normally taken twice a day, at 08:30 hrs and 17:30 hrs. Simultaneously steven screens and anemometers should be made available at the site for measurements of temperature and wind velocities. Several estimations made in India through these experimental steps indicated values several times greater than natural situations with large water bodies and soil regions covered with vegetation etc. In addition some empirical relation/formulae used in this connection are:

(a) *Meyers formula (refer: Garg, S.K., 2007)*

$$E = K_m \left(e_s - e_a\right)\left[1 + \frac{V_9}{16}\right] \quad .....(3.6)$$

where E = Evaporation from water body in mm/day.

e_s = saturation vapour pressure at water surface temperature.

e_a = actual vapour pressure of air in mm of mercury at specific height (9m).

V_9 = monthly mean wind velocity in Km/h at 9m above ground level.

K_m = a coefficient having value of 0.36 for large deep water and 0.50 for small and shallow waters.

(b) *Rohwer's formula(refer: Garg, S.K., 2007)*

$$E = 0.771\,(1.465 - 0.000732\,P_a)\,(0.44 + 0.0733\,V_{0.6})\,(e_s - e_a) \quad .....(3.7)$$

where E, e_s and e_a are same as defined above in the equation 3.6

P_a = mean atmospheric pressure (mm of mercury)

$V_{0.6}$ = mean wind velocity in km/h (0.6 m above ground level).

3.2.3 Soil evaporation and evapotranspiration

Evaporation from a soil surface saturated with water nearly equals the evaporation from free surface water. The evaporation which takes place from the soil grains immediately after the cessation of rains is termed *'soil evaporation'*. The evaporation rates normally are high immediately after the rains, but gradually decrease as the ground surface begins to dry until a constant rate is reached depending upon the water table depth and the climatic conditions. The maximum amount of water the soil can hold against the force of gravity is termed *'field capacity'*. The minimum amount of moisture that is held by soil which is not available for transpiration by vegetation and plants is termed *'wilting point'*. The difference between the field capacity and the wilting point is the *'soil moisture'*. The soil evaporation diminishes when the soil surface dries up gradually and ceases altogether when vapour pressure in the soil equals that of the atmosphere. The soil moisture contents could be estimated by several methods: 1. gravimetric method 2. Lysimeters method and 3. Neutron probes method.

The gravimetric method involves estimations of percentage loss of weight of the soil before and after drying up (P_w) and the density of the dried up soil (d). The percentage of water content (P_v) is obtained by multiplying the P_w parameter with density of the soil.

Soil evaporation estimations are also done through 'Lysimeters'. These are the tanks filled with gravel and water of various sizes. Two tubes are normally provided in these tanks with openings on both sides – one to feed water and the other for measurement of water levels. In this type of experimental setup, the actual evaporation loss (E) is given by the following equation.

$$E = S \times \Delta h \quad .....(3.8)$$

where E = evaporation loss; S = specific yield of the soil and Δh = water level fluctuation.

Sometimes 'neutron probes' could also be used which involves basically estimation of loss in the energy of neutrons emitted from a probe where they are installed in the soils at different depth levels. The losses in the energy levels are proportional to the losses in the water contents of the soils.

3.2.4 Evapotranspiration

Transpiration is the process by which water is absorbed by plants and vegetation through their roots and released into the atmosphere through their leaves in the form of vapour. Large amounts of water is absorbed into the plants and released into the atmosphere than what is required for their consumption and photosynthesis. Therefore transpiration is an important element in the water cycle and water balance estimations. Transpirations are extremely low in the desert and semi-desert areas, but could be of high order in the humid and sub humid areas such as coastal regions. Estimations of transpiration rates alone is a difficult task though some experimental designs using lysimeters are available. In recent years estimations of *'evapotranspiration'* have made great strides which is defined as cumulative loss of evaporation and transpiration. The total water loss at the maximum rate through evaporation from open water surfaces and transpiration by vegetation from an extensive area fully covered by plant growth with unlimited supply of water is termed *'Potential Evapotranspiration'* (PET). The real evapotranspiration occurring in a specific situation in the field is termed *'Actual Evapotranspiration'* (AET) which could be measured by using lysimeters designed for this purpose and sometimes by numerical analysis. Several workers contributed extensively and developed empirical formulae to estimate evapotranspiration values. Thoronthwaite, C.W. (1948) made significant contributions in this direction and gave the following empirical relation.

Thoronthwaite method

Thoronthwaite, C.W. (1948) proposed following formula

$$E = 1.6\left[\frac{10^t}{I}\right]^a \qquad \text{.....(3.9)}$$

where E = Monthly potential evapotranspiration in cm.

t = Mean monthly temperature in ^{0}c.

I = Annual heat index equal to $\sum_{n=1}^{n=12} i_n$ in

where i_n = Mean heat index of the nth month given as $i_n = \left(\frac{t_n}{5}\right)^{1.514}$

$$a = (6.75 \times 10^{-7}\, I^3) - (7.71 \times 10^{-4}\, I^2) + (1.729 \times 10^{-2}\, I) + 0.4923.$$

In order to avoid making tedious calculations, we use Table 3.3 which gives 'i' as a function of the temperature. The monthly indices could be determined for each month. The twelve monthly indices are then added together and a thermal index 'I' is obtained.

Table 3.3 Values of monthly indices used in the Thoronthwaite formula: Monthly index: $i_n=(t_n/5)^{1.514}$

1^0		.0	.1	.2	.3	.4	.5	.6	.7	.8	.9
0^0	.....			0.01	0.01	0.02	0.03	0.04	0.05	0.06	0.07
1^0	.....	0.09	0.10	0.12	0.13	0.15	0.16	0.18	0.20	0.21	0.23
2^0	.....	0.25	0.27	0.29	0.31	0.33	0.35	0.37	0.39	0.42	0.44
3^0	.....	0.46	0.48	0.51	0.53	0.56	0.58	0.61	0.63	0.66	0.69

Table 3.3 ***Contd...***

1^0		.0	.1	.2	.3	.4	.5	.6	.7	.8	.9
4^0	.....	0.71	0.74	0.77	0.80	0.82	0.85	0.88	0.91	0.94	0.97
5^0	.....	1.00	1.03	1.06	1.09	1.12	1.16	1.19	1.22	1.25	1.29
6^0	.....	1.32	1.35	1.39	1.42	1.45	1.49	1.52	1.56	1.59	1.63
7^0	.....	1.66	1.70	1.74	1.77	1.81	1.85	1.89	1.92	1.96	2.00
8^0	.....	2.04	2.08	2.12	2.15	2.19	2.23	2.27	2.31	2.35	2.39
9^0	.....	2.44	2.48	2.52	2.56	2.60	2.64	2.69	2.73	2.77	2.81
10^0	.....	2.86	2.90	2.94	2.99	3.03	3.08	3.12	3.16	3.21	3.25
11^0	.....	3.30	3.34	3.39	3.44	3.48	3.53	3.58	3.62	3.67	3.72
12^0	.....	3.76	3.81	3.86	3.91	3.96	4.00	4.05	4.10	4.15	4.20
13^0	.....	4.25	4.30	4.35	4.40	4.45	4.50	4.55	4.60	4.65	4.70
14^0	.....	4.75	4.81	4.86	4.91	4.96	5.01	5.07	5.12	5.17	5.22
15^0	.....	5.28	5.33	5.38	5.44	5.49	5.55	5.60	5.65	5.71	5.76
16^0	.....	5.82	5.87	5.93	5.98	6.04	6.10	6.15	6.21	6.26	6.32
17^0	.....	6.38	6.44	6.49	6.55	6.61	6.66	6.72	6.78	6.84	6.90
18^0	.....	6.95	7.01	7.07	7.13	7.19	7.25	7.31	7.37	7.43	7.49
19^0	.....	7.55	7.61	7.67	7.73	7.79	7.85	7.91	7.97	8.03	8.10
20^0	.....	8.16	8.22	8.28	8.34	8.41	8.47	8.53	8.59	8.66	8.72
21^0	.....	8.78	8.85	8.91	8.97	9.04	9.10	9.17	9.23	9.29	9.36
22^0	.....	9.42	9.49	9.55	9.62	9.68	9.75	9.82	9.88	9.95	10.01
23^0	.....	10.08	10.15	10.21	10.28	10.35	10.41	10.48	10.55	10.62	10.01
24^0	.....	10.75	10.82	10.89	10.95	11.02	11.09	11.16	10.55	10.62	10.68
25^0	.....	11.44	11.50	11.57	11.64	11.71	11.78	11.85	11.23	11.30	11.37
26^0	.....	12.13	12.21	12.28	12.35	12.42	12.49	12.56	12.63	12.70	12.78
27^0	.....	12.85	12.92	12.99	13.07	13.14	13.21	13.28	13.36	13.43	13.50
28^0	.....	13.58	13.65	13.72	13.80	13.87	13.94	14.02	14.09	14.17	14.24
29^0	.....	14.32	14.39	14.47	14.54	14.62	14.69	14.77	14.84	14.92	14.99
30^0	.....	15.07	15.15	15.22	15.30	15.38	15.45	15.53	15.61	15.68	15.76
31^0	.....	15.84	15.92	15.99	16.07	16.15	16.23	16.30	16.38	16.46	16.54
32^0	.....	16.62	16.70	16.78	16.85	16.93	17.01	17.09	17.17	17.25	17.33
33^0	.....	17.41	17.49	17.57	17.65	17.73	17.81	17.89	17.97	18.05	18.13
34^0	.....	18.22	18.30	18.38	18.46	18.54	18.62	18.70	18.79	18.87	18.95
35^0	.....	19.03	19.11	19.20	19.28	19.36	19.45	19.45	19.61	19.69	19.78
		0.0	0.1	0.2	0.3	0.4	0.5	0.6	0.7	0.8	0.9

(Source: Subrahmanyam, V.P., 1982)

The gross potential evapotranspiration is given by drawing a straight line through the point of convergence and the thermal index (Fig. 3.12). The coordinates of the points on this line give the potential evapotranspiration for each month as a function of the average temperature for the month. This gross potential evaporation is adjusted to allow for the latitude by multiplying the calculated monthly value by a correction factor. The potential evapotranspiration is found by multiplying the values of the gross potential evapotranspiration by the coefficients shown in Table 3.4.

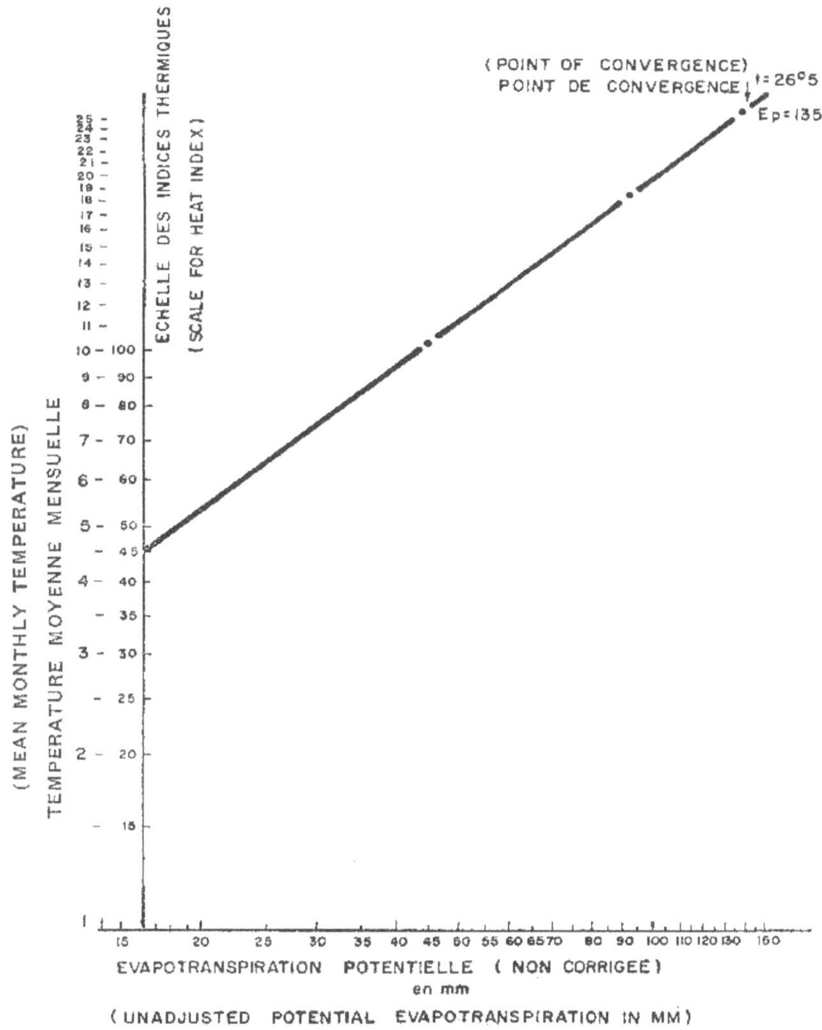

Fig. 3.12 Graphical Solution for unadjusted potential evapotranspiration

Table 3.4 Correction for potential evapotranspiration

Latitude North		Jan.	Feb.	Mar.	Apr.	May	June	July	Aug.	Sept.	Oct.	Nov.	Dec.
20^0	.....	0.95	0.90	1.03	1.05	1.13	1.11	1.11	1.14	1.11	1.02	0.93	0.94
25^0	.....	9.93	0.89	1.03	1.06	1.15	1.14	1.14	1.17	1.12	1.02	0.91	0.91
26^0	.....	0.92	0.88	1.03	1.06	1.15	1.15	1.15	1.17	1.12	1.02	0.91	0.91
27^0	.....	0.92	0.88	1.03	1.07	1.16	1.15	1.18	1.13	1.02	0.99	0.90	0.90
28^0	.....	0.91	0.88	1.03	1.07	1.16	1.16	1.18	1.13	1.02	0.99	0.90	0.90
29^0	.....	0.91	0.87	1.03	1.07	1.17	1.16	1.19	1.13	1.03	0.98	0.90	0.89
30^0	.....	0.90	0.87	1.03	1.08	1.18	1.17	1.20	1.14	1.03	0.98	0.89	0.88
31^0	.....	0.90	0.87	1.03	1.08	1.18	1.18	1.20	1.14	1.03	0.98	0.89	0.88
32^0	.....	0.89	0.86	1.03	1.08	1.19	1.19	1.21	1.15	1.03	0.98	0.88	0.87
33^0	.....	0.88	0.86	1.03	1.09	1.19	1.20	1.22	1.15	1.03	0.97	0.88	0.86
34^0	.....	0.88	0.85	1.03	1.09	1.20	1.20	1.22	1.16	1.03	0.97	0.87	0.86
35^0	.....	0.87	0.85	1.03	1.09	1.21	1.21	1.23	1.16	1.03	0.97	0.86	0.85
36^0	.....	0.87	0.85	1.03	1.10	1.21	1.22	1.24	1.16	1.03	0.97	0.86	0.84

Table 3.4 *Contd...*

Latitude North		Jan.	Feb.	Mar.	Apr.	May	June	July	Aug.	Sept.	Oct.	Nov.	Dec.
37^0	.....	0.86	0.84	1.03	1.10	1.22	1.23	1.25	1.17	1.03	0.97	0.85	0.83
38^0	.....	0.85	0.84	1.03	1.10	1.23	1.24	1.25	1.17	1.04	0.96	0.84	0.83
39^0	.....	0.85	0.84	1.03	1.11	1.23	1.24	1.26	1.18	1.04	0.96	0.84	0.82
40^0	.....	0.84	0.83	1.03	1.11	1.24	1.25	1.27	1.18	1.04	0.96	0.83	0.81
41^0	.....	0.83	0.83	1.03	1.11	1.25	1.26	1.27	1.19	1.04	0.96	0.82	0.80
42^0	.....	0.82	0.83	1.03	1.12	1.26	1.27	1.28	1.19	1.04	0.95	0.82	0.79
43^0	.....	0.81	0.82	1.02	1.12	1.26	1.28	1.29	1.20	1.04	0.95	0.81	0.77
44^0	.....	0.81	0.82	1.02	1.13	1.27	1.29	1.30	1.20	1.04	0.95	0.80	0.76
45^0	.....	0.80	0.81	1.02	1.13	1.28	1.29	1.31	1.21	1.04	0.94	0.79	0.75
46^0	.....	0.79	0.81	1.02	1.13	1.29	1.31	1.32	1.22	1.04	0.94	0.79	0.74
47^0	.....	0.77	0.80	1.02	1.14	1.30	1.32	1.33	1.22	1.04	0.93	0.78	0.73
48^0	.....	0.76	0.80	1.02	1.14	1.31	1.33	1.34	1.23	1.05	0.93	0.77	0.72
49^0	.....	0.75	0.79	1.02	1.14	1.32	1.34	1.35	1.24	1.05	0.93	0.76	0.71
50^0	.....	0.74	0.78	1.02	1.15	1.33	1.36	1.37	1.25	1.06	0.92	0.76	0.70

Penman-Monteith method (Allen, R.G., et al. 1998)

This is an improved version of Penman, H.L. (1948) approach and is also widely adopted for computations of potential evapotranspiration in both humid and arid climatic regions and is given below.

$$ET_0 = \frac{0.408\Delta\left(R_n^{-G}\right) + \gamma\frac{900}{T+273} - u_2\left(e_s - e_a\right)}{\Delta + \gamma(1+0.34u_2)} \qquad .....(3.10)$$

where ET_0 = Evapotranspiration (mm day^{-1})

R_n = net radiation at the crop surface (MJ m^{-2} day^{-1})

G = soil heat flux density (MJ m^{-2} day^{-1})

T = mean daily air temperature at 2 m height (^{0}C)

μ_2 = wind speed at 2 m height (ms^{-1})

e_s = saturation vapour pressure (kPa)

e_a = actual vapour pressure (k)

$e_s - e_a$ = Saturation vapour pressure deficit (kPa)

x = slope of vapour pressure curve (kPa^0 C^{-1})

γ = Psychometric constant (kPa^0 C^{-1})

The above equation requires site location data on air temperature, humidity, radiation and wind speed for PET estimations.

Blaney, H.F. and Criddle, W.D., Method (1950)

The potential evapotranspiration values can be obtained by using the expression mentioned below.

$$V = 25.4\ KF = 25.4\ K\ \Sigma_1^n \qquad .....(3.11)$$

where V= Evaporation or consumptive use in mm

k = Consumptive use crop coefficient for a growing season

F = Sum of the monthly consumptive use factors (f) for a growing period where $$f = P\frac{T}{100}$$

P = Monthly sunshine hours expressed as percentage of yearly sunshine hours in a given locality.

T = Mean air temperature in ^{0}F.

Hargreaves, H.G. (1968)

This method takes into account pan evaporation data through the following equation.

$$E_p = 17.4\ d\ T\ (1.0 = H_n) \quad .....(3.12)$$

where E_p = Class A pan evaporation in mm

d = Monthly day time coefficient, a ratio of day length for the month to 12 hours

T = Average mean monthly temperature in ^{0}C

H_n = Mean monthly relative humidity at noon expressed in decimal form.

Then the evapotranspiration is given by $E_t = K\ E_p$

where E_t = Evapotranspiration in mm; K = Crop coefficient

3.2.5 Run off

A part of the precipitation on the soil after meeting the initial requirements over flows on the surface and joins near by streams and rivers and is termed '*Runoff*'. This could take place through the water that flows on the land surface *'surface runoff'* and also through ground water *'sub-surface runoff'*. A part of the initial rain fall gets stored in the depressions and puddles of the land surface and is termed as *'depression storage'* which gets either evaporated or infiltrated underground at a little later stage. The runoff component occurs when the precipitation amounts exceed the evaporation and infiltration rates on the ground surface as over land flow and meets the nearby stream or river channel. In addition to the surface flow component, a part of the water infiltrating through the soil, could move laterally and join the stream flow and such flow is called, *'sub surface runoff'*. A little later, a part of the infiltrated water would join the water table, contributing to the rise of water table and at the same time creating a flow gradient towards the stream/river, which is termed as *'effluent flow'* or *'base flow'* and such streams are called *'gaining streams'*. Sometimes the streams themselves may loose water, when the stream level is more than the water table level, which is called *'influent stream'*. The influent and effluent seepages take place depending upon the elevations of water table in relation to the stream flow elevations and the prevailing gradients between ground water levels and stream water levels. In a nut shell the stream run off consists of three components.

(i) Direct precipitation over the surface of the stream.

(ii) Surface run off and sub-surface runoff.

(iii) Ground water inflow or base flow.

The stream flow is generally represented in the form of a hydrograph which is a graphical representation of stream flow discharge at a given location against time, where discharge fluctuations could be depicted. The run off hydrograph represents combined discharge of surface run off and also ground water inflow. In order to obtain the surface run off graph, one need to remove the inflow generated from ground water. As discussed earlier the inflow of the system and the effluent stream during the peak stages of the hydrograph could become 'influent' and contribute some amount of stream water to the groundwater system depending upon the built up of water level heads in the stream. This stage however is very small as represented in the schematic diagram (Fig.3.13) and is normally referred as base flow.

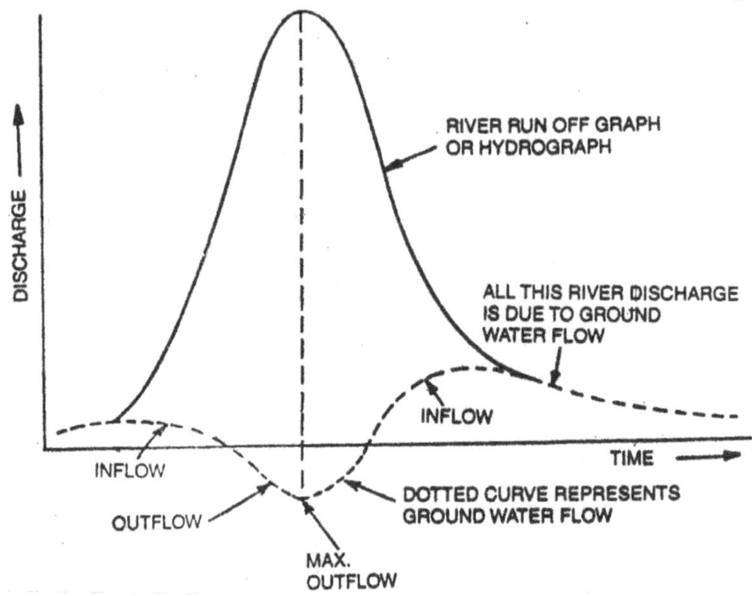

Fig. 3.13 Runoff hydrograph with inflow and outflow situation

As the surface run off gets reduced with time, the water in the stream totally comes from ground water only resulting in steady lowering of water table and at the same time reducing the stream flow as well. A curve drawn between discharge and the time in general referred as *'ground water depletion curve'* or *'base flow curve'*. A schematic diagram of the ground water depletion curve is shown in Fig. 3.14.

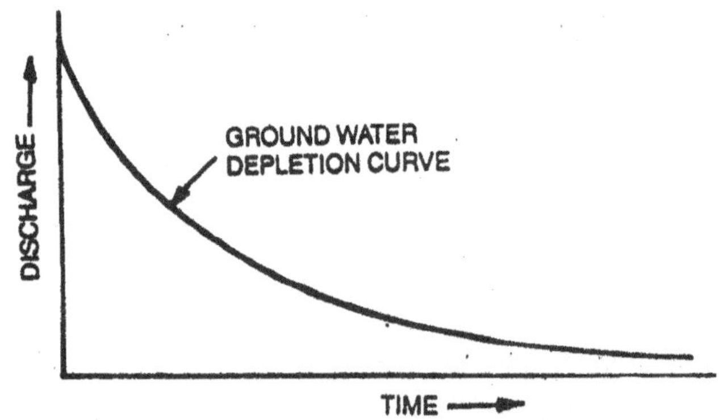

Fig. 3.14 Schematic curve showing the ground water depletion curve

Estimates of surface water run off in a river could be obtained by way of superimposing the ground water inflow curve on the run off hydrograph or connect the starting and recession terminals as shown in Fig. 3.15.

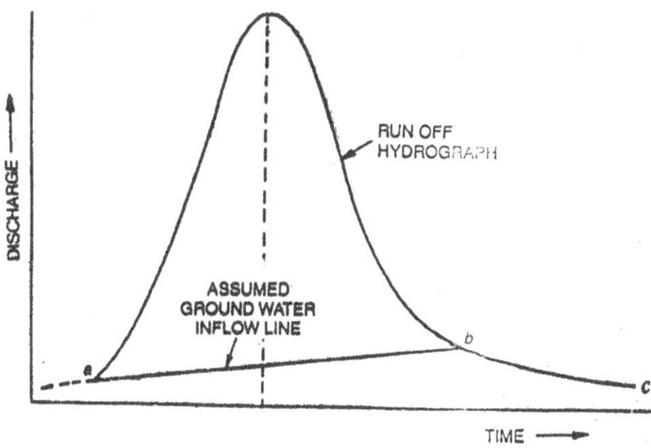

Fig. 3.15 Direct runoff hydrograph with base flow separation

The point 'a' represents the beginning of the accretion of surface run off and 'bc' represents the ceasing point of surface run off and the portion 'bc' of the hydrograph represents nothing but ground water depletion curve. This ground water depletion curve numerically could be expressed by the following equation (Boussinesq, Jr. J., 1877).

$$Q = Q_0 \, e^{-at} \quad .....(3.13)$$

where Q = base flow at any time't';

Q_0 = base flow at time 't_0' and

a = recession constant depending upon terrain morphology /soil characteristics

Annual and Seasonal Runoff

Estimations of annual and seasonal rainfall parameters are essential for long term and regional evaluation of water balance studies. A number of methods in this direction are available to compete monthly runoff but most methods appear the site specific and approximate and cannot be universalized (Garg, S.K., 2007).

Run off coefficient: The volume of runoff can be directly computed by using the following equation.

$$Q = K \times P \quad .....(3.14)$$

where Q = Runoff in cms;

P = Precipitation in cms and

K = a constant depending upon the slope characteristics of the catchment and soil permeability. There are several constraints involved while using this method like slope and soil characteristics which are not incorporated, and is not dependable for long terms evaluations.

Inghlis Formulae: Inglis developed a formula for ghat areas where rainfall exceeds 200 cms or more and non ghat areas where rainfall in less than 200 cms.

(a) Ghat areas: $R = 0.85\,P - 30.50$ .....(3.15)

(b) Non Ghat areas: $R = \dfrac{P - 17.80}{254} \times P$ .....(3.16)

where R = Run off in cms and p is precipitation in cms.

Khosla's formula This formula is expressed as: $R = P - \left[\frac{t - 32}{3.74}\right]$ .....(3.17)

where R = Run off in cms;

P = Precipitations in cms and

T = Mean temperature 0^0 F.

Lacy's Formula This is expressed as: $R = \left[\frac{P}{1+\frac{304.8m}{p.n}}\right]$ .....(3.18)

where R = run off in cm

P = Precipitation in cm.

'm' is a monsoon duration factor as given below.

0.50 : Very short

1.00 : Normal

1.50 : Very long

'n' is a value dependent on catchment characteristics with the following value.

0.25 Flat cultivated black cotton soils

0.60 Flat but partly cultivated

1.00 Average

1.70 Hills and Plains in the insignificant cultivation.

3.45 Hilly slopes without any cultivation.

USDA SCS Curve number method: The following is a brief description as given by Rallison, R.E. (1980). The hypothesis in brief is given by:

$$F / S = Q / P \qquad .....(3.19)$$

where 'F' is the actual retention of precipitation(P) during a rainfall event (inches); 'S' is the potential maximum retention (inches); Q is the direct surface run off (inches).

The above equation could be changed after adding initial abstraction and could be expressed as:

$$F / S = Q / P_a \qquad .....(3.20)$$

where P_a is the rainfall (inches) after the start of run off and $P_a = P - I_a$ where I_a is the initial abstraction (inches). By substituting this value for 'F' and solving for 'Q' one can obtain

$$Q = \frac{(P - I_a)^2}{(P - I_a) + S} \quad \text{for} \quad P_a > I_a \qquad .....(3.21)$$

$$Q = 0 \quad \text{for } P < I_a \qquad .....(3.22)$$

To reduce the number of variables the empirical relationship $I_a = 0.25$ was adopted, which gives the following runoff equation.

$$Q = \frac{(P - 0.2S)^2}{P + 0.8S} \quad \text{for } P > 0.25 \qquad .....(3.23)$$

$$Q = 0 \text{ for } P < 0.25 \qquad .....(3.24)$$

The retention parameter 'S' is related to a curve number CN by the relationship.

$$CN = \frac{1000}{(S + 10)} \qquad .....(3.25)$$

The relationship of run off to rainfall in the curve number method could be expressed in the graphical form as shown in Fig. 3.16.

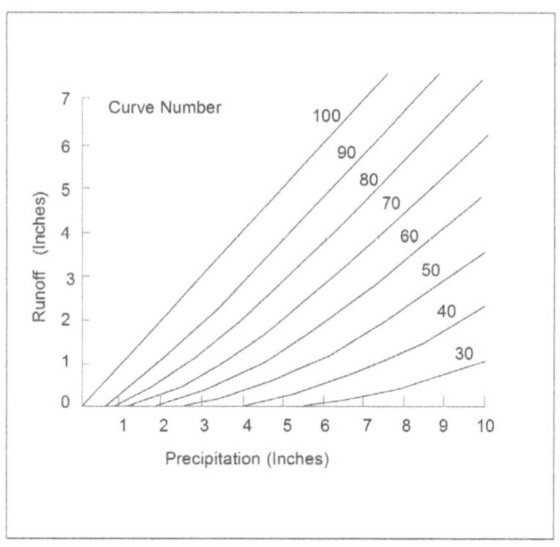

Fig. 3.16 Graphical form of USDA SCA curve number method

The curve numbers incorporated have a range between 100 to 0. The maximum retention 'S' increased exponentially as the curve numbers decrease from 100 as shown in the Table 3.5.

Table 3.5 Maximum potential retention 'S' values for a range of curve numbers

Curve number	S (inches)	S (mm)	Curve number	S (inches)	S (mm)
100	0	0	60	6.67	169.3
95	0.53	13.4	55	8.18	207.8
90	1.11	28.2	50	10.00	254.0
85	1.76	44.8	45	12.22	310.4
80	2.50	63.5	40	15.00	381.0
75	3.33	84.7	35	18.57	471.7
70	4.29	108.9	30	23.33	592.7
65	5.38	136.8			

The curve numbers incorporate the effects such as soil type, topography, cultivation and land use practices on the run off values generated.

3.2.6 Infiltration and soil recharge

Rain water that enters the soil surface is termed 'infiltration'. The extent of infiltration depends on the intensity and duration of rainfall, initial moisture levels of the soil and soil characteristics such as texture, structure, and compaction of the soil. The water initially is absorbed in capillary forces, and subsequently moves under gravity attain full saturation levels of the soil interstices and meets the water table. The maximum rate at which the soil at any given condition is capable of absorbing water is called *infiltration capacity.* Infiltration often begins at a high rate during initial event of rainfall and rapidly decreases to a steady state. Horton, R.E. (1933) gave an empirical formula to measure infiltration rates. The empirical formula says that the infiltration starts at a constant rate 'f_o' and will be decreasing exponentially with time 't'. After sometime when the soil saturation reaches a certain value, the rate of infiltration will level off to the rate 'f_c'. The empirical formula is given as under

$$f_t = f_c + (f_o - f_c)e^{-kt} \qquad .....(3.26)$$

where f_t = infiltration rate at time t

f_o = initial infiltration rate

f_c = constant or final infiltration rate after the soils has been saturated.

K = decay constant specific to the soil.

Infiltration depends upon intensity, duration of rainfall, climate, soil characteristics, vegetative cover land use, initial soil moisture content and water table depth. The soil water recharge could also be estimated by applying the following water balance equation with appropriate cautions wherever required and with certain limitations.

$$\text{Infiltration/soil water recharge} = P - AE - \Delta S - R_o \qquad .....(3.27)$$

where P = precipitation

AE = actual evapotranspiration

ΔS = change in soil water storage and

R_o = run-off.

3.3 SOIL WATER BALANCE ESTIMATIONS FOR SELECTED STATIONS IN DIFFERENT DELTA SYSTEMS

The soil water balance of a basin or a region essentially involves rainfall as the 'input', evapotranspiration as the 'total loss' and the amount of moisture stored in the soil as 'reserve' that could be used subsequently to meet the water needs during the times of water deficit and the 'runoff' that becomes the stream flow and ground water recharge that infiltrates underground and adds to the ground water storage. The concept of water balance was first put forth by Thoronthwaite, C.W. (1948) and later modified by Thoronthwaite, C.W. and Mather, J.R. (1955). The basic water balance equation and the parameters involved are reflected in equation (3.27).

A comparative analysis of rainfall (P) with that of potential evapotranspiration (PET) of a station, one could obtain information on several aspects of water balance parameters such as soil moisture storage, actual evapotranspiration, water surplus, water deficit, runoff and ground water recharge. The concepts of PET and estimation of water balance parameters have been extensively applied to studies related to climatic classifications of a region/watershed and crop water requirements etc., (Subrahmanyam, V.P., 1956; 1970; 1982; Rao, K.N., et al 1971; Kesava Rao, A.V.R., et al 2006; Wani, S.P., et al 2004). The National bureau of soil survey and land use planning (Mandal, C.K., et al

1999) have provided database for evaluation of appropriate water balance parameters for different stations of the coastal deltaic regions of India which are presented and discussed in the following paragraphs.

Evaluation of water balance of a station involves comparative analysis of precipitation (water supply), potential evapotranspiration (water need) and soil water storage. At stations where precipitation is greater than the potential evapotranspiration, the soil remains fully saturated with water, a state of '*water surplus*' occurs. At times when precipitation is less than the potential evapotranspiration, the soil remains partly dry and partly saturated, a state of '*water deficit*' occurs. These values along with soil storage and retention parameters are calculated by using the Thoronthwaite, C.W. and Mather, J.R. (1955) book keeping procedures and methodologies for different basins and would help to derive the 'runoff' and 'recharge' parameters in a qualitative way. The different steps involved in obtaining various water balance parameters are mentioned briefly below (Mandal, C.K., et al 1999; Soil climatic database for crop planning in India: National Bureau of soil survey and land use planning)

Step 1: Estimations PET values for each station based on either Penman, H.L. (1948) or Thoronthwaite, C.W. and Mather, J.R., methods (1955, 1957).

Step 2: Estimations of water quantities either excess (+) or deficit (–) of P over PET where P = Precipitation/rainfall; and PET = Potential evapotranspiration for each month.

Step 3: Computation of accumulated values of P – PET for each month to account for gross potential deficit or gross potential surplus.

Step 4: Computation of storage of soil moisture for each month. The release of soil moisture is an exponential function.

$$\text{STOR} = \text{AWC}.\ \frac{\text{ACC(P} - \text{PET)}}{\text{AWC}}$$

where STOR = Actual storage of soil moisture

AWC = Storage capacity of soil moisture (Field capacity – wilting point)

ACC (P-PET) = Accumulated values of (P – PET) < 0 up to the month considered

This value is always negative (Accumulated Potential Water Loss, APWL)

Step 5: Computation of ΔSTOR

ΔSTOR is the change of actual storage from month to month. When storage remains at capacity level the storage value is zero. When storage reaches values of less than the capacity, the storage is calculated as the subtraction of storage value for the actual month from STOR value of the previous month.

ΔSTOR value can be negative and implies subtraction of water from the storage to be used for evapotranspiration (soil moisture utilization).

ΔSTOR value can be positive and implies infiltration of water in the soil and addition to the soil moisture storage (Soil moisture recharge).

Step 6: When ΔSTOR is negative, the AET = ΔSTOR + P When ΔSTOR is positive, the AET = PET (AET = Actual Evapo Transpiration).

Step 7: DEF = PET – AET, where DEF stands for Water Deficit.

Step 8: SUR = (P – PET) – ΔSTOR, where SUR stands for Surplus water.

Water surplus and water deficit occur in different seasons at many places. Both are significant to indicate the index of moisture effectively, one effecting positively and the other effecting negatively. Thoronthwaite termed this as Moisture index (I_m) and could be computed from the water balance parameters such as water surplus (WS), water deficit (WD) and PET as given below.

1. Aridity index I_a (%) = 100 × WD/PET
2. Humidity index I_h (%) = 100 × WS/PET
3. Moisture index I_m(%) = $I_h - I_a$

A moisture resume of a station is defined by humidity and aridity index values. The humidity index is the percentage ratio of the annual water surplus to the annual potential evapotranspiration. The aridity index is the percentage ratio of annual water deficit to annual water need. The difference between humidity index and aridity index will define the moisture index. The main climatic type of a region is characterized by the moisture index.

Most wet climates have positive I_m values while dry climates have negative I_m values. Thornthwaite, C.W. and Mather, J.R. (1955) have classified climates into three categories namely (i) dry sub humid climate with moisture index values (I_m) varying between 0 and –33.3, (ii) Semi arid climate with moisture index values varying between -33.3 and -66.6 and (iii) Arid climates with I_m values less than –66.6. All the relevant water balance parameters as obtained from the soil conservation database for different coastal delta regions are summarized below based on which a number of graphical analyses are made for each delta system while deriving the approximate runoff and soil recharge values. The Thoronthwaite method of approach takes into account the rainfall, potential evapotranspiration and moisture storage values as basic inputs while computing the water balance of a station. The maximum water holding capacity of the soils (field capacity) depends on texture, structure and vegetative pattern of the soil. By adopting an appropriate field capacity values of soils, precipitation values and potential evapotranspiration values for a year in a station, one can compute the water budget elements such as water deficit, water surplus, surface runoff and soil recharge which help to assess and characterize the water balance situations and recharge potentials in different delta basins under review.

3.3.1 Cauvery, Ponnaiyar and Vaigai delta basins (Tamil Nadu)

The Cauvery, Ponnaiyar and Vaigai are the prominent delta systems in Tamil Nadu from the point of view of water resource potentials and in view of the fact that these river systems are being drained through large sedimentary tracts of the coastal Tamil Nadu region. Ponnaiyar river is disposed north of Cauvery in the districts of Villupuram, Cuddalore and in the Union Territory of Pondicherry. Vaigai delta is disposed south of Cauvery delta in the districts of Ramanathapuram and Sivaganga. Cauvery is the major delta system along the east coast of Tamil Nadu. The Cauvery river after traversing a distance of 800 kms from its source debouches in to Bay of Bengal through different distributaries like Coleroon, Cauvery and Vennar while building up reasonably huge delta system with a span of about 7500 sq. kms and with its apex near Tiruchinapally which is located at an altitude of about 80 m above mean sea level. The delta part is largely distributed in the districts of Tanjore, Thiruvarur, Nagapattinam and in the Union Territory of Karaikal. The river initially branches off into two channels namely Coleroon and Cauvery and further downstream Cauvery gets branched off into two more namely Cauvery and Vettar and then into several subdistributaries. Coleroon remains as the single active river channel carrying the bulk of water and sediment discharge. The upper and middle delta plains are characterized by a number of paleo channels and levee deposits where as the outfall region or the lower delta region near the coast contain tidal flats, creeks, spits, bars and several sand ridges parallel to the coast. The tidal impacts are normally seen up to 20 kms upstream. The delta region is characterized by humid tropical climate with mean maximum and minimum temperatures of the order of 32.2 and 24.6 °C respectively. The relative humidity varies between 80% (wet season) and 65% (dry season).

The Ponnaiyar and Vaigai river deltas are disposed north and south of Cauvery delta respectively and contain several small estuaries in between them. The Ponnaiyar river after traversing a distance of 400 kms debouches into Bay of Bengal with its near Thirukkovilur. It has relatively a small linear delta system with an aerial spread of an order 300 km^2. The delta is linearly disposed along a narrow valley with Tertiary rock massifs located on either side of the river. All the distributaries by and large stand as abandoned river channels. Similarly the Vaigai river after traversing about 260 kms debouches into Bay of Bengal near Mandapam, with its apex at Paramakkudi. The upper delta region of Vaigai contain a number of paleo channels basically disposed south of Vaigai river. The Vaigai delta development appears to have been largely influenced by fluvial and tectonic processes along with sea level oscillations. The delta built up appears to have taken place on either side of the river reflecting the migratory trends of the river. The lower delta region contains a number of tidal flats, mangrove marshes and beach ridges.

The Cauvery, Ponnaiyar and Vaigai delta systems receive over 70% of the rainfall from the northeast monsoon and the rest 30% of rainfall from the southwest monsoon. The average annual rain fall in these delta regions is of the order of 1160 mm with peak rainfall normally taking place in the months of November and December. The observed annual river discharges near the terminal points in respect of Cauvery, Ponnaiyar and Vaigai respectively are 20,950 MCM, 1600 MCM and 770 MCM respectively. The soils in general in all the delta systems are deep to moderate well drained soils with grayish brown and dark brown in color, with clay percentage varying from 25 to 35%. The pH values vary from 7.5 to 9.5 with low nitrogen and with medium levels of phosphates and potash. A brief regional hydrological characters in respect of Cauvery, Ponnaiyar and Vaigai are presented Table 3.6.

Table 3.6 Regional hydrological characters of Cauvery, Ponnaiyar and Vaigai

S.no.	Item	Cauvery	Ponnaiyar	Vaigai
1.	Average annual rainfall	1150-1200 mm	1150-1200 mm	800-900 mm
2.	Altitudinal variations	20 m AMSL	15 m AMSL	15 m AMSL
3.	Average annual temperatures	36.5-27.8°C	42-26°C	31.6-25.5°C
4.	Aereal spreads of the delta	11000 sq. kms	300 sq. kms	2000 sq. kms
5.	Length of the delta	120 kms	40 kms	40 kms
6.	Width of the delta	190 kms	10-20 kms	40 kms
7.	Tidal impacts	40 kms	10 kms	10 kms
8.	Average annual discharge	20,950 MCM	1600 MCM	770 MCM

The isohyetel trends and the hydromet station locations in respect of Ponnaiyar, Cauvery and Vaigai are shown in Fig. 3.17. The relevant water balance parameters for selected number of stations for Cauvery (3), Ponnaiyar (2) and Vaigai (2) delta basins are tabulated Table 3.6 and their graphical trends are shown in Fig. 3.18.

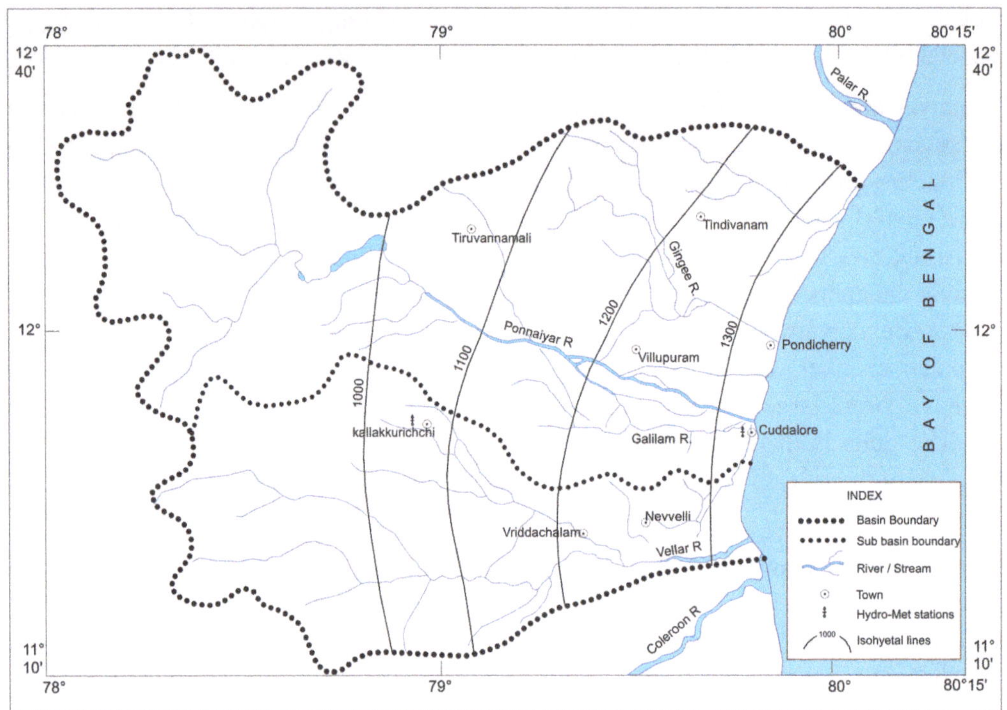

(a) Ponnaiyar

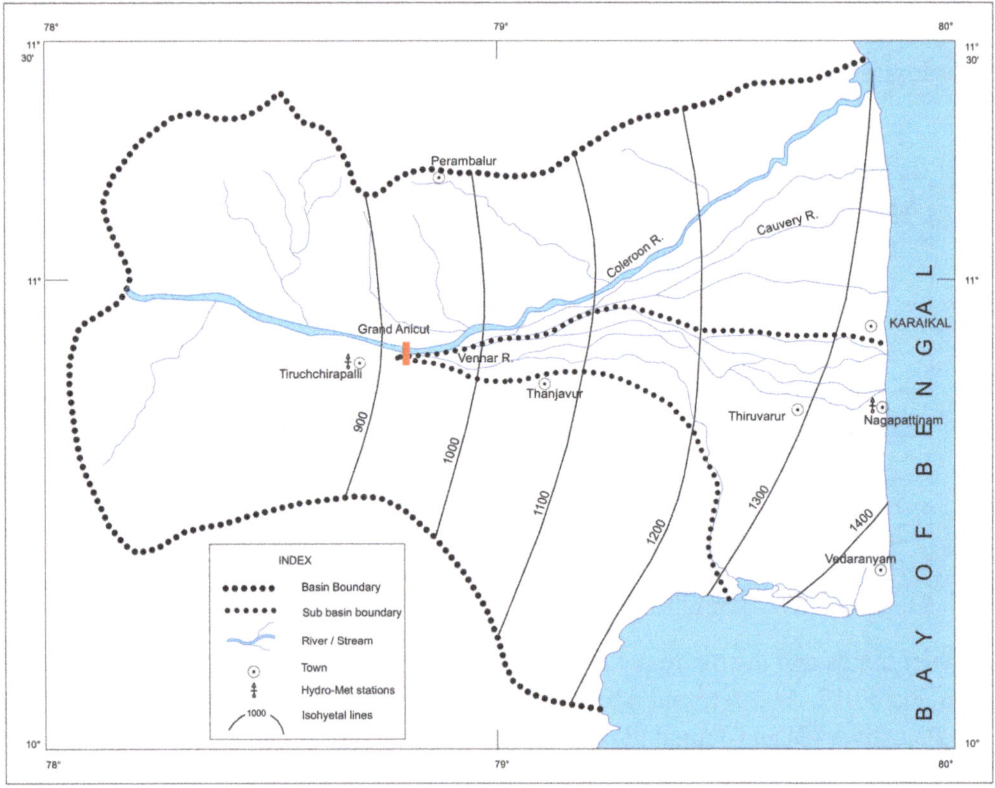

(b) Cauvery

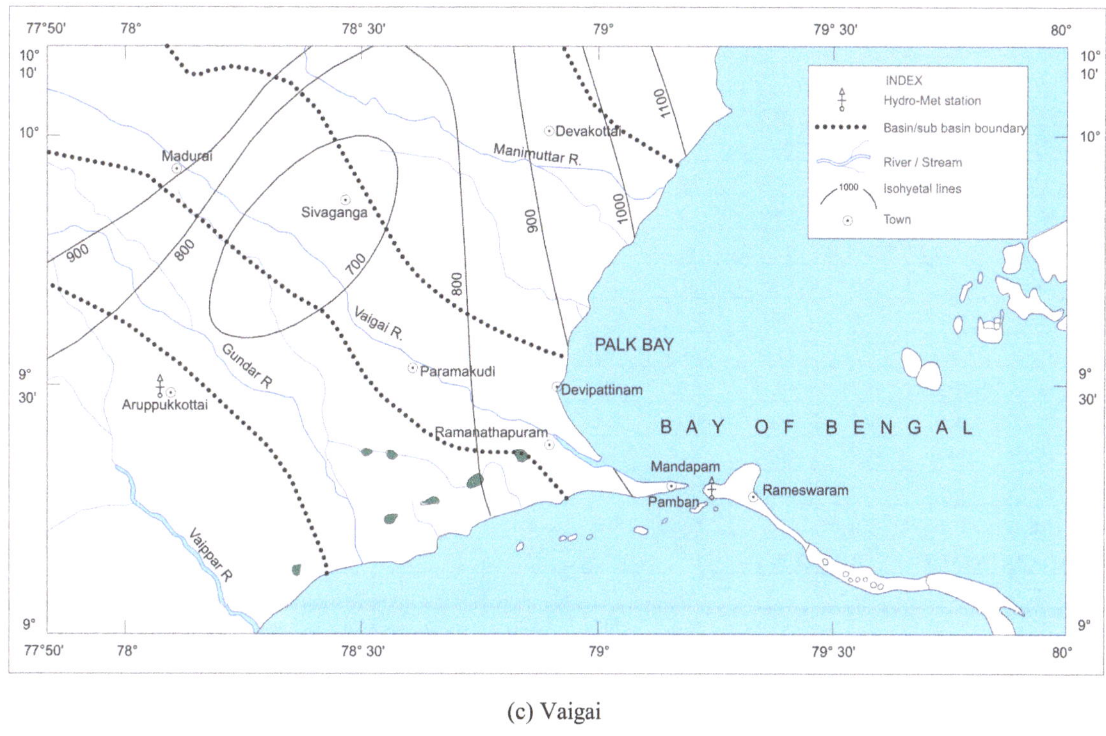

(c) Vaigai

Fig. 3.17 Isohyetel trends and hydromet stations in (a) Ponnaiyar, (b) Cauvery and (c) Vaigai

The Cauvery delta is characterized by a precipitation isohyetel 1400 mm (close to the coast) and 1000 mm (away from the coast). Similarly the Ponnaiyar-Vellar delta regions are bounded by 1200 mm isohyetel (close to the coast) and 1000 mm (away from the coast) and the Vaigai delta region is bounded by 1100 mm (close to the coast) and 700 mm (away from the coast). The precipitation values by and large in all the deltas decrease from coast to the upstream area. In the case of Vaigai delta region is characterized by low precipitation lines around Shivaganga with an increase in the values towards north. The isohyetels in this case cut across the coast with a decrease from north to south.

Seasonal water balance parameters: The seasonal water balance parameters for Cauvery, Ponnaiyar and Vaigai are tabulated in Table 3.7.

Table 3.7 Seasonal water balance parameters for Cauvery and its adjoining regions

Stn : KALLAKURCHI (PONNAIYAR DELTA)											**AWC = 150 mm**		
	J	**F**	**M**	**A**	**M**	**J**	**J**	**A**	**S**	**O**	**N**	**D**	**TOTAL**
P	47.0	14.0	14.0	39.0	81.0	46.0	79.0	136.0	142.0	194.0	175.0	75.0	1042
PE	94.7	126.9	160.1	173.0	191.4	183.5	175.1	173.4	165.3	150.8	111.4	93.7	1799
ΔS	–32.30	–37.42	–14.86	–3.62	–0.94	–0.37	–0.08	–0.02	–0.01	43.17	63.60	–17.15	0
AE	79.3	51.4	28.9	42.6	81.9	46.4	79.1	136.0	142.0	150.8	111.4	92.2	1042
WD	15.4	75.5	131.2	130.4	109.5	137.1	96.0	37.4	23.3	0.0	0.0	1.6	757
WS	0.0	0.0	0.0	0.0	0.0	0.0	0.0	0.0	0.0	0.03	0.0	0.0	0
RO	0.0	0.0	0.0	0.0	0.0	0.0	0.0	0.0	0.0	0.0	0.0	0.0	0
Stn : CUDDALORE (PONNAIYAR DELTA)											**AWC = 150 mm**		
	J	**F**	**M**	**A**	**M**	**J**	**J**	**A**	**S**	**O**	**N**	**D**	**TOTAL**
P	42.0	15.0	27.0	38.0	71.0	35.0	69.0	110.0	113.0	242.0	362.0	183.0	1307
PE	83.2	90.4	143.7	160.1	181.4	183.5	181.4	173.4	158.5	150.8	115.5	98.4	1720

Table 3.7 *Contd...*

Stn : CUDDALORE (PONNAIYAR DELTA)											AWC = 150 mm		
	J	F	M	A	M	J	J	A	S	O	N	D	TOTAL
ΔS	113.97	–45.03	–37.27	–17.64	–7.31	–4.22	–1.32	–0.41	–0.20	90.63	58.80	0.0	150
AE	83.2	60.0	64.3	55.6	78.3	39.2	70.3	110.4	113.2	150.8	115.5	98.4	1039
WD	0.0	30.4	79.4	104.5	103.1	144.3	111.1	63.0	45.3	0.0	0.0	0.0	681
WS	0.0	0.0	0.0	0.0	0.0	0.0	0.0	0.0	0.0	0.6	187.7	84.6	273
RO	44.7	22.3	11.2	5.6	2.8	1.4	0.0	0.0	0.0	0.3	94.0	89.3	272

Stn :TIRUCHIRAPALLI (CAUVERY DELTA)											AWC = 150 mm		
	J	F	M	A	M	J	J	A	S	O	N	D	TOTAL
P	18.0	8.0	8.0	70.0	80.0	34.0	41.0	105.0	108.0	170.0	156.0	71.0	869
PE	95.3	126.9	160.1	173.0	186.9	178.1	181.4	173.4	165.3	150.8	112.0	94.3	1798
ΔS	–30.84	–10.91	–1.78	–0.15	–0.02	–0.01	0.0	0.0	0.0	19.20	44.0	–19.49	0
AE	48.8	18.9	9.8	70.2	80.0	34.0	41.0	105.0	108.0	150.8	112.0	90.5	869
WD	46.5	108.0	150.3	102.9	106.9	144.1	140.4	68.4	57.3	0.0	0.0	3.8	929
WS	0.0	0.0	0.0	0.0	0.0	0.0	0.0	0.0	0.0	0.0	0.0	0.0	0
RO	0.0	0.0	0.0	0.0	0.0	0.0	0.0	0.0	0.0	0.0	0.0	0.0	0

Stn : NAGAPATNAM (CAUVERY DELTA)											AWC = 100 mm		
	J	F	M	A	M	J	J	A	S	O	N	D	TOTAL
P	57.0	25.0	22.0	55.0	56.0	29.0	48.0	62.0	62.0	224.0	458.0	239.0	1337
PE	96.9	105.3	152.2	167.0	181.4	183.5	181.4	173.4	165.3	150.8	113.4	95.9	1767
ΔS	67.10	–37.04	–21.88	–5.51	–1.91	–0.60	–0.12	–0.03	–0.01	73.20	26.80	0.0	100
AE	96.9	62.0	43.9	60.5	57.9	29.6	48.1	62.0	62.0	150.8	113.4	95.9	883
WD	0.0	43.3	108.3	106.5	123.5	153.9	133.3	111.4	103.3	0.0	0.0	0.0	883
WS	0.0	0.0	0.0	0.0	0.0	0.0	0.0	0.0	0.0	0.0	317.8	143.1	461
RO	79.5	39.7	19.9	9.9	5.0	2.5	1.2	0.0	0.0	0.0	174.8	158.9	491

Stn : PUDUKKOTTAI (CAVERY DELTA)											AWC = 100 mm		
	J	F	M	A	M	J	J	A	S	O	N	D	TOTAL
P	39.0	13.0	16.0	48.0	69.0	52.0	73.0	136.0	133.0	158.0	158.0	90.0	985
PE	95.3	126.9	160.1	173.0	186.9	178.1	181.4	173.4	165.3	150.8	112.0	94.3	1798
ΔS	–32.04	–15.03	–1.87	–0.12	–0.01	0.0	0.0	0.0	0.0	7.20	46.0	–4.13	0
AE	71.0	28.0	17.9	48.1	69.0	52.0	73.0	136.0	133.0	150.8	112.0	94.1	985
WD	24.3	98.9	142.2	124.9	117.9	126.1	108.4	37.4	32.3	0.0	0.0	0.2	813
WS	0.0	0.0	0.0	0.0	0.0	0.0	0.0	0.0	0.0	0.0	0.0	0.0	0
RO	0.0	0.0	0.0	0.0	0.0	0.0	0.0	0.0	0.0	0.0	0.0	0.0	0

Stn :ARUPPUKOTTAI (VAIGAI DELTA)											AWC = 100mm		
	J	F	M	A	M	J	J	A	S	O	N	D	TOTAL
P	0.0	8.9	20.5	35.7	76.8	11.4	43.0	64.5	103.5	196.0	152.4	9.9	723
PE	106.8	129.7	160.1	168.5	180.9	175.8	175.1	170.2	160.5	156.3	138.1	108.8	1831
ΔS	–7.45	–1.07	–0.12	–0.01	0.0	0.0	0.0	0.0	0.0	39.70	14.30	–45.35	0
AE	7.5	10.0	20.6	35.7	76.8	11.4	43.0	64.5	103.5	156.3	138.1	55.3	723
WD	99.4	119.7	139.5	132.8	104.1	164.4	132.1	105.7	57.0	0.0	0.0	53.6	1108
WS	0.0	0.0	0.0	0.0	0.0	0.0	0.0	0.0	0.0	0.0	0.0	0.0	0
RO	0.0	0.0	0.0	0.0	0.0	0.0	0.0	0.0	0.0	0.0	0.0	0.0	0

Stn :PAMBAN (VAIGAI DELTA)											AWC= 100 mm		
	J	F	M	A	M	J	J	A	S	O	N	D	TOTAL
P	66.0	19.0	24.0	68.0	24.0	6.0	9.0	12.0	16.0	175.0	308.0	196.0	923
PE	120.0	129.7	152.2	158.5	171.8	167.0	164.7	163.2	157.0	152.2	138.1	120.0	1794
ΔS	58.27	-39.01	-13.91	-3.19	-1.67	-0.39	-0.08	-0.02	0.0	22.80	77.20	0.0	100

Table 3.7 *Contd...*

Stn :PAMBAN (VAIGAI DELTA)										AWC= 100 mm			
	J	F	M	A	M	J	J	A	S	O	N	D	TOTAL
AE	120.0	58.0	37.9	71.2	25.7	6.4	9.1	12.0	16.0	152.2	138.1	120.0	767
WD	0.0	71.7	114.3	87.3	146.1	160.6	155.6	151.2	141.0	0.0	0.0	0.0	1028
WS	0.0	0.0	0.0	0.0	0.0	0.0	0.0	0.0	0.0	0.0	92.7	76.0	169
RO	31.8	15.9	7.9	4.0	2.0	1.0	0.0	0.0	0.0	0.0	51.0	63.5	177

(P = Precipitation, PE = Potential Evapotranspiration, ΔS = Soil moisture storage change, AE = Actual Evaporation, WD = Water Deficiency, WS= Water Surplus and RO = Run Off)

Graphical trends and water balance parameters

The Graphical trends of water balance parameters for different stations of Cauvery,Ponnaiyar and Vaigai deltas are shown below in Fig. 3.18.

The monthly Precipitation amounts in respect of Cauvery, Ponnaiyar and Vaigai deltas vary from 8.0 mm to 458.0 mm; 14.0 mm to 362.0 mm and 0.0 mm to 308.0 mm respectively. The maximum rainfall occurrences are seen during the period of October to November from the northeast monsoon though substantial inputs are also seen during the period June to September from the southwest monsoon. The peak values recorded in different stations are: Cauvery delta: Tiruchirapalli: 170.0 mm; Nagapattinam: 458.0 mm; Pudukkottai: 158.0 mm; Ponnaiyar delta: Kallakurchi: 194.0 mm; Cuddalore: 362.0 mm; Vaigai delta: Aruppukottai: 196.0 mm; Pamban: 308.0 mm. The average annual precipitation levels in respect of Cauvery, Ponnaiyar, Vaigai are 1063.6 mm, 1174.5 mm, 823.0 mm respectively.

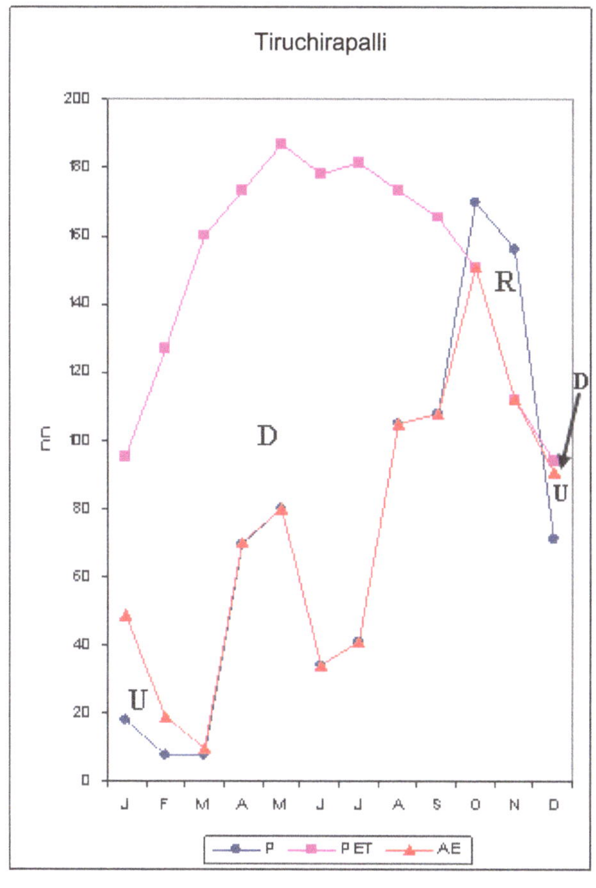

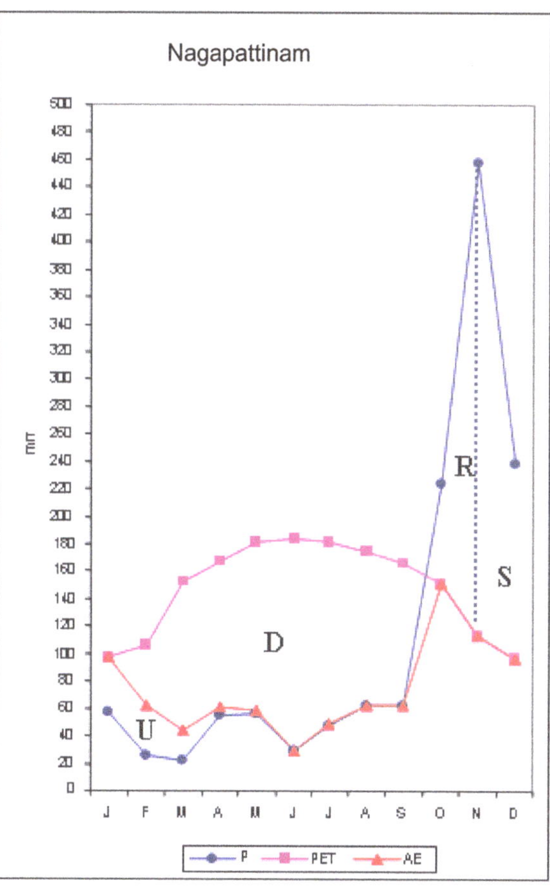

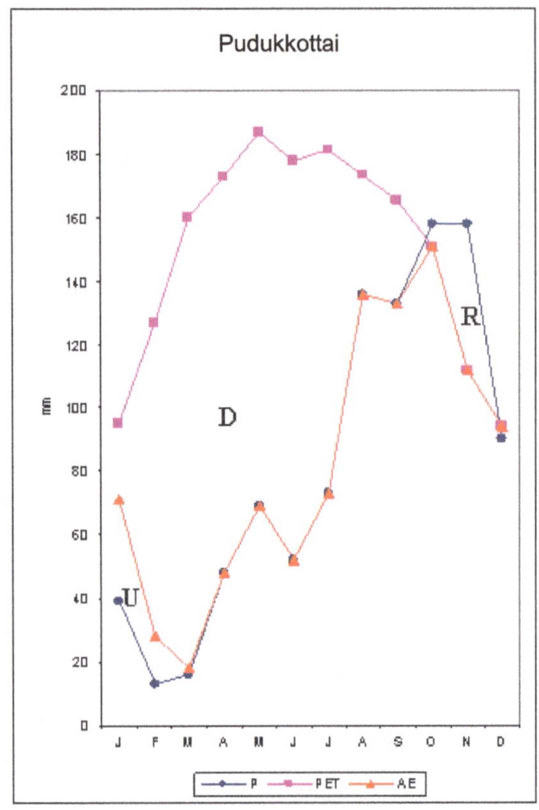
Pudukkottai
mm
D
R
U
J F M A M J J A S O N D
P PET AE

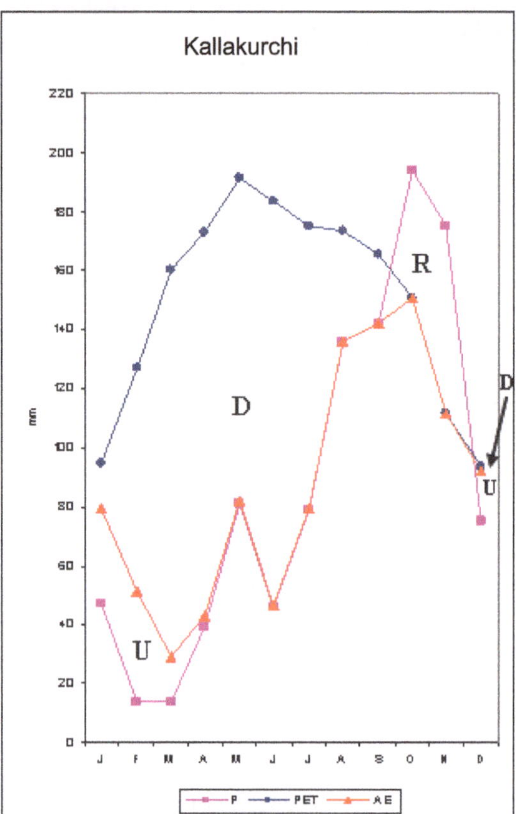
Kallakurchi
mm
D
R
D
U
U
J F M A M J J A S O N D
P PET AE

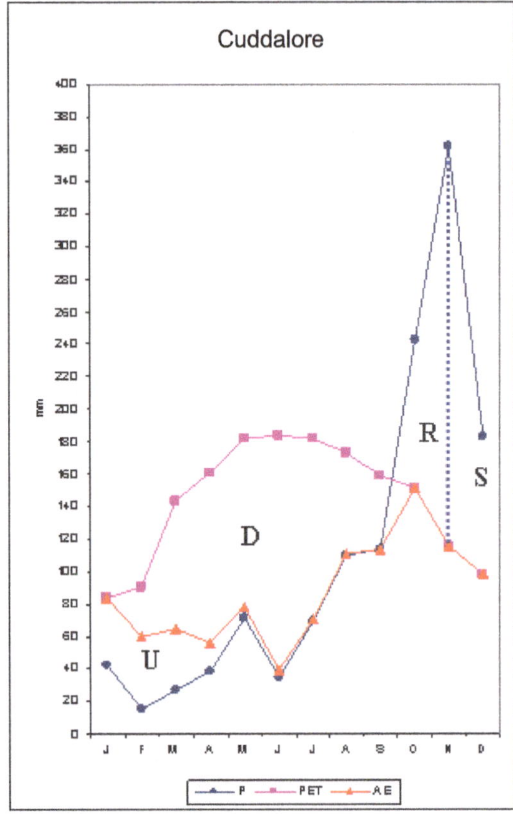
Cuddalore
mm
D
R
S
U
J F M A M J J A S O N D
P PET AE

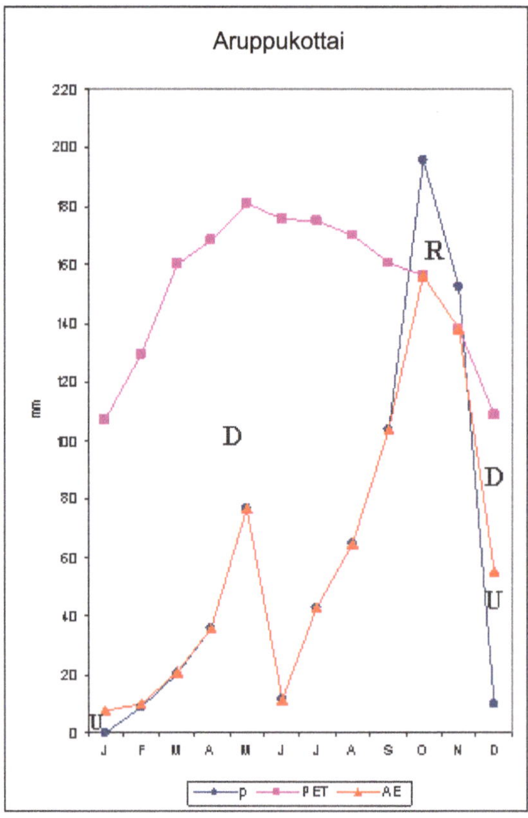
Aruppukottai
mm
D
R
D
U
U
J F M A M J J A S O N D
P PET AE

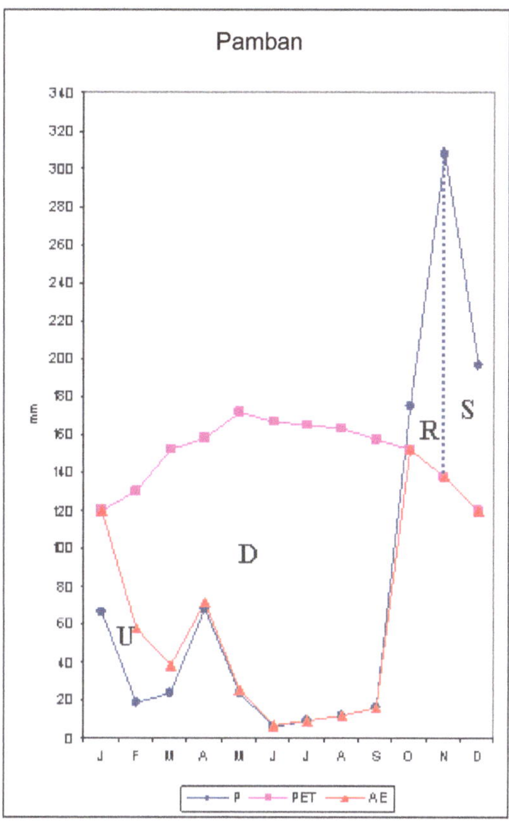

Fig. 3.18 Graphical trends of water balance parameters in respect of Cauvery, Ponnaiyar and Vaigai deltas (R-Recharge accretion in the soil zone, S-Surplus, U-Utilization, D-Water Deficit)

The potential evapotranspiration in Cauvery, Ponnaiyar and Vaigai delta have a range of values varying from 83.2 mm to 186.5 mm. The minimum values normally are recorded in the month of January and the maximum values during the period May to June. The maximum PET values recorded in different stations are: Cauvery delta: Tiruchirapally: 186.9 mm; Nagapattinam: 183.5 mm and Pudukottai: 186.9 mm; Ponnaiyar delta: Kallakurchi: 191.4 mm; Cuddalore: 183.5 mm; Vaigai delta: Aruppukottai: 180.9 mm and Pamban: 167.0 mm. The average annual PET values in Cauvery, Ponnaiyar and Vaigai deltas are: 1787.3 mm, 1759.5 mm and 1812.5 mm respectively.

The water deficiency values in Cauvery, Ponnaiyar and Vaigai deltas in general occur during the period December/January to September and the rest of the periods recorded 'nil' values. The range of water deficiency values are given as under: Cauvery delta: 0.2 mm to 153.3 mm; Ponnaiyar delta: 1.6 mm to 144.3 mm and Vaigai delta: 53.6 mm to 164.4 mm. The peak values of WD recorded in different deltas/stations are: Tiruchirapalli: 144.1 mm; Nagapattinam: 153.9 mm; Pudukkottai: 126.1 mm; Ponnaiyar delta: Kallakurchi: 137.1 mm; Cuddalore: 144.3 mm and Vaigai delta: Arupukottai: 164.4 mm and Pamban: 160.6 mm. The annual average WD recorded for Cauvery, Ponnaiyar and Vaigai are: 875 mm, 719 mm and 1068 mm respectively.

The water surpluses recorded in majority of the stations in these delta systems have shown 'zero' values through out the year, excepting stations like Nagapattinam in Cauvery delta, Cuddalore in Ponnaiyar delta and Pamban in Vaigai delta. The WS recorded in different deltas/stations are noticed during the period November to December and peak values are: Cauvery delta: Nagapattinam: 317.8 mm; Ponnaiyar delta: Cuddalore: 187.7 mm and Vaigai delta: Pamban: 92.7 mm. The average annual WS in Cauvery, Ponnaiyar and Vaigai delta systems are: 154.0 mm; 137 mm; 85 mm respectively.

The seasonal computed runoff values in these regions turn out to be of low order and are graphically shown in Fig. 3.19.

The runoff values for different deltas/stations are recorded during the period starting from November to December with peak values reaching during the month of November for Cauvery, Ponnaiyar and December for Vaigai. The peak RO values in different deltas/stations are: Cauvery: Nagapatnam: 174.8 mm; Ponnaiyar delta: Cuddalore: 94.0 mm; Vaigai delta: Pamban: 63.0 mm. The average annual runoff values for Cauvery, Ponnaiyar and Vaigai deltas are: 163.6 mm, 136.0 mm and 88.5 mm. The estimated annual RO for Cauvery, Ponnaiyar and Vaigai are: 15.3 %, 11.5 % and 10.7 % of the total annual precipitation respectively.

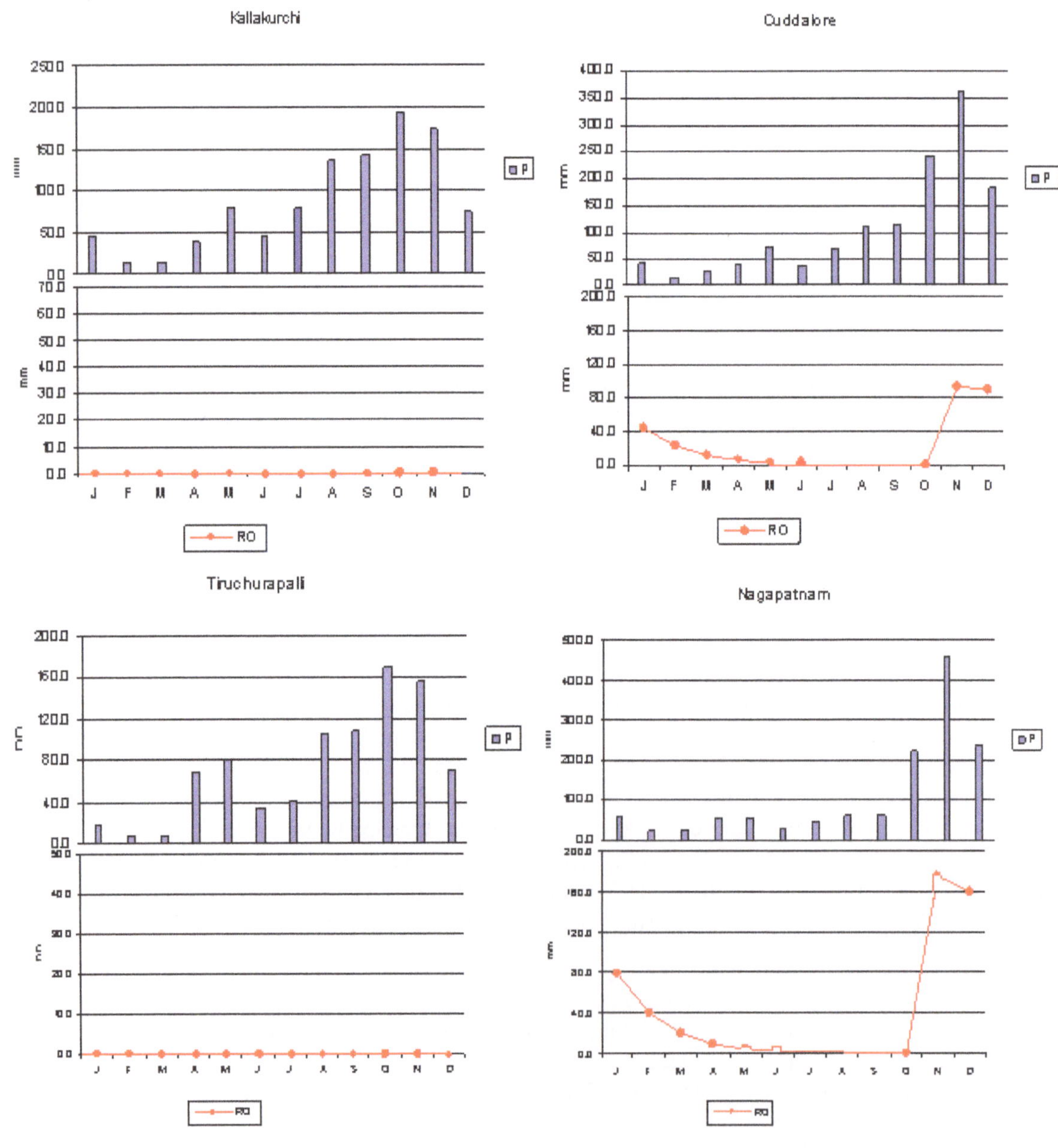

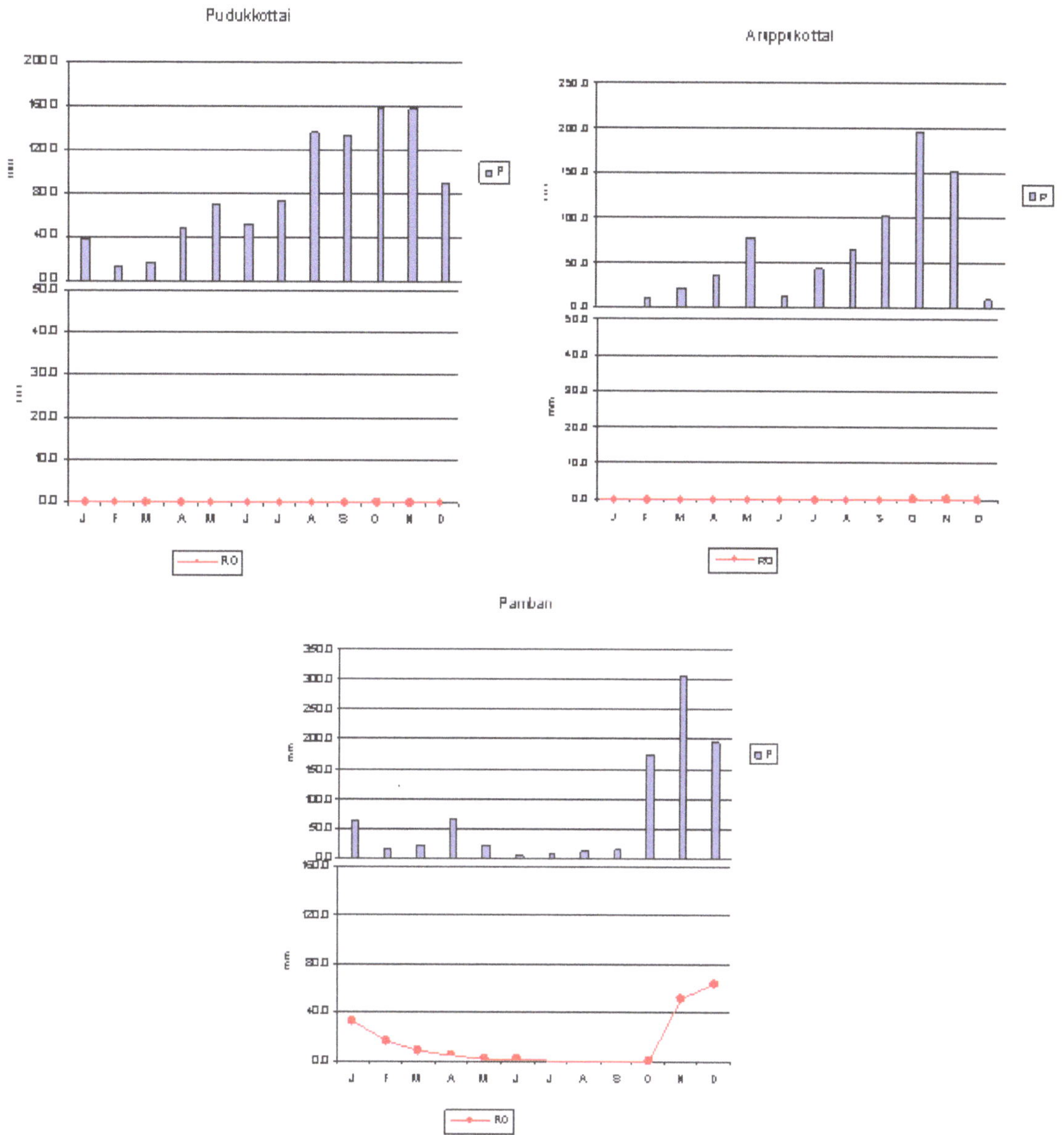

Fig. 3.19 Precipitation-Runoff trends in Ponnaiyar, Cauvery and Vaigai delta systems

Annual climatic water balance and budgetary analyses

The Annual climatic water balance/budgetary estimations for Cauvery, Ponnaiyar and Vaigai deltas are tabulated below in Table 3.8.

Table 3.8 Annual climatic water balance/budgetary estimations

Location	PET (mm)	Rainfall (mm)	Water Surplus (mm)	Water Deficit (mm)	Humidity Index (%)	Aridity Index (%)	Moisture Index (%)	RunOff (mm)
Cauvery Delta								
Tiruchurapalli	1798	869	0	929	0.0	51.7	-51.7	0
Nagapatnam	1766.5	1337	460.9	883.4	26.1	50.0	-23.9	491
Pudukottai	1797.5	985	0	812.5	0.0	45.2	-45.2	0

Table 3.8 ***Contd...***

Location	PET (mm)	Rainfall (mm)	Water Surplus (mm)	Water Deficit (mm)	Humidity Index (%)	Aridity Index (%)	Moisture Index (%)	RunOff (mm)
Ponnaiyar Delta								
Kallakurchchi	1799	1042	0	757	0.0	42.1	– 42.1	0
Cuddalore	1720	1307	272.87	681.3	15.9	39.6	– 23.7	272
Vaigai Delta								
Arupukkottai	1831	723	0	1108	0.0	60.5	– 60.5	0
Pamban	1794	923	169	1028	9.4	57.3	– 47.9	177

3.3.2 Pennar and palar delta basins

Pennar and Palar are medium type delta basins along the east coast located in the south east and north east regions of Andhra Pradesh and Tamil Nadu respectively. The Pennar river after traversing a length of about 560 kms from its source builds up a medium delta in the Nellore district of Andhra Pradesh with an aereal spread of the order of 1470 km^2, with its apex located near Sangham. It has an elevation of about 33 m AMSL and is located at a distance of about 50 kms from the coast line. The average annual discharge near Sangham, the apex point of the delta is about 3235 MCM. The upper and middle portions of the delta between Sangham and Nellore contain a number of paleo channels deposited under fluviotile conditions and the lower region contains a series of sand ridges, lagoonal deposits, spits and bars between Nellore and Mypadu-Isakapalli-Muthur line. The soils in the region could be categorized as well drained lateritic, red loamy, black cotton and alluvial soils. The clay percentages in these soils vary from 50% to 60%. The soils are reported to contain high potash and low to medium levels of phosphates and nitrates with pH values ranging from 7.5 to 8.5 and conductivity values ranging between 500 micro siemens/cm (Inland soils) to 200 micro siemens/cm (outfall coastal soils). The delta region enjoys a subtropical climate with maximum and minimum temperatures of 30^0C and 20.5^0C respectively. The average relative humidity varies between 80% (wet season) to 55% (dry season). The Pennar delta region witnesses more dominantly northeast monsoon during the period November to December (60%) and moderately with southwest monsoon during the period June to September (30%) with an average annual rainfall of about 1070 mm.

The Palar river after traversing for about 340 kms from its source debouches into Bay of Bengal at a point south of Chennai in Tamil Nadu. The Palar river is believed to have its original course in a south easterly direction from its apex point at Walajapet and had built reasonably good delta with an aereal spread of the order about 700 – 800 sq. kms during recent past (Quaternary) and later on changed its course and migrated in south easterly direction and built up a relatively an insignificant delta around south of Chennai with an aereal spread of the order of 100 – 200 sq. kms. The estimated annual average discharge in the river is of the order of 1780 MCM. The peak discharges are normally observed in the months of October and November depending upon monsoonal variations/intensities. The Palar paleo delta basin region at present is drained by several rivulets like Araniar, Kortalaiyar and Cooum. The basin soils are mostly lateritic red loamy type, black and grey soils and thick alluvial soils with clay contents varying from 55% to 65%. The soils in general are rich in lime, potash and magnesium and low in nitrates and phosphates with pH values ranging from 7.0 to 8.5. The average annual maximum and minimum temperatures are 32.9 °C and 24.3 °C respectively. The humidity levels are of the order of 83% (wet season) and 58% (dry season). The basin receives predominantly northeast monsoon rains (60%) and moderate rains through south east monsoon (30%) with an annual average rainfall of the order of 1210 mm. A brief sketch of regional hydrological characters of Pennar and Palar delta basins are tabulated below in Table 3.9.

Table 3.9 Regional hydrological characters of Pennar and Palar basins

S.no.	Item	Pennar delta	Palar delta
1.	Average annual rainfall	1075 mm	1210 mm
2.	Altitudinal variations	0-30 m AMSL	0-15 m AMSL
3.	Average annual temperatures	27.5^{0}C	27.5^{0}C
4.	Aereal spreads of the delta	1470 km^2	500-700 km^2
5.	Length of the delta	45 kms	30 kms
6.	Width of the delta	50 kms	20 kms
7.	Regional Slope of the delta	65 cm/km	50 cm/km
8.	Tidal impacts	7 - 8 kms	2.4 kms.
9.	Average annual discharge	3235 MCM	1780 MCM

Seasonal water balance characteristics

The seasonal relevant water balance characteristics for selected number of hydromet stations in Pennar(4) and Palar(3) delta basin areas are tabulated in Table 3.10 and their graphical trends are shown in Fig. 3.21. The precipitation trend lines and hydromet station locations are shown in Fig. 3.20 (a,b).

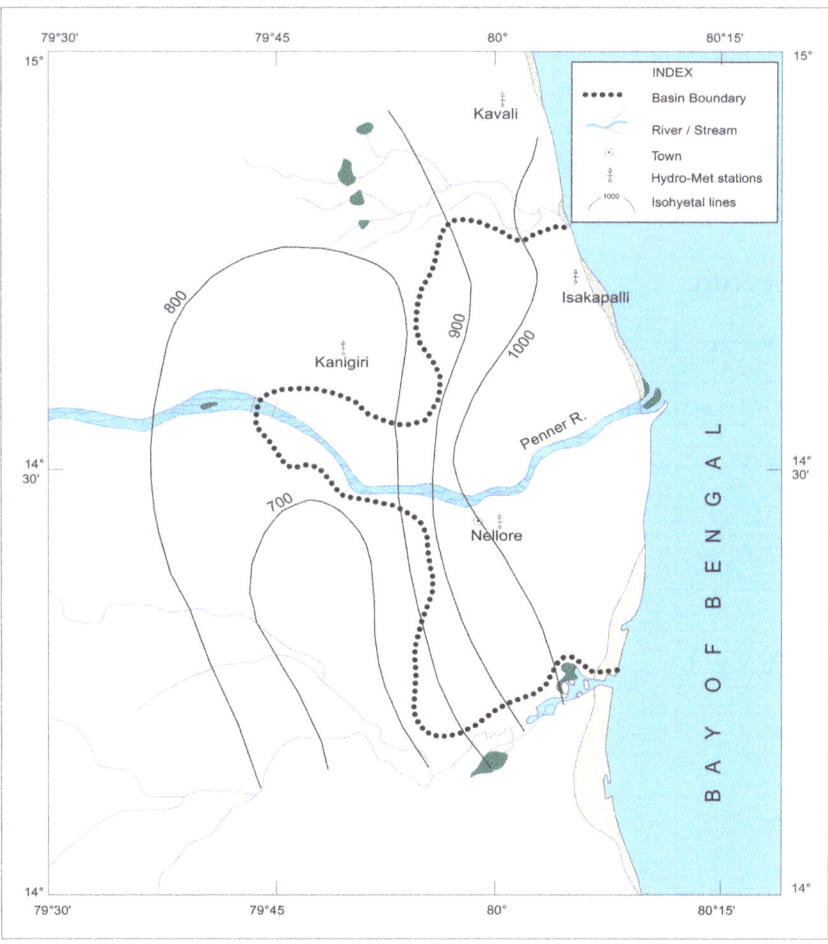

Fig. 3.20(a) Isohyetel trend lines and hydromet stations in Pennar delta

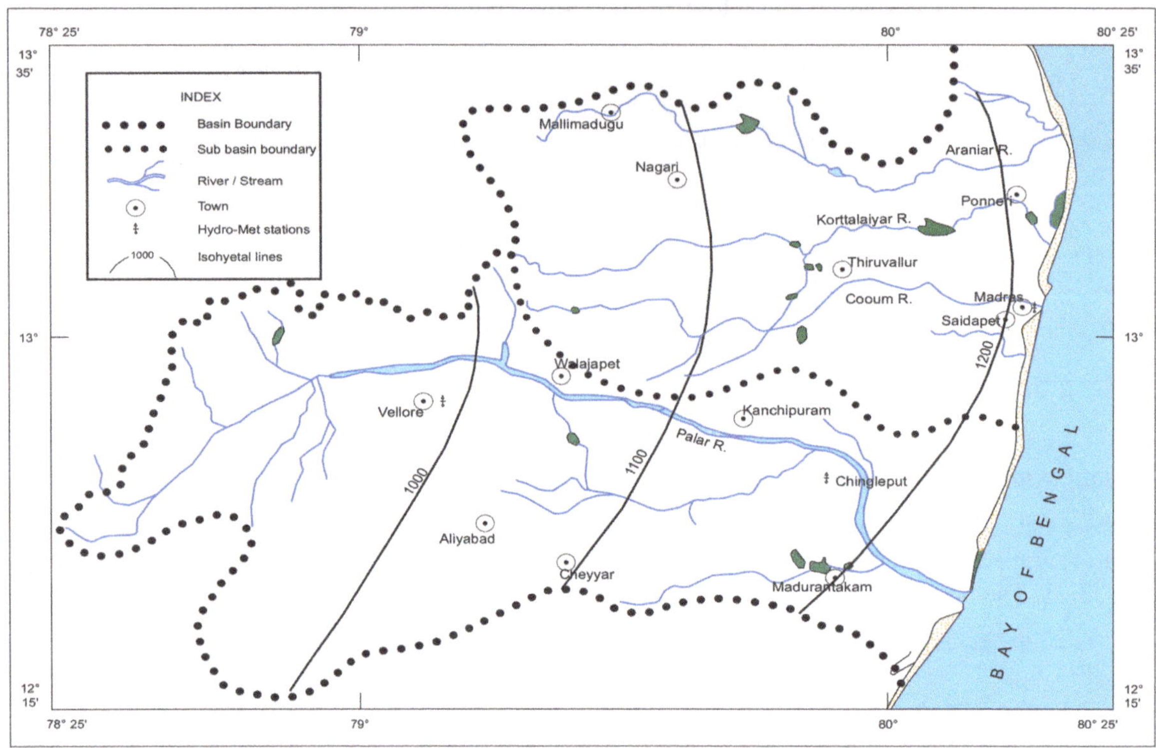

Fig. 3.20(b) Isohyetel trend lines and hydromet stations in Palar delta

Table 3.10 Normal water balance parameters for Pennar and Palar delta basin areas

Stn : ISAKAPALLE (PENNAR DELTA)										**AWC = 150mm**		
	J	**F**	**M**	**A**	**M**	**J**	**J**	**A**	**S**	**O**	**N**	**D**	**TOTAL**
P	23.9	10.7	5.3	10.7	36.6	37.6	57.9	75.7	107.4	283.7	299.0	74.7	1023
PE	111.0	126.0	173.0	184.0	197.0	170.0	152.0	153.0	142.0	122.0	99.0	99.0	1728
ΔS	–56.19	–38.29	–22.27	–7.41	–2.24	–0.69	–0.22	–0.11	-0.03	149.88	0.0	–22.43	0
AE	80.1	49.0	27.6	18.1	38.8	38.3	58.1	75.8	107.4	122.0	99.0	97.1	811
WD	30.9	77.0	145.4	165.9	158.2	131.7	93.9	77.2	34.6	0.0	0.0	1.9	917
WS	0.0	0.0	0.0	0.0	0.0	0.0	0.0	0.0	0.0	11.8	200.0	0.0	212
RO	25.8	12.9	6.5	3.2	1.6	0.0	0.0	0.0	0.0	6.5	103.3	51.6	211
Stn : KANIGIRI (PENNAR DELTA)											**AWC = 150mm**		
	J	F	M	A	M	J	J	A	S	O	N	D	TOTAL
P	12.9	8.1	8.4	20.3	44.5	45.0	80.0	66.3	115.1	160.3	142.0	27.9	731
PE	110.0	124.0	163.0	172.0	194.0	171.0	140.0	146.0	130.0	117.0	107.0	101.0	1675
ΔS	–21.87	–6.88	–1.75	–0.24	–0.03	-0.01	0.0	0.0	0.0	43.30	35.0	–47.52	0
AE	34.8	15.0	10.2	20.5	44.5	45.0	80.0	66.3	115.1	117.0	107.0	75.4	731
WD	75.2	109.0	152.9	151.5	149.5	126.0	60.0	79.7	14.9	0.0	0.0	25.6	944
WS	0.0	0.0	0.0	0.0	0.0	0.0	0.0	0.0	0.0	0.0	0.0	0.0	0
RO	0.0	0.0	0.0	0.0	0.0	0.0	0.0	0.0	0.0	0.0	0.0	0.0	0

Table 3.10 ***Contd...***

Stn : KAVALI (PENNAR DELTA)											**AWC = 150mm**		
	J	F	M	A	M	J	J	A	S	O	N	D	TOTAL
P	25.9	7.1	11.7	11.9	31.2	38.3	65.5	65.5	110.7	283.7	264.9	69.3	986
PE	111.0	126.0	173.0	184.0	197.0	170.0	152.0	153.0	142.0	122.0	99.0	99.0	1728
ΔS	–53.28	–38.20	–20.80	–7.36	–2.29	–0.66	–0.21	–0.11	–0.03	149.88	0.0	-26.94	0
AE	79.2	45.3	32.5	19.3	33.5	39.0	65.7	65.6	110.7	122.0	99.0	96.2	808
WD	31.8	80.7	140.5	164.7	163.5	131.0	86.3	87.4	31.3	0.0	0.0	2.8	920
WS	0.0	0.0	0.0	0.0	0.0	0.0	0.0	0.0	0.0	11.8	165.9	0.0	178
RO	21.6	10.8	5.4	2.7	1.3	0.0	0.0	0.0	0.0	6.5	86.2	43.1	178
Stn : NELLORE (PENNAR DELTA)											**AWC = 150mm**		
	J	**F**	**M**	**A**	**M**	**J**	**J**	**A**	**S**	**O**	**N**	**D**	**TOTAL**
P	32.0	6.9	6.9	9.7	36.6	36.6	70.6	75.9	110.2	266.2	317.5	88.1	1057
PE	111.0	126.0	173.0	184.0	197.0	170.0	152.0	153.0	142.0	122.0	99.0	99.0	1728
ΔS	–57.11	–45.14	–24.94	–8.45	–2.53	–0.78	–0.22	–0.13	–0.04	144.05	5.80	–10.51	0
AE	89.1	52.0	31.8	18.2	39.1	37.4	70.8	76.0	110.2	122.0	99.0	98.6	845.0
WD	21.9	74.0	141.2	165.9	157.9	132.6	81.2	77.0	31.8	0.0	0.0	0.4	883
WS	0.0	0.0	0.0	0.0	0.0	0.0	0.0	0.0	0.0	0.2	212.7	0.0	213
RO	26.6	13.3	6.7	3.3	1.7	0.8	0.0	0.0	0.0	0.1	106.4	53.2	212.0
Stn : VELLORE (PALAR DELTA)											**AWC = 150mm**		
	J	F	M	A	M	J	J	A	S	O	N	D	TOTAL
P	19.0	6.0	8.0	25.0	80.0	77.0	90.0	106.0	152.0	169.0	160.0	77.0	969.0
PE	83.6	90.7	152.2	173.0	191.4	178.1	175.1	173.4	158.5	142.3	97.7	68.9	1685
ΔS	49.92	-29.05	-16.14	-3.70	-0.70	-0.21	-0.07	-0.03	0.0	26.68	62.30	8.10	97.10
AE	68.9	35.1	24.1	28.7	80.7	77.2	90.1	106.0	152.0	142.3	97.7	68.9	971.0
WD	14.7	55.7	128.1	144.3	110.7	100.9	85.0	67.4	6.5	0.0	0.0	0.0	714
WS	0.0	0.0	0.0	0.0	0.0	0.0	0.0	0.0	0.0	0.0	0.0	0.0	0
RO	0.0	0.0	0.0	0.0	0.0	0.0	0.0	0.0	0.0	0.0	0.0	0.0	0
Stn : CHINGLEPUT (PALAR DELTA)											**AWC = 150mm**		
	J	F	M	A	M	J	J	A	S	O	N	D	TOTAL
P	38.0	11.0	14.0	21.0	37.0	52.0	98.0	142.0	160.0	233.0	274.0	135.0	1215.0
PE	95.8	104.4	152.2	167.0	191.4	183.5	181.4	173.4	165.3	150.8	112.4	94.8	1772
ΔS	102.03	-47.29	-32.95	-13.56	-5.29	-1.72	-0.52	-0.13	-0.02	81.65	67.80	0	150
AE	95.8	58.3	47.0	34.6	42.3	53.7	98.5	142.1	160.0	150.8	112.4	94.8	1090.3
WD	0.0	46.1	105.3	132.4	149.1	129.8	82.9	31.3	5.3	0.0	0.0	0.0	682
WS	0.0	0.0	0.0	0.0	0.0	0.0	0.0	0.0	0.0	0.6	93.8	40.2	135
RO	21.8	10.9	5.5	2.7	1.4	0.0	0.0	0.0	0.0	0.3	47.1	43.6	133.2
Stn : MADRAS (PALAR DELTA)											**AWC = 150mm**		
	J	F	M	A	M	J	J	A	S	O	N	D	TOTAL
P	24.0	7.0	15.0	25.0	52.0	53.0	84.0	124.0	118.0	267.0	309.0	139.0	1217
PE	95.8	104.4	152.2	167.0	191.4	183.5	181.4	173.4	165.3	150.8	112.4	94.8	1772
ΔS	92.94	-44.39	-29.10	-11.90	-4.57	-1.73	-0.60	-0.18	-0.13	115.86	33.80	0.0	150
AE	95.8	51.4	44.1	36.9	56.6	54.7	84.6	124.2	118.1	150.8	112.4	94.8	1024
WD	0.0	53.0	108.1	130.1	134.8	128.8	96.8	49.2	47.2	0.0	0.0	0.0	748
WS	0.0	0.0	0.0	0.0	0.0	0.0	0.0	0.0	0.0	0.3	162.8	44.2	207
RO	31.4	15.7	7.9	3.9	2.0	1.0	0.0	0.0	0.0	0.2	81.5	62.8	206

(P = Precipitation, PE = Potential Evapotranspiration, ΔS = Soil moisture storage change, AE = Actual Evaporation, WD = Water Deficiency, WS = Water Surplus and RO = Run Off)

Graphical trends and water balance parameters

The Graphical trends of water balance parameters for different hydro-met stations for Pennar and Palar deltas are shown below in Fig. 3.21.

The Pennar delta basin area is bounded by 1100 mm contour close to the coast, and 700 mm contour away from the coast. A low precipitation loop with 700-800 mm precipitation contours is observed along Gudur - Buchireddipalem axis in the up stream area of the delta. A major portion of the outfall Pennar delta region is characterized by 900-1000 mm contour line. On the other hand the Palar delta region is bounded by precipitation contour lines of 1200 mm and 1000 mm with maximum contour line 1200 mm going close to the coast and the minimum contour line 1000 mm away from the coast. All most all the contour lines go parallel to the coast. A major portion of the Palar delta region is characterized by 1100 mm – 1200 mm contour line.

The south west monsoon impacts are seen early in May itself, and increases gradually and attains peak levels in the months of October and November (North east monsoon) in both Pennar – Palar delta regions. The peak monsoon levels in respect of stations of Pennar and Palar delta regions are: Isakapalle: 299 mm; Kanigiri: 160.3 mm; Kavali: 283.7 mm and Nellore: 317.5 mm; Vellore: 169 mm; Chingleput: 274 mm; Madras: 309 mm. The average annual precipitation levels for Pennar and Palar are 949.2 mm, 1133.6 mm respectively.

The potential evapotranspiration levels in Pennar delta have a range between 99 mm to 197 mm with the minimum and maximum levels during the months of December and May respectively. The peak values attained in different stations are: Isakapalle: 197 mm; Kanigiri: 194 mm; Kavali: 197 mm and Nellore: 197 mm. A minimum value of 99 mm is recorded in all stations except Kanigiri site where the minimum value is 101 mm. Similarly the PET values for Palar delta region register a variation from 68.9 mm to 191.4 mm in all the stations and attain maximum values during the month of May and Minimum values during the month of December. A minimum value 94.8 mm is observed for Chingleput and Madras stations and a value of 68.9 mm for Vellore station. The average annual PET values for Pennar and Palar are 1714.7 mm, 1743.3 mm respectively.

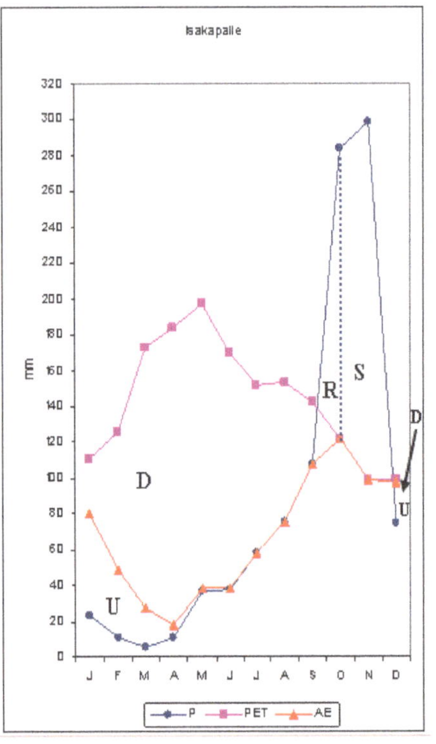

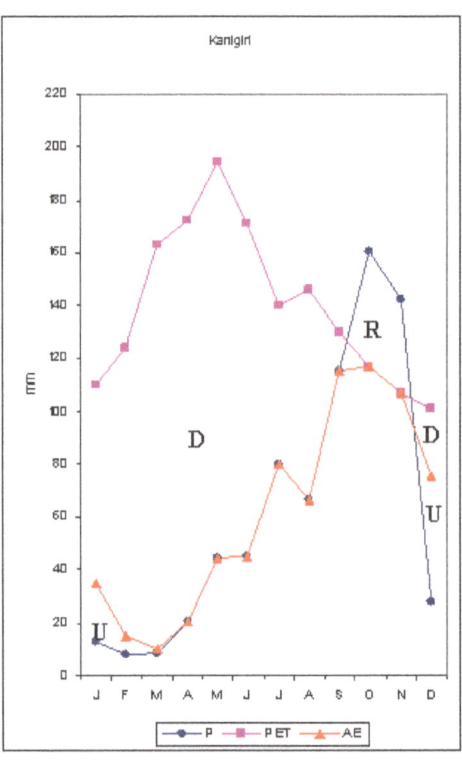

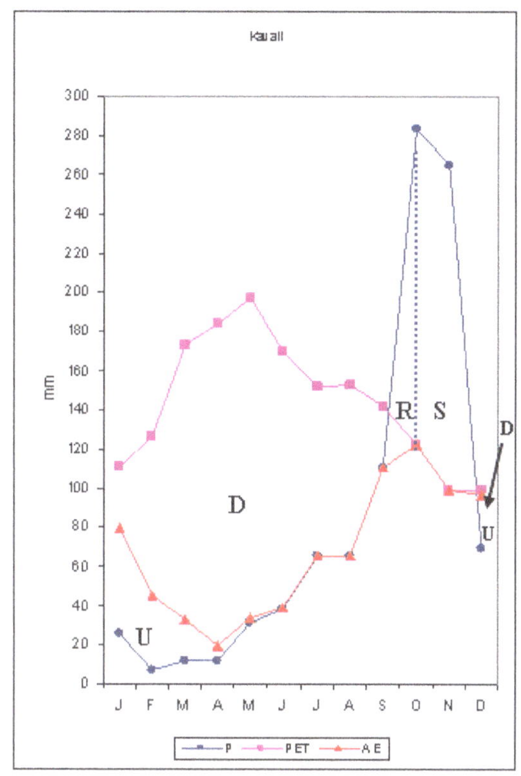
Kavali
mm
R
S
D
D
U
U
J F M A M J J A S O N D
P
PET
AE

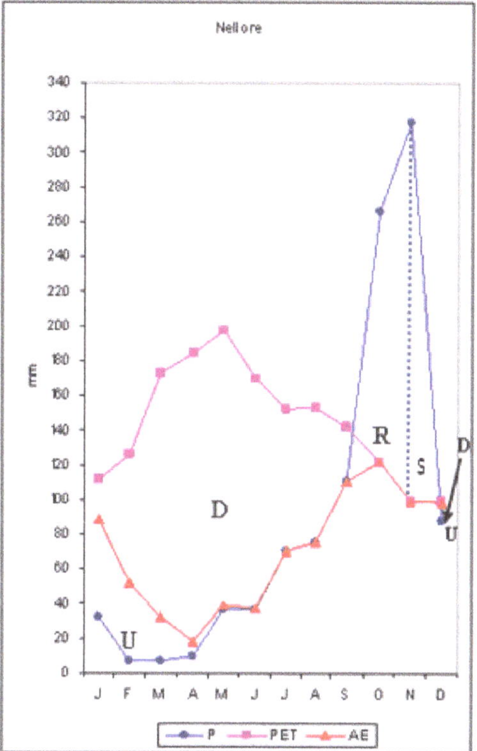
Nellore
mm
R
S
D
D
U
U
J F M A M J J A S O N D
P
PET
AE

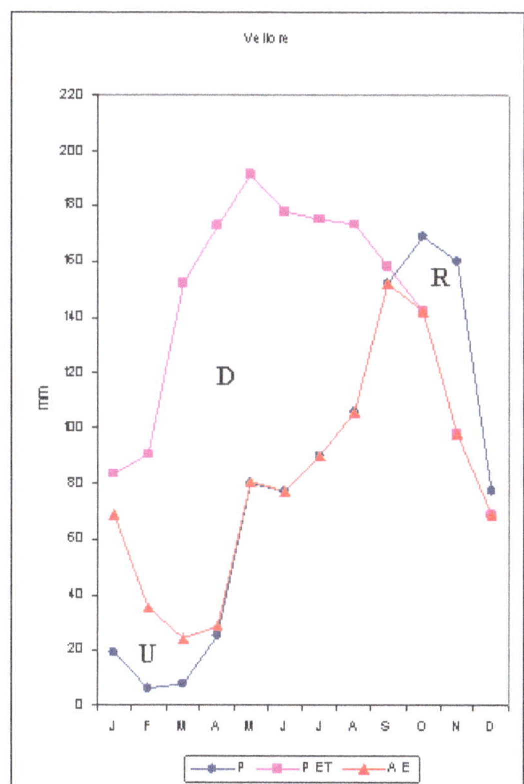
Vellore
mm
R
D
U
J F M A M J J A S O N D
P
PET
AE

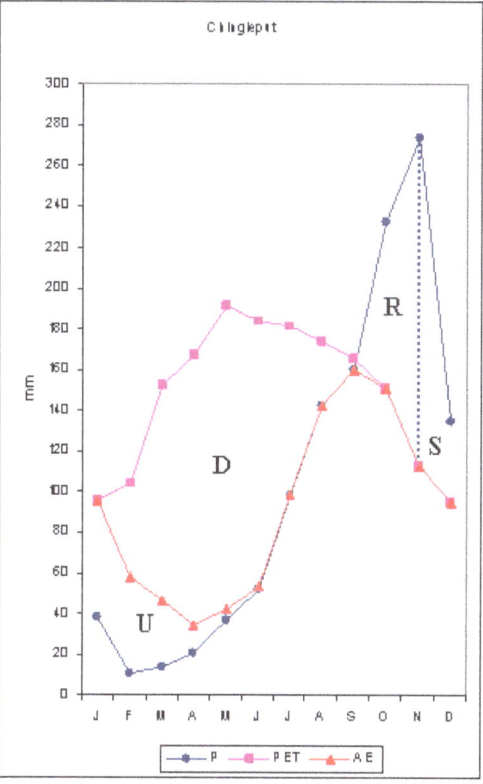
Chingleput
mm
R
S
D
U
J F M A M J J A S O N D
P
PET
AE

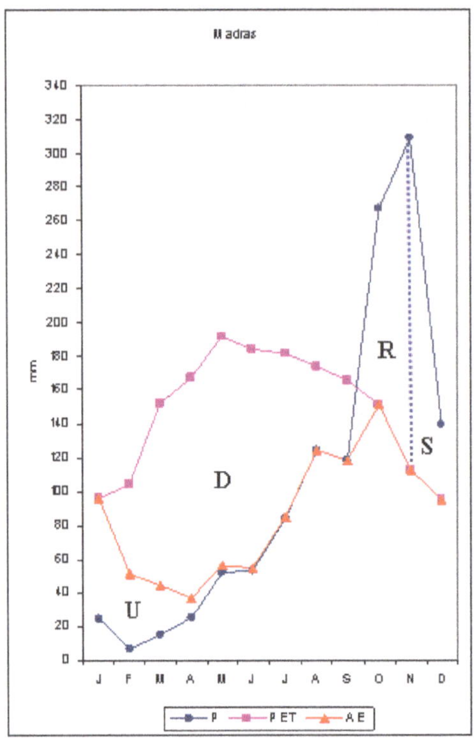

Fig. 3.21 Graphical trends of water balance parameters in respect of Pennar and Palar (R-Recharge accretion in the soil zone, S-Surplus, U-Utilization, D-Water Deficit)

The water deficits in the region are seen through out the year, except in the months of October and November where the values are at 'zero' level. The maximum deficiencies are seen either in the month of April or May and the minimum values in December. The maximum WD values recorded in different stations are: Isakapalle: 165.89 mm; Kanigiri: 152.9 mm; Kavali: 164.7 mm; Nellore: 165.9 mm; Vellore: 144.3 mm; Chingleput: 149.1 mm and Madras: 134.8 mm. The average annual WD for Pennar and Palar are 916 mm, 714 mm respectively.

The water surplus in both the delta systems are observed during the period October, November and December with peak values reaching normally during the month of October. The maximum WS observed in different stations are: Isakapalle: 200 mm; Kavali: 165 mm; Nellore: 212.7 mm; Chingleput: 93.8 mm and Madras: 162.8 mm; Kanigiri in Pennar delta and Vellore in Palar have recorded 'zero' values all through. The average annual WS for Pennar and Palar are 150 mm, 114 mm respectively.

The computed seasonal surface runoff trends for Pennar and Palar delta basin areas are shown in Fig. 3.22.

The runoff values in Pennar delta are recorded during the period starting from October to December with peak values reaching during the month of October. The peak RO values in Pennar delta are: Isakapalle: 103.3 mm; Kanigiri: 0 mm; Kavali: 86.2 mm; Nellore: 106.4 mm. In the case of Palar delta by and large similar trends are recorded with the following peak values: Vellore: 0.0 mm; Chingleput: 47.1 mm; Madras: 81.5 mm. The average annual RO values in Pennar and Palar are 150.25 mm, 113.01 mm respectively. The estimated annual runoff for Pennar and Palar is 15.8 %, 9.9% of the total annual precipitation respectively.

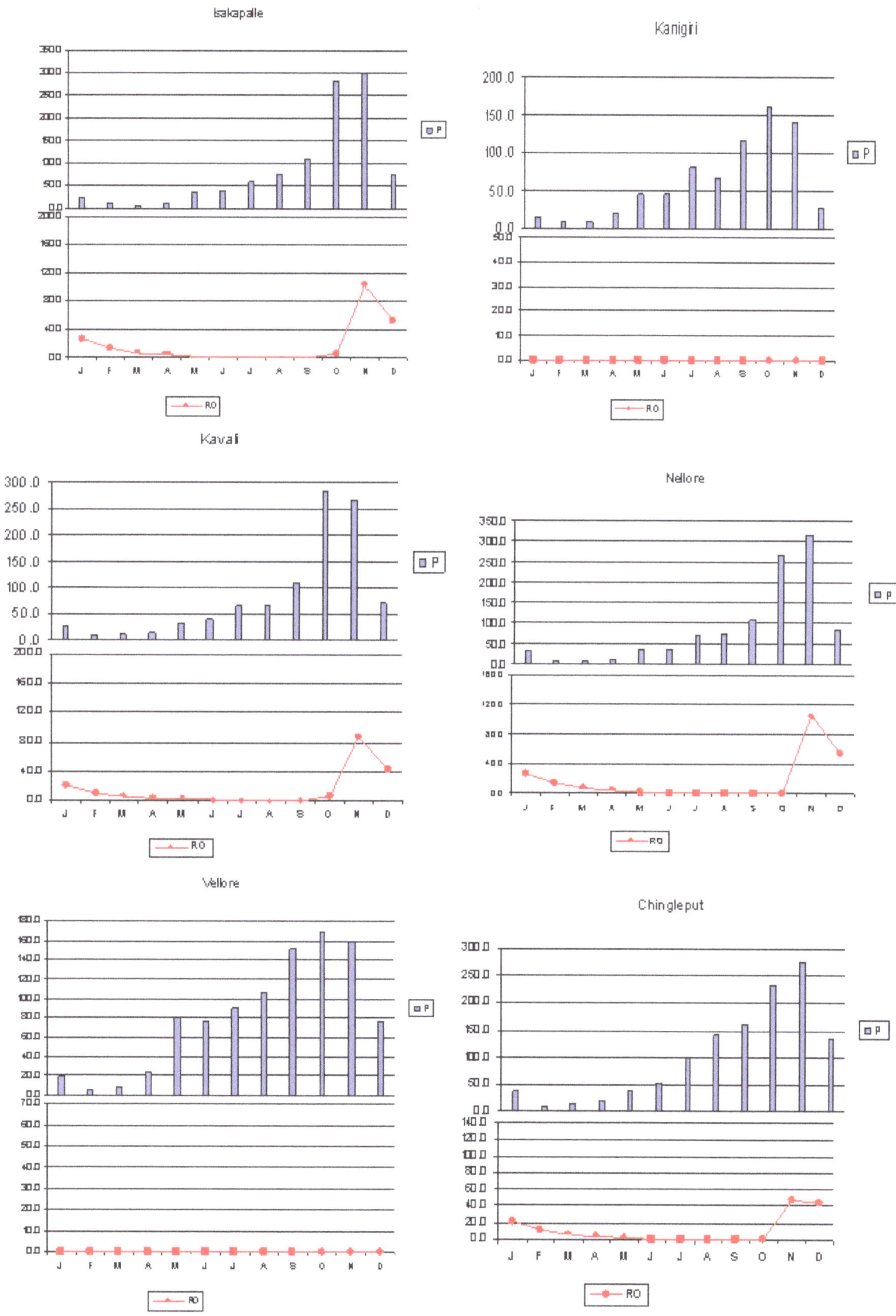
Isakapalle
Kanigiri
Kavali
Nellore
Vellore
Chingleput
P
RO
J F M A M J J A S O N D

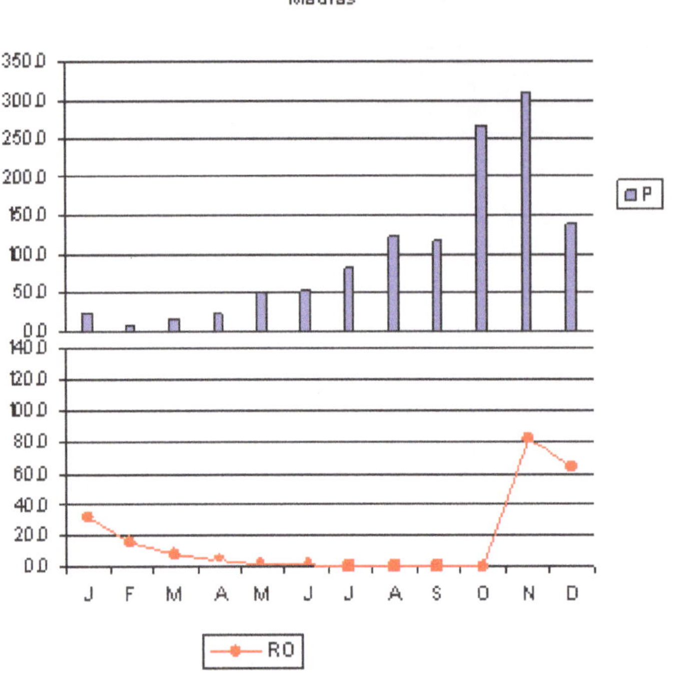

Fig. 3.22 Precipitation-Runoff relations in Pennar- Palar delta regions

Annual climatic water balance budget analysis

The Annual water balance/budgetary estimations for Pennar and Palar deltas are tabulated below in Table 3.11.

Table 3.11 Annual water balance/budgetary estimations for Pennar-Palar delta regions

Location	PET (mm)	Rainfall (mm)	Water Surplus (mm)	Water Deficit (mm)	Humidity Index (%)	Aridity Index (%)	Moisture Index (%)	RunOff (mm)
Pennar Delta								
Isakapalli	1728	1023.2	211.82	916.62	12.3	53.0	–40.8	211
Kanigiri	1675	730.8	0	944.2	0.0	56.4	–56.4	0
Kavali	1728	985.7	177.72	920.02	10.3	53.2	–43.0	178
Nellore	1728	1057.2	212.85	883.65	12.3	51.1	–38.8	212
Palar Delta								
Vellore	1684.9	969	0.02	713.18	0.0	42.3	–42.3	0.03
Chingleput	1772.4	1215	134.55	682.12	7.6	38.5	–30.9	133
Madras	1772.4	1217	207.3	748	11.7	42.2	–30.5	206

3.3.3 Godavari and krishna delta regions

Godavari and Krishna deltas are the major and prominent delta systems along the east coast with aereal spreads of the order of 5500 and 4600 sq. kms respectively. These two twin delta systems support rich agriculture practices round the year and are classified as 'Graneries' of agricultural production in Andhra Pradesh. The Godavari delta with its apex point at Rajahmundry has an elevation of the order of

11 m above mean sea level and branches off into two distributaries namely Vasishta and Gautami in the down stream side. These again further gets branched off into two each namely Vasishta–Vainateyam and Goutami–Nilarevu respectively. The upper and middle delta systems are mainly developed as flood plains with several paleo channels and the lower outfall delta region contain several land form features like spits and bars, tidal flats, levees, tidal channels, creeks and mangroves etc. The impacts of tides are seen up to 50 kms, upstream from the coastline. The Krishna delta with its apex at Vijayawada has an elevation of the order of 19 m above mean sea level, runs as a braided channel for about 60 kms and then branches off into two near Avanigadda namely Krishna and Hamsaladivi paya. Further down stream, Krishna river distributes itself into three more distributaries namely Gollumutha paya, Nadimeru and Krishna river. Most of the river water and sediment discharges are being carried out by main Krishna river and meets the sea near false Divi point. The upper and middle delta plains contain a number of paleo channels and levees, and the outfall region is characterized by spits, bars, tidal channels, sand ridges and mangroves. The tidal impacts are seen up to 40 kms upstream from the coastline.

The soils of Godavari and Krishna deltas are deep moderately well drained with colors varying from grey to brown and sometimes black. The clay contents of the soils in general are of the order of 60-70% with pH values ranging between 7.5 and 8.5. The clay minerals basically are smectites and montimorillonites and are rich in lime, calcium and magnesium with low organic carbon contents. The dominant salts present in the soils are chlorides and sulphates of sodium, magnesium and calcium with rich nutrient values but are deficient in zinc (Zn). The average general maximum and minimum mean temperatures in these delta tracts are 30 °C and 18 °C respectively. The average maximum and minimum relative humidities are 80% (wet season) and 60% (dry season).

Godavari and Krishna receive south west monsoon rainfall during the months of June to September (southwest monsoon) and occasional north-east monsoon during the period of October to December. The average annual rainfall in respect of Godavari and Krishna delta are 1125 mm and 900 mm and with a mean annual temperature of 27.5 °C. The average peak estimated discharges in respect of Godavari and Krishna delta regions are 105,000 MCM and 67,675 MCM respectively. The regional hydrological characters of these two basins at a glance are presented in the Table 3.12 below:

Table 3.12 Regional hydrological characters of Godavari and Krishna delta basins

S.no.	Item	Godavari delta	Krishna delta
1.	Average annual rainfall	1000 mm to 1250 mm	800 mm to 1000 mm
2.	Altitudinal variations	0-20 m AMSL	0-19 m AMSL
3.	Average annual temperatures	27.5 °C	27.5 °C
4.	Areal spreads of the delta	5500 km^2	4600 km^2
5.	Length of the delta	140 kms	120 kms
6.	Width of the delta	80 kms	96 kms
7.	Slope of the delta	20 cm/km	20 cm/km
8.	Tidal impacts	45 kms upstream	40 kms upstream.
9.	Average annual discharge	105,000 MCM	67,675 MCM

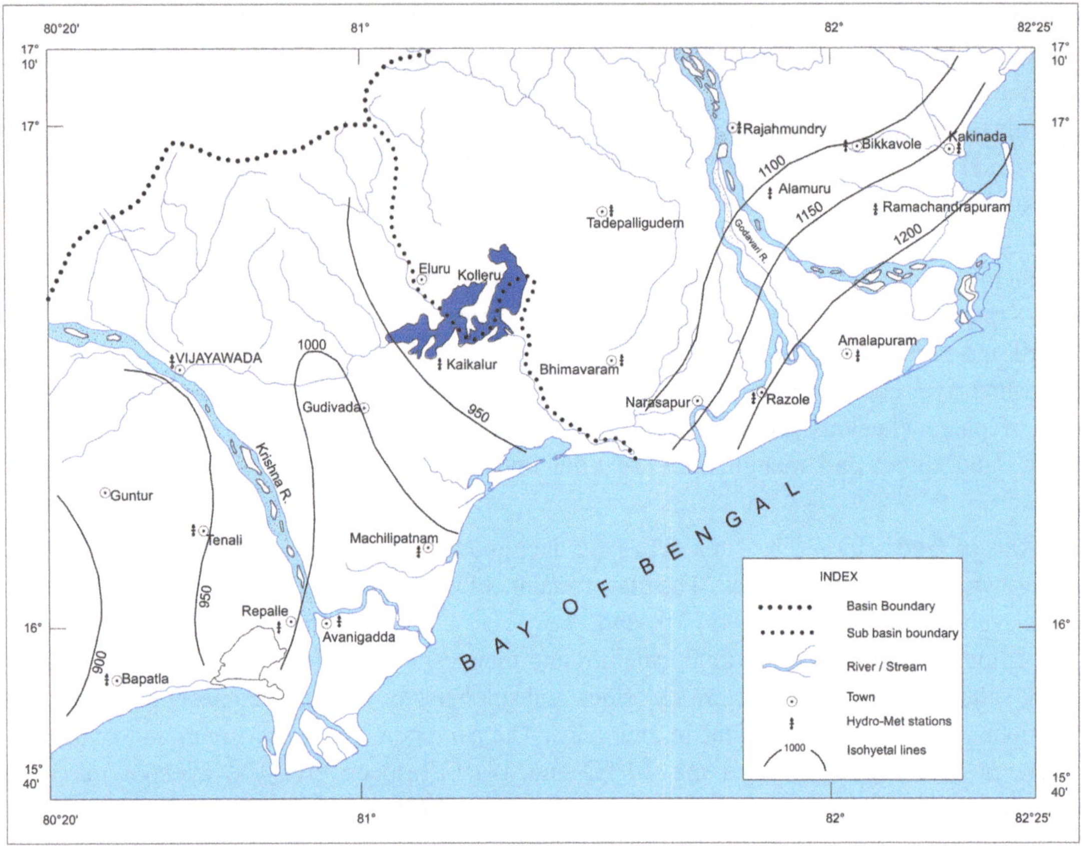

Fig. 3.23 Isohyetel map of Godavari and Krishna delta with selected hydromet stations

The relevant water balance parameters for selected number of hydromet stations namely P(precipitation), PE (PotentialEvapotranspiration), AE (Actual Evapotranspiration), WD (Water Deficit), and WS (Water Surplus) for nine selected stations in Godavari delta and seven selected stations in Krishna delta are presented in Table 3.13 and the graphical trends are shown in Fig. 3.24(a,b,c and d) averaged for 30 year period from 1960 – 1990. The hydromet site locations of selected stations and their precipitation trends are shown in Fig. 3.23.

Seasonal water balance characteristics

Table 3.13 Seasonal water balance characteristics of Godavari and Krishna deltas

Stn : RAJAHMUNDRY (GODAVARI DELTA)										AWC = 200mm		
	J	F	M	A	M	J	J	A	S	O	N	D	TOTAL
P	5.3	16.0	12.5	28.5	53.3	141.2	214.9	179.3	175.3	167.4	58.2	5.3	1057
PE	112.0	123.0	168.0	176.0	194.0	169.0	136.0	137.0	125.0	125.0	116.0	108.0	1689
ΔS	–37.06	–21.79	–16.64	–7.38	–3.42	–0.44	75.99	42.30	50.30	28.50	–50.20	–60.16	0
AE	42.4	37.8	29.1	35.9	56.7	141.6	136.0	137.0	125.0	125.0	108.4	65.5	1040
WD	69.6	85.2	138.9	140.1	137.3	27.4	0.0	0.0	0.0	0.0	7.6	42.5	649
WS	0.0	0.0	0.0	0.0	0.0	0.0	2.9	0.0	0.0	13.9	0.0	0.0	17
RO	1.0	0.0	0.0	0.0	0.0	0.0	1.6	0.8	0.4	7.2	3.6	1.8	16

Table 3.13 *Contd...*

Stn : RAMACHANDRAPURAM (GODAVARI DELTA)										AWC = 150mm		
	J	F	M	A	M	J	J	A	S	O	N	D	TOTAL
P	8.9	12.7	13.2	19.3	44.2	126.0	199.6	174.7	164.1	213.6	120.9	11.4	1108
PE	112.0	123.0	168.0	176.0	194.0	169.0	136.0	137.0	125.0	125.0	116.0	108.0	1689
ΔS	–39.16	–20.63	–12.22	–4.39	–1.50	–0.22	62.94	37.70	39.10	9.60	0.0	–71.22	0
AE	48.1	33.3	25.4	23.7	45.7	126.2	136.0	137.0	125.0	125.0	116.0	82.6	1024.0
WD	63.9	89.7	142.6	152.3	148.3	42.8	0.0	0.0	0.0	0.0	0.0	25.4	665
WS	0.0	0.0	0.0	0.0	0.0	0.0	0.7	0.0	0.0	79.0	4.9	0.0	85
RO	5.6	2.8	1.4	0.0	0.0	0.0	0.4	0.2	0.1	39.5	22.2	11.1	84.0
Stn : ALAMURU (GODAVARI DELTA)											**AWC = 150mm**		
	J	**F**	**M**	**A**	**M**	**J**	**J**	**A**	**S**	**O**	**N**	**D**	**TOTAL**
P	4.1	14.0	11.7	20.1	45.2	135.6	218.9	190.7	179.3	205.5	106.7	5.1	1137
PE	112.0	123.0	168.0	176.0	194.0	169.0	136.0	137.0	125.0	125.0	116.0	108.0	1689
ΔS	–36.42	–17.86	–10.82	–3.81	–1.32	–0.15	82.28	53.70	13.40	0.0	–9.02	–69.98	0
AE	40.5	31.9	22.5	23.9	46.5	135.8	136.0	137.0	125.0	125.0	115.7	75.1	1015
WD	71.5	91.1	145.5	152.1	147.5	33.3	0.0	0.0	0.0	0.0	0.3	32.9	674
WS	0.0	0.0	0.0	0.0	0.0	0.0	0.6	0.0	40.9	80.5	0.0	0.0	122
RO	6.3	3.2	1.6	0.8	0.4	0.2	0.3	0.2	20.5	50.5	25.3	12.6	122
Stn : KAKINADA (GODAVARI DELTA)											**AWC = 150mm**		
	J	**F**	**M**	**A**	**M**	**J**	**J**	**A**	**S**	**O**	**N**	**D**	**TOTAL**
P	4.0	12.0	12.0	22.0	46.0	126.0	218.0	152.0	158.0	307.0	107.0	9.0	1173
PE	112.0	123.0	168.0	176.0	194.0	169.0	136.0	137.0	125.0	125.0	116.0	108.0	1689
ΔS	–37.47	–18.58	–10.97	–3.84	–1.35	–0.20	81.40	15.0	33.0	20.0	–8.74	–68.25	0
AE	41.5	30.6	23.0	25.8	47.4	126.2	136.0	137.0	125.0	125.0	115.7	77.3	1011.0
WD	70.5	92.4	145.0	150.2	146.7	42.8	0.0	0.0	0.0	0.0	0.3	30.8	679
WS	0.0	0.0	0.0	0.0	0.0	0.0	0.6	0.0	0.0	162.0	0.0	0.0	163
RO	10.2	5.1	2.5	1.3	0.0	0.0	0.3	0.2	0.1	81.0	40.5	20.3	162.0
Stn :AMALAPURAM (GODAVARI DELTA)											**AWC = 100mm**		
	J	**F**	**M**	**A**	**M**	**J**	**J**	**A**	**S**	**O**	**N**	**D**	**TOTAL**
P	5.8	11.9	10.7	11.9	46.2	138.7	188.0	185.2	188.5	279.4	153.7	7.9	1227.0
PE	112.0	123.0	168.0	176.0	194.0	169.0	136.0	137.0	125.0	125.0	116.0	108.0	1689
ΔS	–24.04	–8.53	–3.31	–0.70	–0.13	–0.01	51.97	48.0	0.0	0.0	0.0	–63.25	0
AE	29.8	20.4	14.0	12.6	46.3	138.7	136.0	137.0	125.0	125.0	116.0	71.2	972.1
WD	82.2	102.6	154.0	163.4	147.7	30.3	0.0	0.0	0.0	0.0	0.0	36.9	717
WS	0.0	0.0	0.0	0.0	0.0	0.0	0.0	0.2	63.5	154.4	37.7	0.0	256
RO	16.4	8.2	4.1	2.0	1.0	0.0	0.0	0.1	31.8	93.1	65.4	32.7	255.0
Stn :BICCAVOL (GODAVARI DELTA)											**AWC = 150mm**		
	J	**F**	**M**	**A**	**M**	**J**	**J**	**A**	**S**	**O**	**N**	**D**	**TOTAL**
P	7.6	12.5	12.7	15.2	50.5	110.2	204.5	155.5	182.6	199.1	112.3	3.8	1066.0
PE	112.0	123.0	168.0	176.0	194.0	169.0	136.0	137.0	125.0	125.0	116.0	108.0	1689
ΔS	–36.63	–18.99	–11.25	–4.07	–1.31	–0.26	67.95	18.50	57.60	5.40	-3.65	–73.29	0
AE	44.2	31.5	24.0	19.3	51.8	110.5	136.0	137.0	125.0	125.0	116.0	77.1	998.0
WD	67.8	91.5	144.1	156.7	142.2	58.5	0.0	0.0	0.0	0.0	0.1	30.9	692
WS	0.0	0.0	0.0	0.0	0.0	0.0	0.6	0.0	0.0	68.7	0.0	0.0	69
RO	4.3	2.2	1.1	0.0	0.0	0.0	0.3	0.2	0.1	34.4	17.2	8.6	68

Table 3.13 ***Contd...***

Stn :RAZOLE (GODAVARI DELTA)										AWC = 100mm		
	J	F	M	A	M	J	J	A	S	O	N	D	TOTAL
P	4.3	9.7	6.3	8.1	40.4	139.9	199.4	186.7	182.4	261.4	142.5	12.2	1193
PE	112.0	123.0	168.0	176.0	194.0	169.0	136.0	137.0	125.0	125.0	116.0	108.0	1689
ΔS	–25.30	–8.86	–3.37	–0.68	–0.13	0.0	63.37	36.60	0.0	0.0	0.0	–61.63	0
AE	29.6	18.6	9.7	8.8	40.5	139.9	136.0	137.0	125.0	125.0	116.0	73.8	960
WD	82.4	104.4	158.3	167.2	153.5	29.1	0.0	0.0	0.0	0.0	0.0	34.2	729
WS	0.0	0.0	0.0	0.0	0.0	0.0	0.0	13.1	57.4	136.4	26.5	0.0	233
RO	13.9	6.9	3.5	1.7	0.0	0.0	0.0	7.2	32.3	84.4	55.4	27.7	233
Stn :TADEPALLIGUDEM (GODAVARI DELTA)										**AWC = 150mm**		
	J	F	M	A	M	J	J	A	S	O	N	D	TOTAL
P	3.1	13.2	9.9	18.3	48.3	148.3	214.0	199.4	168.4	159.5	72.9	6.9	1062.2
PE	106.0	116.0	150.0	151.0	162.0	150.0	114.0	110.0	105.0	100.0	102.0	99.0	1465
ΔS	–33.19	–16.70	–10.30	–3.92	–1.46	–0.01	98.82	49.0	0.0	0.0	–26.45	–56.69	0
AE	36.3	29.9	20.2	22.2	49.8	148.3	114.0	110.0	105.0	100.0	99.4	63.6	899.0
WD	69.7	86.1	129.8	128.8	112.2	1.7	0.0	0.0	0.0	0.0	2.7	35.4	566
WS	0.0	0.0	0.0	0.0	0.0	0.0	1.3	39.5	63.4	59.5	0.0	0.0	164
RO	6.4	3.2	1.6	0.0	0.0	0.0	0.7	20.1	41.8	50.6	25.3	12.7	163.0
Stn :BHIMAVARAM (GODAVARI DELTA)										**AWC = 150mm**		
	J	F	M	A	M	J	J	A	S	O	N	D	TOTAL
P	4.3	10.4	7.9	12.5	32.8	123.7	205.0	173.0	158.0	179.1	87.1	12.7	1006.0
PE	106.0	116.0	150.0	151.0	162.0	150.0	114.0	110.0	105.0	100.0	102.0	99.0	1465
ΔS	–37.62	–19.60	–11.74	–4.49	–1.70	–0.20	89.95	59.0	0.0	0.0	–14.18	–59.42	0
AE	41.9	30.0	19.6	17.0	34.5	123.9	114.0	110.0	105.0	100.0	101.3	72.1	870.0
WD	64.1	86.0	130.4	134.0	127.5	26.1	0.0	0.0	0.0	0.0	0.7	26.9	596
WS	0.0	0.0	0.0	0.0	0.0	0.0	1.1	4.0	53.0	79.0	0.0	0.0	137
RO	6.7	3.3	1.7	0.0	0.0	0.0	0.6	2.3	27.6	53.3	26.7	13.3	136.0
Stn : VIJAYAWADA (KRISHNA DELTA)										**AWC = 150mm**		
	J	F	M	A	M	J	J	A	S	O	N	D	TOTAL
P	5.6	12.5	9.7	21.8	50.0	119.4	198.4	181.1	147.6	132.8	74.9	5.6	959
PE	125.0	136.0	181.0	197.0	227.0	199.0	148.0	139.0	127.0	121.0	121.0	116.0	1837
ΔS	–21.96	–8.62	–3.81	–0.97	–0.24	–0.04	50.36	42.10	20.60	11.80	–38.55	–50.67	0
AE	27.6	21.1	13.5	22.8	50.2	119.4	148.0	139.0	127.0	121.0	113.5	56.3	959
WD	97.4	114.9	167.5	174.2	176.8	79.6	0.0	0.0	0.0	0.0	7.6	59.7	878
WS	0.0	0.0	0.0	0.0	0.0	0.0	0.0	0.0	0.0	0.0	0.0	0.0	0
RO	0.0	0.0	0.0	0.0	0.0	0.0	0.0	0.0	0.0	0.0	0.0	0.0	0
Stn : KAIKALUR (KRISHNA DELTA)										**AWC = 150mm**		
	J	F	M	A	M	J	J	A	S	O	N	D	TOTAL
P	6.3	9.4	10.2	13.7	40.4	109.5	182.6	169.9	156.5	161.5	77.5	5.6	943
PE	125.0	136.0	181.0	197.0	227.0	199.0	148.0	139.0	127.0	121.0	121.0	116.0	1837
ΔS	–25.40	–11.0	–5.10	–1.50	–0.39	–0.06	34.53	30.90	29.50	40.50	–37.21	–54.77	0
AE	31.7	20.4	15.3	15.2	40.8	109.6	148.0	139.0	127.0	121.0	114.7	60.4	943
WD	93.3	115.6	165.7	181.8	186.2	89.4	0.0	0.0	0.0	0.0	6.3	55.6	894
WS	0.0	0.0	0.0	0.0	0.0	0.0	0.1	0.0	0.0	0.0	0.0	0.0	0
RO	0.0	0.0	0.0	0.0	0.0	0.0	0.0	0.0	0.0	0.0	0.0	0.0	0

Table 3.13 *Contd...*

Stn : MACHILIPATNAM (KRISHNA DELTA)											**AWC = 150mm**		
	J	F	M	A	M	J	J	A	S	O	N	D	TOTAL
P	5.1	15.0	13.5	18.5	39.1	93.7	152.4	162.1	234.4	138.2	18.3	177.0	1067.0
PE	125.0	136.0	181.0	197.0	227.0	199.0	139.0	127.0	121.0	121.0	116.0	148.0	1837
ΔS	–53.32	–28.84	–19.17	–8.32	–3.53	–0.82	49.61	13.0	29.0	107.0	-9.75	–74.86	0
AE	48.1	34.5	24.1	22.1	40.2	93.9	139.0	127.0	121.0	121.0	90.1	148.0	1009.0
WD	76.9	101.5	157.0	174.9	186.8	105.1	0.0	0.0	0.0	0.0	25.9	0.0	828
WS	0.0	0.0	0.0	0.0	0.0	0.0	0.2	0.0	0.0	40.9	17.2	0.0	58.3
RO	4.7	2.4	1.2	0.6	0.0	0.0	0.1	0.1	0.0	20.5	18.8	9.4	58

Stn : AVANIGADDA (KRISHNA DELTA)											**AWC = 150mm**		
	J	**F**	**M**	**A**	**M**	**J**	**J**	**A**	**S**	**O**	**N**	**D**	**TOTAL**
P	7.4	13.2	11.2	14.2	40.9	87.9	154.7	157.0	158.7	222.3	144.8	15.7	1028
PE	125.0	136.0	181.0	197.0	227.0	199.9	148.0	139.0	127.0	121.0	121.0	116.0	1838
ΔS	-41.77	-19.61	-10.49	-3.52	-1.04	-0.23	6.5	18.0	31.70	93.60	0.0	-73.14	0
AE	49.2	32.8	21.7	17.7	41.9	88.1	148.0	139.0	127.0	121.0	121.0	88.8	996
WD	75.8	103.2	159.3	179.3	185.1	111.8	0.0	0.0	0.0	0.0	0.0	27.2	842
WS	0.0	0.0	0.0	0.0	0.0	0.0	0.2	0.0	0.0	7.7	23.8	0.0	32
RO	3.5	1.8	0.9	0.0	0.0	0.0	0.1	0.1	0.0	4.2	14.0	7.0	32

Stn : TENALI (KRISHNA DELTA)											**AWC = 150mm**		
	J	**F**	**M**	**A**	**M**	**J**	**J**	**A**	**S**	**O**	**N**	**D**	**TOTAL**
P	7.4	9.1	9.7	18.5	55.4	106.4	172.5	167.4	157.2	149.1	83.6	10.2	947
PE	111.0	132.0	185.0	196.0	214.0	188.0	152.0	151.0	127.0	121.0	102.0	98.0	1777
ΔS	–20.69	–7.62	–2.43	–0.39	–0.06	0.0	20.49	16.40	30.20	28.10	–16.73	–47.27	0
AE	28.1	16.7	12.1	18.9	55.5	106.4	152.0	151.0	127.0	121.0	100.3	57.5	946
WD	82.9	115.3	172.9	177.1	158.5	81.6	0.0	0.0	0.0	0.0	1.7	40.5	831
WS	0.0	0.0	0.0	0.0	0.0	0.0	0.0	0.0	0.0	0.0	0.0	0.0	0
RO	0.0	0.0	0.0	0.0	0.0	0.0	0.0	0.0	0.0	0.0	0.0	0.0	0

Stn : REPALLE (KRISHNA DELTA)											**AWC = 150mm**		
	J	**F**	**M**	**A**	**M**	**J**	**J**	**A**	**S**	**O**	**N**	**D**	**TOTAL**
P	4.8	14.7	8.9	14.0	45.5	85.9	147.3	151.4	157.7	197.6	128.0	17.0	973
PE	111.0	132.0	185.0	196.0	214.0	188.0	152.0	151.0	127.0	121.0	102.0	98.0	1777
ΔS	–39.98	–19.26	–10.04	–2.73	–0.67	–0.15	0.0	0.28	30.70	76.60	26.0	–60.75	0
AE	44.8	34.0	18.9	16.7	46.2	86.1	147.3	151.0	127.0	121.0	102.0	77.8	973
WD	66.2	98.0	166.1	179.3	167.8	102.0	4.7	0.0	0.0	0.0	0.0	20.3	804
WS	0.0	0.0	0.0	0.0	0.0	0.0	0.0	0.1	0.0	0.0	0.0	0.0	0
RO	0.0	0.0	0.0	0.0	0.0	0.0	0.0	0.1	0.0	0.0	0.0	0.0	0

Stn : BAPATLA (KRISHNA DELTA)											**AWC = 150mm**		
	J	**F**	**M**	**A**	**M**	**J**	**J**	**A**	**S**	**O**	**N**	**D**	**TOTAL**
P	4.8	12.7	8.6	9.9	42.2	75.9	133.9	144.3	168.1	169.2	119.1	13.2	902
PE	111.0	132.0	185.0	196.0	214.0	188.0	152.0	151.0	127.0	121.0	102.0	98.0	1777
ΔS	–30.28	–11.91	–4.66	–0.91	–0.15	–0.03	0.0	0.0	41.09	48.20	17.10	–58.45	0
AE	35.1	24.6	13.3	10.8	42.4	75.9	133.9	144.3	127.0	121.0	102.0	71.7	902
WD	75.9	107.4	171.7	185.2	171.7	112.1	18.1	6.7	0.0	0.0	0.0	26.4	875
WS	0.0	0.0	0.0	0.0	0.0	0.0	0.0	0.0	0.0	0.0	0.0	0.0	0
RO	0.0	0.0	0.0	0.0	0.0	0.0	0.0	0.0	0.0	0.0	0.0	0.0	0

(P = Precipitation, PE = Potential Evapotranspiration, ΔS = Soil moisture storage change, AE = Actual Evaporation, WD = Water Deficiency, WS = Water Surplus and RO = Run Off)

The map indicates that maximum precipitation levels are of the order of 1200 mm observed in the outfall region of Godavari delta along the Razole-Kakinada axis and gradually decreases towards the up stream region to an order of 1100 mm. Similarly the Krishna delta region shows a maximum value trend of the order of 1100 mm in the loop region of Avanigadda-Gudivada and the values get reduced both towards east and west to an order of 900 mm.

Graphical trends and water balance parameters

The Graphical trends of water balance parameters for different hydromet stations of Godavari and Krishna deltas are shown below in Fig 3.24(a,b,c & d).

The monsoon precipitation levels in the Godavari delta region actually starts from the month of June and extend up to the month of October though some minor amounts of precipitation are observed during the rest of the period in a year. In majority of the stations, the peak precipitation levels are seen during the month of October and in few stations during the month of July. The peak levels observed in different stations are: Rajahmundry: 214.9 mm; Ramachandrapuram: 213.6 mm; Alamuru: 205.5 mm; Kakinada: 307.0 mm; Amalapuram: 279.0 mm; Biccavol: 199.1 mm Razole: 261.4 mm; Tadepalligudem: 214.0 mm and Bhimavaram: 205.00 mm. Similarly in Krishna delta the actual monsoon starts from the month of June and extends up to October with minor amounts of rainfall during the rest of the period. The peak monsoon levels are generally observed during the periods July, August, September and October depending upon the station location. The peak values observed in this region are: Vijayawada: 198.4 mm; Kaikaluru: 182.6 mm; Machilipatnam: 234.4 mm; Tenali: 172.5 mm; Repalle: 197.6 mm and Bapatla: 169.0 mm. The overall seasonal values in both the delta systems indicate that precipitation levels reach peak levels in the upland regions early in July and in the outfall region during the month of October. The average annual precipitation levels in respect of Godavari and Krishna deltas are 1114.45 mm and 974.14 mm respectively.

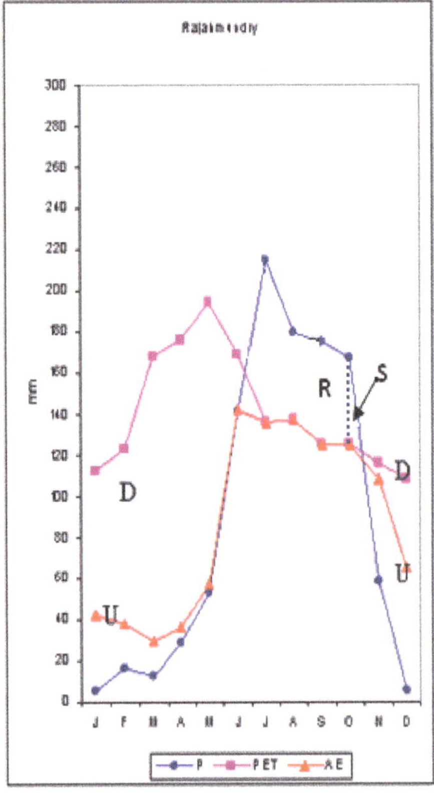

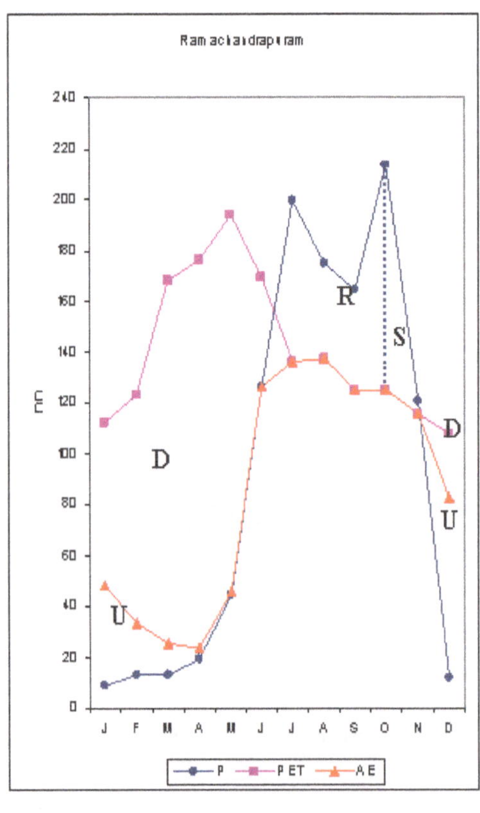

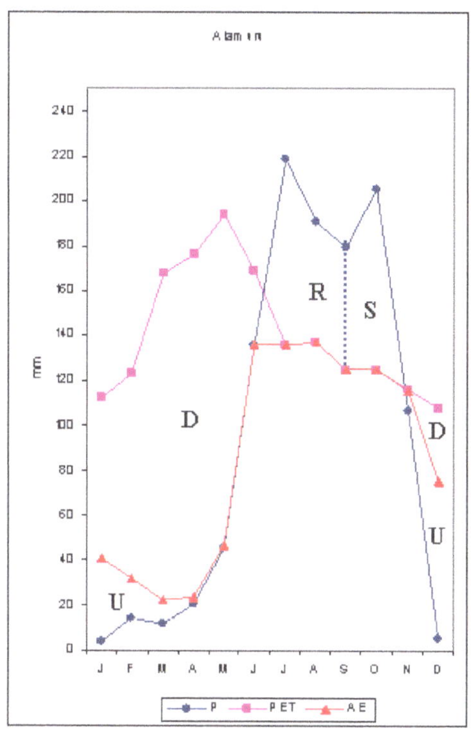

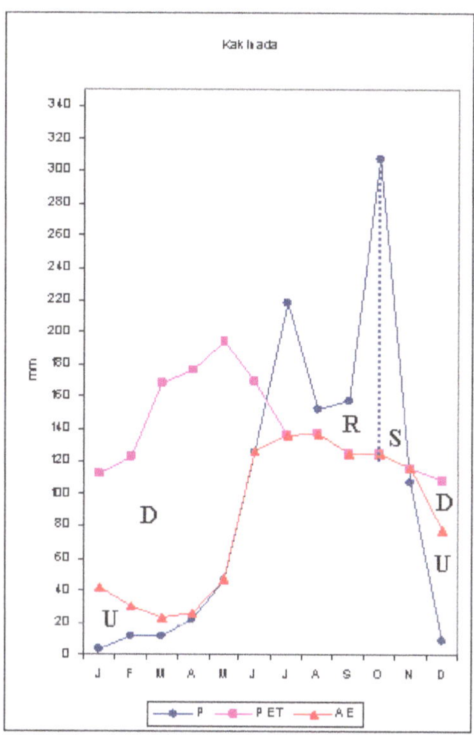

(a) (R-Recharge accretion in the soil zone, S-Surplus, U-Utilization, D-Water Deficit)

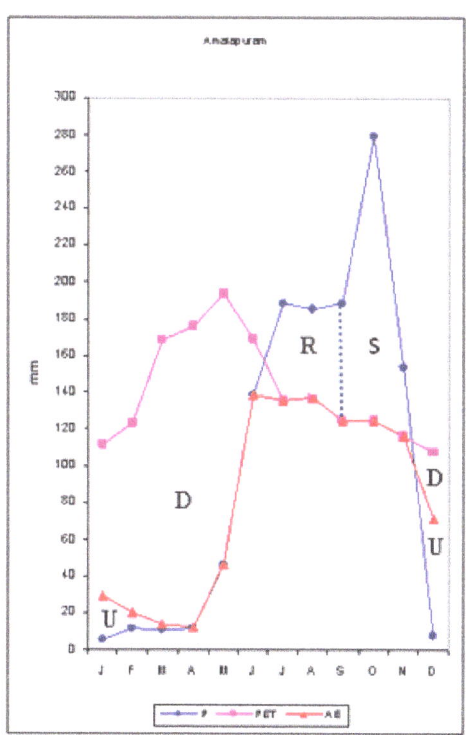

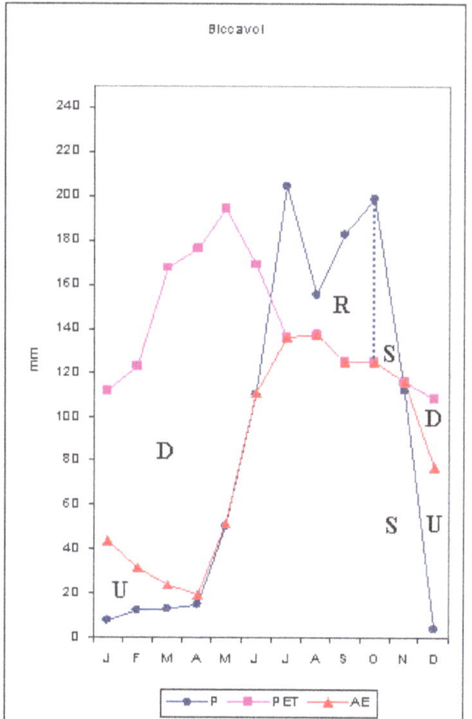

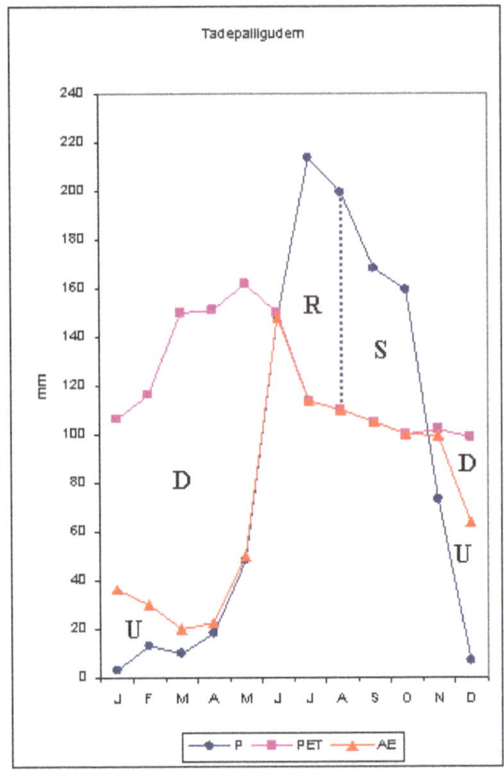

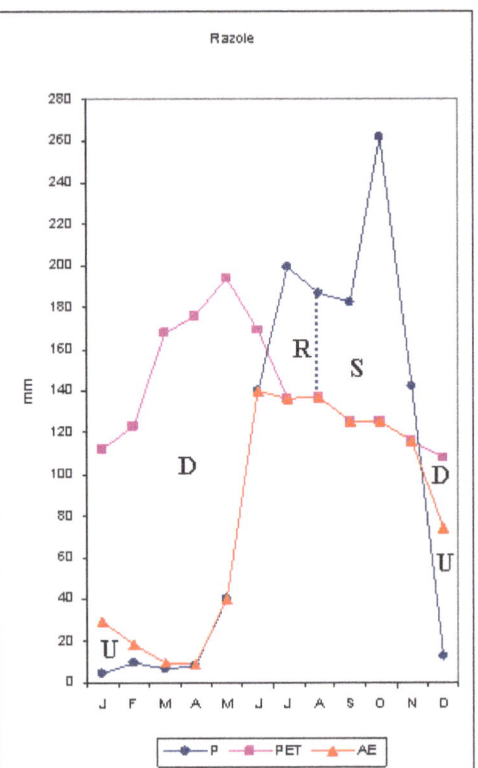

(b) (R-Recharge accretion in the soil zone, S-Surplus, U-Utilization, D-Water Deficit)

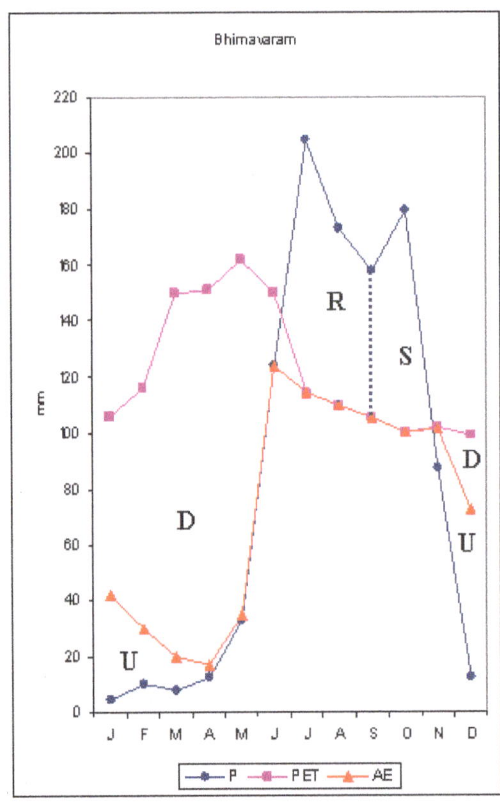

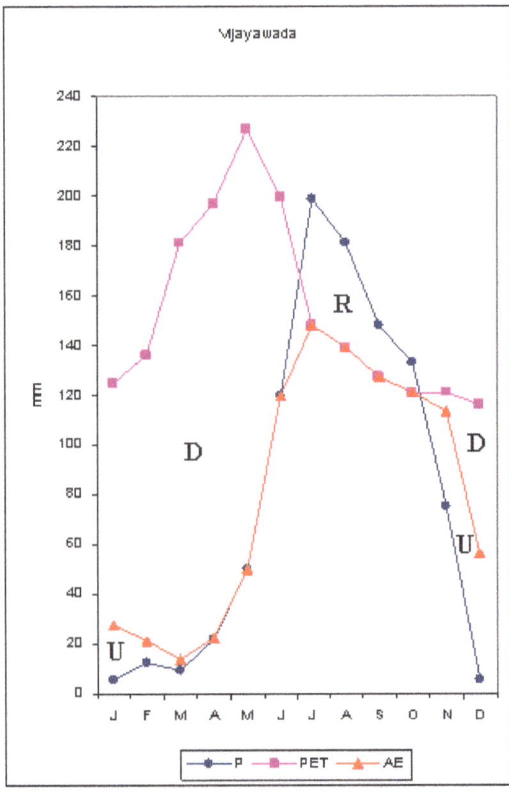

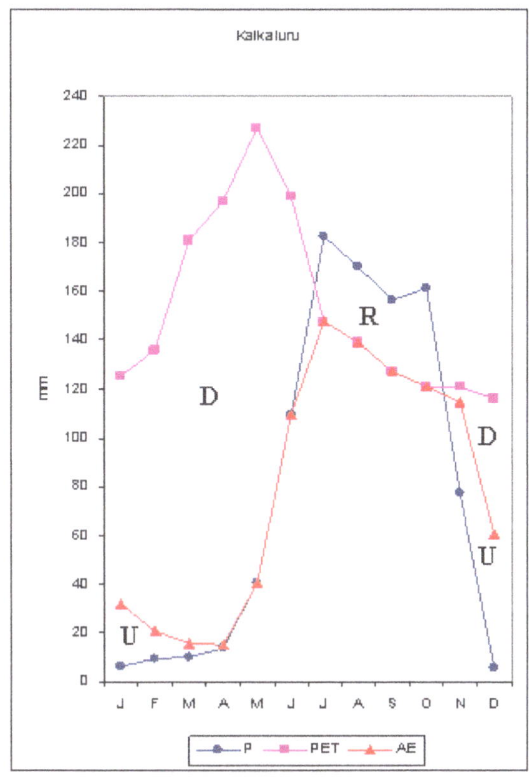

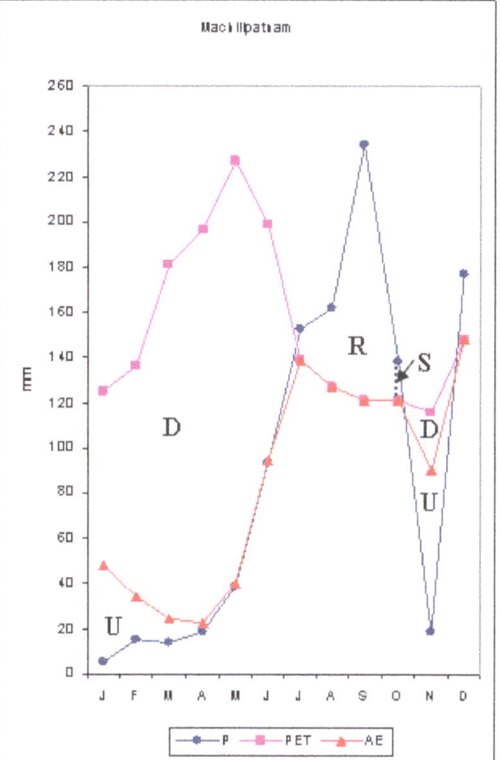

(c) (R-Recharge accretion in the soil zone, S-Surplus, U-Utilization, D-Water Deficit)

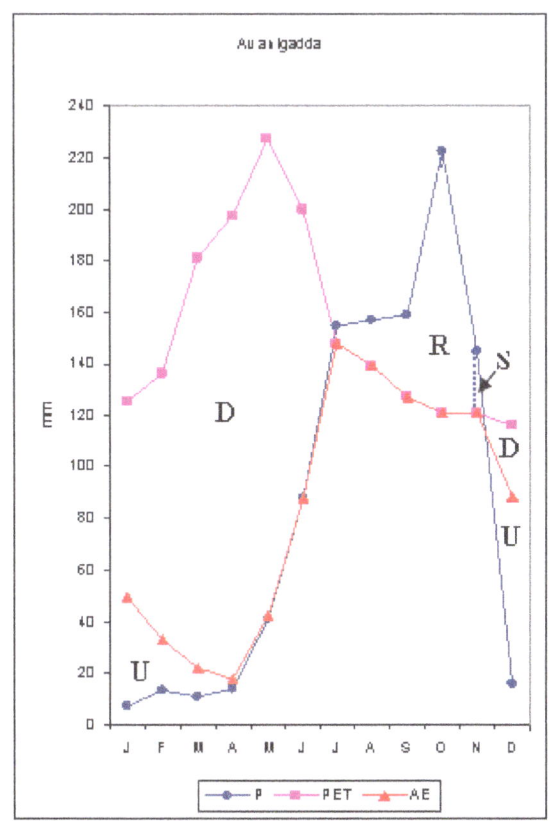

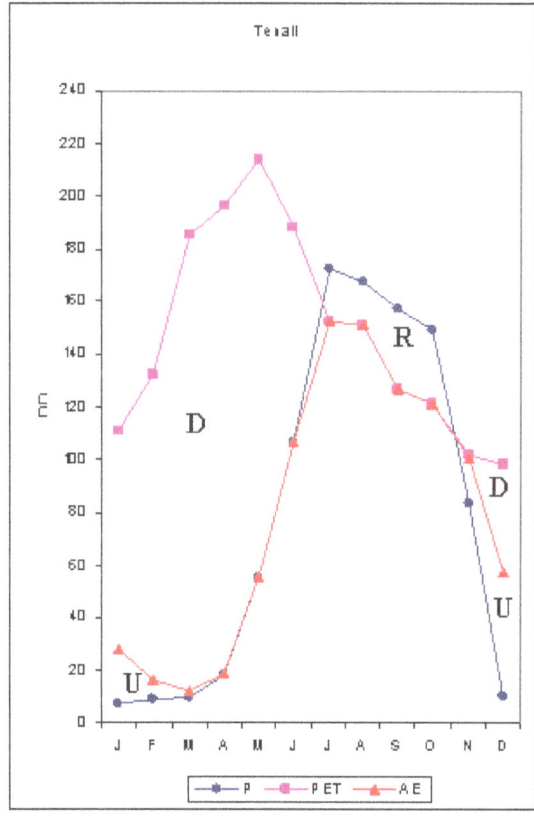

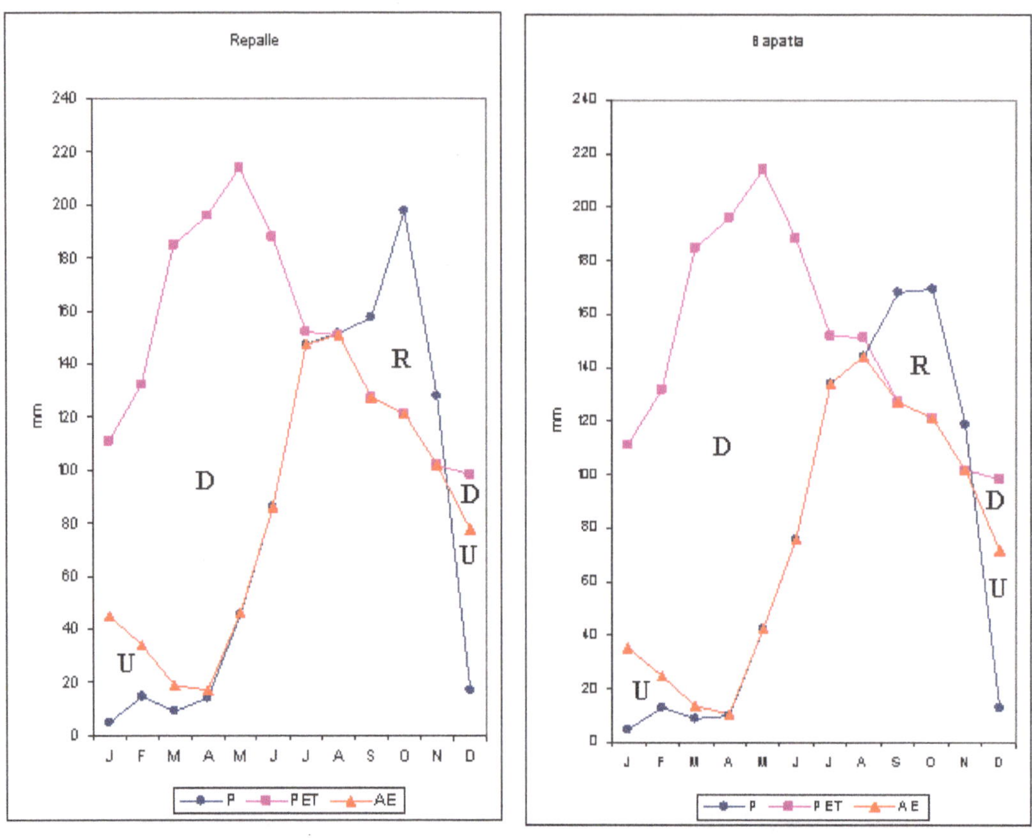

(d) (R-Recharge accretion in the soil zone, S-Surplus, U-Utilization, D-Water Deficit)

Fig. 3.24(a,b,c,d) Graphical trends of water balance parameters in Godavari and Krishna delta regions

The potential evapotranspiration in Godavari delta have a range varying from 99 mm to 194.0 mm. The minimum values generally are observed during the month of December and the maximum values during the month of May. A maximum value of 194 mm is recorded at stations Rajahmundry, Ramachandrapuram, Alamuru, Kakinada, Amalapuram, Biccavol and Rajole. The other stations namely Tadepalligudem and Bhimavaram have recorded a maximum value of 162 mm. The Krishna delta on the other hand had registered a maximum PET values of 227 mm at stations Vijayawada, Kaikaluru, Machilipatnam in the month of May and a minimum values ranging between 125-116 mm during the months of Dec-Jan. The other stations namely Tenali, Repalle and Bapatla have shown maximum value of 214 mm and a minimum value of 98 mm only. A significant feature is that the eastern portion of these deltas has shown relatively high PE values when compared to western flanks. The average annual PET in respect of Godavari and Krishna deltas are 1639.2 mm and 1811.4 mm respectively.

The water deficit in Godavari and Krishna deltas mainly occurs during the period January to May and sometime in December with peak values attaining during the months of April and May. During the rest of the period the values recorded are 'nil'. The peak WD values in the case of Godavari are: Rajahmundry: 140.1 mm; Ramachandrapuram: 152.3 mm; Alamuru: 152.09 mm; Kakinada: 150.16; Biccavol: 150.73 mm; Razole: 167.22 mm; Tadepalligudem: 129.8 mm and Bhimavaram: 134.01 mm. In the case of Krishna delta the peak WD values recorded are: Vijayawada: 176.76 mm; Kaikaluru: 186.21 mm; Machilipatnam: 186.79 mm; Avanigadda: 185.06 mm; Tenali: 177.11 mm; Repalle: 179.27 mm and Bapatla: 185.19 mm. The peak WD values in the case of Krishna delta are marginally higher than those of Godavari delta region. The annual average WD values for Godavari and Krishna are 663 and 850 respectively. The water surplus in both Godavari and Krishna delta takes place only for a few months between July and November in a year depending upon the location and rainfall intensities. The peak values of WS recorded in these regions are: Rajahmundry: 13.9 mm; Ramachandrapuram: 79.0 mm; Alamuru: 80.5 mm; Kakinada: 74.6 mm; Amalapuram: 154.4 mm; Biccavol: 68.7 mm; Rajole: 136.4 mm; Tadepalligudem: 63.4 mm and Bhimavaram: 79.0 mm. In majority of the stations the peak values are observed during the month of October. In the case of Krishna delta WD values recorded at two stations namely Machilipatnam: 40.9 mm and Avanigadda: 23.8 mm. All the rest of the stations significantly reflect nil water surplus values. The annual average WS values for Godavari and Krishna delta systems are 138 mm and 13 mm respectively.

The seasonal computed surface runoff values are graphically shown in Fig. 3.25.

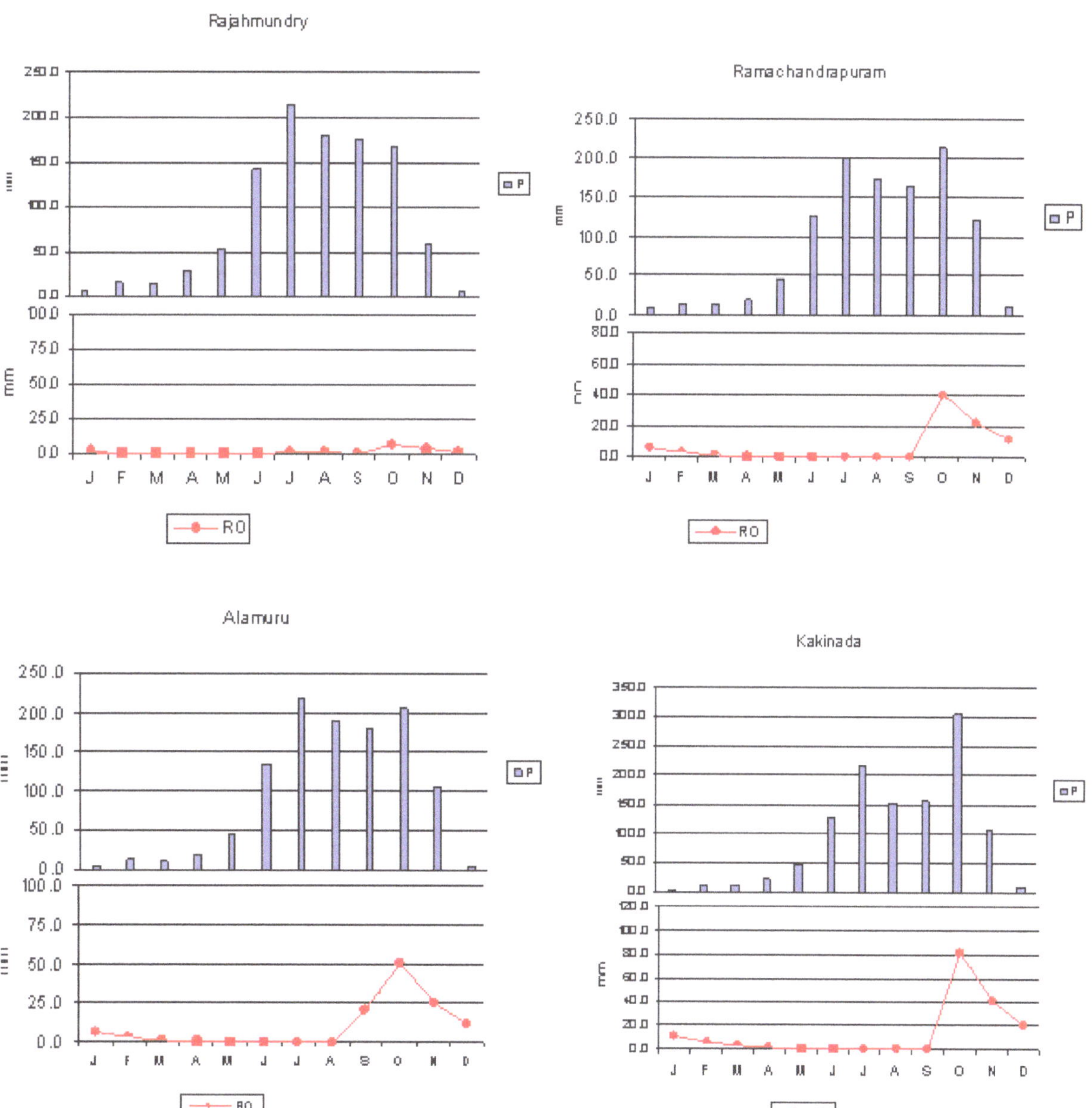

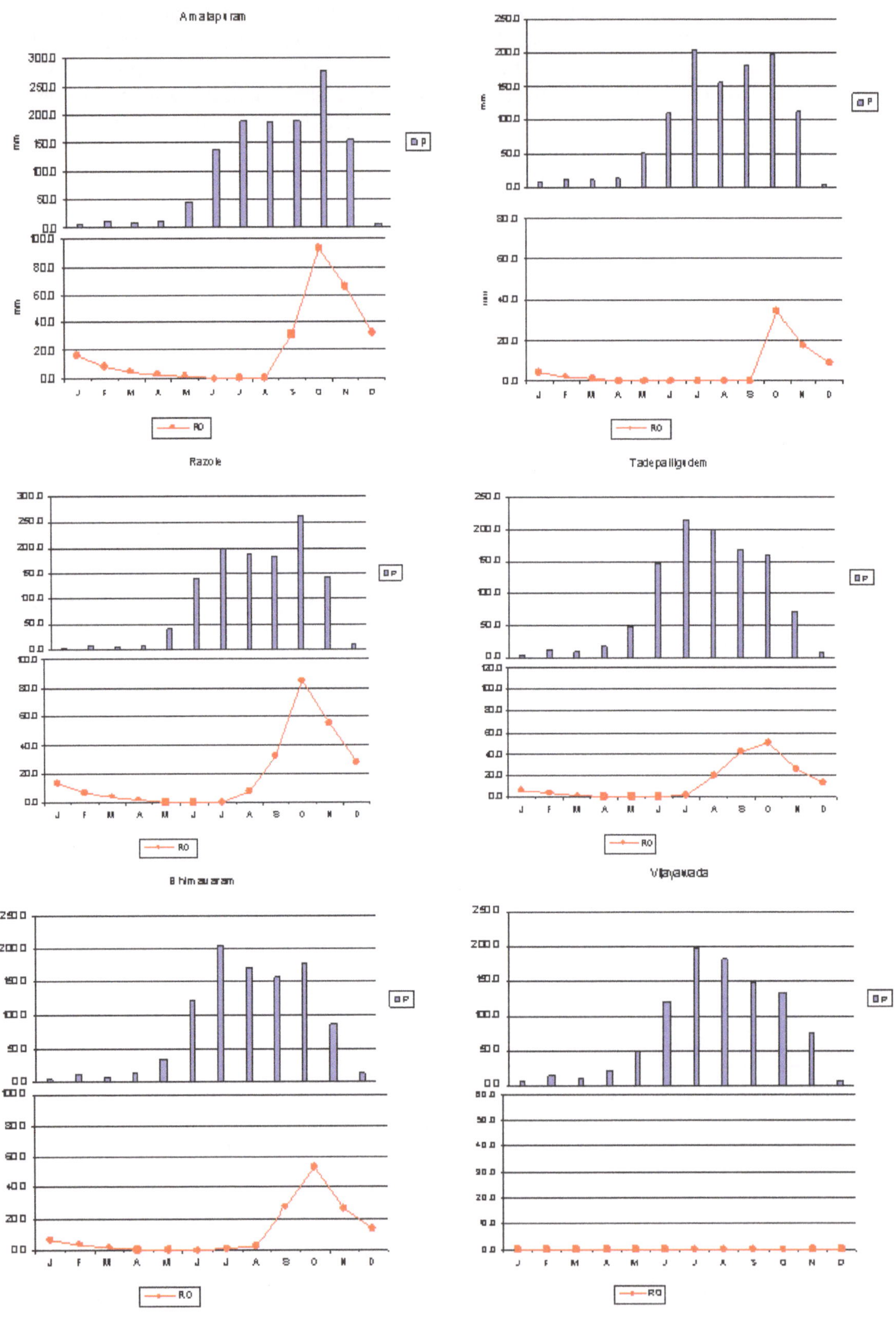

Amalapuram
Biccavolu
Razole
Tadepalligudem
Bhimavaram
Vijayawada
mm
P
RO
J F M A M J J A S O N D

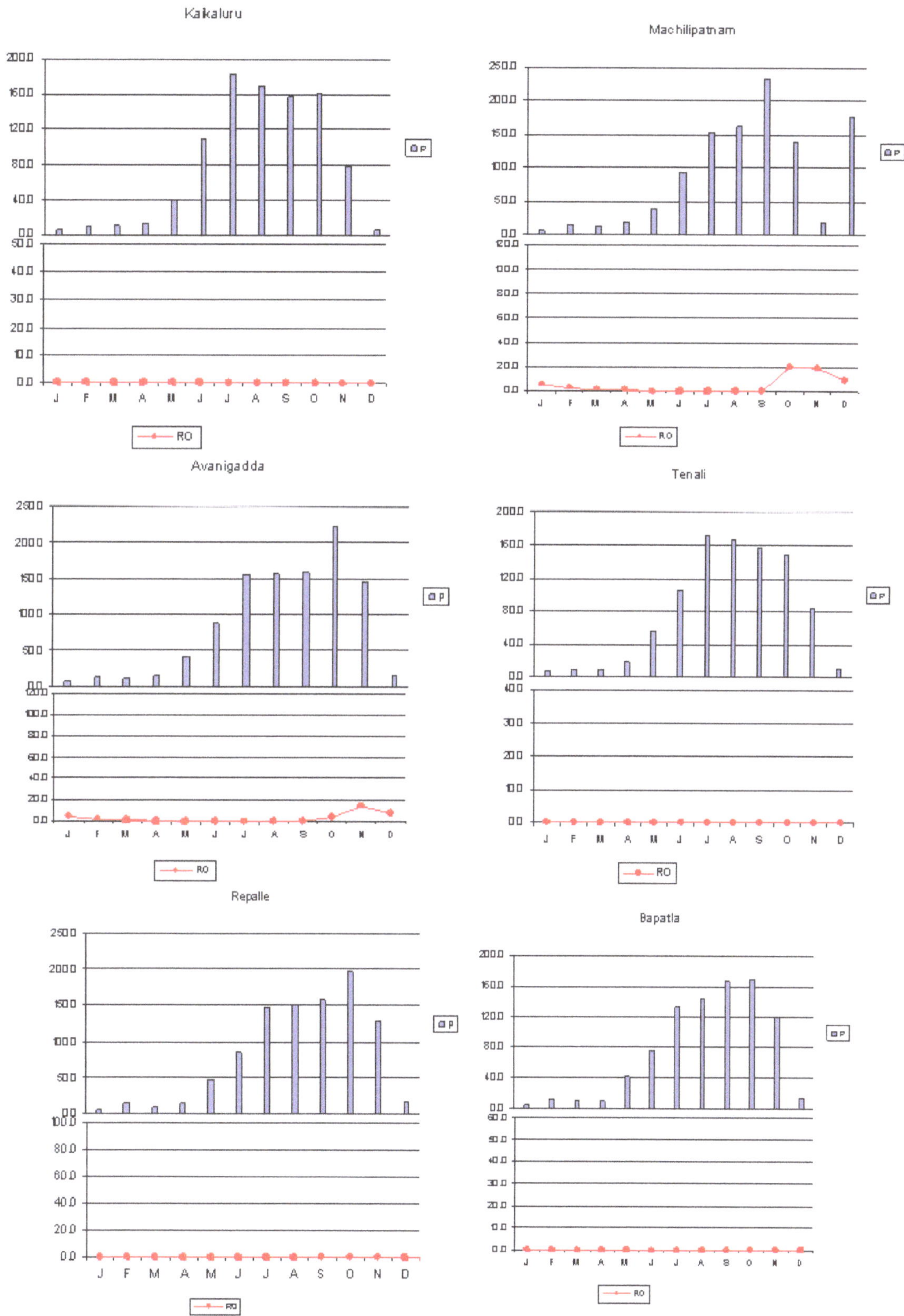

Fig. 3.25 Precipitation- Runoff relations in Godavari and Krishna delta regions

The runoff processes start taking place from the month of July-August attains peak stage during the month of October and gradually gets reduced by the month of March or April. The peak RO values obtained for Godavari delta are Rajhamundry: 7.2 mm; Ramachandrapuram: 39.5 mm; Alamuru: 50.5 mm; Kakinada: 81.0 mm; Amalapuram: 93.1 mm; Biccavol: 34.4 mm; Razole: 84.4 mm; Tadepalligudem: 50.6 mm; Bhimavaram: 53.3 mm. The average annual RO obtained for the Godavari delta is of the order of 137.6 mm which comes to about 12.3% of the total annual average precipitation. In the case of Krishna delta the RO appears to be totally insignificant except at two stations namely Avanigadda and Machilipatnam where the processes start during the month of October and ends up in the month of February to March. The average annual RO which is 12.9 mm appears to be totally insignificant in the case of Krishna delta which comes to an order of 1.3% of total average annual precipitation. This is anamolous and needs to be checked for its relevance.

Annual climatic water balance budget analysis

The Annual climatic water balance budget for Godavari and Krishna deltas are furnished below in Table 3.14.

Table 3.14 Annual water balance parameters in Godavari and Krishna delta regions

Location	PET (mm)	Rainfall (mm)	Water Surplus (mm)	Water Deficit (mm)	Humidity Index (%)	Aridity Index (%)	Moisture Index (%)	RunOff (mm)
Godavari Delta								
Rajahmundry	1689	1057	16.81	649	1.0	38.4	-37.4	16
Ramachandrapuram	1689	1108.6	84.56	664.96	5.0	39.4	-34.4	84
Alamuru	1689	1136	122.02	674.12	7.2	39.9	-32.7	122
Kakinada	1689	1173	162.6	678.6	9.6	40.2	-30.6	162
Amalapuram	1689	1227	255.83	716.93	15.1	42.4	-27.3	255
Biccavol	1689	1066.5	69.25	691.75	4.1	41.0	-36.9	68
Razole	1689	1193.3	233.43	729.13	13.8	43.2	-29.3	233
Tadepalligudem	1465	1062.2	163.68	566.38	11.2	38.7	-27.5	163
Bhimavaram	1465	1006.5	137.15	595.65	9.4	40.7	-31.3	136
Krishna Delta								
Vijayawada	1837	959.4	0.04	877.64	0.0	47.8	-47.8	0.1
Kaikaluru	1837	943.1	0.07	893.97	0.0	48.7	-48.7	0.12
Machilipatnam	1837	1067.3	58.32	828.02	3.2	45.1	-41.9	58
Avanigadda	1837.9	1028	31.7	841.6	1.7	45.8	-44.1	32
Tenali	1777	946.5	0.01	830.51	0.0	46.7	-46.7	0.01
Repalle	1777	972.8	0.12	804.32	0.0	45.3	-45.3	0.12
Bapatla	1777	901.9	0.01	875.11	0.0	49.2	-49.2	0.01

3.3.4 Mahanadi – Brahmini-Baitarni delta basins (Orissa)

Mahandi, Brahmini and Baitarini delta systems constitute a major composite delta system along the east coast with an aereal spread of the order of 13,200 sq.kms and is considered to be a major agricultural production center of Orissa. Mahanadi travels a distance of about 851 kms and debouches into Bay of Bengal. The river spans itself in the form of a delta with its apex near Naraj in Cuttack district which is located at an elevation of 36 m above mean sea level and has several distributaries in its delta frame.

The major distributaries/subdistributaries of Mahanadi delta are Birupa, Mahanadi/Chitrala, Katjhori/Devi and Koyakhai/Daya/Kushubhadra/Bhargavi. The river Brahmini, the second biggest river of Orissa travels a distance of about 800 kms and debouches into Bay of Bengal at the mouth region of Dhamra of Orissa. The delta portion of Brahmini starts from its apex point near Jenapur and gets bifurcated into two distributaries namely Brahmini main and Kharsuan. These two distributaries further down stream join together and further downstream the Birupa a distributary of Mahanadi joins Brahmini, before it meets the sea. Similarly the river Baitarini travels a distance of about 440 kms and debouches into Bay of Bengal, with its apex at Akhudapada which is located at an elevation of 6 m above mean sea level where it gets bifurcated into two arms. One arm of Baitarni joins with Brahmini and further downstream transforms itself as Dhamra river before joining Bay of Bengal. The other arm of Baitarni travels independently joins the sea near Chandbali. All these three rivers Mahanadi, Brahmini and Baitarni from their respective apex centers interlace with each other in the delta portion, make the entire delta complex as a major composite delta system of Orissa .The upper delta region is largely characterized by lateritic soils and admixtures of sands and clays. The mid delta region largely dominated by a number of sub distributaries, inter lacing river channels and paleo channels. The lower delta region has a number of low lying swamps, tidal canals and a series of sand ridges and valleys. The delta region enjoys a typical sub tropical climate with an average maximum and minimum temperatures of the order of 29 °C and 16 °C respectively and with humidity levels ranging between 75% (wet season) and 60% (dry season).

Mahanadi-Brahmini-Baitarni delta mainly receives rainfall from southwest monsoon during the period June to October with occasional showers during November and December. The average annual rainfall is of the order of 1500 mm. The isohyetel map presented in Fig. 3.29 shows two relatively higher precipitation levels – one in Mahanadi delta region with precipitation levels ranging from 1500 mm to 1450 mm and the other towards the outfall region of Brahmini and Baitarni where the trends range from 1600 mm to 1500 mm. A relatively low precipitation loop line with 1450 mm contour line is observed along Kendrapara-Bhadrak axis. The average annual discharges observed at the terminal points of Mahanadi-Brahmini-Baitarni are 66,640 MCM, 18,310 MCM and 5755 MCM respectively.

The soils in these deltaic regions are deep, poorly drained and dark grey in color with clay percentages varying from 40-50%. The dominant clay minerals are illites. The pH of the soils at near surface is about 6.5 and increases with depth to 8.0. The soils are relatively rich in Magnesium and Sodium and contain moderate levels of organic carbon. A brief hydrological characteristics of Mahanadi, Brahmini and Baitarni delta regions are presented in Table 3.15.

Table 3.15 Regional hydrological characters of Mahanadi-Brahmini deltas

S.no.	Item	Mahanadi	Brahmini
1.	Average annual rainfall	1200-1500 mm	1600 mm
2.	Altitudinal variations	About 20-25 m AMSL	About 20-25 m AMSL
3.	Average annual temperatures	26 °C	26 °C
4.	Aereal spreads of the delta	9,500 sq.kms	
5.	Length of the delta	300 kms	300 kms
6.	Average height of the delta	15 m	15 m
7.	Slope of the delta	20 cm/kms	27 cm/kms
8.	Tidal impacts	Up to 30 kms upstream	Up to 30 kms upstream
9.	Average annual discharge	66,640 MCM	18,310 MCM

Seasonal water balance characteristics

The relevant water balance characteristics for selected stations in the region are presented in Table 3.16 and their graphical trends are shown in Fig. 3.27. The locations of the selected stations and their precipitation trends are shown in Fig. 3.26.

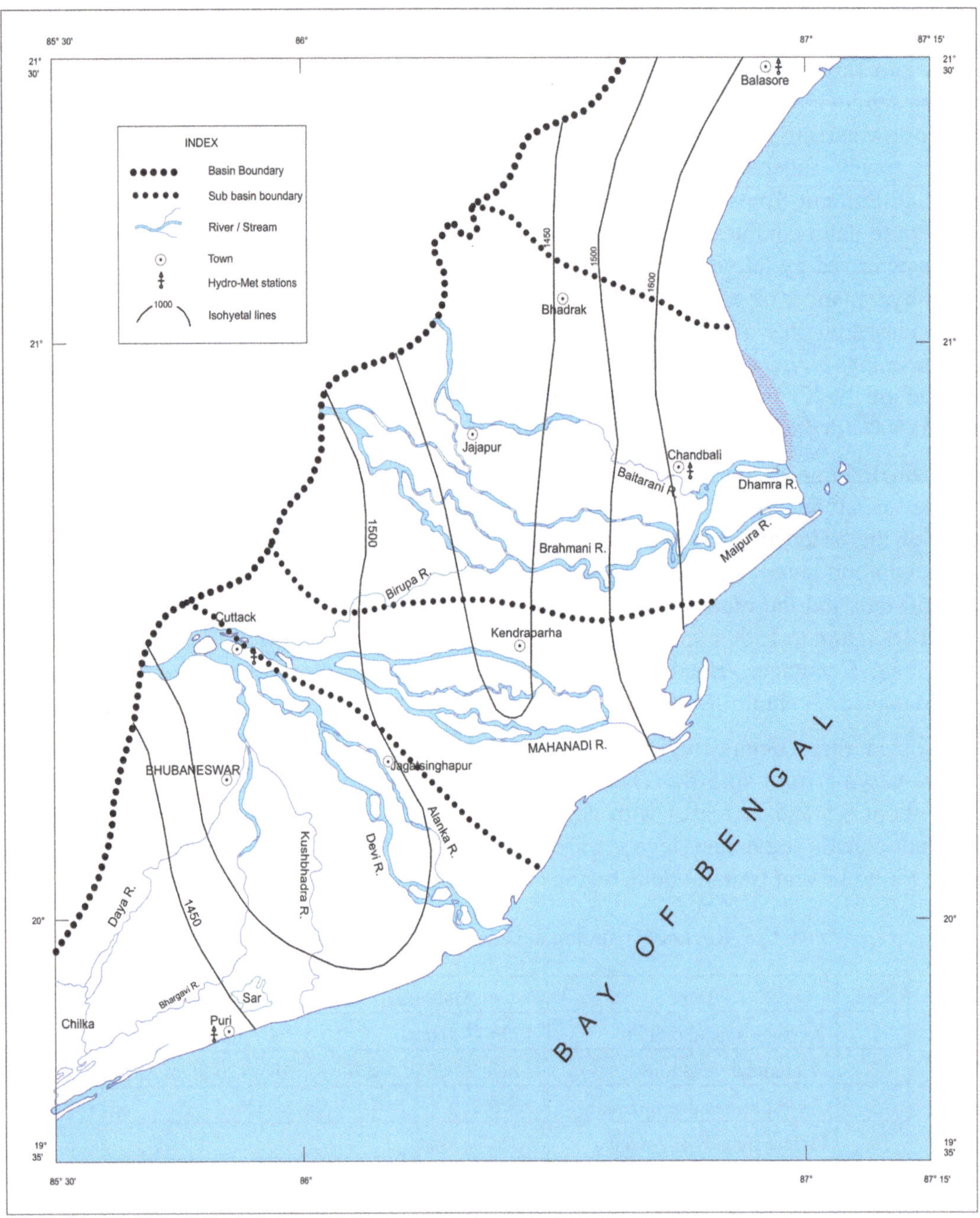

Fig. 3.26 Isohyetel trends and hydromet stations of Mahanadi-Brahmini-Baitarni deltas

Table 3.16 Seasonal water balance characteristics of Mahanadi, Brahmini and Baitarni deltas

Stn : BALASORE (MAHANADI DELTA)										AWC = 200mm		
	J	F	M	A	M	J	J	A	S	O	N	D	TOTAL
P	17.0	31.0	38.0	48.0	98.0	234.0	318.0	335.0	287.0	230.0	45.0	9.0	1690
PE	82.0	102.0	157.0	177.0	219.0	129.0	107.0	101.0	94.0	105.0	88.0	76.0	1437
ΔS	–32.02	–24.91	–26.22	–15.32	–7.68	95.76	95.0	0.0	0.0	0.0	–38.69	–45.92	0
AE	49.0	55.9	64.2	63.3	105.7	129.0	107.0	101.0	94.0	105.0	83.7	54.9	1013
WD	33.0	46.1	92.8	113.7	113.3	0.0	0.0	0.0	0.0	0.0	4.3	21.1	424
WS	0.0	0.0	0.0	0.0	0.0	9.2	116.0	234.0	193.0	125.0	0.0	0.0	677
RO	18.5	9.2	4.6	2.3	1.2	5.1	60.5	147.3	170.1	147.6	73.8	36.9	677
Stn : CHANDBALI (MAHANADI DELTA)											**AWC = 150mm**		
	J	F	M	A	M	J	J	A	S	O	N	D	TOTAL
P	10.0	19.0	26.0	43.0	90.0	201.0	348.0	334.0	264.0	229.0	57.0	7.0	1628.0
PE	88.0	111.0	163.0	194.0	243.0	148.0	122.0	124.0	116.0	115.0	95.0	82.0	1601
ΔS	–28.64	–19.24	–13.62	–5.79	–2.13	51.80	97.0	0.0	0.0	0.0	–33.57	–45.81	0
Stn : CHANDBALI (MAHANADI DELTA)											**AWC = 150mm**		
AE	38.6	38.2	39.6	48.8	92.1	148.0	122.0	124.0	116.0	115.0	90.6	52.8	1026
WD	49.4	72.8	123.4	145.2	150.9	0.0	0.0	0.0	0.0	0.0	4.4	29.2	575
WS	0.0	0.0	0.0	0.0	0.0	1.2	129.0	210.0	148.0	114.0	0.0	0.0	602
RO	16.1	8.0	4.0	2.0	1.0	0.7	64.8	137.4	142.7	128.4	64.2	32.1	602.0
Stn : CUTTACK (MAHANADI DELTA)											**AWC = 200mm**		
	J	F	M	A	M	J	J	A	S	O	N	D	TOTAL
P	10.0	29.0	20.0	27.0	72.0	215.0	355.0	365.0	252.0	168.0	41.0	5.0	1559
PE	88.0	106.0	158.0	180.0	203.0	144.0	113.0	111.0	110.0	115.0	95.0	81.0	1504
ΔS	-33.72	-22.59	-23.97	-12.90	-5.40	65.17	129.0	0.0	0.0	0.0	-47.32	-48.27	0
AE	43.7	51.6	44.0	39.9	77.4	144.0	113.0	111.0	110.0	115.0	88.3	53.3	992.0
WD	44.3	54.4	114.0	140.1	125.6	0.0	0.0	0.0	0.0	0.0	6.7	27.7	512
WS	0.0	0.0	0.0	0.0	0.0	5.8	113.0	254.0	142.0	53.0	0.0	0.0	568
RO	12.7	6.3	3.2	1.6	0.0	3.2	58.1	156.1	149.0	101.0	50.5	25.3	567.0
Stn : PURI (MAHANADI DELTA)											**AWC = 100mm**		
	J	F	M	A	M	J	J	A	S	O	N	D	TOTAL
P	9.0	20.0	14.0	12.0	63.0	187.0	296.0	256.0	258.0	242.0	75.0	8.0	1440.0
PE	109.0	128.0	170.0	168.0	257.0	143.0	136.0	135.0	131.0	133.0	116.0	104.0	1730
ΔS	-16.06	-6.18	-2.50	-0.53	-0.12	43.98	56.0	0.0	0.0	0.0	-33.63	-40.96	0
AE	25.1	26.2	16.5	12.5	63.1	143.0	136.0	135.0	131.0	133.0	108.6	49.0	979.0
WD	83.9	101.8	153.5	155.5	193.9	0.0	0.0	0.0	0.0	0.0	7.4	55.0	751
WS	0.0	0.0	0.0	0.0	0.0	0.0	104.0	121.0	127.0	109.0	0.0	0.0	461
RO	13.5	6.8	3.4	1.7	0.8	0.0	52.0	86.5	106.8	107.9	53.9	27.0	461.0

(P = Precipitation, PE = Potential Evapotranspiration, ΔS = Soil moisture storage change, AE = Actual Evaporation, WD = Water Deficiency, WS = Water Surplus and RO = Run Off)

The precipitation levels in Mahanadi-Brahmini-Baitarni delta complex have a range of variations from a minimum of 5.0 mm to a maximum of 365.0 mm. The monsoon impacts are seen from the month of June and end up by October with peak values reaching during the period July and August in almost all stations. The peak levels observed for different stations are: Balasore: 335 mm; Chandbali: 345 mm; Cuttack: 365 mm and Puri: 296 mm. The average annual precipitation level is 1579.5 mm.

The potential evapotranspiration levels in the region have a range varying from 76.0 mm to 257.0 mm. The minimum values are normally seen during the month of December and the maximum values in the month of May. The maximum PET values observed in different stations are: Balasore: 219 mm; Chandbali: 243 mm; Cuttack: 203 mm and Puri: 257 mm. The average annual PET value in this region is 1568 mm.

The water deficiencies in this delta region are observed during the period November to May and 'zero' values during the period July to October. The maximum deficiencies are seen during the month of May almost in all stations of delta region. The maximum WD values recorded in different stations are: Balasore: 113.68 mm; Chandbali: 150.87 mm; Cuttack: 140.10 and Puri: 193.88 mm. The average annual WD values are of the order of 566 mm. The water surplus is seen during the period June to October, and 'zero' values during the rest of the period. The peak values are normally reached during the month of August in all stations except in Puri where the maximum values are reached in September. The peak WS value recorded are: Balasore: 234 mm; Chandbali: 210 mm; Cuttack: 254 mm and Puri: 127 mm. The annual average WS value in the delta region is 577 mm.

The seasonal computed Runoff values in these regions are graphically shown in Fig. 3.28.

Graphical trends and water balance parameters

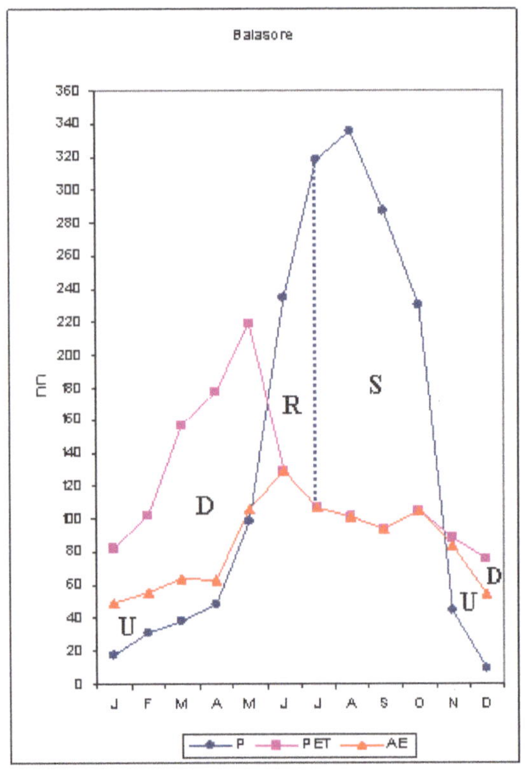

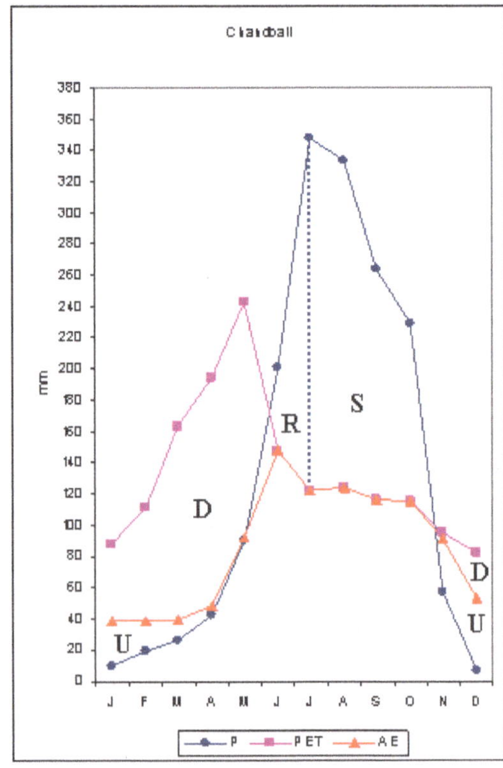

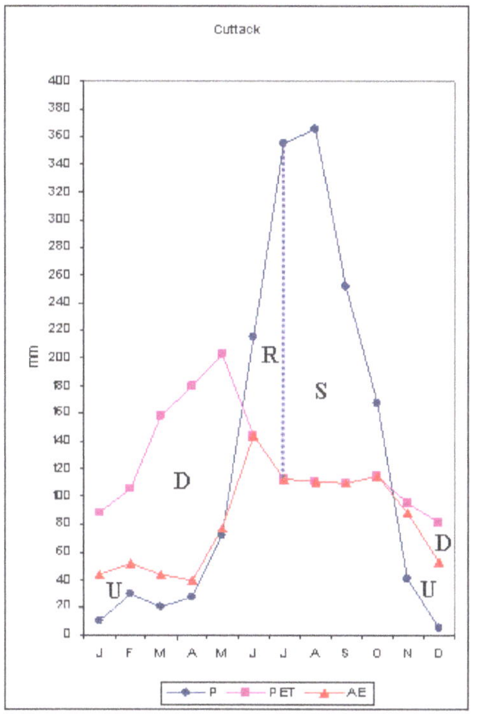

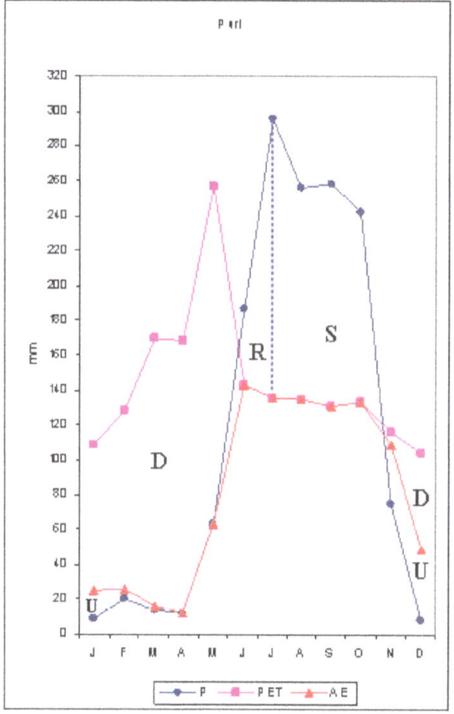

Fig. 3.27 Graphical trends of water balance parameters in Mahanadi-Brahmini-Baitarni deltas and (R-Recharge accretion in the soil zone, S-Surplus, U-Utilization, D-Water Deficit)

The runoff values obtained for different stations in Mahanadi-Brahmini-Baitarni delta systems have a range of variations from zero to a maximum of 170.1 mm. The runoff values normally start in the month of July and attain peak values during the month of August/September and then onwards gradually get reduced to minimum values during the period May to June. The peak RO values obtained in different stations are: Balasore: 170.1 mm; Chandbali: 142.7 mm; Cuttuck: 156.1 mm and Puri: 107.9 mm. The average annual RO value in this delta region is 576.75 mm which is 36.5 % of the total annual precipitation.

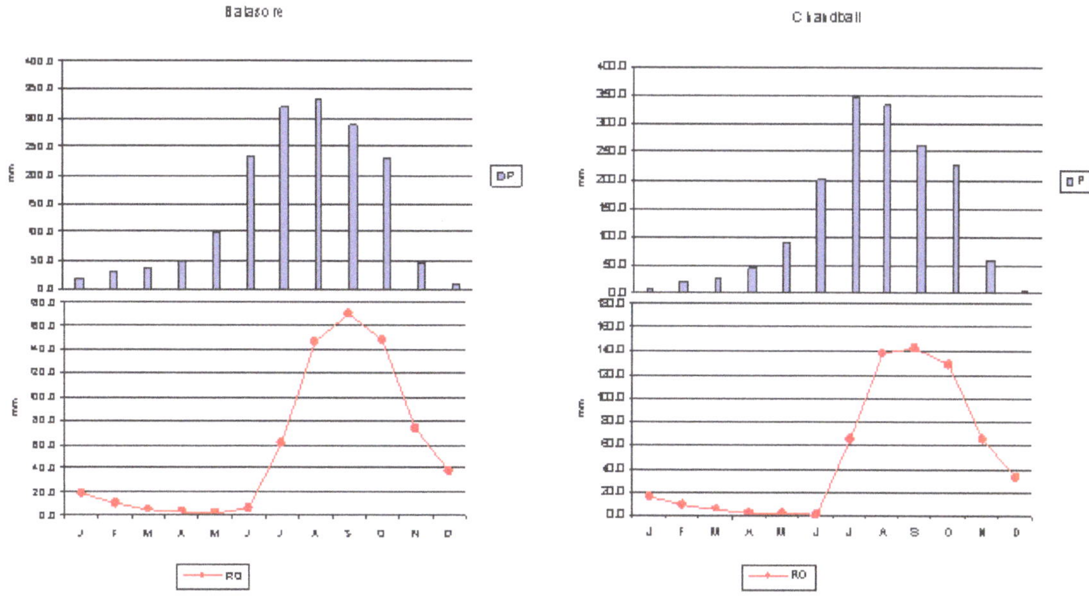

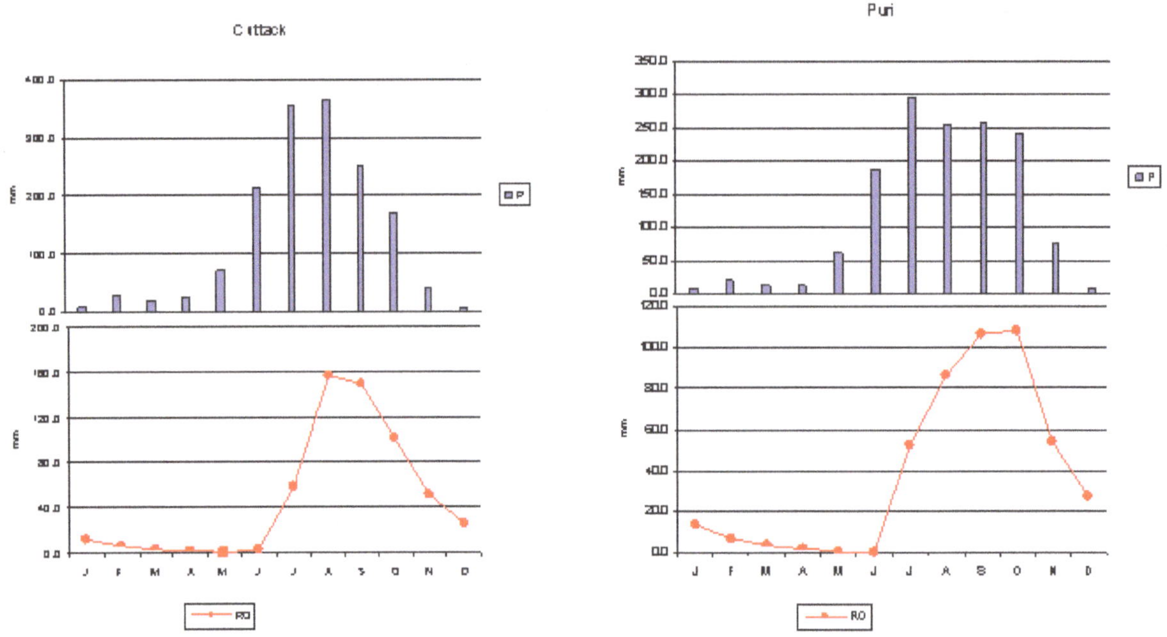

Fig. 3.28 Precipitation-Runoff relations in the composite Mahanadi delta

Annual climatic water balance and budgetary analyses

The Annual climatic water balance/budgetary elements for Mahanadi-Brahmini-Baitarni delta system is tabulated below in Table 3.17.

Table 3.17 Annual water balance parameters in Mahanadi, Brahmini and Baitarni

Location	PET (mm)	Rainfall (mm)	Water Surplus (mm)	Water Deficit (mm)	Humidity Index (%)	Aridity Index (%)	Moisture Index (%)	RunOff (mm)
Mahanadi Delta								
Balasore	1437	1690	677.24	424.24	47.1	29.5	17.6	677
Chandbali	1601	1628	602.2	575.2	37.6	35.9	1.7	602
Cuttuck	1504	1559	567.83	512.83	37.8	34.1	3.7	567
Puri	1730	1440	461.02	751.02	26.6	43.4	-16.8	461

3.3.5 Bhagiradhi – Hoogly delta (West Bengal)

Bhagiradhi – Hoogly river/delta system is one of the main distributary arm of Ganges delta, spread mainly in the southern regions of West Bengal. The river emerges from the apex point at Farrakka, traverses a distance of about 260 kms before joining the Bay of Bengal near the Hoogly estuary mouth area. A number of tributaries like Mayurakshi and Ajay (upper reaches), Damodar and Roopnarayan (middle reaches) and Haldi and Rasulpur (lower reaches) join the main river on its track from the western flanks of the delta. Similarly several small tributaries like Bhairab, Jalangi, Mathabanga and Churin join the main river from the eastern flanks. All the network of several large and small rivers join the main river at different locations along its stretch could make the entire delta complex filled with fluviotile alluvial sediments. The upper and middle delta regions are dominantly characterized by a network of paleo channels, natural levees and flood plains filled with sands and clays largely generated by several tributaries flowing from the western and eastern flanks of the basin area. The lower delta regions are characterized by thick mud flats, tidal inlets, marshes and thick vegetative growth and forest cover. The world's largest mangrove forest 'Sundarbans Mangrove Biosphere reserve' is located in the lower reaches of this delta region. The soils of Bengal delta are deep, poorly drained and grey in color. The clay contents in the soils range between 35% to 50% with mixtures of illites and smectites. The pH values of the soils at near surface range between 4.5 and 5.3 (acidic) and at deeper levels range between 6 to 8 (neutral to alkaline) and contain high Sodium and Magnesium contents with high organic matter. The delta is characterized by sub tropical humid climate with an average maximum temperature of 28 °C and a minimum of 18 °C respectively and with humidity levels ranging between 80% (wet season) and 70%(dry season).

The basin mainly receives rainfall from southwest monsoon during the period June to October and has peak precipitation levels during the months July and August and low to moderate precipitation levels during the rest of the period. The average annual rainfall recorded is of the order of 1500 mm. Though the exact annual discharge figures are not available it is approximately estimated that about 46,800 MCM could takes place through Bhagirathi-Hoogly river into Bay of Bengal. The peak discharge values are normally recorded during the months of August to September. A brief hydrological characteristics of Bhagirathi-Hoogly delta are presented below in Table 3.18.

Table 3.18 Hydrological charactcristics of Bhagiradhi – Hoogly dclta

S.no.	Item	Bhagiradhi-Hoogly
1.	Average annual rainfall	1500 mm
2.	Altitudinal variations	1-23 m
3.	Average annual temperatures	25° to 27.5°C
4.	Aereal spreads of the delta	60,000 sq.km
5.	Length of the delta	260 kms
6.	Width of the delta	
7.	Slope of the delta	8.84 cm/kms
8.	Tidal impacts	100 kms upstream
9.	Average annual discharge	10% of the average discharge of Ganges

Seasonal water balance characteristics

The relevant water balance parameters for the selected stations in the delta region are presented below in Table 3.19 and their graphical trends shown in Fig. 3.30. The locations of the selected stations and their precipitation trends are shown in Fig. 3.29.

Table 3.19 Seasonal water balance characteristics of Bhagiradhi-Hoogly delta region

Stn : CALCUTTA (BAGIRADI-HOOGLY DELTA)										AWC = 200mm		
	J	F	M	A	M	J	J	A	S	O	N	D	TOTAL
P	14.0	24.0	27.0	43.0	121.0	259.0	301.0	306.0	290.0	160.0	35.0	3.0	1583
PE	76.0	96.0	148.0	170.0	171.0	126.0	109.0	106.0	103.0	112.0	89.0	71.0	1377
ΔS	–28.97	–24.09	–25.24	–14.28	–3.56	120.47	67.0	0.0	0.0	0.0	–47.32	–44.01	0
AE	43.0	48.1	52.2	57.3	124.6	126.0	109.0	106.0	103.0	112.0	82.3	47.0	1010
WD	33.0	47.9	95.8	112.7	46.4	0.0	0.0	0.0	0.0	0.0	6.7	24.0	367
WS	0.0	0.0	0.0	0.0	0.0	12.5	125.0	200.0	187.0	48.0	0.0	0.0	573
RO	13.0	6.5	3.3	1.6	0.0	6.9	65.9	133.0	160.0	104.0	52.0	26.0	572

Table 3.19 ***Contd...***

Stn : CONTAI (BAGIRADHI-HOOGLY DELTA)										AWC = 100mm		
	J	F	M	A	M	J	J	A	S	O	N	D	TOTAL
P	16.0	3.0	18.0	19.0	127.0	269.0	314.0	384.0	369.0	272.0	38.0	6.0	1835.0
PE	82.0	108.0	158.0	183.0	244.0	141.0	126.0	124.0	115.0	120.0	94.0	77.0	1572
ΔS	-13.57	-9.43	-3.83	-1.01	-0.16	99.92	0.0	0.0	0.0	0.0	-42.88	-29.04	0
AE	29.6	12.4	21.8	20.0	127.2	141.0	126.0	124.0	115.0	120.0	80.9	35.0	952
WD	52.4	95.6	136.2	163.0	116.8	0.0	0.0	0.0	0.0	0.0	13.1	42.0	620
WS	0.0	0.0	0.0	0.0	0.0	28.1	188.0	260.0	254.0	152.0	0.0	0.0	882.1
RO	23.1	11.6	5.8	2.9	1.4	15.4	101.7	180.9	217.4	184.7	92.4	46.2	883.00

Stn : MIDNAPUR (BAGIRADI-HOOGLY DELTA)										AWC = 200mm		
	J	F	M	A	M	J	J	A	S	O	N	D	TOTAL
P	14.0	30.0	34.0	44.0	109.0	232.0	322.0	336.0	262.0	132.0	36.0	3.0	1554
PE	78.0	98.0	153.0	184.0	207.0	139.0	115.0	109.0	108.0	113.0	87.0	73.0	1464
ΔS	–29.90	–22.86	–25.32	–15.67	–5.99	83.53	107.0	0.0	0.0	0.0	–45.02	–45.77	0
AE	43.9	52.9	59.3	59.7	115.0	139.0	115.0	109.0	108.0	113.0	81.0	48.8	1045
WD	34.1	45.1	93.7	124.3	92.0	0.0	0.0	0.0	0.0	0.0	6.0	24.2	419
WS	0.0	0.0	0.0	0.0	0.0	9.5	100.0	227.0	154.0	19.0	0.0	0.0	509
RO	10.4	5.2	2.6	1.3	0.0	5.2	52.6	139.8	146.9	83.0	41.5	20.7	509

Stn : BURDWAN (BHAGIRATHI-HOOGLY DELTA)										AWC = 200 mm		
	J	F	M	A	M	J	J	A	S	O	N	D	TOTAL
P	11.0	25.0	25.0	46.0	115.0	196.0	314.0	301.0	237.0	107.0	23.0	4.0	1404
PE	71.8	89.6	146.6	177.5	202.9	146.2	129.1	124.5	115.1	117.9	86.9	67.3	1475
ΔS	–26.29	–20.42	–24.40	–14.05	–5.37	40.06	150.20	0.0	0.0	–10.61	–51.79	–37.33	0
AE	37.3	45.4	49.4	60.1	120.4	146.2	129.1	124.5	115.1	117.6	74.8	41.3	1062.0
WD	34.5	44.2	97.2	117.5	82.5	0.0	0.0	0.0	0.0	0.3	12.1	26.0	414
WS	0.0	0.0	0.0	0.0	0.0	9.7	34.7	176.5	121.9	0.0	0.0	0.0	343
RO	6.9	3.5	1.7	0.0	0.0	5.4	20.0	98.3	110.1	55.0	27.5	13.8	342.1

Stn : KRISHNANAGAR (BHAGIRATHI-HOOGLY DELTA)										AWC = 200 mm		
	J	F	M	A	M	J	J	A	S	O	N	D	TOTAL
P	11.0	23.0	31.0	60.0	141.0	252.0	281.0	271.0	205.0	130.0	26.0	3.0	1434
PE	67.0	88.0	144.0	176.0	204.0	127.0	111.0	107.0	103.0	113.0	84.0	64.0	1390
ΔS	–26.94	–23.13	–26.0	–15.26	–5.13	111.15	75.0	0.0	0.0	0.0	–50.35	–39.34	0
AE	37.9	46.1	57.0	75.3	146.1	127.0	111.0	107.0	103.0	113.0	76.4	42.3	1042.2
WD	29.1	41.9	87.0	102.7	57.9	0.0	0.0	0.0	0.0	0.0	7.7	21.7	348
WS	0.0	0.0	0.0	0.0	0.0	13.9	95.0	164.0	102.0	17.0	0.0	0.0	392
RO	7.6	3.8	1.9	0.0	0.0	7.6	51.3	107.7	104.8	60.9	30.5	15.2	392.0

(P = Precipitation, PE = Potential Evapotranspiration, ΔS = Soil moisture storage change, AE = Actual Evaporation, WD = Water Deficiency, WS = Water Surplus and RO = Run Off)

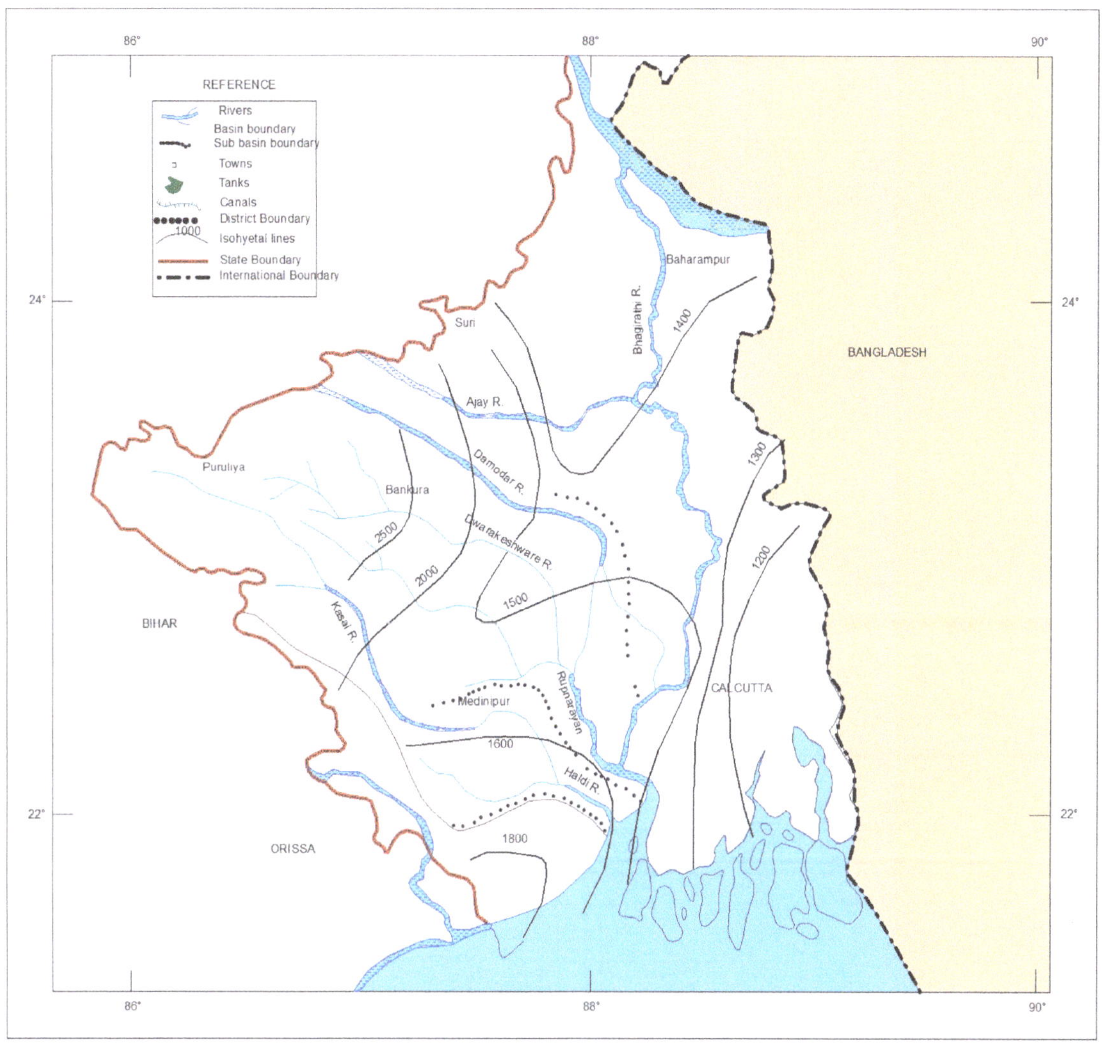

Fig. 3.29 Isohyetel trends and hydromet stations of Bhagiradhi-Hoogly delta

The trends show wide range of variations in the precipitation levels. The western flanks show precipitation levels of the order > 2500 mm where as the eastern flanks show a minimum level <1100 mm. Similarly the southern side shows a range of variations between 1100 mm to 1800 mm and the northern side has a level of the order of 1400 mm and the central region is marked by a level varying between 1400 mm to 1500 mm.

Graphical trends and water balance parameters

The monsoon precipitation levels actually start from the month of June and ends up in the month of October in all most all stations, though some amount of rainfall get registered during the rest of the period. The peak precipitation levels in general are observed in the month of July in the upper delta region and in the month of August for the lower delta region. The peak monsoon levels recorded for different stations are: Culcutta: 306.0 mm; Contai: 384.0 mm; Midnapur: 336.0 mm; Burdwan: 314.0mm and Krishnanagar: 281.0 mm. The average annual precipitation levels for the entire delta region is of the order of 1562 mm.

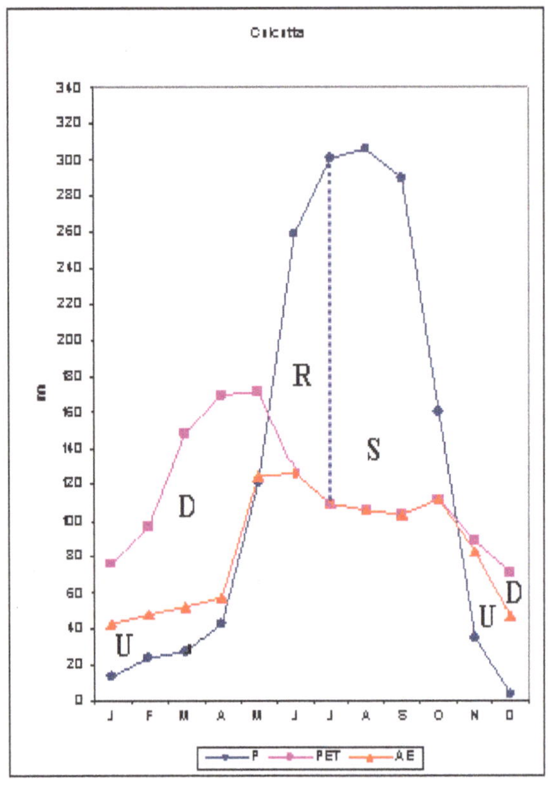
Calcutta
340
320
300
280
260
240
220
200
180
160
140
120
100
80
60
40
20
0
R
S
D
D
U
U
J F M A M J J A S O N D
P PET AE

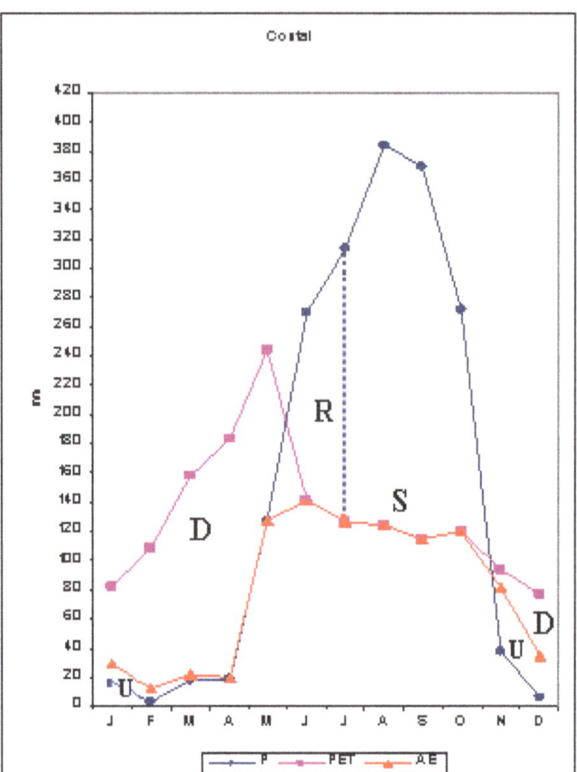
Coastal
420
400
380
360
340
320
300
280
260
240
220
200
180
160
140
120
100
80
60
40
20
0
R
S
D
D
U
U
J F M A M J J A S O N D
P PET AE

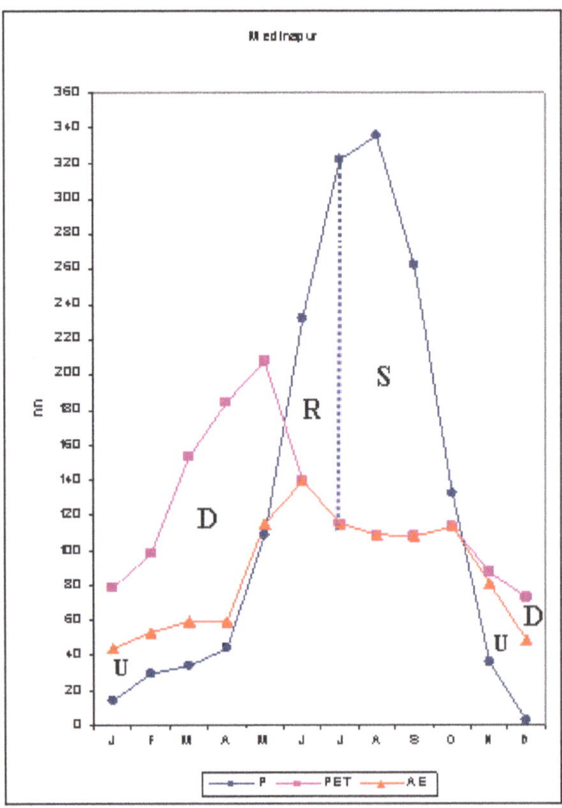
Medinapur
360
340
320
300
280
260
240
220
200
180
160
140
120
100
80
60
40
20
0
S
R
D
D
U
U
J F M A M J J A S O N D
P PET AE

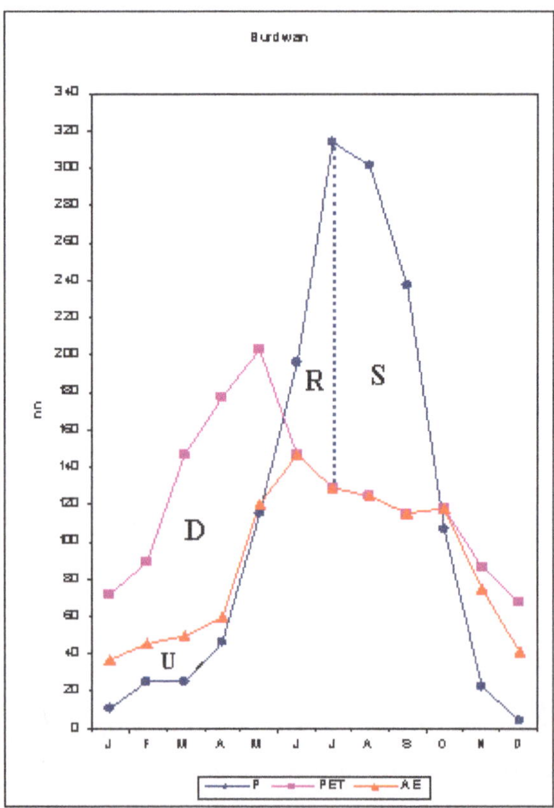
Burdwan
340
320
300
280
260
240
220
200
180
160
140
120
100
80
60
40
20
0
R
S
D
U
J F M A M J J A S O N D
P PET AE

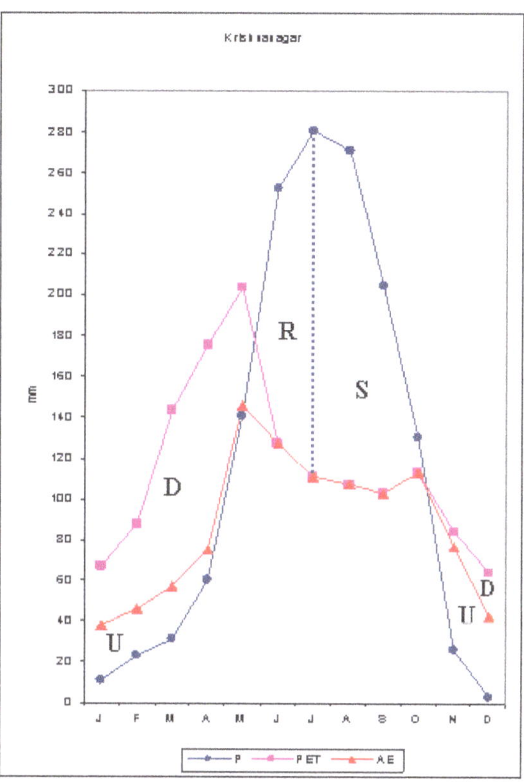

Fig. 3.30 Graphical trends of water balance parameters in Bhagiradhi-Hoogly delta and (R-Recharge accretion in the soil zone, S-Surplus, U-Utilization, D-Water Deficit)

The potential evapotranspiration values have a range varying between 64.0 mm to 244.0 mm. The maximum PET values are observed during the month of May and minimum values during the month of December. The maximum and minimum PET values recorded for different station are seen in the month of May in a year. The maximum PET values recorded for different stations are: Culcutta: 171.0 mm; Contai: 244.0 mm; Midnapur: 207.0 mm; Burdwan: 202.9 mm and Krishnanagar: 204.0 mm. The average annual PET value for this delta region is of the order of 1455.6 mm.

The water deficiencies are observed during the period January to May and again in the month of December and the rest of the period 'zero' values are recorded. The values have range varying from 21.7 mm to 163.0 mm. The peak WD values observed in different stations are: Culcutta: 112.7 mm; Contai: 163.0 mm; Midnapur: 124.3 mm; Burdwan: 117.5 mm and Krishnanagar: 102.7 mm. The average annual WD in this region is of the order of 433 mm.

The water surpluses are recorded during the period June to October only and during the rest of the period the values are at 'zero' level. The WS values have a range varying from 9.5 mm to 227.0 mm. The maximum water surpluses are normally seen in the month of August and the minimum values in the month of July in all stations. The peak WS values recorded in different stations are: Culcutta: 200.0 mm; Contai: 260.0 mm; Midnapur: 227.0 mm; Burdwan: 176.0mm and Krishnanagar: 164.0 mm. The average annual WS value in this delta region is of the order of 540 mm.

The runoff values obtained for different stations in Bhagirathi-Hoogly delta system have a range of variations from 'zero' to a maximum of 217.0 mm. The runoff values normally start in the month of June and attain peak value during the month of September then onwards gradually get reduced and take minimum values during the period August/May. The peak RO values obtained in different stations are: Culcutta: 572 mm; Contai: 883 mm; Midnapur: 509 mm; Burdwan: 342.1 mm and Krishnanagar: 392 mm. The average annual RO value in this delta region is 539.6 mm which is 34.5 % of the total annual precipitation.

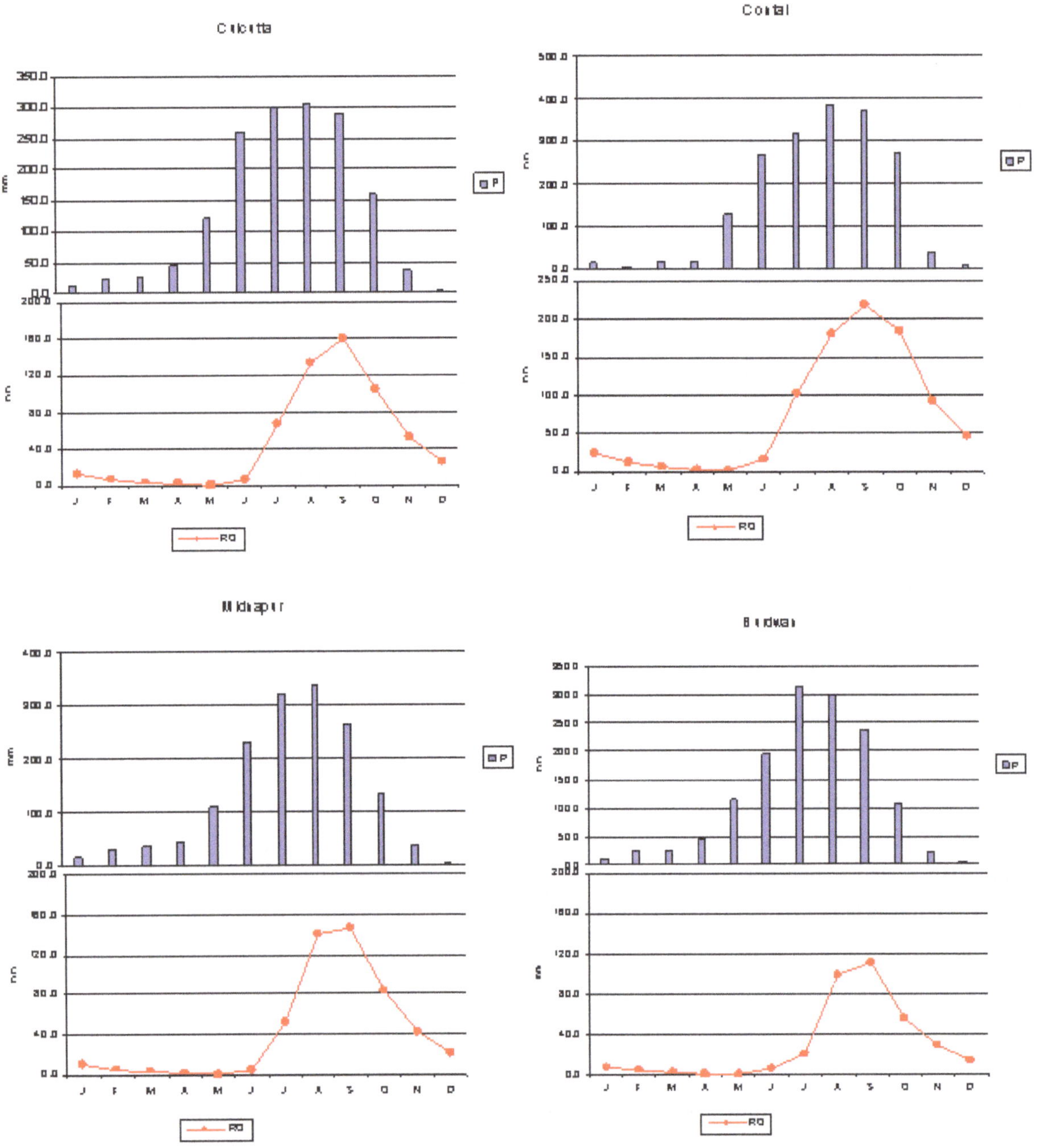

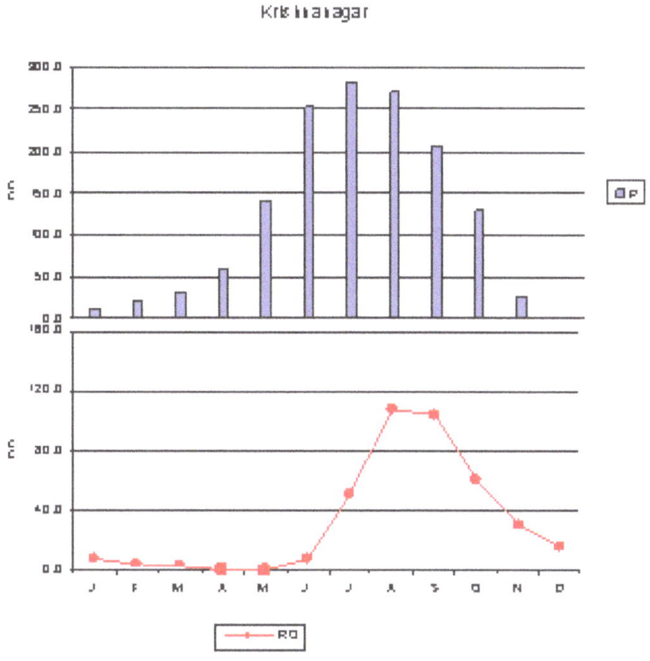

Fig. 3.31 Precipitation-Runoff relations in Bhagiradhi-Hoogly delta

Annual climatic water balance budget analysis

The Annual climatic water balance/budgetary elements for Bhagiradhi-Hoogly delta system is tabulated below in Table 3.20.

Table 3.20 Annual water balance parameters in Bhagiradhi-Hoogly delta region

Location	PET (mm)	Rainfall (mm)	Water Surplus (mm)	Water Deficit (mm)	Humidity Index (%)	Aridity Index (%)	Moisture Index (%)	RunOff (mm)
Bhagirathi-Hoogly								
Culcutta	1377	1583	572.53	366.53	41.6	26.6	15.0	572.00
Contai	1572	1835	882.08	619.08	56.1	39.4	16.7	883.00
Midnapur	1464	1554	509.47	419.47	34.8	28.7	6.1	509.00
Burdwan	1475	1404	343	414	23.3	28.1	-4.8	342.00
Krishnanagar	1390	1434	392	348	28.2	25.0	3.2	392.00

3.3.6 Narmada-Tapti estuary delta systems

Narmada and Tapti are the two major estuary delta systems along the west coast mainly spread in the districts of Bharuch and Surat of Gujarat state. The Narmada traverse a distance of about 1312 kms along a major fault/fracture line (Narmada-Son lineament) and Tapti traverses a distance of about 724 kms (along sub parallel faulted line) from their source areas and debouch into Gulf of Cambay (Arabian Sea) with their apex points at Bharuch and Surat respectively, covering the states of Madhya Pradesh,

Maharashtra and Gujarat. The aerial spread of these delta systems are rather small (< 200 sq.kms) but their flood plains/alluvial coverage is fairly large to the tune of about 5583 sq.kms. The thicknesses of these alluvial sediments increase from an order of 20 m (upstream) to an order of 120 m towards the outfall regions. These two estuary delta systems support reasonably rich agricultural practices in this part of the coast. The soils in the region are well drained black cotton and alluvial soils with their clay percentages ranging between 60 % and 70 %. The soils contain relatively high potash and medium levels of phosphates and nitrates with pH values ranging from 7.0 to 8.0 and conductivity levels ranging between 600 μ mhos/cm to 300 μ mhos/cm. The delta region enjoys a sub tropical climate with an average maximum temperature of 26.6 °C and a minimum temperature of the order of 14.3 °C respectively. The average maximum humidity level is of the order of 85% (wet season) and dips down to 50% (dry season).

The Narmada and Tapti delta/outfall regions have an average rainfall of about 959 mm and 1205 mm respectively, mainly from southwest monsoon during the period June to September. The annual average discharges in respect of Narmada and Tapti are of the order of 40,705 MCM and 17,982 MCM. The peak discharges are normally observed during the months of July to September. A brief sketch of hydrological characters of Narmada and Tapti estuary delta regions are tabulated below in Table 3.21.

Table 3.21 Regional hydrological characters of Narmada and Tapti estuary basins.

S.no	Item	Narmada	Tapti
1.	Average annual rainfall	959 mm	1205 mm
2.	Altitudinal variations	0-25 above mean sea level	0-20 above mean sea level
3.	Average annual temperatures	26.5^0C	26.5^0C
4.	Areal spreads of the delta	< 200 sq.kms (up to the apex points)	< 150 sq.kms (up to the apex points)
5.	Length of the delta	35 kms	20 kms
6.	Width of the estuary	15.0 kms	5.0 kms
7.	Slope of the delta	0.4 m/km (up to the apex points)	0.65 m/km (up to the apex points)
8.	Tidal impacts	10-15 kms	8-10 kms
9.	Average annual discharge	40,705 MCM	17,982 MCM

The precipitation trends for selected hydro-met stations in Narmada and Tapti delta basins are shown in Figures 3.32 and the relevant seasonal water balance parameters are tabulated in Table 3.22 and graphical trends are displayed in Fig. 3.33.

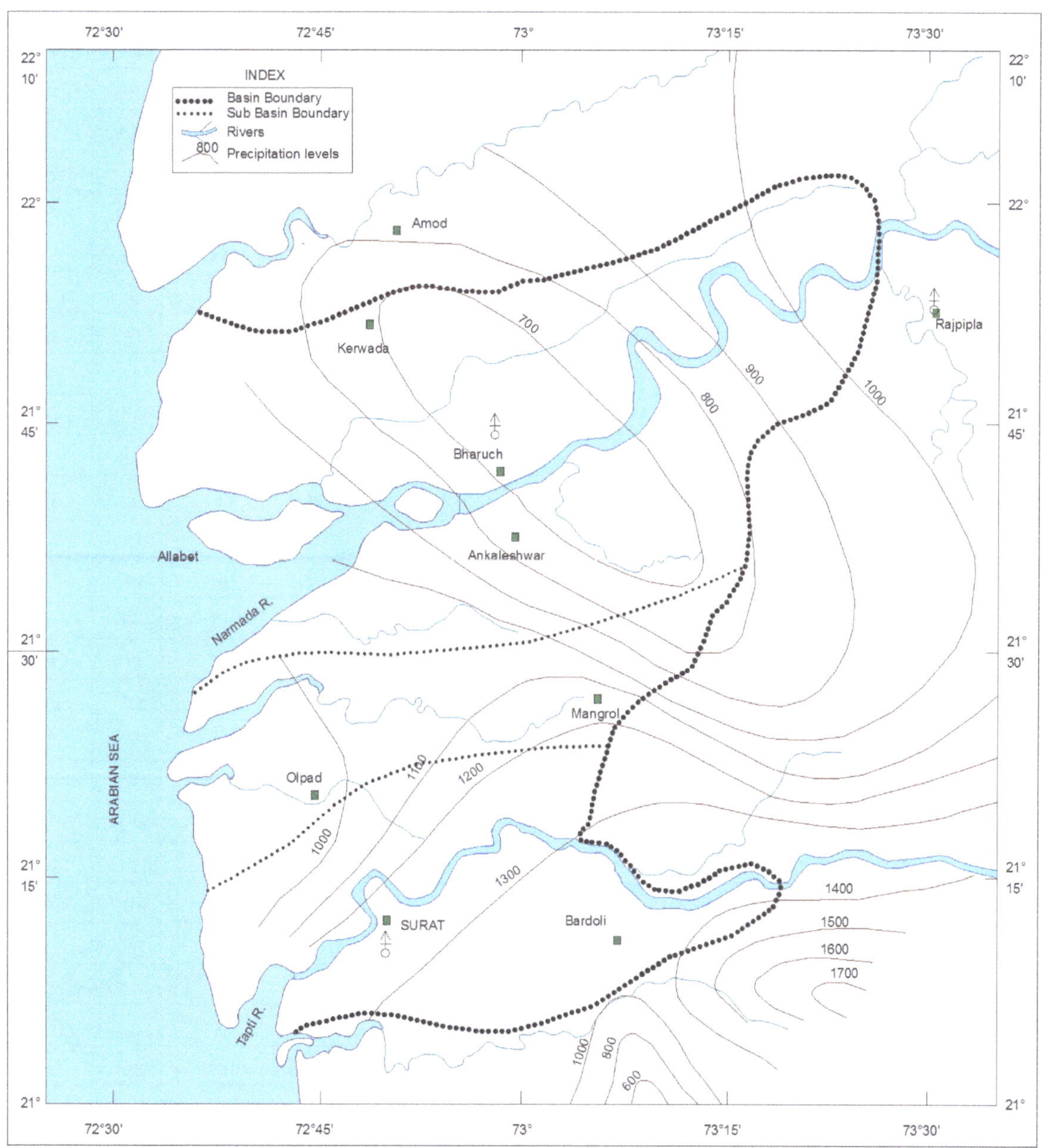

Fig. 3.32 Isohyetel map of Narmada and Tapti with selected hydromet stations locations

Seasonal water balance characteristics

Table 3.22 Seasonal water balance characteristics of Narmada and Tapti delta region

Stn : VADODARA*										AWC = 200mm		
	J	F	M	A	M	J	J	A	S	O	N	D	TOTAL
P	2.0	2.0	1.0	3.0	4.0	115.0	366.0	264.0	186.0	35.0	9.0	0.0	987
PE	82.0	98.0	145.0	181.0	227.0	185.0	120.0	114.0	124.0	130.0	93.0	76.0	1575
ΔS	−18.43	−14.28	−11.90	−6.65	−3.11	−0.45	198.93	0.0	0.0	−75.62	−42.66	-25.83	0

Table 3.22 ***Contd...***

Stn : VADODARA*										AWC = 200mm		
	J	F	M	A	M	J	J	A	S	O	N	D	TOTAL
AE	20.4	16.3	12.9	9.7	7.1	115.5	120.0	114.0	124.0	110.6	51.7	25.8	728
WD	61.6	81.7	132.1	171.4	219.9	69.6	0.0	0.0	0.0	19.4	41.3	50.2	847
WS	0.0	0.0	0.0	0.0	0.0	0.0	47.0	150.0	62.0	0.0	0.0	0.0	259
RO	4.7	2.4	1.2	0.0	0.0	0.0	25.9	87.9	75.0	37.5	18.7	9.4	263
Stn : BHARUCH											AWC = 200mm		
	J	F	M	A	M	J	J	A	S	O	N	D	TOTAL
P	3.6	1.4	0.5	3.3	7.4	145.1	328.7	186.0	166.5	32.7	8.0	0.7	884
PE	82.0	98.0	145.0	181.0	227.0	185.0	120.0	114.0	124.0	130.0	93.0	76.0	1575
ΔS	−17.89	−14.27	−11.83	−6.58	−3.06	−0.28	198.75	0.0	0.0	−77.05	−42.57	−25.22	0
AE	21.5	15.7	12.3	9.9	10.5	145.4	120.0	114.0	124.0	109.8	50.6	25.9	759
WD	60.5	82.3	132.7	171.1	216.5	39.6	0.0	0.0	0.0	20.3	42.4	50.1	816
WS	0.0	0.0	0.0	0.0	0.0	0.0	10.0	72.0	42.5	0.0	0.0	0.0	124
RO	2.6	1.3	0.0	0.0	0.0	0.0	5.5	38.7	40.6	20.3	10.2	5.1	124
Stn : RAJPIPLA											AWC = 200mm		
	J	F	M	A	M	J	J	A	S	O	N	D	TOTAL
P	3.3	1.3	1.7	1.1	7.2	143.2	372.3	248.5	182.6	31.3	9.1	3.0	1005
PE	98.4	115.3	167.1	200.3	237.4	180.1	129.2	118.3	134.1	143.7	109.3	93.9	1727
ΔS	−16.59	−11.85	−8.67	−4.25	−1.70	−0.14	199.35	0.0	0.0	−85.99	−44.93	−25.23	0
AE	19.9	13.2	10.4	5.4	8.9	143.3	129.2	118.3	134.1	117.3	54.0	28.2	782
WD	78.5	102.2	156.7	195.0	228.5	36.8	0.0	0.0	0.0	26.4	55.3	65.7	945
WS	0.0	0.0	0.0	0.0	0.0	0.0	43.8	130.2	48.5	0.0	0.0	0.0	222
RO	4.0	2.0	1.0	0.0	0.0	0.0	24.1	77.1	62.8	31.4	15.7	7.9	226
Stn : SURAT											AWC = 200mm		
	J	F	M	A	M	J	J	A	S	O	N	D	TOTAL
P	1.0	1.0	1.0	5.0	6.0	210.0	448.0	254.0	218.0	43.0	16.0	2.0	1205
PE	100.0	144.9	162.0	186.0	202.0	152.0	109.0	108.0	115.0	142.0	116.0	99.0	1636
ΔS	-17.78	-14.23	-7.48	-3.60	-1.52	58.0	141.08	0.0	0.0	-78.09	-47.97	-28.41	0
AE	18.8	15.2	8.5	8.6	7.5	152.0	109.0	108.0	115.0	121.0	64.0	30.4	758
WD	81.2	129.7	153.5	177.4	194.5	0.0	0.0	0.0	0.0	20.9	52.0	68.6	878
WS	0.0	0.0	0.0	0.0	0.0	0.0	197.9	146.0	103.0	0.0	0.0	0.0	447
RO	7.2	3.6	1.8	0.0	0.0	0.0	108.9	127.4	115.2	57.6	28.8	14.4	465

*(P= Precipitation, PE=Potential Evapotranspiration, ΔS= Soil moisture storage change, AE=Actual Evaporation, WD= Water Deficiency, WS= Water Surplus and RO=Run Off)

*Location is outside the boundary

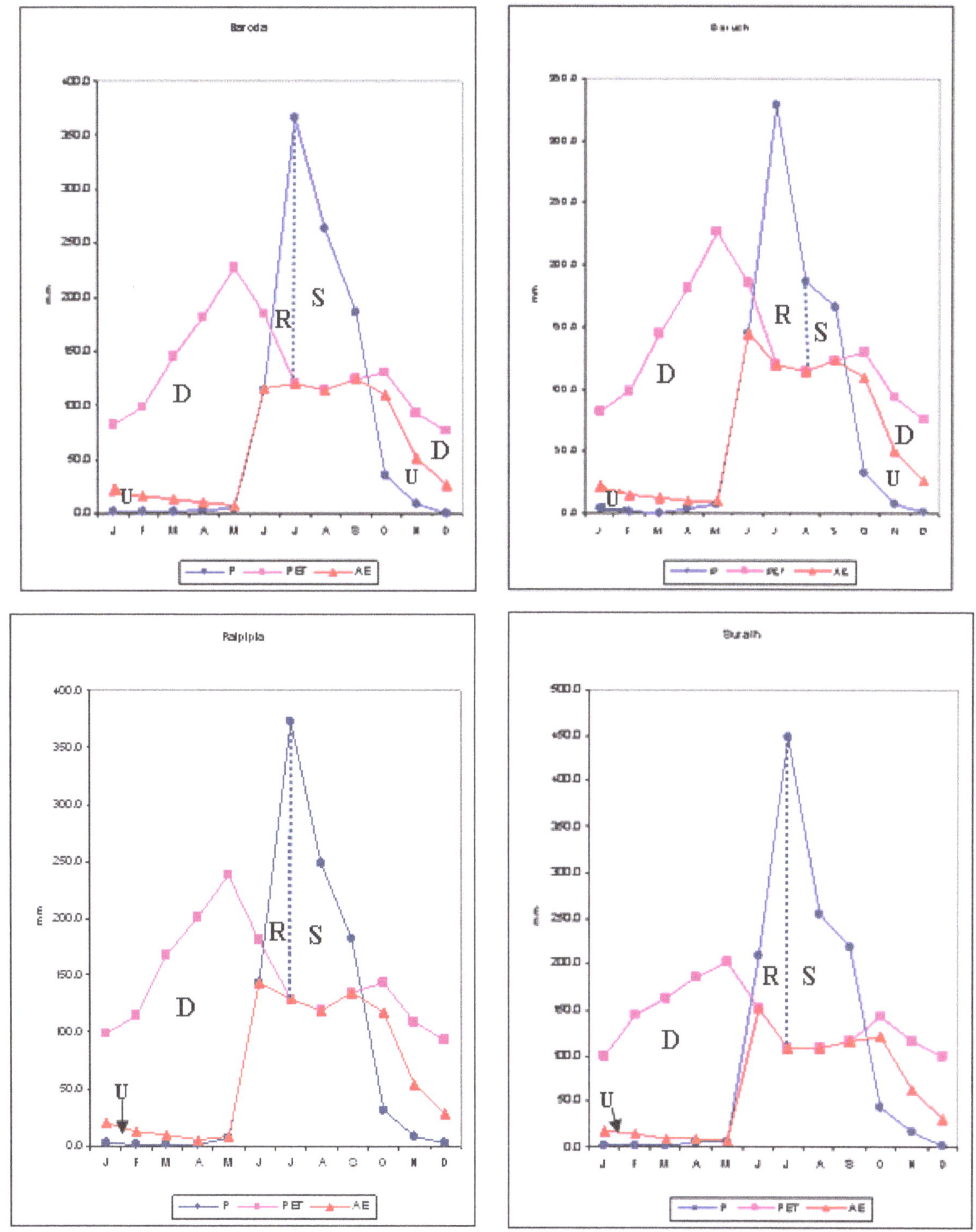

Fig. 3.33 Graphical trends of water balance parameters in Narmada and Tapti delta (R-Recharge accretion in the soil zone, S-Surplus, U-Utilization, D-Water Deficit)

Graphical trends and water balance parameters

The precipitation levels in respect of Narmada have a range of values varying from a minimum of 0.0 mm to a maximum of 372.3 mm and for Tapti with a minimum of 1.2 mm to 448 mm. The monsoon rainfall inputs are observed from June to September with maximum rainfall inputs during the month of July. The peak values recorded in different stations are: Narmada delta region: Vadodara (366.0 mm); Bharuch (328.7 mm); Rajpipla (372.3 mm) and Tapti delta region: Surat (448.0 mm). The average annual precipitation levels in respect of Narmada and Tapti are 959 mm and 1205 mm respectively.

The potential evapotranspiration in Narmada and Tapti delta regions have a range of values varying from 76 to 237.4 mm. The minimum values are normally recorded in the month of December and the maximum values during the month of May. The maximum PET values recorded in different stations are Narmada delta: Vadodara (227 mm); Bharuch (227 mm); Rajpipla (237.4 mm) and Tapti delta: Surat (202 mm). The average annual PET values in Narmada and Tapti delta are 1626 mm and 1636 mm respectively.

The water deficit values in Narmada and Tapti deltas in general occur during the period January to June and then from October to December and the rest of the periods nil values. The range of variations of water deficit values in Narmada-Tapti delta are as follows :19.4 mm to 228.5 mm; Tapti delta: 20.9 mm to 164.5 mm. The peak values of WD recorded in different stations are: Narmada delta: Vadodara (219.9 mm); Bharuch (216.5 mm); Rajpipla (228.5 mm) and Tapti delta: Surat (164.5 mm). The annual average WD recorded for Narmada and Tapti region are 869 mm and 878 mm respectively.

The water surpluses recorded in all the stations in both the delta systems have shown 0 values excepting during the months of July, August and September. The peak WS values recorded in different stations are: Narmada delta: Vadodara (150 mm); Bharuch (72.0 mm), Rajpipla (130.2 mm) and Tapti delta: Surat (197.9 mm). The average annual WS in respect of Narmada and Tapti are 202 mm and 447 mm respectively. The annual average seasonal computed Run Off and Recharge values for Narmada delta: 204 mm and 264 mm and Tapti delta region: 465 mm and 230 mm respectively. The monthly variations of Runoff and Recharge parameters are graphically shown in Figures 3.34.

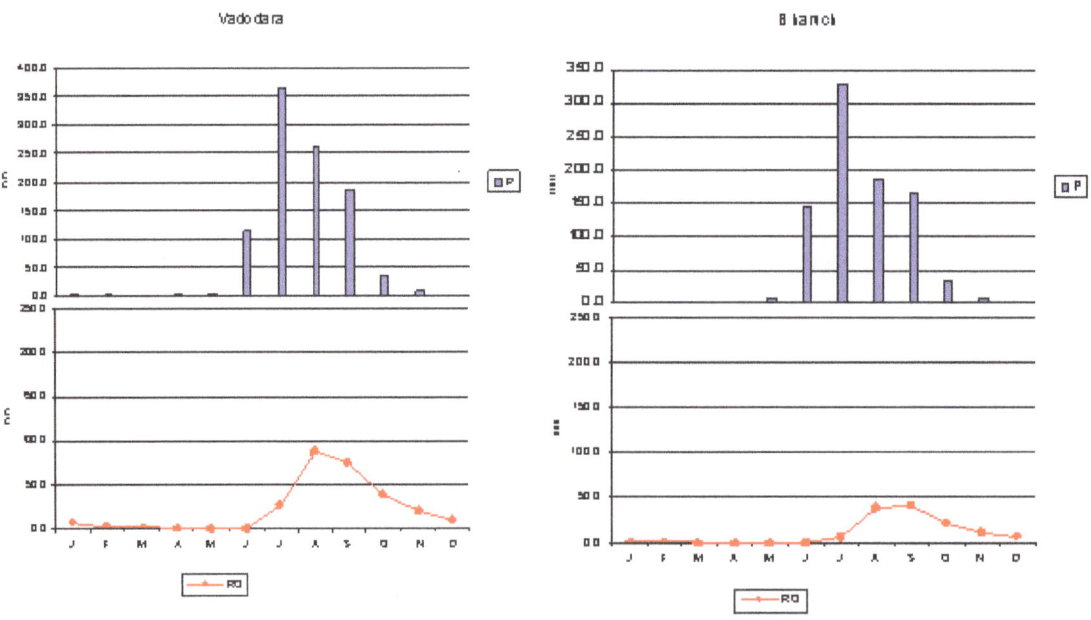

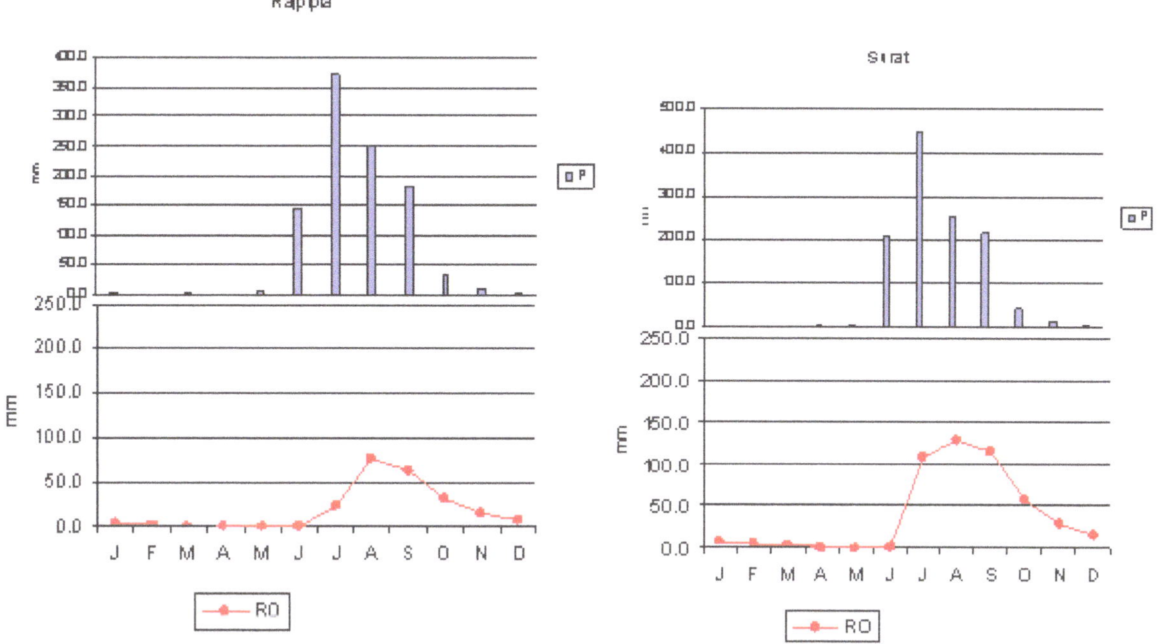

Fig. 3.34 Precipitation-Runoff relations in Narmada-Tapti delta region

The Runoff values for Narmada and Tapti are recorded during the period starting from June to December with peak values reaching during the month of August. The peak RO values for Narmada estuary: Vadadora (87.9 mm), Bharuch (40.6 mm), Rajpipla (77.1 mm) and Tapti estuary: Surat (127.4). The average annual Runoff for Narmada is 204 mm. and for Surat 465 mm. The estimated annual RO for Narmada and Tapti are 21% and 38.6 % of the total annual precipitation respectively.

Annual climatic water balance budget analysis

The Annual climatic water balance budget for Narmada and Tapti delta is furnished below in Table 3.23.

Table 3.23 Annual water balance parameters in Narmada-Tapti delta region

Location	PET (mm)	Rainfall (mm)	Water Surplus (mm)	Water Deficit (mm)	Humidity Index (%)	Aridity Index (%)	Moisture Index (%)	RunOff (mm)
Narmada-Tapti Delta								
Baroda	1575	987	259.07	847.07	16.4	53.8	-37.3	263.00
Baruch	1575	883.9	124.45	815.55	7.9	51.8	-43.9	124.00
Rajpipla	1727	1004.6	222.45	944.95	12.9	54.7	-41.8	226.00
Surat	1635.9	1205	446.92	877.82	27.3	53.7	-26.3	465.00

CHAPTER 4

ASSESSMENT OF FRESH GROUNDWATER RESOURCES IN THE COASTAL DELTA SYSTEMS

Resource estimation/evaluation studies form an integral part of the groundwater systems and demand clear understanding of parameters like coefficient of transmissivity (T), coefficient of storage/storativity (S), leakage coefficient and aquifer thicknesses which are obtained normally through rigorous drilling procedures, subsurface aquifer mapping, hydraulic tests and graphical analysis. Accurate estimations of these parameters will help obtaining correct assessment of the resource potentials in the groundwater systems. Over the last few decades a number of workers have carried out extensive studies on these aspects. Theim, G., 1906; Theis, C.V., 1935; Jacob, C.E., 1940; Chow, V.T., 1952; Cooper, H.H. and Jacob, C.E., 1946; De Glee, G.J., 1930; Hantush, M.S., 1956, 1960, 1962, 1964; Walton, W.C., 1962, 1970; Lohman S.W. 1972; Boulton, N.S., 1954, 1963; Wenzel, L.K., 1942; Kruseman, G.P., et al 1970 and Ferris, J.G., et al., 1962; have contributed immensely towards development of appropriate methodologies to understand and estimate the hydraulic parameters as correctly as possible incorporating appropriate boundary conditions. A brief account of aquifer properties, flow equations and selected methodologies pertinent to coastal aquifer systems are dealt below in the following paragraphs.

4.1 GROUNDWATER OCCURENCES AND HYDRAULIC PROPERTIES

4.1.1 Groundwater occurrence and movement

The occurrence and movement of subsurface water depends upon several factors like lithology, texture, structure, gradient, permeability and storage capacity etc. The sub surface geological formations which can hold and transmit water are classified as *aquifers*. These aquifers could occur in different forms like (a) confined aquifers, (b) semi confined aquifers/leaky confined aquifers, (c) unconfined aquifers and (d) semi unconfined aquifers.

Confined aquifers: Confined aquifer is a completely saturated system bounded by impervious layers on the upper and lower sides. The hydrostatic pressure existing in these aquifers is normally higher than that of the atmospheric pressure and the water level in the wells stands above the saturation level in the aquifers.

Semi confined aquifers: A semi confined aquifer is also totally a saturated system with semi pervious layers on the top and impervious or semi pervious layers below. These aquifers when subjected to hydraulic stresses will have hydraulic linkages from the top semi pervious layers.

Unconfined aquifers: These are permeable near surface aquifers immediately below the ground soil surfaces and are partially filled with water with free water surface under atmospheric pressure. The water levels in this type of aquifers exist by and large at the same saturation level. These aquifers are also called phreatic aquifers.

Semi unconfined aquifers: These are the aquifers which have characters intermediate to semi confined and unconfined aquifers. The characterization of above aquifer systems are presented in Figure 4.1.

These aquifers when subjected to hydraulic stresses (pumping tests) would give rise to different patterns of time-drawdown curves reflecting the inherent hydraulic properties of the systems. The trend diagrams plotted on a double logarithmic graph for different aquifer types are shown in Figure 4.2.

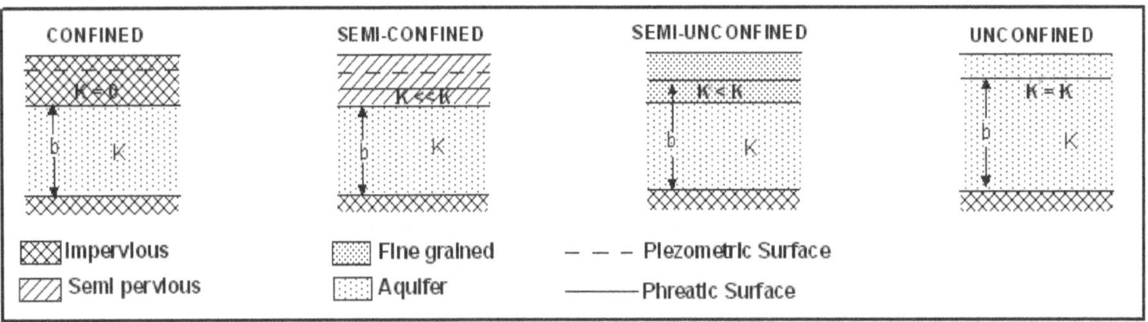

Fig. 4.1 Characteristics of different aquifer systems

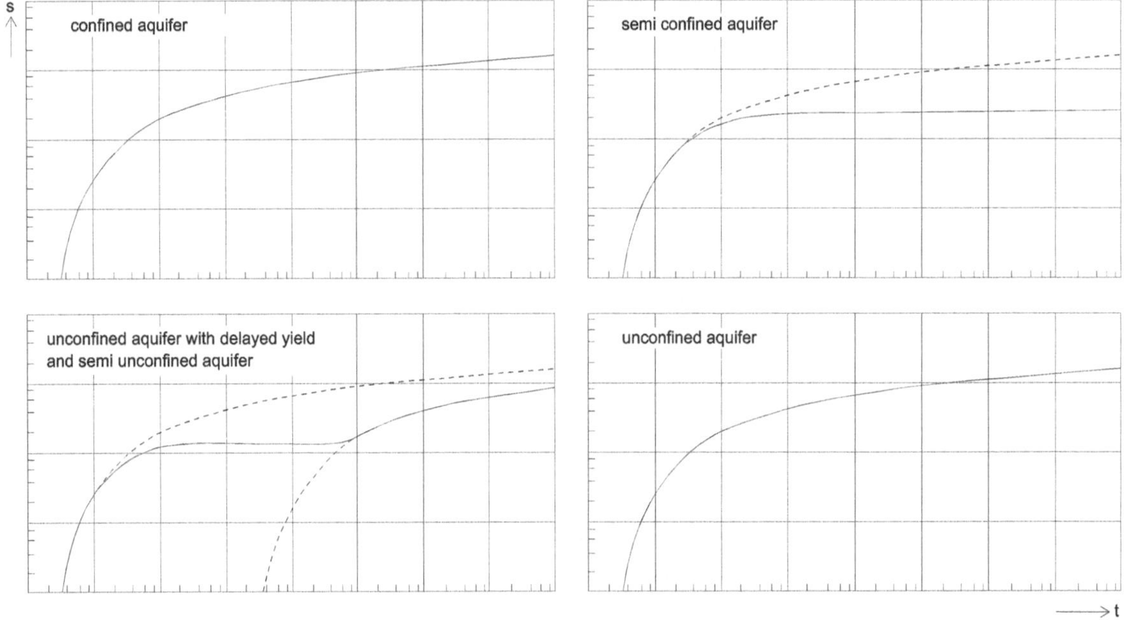

Fig. 4.2 Typical time-draw down curves for different types of aquifers

(Source: Kruseman, G.P. & De Ridder, N.A., 1970)

4.1.2 Hydraulic properties

An understanding of the hydraulic properties of different types of aquifers is most basic for the estimation of groundwater resources and is therefore briefly described below.

Hydraulic conductivity (K): The hydraulic conductivity or permeability of an aquifer system is defined as the amount of water that flows over a unit cross sectional area under unit hydraulic gradient. It has the dimensions of velocity (LT^{-1}) and generally expressed in 'm/day'. The hydraulic conductivity depends upon both fluid and aquifer properties.

Coefficient of transmissivity: The coefficient of transmissivity is defined as the rate of flow of water at the prevailing field temperature under a unit hydraulic gradient through a vertical strip of aquifer of unit width and extending through the entire saturated thickness of the aquifer. It is a product of permeability and the thickness of the aquifer normally expressed as: T = Kb where 'b' is the thickness of the aquifer and 'K' is the permeability or hydraulic conductivity of the aquifer. Transmissivity has the dimensions of L^2T^{-1} and is generally expressed in 'm^2/day'.

Coefficient of storage or storativity and specific yield: The storativity or coefficient of storage of an aquifer is defined as the volume of water released from a vertical column of aquifer of a unit cross sectional area under a unit hydraulic gradient. The first term is attributed to the compressibility of the aquifer skeleton and the second term is related to the expansibility of water. The storage coefficient 'S' is given by

$$S = b\,S_s \qquad \text{.....(4.1)}$$

where 'b' is thickness of the aquifer and 'S_s' is the specific storage.

This is a dimensionless parameter and the value for confined aquifers ranges between 10^{-4} and 10^{-6}. In unconfined aquifers, the storage coefficient becomes *'specific yield'* without considering either the compressibility of the aquifers or to its fluid. The specific yield is defined as the ratio of volume of water that an aquifer yields by gravity to its own volume. In other words it is the effective porosity of the aquifer. The values normally range between 10^{-1} and 10^{-3}.

Hydraulic diffusivity: The hydraulic diffusivity is defined as the ratio of transmissivity and storativity for a given aquifer and is given by:

$$D = \frac{T}{S} = \frac{K}{S_s} \qquad \text{.....(4.2)}$$

where D is the diffusivity expressed in m^2/day and has the dimensions of L^2/T^{-1}.

Hydraulic resistance: This is called reciprocal leakage coefficient or the resistance against vertical flow and is a property of semi confined aquifers/leaky aquifers. It is the ratio of the vertical saturated thickness of semi pervious layer 'b' to its vertical hydraulic conductivity 'k'. It characterizes the vertical movement of water through leakage. It has the dimension of 'Time'.

Leakage factor: The leakage factor $L = \sqrt{T.C} = \sqrt{T/_{k'/b'}}$. It determines the distribution of leakage into the semi confined aquifer. The factor 'L' has the dimension of length and is generally expressed in meters. High values of 'L' indicate a great resistance of the semi pervious strata to flow as compared with the resistance of the aquifer. The factor 'L' has the dimension of length and is expressed in meters.

Drainage factor: The drainage factor $B = \sqrt{KD/\alpha S_y}$ which is normally noticed in unconfined aquifers. Large values of 'B' indicate fast drainage. The drainage factor has the dimension of length and is expressed in meters. If $B = \infty$, the yield is instantaneous with the lowering of the water table and then the aquifer is unconfined without delayed yield. The coefficient $1/\infty$ is called 'Boulton delay index' which is an empirical constant while 'S_y' is the specific yield after a long time of pumping.

4.1.3 Darcy's law/groundwater flow and velocity

The fundamental and basic equation governing the groundwater flow is *'Darcy's equation'*. It states that the flow rate through porous medium is proportional to the head loss and inversely proportional to the length of the flow path and is expressed as follows

$$Q = -KA\frac{h_1 - h_2}{L} \text{ or } Q = -KA\frac{dh}{dL} \text{ or } V = \frac{Q}{A} = -K\frac{dh}{dL} \quad .....(4.3)$$

where

Q = flow rate

k = constant depending upon proportion of the porosity and permeability of the medium

I = hydraulic gradient $\frac{h_1 - h_2}{L}$

A = area perpendicular to the flow direction

h_1 and h_2 = water levels/heads at two points in the direction of flow

L = length of the flow path between two points h_1 and h_2

V = Darcy velocity or specific discharge

It is also important to understand the limitations of its application and its validity. The Darcy's law could be applied when the flow is laminar. Renolds number which is a dimensionless ratio of inertia to viscous forces serves as a criteria to recognize the laminar and turbulent flows which is expressed as:

$$N_R = \frac{\rho VD}{\mu} \quad .....(4.4)$$

where 'ρ' is fluid density, 'V' is velocity, 'D' is diameter of the pipe and 'μ' is viscosity of the fluid

Several experiments have shown that Darcy's law is valid when $N_R \leq 1$ when the flow is laminar and become turbulent when $N_R \geq 1000$.

4.2 DIFFERENTIAL EQUATIONS GOVERNING THE GROUNDWATER FLOW

An approximate partial differential equation controlling the groundwater flow in an unsteady state for a homogeneous and isotropic aquifer is given by

$$\frac{d^2h}{dx^2} + \frac{d^2h}{dy^2} + \frac{d^2h}{dz^2} = \frac{S_S}{K} \cdot \frac{dh}{dt} \quad .....(4.5)$$

where 'S_s' is specific storage, 'K' is permeability, 'h' is hydraulic head and 't' is time.

The above equation is the same as the fundamental equation in applied physics and is known as *'diffusion equation'*. The same form of equation appear in the theory of unsteady state flow of heat and electricity from which solutions to many groundwater flow problems have been obtained.

The same equation could be rewritten in the following format.

$$\frac{d^2h}{dx^2}+\frac{d^2h}{dy^2}+\frac{d^2h}{dz^2}=\frac{S}{T}\cdot\frac{dh}{dt} \qquad .....(4.6)$$

where 'S' is the storage coefficient and 'T' is the transmissivity

The above equation under steady state flow conditions becomes:

$$\frac{d^2h}{dx^2}+\frac{d^2h}{dy^2}+\frac{d^2h}{dz^2}=0 \qquad .....(4.7)$$

This is also known as basic *'laplace equation'* which governs the steady state groundwater flow in a homogeneous and isotropic aquifer. The above equation under non steady state radial flow situation in polar coordinates is written as:

$$\frac{d^2h}{dr^2}+\frac{1}{r}\cdot\frac{dh}{dr}=\frac{S}{T}\cdot\frac{dh}{dt} \qquad .....(4.8)$$

where 'r' is the distance from the well to the point of observation.

4.2.1 Steady state radial flow in a confined aquifer

A steady radial flow to a well for an aquifer fully penetrating a confined aquifer at a distance 'r' from the axis of the well as shown in Fig. 4.3 can be expressed in the following way

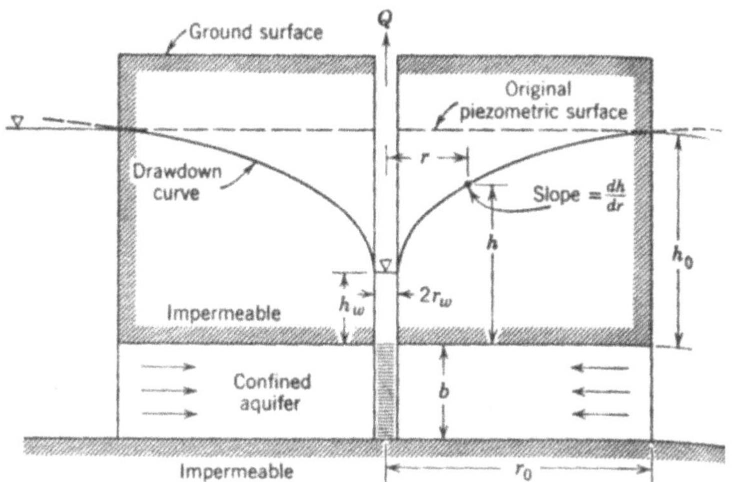

Fig. 4.3 Steady radial flow to a well fully penetrating a confined aquifer

$$Q = AV = -2\pi rKb.\frac{dh}{dr} \qquad .....(4.9)$$

where 'Q' is rate of flow; Kb = transmissivity

Rearranging and integrating for the boundary conditions at the well, $h = h_w$ and $r = r_w$ at the edge of equipotential boundary, $h = h_0$ and $r = r_0$, yields

$$h_0 - h_w = \frac{Q}{2\pi Kb}\ln\frac{r_0}{r_w} \qquad .....(4.10)$$

or
$$Q = \frac{2\pi Kb(h_0 - h_w)}{\ln(r_0 / r_w)} \quad .....(4.11)$$

This is known as the equilibrium or Thiem's equation. The drawdown 's' at a radial distance 'r' from the well is given as

$$s = h_0 - h_w = \frac{Q}{2\pi Kb} \ln \frac{r_0}{r_w} \quad .....(4.12)$$

4.2.2 Steady state flow in an unconfined aquifer

An equation for steady radial flow to a well in an unconfined aquifer can also be derived with the help of Dupuit's (1863) assumption as shown below in Fig. 4.4, where the well completely penetrates an unconfined aquifer and the well discharge 'Q' is expressed as follows:

$$Q = -2\pi rKh.\frac{dh}{dr} \quad .....(4.13)$$

where 'h' denotes the water table above the base of the unconfined aquifer boundary at any distance 'r' from the well.

The above equation when integrated between the limits $h = h_w$ at $r = r_w$ and $h = h_o$ at $r = r_o$ yields

$$Q = \pi K \frac{h_0^2 - h_w^2}{\ln(r_0 / r_w)} \quad .....(4.14)$$

or
$$K = \frac{Q}{\pi(h_0^2 - h_w^2)} \ln(r_0 / r_w) \quad .....(4.15)$$

The above equation fails to account for drawdown curve near the well because of large vertical flow components. However for small draw downs and large aquifer thicknesses, the estimates of 'K' may be reasonably good.

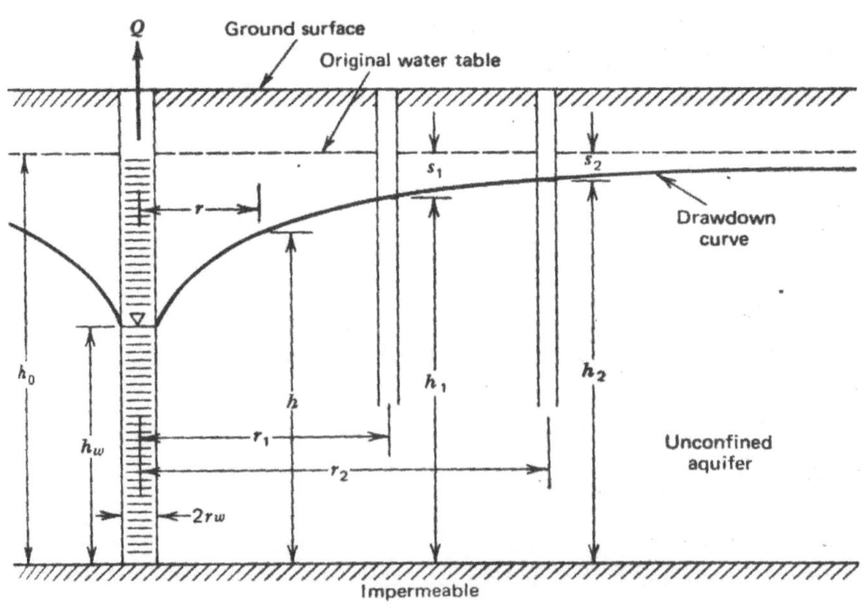

Fig. 4.4 Radial flow to a well penetrating an unconfined aquifer

4.2.3 Non-steady state flow in confined aquifers

A well penetrating an extensive confined aquifer, when pumped is governed by differential equation in plain polar coordinates as

$$\frac{d^2h}{dr^2}+\frac{1}{r}.\frac{dh}{dr}=\frac{S}{T}.\frac{dh}{dt} \quad .....(4.16)$$

where 'h' is the hydraulic head, 'S' is the storage coefficient, 'T' is the transmissivity, 't' is the time and 'r' is the distance from the well.

Theis, C.V. (1935) obtained a solution for the above equation based on the analogy between groundwater flow and heat conduction and is expressed as

$$s=\frac{Q}{4\pi T}\int_{u}^{\infty}\frac{e^{-u}}{u}du \quad .....(4.17)$$

and

$$u=\frac{r^2S}{4Tt} \quad .....(4.18)$$

where 's' is the draw down in observation well at a distance r from pumped well, T is the Transmissivity of the aquifer, Q is the constant discharge, S is the storage coefficient, 't' is the time.

The above equation traditionally called 'Theis non equilibrium equation'. The exponential integral of equation 4.17 is given as infinite series and is called well function W(u) which is represented as

$$W(u)=\left[-0.5772-\ln u+u+u-\frac{u^2}{2.2!}+\frac{u^3}{3.3!}-\frac{u^4}{4.4!}+......\right] \quad .....(4.19)$$

The values of W(u) for values of u from 10^{-15} to 1 are estimated by Wenzel (1942) and could be refered in any standard books (Todd, D.K., 1980; Walton, W.C., 1970; Kruseman, G.P. and De Ridder, N.A., 1970). The ultimate equation of the non equilibrium equation of this could be written as

$$s=\left(\frac{Q}{4\pi T}\right)W(u) \quad .....(4.20)$$

or could be rewritten as

$$\frac{r^2}{t}=\left(\frac{4T}{S}\right)u \quad .....(4.21)$$

The above equation implies that the relation between W(u) verses 'u' is similar to S and r^2/t values obtained in a well test programme and therefore helps to obtain 'S' and 'T' through a graphic method of superimposition.

A plot of W(u) vs u known as type curve could be plotted on a logarithmic paper and similarly the values of drawdown 'S' could be plotted against r^2/t on another logarithmic paper and superimpose both the graphical trends, keeping the axes parallel and obtain a matching curve point as shown in Fig. 4.5. The values W(u), u, S and r^2/t, thus obtained will help determining the 'T' and 'S' parameters for the aquifer system under testing, by adopting the above equations 4.20 and 4.21.

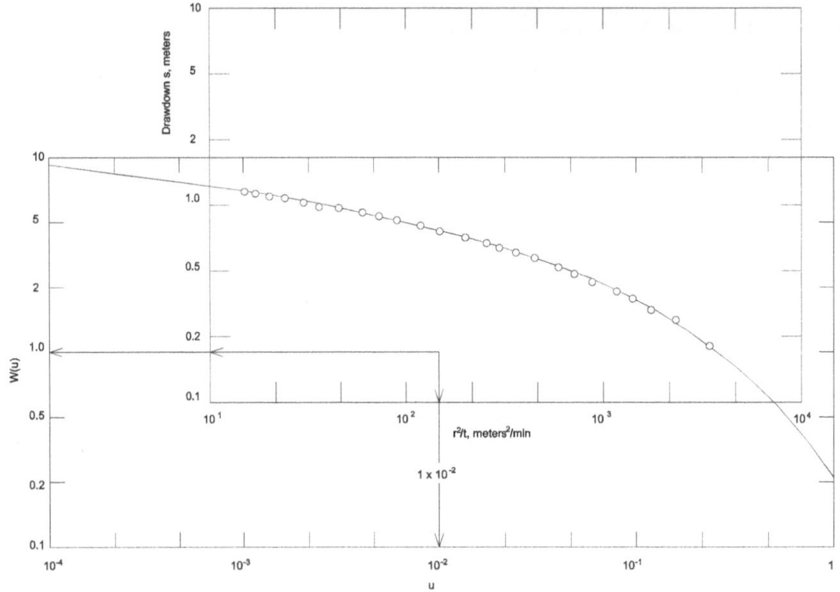

Fig. 4.5 Method of superposition for solution of the non-equlibrium equation

Field example (test site at Kansa well field): A pumping test was organized at Kansa well field (Orissa) for a period of 48.5 hours with a constant discharge of 43.14 m^3/h. The static water levels in test well and observation well are respectively 1.460m below ground level and 1.215m below ground level. Systematic measurements of drawdown were made in both test well and observation well during the pumping and recovery period. The maximum drawdown values for test and observation wells are 19.695 m and 5.775 m respectively. The time-drawdown data obtained at observation well is superimposed on Theis type curve plot of W(u) vs 1/u and is shown in Figure 4.6.

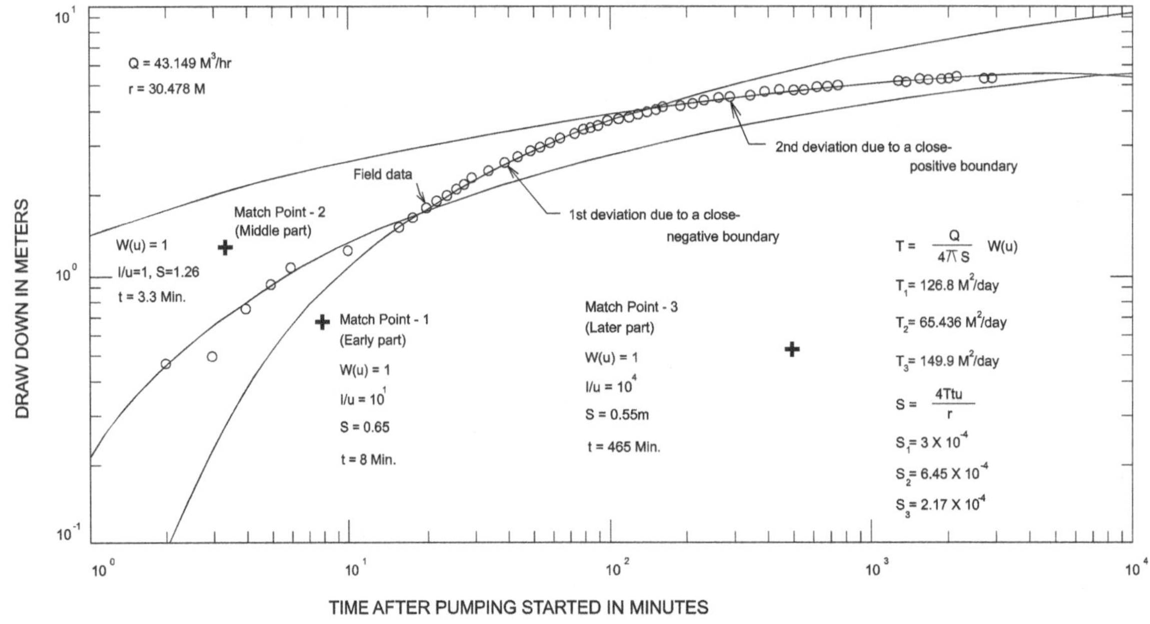

Fig. 4.6 Time draw down test analysis (Theis method)

(Source: Radhakrishna, I. & Venkateswarlu, T., 1980)

The graph shows three different segments in the time-drawdown curve with three different match points. The transmissivity and storage coefficient values are determined for the three different match points using the following equations.

$$T = \frac{Q}{4\pi s} W(u) \qquad .....(4.22)$$

$$S = \frac{4Tu}{r^2/t} \qquad .....(4.23)$$

Matchpoint 1, Early part of the curve: $T = 126.8\ m^2/day$

0-16 min: $S = 3 \times 10^{-4}$

Matchpoint 2, Middle part of the curve: $T = 65.436\ m^2/day$

16-150 min: $S = 6.45 \times 10^{-4}$

Matchpoint 3, Later part of the curve: $T = 149.9\ m^2/day$

150 min: $S = 2.17 \times 10^{-4}$

Cooper – Jacob method: Cooper, H.H. and Jacob, C.E. (1946) have shown that for small values of 'r' and large values of 't' the 'u' value becomes small and the series in terms of equation 4.18 becomes negligible. The drawdown can therefore be expressed as

$$s = \frac{Q}{4\pi T}\left[-0.5772 - Ln\frac{r^2S}{4Tt}\right] \qquad .....(4.24)$$

This gets reduced to

$$s = \frac{2.30Q}{4\pi T}\log\frac{2.25Tt}{r^2S} \qquad .....(4.25)$$

This implies that a plot of 's' vs logarithem of 't' forms a straight line. Projecting the line to $s = 0$, where $t = t_0$, the equation becomes

$$0 = \frac{2.30Q}{4\pi T}\log\frac{2.25Tt_0}{r^2S} \qquad .....(4.26)$$

The above equation follows that $\frac{2.25Tt_0}{r^2S} = 1$

This amounts to say that

$$S = \frac{2.25Tt_0}{r^2} \qquad .....(4.27)$$

The value of 'T' can be obtained by noting that its $t/t_0 = 10$, then $\log t/t_0 = 1$ and therefore by replacing s by Δ s, where Δ s is the drawdown over a log cycle, the equation 4.22 becomes

$$T = \frac{2.30Q}{4\pi\Delta s} \qquad .....(4.28)$$

By taking this procedure one can determine the 'T' and 'S' under the conditions mentioned above.

Field example (test site at Kansa well field): A time drawdown plot of the field data of Kansa well with the same discharge and drawdown values as shown above for Kansa well is adopted and the analysis are reflected in Figure 4.7. The method is applicable for small values of 'r' and large values of 't'. The plot shows two major slopes before the system attains a steady state condition.

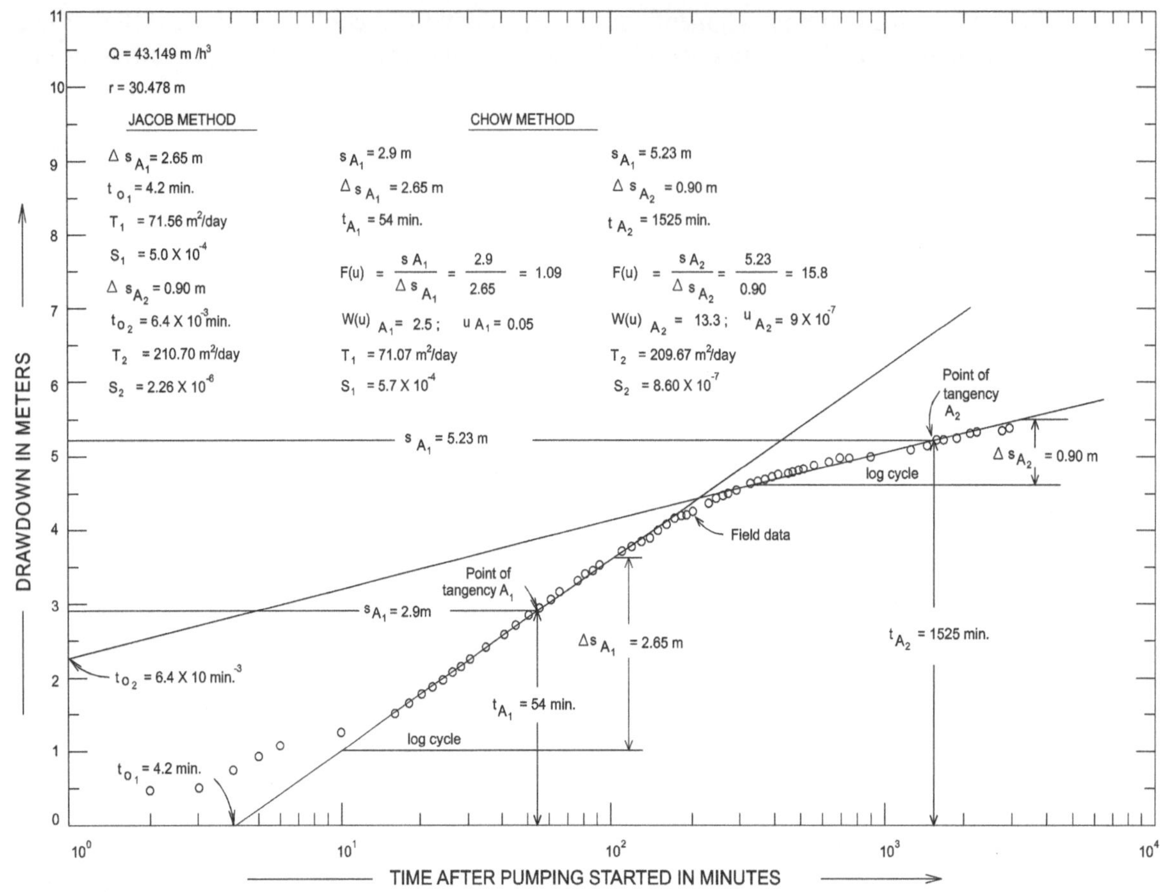

Fig. 4.7 Time draw down test analysis (Jacob and Chow method)

(Source:Radhakrishna,I. & Venkateswarlu, T., 1980)

The T and S values for these two different slopes are given below

1st slope: T = 71.56 m^2/day

16-150 min: S = 5.0×10^{-4}

2nd slope: T = 210.70 m^2/day

150 min and beyond: S = 2.26×10^{-6}

Chow method: Chow, V.T. (1952) developed a method which has the advantage of avoiding this type curve and also restrictions imposed on Jacob's method. Chow introduced a function to find out the values of W(u) and 'u' corresponding to the drawdown 's' values measured at time 't'. Chow has represented the values of F(u) for different values of W(u) and 'u' as shown in Fig. 4.8.

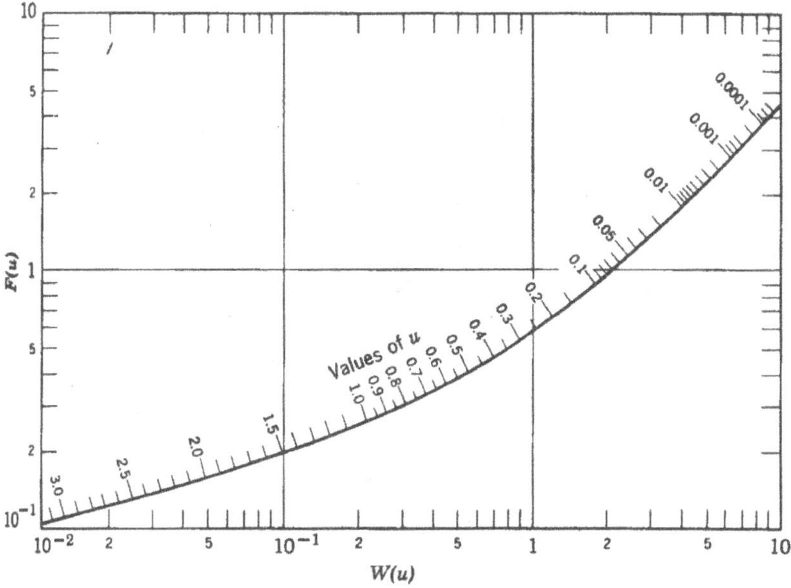

Fig. 4.8 Relation among F(u), W(u), and u

The method involves measurement of draw down values in an observation well in the vicinity of pumping well and plotting the data in the same manner as in the case of Cooper and Jacob method on a semi logarithmic paper and selecting an arbitrary point on the time-draw down curve and note the corresponding draw down (s) and time (t) component values. The values of F(u) for the arbitrary point is given by

$$F(u) = \frac{s}{\Delta s} \quad \text{.....(4.29)}$$

where Δs is the slope of the tangent at the arbitrary point. The corresponding values of W(u) and u could be derived from the nomograph given above and evaluate the 'T' and 'S' values.

Field example (test site at Kansa well field): A method of analysis is made similarly by Kansa well field by adopting Chow, V.T. (1952) procedure which is not restricted to small values of 'r' and large values of 't'. The T and S values are obtained through this method are given below

1st slope: $T = 71.07\ m^2/day$

16-150 min: $S = 5.7 \times 10^{-4}$

2nd slope: $T = 209.67\ m^2/day$

150 min and beyond: $S = 8.6 \times 10^{-7}$

Theis recovery method: Theis has shown that when a well pumped for a known period of time and when it was shutdown, the recovery drawdown thereafter will be identically the same or a mirror image. The residual draw down s' thus obtained is given as:

$$s' = \frac{Q}{4\pi T}\left[W(u) - W(u')\right] \quad \text{.....(4.30)}$$

where $$u = \frac{r^2S}{4Tt} \text{ and } u' = \frac{r^2S}{4Tt'}$$ .....(4.31)

A schematic diagram showing the water level changes with time during and after pumping is shown in Fig. 4.9 where t and t' are defined. For small values of r and large values of t', the well function could be approximated by the first two terms of the equation 4.19, so that the equation 4.30 could be written as

$$s' = \frac{2.30Q}{4\pi T} \log t/t'$$ .....(4.32)

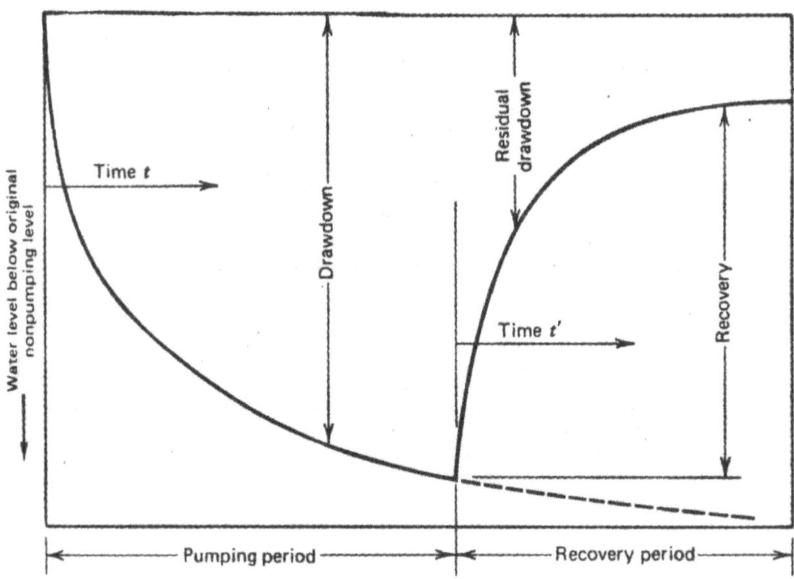

Fig. 4.9 Draw down and recovery curves in an observation well near a pumping well

Which means a graphical plot of residual draw down s' vs. log of t/t' results in a straight line, so that the residual draw down over a log cycle of t/t', facilitates derivation of T values by the following equation

$$T = \frac{2.30Q}{4\pi\Delta s'}$$ .....(4.33)

Note: The storativity values (S) can't be determined by this method of analysis.

Field example (test site at Kansa well field): The values obtained through the Theis recovery plots for the Kansa well field site are given in Figure 4.10. The T values of both main well and observation well are represented by a value of 65.39 m^2/day.

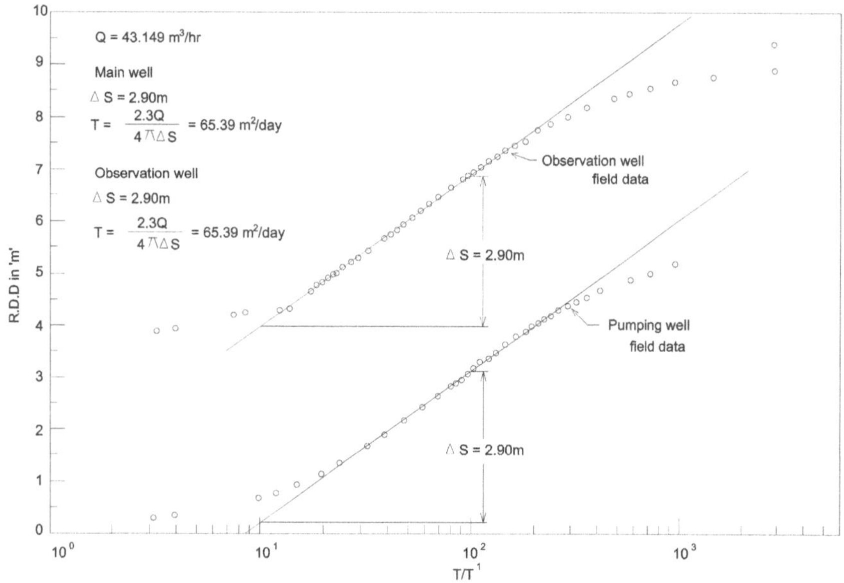

Fig. 4.10 Residual drawdown vs t/t'

(Source:Radhakrishna, I. & Venkateswarlu, T., 1980)

4.2.4 Non-steady state radial flow in a semi confined aquifer

A semi confined aquifer with an overlying leaky aquifer (aquitard) when pumped is shown in Fig. 4.11. The discharge of water is supplied from storage within the aquifer and also through leakage through the aquitard. Lowering of piezometric head in the aquifer through pumping creates a hydraulic gradient within the aquitard and then consequently groundwater migrates vertically downward in the aquifer. The quantity of water moving downward is proportional to the difference between the water table and the pieziometric head. Steady state flow is possible and the equilibrium will be established when the discharge rate equals the recharge rate of vertical flow into the aquifer. In such situation if k' is the coefficient of permeability of the aquitard, b' is the thickness of the aquitard, $L^2 = \left(T/_{k'.b'}\right)$ where L is termed as leakage factor.

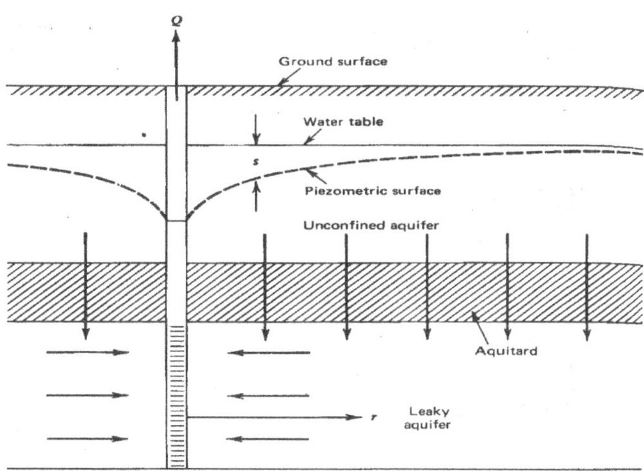

Fig. 4.11 Well pumping from a leaky aquifer

The flow equation for such a system could be written as:

$$\frac{d^2s}{dr^2} + \frac{1}{r}\frac{ds}{dr} - \frac{s}{L^2} = \frac{S}{T}\frac{ds}{dt} \qquad \text{.....(4.34)}$$

where L is the leakage factor; s is the drawdown; T is the transmissivity and 'r' is the distance of observation well from pumping well.

The drawdown solution to the above equation as presented by Hantush (1955) is described below:

$$s = \frac{Q}{4\pi T}\int_u^\infty \frac{1}{u}\exp\left(-u - \frac{r^2}{4L^2u}\right) du \qquad \text{.....(4.35)}$$

or

$$s = \frac{Q}{4\pi T} W\left(u, \frac{r}{L}\right) \qquad \text{.....(4.36)}$$

where

$$u = \frac{r^2S}{4Tt} \qquad \text{.....(4.37)}$$

The quantity r/L is given by

$$\frac{r}{L} = \frac{r}{\sqrt{T/(K'/b')}} \qquad \text{.....(4.38)}$$

where s is the drawdown, r is the distance from pumped well to observation well; Q is the discharge; t is the time since pumping started; T is the coefficient of transmissivity; S is the coefficient of storage; K' is the coefficient of vertical permeability of the aquitard; and b' is the thickness of the aquitard

The values of the function W(u, r/L) were given by Hantush, M.S. (1956), it is to be noted that the equation has the same form of Theis equation (4.33), when in a confined aquifer $K' \to 0$ so that $L \to \infty$ and $r/L \to 0$. Employing this analogy Walton prepared a set of family type curves for W(u, r/L) as shown in Fig. 4.12.

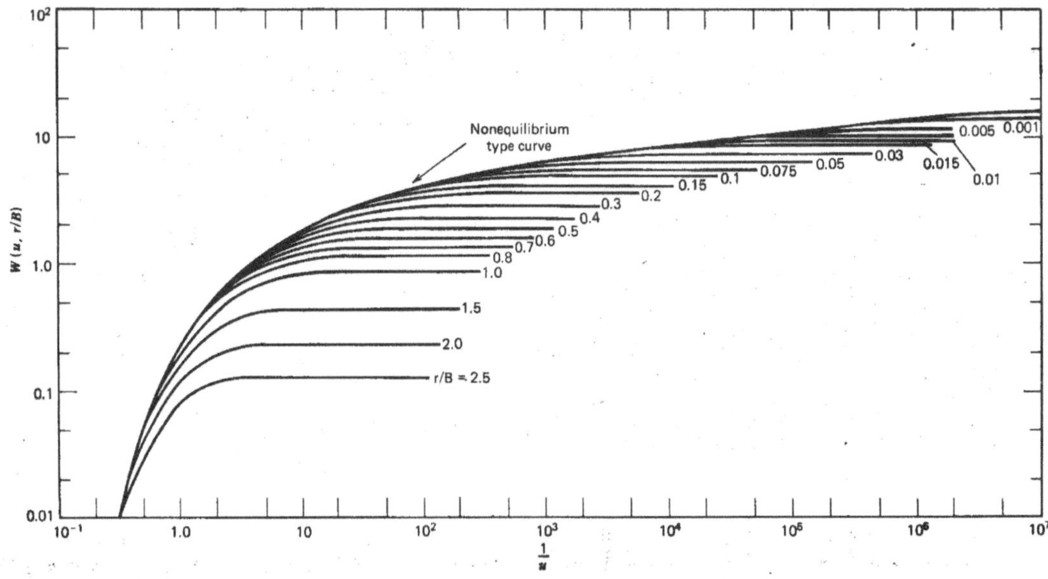

Fig. 4.12 Walton type curves for analysis of pumping test data to evaluate S and T parameters in leaky aquifers

Here the values of W(u, r/L) could be plotted against $\frac{1}{u}$ for various values of $\frac{r}{L}$ and prepare a set of type curves. Again in another logarithmic paper of the same scale plot the values of 's' vs 't' and superimpose both the sheets, keeping the axes parallel. A match point could be selected where most of the data points fall on one sector of the type curve. The values of W(u, r/L), $\frac{1}{u}$, 's' and 't' could be obtained. Then subsequently 'T' and 'S' could be evaluated by deploying the above equations and at the same time the value of k' could also be derived when b' value is known from the field data.

Field example (test site at Kansa well field): Walton, W.C., 1962 based on Hantush, M.S., 1956 well function values for leaky aquifers developed type curve graphical analysis for different values of 'r/L'. The time-draw down data for Kansa well is superimposed on the family of Walton's type curve as shown in Fig. below.

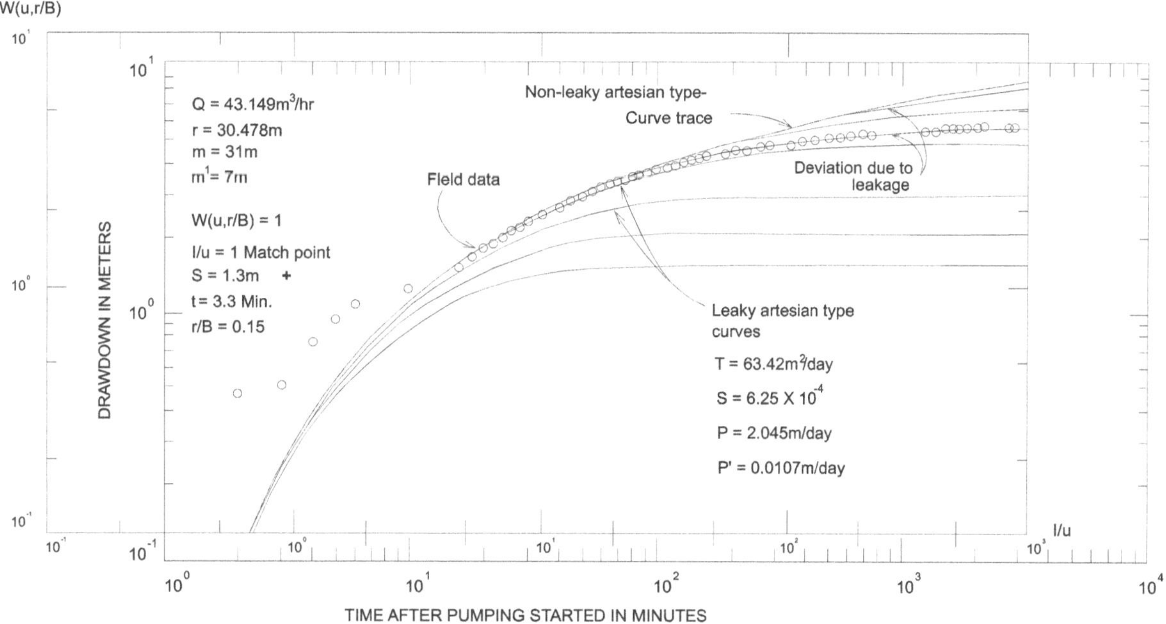

Fig. 4.13 Time draw down test analysis superimposed on Walton family type curves

(Source: Radhakrishna, I. & Venkateswarlu, T., 1980)

The equations adopted in this method given by Hantush, M.S. and Jacob, C.E. (1955) are as follows:

$$T = \frac{Q}{4\pi T} W\left(u, \frac{r}{L}\right) \qquad .....(4.39)$$

$$S = \frac{4Tt}{r^2} u \qquad .....(4.40)$$

The plotted points all along the type curve with r/L = 0.150, an appropriate match point is selected on the curve. The coordinates for W(u, r/L), 1/u and the corresponding values of 's' and 't' are recorded. The parameters T and S calculated using the equations 4.36, 4.37 are given below.

$T = 63.42\ m^2/day$

$S = 6.25 \times 10^{-4}$

Since the value of r = 30.478 meters and r/L = 0.15, it follows that the leakage factor L = 30.478/0.15 = 203.18 meters. Therefore, the hydraulic resistance is given by

$L^2/T = 650.932$ days

Thus the 'leakage coefficient' (p`/m`) of the system is 0.00153 days.

Field example2 (Hantush method): Hantush, M.S. (1956) developed a method using the inflection point of the time drawdown curve on a single logarithmic paper. However to determine the inflection point a steady state drawdown is required to be reached, since the well at Kansa shows by and large attainment of steady state condition and therefore the Hantush inflection point method is shown in Figure 4.14.

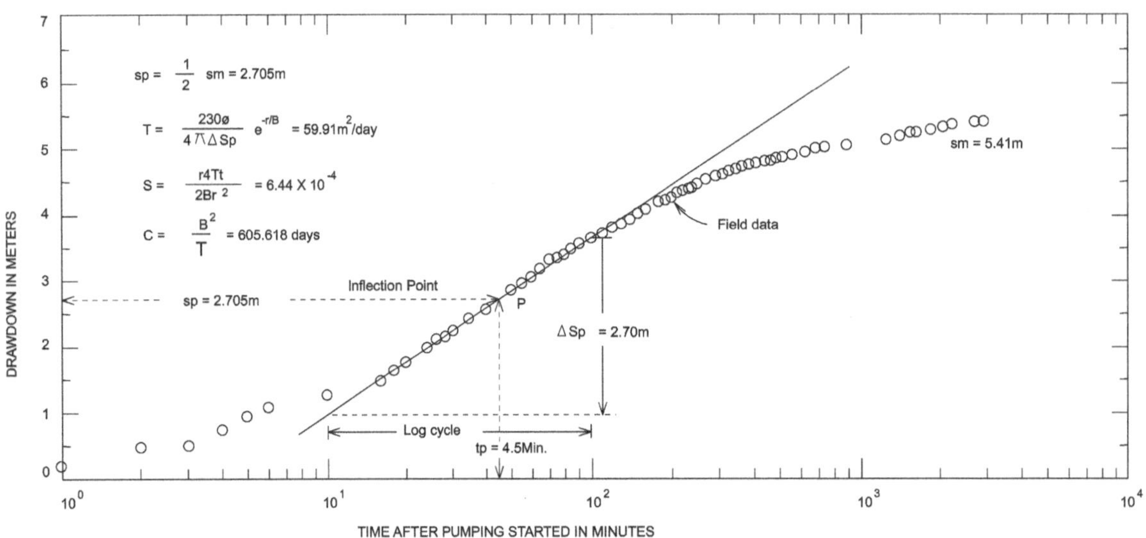

Fig. 4.14 Time draw down test analysis with inflection point (Hantush method)

(Source:Radhakrishna, I. & Venkateswarlu, T., 1980)

The approach basically involves location of inflection point on time drawdown graph and taking the corresponding time component and Δs_p values for substitution in the following formulae. A time draw down curve at inflection point P hold the following relations

$$s_p = 1/2\, s_m \qquad .....(4.41)$$

where s_p = drawdown at the inflection point and s_m= steady-state drawdown.

$$u_p = \frac{r^2\, s}{4\, T\, t_p} = \frac{r}{2\, L} \qquad .....(4.42)$$

and u_p = time at inflection point

$$\Delta s_p = \frac{2.30\, Q}{4\, \pi\, T}\, e^{-r/L} \qquad .....(4.43)$$

where Δs_p = drawdown difference per log cycle of time, and

$$2.3\frac{s_p}{\Delta s_p} = e^{r/L}\, K_o\, (r/L) \qquad .....(4.44)$$

The r/L value is obtained by comparing the functional values given by Hantush, M.S. (1956). In the present case, r/L value is 0.16, and since r = 30.478 meters, it follows that

$$L = \frac{r}{0.16} = 190.480 \text{ meters and } e^{-r/L} = e^{-0.16} = 0.852$$

Substituting these values in equation 5.40, 5.41 and 5.39, the values of T and S come out to be 59.91 m^2/day and 6.44×10^{-4}, respectively. The factor C = 605.618 days and the coefficient of leakage for the system = 0.00165 days.

4.2.5 Non-steady state radial flow in unconfined or semi-unconfined aquifer with delayed yield

In an unconfined aquifer, the water is derived from storage through (i) gravity drainage (ii) compaction of the aquifer and (iii) expansion of water. A pumping test organized in unconfined aquifers reveal that water released from storage instantaneously does not satisfy the non steady state equations, but instead shows delayed yield phenomenon. The flow equation for such a situation as presented by Boulton, N.S. (1954, 1963) is given below.

$$T\left[\frac{d^2s}{dr^2} + \frac{1}{r}\frac{ds}{dr}\right] = S\frac{ds}{dt} + D_i\ S_y\int_0^t \frac{ds}{dt}\exp\left[D_i(t-\tau)\right]d\tau(t=\tau) \qquad \text{.....(4.45)}$$

where Di is the reciprocal of delayed yield index. The integral term is the rate of delayed gravity drainage per unit area at time 't'.

The drawdown solution of the above equation is analogous to that for a semi confined condition and in general is represented as

$$s = \frac{Q}{4\pi T} W(u_{Ay}, {}^{r}\!/\!_{B}) \qquad \text{.....(4.46)}$$

During the initial time of pumping the aquifer responds in a similar way as that of a confined aquifer and the drawdown solution can be written as:

$$s = \frac{Q}{4\pi T} W(u_A, \frac{r}{B}) \text{ and } u_A = \frac{r^2 S_A}{4Tt} \qquad \text{.....(4.47)}$$

where S_A is effective early storage coefficient.

This equation represents the first segment of early part of time drawdown curve which is called 'A' type curve. Further the slope of the time drawn curve decreases because of the gravity drainage. Again after sometime (ranging from few minutes to few days) the draw down increases and a similar trend is achieved as seen in the early part of the time drawn curve which is called 'Y' type curve. The drawdown solution then becomes:

$$s = \frac{Q}{4\pi T} W(u_Y, {}^{r}\!/\!_{B}) \text{ and } u_Y = \frac{r^2 S_Y}{4Tt} \qquad \text{.....(4.48)}$$

where S_y is specific yield of the aquifer.

Similar to 'leakage factor' described earlier, Boulton introduced the term 'drainage factor' 'B' which is expressed as:

$$B = \sqrt{{}^{T}\!/\!_{\alpha S_y}} \qquad \text{.....(4.49)}$$

where α is the reciprocal of delay index (D_i)

Boulton, N.S. (1963) developed a series of values of the functions $W(u_A, r/B)$ and $W(u_Y, r/B)$ for different values of $1/u_A$ and $1/u_Y$, which could be converted into a graphical format by plotting $W(u_{AY}, r/B)$ vs $1/u_A$ and $1/u_Y$ as shown in Fig. 4.15.

The field data (time-drawdown data) could be plotted on a logarithmic paper on the same scale as that of the type curve $W(u_{AY}, r/B)$ vs $1/u_A$ and $1/u_Y$ for different values of r/B. A suitable match point be made between early parts of both the curves keeping the axes of both curves parallel. Substituting the values in the equation 4.38 the T and S_A values could be obtained. A similar match could be made to the later parts of the curve by applying the equation 4.39. Subsequently the delay index of the system could also be estimated by using the equation 4.40.

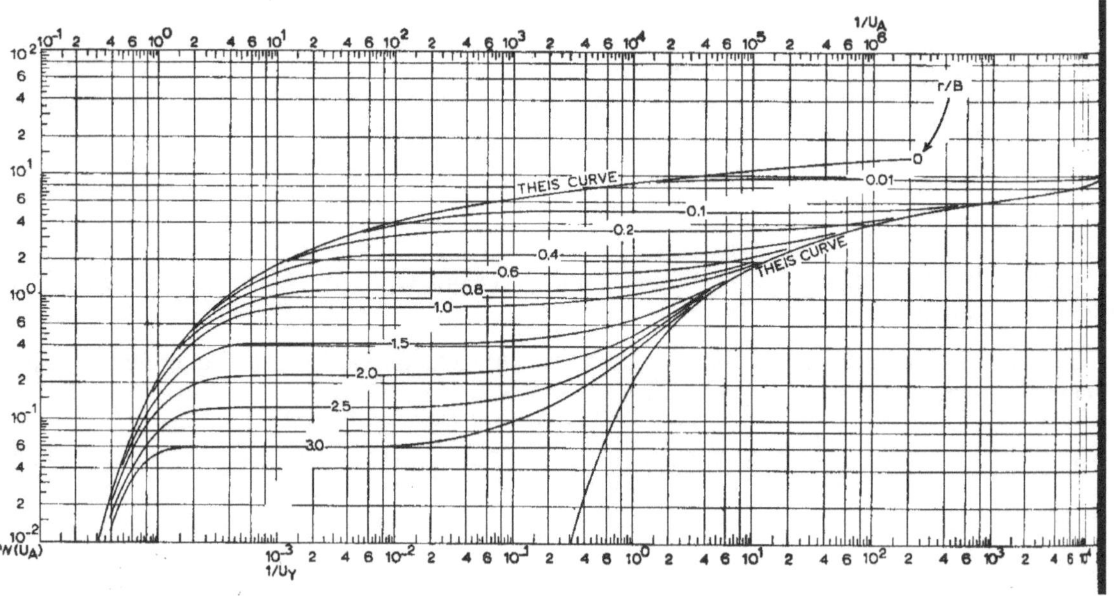

Fig. 4.15 Family of Boulton type curves: W(uA, r/B) vs 1/ uA and W(uY, r/B) Vs 1/ uY for different values of r/B

Field example: A pumping test was organized at Bhagiradhipet well site in Lower Maner basin (Andhra Pradesh) where a test well and an observation well are constructed located 12.7 m apart from each other. The test was conducted with a constant discharge of 9.0 m^3/hour and the maximum draw downs recorded at the main and observation wells are respectively 13.0 m and 0.275 m. The drawdown response observed when superimposed on time drawdown curve is found to be matching well with the Boulton type of curve as shown in Fig. 4.16.

The left hand portion of the time-drawdown curve was superimposed on the family of Boulton type curves and adjusted for a suitable match point. A suitable matching curve was taken with value of r/B at 1.5. A match point was chosen with the following coordinates.

$$1/uA = 0.1;\ W(uA, r/B) = 1;\ s = 0.42\ m \text{ and } t = 3.45\ min.$$

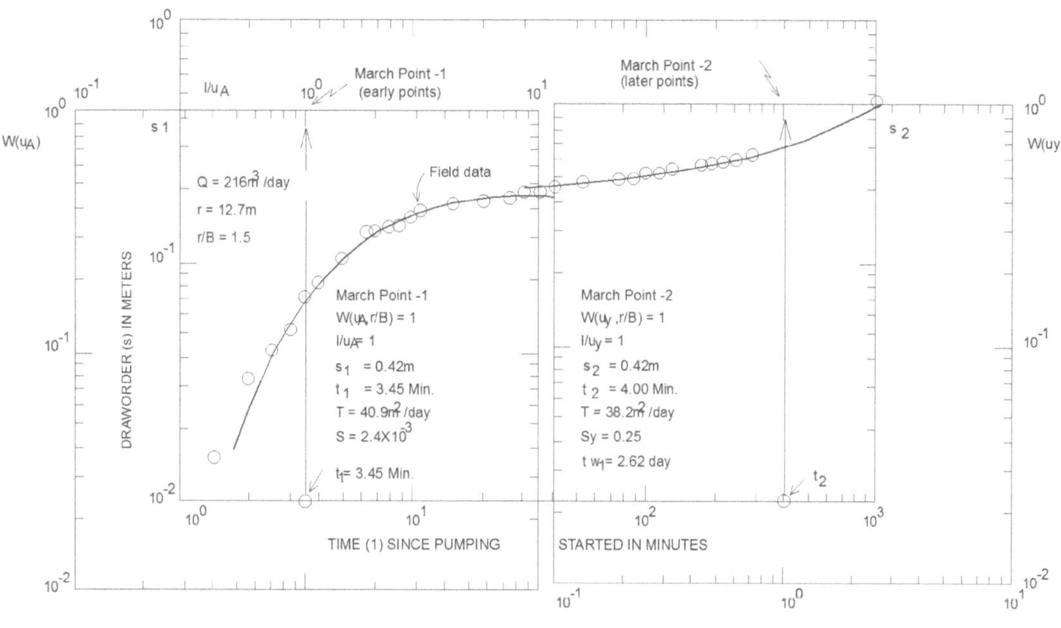

Fig. 4.16 Time draw down test data superimposed on Boulton family type curves

(Source: Radhakrishna, I, et al 1979)

Substituting these values in equation 4.38, we have T = 40.90 m^2/day and $S_A = 2.4 \times 10^{-3}$. Again the right hand portion of the curves was superimposed on the right hand portion of the type curve. A second match point was chosen. This point is characterized by the following coordinates:

$$1/u_A = 1;\ W(u_A, r/B) = 1;\ \ s = 0.45 \text{ m and } t = 400 \text{ min.}$$

Substituting these values in equation 4.39, we have T = 38.2 m^2/day and $S_A = 0.25$.

It follows that B = r/(r/B) = 1.50 and α = 72.71 and t_{Wt} = 2.62 days where t_{Wt} is the time at which delayed gravity drainage stops.

4.2.6 Partially penetrating wells

A partially penetrating well is one whose length of construction of well is less than the thickness of the aquifer. In such cases the flow equation of a fully penetrating confined aquifer equation has to be modified. The drawdown in a partially penetrating well will be more than the drawdown noticed for a fully penetrating well for same amount of discharge. In other words the discharge of a partially penetrating well will be less than that of a fully penetrating well for the same drawdown. This situation is shown in Fig. 4.9 where if $Q_p = Q$ then $s_p > s$; and if $s_p = s$ then $Q_p < Q$ where Q = well discharge; s = drawdown and prefers to the subscript of the partially penetrating well.

The drawdown of a partially penetrating well is expressed as:

$$s_p = s + \Delta s \qquad .....(4.50)$$

where Δs refers to the additional drawdown as a result of partial penetration under steady state condition. The above equation can be rewritten as

$$\Delta s = \frac{Q_p}{2\pi T} - \frac{1-p}{p} \ln \frac{(1-p)h_s}{r_w} \qquad .....(4.51)$$

where T = transmissivity; p = penetration fraction; $p = h_s/b$ and h_s and r_w are shown in Fig. 4.17(a). The above equation is valid when p > 0.20 then the above equation could be re-written for the well screens located in the central portion of the aquifer as in Fig. 4.17(b).

$$\Delta s = \frac{Q_p}{2\pi T}\frac{1-p}{p}\ln\frac{(1-p)h_s}{2r_w} \qquad .....(4.52)$$

The above equation could be modified for an unconfined aquifer as

$$\Delta s\, 2h_w = \frac{Q_p}{\pi K}\frac{1-p}{p}\ln\frac{(1-p)h_s}{r_w} \qquad .....(4.53)$$

where h_w is the saturated thickness at the well face with full penetration of the aquifer.

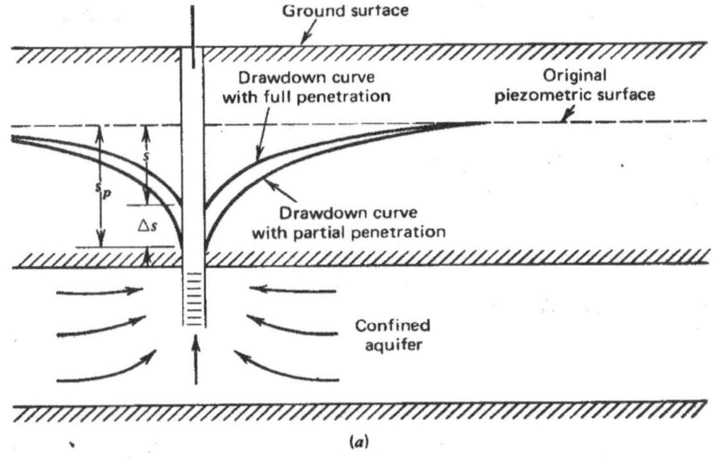

Fig. 4.17(a) Effect of partially penetrating well on drawdown

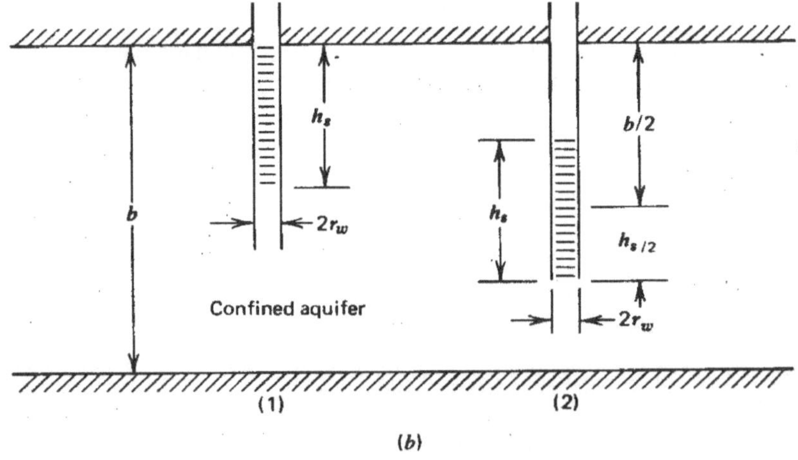

Fig. 4.17(b) Two configurations of partially penetrating wells

4.2.7 Large diameter wells

It so happens in many areas, that proper tube wells for testing purposes are not available, and several shallow and deep large diameter wells are available, and one may require to evaluate the aquifer parameters. In such cases, the most appropriate method one can adopt the method presented by

Papadopulos, I.S. and Cooper, H.H. (1967), which is detailed below. The drawdown equation for such a large diameter well is shown below in equation 4.54 and also diagrammatically shown in Fig. 4.18. The empirical equation has the following assumptions: (i) the aquifer has an infinite aereal extent, (ii) the aquifer is homogeneous, isotropic and of uniform thickness and confined, (iii) the flow to the well is non steady state and (iv) the well diameter is significantly large and storage has to be considered.

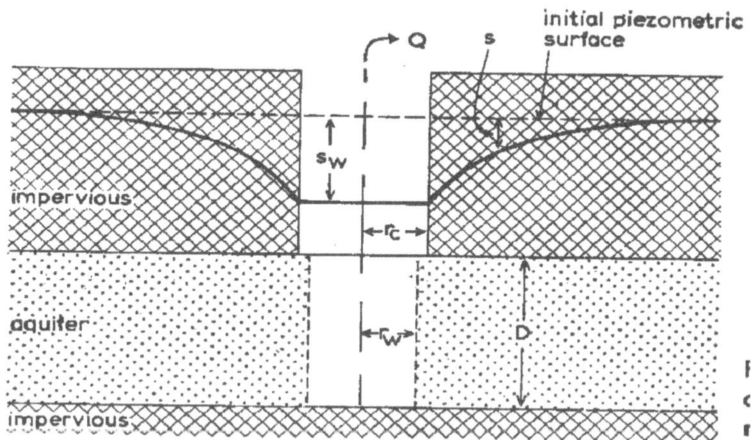

Fig. 4.18 Schematic cross-section of a confined aquifer pumped by a large diameter well

(Source: Kruseman, G.P., 1970)

$$s_w = \frac{Q}{4\pi T} F(u_w, \beta) \qquad .....(4.54)$$

where $F(u_w, \beta)$ is a function for which numerical values are given by Papadopulos and Cooper (1967)

$$u_w = \frac{r_w^2 S}{4Tt} \qquad .....(4.55)$$

and

$$\beta = \frac{r_w^2 S}{r_c^2} \qquad .....(4.56)$$

The index 'w' stand for 'at the pumped well' and r_c = radius of the unscreened part of the well.

Procedure:

(a) Develop a family type curves by way of plotting the values of $f(u_w, \beta)$ vs $1/u_w$ for different values of β (Fig. 4.19) as given by Papadopulos, I.S. and Cooper, H.H. (1967) in a double log paper.

(b) Plot the values of s_w vs t in another graph paper as obtained in the well test and superimpose the data curve on the type curve solution, and obtain a good match point of the observed data curve.

(c) Choose an arbitrary point 'A' on the superimposed sheet and note the values of $f(u_w, \beta)$, $1/u_w$, s_w and t and also note the value of β, the best fit of observed and type curve. Substitute the values obtained in the equation 4.45 together with the observed values of 'Q' at the testing site and estimate the values of 'T'. Subsequently by way of introducing the values of r_w, $1/u_w$, t and T in the equation 4.46 or by introducing the values of r_c, r_w and β in equation 4.47, the 'S' values could be estimated.

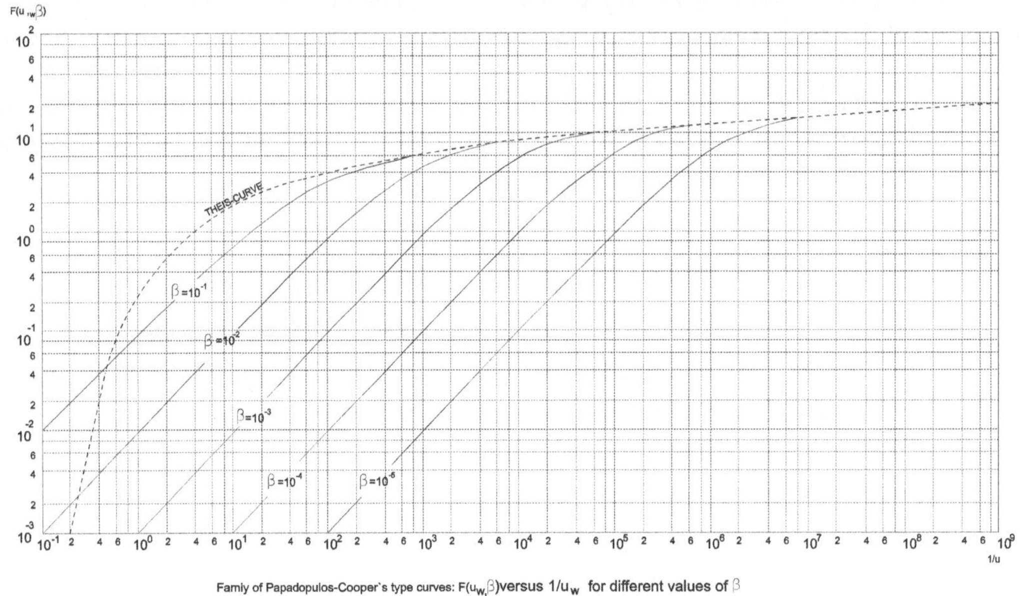

Fig. 4.19 Family of Papadopulos-Cooper's type curves.F/u, ß Vs 1/u, for different values of ß

Test data example: A large diameter well with a radius 1.75 meter was tested with a discharge of 1.173 m^3/min and obtained a drawdown of 0.515 m at the end of 360 min. A graphical plot of S vs T superimposed on a Papadopulos-Cooper type curves are shown in Figure 4.20 corresponding to β value of 10^{-2}. The point values for function $f(u_w, \beta)$, $1/u_w$, s and t are respectively 10, 10^3, 0.56 meters, and 5.4 min. The T and S values obtained using the equations of 4.45 and 4.46.

T = 2399 m^2/day; S = 0.01

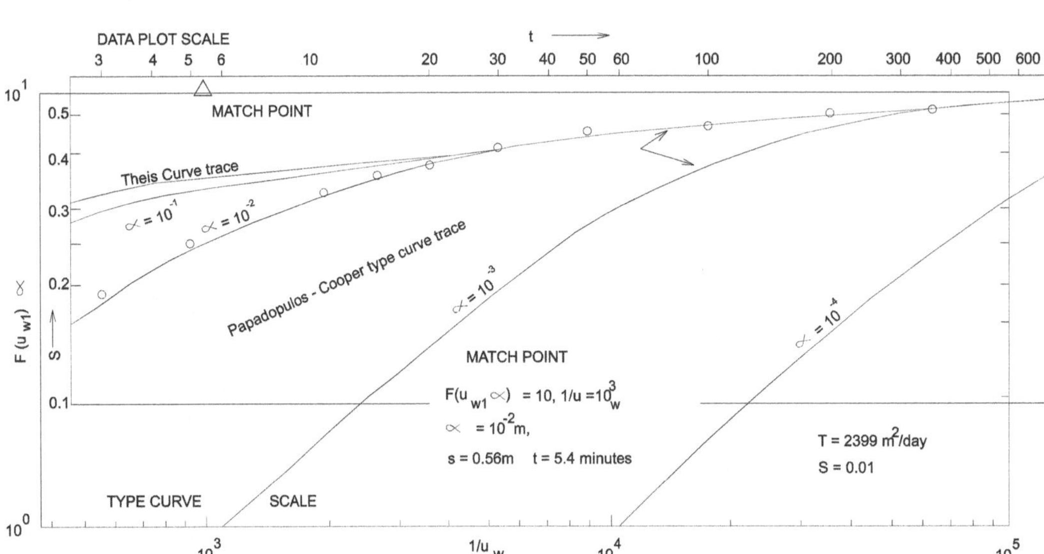

Fig. 4.20 Time-draw down test analysis in large diameter well superimposed with Papadopulos-Cooper type curves

(Source: Karanth, K.R., 1987)

4.3 METHODOLOGIES RELATED TO THE ESTIMATION OF RESOURCE POTENTIALS

Estimation/assessment of fresh groundwater resources in a coastal basin involves (i) accurate estimates of rainfall inputs, (ii) land use and cropping patterns including estimates of evaporation, evapotranspiration and infiltration data, (iii) salinity impacts both surface and deep aquifer systems, (iv) S and T parameters for near surface and deep aquifer systems, (v) canals and tank infiltration parameters and (vi) the regional draft components for different basins are mentioned in the chapter-1 earlier. A brief description on the application of these inputs leading to the assessment of resource potentials are detailed below.

Groundwater balance equation: A groundwater balance equation considering various inflow and outflow components could be written as follows:

$$R_i + R_c + R_r + R_t + S_i + I_g = E_t + T_p + S_e + O_g + \Delta S \qquad .....(4.57)$$

where R_i = Recharge from rainfall; R_c = recharge from canals; R_r = recharge from irrigation returns; R_t = recharge from tanks; S_i = influent seepage from rivers; I_g = inflow from adjacent basins; E_t = evapotranspiration; T_p = groundwater draft; S_e = effluent seepage to rivers; O_g = outflow to other basins; ΔS = change in groundwater storage.

Estimations of all the above parameters involve dedicated team and analyses involving different specialists groups to work out an integrated/holistic estimate of water balance in a basin. Some methods in this regard for assessment programmes in groundwater are described below.

4.3.1 Empirical methods

Chaturvedi, R.S. (1973) derived an empirical relationship to arrive at the recharge components for alluvial system of aquifers which is expressed below.

$$R = 2.0\,(P - 15)^{0.4} \qquad .....(4.58)$$

and subsequently modified by UP irrigation research institute, Roorkee as:

$$R = 1.35\,(P - 14)^{0.5} \qquad .....(4.59)$$

where R = net recharge due to precipitation in inches; P = annual precipitation in inches

The above formula has been widely used for initial assessment based on rainfall input to Groundwater systems of sedimentary terrains.

4.3.2 Soil water balance methods

The soil water balance approach has already been described in the earlier chapter III. The approach basically reflects on the vertical recharge that could probably take place after meeting all the soil water balance parameters and is expressed as:

$$I = P - AE - \Delta S - R_o \qquad .....(4.60)$$

where I = Infiltration; P = Precipitation, AE = actual evapotranspiration; ΔS = change in soil water storage and R_o = Runoff

In addition, several nuclear techniques are widely used such as natural Tritium technique and injection Tritium techniques (Athavale, R.N., et al, 1980). Tritium is the radio isotope of hydrogen with half life of 12.43 years, and could be treated as a tracer for soil moisture movement. The environmental Tritium method basically involves tracing of Tritium peaks in the soil profiles, caused as a result of nuclear tests in atmosphere. But these methods have less importance with time. In view of its diffusion over a period of time alternatively Tritium injection method is being used to trace the movement of

Tritium which is artificially injected in a particular soil of a region, and monitor its movement in the soil (Zimmerman, U., et al 1967) through the already available known and laboratory procedures (Rangarajan, R., et al 2000).

4.3.3 Water level fluctuations/groundwater storage method

A systematic monitoring of water levels in observation wells and rigorous practices of estimations of specific yields/storage coefficient values would help estimating the groundwater recharge in a specific terrain. Change in water levels between the beginning of monsoon and the end of monsoon in a year and similarly the change in water levels at the end of monsoon and the beginning of monsoon which help in estimating the recharge and draft components in a basin by using the following equation:

$$R = h \times S_y \times A + D_G \qquad .....(4.61)$$

where R = Recharge; h = rise in water level in the monsoon season; S_y = specific yield; A = area for computation of recharge and D_G = gross Groundwater draft.

The specific yield values should be normally computed through resource hydrological tests described earlier in these chapters.

4.3.4 Inflow and outflow estimation in a basin

The inflow and outflow in a given subsurface phreatic or a deep confined aquifer could be made by using the Darcy's law and by way of preparing water levels/piezometric contain maps as the case may be depending upon the type of aquifer of influent, and also by way of estimating the 'Transmissivity' of the aquifer system through rigorous hydrological pumping test analyses. The equation governing such flow is given by:

$$Q = T.I.L \qquad .....(4.62)$$

where Q = Quantity of Groundwater flow; T = transmissivity of the aquifer; L = Length of the medium against which the flow is considered; I = Hydraulic gradient

This facilitates estimations of groundwater flow quality from one region to another and in different regions.

4.3.5 Recharge from canal seepages

In most coastal deltaic regions and in major inland valley fills where canal net works densities are high, the canal seepage into the groundwater system is relatively very high, and play a very significant role in recharging the aquifers. These canal seepage losses could be empirically calculated using the following equation for lined channels.

$$S = 1.25 \times Q^{2.56} \qquad .....(4.63)$$

where S = seepage loss in cusecs per million square foot of wetted perimeter and Q = Discharge in cusecs in the channel.

In the case of unlined channels the seepage losses may go up by four to five times depending upon the texture of the soil bed.

Alternately another equation could also be used as given below for unlined canals.

$$S = \frac{C}{200}(B + D)^{2/3} \qquad .....(4.64)$$

where S = seepage losses in cumecs/km; B, D = bed widths and depths of the channel; C = Constant (Normally taken as 1.0 for intermittent running channels and 0.75 for constant running channels).

4.3.6 Nuclear methods

Nuclear techniques of late are being used widely for estimations of groundwater recharge and also to determine the ages groundwaters in different aquifers underground. The methods commonly deployed in the Indian context are Tritium tagging (injection) method, Environmental tritium method, Groundwater dating methods.

Tritium tagging (injection) method: This method is based on the assumption that the soil moisture moves downward discrete layers underground (Piston Flow Model) under the gravitational force (Zimmermann, U., et al., 1967; Munnich, K.O., 1968). Tritium is normally injected at a suitable point or a series of points below the root zone in a basin, preferably before the onsets of monsoon and subsequently after the monsoon, the core samples are obtained at different depth levels and the tritium concentrations are measured. A graphical representation of tritium concentrations at different depth levels will indicate movement and peak concentrations in a particular soil profile. The vertical recharge could be estimated by measuring the displacements of soil moisture contents and applying the following formula.

$$W = \frac{d_d}{(100+d_d)} p_w h \qquad \text{.....(4.65)}$$

where W= recharge or amount of water added per sq.cm of soil; d_d= dry weight per cent of soil moisture; p_w= wet bulk-density of soil, in situ; h= distance by which the tracer is displaced

(This method has been applied in several parts of the country: Athavale, R.N., et al, 1980; Datta, P.S., et al, 1980; Goel, P.S., et al, 1975; Rangarajan, R. and Athavale, R.N., 2000).

Environmental tritium method: The natural concentration of tritium in the atmosphere produced through interaction of cosmic rays with nitrogen nuclei is of the order of 5 to 10 tritium units (Lal, D. and Peters, B., 1962). These values are supposed to have been triggered after a series of bomb blasts after 1952 by several countries (U.S.A and U.S.S.R), and reached peak values by 1963. These bomb afflicted tritium through the atmospheric rain is believed to have been infiltrated into the soil layers in different basins subsequently. The recharge through processes could be estimated by adopting the different approches (i) peak method (ii) integrated method (Munnich, K.O., 1968). The peak method involves identification of the peak tritium value at some depths. The amount of soil moisture stored above the peak in relation to the precipitation since 1963 gives the measure of recharge by adopting the following formula.

$$W = \frac{100S_m}{P_t} \qquad \text{.....(4.66)}$$

where W= recharge as percentage of rainfall; S_m= total soil moisture (in cm) in the soil column down to the depth at which 1963 peak is Located; P_t= total precipitation (in cm) since 1963 to the time of investigation

The second approach named integrated method: The total tritium fall out in the rain since 1952 to the date of sample collection is computed by summing up the products of yearly rainfall and average tritium concentration and the amount of tritium upto to depth level where the concentration is insignificant. The estimations could be made by adopting the following formula.

$$W = \frac{\sum a_x m_x}{\sum A_i P_i} 100 \qquad \text{.....(4.67)}$$

where a_x = tritium concentration of soil segment x; m_x= moisture content, in cm, of soil segment x; A_i = tritium concentration of precipitation in month or year; P_i= precipitation, in cm in months/years.

The approach becomes no longer applicable because majority of the tritium peaks of 1963 must have been mixed with the water table and got diluted. However this method has been applied during 70's in several parts of the country, but lost its validity subsequently (Athavale, R.N. and Murti, C.S., 1980; Sukhija, B.S., et al, 1973).

Dating of groundwater: Radio active tracers used for estimating the age of groundwater are ^{32}Si (half-life 400 ± 100 years), ^{14}C (half-life 5740 ± 30 years), and tritium (half-life 12.4 years). Estimation of groundwater age is based on the principle of radio active decay as reflected in the equation given below

$$C_t = C_0 e^{-t}\lambda \qquad \text{.....(4.68)}$$

$$T\tfrac{1}{2} = \frac{\ln 2}{\lambda} \qquad \text{.....(4.69)}$$

where C_t = activity at time t; C_0 = activity at time t_0; λ = radioactive decay constant and $T\tfrac{1}{2}$ = half life. The age of groundwater could be estimated by determining the C_t of a radio isotope for a known radio active decay constant. The groundwater recharge rate could be estimated by using the following formula (Vogel, J.C., 1967).

$$R = 0.69\frac{\theta h}{t_{\frac{1}{2}}} \qquad \text{.....(4.70)}$$

where, R = recharge rate, in m/yr; θ = porosity of aquifer; h = total depth of aquifer in m and $t\tfrac{1}{2}$ = characteristic age at half the depth of the aquifer.

Bhattacharya, S.K., et al (1979) have used ^{32}Si and ^{14}C measurements on groundwater samples from deep confined aquifers of Mehshana district of Gujarat and Jaisalmer district of Rajasthan with residence times of 500 years and 500-3000 years respectively, Similarly Sukhija, B.S., et al (1996) reported using ^{14}C isotope ages for deep groundwater systems for cuddalore sandstones of Neyveli lignite mining area to be between 90-30,000 from recharge areas to discharge areas.

4.4 EVALUATION OF FRESH GROUNDWATER RESOURCE POTENTIALS IN DIFFERENT COASTAL DELTA COMPLEXES AND THEIR ADJOINING ENVIRONMENTS

Coastal delta systems are the prominent land forms endowed with rich surface (high density canal networks and irrigation tanks) and groundwater resources (dugwells, filter points, shallow and deep tubewells) supporting large agriculture outputs in both kharif and rabi seasons. Over the last two or three decades a number of exploratory endeavours have been carried out by both central and state agencies covering all most all deltaic tracks of India. These exploratory endeavours have thrown up important light on the lithological framework, aquifer distribution systems and resource potentials in both shallow and deep levels and also helped in delineation of saline-fresh water bodies. These exploratory studies have also revealed that the fresh water systems in the shallow regions by and large are restricted to paleo channels, levee deposits and beach ridges of Holocene and Quaternaries. The deep fresh water systems occur largely in the Tertiaries and Mesozoics with varying thicknesses and permeabilities. The regional hydrogeological framework of these delta systems are summarized by Das, S. (1991). Majority of these fresh water systems are associated with saline water bodies at various levels which could be resolved through deep borehole geophysical studies Radhakrishna, I., (1981) and Radhakrishna, I., et al (1990). Adoptation of required pumping strategies and evaluation of resource potentials in each of the delta systems demands understanding of S and T parameters. Resource evaluations/water balance studies are considered scientifically valid and rational and data bases should be developed basin/watershed wise and workout appropriate optimization schemes and management strategies. A first order fresh water resource assessment in each of the delta basins are presented in the following paragraphs.

4.4.1 Palar-Araniyar-Korattalaiyar delta basin

Palar is one of the largest medium level interstate river with a stretch of about 348 km covering states of Karnataka, Andhra Pradesh and Tamil Nadu and terminates at a point 5 km south of Chennai, with an annual average discharge potential of the order of 1780 MCM. The river waters in the upstream and in the middle stretches are heavily used for industries and irrigation practices and are considered as one of the most polluted river basins of Tamilnadu. It has several anicuts constructed along its course and its tributaries, the largest one being 25 km downstream of Vellore, with an irrigation potential of the order of 33,600 hectares.

The present Palar river near the coastal region flows in south eastern direction from the delta apex point at Walajapet. The river is believed to have had its earlier course in a north eastern direction from its apex point during the Quaternary times and built up a good delta complex with thick sequences of alluvial sediments around Minjur-Panjetti area which are acting as good repositaries for groundwater resources. This terrain is now being drained by three important rivers namely Araniyar, Korattalaiyar and Cooum rivers, which serve as good recharge source for this Paleo Palar delta complex. Araniyar, Korattalaiyar and Cooum rivers of the basin area are relatively short stretched with their lengths varying from 130 km (Korattalaiyar) to 90 km (Cooum). Araniyar has an intermediate length of about 110 kms. Both Korattalaiyar and Cooum take their origin from the adjacent high lands of eastern ghats with in the state while Araniyar take its origin in the high land region of Chittoor (Andhra Pradesh) close to the interstate boundary. Though all these rivers are short stretched and ephemeral in nature, play a very important role in the augmentation of surface and groundwater resources by way of charging the innumerable number of major and minor surface tank areas and also the near surface/shallow groundwater systems in their sub basin areas. The southern part of the basin area is totally occupied by Palar and its tributaries, which flow in a south easternly direction from its delta apex point at Walajapet and join the coast south of Chennai city. Cheyyeru is an important tributary to Palar which joins the main river near Walajabad located close to the coastal boundary. A prominent hydrological feature in the basin area is the existence of a long coastal canal (Buckingham canal) stretched from the southern tip of Pulicat Lake in the north to Chennai city in the south. The main Palar basin and Araniyar – Korattalaiyar – Cooum sub basins (AKC basin) boundaries with flow trends of the rivers are presented in Fig. 4.21. The outlined basin area covers an area of 17,900 sq.km stretched mainly in the districts of Thiruvallur (100%), Kanchipuram (88%), Chennai (100%) and Vellore (85%) and partly in Tiruvannamalai (48%).

The annual average precipitation levels and their trends are shown in Fig. 4.21. The average annual rain fall in the basin is about 1135 mm, with variations from 1200 mm (towards the coast) and 1000 mm (away from the coast). The basin receives over 60% of the rainfall from northeast monsoon spread in the months of October, November and December and the rest 40% of the rainfall comes from southwest monsoon during the period July, August and September. The delta region has characteristic tropical climate with mean summer temperatures of 32.9 °C (March to June) and with a mean winter temperature of 24.3 °C (November to February). The maximum (November to January) and minimum (May to June) relative humidities respectively are of the order varying from 84% to 58%. The average annual potential evapotranspiration is of the order of 1743 mm which is more than the average annual precipitation level. There are about 5622 tanks and about 619 no. of canals with a total irrigation potential of the order of 1,75,982 hectares. Similarly in addition there are about 2,47,380 open wells and about 43,383 tube wells with an irrigation potential of the order of 2,61,899 hectares. The percentage of irrigation potential from groundwater is about 60%. The proportionate district wise irrigation potentials in the AKC basin are reflected in the Table 4.1.

Table 4.1 Surface and groundwater irrigation potentials in Palar-Araniyar-Korattalaiyar delta region

District (%)	Surface Irrigation						Groundwater Irrigation					Gross Irrigation (ha)
	Tanks (ha)		Canals (ha)		Other sources (ha)	Total (ha)	Tubewells (ha)		Dug Wells (ha)		Total (ha)	
	Number	Area Irrigated (ha)	Number	Area Irrigated (ha)			Number	Area Irrigated (ha)	Number	Area Irrigated (ha)		
Tiruvallur (100%)	1895	23936	17	3676	1619	**29231**	18855	67544	19457	32667	**100211**	**129442**
Vellore (85 %)	1152	17904	513	371	19067	**37343**	3161	9789	103397	59397	**69186**	**106529**
Kanchipuram (88 %)	1709	53444	18	466	28710	**82621**	21367	10706	49613	41582	**52288**	**134908**
Tiruvannamalai (48 %)	866	6778	69	1144	18866	**26788**	0	0	74913	40214	**40214**	**67002**
Total	**5622**	**102063**	**617**	**5658**	**68262**	**175982**	**43383.4**	**88038.68**	**247380**	**173860.3**	**261899**	**437881.4**

(Source: Department of Economics & Statistics, Govt. of Tamil Nadu (2005-06))

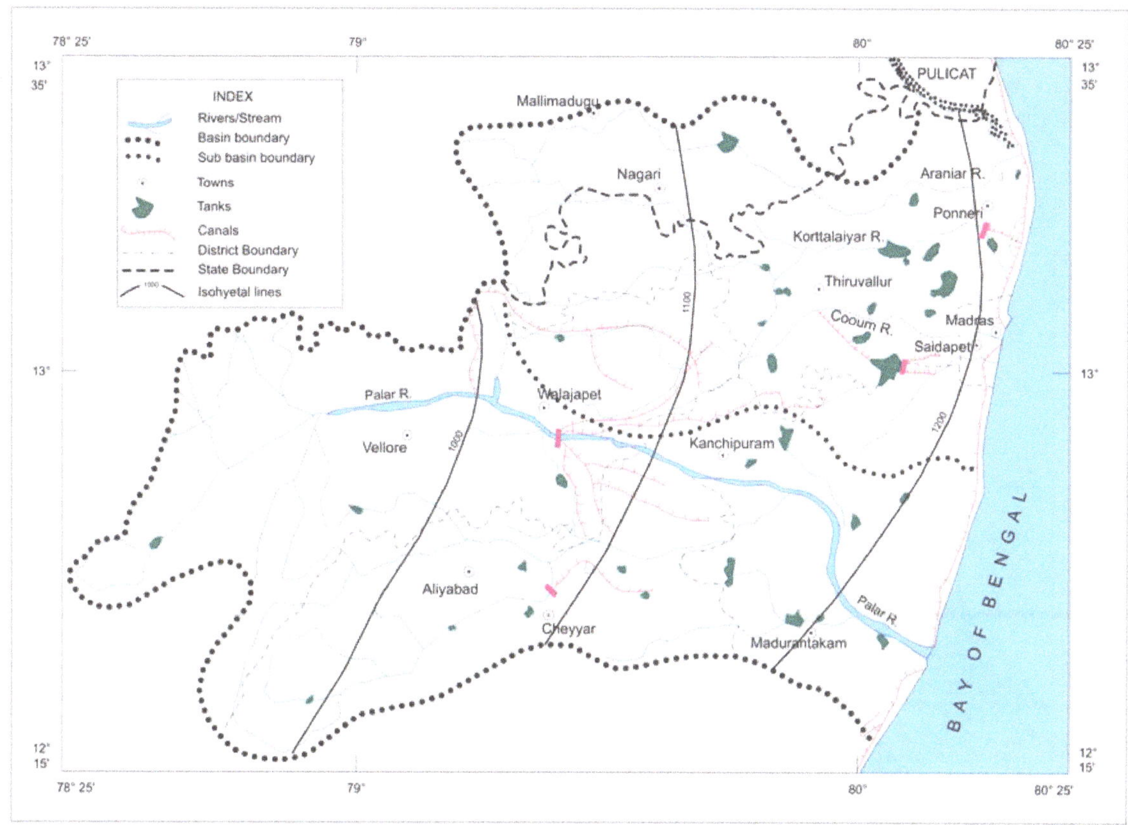

Fig. 4.21 Drainage & Isohyetels in Palar-Araniyar- Korattalaiyar delta region

The total irrigation potential in the basin through surface and groundwater resources comes to an order of 4,37,881.4 hectares.

Regional hydrogeological characteristics and system parameters

A regional geological map within the frame work of the basin boundaries is shown in Fig. 4.22. The different geological formations include Archeans, Precambrians, Mesozoics, Tertiaries, Quaternaries and Recent. Several exploratory wells were drilled by both central and state agencies up to depths varying between 300 to 800 m. The deepest exploratory boreholes drilled in the region are at Veerapuram and Puduvoyal up to depths of 790 m and 745 m below ground level respectively where most of the lithological units encountered belong to Tertiaries, Mesozoics and Archeans with intrusions.

The Archeans occupy more than 70% of the basin area spread out in the western and northwestern regions. In the main Palar region Archeans are exposed in the western flanks and also occur as large *inliers* in the central and out fall region of the basin.The Archeans by and large in the entire region are represented by granites, gneisses, schists and charnockites with intrusions of dolerite dykes and pegmatite veins. Groundwater occurrences in these rock types are mainly restricted to the top 10 to 20 meters of weathered zones and sometimes in fractured and fissured portions of the rocks extending up to depth levels varying between 40 m to 60 m below ground level. The depth of weathering and intensity of fracturing etc are factors related to the stream density and morphotopography of the terrain. Groundwater in these rock types occurs in both unconfined and semi confined conditions with water levels varying from 3.0 to 18.0 m below ground level (pre monsoon) and 1.0 to 15.4 m below ground level (post monsoon). Similarly the piezometric surface water levels of deeper fractured system have a range varying from 3.78 to 21.70 m below ground level (pre monsoon) and 2.1 to 8.5 m below ground level (post monsoon). Large diameter wells with yields varying from 60 to 500 lpm are common, but at times shallow tube wells up to depths varying between 30 to 60 m below ground level are also constructed depending upon depth, intensity of fracture zones and their saturation levels. The transmissivity levels of these aquifer systems are of the order varying from 10 m^2/day to 60 m^2/day.

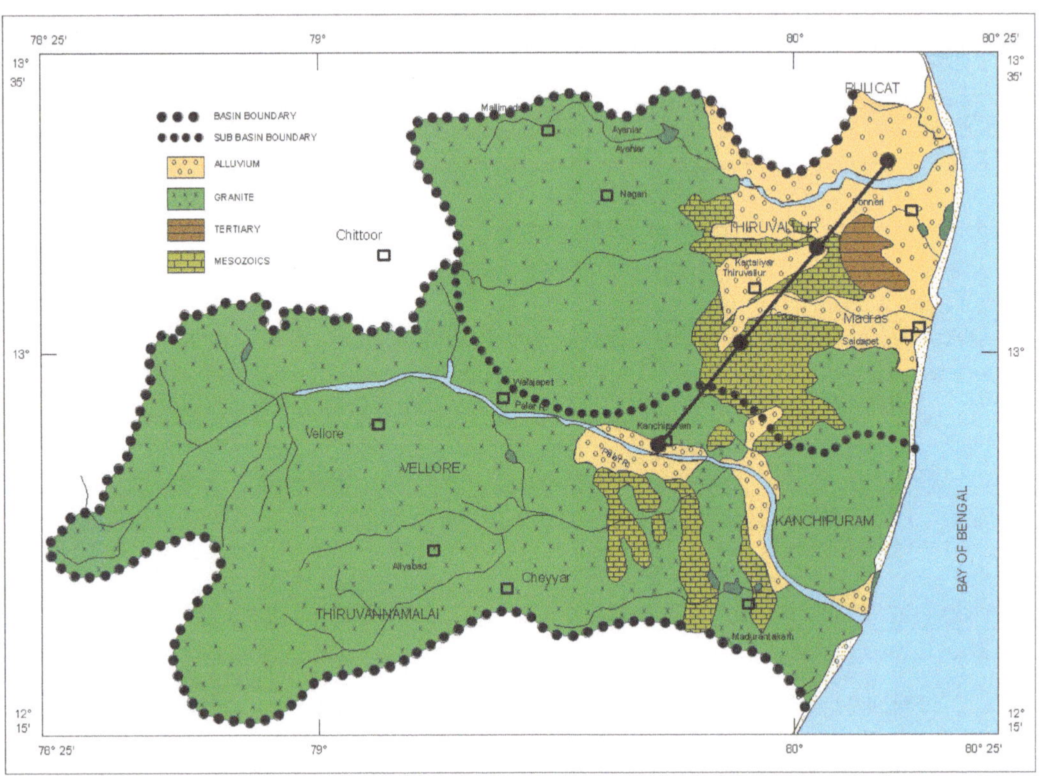

Fig. 4.22 A regional geological map of Palar-Araniyar-Korattalaiyar delta basin

Mesozoics, Tertiaries, Quaternaries and Recent constitute important and widely spread out sedimentaries in the basin. Mesozoics are represented by lower Jurassics (Sriperumbudur beds) and lower Cretaceous sediments (Satyavedu conglomerates). Sriperumbudur beds constitute essentially shales and sandstones with intercalations of clays and the Satyavedu beds are represented by ferruginous sandstones, conglomerates and sometimes pebbles and boulders. These rocks in general are hard and compact with low permeabilities. The Mesozoics outcrop extensively along the western flanks and in the central portions of the basin area and lie unconformably over the Archeans. They lie underneath the Quaternary alluvial sediments and Tertiaries in the central and outfall regions of the Araniyar – Korattalaiyar sub basins and extend to the outfall regions. Some of the exploratory wells in this group of sediments indicated rather low yields with discharges varying from 60 to 350 lpm with low to medium level transmissivities varying from 5 to 300 m^2/day. Some of the deep exploratory wells drilled indicated thicknesses of the order varying from 500 to 700 m but with insignificant yield characteristics.

The Tertiary group of rocks are represented by friable brownish, red feldspathic sandstones, quartz grits and overlies the Mesozoics. These are exposed in the northern and central regions of the Araniyar – Korattalaiyar – Cooum river basins as reflected in the map and extend regionally underneath the alluvial sediments up to the coastal boundaries and beyond. Exposures of Canjeevaram gravels (Pliomiocene) consisting of gravels, pebbles and grits occur north of Kancheepuram near the border of

AKC sub basin boundary. The thicknesses of these sandstones in this region are relatively small varying between 10-20 meters with good yield potentials of the order varying between 1000 lpm to 2500 lpm and with the transmissivities varying from 1000 to 4000 m^2/day. In general they are seen occurring underneath the alluvial systems in confined situations. The storativity for these confined systems are of the order varying from 1.2×10^{-3} to 8×10^{-4}. These aquifer systems often combinedly tapped either with the underlying Mesozoics or with overlying Quaternary alluvial sediments and the hydraulic data individually for the Tertiary sediments as such are not available in this region.

Quaternaries and Recent alluvials are extensively developed over many parts of the basin, along the present river courses, paleo palar river course (Araniyar – Korattalaiyar – Cooum sub basin) and along the narrow coastal tracks with thicknesses ranging from 50 to 60 m below ground level. These are mostly unconsolidated fine to coarse grained sands, gravels, pebbles and clays. These form potential aquifers and support large number of filter points, dug wells and dug cum bore wells and shallow tube wells. Large scale pumping well fields in theses tracks have come up over the last several years with large scale groundwater pumpage in the AKC region namely: (i) Minjur well field (ii) Panjetty well field and (iii) Tamarapakkam well field. Groundwater in these well fields occurs both in the water table and confined conditions. The wells constructed in these well fileds have yields varying between 1200 lpm and 2500 lpm at relatively small drawdowns of the order of 6 to 8 m. These aquifer systems have transmissivity values ranging from 1000 to 4500 m^2/day and storativity values ranging from 5×10^{-2} to 5×10^{-3}. Similarly the Quaternary alluvial sediments in the Palar basin have thicknesses varying from 10 to 30 m below ground level, but their distribution mostly is restricted to the region adjoining the river banks and the coastal zone. The yields are of the order of 300 to 750 lpm at low draw downs ranging from 3 to 5 meters with transmissivity values varying from 550 m^2/day to 4500 m^2/day. Some of the promising potential zones identified in this sub basin are (i) Poini-Palar confluence (ii) Chingleput and (iii) Paiaganallur area.

A generalized schematic diagram showing the geological units in this basin is shown in Fig. 4.23.

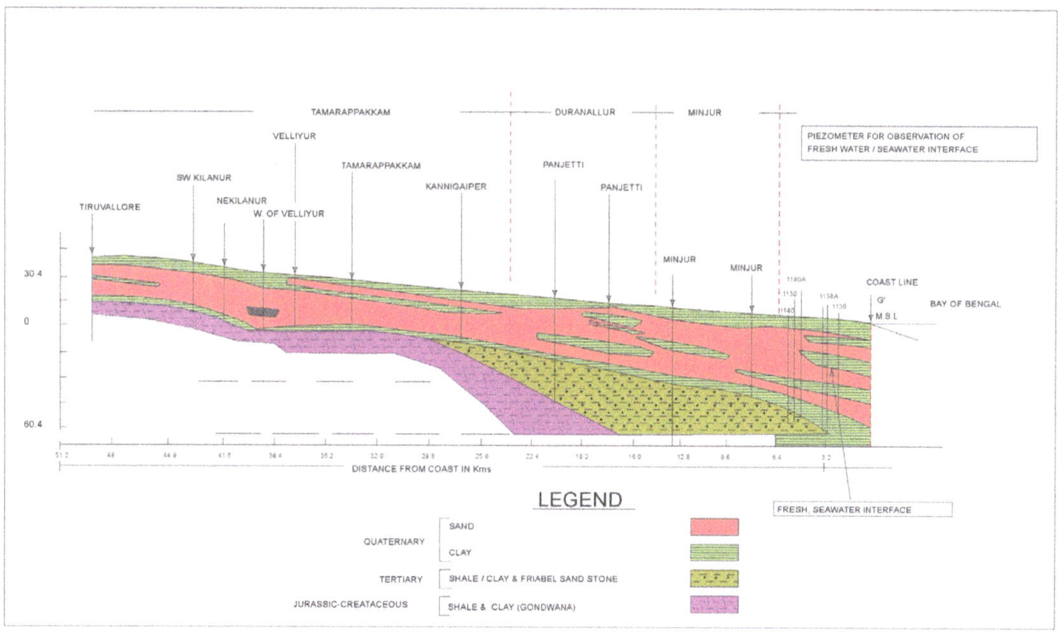

Fig. 4.23 A schematic geological section across Palar-Araniyar-Korattalaiyar delta basin

(Source: CGWB, 1998)

A regional geological section along Kanchipuram to Puduvoyal cutting across all the geological units is shown in Fig. 4.24 covering a distance of over 73 km with intermediate boreholes at Melnalathur and Veerapuram. The depths of these boreholes range between 300 to 830 m below ground level. The lithological units identified across the section are Quaternaries, Tertiaries, Mesozoics and Archeans. The crystalline basement is seen occurring at a depth of 780 m at Veerapuram.

The relevant hydrological parameters for selected exploratory wells as shown in Fig. 4.25 in different sub basins are presented in Table 4.2 (Data source: CGWB, 1998 and UNDP, 1975).

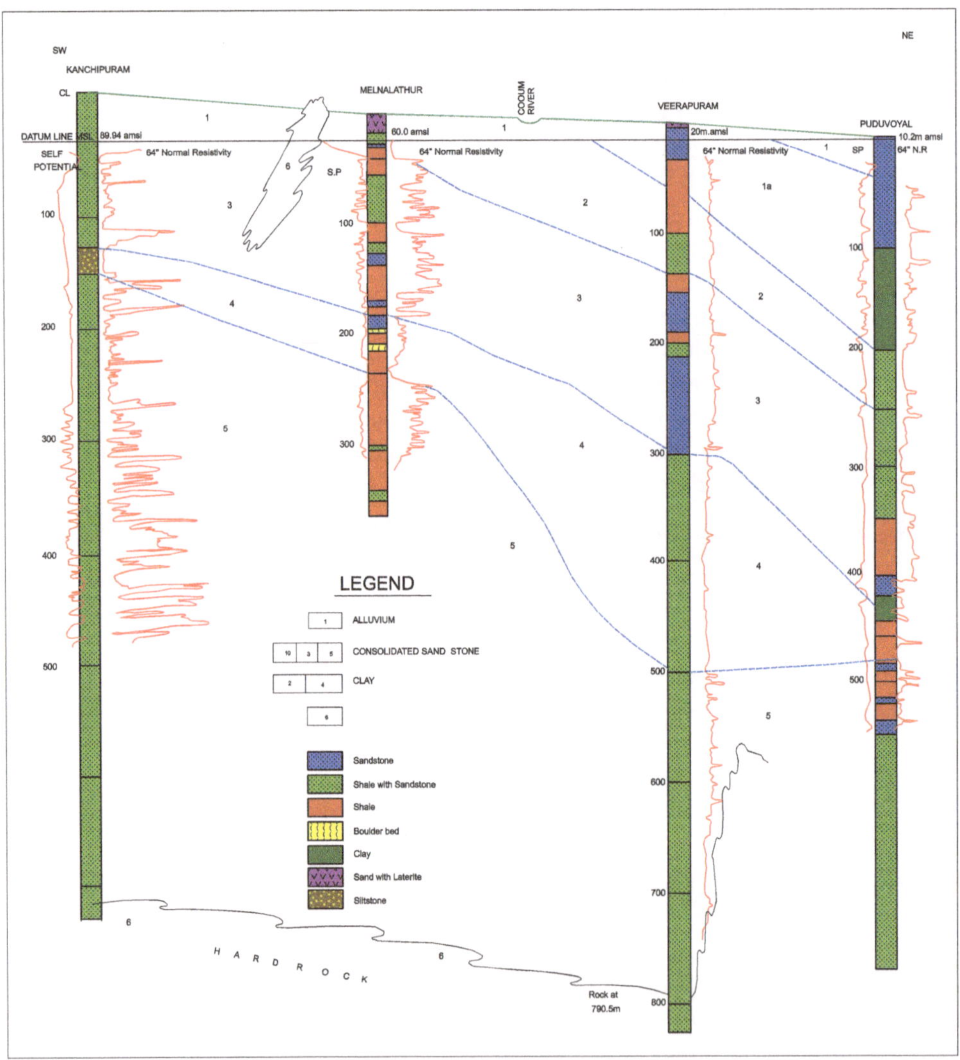

Fig. 4.24 A hydrogeological cross section across Kanchipuram to Puduvoyal in palar delta

(Source:CGWB, 1998)

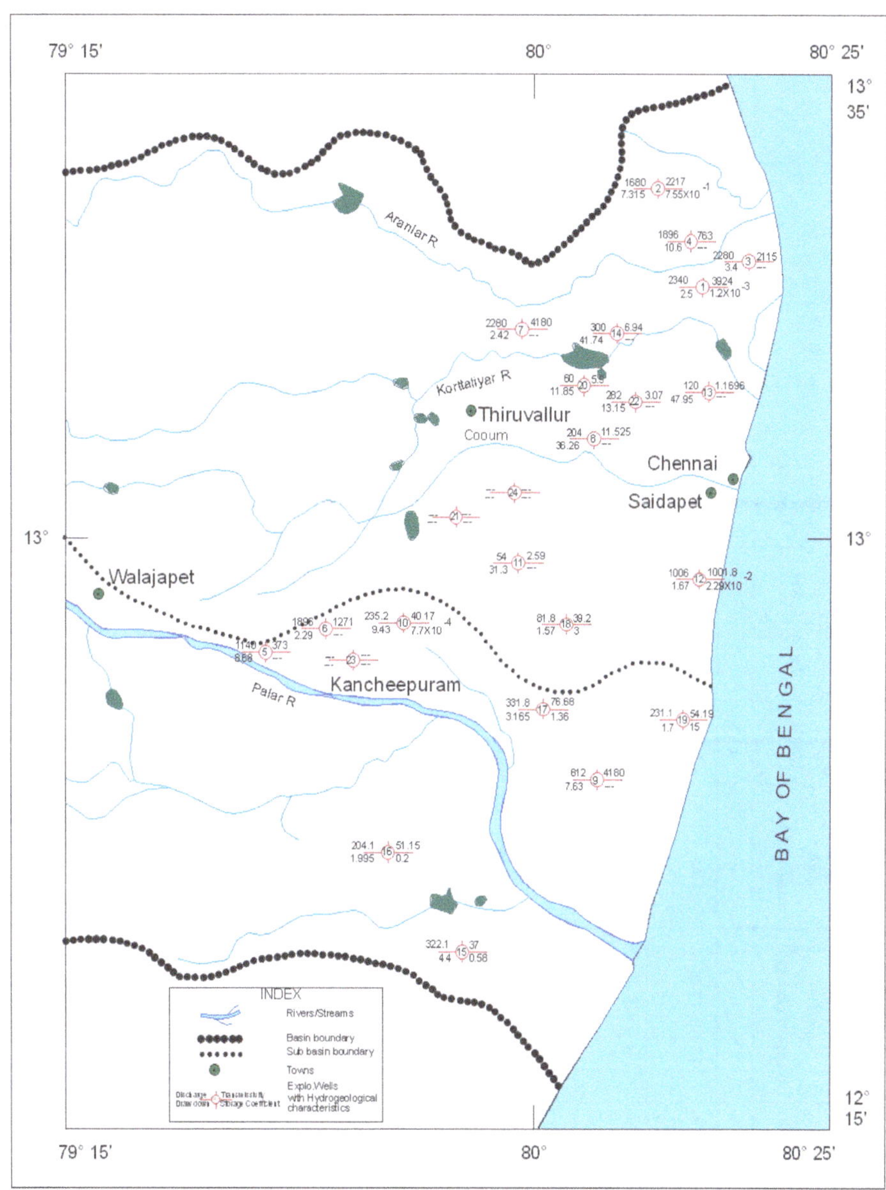

Fig. 4.25 Location map of selected exploratory wells in Palar-Araniyar-Korattalaiyar delta basin

Table 4.2 Regional hydrogeological parameters for selected exploratory wells in Palar delta region

S.No	Location	Depth drilled (m) / Constructed (mbgl)	Depth Range of Aquifers tapped (mbgl)	Thickness of the zones (m)	SWL (mbgl)	Discharge (lpm) / Drawdown (m)	Specific Capacity (lpm/m)	T (m^2/day)	S	Remarks
1	2	3	4	5	6	7	8	9	10	11
1	Yellareddi Kandigai	90.24	17.00-33.00	16	5.5	2340	936	3924	1.2×10^{-3}	Sand, Sandstone shale &
	(13°18'30'';80°13'00'')	36.4				2.5				quartzite
2	Panjetti	63	9.1-35.00	27.9	4.9	1680	230	2217	7.55×10^{-4}	Sand, clay sandstone and
	(13°17'15'';80°09'00'')	43				7.315				shale
3	Neidavoyal	91.43	4.6-49.00	32.4	1.5	2280	671	2115	---	sand, clay quartzite shale
	(13°18'20'';80°16'30'')	5.2				3.4				
4	Parikkipattu	455.66	5.63-42.3	30.5	1.8	1896	178	763	---	Sand, clay sandstone and
	(13°21'30'';80°12'00'')	45.26				10.6				shale
5	Musaravakkam	124.82	19.00-25.00	6	2	1140	130	373	---	Alluvium clay sandstone
	(12°51'00'';79°36'00'')	27.43				8.68				and shale
6	Kilambi	150.9	20.42-28.65	8.23	7.46	1896	828	1271	---	Alluvium clay sandstone
	(12°51'00'';79°39'00'')	31				2.29				and quartzite
7	Velappakam	135.33	12.59-24.3	11.71	9.25	2280	942	4180	---	Alluvium, clay sandstone
	(13°13'30'';79°50'30'')	24.3				2.42				
8	Avadi	200.4	13.48-155.4	71.76	5.38	204	5.6	11.525	---	Lateritic soil clay,
	(13°07'00'';79°50'30'')	155.4				36.26				claystone sandstone and shale
9	Oragadam	300.1	174.0-285.0	44	18.82	612	80	158.4	---	Lateritic sand, gravel, glay
	(13°19'30'';80°06'50'')	288				7.63				sandstone
10	Enathur	365.01	126.0-341.0	82	13.3	235.2	24.94	40.17	7.7×10^{-4}	Clay sand, sandstone and
	(12°52'00'';79°44'15'')	344				9.43				shale
11	Sriperumbuduru	217.96	90.00-209.00	19	1.74	54	1.757	2.59	---	Clay, shale sandstone granite
	(12°58'00'';79°57'00'')	213				31.3				

Table 4.2 ***Contd…***

S.No	Location	Depth drilled (m) / Constructed (mbgl)	Depth Range of Aquifers tapped (mbgl)	Thickness of the zones (m)	SWL (mbgl)	Discharge (lpm) / Drawdown (m)	Specific Capacity (lpm/m)	T (m^2/day)	S	Remarks
1	2	3	4	5	6	7	8	9	10	11
12	Porur	143.5	9.00-25.00	12	6.5	1006	602.3	1001.8	2.29×10^{-2}	Clay, sand gravel, shale and sandstone charnockite
	(13°02'20";80°08'30")	28				1.67				
13	Almadi	309.15	77.00-281.1	41	1.74	120	2.50	1.1696	---	Alluvium clay, sand stone and shale
	(13°12'00";80°17'30")	284.18				47.95				
14	Puduvoyal	764.2	206.6-625.3	172	10.24	300	7.18	6.94	---	Alluvium clay, sand-stone & carbonaceous shale
	(13°19'30";80°06'50")	629.38				41.74				
15	Madurantakam	7.9	…	…	3.2	322.1	73.20	37	0.58	Charnockite Jointed
		…				4.4				
16	Uttiramerur	10.5	…	…	4.73	204.1	102.3	51.15	0.2	Charnockite Jointed
		…				1.995				
17	Singaperumalkovil	10	…	…	2.45	331.8	104.83	76.68	1.36	Charnockite Jointed
		…				3.165				
18	Tirupporur	8	..	…	3.76	81.8	52.10	39.2	3	Charnockite Jointed
						1.57				
19	Padappai	8.25	…	…	4.51	231.1	135.92	54.19	15	Charnockite Jointed
		…				1.7				
20	Veerapuram	829.7	…	…	1.5	60	5.06	5.9		
	(13°10'30";80°05'30")	…				11.85				
21	Moppedu	30.45	12.0-27.0	15	2.72	….	….	…	…	
	(13°01'00";79°54'00")	30								
22	Arakkambedu	601	….	…	12.15	282	21.44	3.07	…	
	(13°10'05";80°06'10")	…				13.15				

Regional groundwater flow net analysis

A regional groundwater flow trend map across different profiles for both near surface and deep aquifer systems are provided in the Figure 4.26. The groundwater flow parameters across different lengths in different sub basins are provided in Table 4.3.

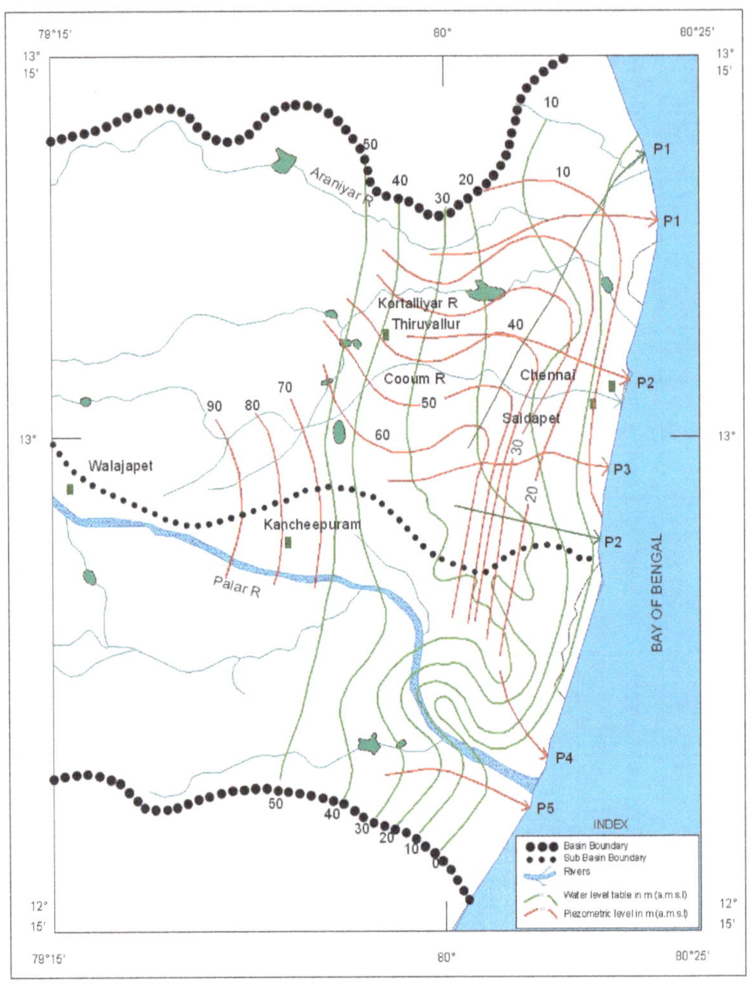

Fig. 4.26 Regional groundwater flow map for near surface and deep aquifer systems of Palar-Araniyar-Korattalaiyar delta basin

Table 4.3 Groundwater flow parameters across different profiles in Palar-Araniyar-Korattalaiyar delta basin

S.No	Basin	Profile	Avg.Transmissivity (T) (m^2/day)	Gradient (I) (m/km)	Length of the Coast (L) (km)	Total flow (Q) (MCM)
Near Surface Aquifers						
1	AKC	P1	1500	1.03	40	22.56
		P2	1500	1	25	13.69
		P3	1000	1.03	22	8.27
2	Palar	P1	1500	3	41	67.34
		P2	1500	2	22	24.09
Deep Aquifers						
1	AKC	P1	2000	1.14	55	45.77
		P2	1000	4	30	43.8

The total estimated groundwater flow across the whole coastal margins of AKC and Palar basins is of the order of 225.52 MCM which comes to about 6.3% of the dynamic potentials of the basin.

Seasonal fluctuation in selected observation stations: The seasonal water level fluctuations for a 5 year period (1980-85) at selected observation stations are shown in Figure 4.27 in different geological horizons as reflected in Table 4.4. The average fluctuation in the granitic region is about 4.558 m and in the coastal alluvial/sedimentary tracks is about 3.30 m.

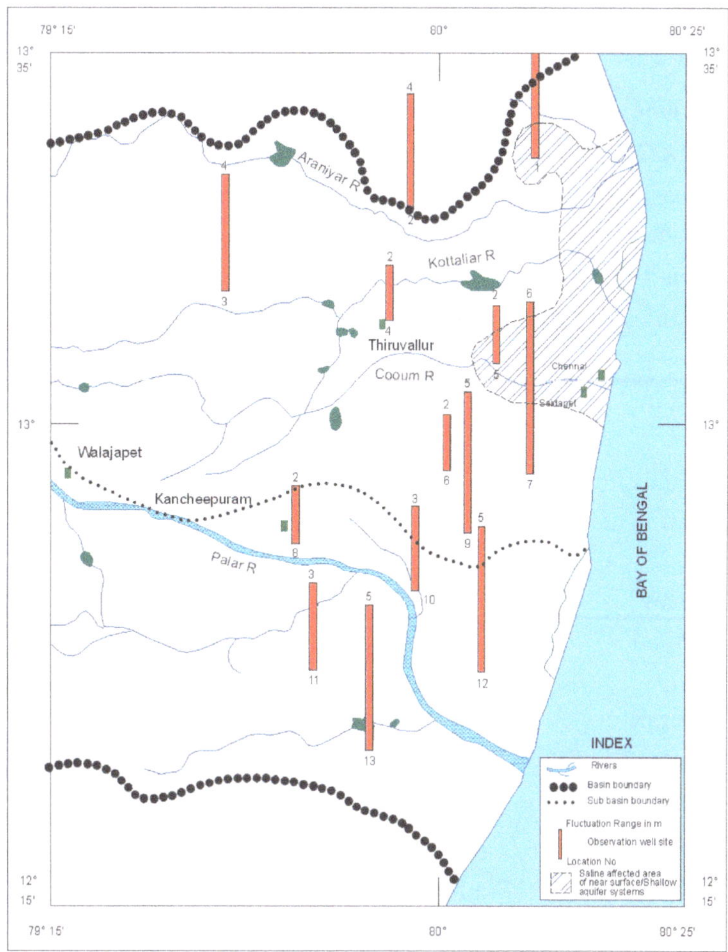

Fig. 4.27 Regional groundwater level fluctuation trends in Palar delta basin (Tamil Nadu)

Table 4.4 Seasonal fluctuations in selected observation stations

S.No	Name of the Site	Geological Horizon	Avg. Fluctuation (m) for a 5 Year period (1980-85)
1.	Gummidipoondi	Teriary	6.03
2.	Uttukottai	Teriary	4.11
3.	Tiruttani	Granite	4.23
4.	Thiruvallur	Teriary	1.68
5.	Avadi	Teriary	1.63
6.	Sriperumbudur	Mesozoics	2.32
7.	Tambaram	Granite	5.73

Table 4.4 ***Contd...***

S.No	Name of the Site	Geological Horizon	Avg. Fluctuation (m) for a 5 Year period (1980-85)
8.	Kanchipuram	Mesozoics	2.28
9.	Padappai	Granite	4.78
10.	Changalpattu	Granite	3.17
11.	Uttiramerur	Mesozoics	3.01
12.	Tirukkalukkundram	Granite	4.88
13.	Madurantakam	Mesozoics	5.4

Resource potentials in Palar-Araniyar-Korattalaiyar basins

Resource potentials in AKC (Araniyar-Korattalaiyar-Cooum) basin

- Area of the AKC basin : 7200 sq.km
- Coverage of Saline/brackish area : 700 sq.km
- Coverage of Alluvials : 2300 sq.km
- Coverage of Mesozoics & Tertiaries : 900 sq.km
- Coverage of Archeans : 4000 sq.km
- Dynamic potentials in Archeans :

(i) Fluctuation method : 640 MCM

- Dynamic potentials for sedimentaries (excluding saline track) :
- Rainfall Infiltration method

(i) Alluvials (fresh) : 317.11 MCM

(ii) Mesozoics+Tertiaries : 178.37 MCM

- Fluctuation method

(i) Alluvials : 1246.44 MCM

(ii) Mesozoics + Tertiaries : 269.25 MCM

- Average dynamic potentials

(i) Alluvials : 798.55 MCM

(ii) Mesozoics+Tertiaries : 224.19 MCM

- Total dynamic potentials in AKC basin (640 + 798.55 + 224.19) : 1662.74 MCM
- Resource potentials in deep confined aquifer systems(static) (Area × Thickness × S.Coefficient):

(i) Alluvials ($2300 \times 30 \times 5 \times 10^{-3}$) : 345 MCM

(ii) Mesozoics + Tertiaries ($3200 \times 30 \times 2.5 \times 10^{-4}$) : 24 MCM

(iii) Archeans ($4000 \times 10 \times 5 \times 10^{-3}$) : 200 MCM

- Total estimated resource potentials in deep confined aquifer systems (345 + 24 + 200) : 569 MCM

Note: 0-100 m is the depth range considered for estimations of dynamic & deep aquifer potential system in the basin

Resource potentials in Palar basin

- Area of the Palar basin : 10,700 sq.km
- Coverage of Archeans : 9,650 sq.km
- Coverage of Alluvials : 600 sq.km

- Coverage of Mesozoics & Tertiaries : 450 sq.km
- Dynamic potentials in Archeans:

(i) Fluctuation method : 1544 MCM

- Dynamic potentials for sedimentaries:

(i) Rainfall Infiltration method
  (a) Alluvials : 118.9 MCM
  (b) Mesozoics+Tertiaries : 89.1 MCM

(ii) Fluctuation method
  (a) Alluvials : 360 MCM
  (b) Mesozoics+Tertiaries : 155.2 MCM

(iii) Average dynamic potentials
  (a) Alluvials : 239.46 MCM
  (b) Mesozoics+Tertiaries : 122.21 MCM

- Total dynamic potentials for Palar basin (1544 + 239.46 + 122.21) : 1905.67 MCM
- Resource potentials in deep confined aquifer systems (static) (Area × Thickness × S.Coefficient) :

(i) Alluvials ($600 \times 20 \times 5 \times 10^{-3}$) : 60 MCM
(ii) Mesozoics+Tertiaries ($1050 \times 30 \times 2.5 \times 10^{-4}$) : 7.87 MCM
(iii) Archeans ($9650 \times 10 \times 5 \times 10^{-3}$) : 482.5 MCM

- Total estimated resource potentials for deep confined aquifer system (60 + 7.87 + 482.5) : 550.37 MCM
- Total estimated dynamic + static potentials in AKC sub basin + Palar sub basin (2231 + 2455) : 4686 MCM

Note: 0-100 m is the depth range considered for estimations of dynamic & static potentials in this basin

4.4.2 Ponnaiyar – Vellar delta basins

Ponnaiyar and Vellar delta basins in Tamil Nadu constitute very important river basins from the water and mineral resources point of view. Ponnaiyar is an inter-state river with a stretch of about 396 km and with an average annual discharge potential of about 1600 MCM. It has a relatively short and linearly developed delta system with its apex near Thirukovilur. In the vicinity of Ponnaiyar delta region, further south, a short stretched river namely Gadilam emerges, flows nearly parallel to Ponnaiyar river and gets interlaced with Ponnaiyar in the outfall regions. The Ponnaiyar together with Gadilam forms a potential alluvial track in this part of the basin with good stretches of flood plains and thick alluvial deposits. The southern part of the basin is occupied by Vellar river, another prominent river but without any significant delta build up. It is relatively short stretched coastal river emerges within the state boundaries and with a length of over 193 km. It has an annual discharge potential of the order of about 850 MCM. Manimukthanadi is an important tributary to Vellar, and joins the main river at the middle stage of the river course from the north. The Vellar together with Manimukthanadi forms good flood plains with build up of thick alluvial sediments. A regional drainage map with precipitation levels and with basin boundaries is presented in Fig. 4.28. These two coastal river basins act as very important resource systems with both surface and groundwater potentials. A significant feature in this part of the coast is the occurrence of 'Lignite coal' in a big way in the inter stream area of Ponnaiyar and Vellar, where large scale continuous pumping is in operation from the aquifers underneath the Lignite belt to facilitate open cast mining operations.

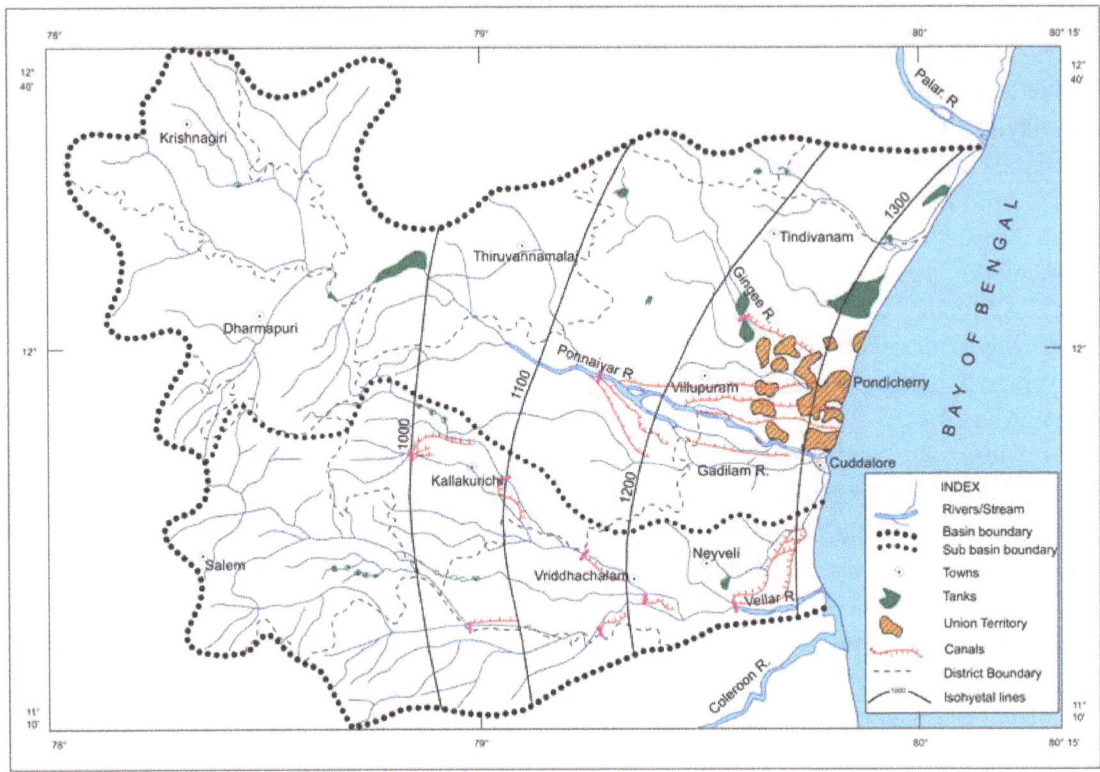

Fig. 4.28 Drainage and Isoheytal map of Ponnaiyar-Vellar delta basin

The outlined basin area covers approximately 22,100 sq.km mainly spread over in the districts Villupuram (100%), Cuddalore (82%) and partly in the districts of Thiruvannamalai (52%), Perambalur (37%), Salem (47%), Dharmapuri (42%), Krishnagiri (33%) and Kanchipuram (13%). The annual average precipitation is of the order of 1175 mm with variations from 1300 mm (towards the coast) and 1000 mm (away from the coast). The basin area in general is characterized by tropical climate and receives about 65% of the rainfall from northeast monsoon spread over in the months of October to December and rest 35% comes from the southwest monsoon during the period July to September. The average maximum and minimum temperatures in the basin area are about 32.4 °C and 23.9 °C respectively. The peak temperatures are normally recorded during the period April to May and the lowest temperatures are recorded during the months December to January. The average relative humidity is of the order varying between 83% (wet period) and 60% (dry period). It is also observed that the western flanks of the basin have lesser range of humidities when compared to the eastern coastal margins. The average annual potential evapotranspiration is of the order of 1760 mm which is remarkably higher than the average mean precipitation level (see chapter-3 for details). Both surface and groundwater resources in the basin are well developed through surface tanks, canals, tube wells and irrigation dug wells. There are about 5,521 no. of irrigation tanks and 861 no. of canals contributing to an irrigation potential of the order of 2,29,612 hectares. Similarly there are about 3,84,918 dug wells and about 61,866 tube wells contributing to the irrigation potential of the order of 4,29,065 hectares from groundwater. The cumulative irrigation potentials in the basin is of the order of 6,58,677 hectares. The percentage of irrigation potential from groundwater is about 65%. A break up of irrigation potentials in the basin area (proportionate district wise) are presented in Table 4.5.

Table 4.5 Surface and groundwater irrigation potentials in Ponnaiyar-Vellar delta basin

District	Surface Irrigation						Groundwater Irrigation					Gross Irrigation (ha)
	Tanks (ha)		Canals (ha)		Other sources (ha)	Total (ha)	Tubewells (ha)		Dug Wells (ha)		Total (ha)	
	Number	Area Irrigated (ha)	Number	Area Irrigated (ha)			Number	Area Irrigated (ha)	Number	Area Irrigated (ha)		
Cuddalore (82%)	487	6549	221	40185	189	**46923**	25163	93489	9236	7838	**101327**	**148249**
Villupuram (100%)	2085	61205	196	6648	38146	**105999**	23454	54525	164871	120661	**175186**	**281185**
Perambalur (37%)	332	3748	257	3671	28	**7447**	5228	8465	16281	12422	**20888**	**28335**
Salem (47%)	257	803	37	987	10202	**11992**	4467	3680	52415	40577	**44258**	**56250**
Kanchipuram (13%)	252	7895	3	69	4241	**12205**	3157	1582	7329	6143	**7724**	**19930**
Dharmapuri (42%)	389	2722	36	2874	4913	**10510**	305	864	31783	21546	**22410**	**32920**
Krishnagiri (33%)	781	3337	36	1597	582	**5516**	92	8	21848	13699	**13707**	**19223**
Tiruvannamalai (52%)	938	7343	75	1239	20438	**29020**	0	0	81155	43566	**43566**	**72586**
Total						**229612**					**429065**	**658677**

Source: Department of Economics & Statistics, Govt.. of Tamil Nadu (2005-06))

Regional hydrogeological characteristics and system parameters

A regional geological map of the basin is presented in Fig. 4.29. A large chunk of western and north western region is occupied by Archeans and the rest of the area along the eastern margins by sedimentaries namely Mesozoics, Tertiaries, Quaternaries and Recent. The Archeans are basically represented by granites, gneisses and charnokites with intrusions of dolerite dykes and pegmatitic veins. Groundwater in these rock types occur largely in unconfined and semi confined situations and are restricted to the top 15 to 20 meters of the weathered zone and sometimes extend up to 60 m in the fractured and fissured portions of the rock matrix. Dug wells and dug cum bore wells with discharges varying from 60 to 350 lpm are in operation in most parts of the granitic terrain. The water levels in these tracks vary from 0.74 to 10 m below ground level (pre monsoon) and 0.60 to 4.5 m below ground level (post monsoon). Similarly the piezometric surface levels for deep fractured aquifer zones vary from 2.8 to 11.50 m below ground level (pre monsoon) and 0.5 to 6.5 m below ground level (post monsoon). The estimated transmisivities are of the order varying from 1 m^2/day to 25 m^2/day depending upon the thicknesses of the saturated zones and nature of the aquifer material.

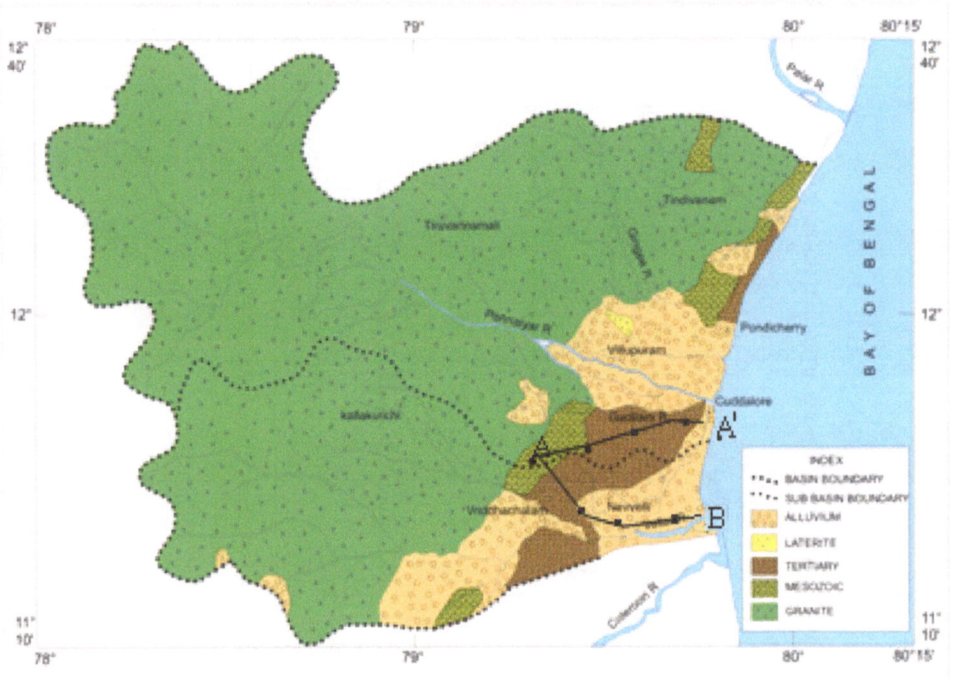

Fig. 4.29 Regional geological map of Ponnaiyar delta basin

The Mesozoics are represented by Cretaceous sediments mainly exposed over a width of 3-5 kms trending in NNE-SSW direction along the western flanks of the sedimentary basin. These are fossiliferrous, calcareous sandstones and siliceous limestone regionally widespread at shallow depth levels in the western flanks and gradually increase in depth levels towards the central and outfall regions (40 to 450 m below ground level). The groundwater in these rock types near the western margins occur in unconfined and semi confined conditions at shallow depth levels and gradually change over to deep confined conditions in the central and outfall regions of the sedimentary cover. These are rather low yielding aquifers with discharges varying from 300 to 800 lpm and with transmissivities varying from 50 to 200 m^2/day.

The Tertiaries in the basin area are represented by the lower Gopuram formations of Eocene age and the upper Cuddalore formations of Plio-Miocene age. The Gopuram formations are seen overlying the Cretaceous group of rocks along the western flanks of the sedimentary basin and sometimes occur as *inliers* within the overlying Plio-miocene sedimentary rocks close to western margins and are seen at deeper levels below Cuddalores invariably at many places covering the entire sedimentary basin. They are largely represented by algal limestones, claystones, and argillaceous sandstones. The upper Cuddalore formations are largely represented by ferruginous and feldspathic sandstones, grits and clays of various colours and textures. These are capped by laterites at several places. The main occurrences of these rocks are seen around Cuddalore and Panrutti and along large tracks in between Vridhachalam and Cuddalore. They are also seen abundantly in the Vellar basin area in the vicinity of Parur, Meppuliyur and Kuttadi areas. They attain thicknesses of the order of 100 to 300 meters as we move from west to east of the sedimentary basin. High discharge wells with yields varying from 200 to 4000 lpm are designed with transmissivities varying from 150 m^2/day to 4000 m^2/day and sometimes transmissivities of the order of 6000 to 8000 m^2/day depending upon the thicknesses of the aquifer zones. High discharge well fields are being designed in this group of rocks particularly around Neyveli, Sathiathope, Cuddalore, Panrutti, Valiyamdevi, and Chidambaram. Many of these well field areas are characterized by high pressure heads, sometimes with artesian conditions. A high discharge well field around Neyveli mining area with discharges of the order varying between 2000 to 3500 lpm are in operation to facilitate mining of Lignite.

Quaternary and Recent alluvial sediments are quite wide spread in the basin represented mostly by coarse to fine sands, silts and clays. They cover wide area in the Ponnaiyar, Gadilam and Vellar basin areas and attain thicknesses of the order varying between 20 to 40 m. A large number of dug wells, filter points and shallow tube wells are in operation in this belt with designed discharges of the order varying between 1200 to 2000 lpm and with transmissivities of the order of 4000 to 8000 m^2/day.

Two subsurface geological configuration sections involving different lithological units supported by electrical logs are presented in Figs. 4.30(a) and 4.30(b). The section A-A^{l} covers a length of about 52 km from Kopuvanur (North West) to Chavadi (North East), with depths of the bore holes ranging between 240 to 750 m below ground level. The section depicts the availability of considerable thicknesses of Cretaceous sediments in the western part of the region with extensions to the north eastern parts below Tertiary beds. The Tertiaries/Cuddalores have an average thickness varying between 140 to 440 m with thicknesses increasing from northwest to northeast and also with granular facies dominating towards the outfall regions. These horizons constitute potential well fields along this track. The large alluvial sediments are predominantly seen in the top levels of the section east of Naduveerapattu. The middle levels of the section with high thicknesses ranging from 114 to 435 m constitute potential Groundwater regions of Cuddalore formations. The bore hole section at Kopuvanur does not represent any Tertiary beds, but appear belong to Mesozoics. The lower level beds are dominantly rich with shales and clay stones and probably represent Mesozoics. These beds hold relatively low Groundwater yielding characteristics.

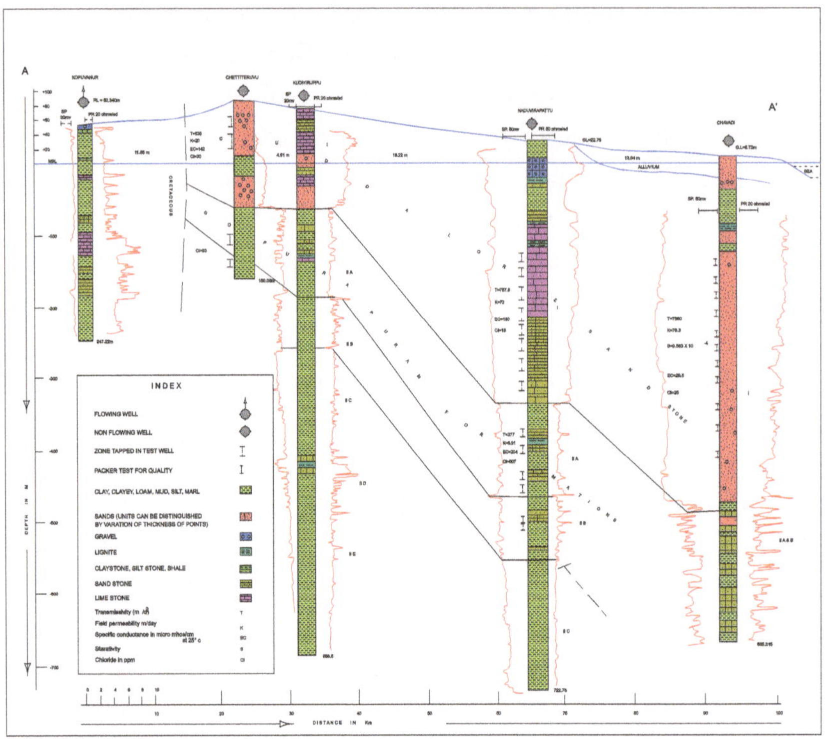

Fig. 4.30(a) Lithological section across profile AA' in ponnaiyar delta basin

(Source: CGWB, 1998)

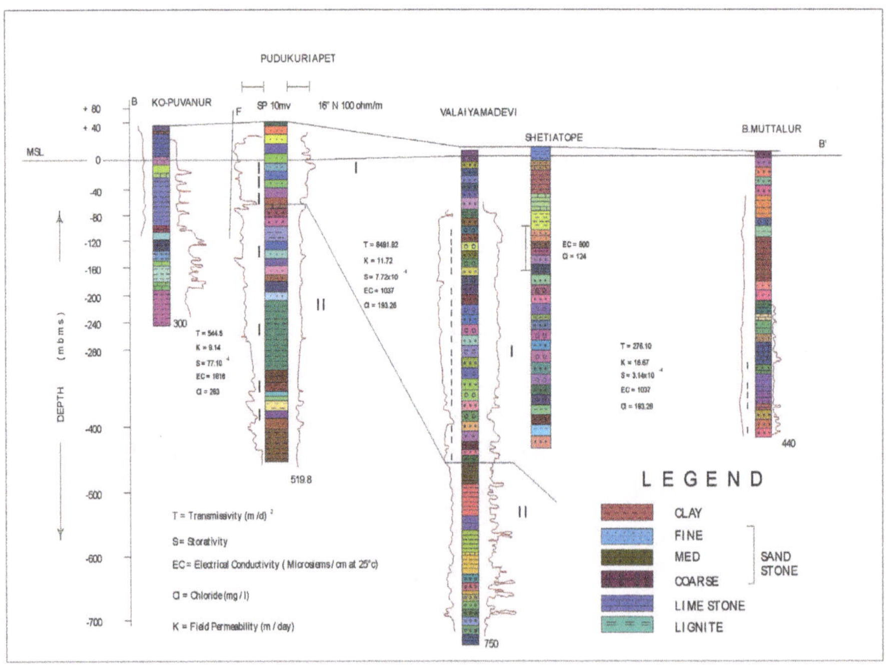

Fig. 4.30(b) Lithological section across AB in ponnaiyar basin

(Source: CGWB, 1998)

The second section A-B represents a track from northwest to southeastern side of the basin area from Kopuvanur to B.Muttalur, transecting the bore holes at Pudukuraipet, Valaiyamadevi and Shatiathope, with depth of the bore holes ranging between 300 to 750 m below ground level. Valaiyamadevi represents deepest bore hole with a depth of 750 m and Kopuvanur represents the shallow bore hole with a depth of 300 meters. The stratigraphic situation by and large is similar to section above, but with large thickness variations and with marine facies dominating towards the out fall region. As remarked earlier the near surface and middle level Tertiary systems constitute potential Groundwater well fields. The lower Mesozoic (Cretaceous) layers are characterized by low yields but are found with high artesian conditions.

A Summarized account of hydrological parameters for selected number of sites as shown in the map (Fig. 4.31) of the basin area are presented in Table 4.6 (Data source: CGWB, 1998 and UNDP, 1974).

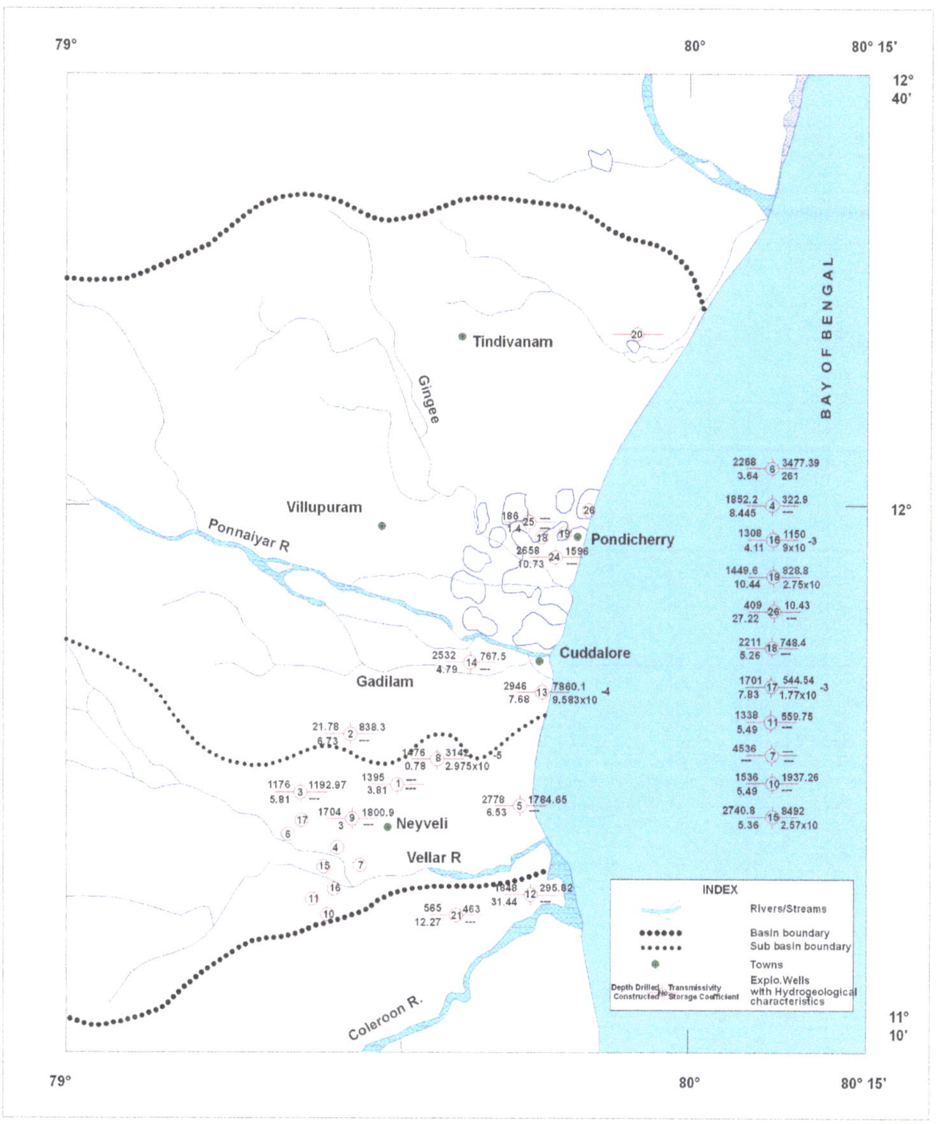

Fig. 4.31 Locations of selected well sites in Ponnaiyar and Vellar delta basins

Table 4.6 Regional hydrogeological parameters in the Ponnaiyar – Vellar delta basin

S.No	Location	Depth drilled (m) / Constructed (mbgl)	Range of Aquifers tapped (mbgl)	Thickness of the zones (m)	SWL (mbgl)	Discharge (lpm) / Drawdown (m)	Specific Capacity (lpm/m)	T (m^2/day)	S	Remarks
1	2	3	4	5	6	7	8	9	10	11
1	Adimanakuppam	457.2	85.17-115.0	22.5	53.2	1395	366.14	..	…	Sand, Clay and sand stone
	(11°37'37";79°26'00")	115.2				3.81				
2	Chettiteruvu	241.46	19.82-68.62	42.7	8.71	2178	323.5	838.3	--	Sand with clay, cuddalore sandstone with shale
	(11°40'00";79°24'30")	71.63				6.73				
3	Ragavankuppam	304.2	50.93-62.32	11.4	49.98	1176	202.33	1192.97	--	Sand with clay sandstone
	(11°35'00";79°22'00")	65.23				5.81				
4	Mettukuppam	195.07	157.50-192.07	24	2.3	1852.2	219.33	322.9	--	Sand with clay Cuddalore sandstone
	(11°31'00";79°31'30")	194.57				8.445				
5	Alapakkam	457.2	25.30-71.98	34.5	0.98	2778	425.46	1784.65	--	Sand with clay and gravel, sandstone
	(11°36'15";79°43'45")	74.9				6.53				
6	Budamur	210.67	59.75-84.14	24.4	18.67	2268	623.07	3477.39	261	Sandy clay, sandstone
	(11°30'30";79°20'00")	86.56				3.64				
7	Shatiatope	457.31	184.75-221.3	36.59	17.43	4536	…	….		Clay sand, sandstone with lignite
	(11°26'00";79°32'00")	224.39								
8	Kadambuliyur	215.39	86.85-140.20	38.15	50.6	1476	1892	3142	2.975×10^{-5}	Sand and clay, Cuddalore sandstone
	(11°42'00";79°33'00")	143.26				0.78				

Table 4.6 *Contd…*

S.No	Location	Depth drilled (m) / Constructed (mbgl)	Range of Aquifers tapped (mbgl)	Thicknes s of the zones (m)	SWL (mbgl)	Discharge (lpm) / Drawdown (m)	Specific Capacity (lpm/m)	T (m^2/day)	S	Remarks
1	2	3	4	5	6	7	8	9	10	11
9	Vadakkuthu	247.8	158.50-195.12	30.5	21.53	1704	568	1800.9	--	Sand, clay and Cuddalore sandstone, lignite
	($11^{0}36^{|}00^{||};79^{0}33^{|}30^{||}$)	198.12				3				
10	Palayamkotta	229.72	145.75-195.12	43.4	6.49	1536	279.7	1937.26	--	Sand, gravel, clay, Cuddalore sandstone
	($11^{0}21^{|}30^{||};79^{0}28^{|}50^{||}$)	198.12				5.49				
11	Ayyampettai	457.2	134.14-164.63	24.4	24.38	1338	243.7	559.75	--	Sand, gravel and glay Cuddalore sandstone
	($11^{0}22^{|}00^{||};79^{0}25^{|}30^{||}$)	167.64				5.49				
12	B.Mutlur	430	312.78-427.41	45.3	5.74 (m agl)	1848	58.77	295.82	--	Sand limestone, claystone and sandstone
	($11^{0}28^{|}20^{||};79^{0}43^{|}00^{||}$)	427.4				31.44				
13	Chavadi	670.5	141.86-415.23	31	0.85	2946	383.5	7860.1	9.583×10^{-4}	Sand and clay and gravel Cuddalore sandstones with pyrite nodules
12	B.Mutlur	430	312.78-427.41	45.3	5.74 (m agl)	1848	58.77	295.82	--	Sand limestone, claystone and sandstone
	($11^{0}28^{|}20^{||};79^{0}43^{|}00^{||}$)	427.4				31.44				
13	Chavadi	670.5	141.86-415.23	31	0.85	2946	383.5	7860.1	9.583×10^{-4}	Sand and clay and gravel Cuddalore sandstones with pyrite nodules
	($11^{0}45^{|}450^{||};79^{0}45^{|}00^{||}$)	417.76				7.68				

Table 4.6 *Contd…*

S.No	Location	Depth drilled (m) / Constructed (mbgl)	Range of Aquifers tapped (mbgl)	Thickness of the zones (m)	SWL (mbgl)	Discharge (lpm) / Drawdown (m)	Specific Capacity (lpm/m)	T (m^2/day)	S	Remarks
1	2	3	4	5	6	7	8	9	10	11
14	Naduveerapattu	370.3	153.68-344.96	106.4	20.33	2532	528.6	767.5	--	Cuddalore sandstone with clay intercalations
	(11°44'15";79°37'30")	348				4.79				
15	Valaiyamadevi	750	128.53-471.99	178.14	8.15	2740.8	511.3	8492	2.57×10^{-1}	Sand and clay, Cuddalore sandstone
	(11°28'00";79°30'00")	475				5.36				
16	Srimushnam	611	55.00-299.00	90	23.25	1308	318.2	1150	9×10^{-3}	Cuddalore sandstone with clay intercalations
	(11°24'04";79°24'22")	302				4.11				
17	Pudukuraiyapet	519.8	65.85-449.02	88.45	32.57	1701	217.24	544.54	1.77×10^{-3}	Sandstone clay sahle, calcareous sandstone with intercalations of limestone
	(11°31'15";79°21'10")	452				7.83				
18	Lake Estate	450	76.40-374.00	69.18	24.09	2211	420.34	748.4	--	Clay, sandstone, Limestone and shale
	(11°57'00";79°45'30")	377				5.26				
19	Auroville	450	311.08-393.33	94.91	37.98	1449.6	138.9	828.8	2.75×10^{-3}	Sand, calcareous sandstone limestone shale and chamockite
	(12°01'00";79°50'00")	397.68				10.44				
20	Chunampet									

Table 4.6 *Contd...*

S.No	Location	Depth drilled (m) Constructed (mbgl)	Range of Aquifers tapped (mbgl)	Thickness of the zones (m)	SWL (mbgl)	Discharge (lpm) Drawdown (m)	Specific Capacity (lpm/m)	T (m^2/day)	S	Remarks
1	2	3	4	5	6	7	8	9	10	11
21	Annamalainagar				11.90 (m agl)	565	46.04	463		
						12.27				
22	Kilanuvampatti				…	546	51.02	167		
						10.7				
23	Lalpettai				12.32 (m agl)	1610	156	197.5		
						10.32				
24	Moolakulam	600.06	58.0-174.9	61.9	8.535	2658	248	1596	…	
		178				10.73				
25	Karasur	336.4	90.9-285.9	77.7	18.5	186	132.8	….	…	
		287.9				1.4				
26	Alankuppam	253	40.62-200.3	38	9.35	409	15.02	10.43	….	
		204				27.22				

Regional groundwater flow net analysis: A regional groundwater flow trend map across different profiles for both near surface and deep aquifer systems are provided in the Fig. 4.32. The Groundwater flows across different lengths of different sub basins are provided in Table 4.7.

Table 4.7 Groundwater flow parameters across different profiles in Ponnaiyar-Vellar delta basin

S.No	Basin	Profile	Avg.Transmissivity (T) (m^2/day)	Gradient (I)(m/km)	Length of the Coast (L)(km)	Total flow (Q) (MCM)
Near Surface Aquifers						
1.	Ponnaiyar	P1	2000	1.14	35	11.65
		P2	2000	0.94	35	9.6
		P3	2000	1.25	25	9.12
2.	Vellar	P1	2000	1.25	20	7.3
Deep Aquifers						
1.	Ponnaiyar	P1	1500	1.94	35	44.61
		P2	1500	1.5	25	24.63
2.	Vellar	P1	1500	1.38	20	18.13

The total estimated groundwater flow across the whole coastal margins of Ponnaiyar and Vellar basins is of the order of 125.04 MCM which comes to about 2.5 % of the dynamic potentials of the basin.

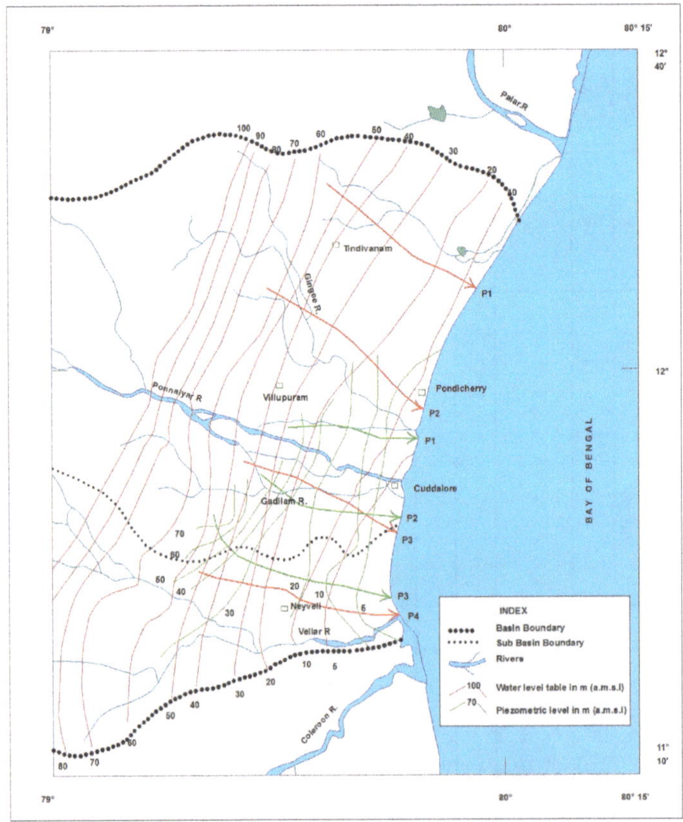

Fig. 4.32 Groundwater flow map for near surface and deep aquifer systems in Ponnaiyar-Vellar delta basin

Seasonal water level fluctuations: The seasonal water level fluctuations for a 5 year period (1980-85) at different site locations in different geological horizons are shown in the Table 4.8 and the site locations are reflected in the Figure 4.33. The average fluctuation in the granitic region is about 3.385 m and in the coastal alluvial/sedimentary tracks is about 4.85 m.

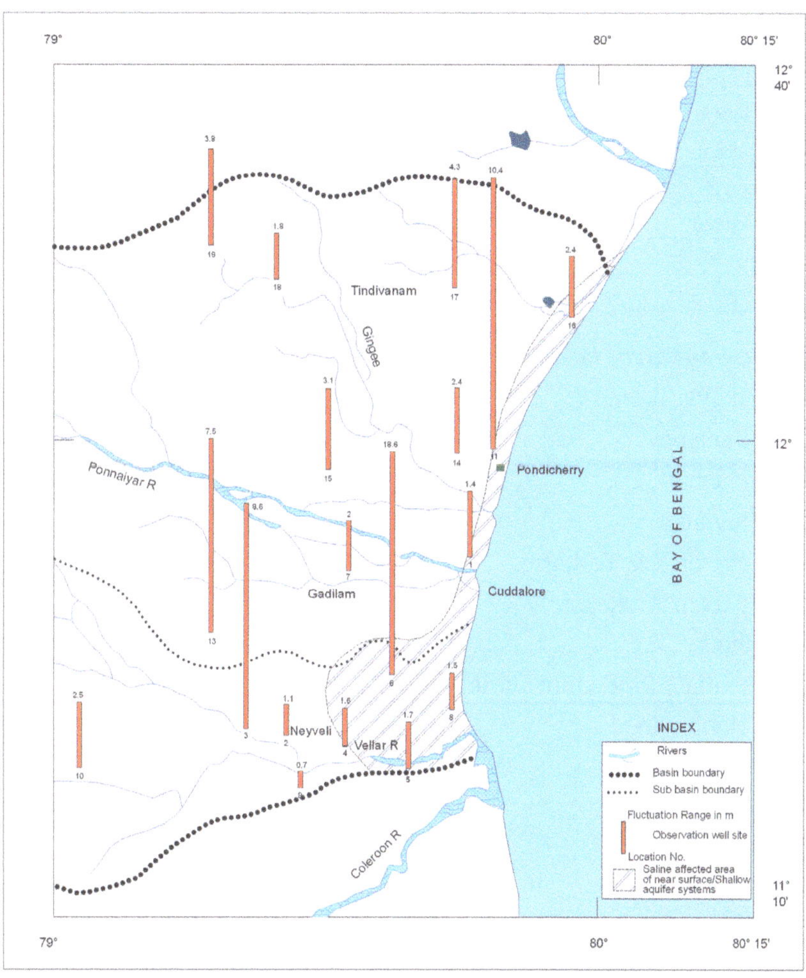

Fig. 4.33 Network of observation stations and their fluctuations in Ponnaiyar-Vellar delta basin

Table 4.8 Seasonal fluctuations in selected observation stations

S.No	Name of the Site	Geological Horizon	Avg. Fluctuation (m) for a 5 Year period (1980-85)
1.	Cuddalore	Alluvium	2.89
2.	Neyveli	Tertiary	1.11
3.	Virudhachalam	Alluvium	8.69
4.	Sethiatope	Alluvium	1.61
5.	Chidambaram	Alluvium	1.71
6.	Kurinjipadi	Tertiary	8.61
7.	Panrutti	Alluvium	1.95
8.	Puduchattram	Alluvium	1.52

Table 4.8 *Contd...*

S.No	Name of the Site	Geological Horizon	Avg. Fluctuation (m) for a 5 Year period (1980-85)
9.	Srimushnam	Tertiary	0.75
10.	Tholudur	Granite	2.55
11.	Auroville	Alluvium	10.43
12.	Kallakurichi	Granite	4.38
13.	Ulundurpet	Mesozoic	7.51
14.	Lake Estate	Tertiary	2.47
15.	Villupuram	Alluvium	3.16
16.	Markanam	Alluvium	2.41
17.	Tindivanam	Granite	4.31
18.	Gingee	Granite	1.86
19.	Avalurpet	Granite	3.81

Resource potentials in Ponnaiyar basin

- Area of the Ponnaiyar basin : 14,600 sq.km
- Coverage of Saline area : 700 sq.km
- Coverage of Archeans : 11,850 sq.km
- Coverage of Alluvials : 1,050 sq.km
- Coverage of Mesozoics & Tertiaries : 1,700 sq.km
- Dynamic potentials in Archeans :

(i) Fluctuation method : 2073.7 MCM

- Dynamic potentials for Sedimentaries (excluding saline track):

(i) Rainfall Infiltration method
 (a) Alluvium : 156.07 MCM
 (b) Mesozoics+Tertiaries : 270.53 MCM

(ii) Fluctuation method
 (a) Alluvium : 375 MCM
 (b) Mesozoics+Tertiaries : 1423.5 MCM

(iii) Average dynamic potentials
 (a) Alluvium : 265.53 MCM
 (b) Mesozoics+Tertiaries : 874.01 MCM

- Total dynamic potentials in Ponnaiyar basin is (2073.7 + 265.53 + 874.01 : 3213.43 MCM
- Resource potentials in deep confined aquifer systems (static) (Area × Thickness × S.Coefficient) :

(i) Alluvials ($1050 \times 30 \times 8 \times 10^{-3}$) : 252 MCM

(ii) Mesozoics+Tertiaries ($2750 \times 100 \times 2.5 \times 10^{-4}$) : 68.75 MCM

(iii) Archeans ($11850 \times 10 \times 5 \times 10^{-3}$) : 592.5 MCM

- Total estimated Resource potentials in deep confined aquifer systems

252 + 68.75 + 592.5 = 913.25 MCM

Note: 0-200 m is the depth range considered for estimations of near surface (dynamic) & deep confined system (static) potentials in this basin.

Resource potentials in Vellar basin

- Area of the Vellar basin : 7,500 sq.km
- Coverage of Saline area : 700 sq.km
- Coverage of Archeans : 5,150 sq.km
- Coverage of Alluvials : 1,600 sq.km
- Coverage of Mesozoics & Tertiaries : 750 sq.km
- Dynamic potentials in Archeans :

(i) Fluctuation method : 643.7 MCM

- Dynamic potentials for Sedimentaries (excluding saline track) :

(i) Rainfall Infiltration method
   (a) Alluvials : 208.10 MCM
   (b) Mesozoics+Tertiaries : 135.26 MCM

(ii) Fluctuation method
   (a) Alluvials : 510 MCM
   (b) Mesozoics+Tertiaries : 299 MCM

(iii) Average dynamic potentials :
   (a) Alluvials : 359.04 MCM
   (b) Mesozoics+Tertiaries : 217.13 MCM

- Total dynamic potentials in the velar sub basin is (643.7 + 359.04 + 217.13) : 1219.92 MCM
- Resource potentials in deep confined aquifer systems (static) (Area × Thickness × S.Coefficient) :

(i) Alluvials ($1600 \times 40 \times 8 \times 10^{-3}$) : 512 MCM

(ii) Mesozoics+Tertiaries ($2350 \times 120 \times 2.5 \times 10^{-4}$) : 70.5 MCM

(iii) Archeans ($5150 \times 10 \times 5 \times 10^{-3}$) : 257.5 MCM

- Total estimated resource potentials in deep confined aquifer systems (512 + 58.75 + 257.5) : 840 MCM
- Total estimated (dynamic + deep confined system) potentials in Ponnaiyar + Vellar sub basins : 6185.74 MCM

Note: 0-300 m is the depth range considered for estimations of near surface (dynamic) & deep seated aquifer system (static) potentials in this basin.

4.4.3 Cauvery coastal delta basin

Cauvery delta is the second biggest delta system with an area of the order of 11,000 sq.km along the east coast of India and is considered as the agricultural rich granary of south India. The Cauvery river has a stretch of about 800 km and debouches into Bay of Bengal through several distributaries with an estimated annual discharge of the order of about 20,950 MCM and has built up a major fan shaped delta system. The delta has its apex at Tiruchinapally where the river bifurcates into two: the northern one is the Coleroon and the southern one remains as Cauvery. Further downstream after the Grand anicut, Cauvery bifurcates into two namely Cauvery and Vennar with several distributaries downstream. The entire Cauvery delta region could be broadly divided into three sub basins: I. Cauvery sub basin, II. Vennar sub basin and III. New delta region. The Mettur dam located about 200 km upstream of Tanjavur forms the main source for surface water irrigation through a network of several distributary canals spread over the entire delta region. The surface water supplies however in recent years have become highly erratic and non sustainable in view of the erratic monsoon conditions in the area affecting drastically the surface irrigation potentials in the downstream delta area. As a consequence Groundwater exploitation in this region has been intensified, and several deep and shallow well structures have come into operation utilizing the available fresh groundwater resources in this part of the delta.

A regional drainage map of Cauvery delta basin and its adjoining regions with isohyetals is presented in Fig. 4.34. The total basin area considered is about 4,750 sq.km covering primarily the districts of Tanjore (100%), Tiruvarur (100%) and Nagapattinam (100%) and partially covering the districts of Cuddalore (17%), Pudukottai (45%), Perambaluru (62%), Tiruchurapally (86%), Namkal

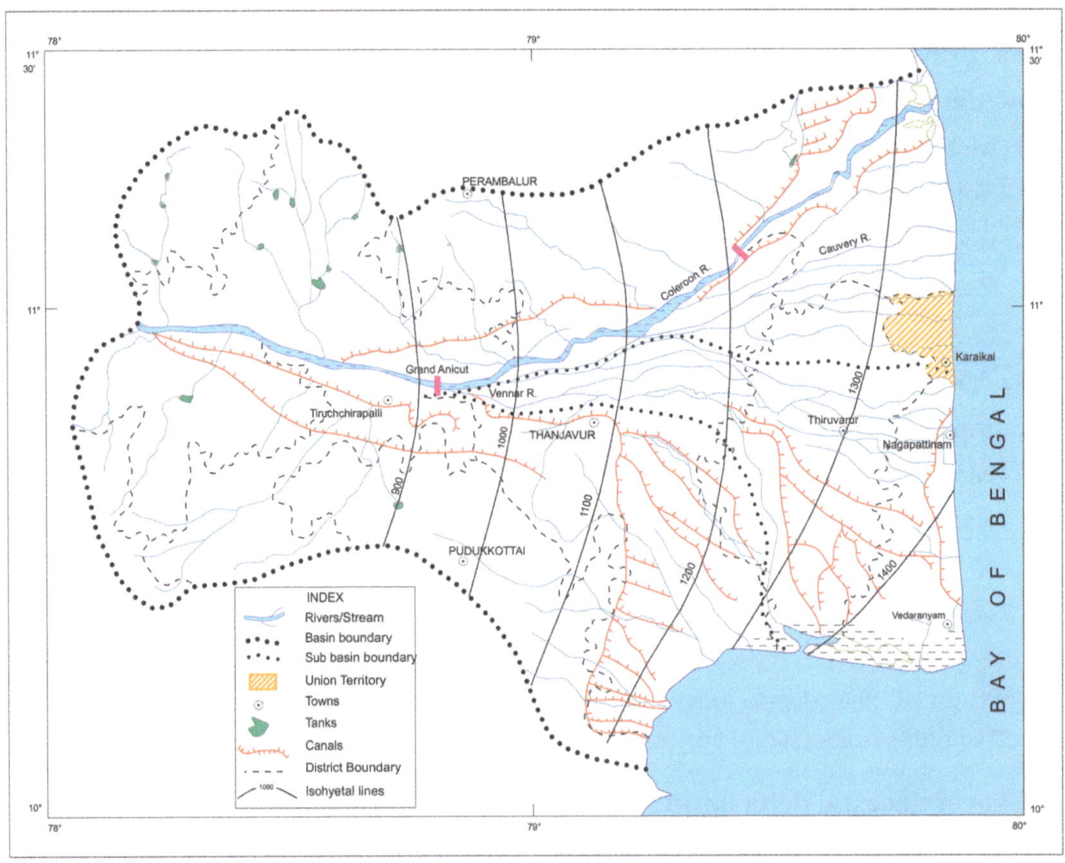

Fig. 4.34 Drainage and isohyetals in the Cauvery delta (Tamil Nadu)

Table 4.9 Surface and groundwater irrigation potentials in Cauvery delta basin

District	Surface Irrigation						Groundwater Irrigation					Gross Irrigation (ha)
	Tanks (ha)		Canals (ha)		Other sources (ha)	Total (ha)	Tubewells (ha)		Dug Wells (ha)		Total (ha)	
	Number	Area Irrigated (ha)	Number	Area Irrigated (ha)			Number	Area Irrigated (ha)	Number	Area Irrigated (ha)		
Thanjavur	428	500	25	155378		**155878**	38035	37279	54032	513	**37792**	**193670**
Nagapattinam	0	0	19	252482	1539	**254021**	18290	4385	9599	844	**5229**	**259250**
Thiruvarur	0	0	13	147564		**147564**					**17798**	**165362**
Cuddalore	101	1358	46	8331	39	**9728**	5217	19382	1915	1625	**21007**	**30735**
Perambalur	556	6281	431	6152	47	**12479**	8761	14185	27282	20816	**35001**	**47480**
Pudukottai	2453	32294	13	4806	610	**37709**	3771	8506	17301	3817	**12323**	**50032**
Tiruchurapalli	1514	7560	116	27863	10146	**45570**	7730	9444	82000	49779	**59223**	**104793**
Namkal	75	232	1	1425	5749	**7406**	1492	6302	22029	11855	**18157**	**25562**
Karur	128	288	11	7735	1288	**9312**	849	2490	23774	15747	**18237**	**27549**
Total	**5255**	**48513**	**674**	**611735**	**19418**	**679666**	**84144**	**101972**	**237932**	**104996**	**224766.2**	**904432.5**

(Source: Department of Economics & Statistics, Govt. of Tamil Nadu 2005-06)

(29%) and Karur (48%). The average annual rainfall in the delta region is of the order 1200 mm with maximum value of 1400 mm (close to the coast) and minimum value of 800 mm (away from the coast). About 65% of the precipitation comes from the northeast monsoon spread over the periods of October, November and December and the rest of the rainfall comes from southwest monsoon during the period July, August and September. The delta area enjoys a tropical climate with average maximum temperature of about 36 °C and a minimum temperature of about 22 °C. The relative humidity ranges between 85% (wet season) and 70% (dry season). The annual average potential evapotranspiration is about 1787 mm which is substantially high order 1.59 times to the average precipitation level (refer chapter-III). Both surface water through canals and tanks and the groundwater through tubewells, dugwells and filterpoints in the Cauvery delta region are intensely utilized. There are approximately 5,255 tanks and 674 canals contributing to an irrigation potential of the order of 6,79,666 hectares. In addition there are about 84,144 tube wells and 2,37,932 dug wells contributing to an irrigation potential of the order of 2,24,766 hectares. The combined gross irrigation potentials is of the order of 9,04,433 hectares. The percentage of groundwater utilization as per the existing data is about 25% to the total irrigation potential created. The proportionate district wise breakup of surface and groundwater irrigation potentials in the basin area are presented in Table 4.9.

Regional hydrogeological characteristics and system parameters

Cauvery delta region is one of the most intensely studied basin through several exploratory endeavours both by central and state agencies and through UNDP programmes in different phases. A regional geological map of the basin is presented in the Fig. 4.35. The geological units out cropping in the basin area are Archeans, Mesozoics, Tertiaries, Quaternaries and Recent.

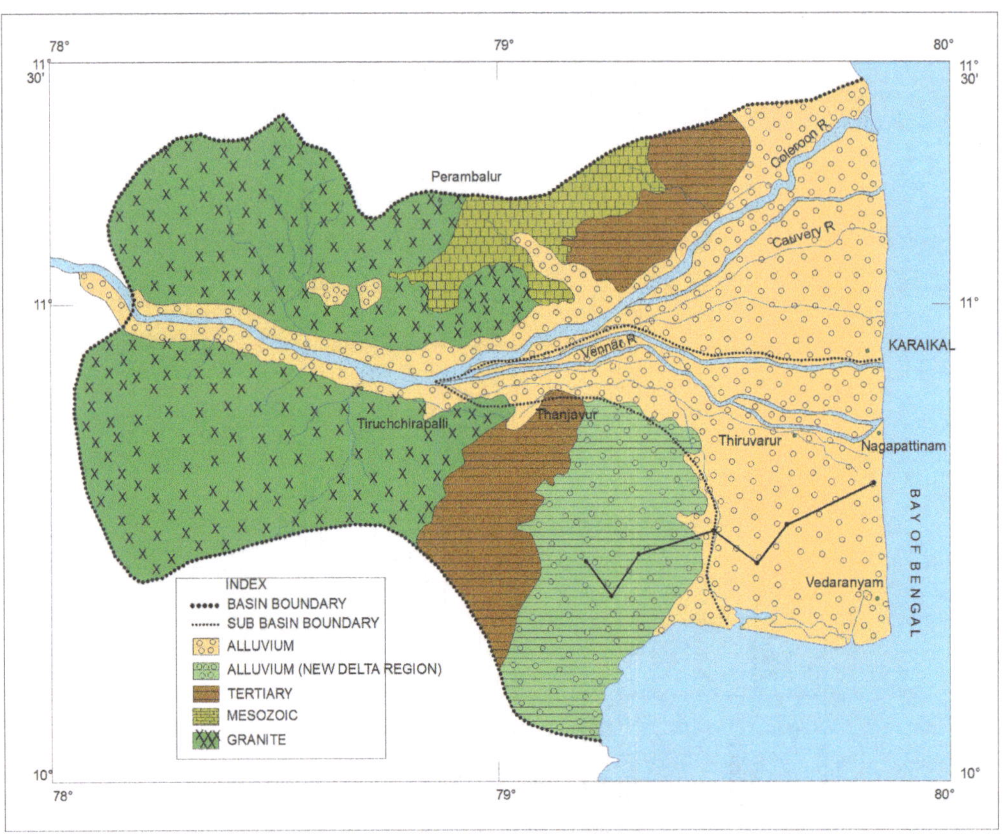

Fig. 4.35 Regional Geological map of Cauvery delta (Tamil Nadu)

Archeans are the oldest group of rocks exposed in the upper regions of the basin. They constitute mainly granites, gneisses, schists, and charnockites with intrusions of dolerite dykes and pagmatitic veins. The top 10 to 20 m are normally found to be weathered and below 60 m are found to be fractured and fissured. Dug wells and dug cum bore wells with yields ranging from 30 to 160 lpm are common in this terrain. These aquifers have a transmissivity values ranging between 5 to 10 m^2/day.

Mesozoics are represented by Cretaceous group of rocks which occur as narrow belts lying unconformably over the Archeans. In general these units contain yellow calcareous sandstones, clays, marls and silt stones believed to have been deposited under marine conditions. These rock types outcrop as narrow belts in the upstream areas and are regionally extensive and well spread out underneath the Tertiaries in the deltaic track and attain reasonably high thicknesses of the order varying between 200 to 400 m in the central and outfall regions of the delta. These aquifers in the upstream regions occur in both confined and unconfined situations with yields varying from 300 to 400 lpm and with transmissivities values of the order varying from 50 to 100 m^2/day.

Tertiaries are represented by Eocene, Miocene and Pliocene sediments. The Eocene sediments are the oldest of the sedimentaries and are not exposed any where in the basin terrain but are encountered in the deep bore holes at depths varying between 140 to 450 m below ground level. These include intercalated beds of silts, clays, shales and sandy clays and occasionally limestones. These lithological data available through deep bore holes of the region indicate that the western region of the basin around Orthanadu exhibits rather coarse to medium grained sediments which gradually become fine grained towards eastern parts. Groundwater in these units occur in deep confined conditions with piezometric levels varying between 2.25 to 4 m below ground level and with their yields varying between 300 to 600 lpm. The estimated transmissivities for these aquifers vary between 1500 to 1800 m^2/day. The Miocene sediments (Cuddalores) contain the most dominant and potential group of aquifers with several well fields under operation covering the entire delta. The sediments are sub divided into two units namely (i) The lower Miocene beds/Orthanadu beds and (ii) Upper Miocene flowing zone. The lower Miocene aquifers or Orthanadu beds are wide spread over the entire delta region. The region probably had taken a delta shape during the deposition of these beds or during early Miocene times. The outcropping widths vary from 4 to 5 km in the region south of Tanjore. They are coarse grained near the outcrop area and gradually turns out to be fine grained towards outfall area of the basin. The thicknesses of these beds vary between 30 to 70 m. The groundwater in these units occur in deep confined conditions with their piezometric levels varying between 2 to 7.5 m above ground level (Artesian flows). These aquifers are normally encountered at depth levels varying between 150 to 350 m below ground level. The yields in these aquifer systems vary between 600 to 1200 lpm and the transmissivities vary between 250 to 1500 m^2/day. The average estimated storativity for these aquifers is of the order of 5×10^{-4}. The upper Miocene zone which is the main flowing zone contain coarse and fine sediments and are distributed abundantly in the Vennar sub basin and in the new delta region. These sediments have an average thickness of the order of 35 m and are encountered in the boreholes at depths varying between 100 to 250 m. The yields of these aquifers are of the order of 500 to 1000 lpm and piezometric levels stand at 1 to 6 m above ground level. The lower Orthanadu and upper flowing aquifers are often interconnected hydrologically. The artesian flow encountered is a result of interconnectivity pressure heads prevailing in these geological units. The transmissivities in the units vary between 200 m^2/day to 1400 m^2/day.

The Pliocene sediments which constitute the youngest group of sediments of the Tertiary group of rocks lie conformably over the Miocene sediments. These are exposed abundantly in the new delta area and around the Grand anicut area in the upstream. These are essentially sands and sandy clays and gravels which form as flat beds and elongated depressions blanketed with lateritic rocks and sometimes with wind blown sands. They are also seen underlying the Quaternaries in the sub basin regions of the Cauvery and Vennar. These are essentially deposited under fluviatile conditions of the delta and turn out to be estuarine and marine near the coastal margins. The thicknesses of these sediments have a range varying between 50 to 75 m and yields of the order varying between 150 to 450 lpm. The estimated transmissivites for these aquifers is of the order varying between 500 to 1300 m^2/day and an average storativity value of the order of 4×10^{-4}.

Quaternaries and Recent form the youngest group of sediments in the deltaic track. They are mostly fluviatile sediments but tend to become marine towards the outfall regions of the coast. The average thickness of these sediments in the upstream area is about 12 m and increases to an order varying between 40 to 60 m in the outfall region of the basin. The aquifer yields in these systems vary from 500 to 750 lpm and transmissivities range from 1200 to 1500 m^2/day.

A subsurface hydrogeological section depicting various lithological units and their depth levels from west (Nambivoyal) to east (Prathaparamapuram) is presented in Fig. 4.36. The section displays the thickness variations of the lithological units from Recent to Eocene. The upstream areas of the section displays relatively large thicknesses of beds but as we move towards the outfall region the thick granular zones become bifurcated into several alternate beds of sands and clays in different geological horizons. In general all the beds display a gradual dipping trend towards the coastal margin.

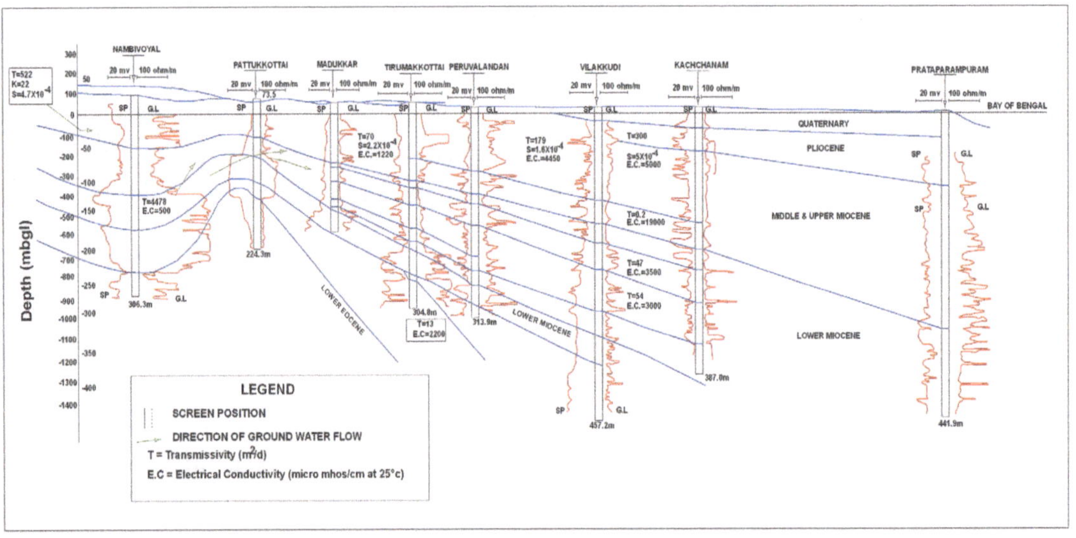

Fig. 4.36 Lithological and electrical log sections in the southern part of Cauvery delta

(Source: CGWB, 2005)

A Summarized account of hydrological parameters for selected number of sites as shown in Fig. 4.37 are presented in Table 4.10 (Data source: CGWB, 1998 and UNDP, 1974).

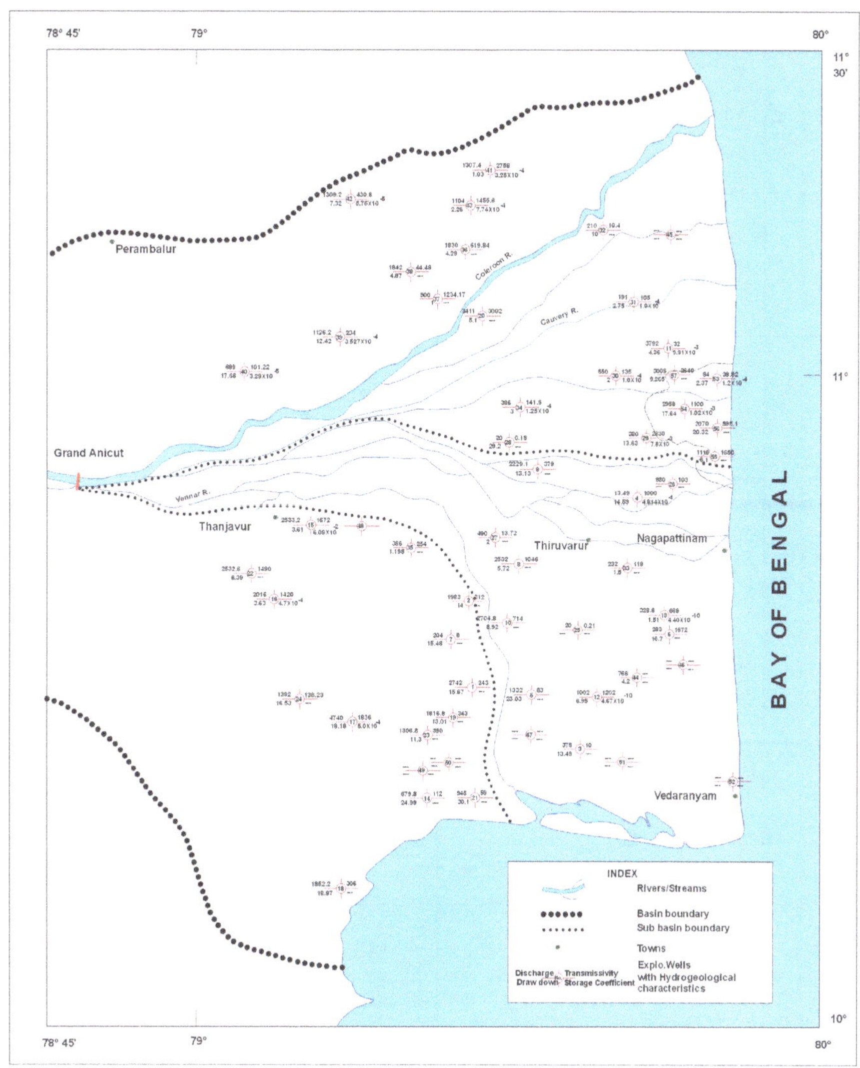

Fig. 4.37 Location map of selected exploratory well sites in Cauvery delta basin

Table 4.10 Hydrogeological parameters in the Cauvery delta basin

S.No	Location	Depth drilled(m) / Constructed (mbgl)	Range of Aquifers tapped (mbgl)	Thickness of the zones (m)	SWL (mbgl)	Discharge (lpm) / Drawdown (m)	Specific Capacity (lpm/m)	T (m^2/day)	S	Remarks
1	**2**	**3**	**4**	**5**	**6**	**7**	**8**	**9**	**10**	**11**
1	Adichchapuram	280.26	33.53-205.74	91.96	1.520 (magl)	2742	174.9	243	--	Sand with clay and sandstones
	($10^0 30' 00''$; $79^0 30' 00''$)	207.26				15.67				
2	Moovanallur	307.84	35.05-306.32	121	2.02	1983	141.6	312	--	Sand with clay and carbonaceous shale and Tertiary sandstones
	($10^0 41' 00''$; $79^0 26' 00''$)	307.84				14				
3	Pinnathur	305.1	167.64-196.12	19.34	4.85 (magl)	378	28	10	--	Sand, clay, sandstone and shale
	($10^0 27' 25''$; $79^0 34' 00''$)	198.12				13.49				
4	Papancheri	271.27	218.85-258.17	39.32	4.570 (magl)	2274	155.7	1000	4.814×10^{-4}	Sand, clay sandstone & limestone
	($10^0 48' 30''$; $79^0 50' 30''$)	259.69				14.59				
5	Thirumakottai	305.4	48.07-251.59	67.47	0.26 (magl)	1332	57.83	83	--	Sand, clay and sandstone
	($10^0 31' 55''$; $79^0 27' 20''$)	251.25				23.03				
6	Prataparamapuram	458.11	242.32-297.18	42.67	3.40 (magl)	283	17	1672	--	Sand with clay sandstone and limestone
	($10^0 40' 30''$; $79^0 49' 30''$)	298.7				16.7				

Table 4.10 ***Contd...***

S.No	Location	Depth drilled(m) / Constructed (mbgl)	Range of Aquifers tapped (mbgl)	Thickness of the zones (m)	SWL (mbgl)	Discharge (lpm) / Drawdown (m)	Specific Capacity (lpm/m)	T (m^2/day)	S	Remarks
1	2	3	4	5	6	7	8	9	10	11
7	Kantikudikadu	398.07	176.78-268.22	54.82	2.28	204	13.17	8	--	Sandstone with clay and shale (Abandoned)
	($10^0 38' 47''$; $79^0 23' 41''$)	275.84				15.48				
8	Adanurtatti	456.89	24.38-301.75	55.46	0.62	2532	442	1046		Sand with clay & sandstone and limestone
	($10^0 46' 10''$; $79^0 23' 10''$)	303.77				5.72				
9	Thiruvarur (Kattur)	457.2	190.50-274.32	57.91	2.43 (magl)	2229.1	169.7	379		Sand, clay and sandstone
	($10^0 46' 35''$; $79^0 34' 05''$)	207.84				13.13				
10	Maravakkadu (Mannargudi)	465.61	190.54-294.14	67.45	3.198 (magl)	2704.8	303	714		Sand, clay and sandstone
	($10^0 37' 35''$; $79^0 27' 30''$)	294.78				8.92				
11	Nalladai	376.24	41.15-280.42	60.95	0.27 (magl)	3792	870	32	9.91×10^{-3}	Sand with clay, sandstone and limestone
	($11^0 00' 45''$; $79^0 44' 10''$)	283.46				4.36				
12	Thiruthuraipoondi	666.2	500.00-524.16	24.16	12.30 (magl)	1002	144.3	1202	4.67×10^{-10}	Sandstone claystone and limestone with sand and clay
	($10^0 31' 45''$; $79^0 39' 00''$)	524				6.95				
13	Thevur	687	278.00-358.00	27	2.910 (magl)	328.8	217.8	669	4.40×10^{-10}	Sand, clay & sandstone
	($10^0 43' 00''$; $79^0 44' 00''$)	365				1.51				
14	Rajankadu	228.6	30.68-196.73	38.64	2.06	679.8	27.21	112		Sandstone with clay and shale
	($10^0 23' 00''$; $79^0 23' 40''$)	198.45				24.99				

Table 4.10 ***Contd…***

S.No	Location	Depth drilled(m) / Constructed (mbgl)	Range of Aquifers tapped (mbgl)	Thickness of the zones (m)	SWL (mbgl)	Discharge (lpm) / Drawdown (m)	Specific Capacity (lpm/m)	T (m^2/day)	S	Remarks
1	2	3	4	5	6	7	8	9	10	11
15	Kattuthottam	245.05	46.17-134.56	60.98	3.4	2533.2	701.7	1672	6.09×10^{-2}	Sand with clay
	($10^0$46'35"; $79^0$10'10")	136.09				3.61				
16	Ichankottai	304.8	58.48-167.94	82.29	10.37	2016	555.3	1420	4.7×10^{-4}	Sand with clay & gravel
	($10^0$39'30"; $79^0$09'20")	169.77				3.63				
17	Nambivoyal	306.11	149.35-243.84	54.19	10.06 (magl)	4740	260.73	1836	5.0×10^{-4}	Sandstone with clay
	($10^0$29'00"; $79^0$15'20")	246.88				18.18				
18	Udainadu	309.37	111.25-163.07	45.72	0.91 (magl)	1852.2	98.15	306	--	Sand with clay
	($10^0$13'25"; $79^0$15'55")	166.12				18.87				
19	Madukkur	304.8	24.38-207.26	55.71	0.34 (magl)	1816.8	139.6	343	--	Sand with clay
	($10^0$28'00"; $79^0$23'55")	210.31				13.01				
20	Pattam	305.71	124.05-194.15	70.1	9.07 (magl)	3411	668.8	3002	--	Sand, clay gravel & sandstone
	($11^0$06'20"; $79^0$27'10")	197.51				5.1				
21	Othankadu	304.8	38.10-111.25	59.43	1.05	945	31.46	59	--	--
	($10^0$20'22"; $79^0$27'30")	112.77				30.1				
22	Marungulam	243.84	46.63-109.40	41.99	16.08	2532.6	396.64	1490	--	Sand with clay and limestone
	($10^0$40'52"; $79^0$30'26")	112.17				6.39				
23	Veppankulam	243.84	68.12-185.10	36.11	3	1306.8	115.74	390	--	Sand with clay and calcareous sandstone
	($10^0$28'00"; $79^0$22'30")	187.15				11.3				

Table 4.10 *Contd…*

S.No	Location	Depth drilled(m) / Constructed (mbgl)	Range of Aquifers tapped (mbgl)	Thickness of the zones (m)	SWL (mbgl)	Discharge (lpm) / Drawdown (m)	Specific Capacity (lpm/m)	T (m^2/day)	S	Remarks
1	2	3	4	5	6	7	8	9	10	11
24	Tiruvonam	304.8	102.70-127.10	24.4	2.44	1392	84.15	138.23	--	Clay and sand
	(10°30'00"; 79°09'30")	130.14				16.53				
25	Vilakkudi	152	124.97-148.03	23.06	-12.68	20	…	0.21	---	Pliocene
	(10°34'45";79°43'20")	149.38				---				
26	Kilvalur	397.76	252.98-280.4	27.42	5.95	980	…	1043	…	Miocene
	(10°45'40";79°43'20")	283.47				..				
27	Adangudi	189.5	166.23-182.63	16.4	1.59	490	245	13.72	…	Miocene
	(10°43'40";79°30'00")	182.63				2				
28	Kadmbangudi	59.15	41.18-45.75	4.57	0.56	20	0.68	0.18	…	Miocene
	(10°55'30";79°30'40")	48.8				29.2				
29	Kovilkadankudi	65.08	45.29-63.44	12.17	-2.56	380	27.8	2830	7.8×10^{-3}	Pliocene
	(10°57'00";79°43'15")	63.44				13.63				
30	Kiranur	152.4	109.6-140	30.4	0.478	550	275	135	3.3×10^{-4}	Miocene
	(11°00'05";79°40'00")	146.54				2				
31	Pattamangalam	153.93	124.34-148.73	45.19	-0.54	191	69.45	105	1.0×10^{-4}	Miocene
		151.52				2.75				
32	Pattavarthi	184.41	145.46-175.97	18.19	1.99	210	21	10.4	…	Miocene
		178.99				10				
33	Sikar	38.1	22.65-27.48	4.83	-1.63	232	154.6	119	…	Quaternary
	(10°42'15";79°40'20")	32.62				1.5				

Table 4.10 ***Contd…***

S.No	Location	Depth drilled(m) Constructed (mbgl)	Range of Aquifers tapped (mbgl)	Thickness of the zones (m)	SWL (mbgl)	Discharge (lpm) Drawdown (m)	Specific Capacity (lpm/m)	T (m^2/day)	S	Remarks
1	2	3	4	5	6	7	8	9	10	11
34	Sivanaragaram	153.92	137.17-149.38	12.21	0.01	386	128.6	141.5	1.25×10^{-4}	Miocene
		150.91				3				
35	Vellikkadu	153.2	62.41-117.72	37.3	2.605	386	323	254	…	Miocene
	($10^0 32^{\|} 40^{\|\|}; 79^0 33^{\|} 20^{\|\|}$)	123.34				1.195				
36	Jayamkondacholapuram	250.24	118.87-137.16	18.29	15.04	1830	426.57	619.84		cuddalore sandstone
		140.2				4.29				
37	Kodangudi	231.34	80.77-131.06	36.58	4.29 (magl)	900	900	1234.17		cuddalore sandstone
		131.16				1				
38	Tularankuruchi	457.2	54.86-240.79	36.6	27.04	1842	378.23	44.48		cuddalore sandstone with clay, shale and lignite
		243.79				4.87				
39	Vikramangalam	376.8	139.00-288.00	116	6.1	1126.2	90.66	234	3.527×10^{-4}	sandstone with clay charnockite
		291				12.42				
40	Venkatanarayanapuram	728.5	99.00-360.00	90	16.89	689	38.97	101.22	3.29×10^{-5}	cuddalore sandstone with clay inter-calations
		363				17.68				
41	Kaduvetti	610	138.0-528.0	132	19.75	1307.4	1269.7	2758	3.25×10^{-4}	cuddalore sandstone with clay intercaltions
		531				1.03				

Table 4.10 ***Contd…***

S.No	Location	Depth drilled(m) / Constructed (mbgl)	Range of Aquifers tapped (mbgl)	Thickness of the zones (m)	SWL (mbgl)	Discharge (lpm) / Drawdown (m)	Specific Capacity (lpm/m)	T (m^2/day)	S	Remarks
1	2	3	4	5	6	7	8	9	10	11
42	Cholankurichi	567	165.0-315.0	78	25.38	1309.2	178.8	430.8	5.75×10^{-5}	cuddalore sandstone with clay intercaltions
		318				7.32				
43	Jayankondam	628	224.0-520.0	138.89	34.44	1104	488.49	1455.6	7.74×10^{-6}	cuddalore sandstone with clay intercaltions
		523				2.26				
44	Kokaladi(5217)	457.2	50.9-161.2	43	3.5	768	182.8			Abandoned
		161.23				4.2				
45	Sirkazhi(5201)	448.06								Abandoned
46	Thalanayar(5213)	359.1								Abandoned
47	Uppur(5218)	632	201.0-415.0							Abandoned
48	Karunthattankudi (3603)	334.06								Abandoned due to non-encountering of any granular zones
49	Turavikkadu(3611)	129.23								Abandoned PY
50	Vallam(3612)	182.57								Abandoned due to absence of granular zones
51	Voimedu(5221)	711								Abandoned
52	Vedaranyam(5222)	489.7								Abandoned

Table 4.10 ***Contd…***

S.No	Location	Depth drilled(m) / Constructed (mbgl)	Range of Aquifers tapped (mbgl)	Thickness of the zones (m)	SWL (mbgl)	Discharge (lpm) / Drawdown (m)	Specific Capacity (lpm/m)	T (m^2/day)	S	Remarks
1	2	3	4	5	6	7	8	9	10	11
53	Kottucheri	225	167.9-222.0	54.1	2.86	84	35.44	38.82	1.2×10^{-4}	cs226-229
		224				2.37				
54	Tirunallar	380.7	261.6-380.0	73.35	5.6	2958	167.5	1100	1.02×10^{-3}	cs275-258
		382				17.64				
55	Polagam	325.4	195.0-285.0	69.9	8.65	1116	182.9	1650		
		287.1				6.1				
56	Thalatheru	220	120.9-211.0	84.1	0.6	2070	101.8	595.1		
		213				20.32				
57	Nallathur	329.8	170.6-323.8	77.3	5.1	3006	324.4	2640		
		325.8				9.265				

Regional groundwater flow net analysis: A groundwater flow trend map across different profiles for both near surface and deep aquifer systems is provided in the Figure 4.38. The groundwater flows across different lengths of different sub basins are provided in Table 4.11.

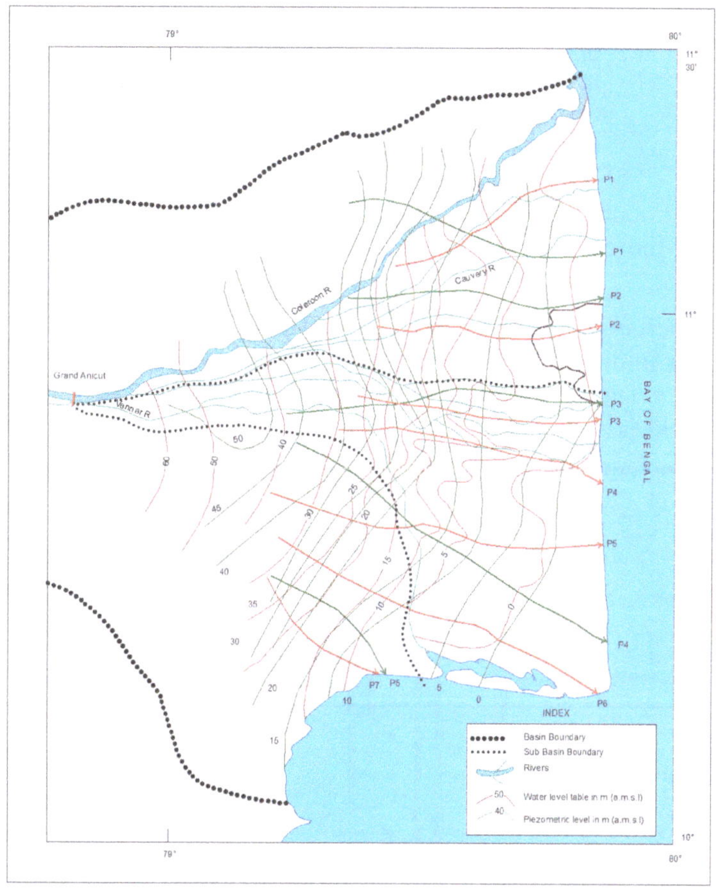

Fig. 4.38 Groundwater flow map for near surface and deep aquifer systems in Cauvery delta

Table 4.11 Groundwater flow across different profiles in Cauvery delta

S.No	Basin	Profile	Avg. Transmissivity (T) (m^2/day)	Gradient (I) (m/kms)	Length of the Coast (L) (kms)	Total flow (Q) (MCM)
Near Surface Aquifers						
1	Cauvery & it's adjacent basins	P1	2500	0.63	40	22.81
		P2	2500	0.58	23	12.2
		P3	2500	0.43	17	6.6
		P4	2500	0.54	13	6.44
		P5	2500	0.60	25	13.68
		P6	2500	0.54	35	17.35
		P7	2500	0.69	44	27.68

Table 4.11 *Contd...*

S.No	Basin	Profile	Avg. Transmissivity (T) (m^2/day)	Gradient (I) (m/kms)	Length of the Coast (L) (kms)	Total flow (Q) (MCM)
Deep Aquifers						
1	Cauvery & it's adjacent basins	P1	1500	0.61	42	13.93
		P2	1500	0.45	17	4.23
		P3	1500	0.69	38	14.4
		P4	1500	0.80	33	14.45
		P5	1500	1.32	38	27.37

The total estimated groundwater flow across the whole coastal margins of Cauvery basin and its adjoining regions is of the order of 181.14 MCM which comes to about 3.25 % of the dynamic potentials of the basin.

Seasonal water level fluctuations: The seasonal water level fluctuations for a 5 year period (1980-85) at different site locations in different geological horizons are shown in the Table 4.12 and are reflected in the Figure 4.39. The average fluctuation in the granitic region is about 3.55 m and in the coastal alluvial/sedimentary tracks is about 3.32 m.

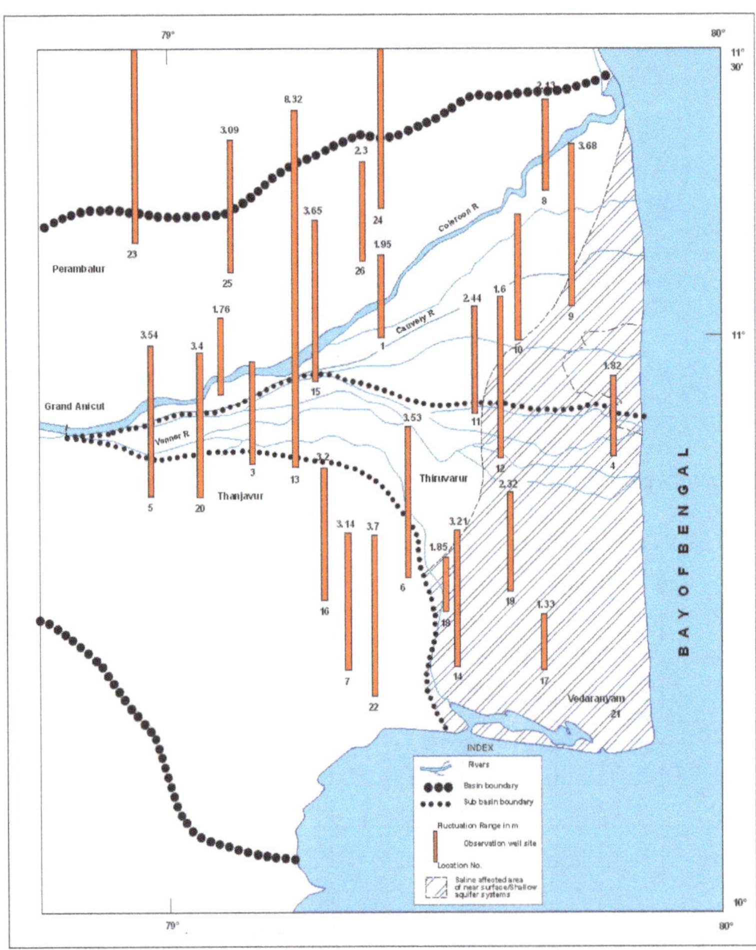

Fig. 4.39 Network of observation wells and their fluctuations in Cauvery delta

Table 4.12 Seasonal water level fluctuations in Cauvery delta

S.No	Name of the Site	Geological Horizon	Avg. Fluctuation (m) for a 5 Year period (1980-85)
1	Kumbakonam	Alluvium	1.96
2	Thiruvaiyaru	Alluvium	1.77
3	Thanjavur	Tertiary	2.27
4	Nagapattinam	Alluvium	1.83
5	Sengipatti	Granite	3.55
6	Ullikottai	Alluvium	3.53
7	Pattukottai	Alluvium	3.14
8	Sirkali	Alluvium	2.13
9	Nalladai	Alluvium	3.69
10	Peralam	Alluvium	2.87
11	Nannilam	Alluvium	2.44
12	Tiruvarur	Alluvium	3.69
13	Salivamangalam	Alluvium	8.33
14	Mothupet	Alluvium	3.21
15	Papanasanam	Alluvium	3.65
16	Orathanadu	Alluvium	8.21
17	Thagattur	Alluvium	1.33
18	Tirumakkottai	Alluvium	1.35
19	Thiruthuraipoondi	Alluvium	2.33
20	Vallam	Tertiary	3.40
21	Vedaranyam	Alluvium	1.10
22	Athiramapatinam	Alluvium	3.72
23	Perambalur	Mesozoic	4.58
24	Jayamkondancholapuram	Tertiary	7.21
25	Ariyalur	Mesozoic	3.09
26	T.palur	Tertiary	2.31

Resource potentials in Cauvery basin

- Area of the Cauvery basin : 21,000 sq.km
- Coverage of Saline affected area : 3365 sq.km
- Coverage of Archeans : 8,450 sq.km
- Coverage of Alluvials : 10,100 sq.km
- Coverage of Mesozoics : 1,200 sq.km
- Coverage of Tertiaries : 1,250 sq.km
- Dynamic Potentials in Archeans :

(i) Fluctuation method : 1478.7 MCM

- Dynamic potentials for Sedimentaries (excluding the saline affected region) :
- Rainfall Infiltration method

(i) Alluvials : 1368.83 MCM

(ii) Mesozoics : 243.89 MCM

(iii) Tertiaries : 254.05 MCM

- Fluctuation method

(i) Alluvials : 3131.78 MCM

(ii) Mesozoics : 444 MCM

(iii) Tertiaries : 473.75 MCM

- Average dynamic potentials

(i) Alluvials : 2250.3 MCM

(ii) Mesozoics : 343.9 MCM

(iii) Tertiaries : 363.9 MCM

- Total dynamic potentials in Cauvery basin is (1478.7 + 2250.3 + 343.9 + 363.9) : 4436.89 MCM
- Resource potentials in deep confined aquifer systems (static) (Area × Thickness × S.Coefficient) :

(i) Alluvials ($7219 \times 30 \times 8 \times 10^{-3}$) : 1732.56 MCM

(ii) Tertiaries ($8469 \times 100 \times 2.5 \times 10^{-4}$) : 211.71 MCM

(iii) Tertiaries (underneath the saline area) ($2881 \times 50 \times 2.5 \times 10^{-4}$) : 36.01 MCM

(iv) Mesozoics ($12550 \times 100 \times 2.5 \times 10^{-5}$) : 31.37 MCM

(v) Archeans ($8450 \times 10 \times 5 \times 10^{-3}$) : 422.5 MCM

- Total estimated resource potentials in deep confined aquifer systems : 2434.17 MCM
- Total estimated (dynamic + deep confined system) potentials in Cauvery basin : 6871.06 MCM

Note: 0-400 m is the depth range considered for estimations of dynamic & static potentials in this basin

4.4.4 Vaigai delta and its adjacent basins

Vaigai delta basin located further south of Cauvery (E.long: $77^0 50^|$ to $80^0 00^|$ and $10^0 10^|$ to $09^0 00^|$)) is considered to be an important medium river basin covering the districts of Ramanathapuram (93%) and Shivaganga (82%) mainly and partially the districts of Virudanagar (16%), Madurai (42%) and Pudukkottai (11%). Vaigai has a stretch of about 258 km and debouches into Bay of Bengal with an annual discharge of the order of 770 MCM. In the immediate vicinity of the Vaigai two short stretched rivers also exist namely Manimuttar towards north and Gundar towards south and have relatively small delta systems built near the mouth areas. These two rivers have short stretch lengths of the order of 125 km and 146 km with discharges of the order of 310 MCM and 480 MCM respectively. The Vaigai river and its delta track is a linearly disposed one with its apex near Madurai. It has an aereal extension of the order of about 2,500 sq.km with two or three delta stages for Vaigai with different apex points, the first one at Madurai; second one at Tirupuvanam and the third one at Paramakkudi. In all these three stages of delta formation reasonably good built up of alluvial flood plains could be seen with increasing thicknesses towards the outfall regions. All these three rivers have contributed quite substantially to the built up of interlaced fluvial systems with thicknesses varying between 20 to 40 m. These delta systems are endowed with numerous lakes and tanks with raised uplands and with heights varying between 1 to 3 m giving the delta a typical bird foot shape. The delta system is characterized mainly by flood plains, fluvial sands and clays disposed mainly towards the southern flanks of the Vaigai river and with a number of paleo channels distributed on either side of the river. The Manimuttar river located north of Vaigai also carries with similar morphological trends but Gundar river located south of Vaigai has little or no significant delta.

A regional drainage map of Vaigai, Manimuttar and Gundar river/delta systems with isohyetal trends marked is shown in Fig. 4.40. The average annual precipitation levels in the basin area is of the order of 823 mm. Over 60% of the annual rainfall is received from northeast monsoon during the period October to December and over 25% of the annual rainfall comes from southwest monsoon during the period June to September. The rest of the 15% is spread over in the period from January to May. The precipitation trends presented in the basin show a remarkable low (700 mm) in the central portion of the basin with a gradual increase towards the northeast (1100 mm) and southwest (900 mm). The basin area traditionally is considered drought prone region with low rainfall conditions. The average mean summer temperature is about 30 °C and winter mean temperature is about 25.5 °C and with humidity levels ranging between 84% (wet season) to 67% (dry season). The average annual potential evapotranspiration in the basin area is of the order of 1813 mm which comes to nearly 2.2 times of average precipitation levels of the basin which characterizes the drought proneness of the region.

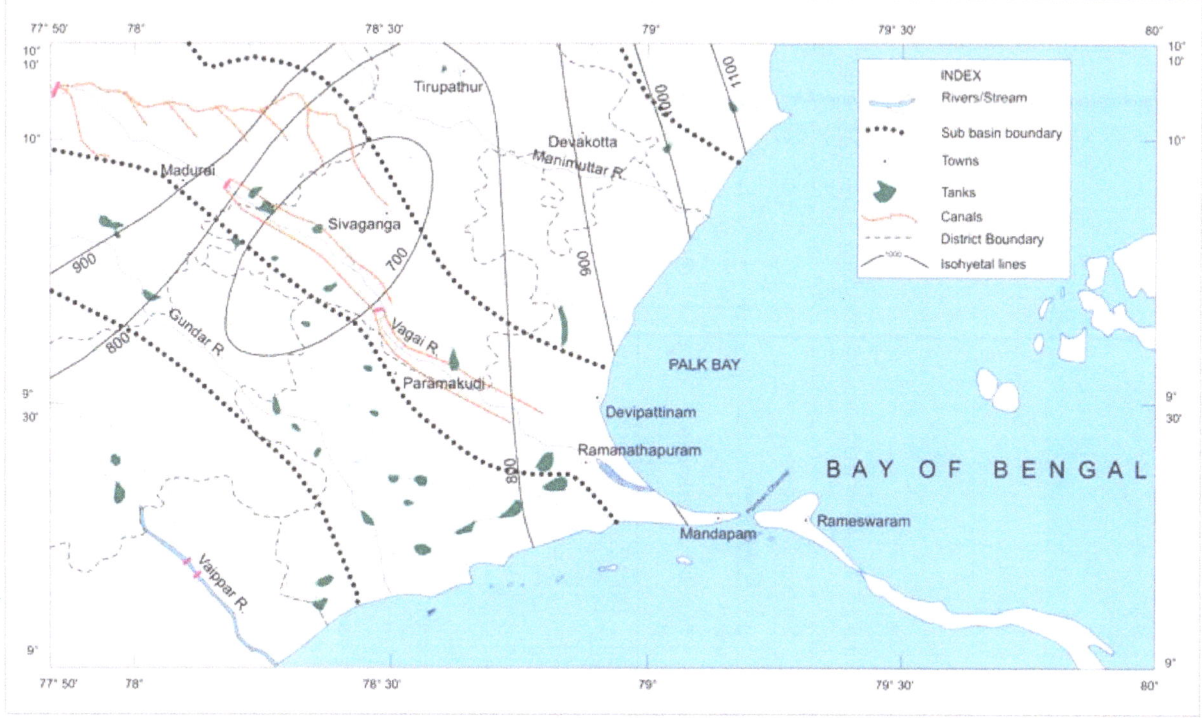

Fig. 4.40 Drainage and isohyetal lines in Vaigai and its adjoining deltas

All the three rivers namely Vaigai, Manimuttar and Gundar are ephemeral in nature and the water flows are observed only during monsoon and dry during the rest of the seasons. There are approximately 7,311 tanks and 37 no's canal systems contributing to surface irrigation potential of the order of about 1,52,188 hectares. Similarly there are approximately 2,325 tube wells and 48,107 dug wells contributing to the irrigation potential of the order of 48,152 hectares. The proportionate district wise breakup of surface and groundwater irrigation potentials in the basin area is presented in the Table 4.13.

Table 4.13 Surface and groundwater irrigation potentials in Vaigai, Manimuttar and Gundar delta region

District (%)	Surface Irrigation						Groundwater Irrigation					Gross Irrigation (ha)
	Tanks (ha)		Canals (ha)		Other sources (ha)	Total (ha)	Tubewells (ha)		Dug Wells (ha)		Total (ha)	
	Number	Area Irrigated (ha)	Number	Area Irrigated (ha)			Number	Area Irrigated (ha)	Number	Area Irrigated (ha)		
Ramanathapuram (93%)	1575	53042	0	0	0	**53042**	288	412	7194	10295	**10707**	**63749**
Sivaganga (82%)	4027	59736	0	0	0	**59736**	890	1805	14127	11439	**13244**	**72980**
Madurai (42%)	961	7689	34	15534	2742	**25965**	225	266	22556	15520	**15787**	**41751**
Virudunagar (16%)	148	1037	0	0	3191	**4228**	0	0	0	5402	**5402**	**9630**
Pudukottai (11%)	600	7894	3	1175	149	**9218**	922	2079	4229	933	**3012**	**12230**
Total						**152188**					**48152**	**200340**

Regional hydrogeological characteristics and system parameters

A regional geological map with different outcropping geological units is presented in Fig. 4.41 which includes Archeans, Mesozoics, Tertiaries, Quaternaries and Recent. Archeans outcrop mainly in the upper part of the basin area represented by granites and gneisses with intrusions of pegmatites and quartz veins. The top near surface weathered fractured and fishered portions of the granites and gneisses have a range of thicknesses varying from 10 to 30 m and are normally suitable for groundwater development through dugwells and dug cum borewells with yields varying from 120 to 300 lpm at drawdowns varying from 6 to 8 m. The average transmissivity for these systems range between 1 to 10 m^2/day and the storage coefficient values have a range between 1.3×10^{-3} to 8×10^{-3}.

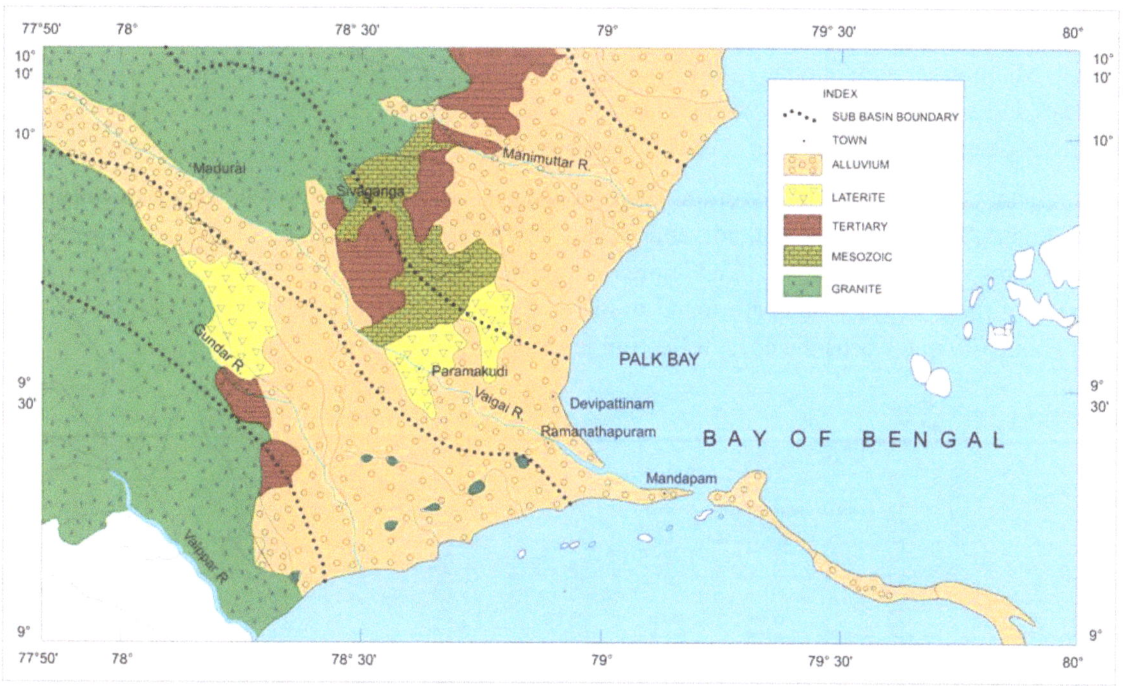

Fig. 4.41 Regional geological map of Vaigai delta (Tamil Nadu)

The Mesozoics are represented by lower and upper Cretaceous group of rocks exposed in the upper half of the sediment region overlying the Archean basement complex unconfirmbly. At the same time they are found to extend all along the coastal region at different depth levels varying from 116 to 890 m below ground level. It is estimated that the lower Cretaceous group of rocks have a maximum thickness of the order of 535 m and the upper Cretaceous rocks have a maximum thickness of the order of 136 m. The lower Cretaceous rocks mainly consist of sandstones, shales and fine grained gray limestones. The upper units contain sandstones, shales and flesh red limestones and pink silty sandstones. Groundwater in these units occurs in both confined and semi confined conditions in the upper outcrop regions and as deep confined systems in the outfall regions. These deep confined systems sometimes turn out to be artesian with heads ranging from 6.00 to 13.00 m above ground level with yields ranging from 400 to 1000 lpm at draw downs of 1.2 to 50 m. The transmissivities for these systems have a range varying from 50 m^2/day to 500 m^2/day and storativity values have a range varying between 5×10^{-4} to 2×10^{-5}.

The Tertiary units overlying the Cretaceous sediments could be grouped into two units: 1. The lower Eocene-Paleocene formations consisting of clays and shales with layers of sandstones with thicknesses varying from 40 to 130 meters. In general they are not seen out cropping any where but are encountered at depths varying from 170 to 270 m below ground level in the outfall areas 2. The upper unit namely Plio-Miocene sediments (Cuddalores) are seen out cropping in the central part of the basin, and are also encountered at depths varying from 120 to 350 m below ground level, with thicknesses ranging from 100 to 280 m. They consist of coarse grained friable feldspathic sandstones, mottled grits and clays. Groundwater in these formations too occur in both unconfined to semi confined conditions near the outcrop areas and in deep confined conditions in the outfall areas. The estimated discharges from these aquifer units are of the order varying from 500 to 3000 lpm with transmissivities ranging between 50 m^2/day to 3500 m^2/day.

The Quaternaries and Recent sediments are fairly wide spread in the flood plains and alluvial tacks of Vaigai, Manimuttar and Gundar with thicknesses varying from 20 to 30 m. They basically consist of sand, clays and sometimes gravels with yields 60 to 700 lpm at drawdowns varying from 5 m to 12 m. The laterites occur as cap rocks over the sedimentaries and are largely suited for dugwell structures.

A regional hydrogeological section from west to east covering the boreholes Kappalur, Devandaduvu, Andavurani, Pandugudi and Tondi is presented in Fig. 4.42. The section displays prominent presence of Miocene (Cuddalores), and Eocene and also Cretaceous sediments. The thicknesses of these Tertiary units are found to increase from an order of about 120 m in the west to an order of about 400 m towards the coast reflecting increasing potentials from west to east.

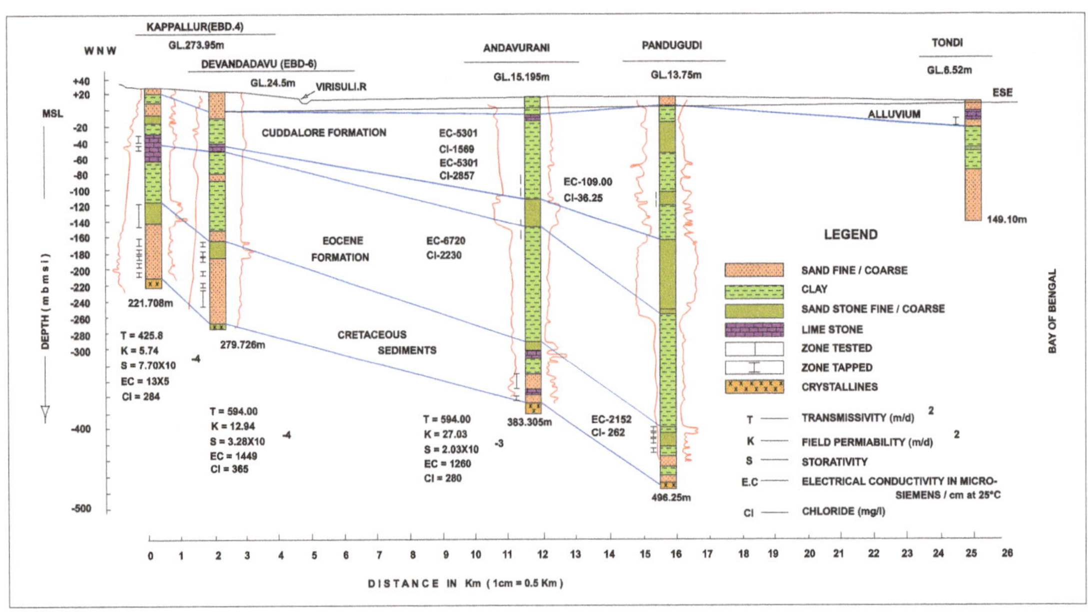

Fig. 4.42 Hydrogeological section across Kappallur and Tondi in Vaigai delta basin

(Source: CGWB, 1998)

The hydrogeological parameters for selected boreholes in the basin area is presented below in Table 4.14 and site locations shown in Fig. 4.43 (Data source: CGWB, 1998 and 2002).

Table 4.14 Hydrogeological parameters in the Vaigai delta basin

S.No	Location	Depth drilled (m) / Constructed (mbgl)	Depth Range of Aquifers tapped (mbgl)	Thickness of the zones (m)	SWL (mbgl)	Discharge (lpm) / Drawdown (m)	Specific Capacity (lpm/m)	T (m^2/day)	S	Lithology/ Remarks
1	2	3	5	6	7	8	9	10	11	12
1	Tiruvadanai	389	351.00-375.00	42	13.25 magl	180	3.6	63.95	1280	Sand, clay and tertiary sandstone, homblende gneiss
	($09^0 45' 00''$; $78^0 54' 30''$)	382				50				
2	Pandukudi	451	416.0-447.0	17	6.00 (magl)	240	4.76			Sand, clay, calcareous sandstone, limestone
	($09^0 48' 00''$;$78^0 50' 30''$)	450				50.37				
3	Melpanaiyur	669.5	564.0-660.0	65	10 (magl)					Sand,clay, sandstone and limestone shale with shell fragments.
	($09^0 43' 45''$;$78^0 52' 45''$)	663								cement sealing from 490 to 500 mbgl flowing well
4	Kamudakudi	37	12.00-30.00		5.575	60				Sand admixed with clay and sandstone.
	($09^0 33' 30''$;$78^0 32' 30''$)	33				….				(very low yield)
5	Puliyur	500	440.50-482.21	35.71	6 (magl)	60				sand admixed with clay, limestone, fossiliferrous sandstone, weathered gneiss
	($09^0 49' 45''$;$79^0 04' 00''$)	485.2				…..				
										flowing well, cement sealing techniques were adopted, between 325 and 335 mbgl

Table 4.14 ***Contd…***

S.No	Location	Depth drilled (m) / Constructed (mbgl)	Depth Range of Aquifers tapped (mbgl)	Thickness of the zones (m)	SWL (mbgl)	Discharge (lpm) / Drawdown (m)	Specific Capacity (lpm/m)	T (m^2/day)	S	Lithology/ Remarks
1	2	3	5	6	7	8	9	10	11	12
6	Srikambaiyur	478	421.91-440.00	18.09	6	30				Sand admixed with clay, limestone, weathered gneissic granite cement sealing between 410.00 and 415.00 mbgl
	($09^0 52^| 30^{||}$;$79^0 02^| 30^{||}$)	443				....				
7	Melperungarai	57.73	37.00-49.00	12	12.17	90	5.23	7.064		
	($09^0 34^| 00^{||}$;$78^0 26^| 00^{||}$)	52				17.19				
8	Mandalamanickam	26	14.00-20.07	6	7.4	21.5				Sand admixed with kankar and clay, weathered gneiss
	($09^0 29^| 00^{||}$;$78^0 20^| 35^{||}$)	23.41								
9	Pillaiyarkulam	152.16	22.50-40.50	18	5.565	270	10.3	603	--	Sand and clay with limestone, gametiferrous gneiss
	($09^0 10' 30''$; $78^0 27' 00''$)	43.5				2.615				
10	A.Poonavisal	290.66	87.00-156.00	42	4.845	...	--	340	--	Sand with clay, fossiliferous sandstone and shale
	($09^0 15' 00''$; $78^0 30' 00''$)	159				23.155				cs77-82
11	Karisalkulam	68.85	10.00-37.12	21	7.98	270	223.1	3616.84	--	Sand and clay, weathered granite
	($09^0 18' 00''$; $78^0 24' 30''$)	40.15				1.21				
12	Mangalakkudi	370	312.00-360.00	36	5.9	2100	--	--	--	Sand, clay and Cuddalore sandstone
	($09^0 52' 15''$; $78^0 57' 35''$)	362.9								

Table 4.14 *Contd…*

S.No	Location	Depth drilled (m) / Constructed (mbgl)	Depth Range of Aquifers tapped (mbgl)	Thickness of the zones (m)	SWL (mbgl)	Discharge (lpm) / Drawdown (m)	Specific Capacity (lpm/m)	T (m^2/day)	S	Lithology/ Remarks
1	2	3	5	6	7	8	9	10	11	12
13	C.K.Mangalam	405	346.00-362.00	16	4.04	690	23.22	---	--	Sand, clay and Cuddalore sandstone (fossiliferrous) and granite
	(09°47'07" ; 78°55'33")	365				29.71				
14	Sakkottai	152.4	22.86-54.34	31.48	6.1	189	9.25	--	--	Sand and clay tertiary sandstones
	(10°05'30" ; 78°51'30")					20.42				
15	Devakottai	180.63	109.73-138.08	23	4.14	318	52.1	--	--	Sand and clay, tertiary sandstone
	(09°57'30" ; 78°49'35")	140				6.1				
16	Kannangudi	263	184.16-253.00	68.84	6.700 (magl)	2310	84.6	552.5	--	Sand with clay sandstone granite
	(09°56'00" ; 78°57'00")	253				27.3				
17	Maviduthikottai	173	50.16-169.53	50	1.86	1200	57.7	96 to 113	7.6×10^{-5}	Sand and clay, sandstone basement rock
	(09°54'30" ; 78°58'20")	171.16				20.81				
18	Sattanikottai	215.5	40.00-197.85	68.5	7.498)	1740	142.4	233.85	3.59×10^{-4}	Sand & clay sandstone caleareous material gneisses
	(09°56'15" ; 78°54'45")	...			(magl	12.357				
19	Kappalur	249.1	59.00-238.00	96	8.115	1740	233	425.8	1.70×10^{-4}	Sand with clay argillaceous sandstone calearous material, granites, gneisses
	(09°57'00" ; 78°56'00")	241				7.467				

Table 4.14 *Contd…*

S.No	Location	Depth drilled (m) Constructed (mbgl)	Depth Range of Aquifers tapped (mbgl)	Thickness of the zones (m)	SWL (mbgl)	Discharge (lpm) Drawdown (m)	Specific Capacity (lpm/m)	T (m^2/day)	S	Lithology/ Remarks
1	2	3	5	6	7	8	9	10	11	12
20	Neyvayal	316	221.00-308.19	56	4.27	1920	82.47	103.57	5.54×10^{-4}	Sand and clay sandstone limestone argillaceious sandstone
	(09⁰52'00" ; 78⁰53'20")	311.19				23.28				
21	Amaravathipud ur	299.07	86.00-185.83	55.61	11.55	249	7.94	4.6	--	Sand and clay, sandstone with shale
	(10⁰01'00" ; 78⁰46'00")	188.83				31.33				
22	Devandathavu	296.25	192.00-271.00	51	13.242 (magl)	4500	177	594.66	3.28×10^{-4}	Sand clay tertiary sandstones limestone gneisses
	(09⁰55'36" ; 78⁰57'12")	…				25.3				
23	Nachankulam	67.1	30.00-49.9	20	6.52	48	1.73			sand with clay, sandstone basement rock
	(095700;785900)	52.92				27.7				
24	Andavurani	386.5	347.0-375.0	22	7.6	1260	--	480	2.722×10^{-5}	
		378								
25	Talakal	100.75	80.0-92.0	12	8.83	210	6.7	80.95		
		95				30.97				
26	Somanathnagar	127	28.0-95.0	24	6.69	255	19.78	57		
		98				12.89				
27	Pulvoyakulam	106.38	82.0-100.0	15		60	--			cs77-80.0
		103								
28	Rajasinga mangalam	304.8								Abandoned
29	Tharikomban	750	55.0-485.0	38						Abandoned

Table 4.14 ***Contd…***

S.No	Location	Depth drilled (m) Constructed (mbgl)	Depth Range of Aquifers tapped (mbgl)	Thickvness of the zones (m)	SWL (mbgl)	Discharge (lpm) Drawdown (m)	Specific Capacity (lpm/m)	T (m^2/day)	S	Lithology/ Remarks
1	2	3	5	6	7	8	9	10	11	12
30	Naranamangalam	304.8								Abandoned
31	Ariyanendal	388.6	352.0-359.0	7						Abandoned based on E-log
		360								
32	Tinaikulam	201			3.9	102	5.5			Abandoned
		….				18.35				
33	Pullangudi	369.72			1.52	150				Abandoned
34	Palangulam	384.4								Abandoned based on E-log results
35	Ettivoyal	301.26	38.6-170.0	24.5						Abandoned
		….								
36	Kilkottai	368.2	249.7-256.7							Abandoned
37	Salaigram	302.54								Abandoned
38	Periyakkudi	777								Abandoned based on E-log results
39	Uchipuli	195								Abandoned due to poor concentration of brine water Flowing well
40	Valinokkam	101	30.0-90.0	36		120				BRINE WATER
		93								
41	Morekulam	100	70.0-95.0			1110				BRINE WATER
		98		11	3.5					

Table 4.14 *Contd…*

S.No	Location	Depth drilled (m) / Constructed (mbgl)	Depth Range of Aquifers tapped (mbgl)	Thickness of the zones (m)	SWL (mbgl)	Discharge (lpm) / Drawdown (m)	Specific Capacity (lpm/m)	T (m^2/day)	S	Lithology/ Remarks
1	2	3	5	6	7	8	9	10	11	12
42	Thangachimadam	200	4.00-25.00	15	4.37	1080	?	223.2	--	BRINE WATER
	($09^0 12' 03"$; $79^0 15' 18"$)	28				(by compressor) 10 lps (APT) 1.45				
43	Theertandatanam	200	53.00-73.00	16	3.15	600	184.6	316	--	BRINE WATER
	($09^0 40' 30"$; $78^0 04' 50"$)	62				3.25				
44	Kallur	450	332.0-356.0	24	4.65(agl)	1680	54.6			free flowing well cement sealed at the bottom
		360				31.11				(deposit well)
45	Orasur	162	85.0-154.0	45		1020				Deposit Well
		157								

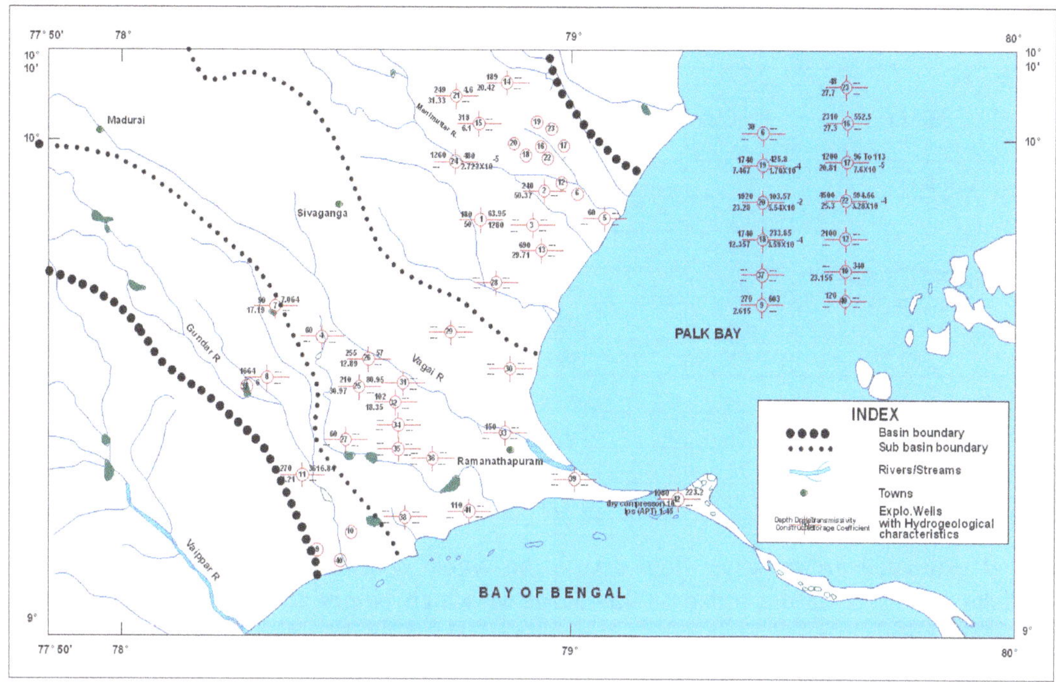

Fig. 4.43 Selected exploratory well sites in Vaigai and its adjacent delta basin

Regional groundwater flow net analysis: A groundwater flow trend map across different profiles for both near surface and deep aquifer systems are provided in the Figure 4.44. The groundwater flows across different lengths in different sub basins are shown in Table 4.15.

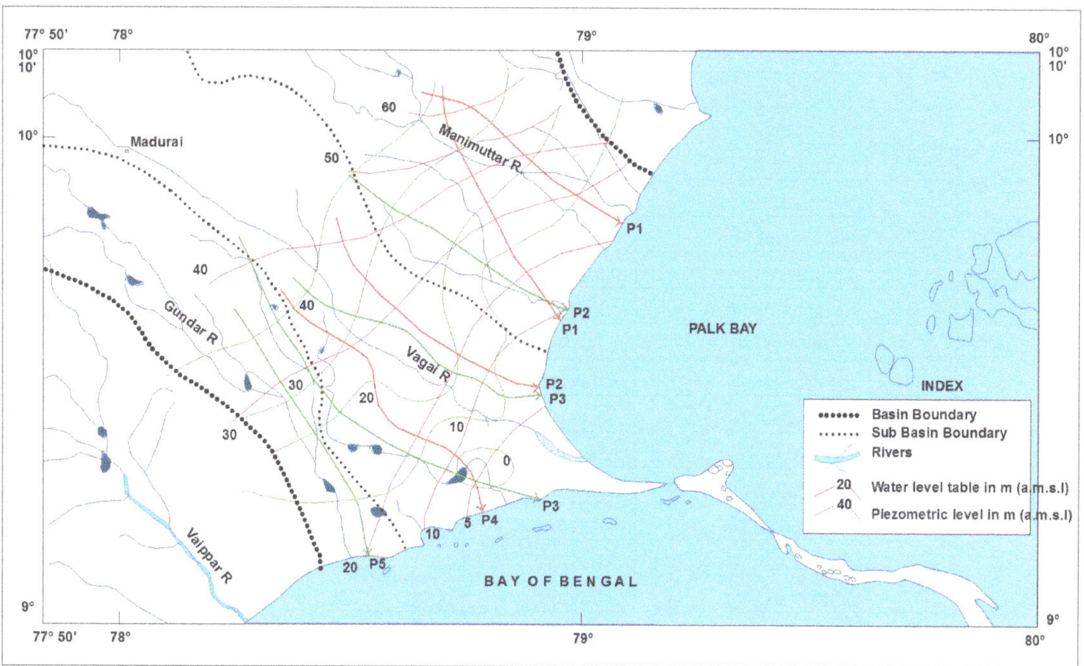

Fig. 4.44 Groundwater flow map for near surface and deep aquifer systems in Vaigai delta

Table 4.15 Groundwater flows across different profiles in Vaigai delta

S.No	Basin	Profile	Avg. Transmissivity (T) (m^2/day)	Gradient (I) (m/Kms)	Length of the Coast (L) (kms)	Total flow (Q) (MCM)
Near Surface Aquifers						
1	Vaigai and It's adjacent basins	P1	500	0.71	30	3.88
		P2	500	0.75	20	2.73
		P3	500	0.54	25	2.46
		P4	500	0.73	30	3.99
		P5	500	0.67	30	3.66
Deep Aquifers						
1	Vaigai and It's adjacent basins	P1	1000	0.75	45	12.31
		P2	1000	0.6	30	6.57
		P3	1000	0.45	60	9.85

The total estimated groundwater flow across the whole coastal margins of Vaigai delta and its adjacent delta basins is of the order of 45.45 MCM which comes to about 3.6% of the dynamic potentials of the basin.

Seasonal water level fluctuations: The seasonal water level fluctuations for a 5 year period at different site locations as shown Fig. 4.45 in different geological horizons are given in Table 4.16. The average fluctuation in the coastal alluvial/sedimentary tracks is about 2.20 m.

Table 4.16 Seasonal water level fluctuations in Vaigai delta

S.No	Name of the Site	Geological Horizon	Avg. Fluctuation (m) for a 5 Year period (1990-95)
1	Tiruvegampattu	Alluvium	2.14
2	Tiruvadanai	Alluvium	1.53
3	Tondi	Alluvium	1.37
4	R.S.Mangalam	Alluvium	2.42
5	Solandur	Alluvium	2.70
6	Saligram	Laterite	3.69
7	Nainarkoil	Alluvium	2.23
8	Paramakudi	Alluvium	2.09
9	Somanathanagar	Alluvium	3.20
10	Kamuthakudi	Alluvium	1.67
11	Parthibanur	Alluvium	2.52
12	Kamudhi	Tertiary	3.28
13	Peraiyur	Alluvium	5.83
14	Udaikulam	Alluvium	2.16
15	Ariyanendal	Alluvium	2.33
16	Bogalur	Alluvium	1.64
17	Mudukulathur	Alluvium	1.58
18	Melsirupothu	Alluvium	3.45
19	U.Kosamangai	Alluvium	1.90
20	Devipattinam	Alluvium	3.15
21	Ramanathapuram	Alluvium	1.80

Table 4.16 ***Contd...***

S.No	Name of the Site	Geological Horizon	Avg. Fluctuation (m) for a 5 Year period (1990-95)
22	Perungulam	Alluvium	1.54
23	Uchipulli	Alluvium	1.04
24	Nadumanaikadu	Alluvium	1.55
25	Mandapam	Alluvium	1.40
26	Thangachimadam	Alluvium	1.59
27	Rameswaram	Alluvium	0.70
28	Kilakkarai	Alluvium	2.48
29	Erwadi	Alluvium	1.72
30	Sikkal	Alluvium	2.43
31	Sayalkudi	Alluvium	1.11
32	Perunali	Alluvium	2.27

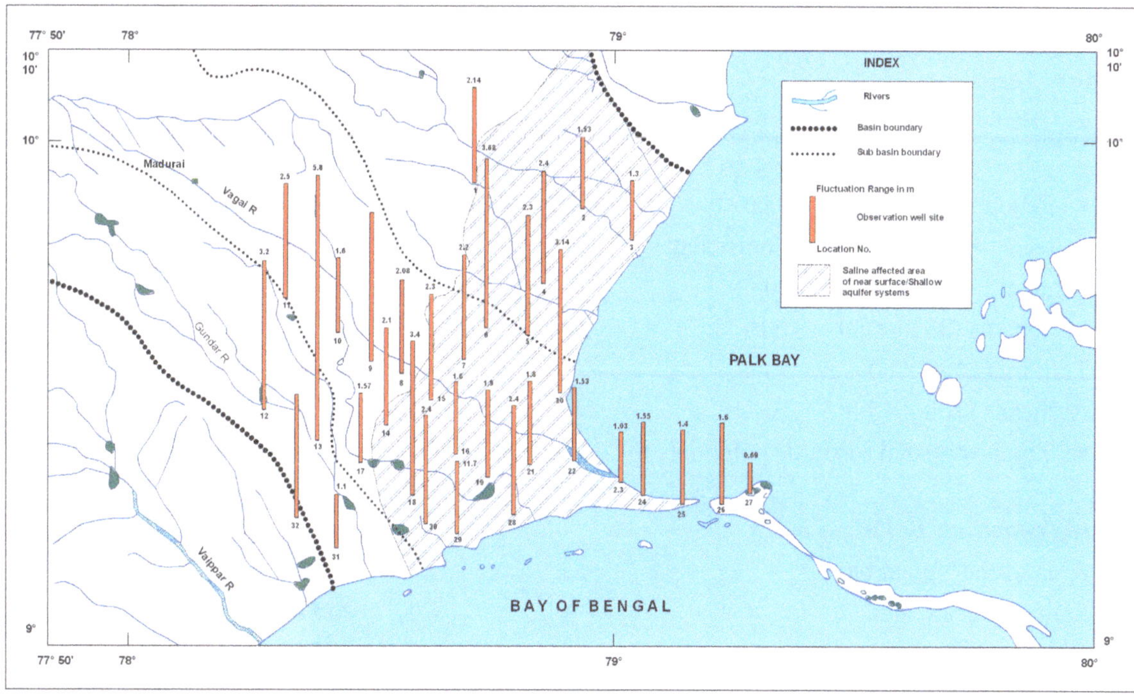

Fig. 4.45 Network of observation well stations and their fluctuations in Vaigai delta

Resource potentials in Manimuttar basin

- Area of the Manimuttar basin : 3600 sq.km
- Saline affected area : 2030 sq.km
- Coverage of alluvial : 2000 sq.km
- Coverage of Archeans : 700 sq.km
- Coverage of Laterites : 90 sq.km
- Coverage of Tertiary : 450 sq.km
- Coverage of Mesozoics : 360 sq.km
- Dynamic potentials in Archeans

(i) Fluctuation method : 10.5 MCM

- Dynamic potentials for sedimentaries not affected by salinity:

(i) Rainfall infiltration method :
  (a) Alluvials : 23.32 MCM
  (b) Mesozoics : 42 MCM
  (c) Tertiaries : 70 MCM

(ii) Fluctuation method :
  (a) Alluvials : 56.25 MCM
  (b) Mesozoics : 60.75 MCM
  (c) Tertiaries : 101.25 MCM

(iii) Average dynamic potentials
  (a) Alluvial : 39.75 MCM
  (b) Mesozoics : 51.36 MCM
  (c) Tertiaries : 85.6 MCM

- Total estimated dynamic potentials in Manimuttar basin
  (10.5 + 39.75 + 51.36 + 85.6) : 187.24 MCM
- Resource potentials in deep confined aquifer systems (static)
  (Area × Thickness × S.Coefficient) :

(i) Alluvial ($2000 \times 50 \times 8 \times 10^{-4}$) : 80.0 MCM
(ii) Mesozoics ($2810 \times 80 \times 3 \times 10^{-4}$) : 67.4 MCM
(iii) Tertiary ($2450 \times 80 \times 3 \times 10^{-4}$) : 58.8 MCM
(iv) Archeans ($700 \times 15 \times 5 \times 10^{-3}$) : 52.5 MCM

- Total estimated resource potentials in deep confined aquifer systems
  : 258.7 MCM

Resource potentials in Vaigai basin

- Area of the Vaigai basin : 4010 sq.km
- Saline affected area : 1170 sq.km
- Coverage of Alluvium : 1900 sq.km
- Coverage of Laterites : 290 sq.km
- Coverage of Tertiary : 270 sq.km
- Coverage of Mesozoics : 200 sq.km
- Coverage of Archeans : 1350 sq.km
- Dynamic potentials in Archeans :

(i) Fluctuation method : 20.25 MCM

- Dynamic potentials for Sedimentaries not affected by salinity :

(i) Rainfall infiltration method
  (a) Alluvials : 155.45 MCM
  (b) Laterites : 10.10 MCM
  (c) Tertiaries : 41.97 MCM
  (d) Mesozoics : 24.09 MCM

(ii) Fluctuation method

(a) Alluvials : 375 MCM

(b) Laterites : 1.95 MCM

(c) Tertiaries : 60.75 MCM

(d) Mesozoics : 34.87 MCM

(iii) Average dynamic potentials

(a) Alluvials : 265.22 MCM

(b) Laterites : 6.03 MCM

(c) Tertiaries : 51.36 MCM

(d) Mesozoics : 29.48 MCM

- Total estimated dynamic potentials in vagai basin (20.25 + 265.22 + 6.03 + 51.36 + 29.48) : 372.34 MCM
- Resource potentials in deep confined aquifer systems (static) (Area x Thickness x S.Coefficient) :

(i) Alluvial ($1900 \times 50 \times 8 \times 10^{-4}$) : 76.0 MCM

(ii) Mesozoics ($2370 \times 80 \times 3 \times 10^{-4}$) : 56.8 MCM

(iii) Tertiary ($2170 \times 80 \times 3 \times 10^{-4}$) : 52.8 MCM

(iv) Archeans ($1350 \times 15 \times 5 \times 10^{-3}$) : 101.2 MCM

- Total estimated resource potentials in deep confined aquifer systems : 286.2 MCM

Resource potentials in Gundar sub basin

- Area of the Gundar basin : 5030 sq.km
- Saline affected area : 800 sq.km
- Coverage of Alluvials : 2250 sq.km
- Coverage of Laterites : 350 sq.km
- Coverage of Tertiaries : 130 sq.km
- Coverage of Mesozoics : 1200 sq.km
- Coverage of Archeans : 1100 sq.km
- Dynamic potentials in Archeans :

(i) Fluctuation method : 16.5 MCM

- Dynamic potentials for Sedimentaries:

(i) Rainfall infiltration method

(a) Alluvials : 225.40 MCM

(b) Laterites : 54.41 MCM

(c) Tertiaries : 20.21 MCM

(d) Mesozoics : 18.65 MCM

(ii) Fluctuation method

(a) Alluvials : 543.75 MCM

(b) Laterites : 10.53 MCM

(c) Tertiaries : 29.25 MCM

(d) Mesozoics : 270 MCM

(iii) Average dynamic potentials

(a) Alluvials : 384.57 MCM

(b) Laterites : 32.47 MCM

(c) Tertiaries : 24.73 MCM

(d) Mesozoics : 228.27 MCM

- Total estimated dynamic potentials in Gundar basin (16.5 + 384.57 + 32.47 + 24.73 + 228.27) : 686.54 MCM
- Resource potentials in deep confined aquifer systems (static) (Area x Thickness x S.Coefficient) :

(i) Alluvial ($2250 \times 50 \times 8 \times 10^{-4}$) : 90.0 MCM

(ii) Mesozoics ($3580 \times 80 \times 3 \times 10^{-4}$) : 85.9 MCM

(iii) Tertiaries ($2380 \times 80 \times 3 \times 10^{-4}$) : 57.12 MCM

(iv) Archeans ($1100 \times 15 \times 5 \times 10^{-3}$) : 82.5 MCM

- Total estimated Resource potentials in deep confined aquifer systems : 315.54 MCM

Note: Over 60% of the groundwater systems in the Tertiaries and Mesozoics in this part of the basin appear to be brackish. Therefore the total freshwater potentials approximately should be the order of 126.2 MCM

- Total estimated (dynamic + deep confined systems) potentials in Vaigai & and its adjacent basins : 2106.56 MCM

Note: 0-400 m is the depth range considered for estimations of dynamic & deep confined system potentials in this basin.

4.4.5 Pennar delta region

Pennar delta region is located north of Palar basin covers an area of about 1470 sq.kms in the Nellore district of Andhra Pradesh. The river has its origin in the Chennakeshava hills of Karnataka and stretches of about 597 km and debouches into Bay of Bengal with an annual discharge of about 3,338 MCM. The delta appears to have been evolved in two stages with its first apex near Sangam and the second apex near Nellore. Both these stages had helped to build up a moderate delta system with alluvial sediments having thicknesses of the order ranging between 40 to 60 m below ground level. The delta region has a gentle slope up to the coastal boundary. It is a single channel river up to the coastal boundary and does not have any distributaries. The delta region is characterized by three distinct features, the erosional (upper delta region), fluvial (central delta region) and estuarine/marine (coastal region). A number of paleo channels (5) and three distinct strand lines could be recognized in this part of the delta reflecting the progradation processes in the recent times (Kameswara rao, K. 1991).

A drainage map covering the delta region with the marked basin boundaries along with isohyetal trends is presented in fig 4.46 covering about 13.5% of the Nellore district. The average annual precipitation trends in Pennar delta is about 913 mm with a maximum precipitation level of the order of about 1000 mm (near to the coast) and a minimum trend level of the order of about 700 mm (away from the coast). Over 60% of the precipitation is received through northeast monsoon during the period October, November and December and about 30% is received through southwest monsoon during the period June to September and the rest 10% is spread over the year sporadically. The average peak summer temperature is about 33.5°C and the average minimum winter temperature is about 24.5 °C. The humidity levels in the region vary from 84% (wet season) and 57% (dry season). The average evapotranspiration level is of the order of 1710 mm which is about 1.87 times of the precipitation level. The delta region is rich in both surface and groundwater potentials represented by good network of canal systems, reservoirs/tanks, tube wells, dug wells and filter points. Sangam and Nellore are the major irrigation anicut irrigation reservoirs where major canal systems emerge. The combined surface irrigation potential in the carved out basin area is of the order of about 50,197 hectares and the combined groundwater irrigation potentials is of the order of about 15,556 hectares. The ratio of irrigation potentials between surface and Groundwater is 3:1. The proportionate district irrigation potentials of both surface and groundwater potentials are presented below in Table 4.17

Table 4.17 Surface and groundwater irrigation potentials in Pennar delta basin

District	Surface Irrigation by					Groundwater Irrigation by			Gross Irrigation (ha)
	Tanks (ha)	Canals (ha)	Lift Irrigation (ha)	Other sources (ha)	Total (ha)	Tube wells (ha)	Dug Wells (ha)	Total (ha)	
Nellore (13.5%)	1746	47093	873	485	50197	14800	756	15556	**65753**

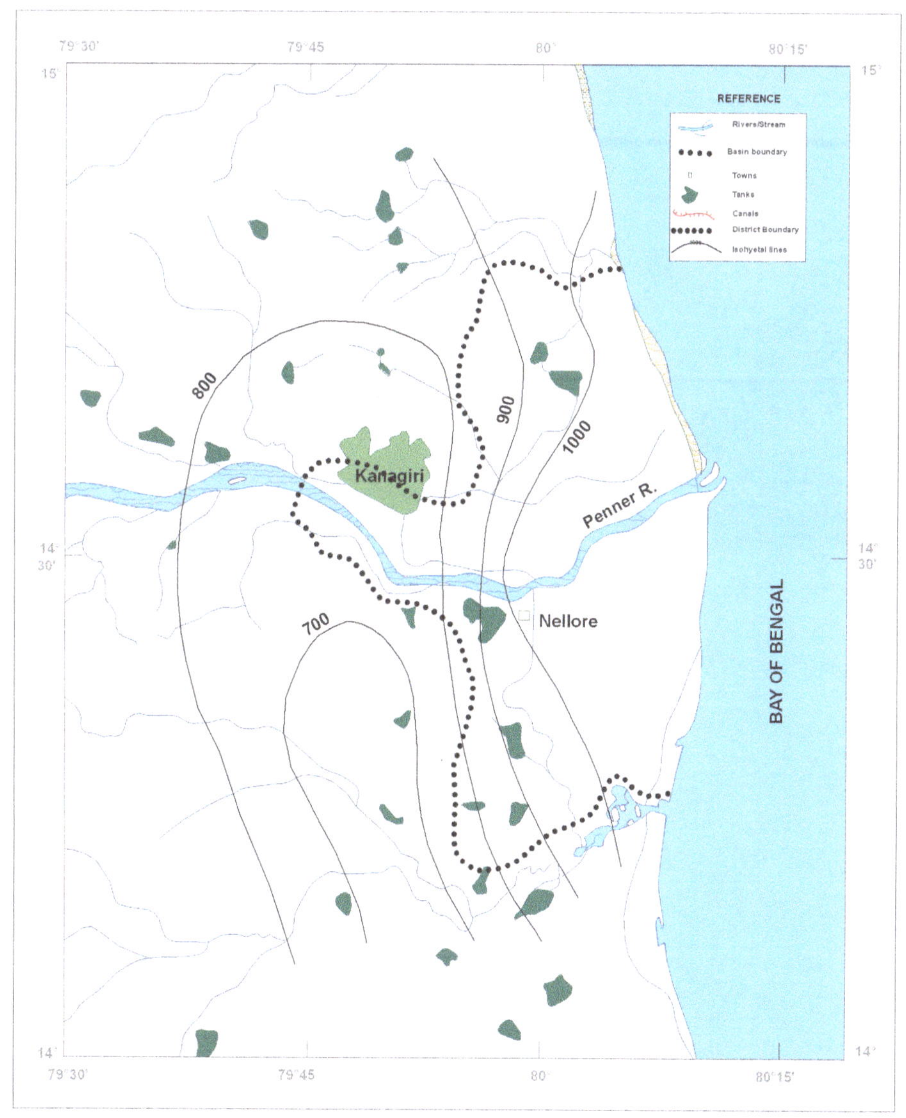

Fig. 4.46 Drainage and isohyetal level trends in Pennar delta (Andhra Pradesh)

Regional Hydrogeological characteristics and system parameters

A regional geological map of Pennar delta basin is shown in Fig. 4.47. The geological units in the basin include Archeans, Tertiaries (below Laterites) and Quaternaries and Recent. Archeans occur across the upper boundary region of the basin. These are basically granites and gneisses with intrusions. The aquifers in these rock units occur in unconfined (weathered portions of the granitic and gneissic region) and semiconfined (fractured and fissured portions) up to depth levels varying between 20 to 40 m below ground level. In general dugwell structures are suitable with discharges or yields varying between 300 – 500 lpm. These aquifers have transmissivity values varying between 1 to 10 m^2/day. The Tertiaries occur in upstream area as thin beds overlained by laterites and in the downstrem side. These Tertiary beds are overlained by Quaternary alluvium. Majority of the Tertiary aquifers in the outfall region appear to be brackish and are not suitable for irrigation through tubewell structures. The laterites in the upstream area however are suitable for groundwater structures with yields ranging from 500 to 600 lpm. The Quaternary alluvium consists of sands, gravels and clays covers a large chunk of the delta region and has thicknesses varying between 60 to 80 m. The yields of the Quaternary aquifers have a range of

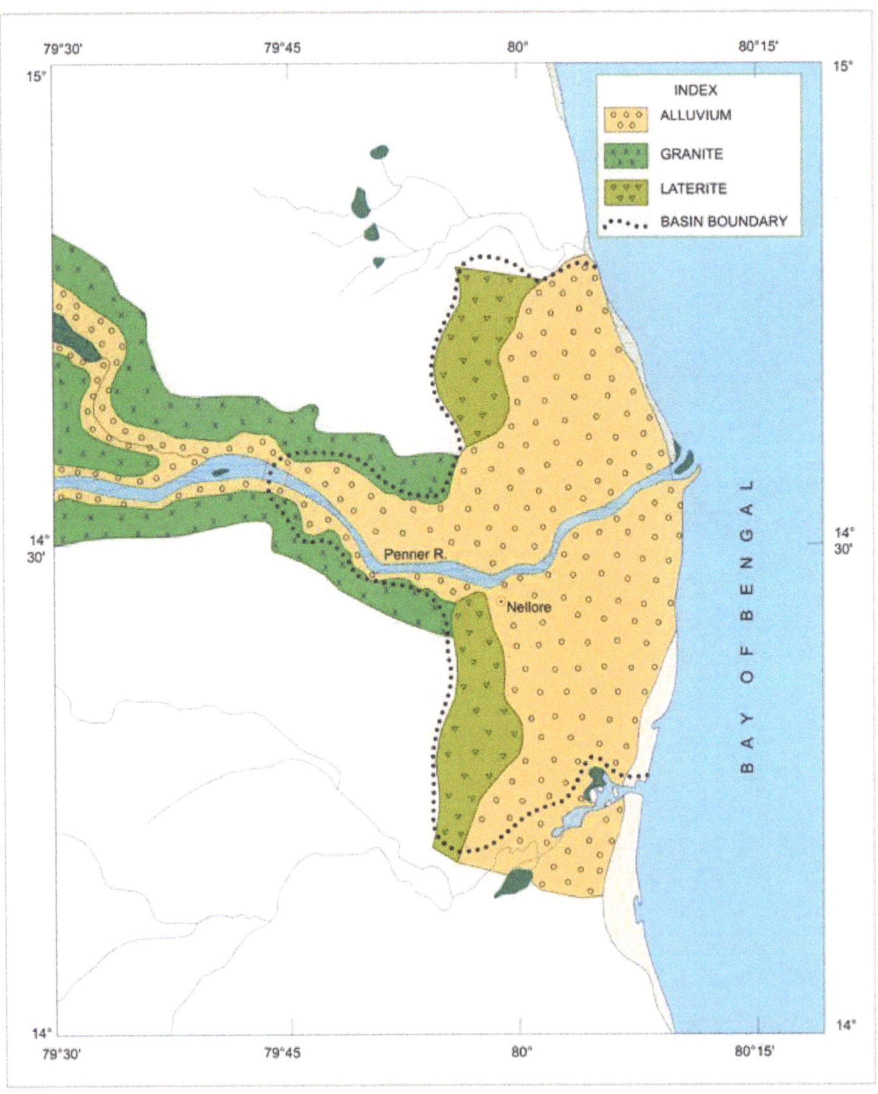

Fig. 4.47 Regional geological map of Pennar delta basin (Andhra Pradesh)

variation from 600 to 2000 lpm and with transmissivity values varying between 1500 to 2500 m^2/day and the storage coefficient between 2×10^{-4} to 8×10^{-4}. The boulder beds which occur between 60 to 80 m below ground level observed in several deep bore holes appear to be the marker horizons separating Quaternaries and Tertiaries. Quaternaries in general are represented by dominating sand facies which act as good aquifer systems and the Tertiaries are represented by dominating clay and silt facies with relatively poor yields and brackish water conditions. A hydrogeological section representing mostly the Quaternary sediments is presented in Fig. 4.48.

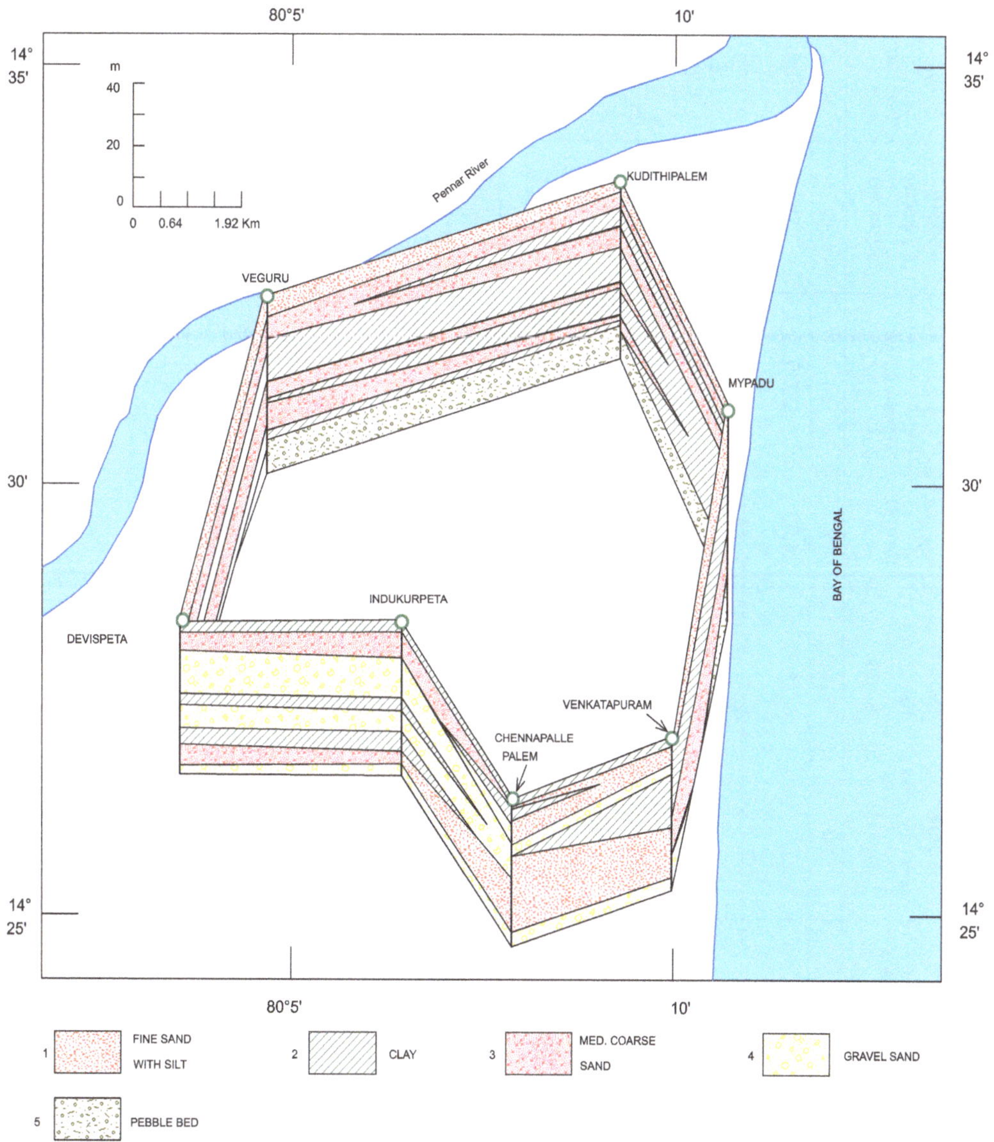

Fig. 4.48 Fence diagram showing the lithological correlations in pennar delta

(Source: Radhakrishna, I.,1991)

The hydrological parameters for selected boreholes in the basin are presented in Table 4.18 and the site locations are shown in Fig. 4.49 (Data source: Radhakrishna, I. and Krishna, G.S.R., 1991; CGWB, 2001).

Table 4.18 Hydrological parameters in the Pennar delta basin

S.No	Location	Depth drilled (m) Constructed (mbgl)	Aquifer zones tapped (mbgl)	Thickness of the zones (m)	SWL (mbgl)	Discharge (lpm) Drawdown (m)	Specific Capacity (lpm/m)	T (m^2/day)	S	Remarks
1	2	3	4	5	6	7	8	9	10	11
1	Mypadu	…		25.91	1.83	2075	324.7	2100	--	Recent alluvium
		…				6.39				
2	Kuditipalem	…		26.75	1.98	2298	371.8	1800	7.4×10^{-4}	-- do --
		…				6.18				
3	Veguru	…		21.28	4.4	2003	326.2	1320	7.5×10^{-4}	-- do --
		…				6.14				
4	chinnapallepalem	…		19.82	1.95	532	32.3	2626	…..	--do--
		…				16.47				
5	Kandalipadu	47			4.14	630	93.19		…	Abandoned due to poor quality
		….				6.76				
6	Suryapalem	8.3			2.59	250	297.6	2200	….	
		….				0.84				
7	Nellipudi	24.34			2.33	458	119.89		….	
		….				3.82				
8	Bhimavaram	12.8			3.73	--	--		….	Well not tested due to poor yield
		….				--				

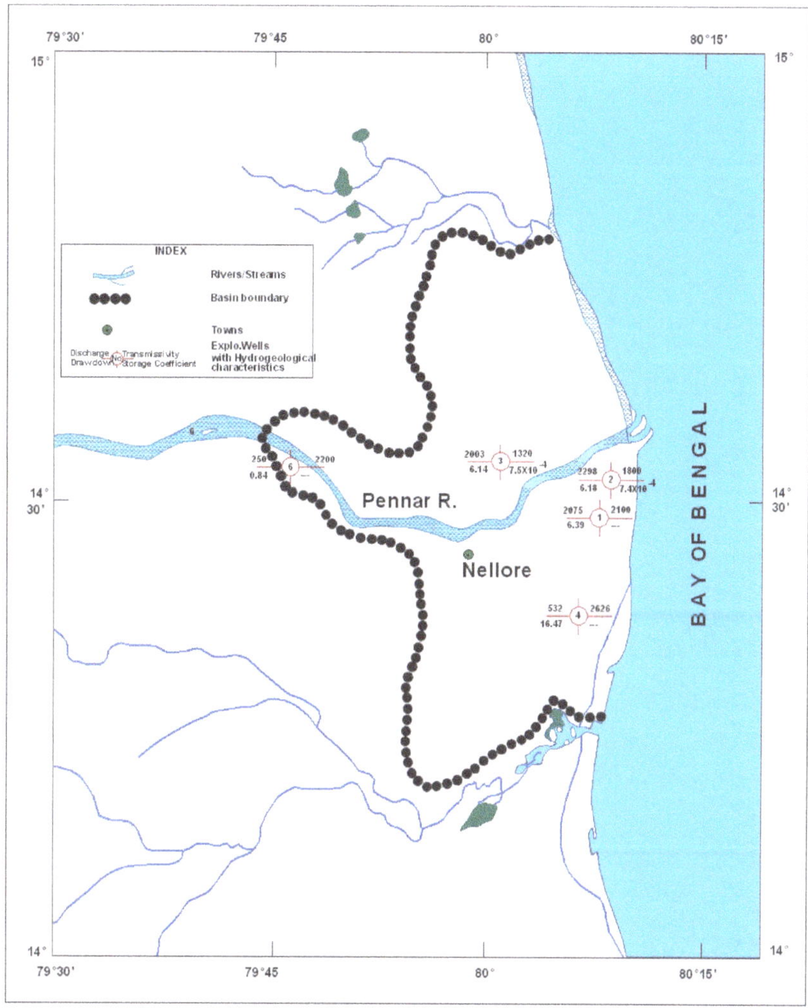

Fig. 4.49 Selected exploratory well sites in Pennar delta (Andhra Pradesh)

Regional groundwater flow net analysis: A Groundwater flow trend map across different profiles for near surface aquifer system is provided in the Fig. 4.50. The groundwater flows across the basin are provided in Table 4.19.

Table 4.19 Groundwater flow parameters across different profiles in Pennar delta

S.No	Basin	Profile	Avg.Transmissivity (T) (m^2/day)	Gradient (I) (m/kms)	Length of the Coast (L) (kms)	Total flow (Q)
1	Pennar	P1	2000	0.90	25	16.43
		P2	2000	0.87	33.5	21.28

*Total flow for near surface system along the coast line is worked out to be an order of 37.7 MCM

The data for deep system is not available. However assuming the same gradient levels for deeper systems and a lesser average transmissivity of an order of 1000 m^2/day, an approximate estimation of groundwater flow across the same length of the coast is of the order of 18.58 MCM. The total estimated combined groundwater flow across the coastal region is of the order of 56.28 MCM which comes to about 9.5% of the dynamic potentials of the basin.

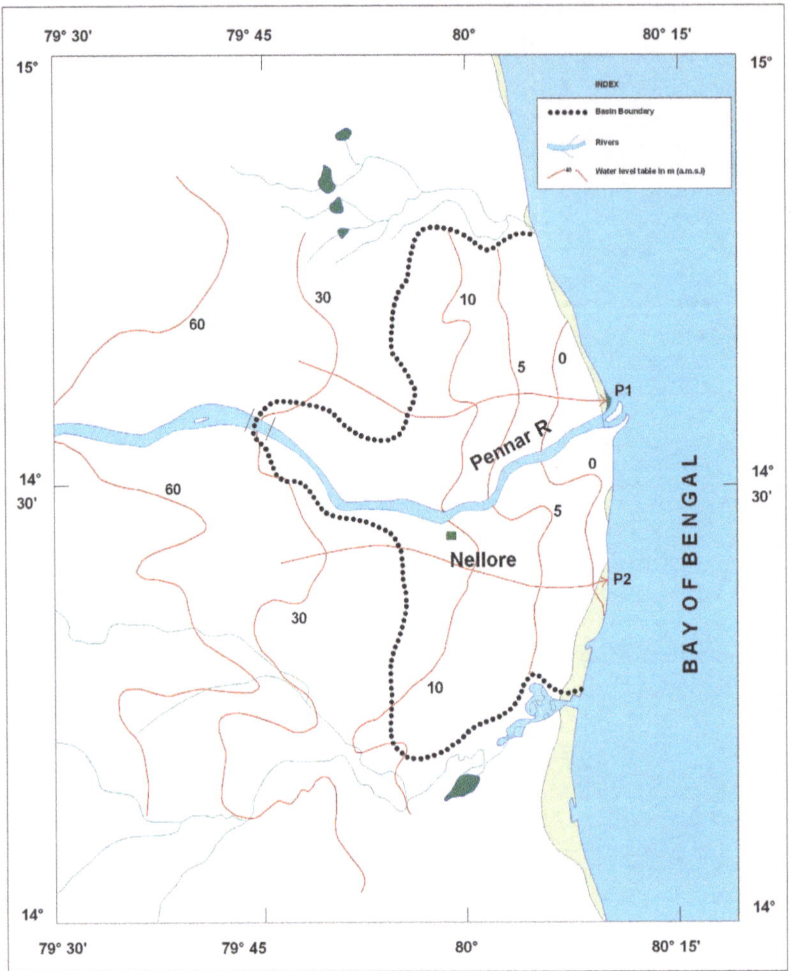

Fig. 4.50 Regional groundwater table map in Pennar delta (Andhra Pradesh)

Seasonal water level fluctuations: The seasonal water level fluctuations for a 5 year period at different site locations in different geological horizons are shown in the Table 4.20 below and are reflected in the Fig. 4.51. The average fluctuation in the coastal alluvial/sedimentary tracks is about 3.58 m.

Table 4.20 Seasonal water level fluctuations in pennar delta

S.No	Name of the Site	Geological Horizon	Avg. Fluctuation (m) for a 5 Year period (1980-85)
1	Jagadevipeta	Alluvium	3.87
2	Somasila		3.21
3	Talamanchi	Alluvium	2.07
4	Veguru	Alluvium	4.81
5	Graddagunta		2.69
6	Vakadu		3.46
7	Thummuru		5.92
8	Sullurpet		0.92

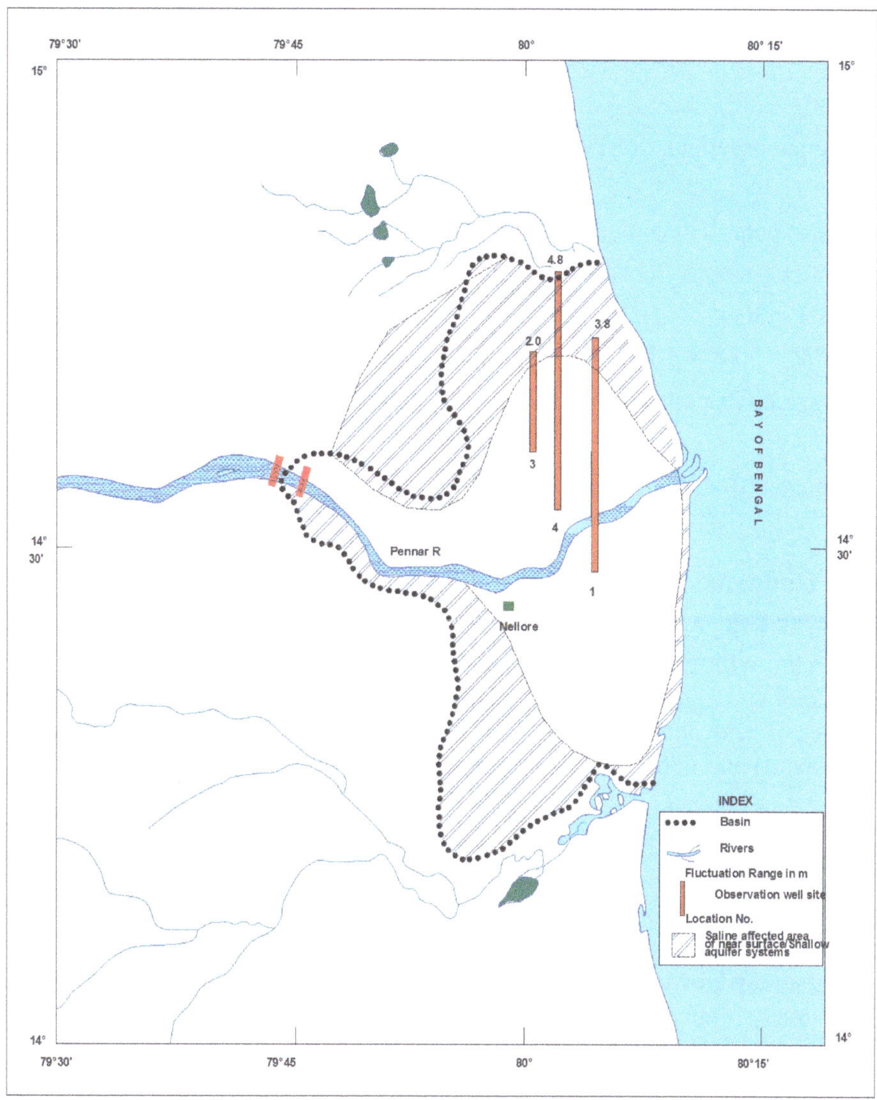

Fig. 4.51 Network of observation well stations and their fluctuations in Pennar delta (Andhra Pradesh)

Resource potentials in Pennar basin

- Area of the Pennar delta basin and its adjacent regions : 1777 sq.km
- Coverage of Alluvium : 1403 sq.km
- Coverage of Laterites : 374 sq.km
- Dynamic potentials for sedimentaries (excluding saline affected area)

(i) Rainfall infiltration method

(a) Alluvium : 228.13 MCM

(b) Laterites : 60.81 MCM

(ii) Fluctuation method

(a) Alluvium : 841.8 MCM

(b) Laterites : 46.75 MCM

(iii) Average dynamic potentials

(a) Alluvium : 534.96 MCM

(b) Laterites : 53.78 MCM

- Total dynamic potentials for Pennar basin is (534.96 + 53.78) : 588.75 MCM

- Groundwater potentials in deep confined systems (static) (Area × Thickness × S.Coefficient)

(i) Alluvials ($1403 \times 30 \times 2 \times 10^{-4}$) : 8.418 MCM

(ii) Tertiaries ($1777 \times 50 \times 8 \times 10^{-4}$) : 71.08 MCM

- Total estimated Groundwater potentials in deep confined systems : 80 MCM

- Total estimated (dynamic + deep confined systems) potentials in Pennar delta basin is : 668.25 MCM

4.4.6 Krishna - Godavari delta region

Godavari and Krishna are the major delta systems along the east coast with an aereal spread of the order of 6322 and 5476 sq.km with catchment areas of the order of 312812 and 258948 sq.kms respectively. Both these rivers take their origin in the western ghats traverse a distance of about 1465 km (Godavari) and 1400 km (Krishna) and debouch into Bay of Bengal with an annual discharge of the order of 105000 MCM (Godavari) and 67675 MCM (Krishna).

Godavari and Krishna deltas have their apex points near Rajahmundry and Vijayawada respectively and spread in the delta frame with a number of distributaries. Both the delta areas have gentle slope characteristics of the order varying between 0.1 to 0.2 m/km up to the coast line and with spread out lengths of the order of 150 km (Godavari) and 125 km (Krishna) from apex points to the coast line. The Godavari river runs as a single channel from the apex up to 6 kms downstream and bifurcates into two major distributaries namely Gowthami and Vasista. Vasista further downstream gets bifurcated into two near Gannavaram namely Vainatheyam and Vasista. Similarly the Gowthami also gets bifurcated into two namely Neelarevu and Gowthami. All these subdistributariesin the outfall region show strong sinusoidal and meandering characteristics. The Krishna river on the other hand runs as a single channel from apex point up to 60 km downstream and gets branched off into two near Avanigadda namely Hamsaladeevipaya (northern arm) and Krishna. The Hamsaladeevipaya is almost dormant with insignificant sediment and water discharge. The Krishna river further downstream branches off again into three distributaries namely Gollumuthapaya, Nadimeru and the main Krishna river. Here again most of the sediment and water discharges take place through the main Krishna river which meets the sea near 'falsedivi' point. Both the delta systems have characteristic geomorphic features like flood plains, paleo channels and natural levees in the upper regions of the deltas and a number of tidal channels, marshes, lagoons and beach sand ridges in the lower delta plains. The evolutions of both Godavari and Krishna have taken place essentially during Holocene times with a number of evolutionary stages as described and discussed earlier in the chapter-II.

The Godavari and Krishna delta basin boundaries are presented in Figure 4.52 where the drainage lines, canal density network and annual precipitation lines are shown. The Godavari delta area presented in the map represents 42.6% of East Godavari area and 81.16% of West Godavari area. Similarly the Krishna delta area constitutes about 70% of Krishna district and 33.33% of Guntur district. The annual average precipitation in respect of Godavari and Krishna deltas are of the order of 1150 and 950 mm respectively. About 65 % of the precipitation in these delta systems comes from south west monsoon and the rest 35% through the north east monsoon. The maximum and minimum precipitation trend line variations observed in the case of Godavari delta are 1200 mm (close to the coast) and 1100 mm (away from the coast line). Similarly the maximum precipitation trend line in respect of Krishna delta is 1000 mm (close to the coast) and 900 mm (away from the coast line). The average mean summer temperature in the Godavari delta region is about 36 °C and the mean winter temperature is about 19 °C. Similarly in the case of Krishna delta the mean summer temperature is about 37 °C and the mean winter temperature is about 19 °C. The humidity levels in both the systems are of the order of 80% (wet season) and 70% (dry season). The average annual potential evapotranspiration in Godavari and Krishna delta basins are of the order of 1639 mm and 1811 mm respectively which comes to nearly 1.5 and 1.9 times of the average precipitation levels of the two basins respectively.

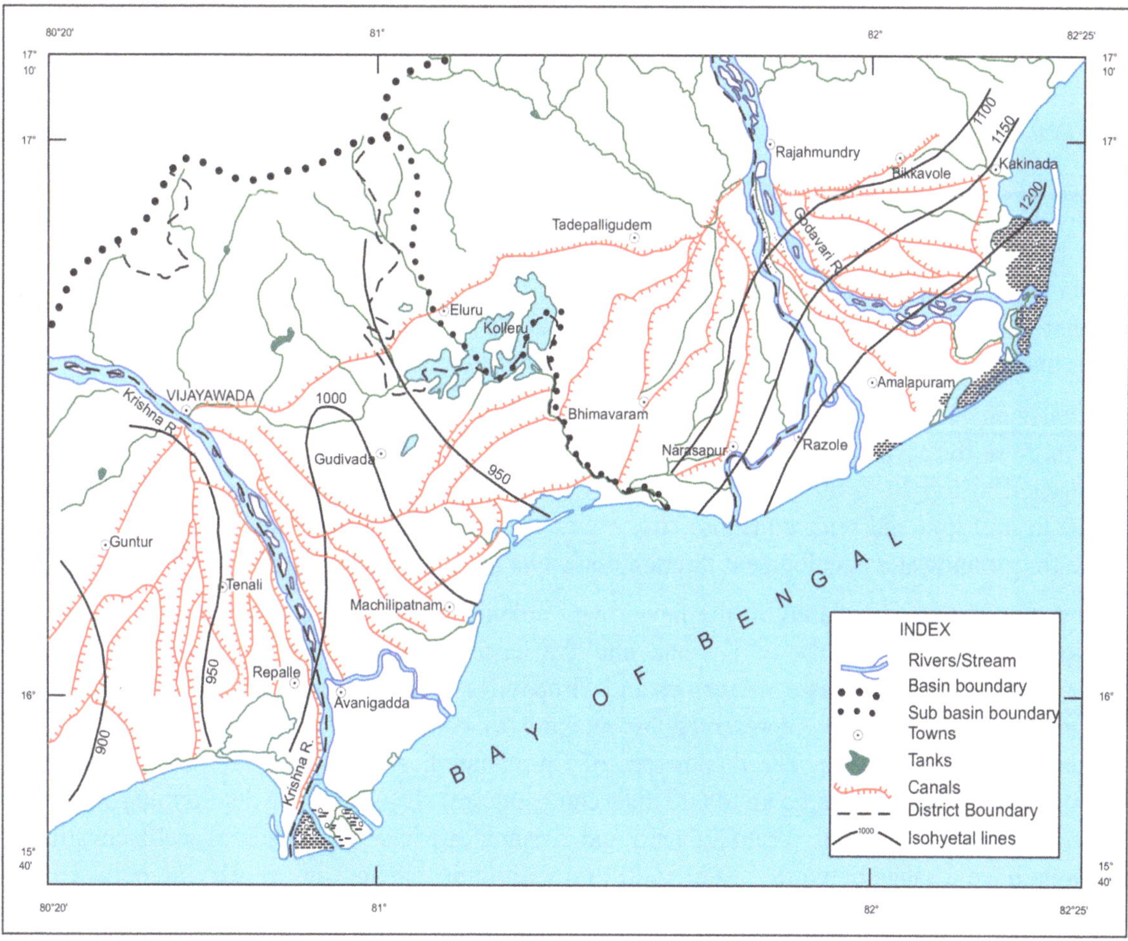

Fig. 4.52 Drainage map with isohyetal trends in Krishna-Godavari deltas (Andhra Pradesh)

The surface and groundwater resources play important role in the agricultural development in these delta systems. Intensive canal network systems have come into operation in both the delta systems through major barrage/dam projects in the upstream areas. The proportionate district wise breakup of surface and groundwater irrigation potentials through a number of canal networks and tube wells/filter points are provided in the Table 4.21.

Table 4.21 Surface and groundwater irrigation potentials in Krishna and Godavari delta region

District	Surface Irrigation by					Groundwater Irrigation by			Gross Irrigation (ha)
	Tanks (ha)	Canals (ha)	Lift Irrigation (ha)	Other sources (ha)	Total (ha)	Tube wells (ha)	Dug Wells (ha)	Total (ha)	
Krishna (70%)	18754	152808	...	9686	181248	24769	5658	30427	**310267**
Guntur (33%)	108	131826	4723	86	136743	19527	966	20493	**157236**
East Godavari (43 %)	9695	74608	...	4468	88771	24448	217	24665	**202500**
West Godavari (81%)	19096	156679	...	6484	182258	109033	6038	115071	**507734**

(Source: Department of Economics and statistics, Govt. of Andhra Pradesh, 2005-06)

Regional hydrogeological characteristics and system parameters

A regional geological map of Godavari and Krishna delta systems and their adjoining environments is presented in Fig. 4.53. Both these delta systems by and large are characterized by similar geological and hydrological environments with minor variations. The outlined basin area displays several geological units namely Archeans, Gondwanas (Chintalpudis, Gollipallies and Tirupaties), Traps, Tertiaries, Quaternaries and Recent. The Archeans, Gondwanas and Tertiaries occupy the upland areas of the delta systems and Quaternaries and Recent sediments occupy the delta frames.

Archeans in general are represented by granites, gneisses, charnokites and khondalites. Groundwater occurs in these rock types in both unconfined and semi confined situations. Majority of these aquifers are restricted to top 20 to 60 m weathered and fractured layers. The yield characters in general vary from 10 to 450 lpm and with transmissivity values varying from 1 to 30 m^2/day. These systems are suitable for groundwater development through dug wells and dug cum bore wells.

Gondwanas occur abundantly in the interstream upland areas of Krishna and Godavari basins and also along the western flanks of Krishna and the eastern flanks of Godavari. These are mainly represented by Chinthalpudies, Gollapallies and Tirupathies. Chinthalpudies are the oldest sedimentary rocks in the region and are represented by medium to coarse grained sandstones and sometimes intercalated with clay stones. The thicknesses of Chinthalpudies sandstones are of the order varying from 60 to 305 m but at times go up to 600 m (Achuthapuram). They are seen outcropping widely in the interstream upland regions of West Godavari and Krishna districts. The yields of wells constructed in these units have a range between 420 to 1680 lpm with their transmissivities varying between 50 and 1450 m^2/day. Overlying these Chinthalpudies are Gollapallies represented by fine grained sandstones and shales. The outcrops of Gollapallies are dominantly seen in the interstream upland areas of Krishna and West Godavari districts. The wells in general are constructed up to depth levels varying from 75 to 120 m with yields varying from 240 to 960 lpm. The transmissivities of these aquifers have a range varying between 240 to1050 m^2/day. The Tirupathi sandstones again are widely noticed in the interstream upland areas of Krishna and Godavari and appear to pinch out towards east and west. The depth range of wells in these units varies between 100 to 300 m with discharges from 240 to 2700 lpm and with transmissivities varying from 80 to 850 m^2/day.

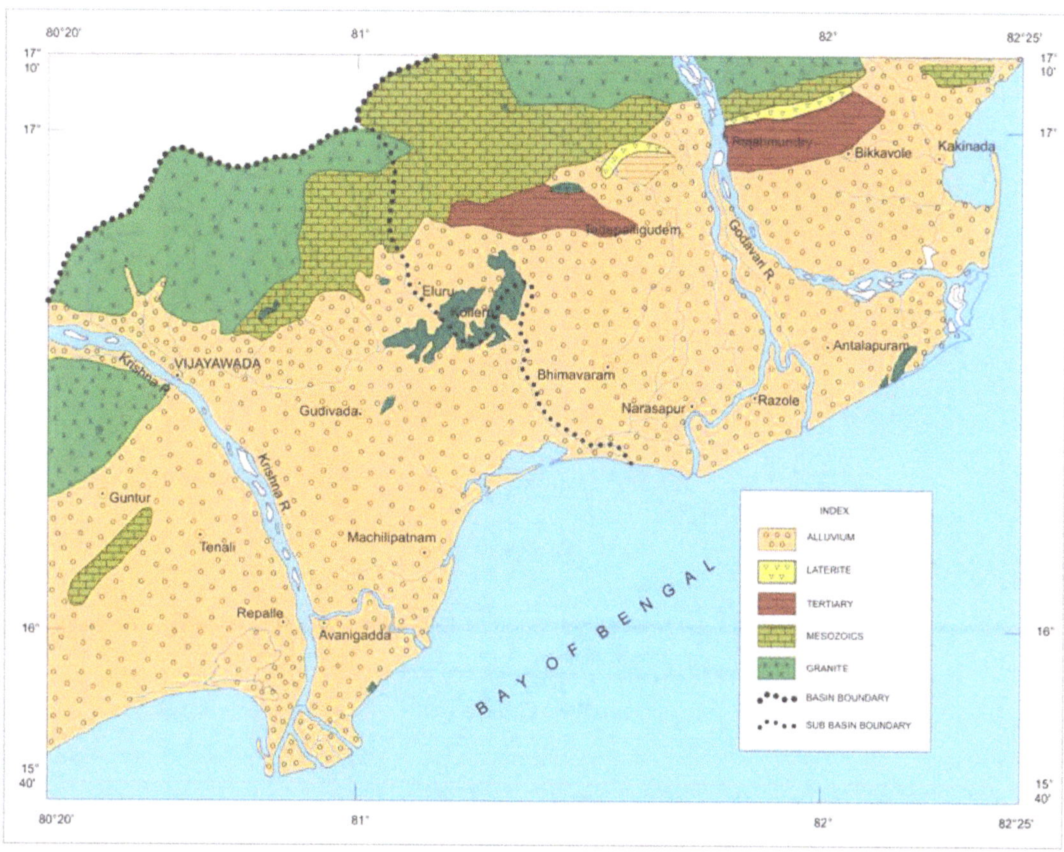

Fig. 4.53 A Regional Geological map of Krishna - Godavari delta basin (Andhra Pradesh)

The Tertiary formations are seen abundantly outcropping in the upland regions of Godavari delta with wide outcrop margins and gradually pinch towards the east and also towards west. These are mainly represented by medium to coarse grained sandstones with thicknesses varying from 10 to 400 m. The depth of wells constructed in these units range between 30 to 600 m with discharges varying from 900 to 2100 lpm and with transmissivity values varying from 60 to 900 m^2/day.

Quaternaries and Recent are the youngest group of sediments and occupy delta frames of both Godavari and Krishna. Thicknesses of Quaternary and Recent alluvial strata vary widely from a depth range of 40 m to an order of about 200 m or even more. These are basically represented by sands, clays and gravels. A major part of the delta aquifer systems are found to have become either brackish or saline excepting the places controlled by paleo channels, lobes etc, where the aquifer systems at shallow depths are found to be fresh. These freshwater bearing shallow aquifer systems are more predominantly seen in the central and eastern delta systems of Godavari delta and sparsely in the western parts of Godavari delta and in the central parts of Krishna delta. The boundaries of these freshwater systems in the delta frames of both Krishna and Godavari have been marked and discussed elsewhere (Chapter-V). Thousands of filter points and shallow tube wells have come in to operation in these shallow freshwater regions of both Krishna and Godavari deltas. The depths of these filter points/shallow tube wells have a range varying between 20 to 60 m and with yields of the order varying between 90 to 1500 lpm and with transmissivity values ranging between 800 to 2500 m^2/day.

A fence diagram showing lithological variations in respect of Godavari delta and their adjoining sedimentary regions is shown in Fig. 4.54 (a, b).

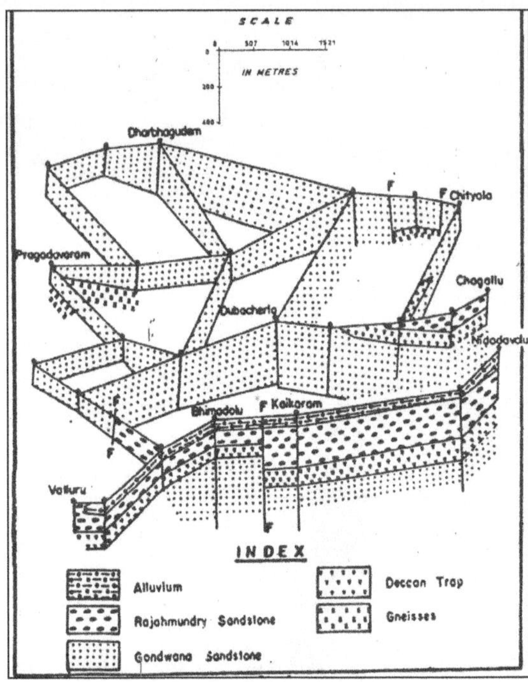

(a) West Godavari

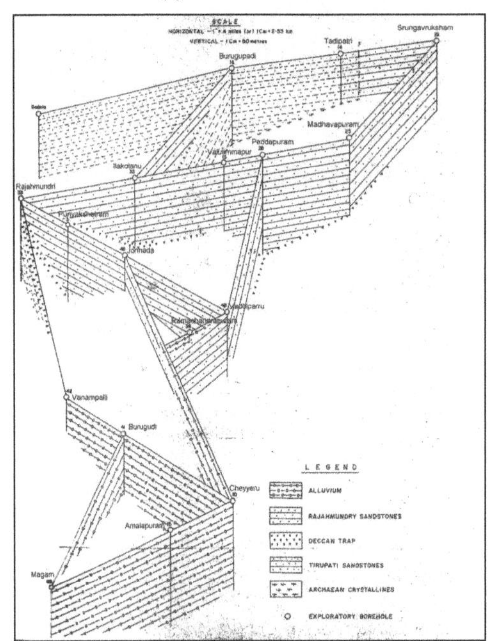

(b) East Godavari

Fig. 4.54 Fence diagram showing lithological variations in West and East Godavari districts of Godavari delta (Andhra Pradesh)

(Source: CGWB, 1995, 1999)

The hydrogeological parameters for selected boreholes in respect of Godavari and Krishna deltas and their adjoining sedimentary regions are presented in Table 4.22 and site locations shown in Fig. 4.55 (Data source: CGWB, 1999; CGWB, 1995 and Radhakrishna, I., 2001; CGWB, 1995; CGWB, 2006; Raju, M.B., et al., 1982).

Table 4.22 Hydrogeological parameters in Godavari, Krishna deltas and its adjoining reigons

S. No	Location	Depth drilled (m) / Constructed (mbgl)	Range of Aquifer zones tapped (mbgl)	Thickness of the zones (m)	SWL (mbgl)	Discharge (lpm) / Drawdown (m)	Specific Capacity (lpm/m)	T (m^2/day)	S	Remarks
1	2	3	4	5	6	7	8	9	10	11
Krishna District										
1	Sitaramapuram	153.9	…	30	2.55	2115	350.7	1284	…	Production well
		142.2				6.03				
2	Narasannapalem	129.85	…	41	3.73	3047	713.5	1807	….	Production well
		115.82				4.27				
3	Peddaparapudi	167	…	24	….	1320	591.9	2303		Production well
		60.9				2.23				
4	Tallavalli	121	…	41	….	810	45	….	…	Production well
		104.54				18				
5	Viravalli	125	…	34	7.74	950	109.4	299	…	Production well
		122.02				8.68				
6	Telaprolu	226	…	45	1.76	2744	191.6	799	…	Production well
		150				14.32				
7	Thotlavalluru	437.35	…	51	5.13	3060	938.3	5557	…	Production well
		153				3.261				
Guntur District										
8	Musalipadu	305.56	…	48	2.23	2089	566.12	18950	…	
		…				3.69				
9	Anumarlapudi	47	…	…	3.52	1076	123.6	471.2		
		…				8.7				
10	Sokur	65	…	….	….	1269	151.9	440	….	
		…				8.35				
11	Vadlamudi	65	…	….	….	232.6	15.6	26.18		
		…				14.88				
12	Guntur	41	…	….	2.85	120				
		…				…				

Table 4.22 *Contd…*

S. No	Location	Depth drilled (m) / Constructed (mbgl)	Range of Aquifer zones tapped (mbgl)	Thickness of the zones (m)	SWL (mbgl)	Discharge (lpm) / Drawdown (m)	Specific Capacity (lpm/m)	T (m^2/day)	S	Remarks
East Godavari										
13	Razole	157.07	79.2-153.0	36.91	3.81	2403	419.37	2261		--
		156.05				5.73				
14	Dowleswaram	273.41	75.48-201.02	83	12.55	2748	546.32	2240		Basalt touched at 271.58 m bgl
		203.6				5.03				
15	Punyakshetram	304.8	40.55-221.59	39.61	11.03	2381	360.2	3168		Basalt at 229.4 mbgl
		221.6				6.61				
16	Peria ramachandrapuram	305.1	65.0-228.29	67	6.79	2184	373.3	1537		--
		231.03				5.85				
17	Balavram	167.75	136.55-160.32	23.7	28.98	2302	469.7	1747		
		163.37				4.9				
18	Kotur	297.7	63.0-140.0	46	11.21	283	32	36.38		--
		147.73				8.83				
19	Lakshminarasapuram	620	145.0-470.65	150	6.94	1512	61.7	665.76		
		473.65				24.49				
20	Rajahmundry	302.17	55.0-297.0	59	18.9	1323	158.8	580.89		
		300				8.33				
21	Alamuru	200.5	64.0-190.0	39	3.1	37	154.1	208.8		Air test
		193				0.24				
22	Ramachandrapuram	30	9.0-28.0	13	0.95	60	857.1	--		--
		-				0.07				
23	Bikkavolu	120.14	42.0-104.0	29	3.645	720		1268.4		
		115				2.24	321.4			
24	Mallavaram	289.7	36.57-214.29	51.5	1.07	1736	266.25	752		--
		215.81				6.52				

Table 4.22 *Contd...*

S. No	Location	Depth drilled (m) / Constructed (mbgl)	Range of Aquifer zones tapped (mbgl)	Thickness of the zones (m)	SWL (mbgl)	Discharge (lpm) / Drawdown (m)	Specific Capacity (lpm/m)	T (m^2/day)	S	Remarks
25	Tatiparti	153.01	52.02-115.88	47.7	6.4	332	9.29	17.2		Crystalline contact encountered at 153.01 m bgl
		118.92				35.7				
26	Burugupudi	228.9	28.95-135.33	77	23.72	321	15.45	39.5		Crystallines encountered at 227.99 m bgl
		136.12				20.77				
27	Ravikampadu	91.5	34.5-89.12	30.55	11.62	106	7.68	31.8		Khondalite basement encountered at 88.54 m bgl.
		90.62				13.8				
28	Ambajipet	27	14.0-20.00	6	2.8	775.8	674.6	5925		
		22				1.15				
29	Srikrishna patnam	194.16	49.00-119.05	88	45.46	227	111.8	2774		Basalt encountered at 191.72-194.16
		137.21				2.03				
30	Vanapalli	188	32.0-70.00	38	2.6	660	132	6215		--
		73				5				
31	Jonnada	200	40.0-109.0	47	5.9	870	813	764.7		--
		112				1.07				
32	Dangeru	268	35.0-112.7	46						Abandoned due to high salinity
		217								
33	Cheyyeru	272.8	12.0-191.0	137						Abandoned due to high salinity
34	Peddada	248.7	27.0-230.0	110						Abandoned due to high salinity
	West Godavari									
35	Nallajerla	293.33	77.11-291.5	163	2.87	2642	230.5	553.5	3.018×10^{-4}	
						11.46				

Table 4.22 ***Contd…***

S. No	Location	Depth drilled (m)	Range of Aquifer zones tapped (mbgl)	Thickness of the zones (m)	SWL (mbgl)	Discharge (lpm)	Specific Capacity (lpm/m)	T (m^2/day)	S	Remarks
		Constructed (mbgl)				Drawdown (m)				
36	Devarapalli	121	30.0-48.00	18	3.93	795	112.9	3798		
		51				7.04				
37	Bhimadolu I	493	360.9-490.0	89	22.24	3710	1.5	916	-	
						2471				
38	Dubacherla	278.14	48.00-347.00	214	35.66	1748	310.48	846	--	
						5.63				
38	Dubacherla	278.14	48.00-347.00	214	35.66	1748	310.48	846	--	
						5.63				
39	Jaganathapuram	248	103.00-248.00	100	--	3300	0.825	3000	1.2×10^{-3}	
						3998				
40	Chagallu	59.45	36.3-55.2	19	19.8	972	95.29	848	--	
						10.2				
41	T.Narsapuram	231.65	77.00-202.00	73	13.72	1448	83.79	125.68		
						17.28				
42	Borrampalem	271.88	58.52-268.63	130	1.5	9.74	0.33	85.11	--	
						29.41				
43	Kaikaram	565	99.18-561.52	240	5.29	7683	401.4	2644	8.5×10^{-5}	
						19.14				
44	Vempadu	252.37	70.71-250.3	179	1.62	1932	125.45	408.2	3.986×10^{-4}	
						15.4				
45	Pulla	554	60.00-544.00	177	1.368	5250	1138.8	3540	5.35×10^{-4}	
						4.61				
46	V.Ramannagudem	124.05	24.94-121.0	34	19.35	1879	311.09	1992		
						6.04				
47	Yadavolu	292.61	166.00-315.00	149	2.37	1876	135.7	309.5	1.309×10^{-2}	
						13.82				
48	Kavaluru	148.45	61.30-143.6	70	2.1	2437	290	1897	--	
						8.4				

Table 4.22 ***Contd…***

S. No	Location	Depth drilled (m) Constructed (mbgl)	Range of Aquifer zones tapped (mbgl)	Thickness of the zones (m)	SWL (mbgl)	Discharge (lpm) Drawdown (m)	Specific Capacity (lpm/m)	T (m^2/day)	S	Remarks
49	Gunnampalli	129	20.00-124.00	67	0.49	1522	228.87	394.4	-	
						6.65				
50	Chinayegnapuram	390	51.00-387.00	135	3.09	6000	310.9	903	2.4×10^{-3}	
						19.298				
51	Veerampalem	280	136.00-277.00	131	8.33	35.58	1.586	137.28	--	
						22.43				
52	Gavaravaram	146.78	71.77-143.74	72	15.2	1493	147.82	648.45	-	
						10.1				

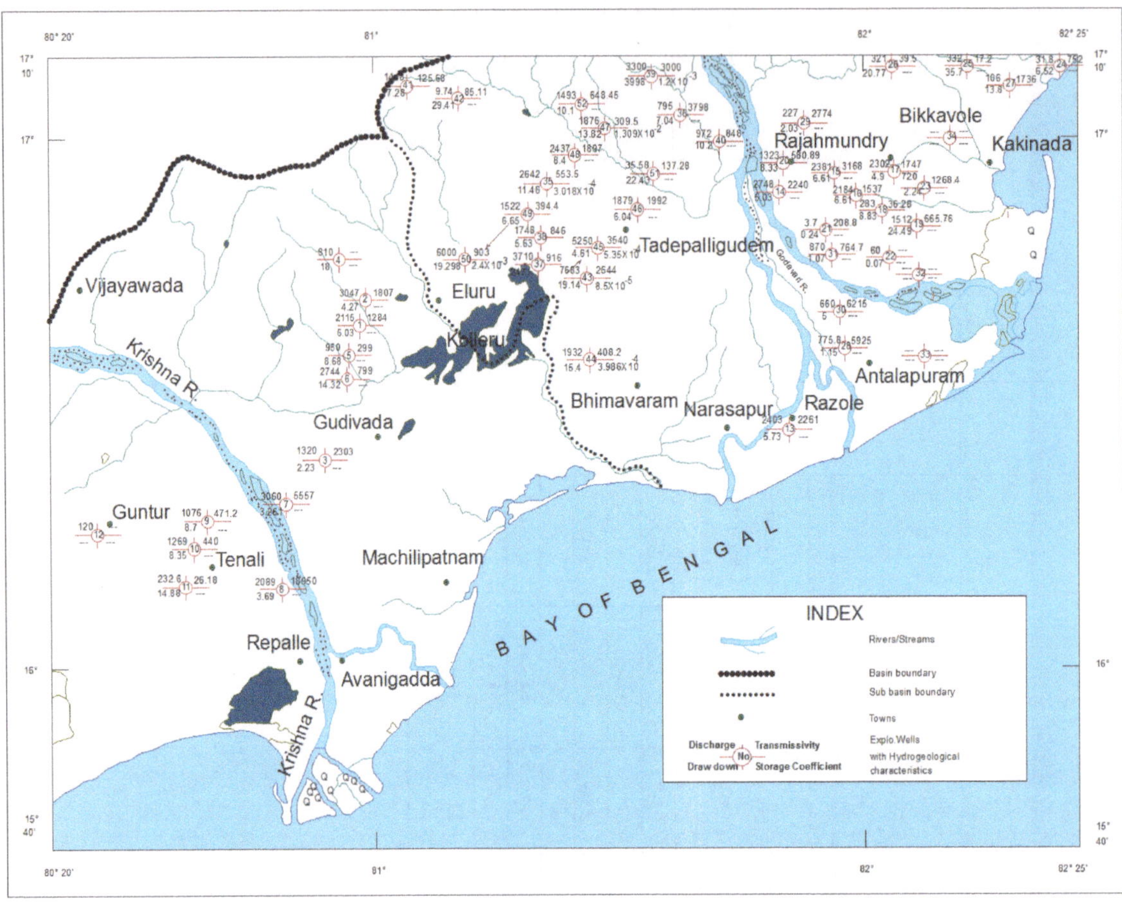

Fig. 4.55 Exploratory well site locations in Krishna – Godavari deltas (Andhra Pradesh)

Regional groundwater flow net analysis: A regional groundwater flow trends across different profiles for near surface aquifers in respect of Krishna and Godavari are presented in Fig. 4.56. The groundwater flows across different lengths in these two basins are provided in Table 4.23.

Table 4.23 Groundwater flow parameters across different profiles in Krishna-Godavari delta basins

S.No	Basin	Profile	Avg.Transmissivity (T) (m^2/day)	Gradient (I) (m/Km)	Length of the Coast (L) (Kms)	Total flow(Q) (MCM)
Near Surface Aquifers						
1	Krishna	P1	1800	1.16	39	29.72
		P2	1800	0.43	51	14.41
2	Godavari	P3	2000	0.56	27	11.04
		P4	2000	0.42	31	9.5
		P5	2500	1.57	47	67.33

The total estimated groundwater flow across the whole coastal margins of Krishna-Godavari basins is of the order of 132 MCM which comes to about 3.38% of the dynamic potentials of the basin.

* The deeper system parameters could not be presented in view of their high contamination levels with brackish/saline waters.

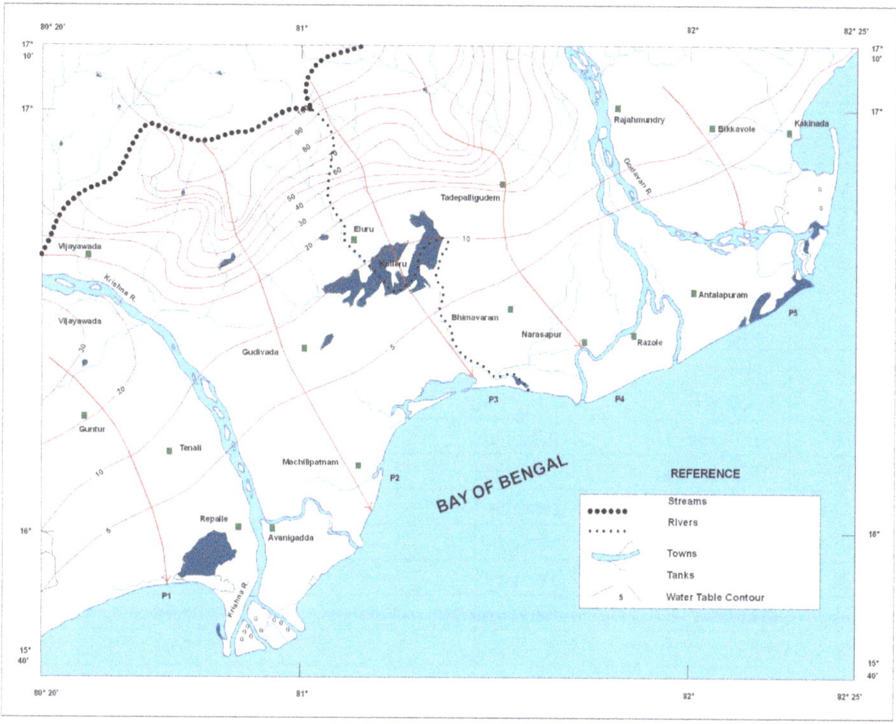

Fig. 4.56 Regional groundwater table map in parts of Krishna – Godavari deltas (Andhra Pradesh)

Seasonal water level fluctuations: The seasonal water level fluctuations for a 5 year period at different site locations in different geological horizons are shown in the Table 4.24 and their locations are reflected in the Figure 4.57. The average fluctuation in the coastal alluvial/sedimentary tracks is about 3.22 m and for Archeans is about 5.18 m.

Table 4.24 Seasonal water level fluctuations in Krishna-Godavari deltas

S.No	Name of the Site	Geological Horizon	Avg. Fluctuation (m) for a 5 Year period (1975-80) (m)
Krishna Delta			
1	Guntur	Alluvium	2.48
2	Bapatla	Alluvium	0.94
3	Ponnur	Alluvium	2.61
4	Mangalagiri	Alluvium	4.54
5	Machilipatnam	Alluvium	2.46
6	Vijayawada	Alluvium	9.67
7	Avanigadda	Alluvium	4.33
8	Challapalli	Alluvium	2.18
9	Vuyyuru	Alluvium	2.45
10	Gudivada	Alluvium	2.00
11	Pedana	Alluvium	2.18
12	Kondapalli	Alluvium	4.50
13	Nuzvidu	Granite	10.15
14	Tiruvuru	Mesozoics	9.98

Table 4.24 *Contd...*

S.No	Name of the Site	Geological Horizon	Avg. Fluctuation (m) for a 5 Year period (1975-80) (m)
Godavari Delta			
1	Kakinada	Alluvium	2.00
2	Gannavaram	Alluvium	3.46
3	Yanam	Alluvium	2.02
4	Amalapuram	Alluvium	1.16
5	Razole	Alluvium	2.58
6	Bikkavolu	Tertiary	1.98
7	Kotipalli	Alluvium	1.96
8	Tallarevu	Alluvium	2.06
9	Kottapeta	Alluvium	3.56
10	Samlkota	Alluvium	4.38
11	Bhimavaram	Alluvium	0.98
12	Gopalapuram	Granite	2.62
13	Dwaraka Tirumala	Mesozoics	0.94
14	Gollagudem	Tertiary	10.34
15	Akivedu	Alluvium	1.59
16	Eluru	Alluvium	1.91
17	Narasapur	Alluvium	1.46
18	Kannapuram	Granite	2.76
19	Jangareddygudem	Mesozoics	4.08
20	Dharmajigudem	Alluvium	5.03
21	Attili	Alluvium	1.13

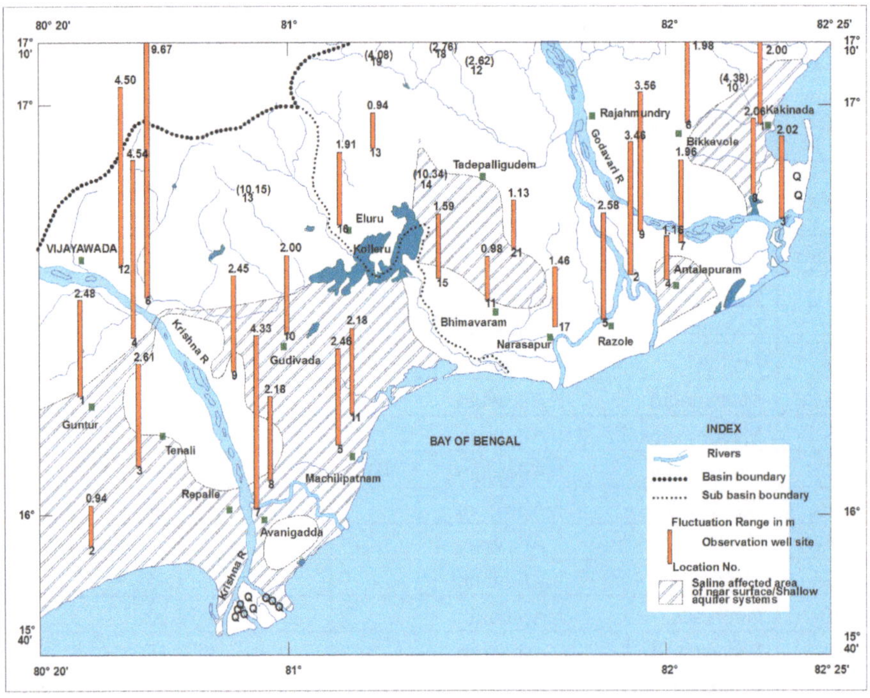

Fig. 4.57 Network of observation well stations and their fluctuation levels in Krishna-Godavari deltas (Andhra Pradesh)

Resource potentials in Krishna delta basin

- Area of the Krishna delta basin and its adjacent sedimentaries : 9900 sq.km
- Saline affected Area : 5707 sq.km
- Coverage of Alluvium : 7350 sq.km
- Coverage of Mesozoics : 650 sq.km
- Coverage of Granites : 1900 sq.km
- Dynamic potentials for alluvials and sedimentaries (excluding saline affected area)

(i) Rainfall infiltration method

(a) Alluvium : 327.00 MCM

(b) Mesozoics : 75.47 MCM

(ii) Fluctuation method

(a) Alluvium : 1536.97 MCM

(b) Mesozoics : 72.02 MCM

(iii) Average dynamic potentials

(a) Alluvium : 932 MCM

(b) Mesozoics : 73.75 MCM

- Dynamic potential for Granitic area by fluctuation method : 95 MCM
- Total dynamic potentials for Krishna basin is (932 + 73.75 + 95) : 1100.75 MCM

Note: The relatively high recharge levels through fluctuation method could be due to the recharge through high intensity canal networks and irrigation returns.

- Resource potentials in deep confined aquifer systems (static)

(Area × Thickness × S.Coefficient)

(i) Mesozoics ($430 \times 100 \times 2.5 \times 10^{-4}$) : 10.75 MCM

(ii) Granites ($1900 \times 10 \times 1 \times 10^{-3}$) : 19 MCM

- Total estimated resource potentials in deep aquifers restricted to the basin boundaries considered

: 29.75 MCM

Note: The deep alluvial aquifer systems in the delta basin area are totally either brackish or saline and therefore can't be considered for estimations of fresh water resource potentials.

Resource potentials in Godavari delta basin

- Area of the Godavari delta basin and its adjacent regions : 10,850 sq.km
- Saline affected Area : 1600 sq.km
- Coverage of Alluvium : 7300 sq.km
- Coverage of Mesozoics : 1800 sq.km
- Coverage of Tertiaries + Laterites : 1000 sq.km
- Coverage of Granites : 750 sq.km
- Dynamic potentials for sedimentaries (excluding saline affected area)

(i) Rainfall infiltration method
  (a) Alluvium : 1128.97 MCM
  (b) Mesozoics : 356.51 MCM
  (c) Tertiaries : 198.06 MCM
(ii) Fluctuation method
  (a) Alluvium : 3078 MCM
  (b) Mesozoics : 301.5 MCM
  (c) Tertiaries : 616 MCM
(iii) Average dynamic Potentials
  (a) Alluvium : 2103 MCM
  (b) Mesozoics : 329.0 MCM
  (c) Tertiaries : 407.03 MCM
(iv) Dynamic potential for Granitic area by fluctuation method : 10.08 MCM

- Total dynamic potentials for Godavari delta basin is
  (2103 + 329.0 + 407.03 + 10.08) : 2849.88 MCM

Note: The relatively high recharge levels through fluctuation method could be due to the recharge through high intensity canal networks and irrigation returns.

- Static potentials (Area × Thickness × S.Coefficient)

(i) Mesozoics ($1800 \times 100 \times 2.5 \times 10^{-4}$) : 45 MCM
(ii) Tertiaries ($1400 \times 100 \times 6 \times 10^{-4}$) : 84 MCM
(iii) Granites ($750 \times 10 \times 1 \times 10^{-3}$) : 7.5 MCM

- Total estimated static Groundwater potentials : 136.5 MCM
- Total estimated (dynamic + deep confined systems) potentials in Krishna-Godavari delta basins : 4116.88 MCM

4.4.7 Mahanadi-Brahmini-Baitarni delta basins

Mahanadi, Brahmini and Baitarni form the largest composite delta system along the east coast of India in the state of Orissa with an areal spread of the order of about 13,200 sq.km. Mahanadi, Brahmini and Baitarni have stretches of about 850, 800 and 360 km respectively from their origins and debouch into Bay of Bengal with an annual combined discharge of the order of about 84,950 MCM. Mahanadi near Naraj (apex point) distributes itself as a fan shaped delta through a number of distributaries and sub distributaries namely Birupa, Mahanadi/Chitrala, Katjhori/Devi and Koayakhai/Daya/Kushubadra/ Bhargavi. Similarly Brahmini and Baitarni also have apex points near Jeenapur and Akhudapada respectively from where they get bifurcated into sub distributaries further down stream side and get interlaced with each other near the coast. There are several tidal inlets that make their way from the shore line to several interior regions, sometime up to the central region of the delta, thereby reducing the freshwater outflow into the composite delta complex. The delta complex in general is a flat terrain with gentle slope towards the Bay of Bengal and with altitudes varying between 6 to 15 m amsl in the upper and the central parts of the region and with a set of parallel sand dune ridges along the coast, attaining heights varying from 6 to 8 m above mean sea level. A typical characteristic geomorphic feature of the composite Mahanadi-Brahmini-Baitarni delta complex is the formation of long narrow islands in the central part of the delta which appear to be the manifestation of the subsurface structure in the region. A drainage map with carved out coastal boundary with different sub basins is presented in Fig 4.58 covering the districts of Bhadrak (100%), Jajpur (80%), Jagatsinghpur (100%), Kendrapara

(100%), Cuttuck (53%), Puri (100%) and Khorda (35%). The different sub basins marked are 1) Brahmini-Baitarni, 2) Mahanadi and 3) Koyakhai-Katjhori. Each of these sub basins exhibit different hydrological, hydrogeological and hydrogeochemical characteristics. The entire delta region including the identified sub basins could be classified into upper, central and lower regions based on soils, morphology and drainage patterns. The upper delta region lies between Naraj (Mahanadi), Jeenapur (Brahmini) and Akhudapada (Baitarni) and the down stream line that joins the further bifurcation of Birupa, Mahanadi, Koyakhai, Brahmini and Baitarni rivers. This area is characterized by coarse to medium sandy facies with little or no admixtures of clays and silts. The central delta region encompasses the area between the sub distributary delta systems and the upper part of the tidal delta reaches. This region is characterized by a number of lenticular islands that developed as a result of bifurcation and rejoining of the distributary river systems. The lower delta region adjoining the coast line is characterized by a number of low lying swamps, tidal canals and a series of sand ridges and valleys.

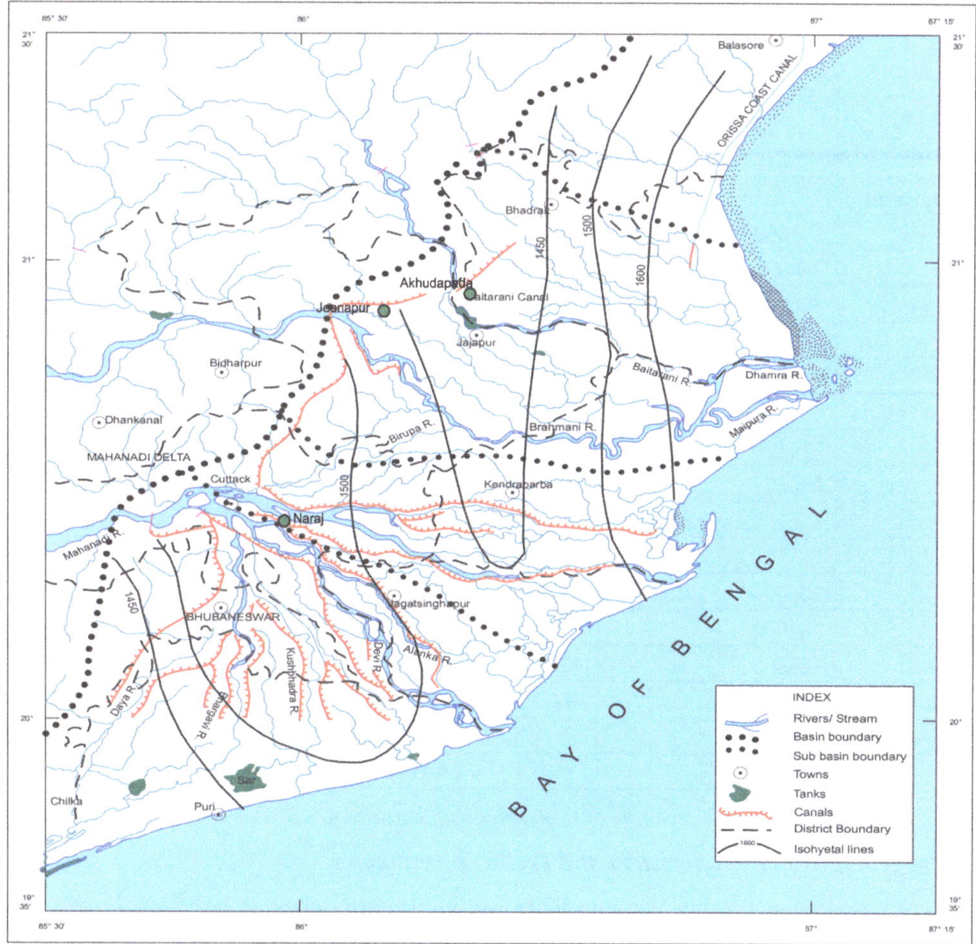

Fig. 4.58 Drainage map with isohyetal trend lines in Mahanadi-Brahmini-Baitarni delta complex (Orissa)

The annual average precipitation in the composite delta system is of the order of 1579 mm with variation levels ranging from 1600 mm (close to the coast) and 1450 mm (away from the coast). Over, 80% of the rainfall is received from the southwest monsoon spread over the period from June to September and the rest of the precipitation generally comes through northeast monsoon during the period of October to December. The delta region has typical tropical climate with mean summer temperature of 37 °C (March to June) and with a mean winter temperature of 22 °C (November to February). The humidity values in the delta range from 90% (wet season) and 25% (dry season). The average annual potential evapotranspiration is of the order of 1568 mm which is marginally more than the average annual precipitation levels (refer chapter-III).

Mahanadi-Brahmini-Baitarni coastal delta complex is rich in both surface and Groundwater resources with good network of canals, tanks, dug wells and tubewells. The canal networks, tanks and lift irrigation schemes together contribute to an irrigation potential of the order of 9,26,545.45 hectares. Similarly the irrigation potential developed through different groundwater structures is of the order of 81,417.3 hectares. The percentage of irrigation potentials developed through groundwater is about 11.3%. The sub basin wise/proportionate district wise surface and groundwater irrigation potentials in the composite Mahanadi-Brahmini-Baitarni deltas are presented in Table 4.25.

Table 4.25 Surface and groundwater irrigation potentials in the Mahanadi-Brahmini-Baitarni delta regions

District	Surface Irrigation			Groundwater Irrigation	Gross Irrigation (ha)
	Major and Medium (Canals + Tanks) (ha)	Minor Flow (ha)	Total (ha)	Minor sector/ Groundwater (ha)	
Brahmani-Baitarni					
Bhadrak (100%)	103000	6155	109155	7846	117001
Jajpur (80%)	76000	10710.4	86710.4	11390.4	98100.8
Kendraparha (46%)	50600	63.94	50663.9	12919.1	63583
			246529	**32155.5**	**278685**
Mahanadi					
Kendraparha (54%)	59400	75.06	59475.1	15165.9	74641
Cuttack (33%)	210870	7122.39	217992	9067.41	227060
Jagatsinghpur (54%)	28080	582.12	28662.1	845.64	29507.8
			306130	**25078.95**	**331209**
Daya					
Khorda(35%)	24500	7248.85	31748.9	3893.75	35642.6
Puri(100%)	184000	2136	186136	14089	200225
Cuttack(20%)	127800	4316.6	132117	5495.4	137612
Jagatsinghpur(45%)	23400	485.1	23885.1	704.7	24589.8
			373887	**24182.85**	**398069**
Total			**926545**	**81417.3**	**1007963**

(Source: Directorate of Economics and Statistics, Orissa 2007-08)

Regional hydrogeological characteristics and system parameters

A regional geological map of the Mahanadi-Brahmini-Baitarni coastal delta region is presented in Fig. 4.59. The geological units in and around the basin are Archeans, Mesozoics, Tertiaries, Quaternaries and Recent. The Archeans outcrop in the upland regions of Puri and Cuttack districts regions and are represented by granites, gneisses, chornakites and khondalites. Groundwater in these rock aquifers occur in both unconfined and semi-confined conditions restricted to the shallow weathered and fractured portions up to a depth range varying from 30-50 m and with yields ranging between 180-250 lpm with drawdowns varying between 8-15 m. The transmissivity and storage coefficients for these rock aquifers have a range varying between 2-12 m^2/day and storativity are in the order varying between 1×10^{-3} to 5×10^{-3}.

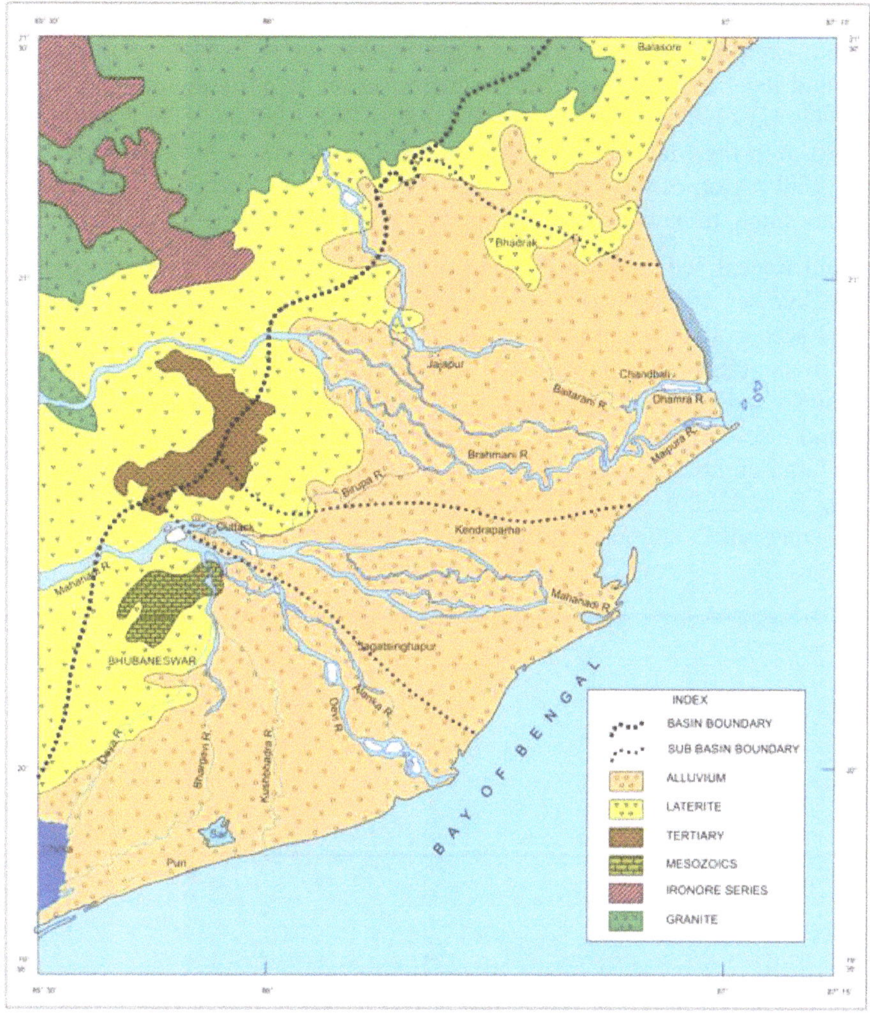

Fig. 4.59 Regional geological map of Mahanadi-Brahmini-Baitarni delta complex (Orissa)

The Mesozoics are represented by Athagarh sandstones consisting of sandstones, shales and claystones. Several exploratory wells were drilled in these tracks up to depths varying between 15-120 m below ground level. The general yield characteristics for these sandstone aquifers in the vicinity of the basin is of the order varying between 120-1800 lpm for drawdowns varying from 15-20 m and with an order of transmissivity varying from 80 m^2/day to 450 m^2/day.

The Tertiary group of rocks are widely distributed over the coastal delta basin both as surface outcrops and in subsurface. These are represented by Plio-Miocene and Miocene sediments with extensive thicknesses and occur below Quaternaries. The Plio-Miocene, Miocene sediments largely consist of fine to medium sandstones with intercalatious of thin layers of clays and silts and sometimes fossiliferrous limestones. These sediments are spread out in the central and lower delta regions of the basin and have huge thicknesses of the order varying from 60 to 600 m. Maximum thicknesses of these sediments are seen in the central and outfall regions of Mahanadi delta and get reduced on either side of Mahanadi delta. The lithological variations available through several exploratory and production well programs indicate a dominance of clay facies towards the outfall region and sandy facies towards the central and upper delta regions. These sediments constitute most potential aquifer systems in the delta region with yields varying from 2500 to 4500 lpm at drawdowns varying from 8 to 12 m and with transmissivities varying from 800 m^2/day to 4000 m^2/day. A large portion of these aquifers in the outfall regions of Mahanadi-Brahmini-Baitarni have become brackish up to depth levels varying between 60 to 300 m below ground level below which fresh waters are recorded with reasonably high yields. The interface depth levels vary between 60 to 150 m in the outfall regions of Brahmini and Baitarni delta and from 60 to 300 m in the Mahanadi delta outfall region. In the southwest extensions of Mahanadi delta and in the outfall region of the sub basin Koyakhai-Katjhori where the entire stretch of the Plio-Miocene aquifers are found to have become either saline or brackish.

Quaternaries and Recent sediments occur as sands, clays and gravels overlying the Plio-Miocene sediments with thicknesses varying from 40 to 120 m below ground level. Maximum thicknesses of these sediments are seen in the upper regions of the delta sub basin and become thin down towards the outfall region. Aquifers in this unit occur in both unconfined and semi confined situations with potential patches of freshwater resources occurring in the upper delta regions of Mahanadi, Birupa, Brahmini and Baitarni regions and also in the downstream regions/outfall regions of Koyakhai and Katjhori (Puri, Konark coast). A large number of filter points and shallow tube wells are in operation in these units with yields varying from 2500 to 4500 lpm at rather small drawdowns ranging from 4 to 6 m and with relatively high transmissivities ranging from 2500 to 4500 m^2/day.

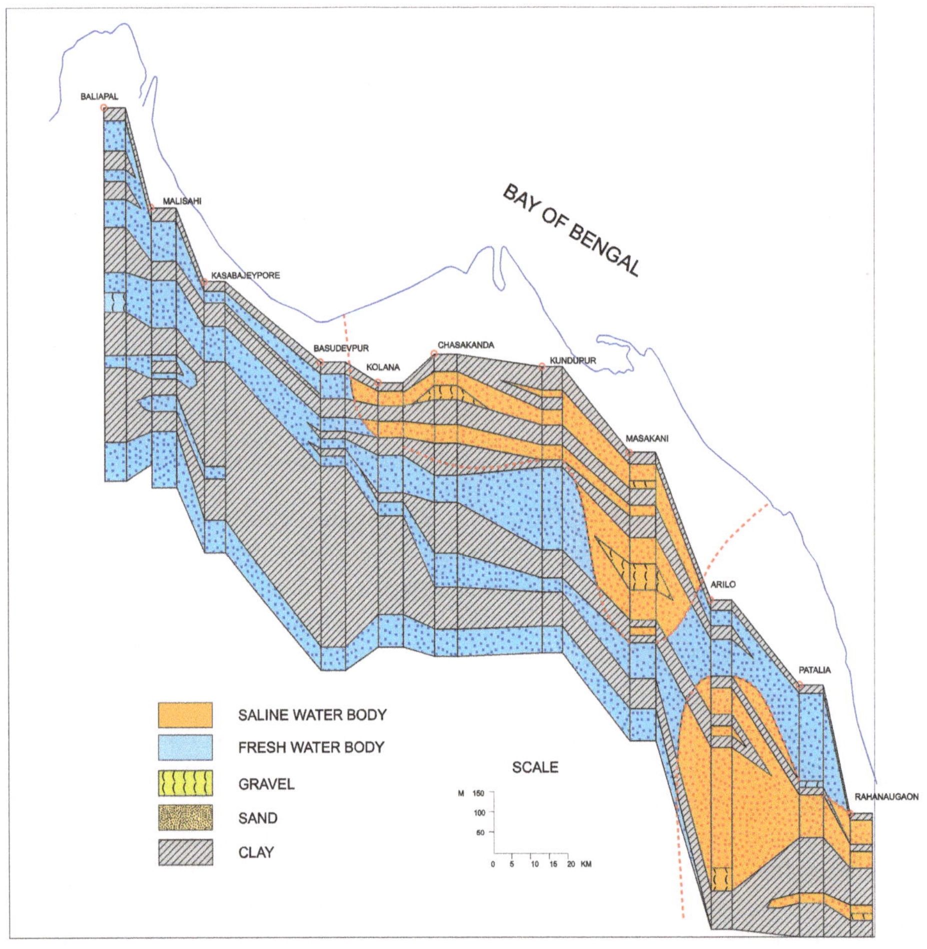

Fig. 4.60 A fence diagram showing lithological and water quality variations in Mahanadi-Brahmini-Baitarni delta complex (Orissa) parallel to the coastline

A regional subsurface lithological variations map reflecting the sand clay facies based on data obtained through several exploratory and production well programs is incorporated in Fig. 4.60. Also a few selected hydrogeological sections along and across the basin area are also presented in Fig. 4.61.

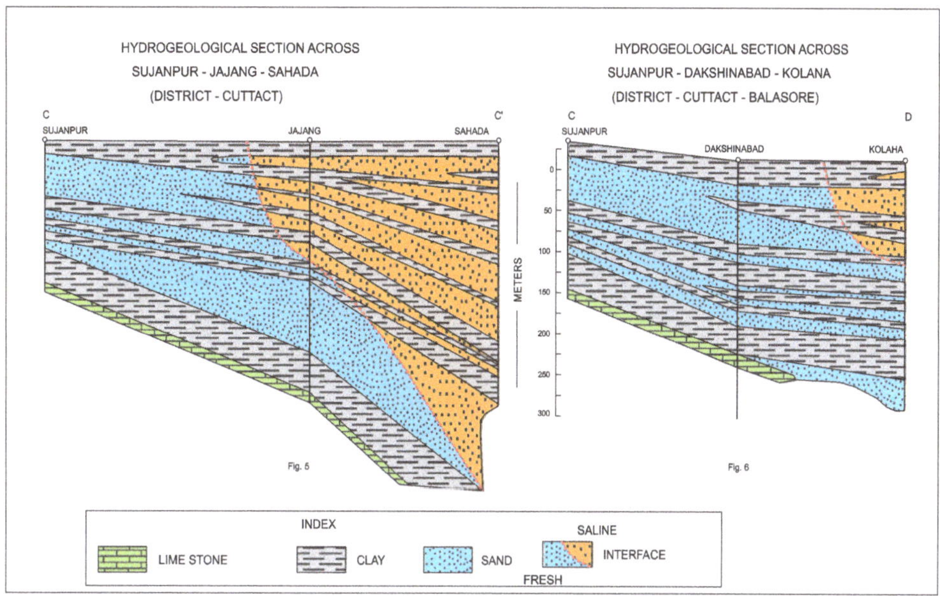

(Source: Radhakrishna, I. & Dutt, D.K., 1976)

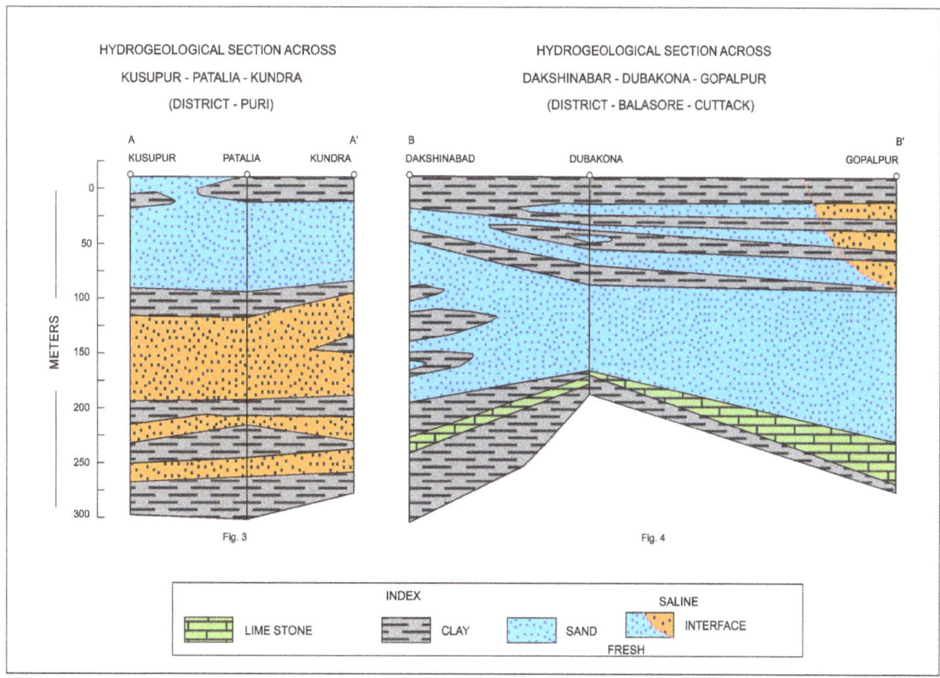

(Source: Radhakrishna, I. &Dutt, D.K., 1976)

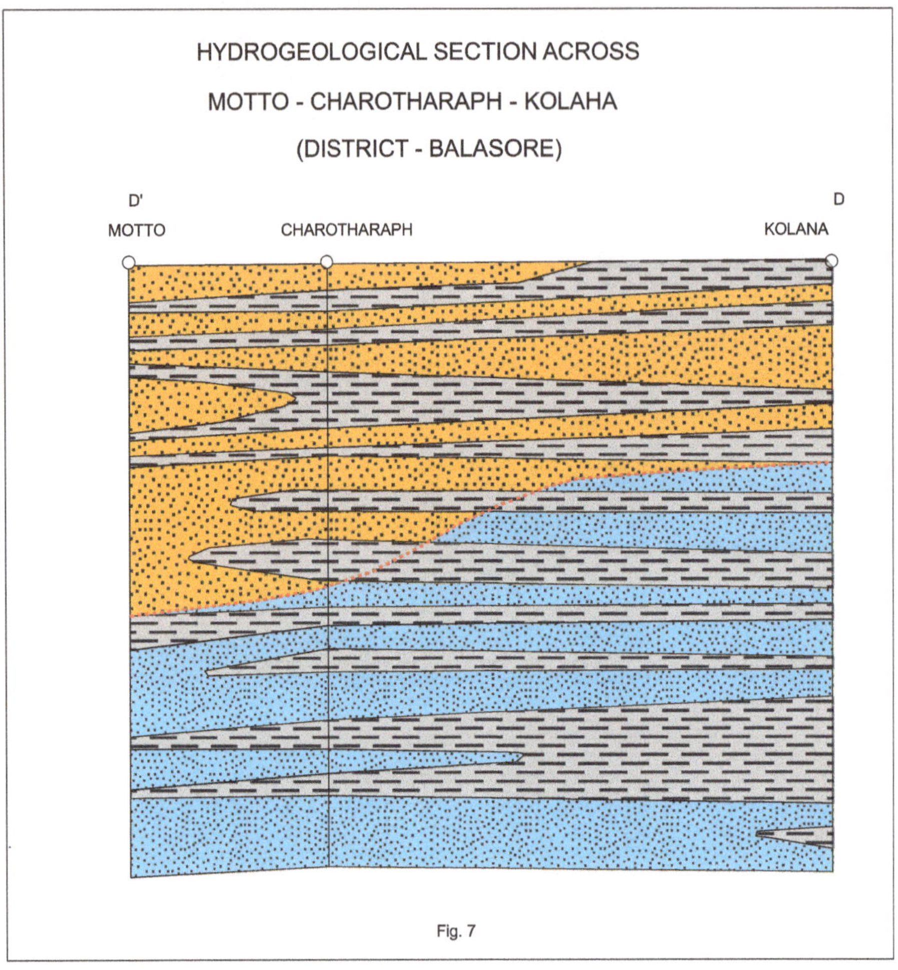

Fig. 4.61 Hydrogeological sections across different sections in Mahanadi-Brahmini-Baitarni delta complex (Orissa)

(Source: Radhakrishna, I. &Dutt, D.K., 1976)

A distinct feature of these sub surface sections is that the transverse section (perpendicular to the coast) show gently dipping nature of beds with thicknesses enlarging towards the coast and the sections drawn parallel to the coast exhibit horizontal nature of beds. The sections along Kusupur-Kundra (south Mahanadi delta region) shows abundant granular zones up to 180 m below which clay facies prevail. The section drawn along Sujanpur-Sahada (Mahanadi delta region) characterizes typical continental shelf facies particularly along the bottom beds. The section also characterizes with a series of alternate beds of clays and sands with a clear domination of sand facies up to 250 m below ground level below which grey/green facies develop followed by beds of arenaceous limestones. Another section drawn between Sujanpur-Kolaha (Brahmini-Baitarni region) demonistrates a similar situation but with differing sand clay facies with depth. The upper half of the section is characterized by dominant sand facies and in the lower half by clay facies. Similarly another section drawn between Dakshinabad and Gopalpur (Brahmini-Baitarni region) also displays a system of alternate beds of sands and clays up to a depth of 75 m below ground level below which thick potential granular zones exist followed by green clays and arenaceous limestones. The beds here show opposite dip slopes on either side of Dubakona site. Another section between Kolaha-Motto (Baitarni region) demonstrates thick granular material with definite increase of sand facies towards the coast and clay facies inland. All the sections presented above in different sub delta basins in this part of the coast present a complex situation of sand clay facies with wide variations in their thicknesses. These thickness variations perhaps could be better comprehended through an isopach map presented in Fig. 4.62 spread over the entire delta complex. The map demonstrates thinning of granular zones towards up stream side and thickening towards coastal margin.

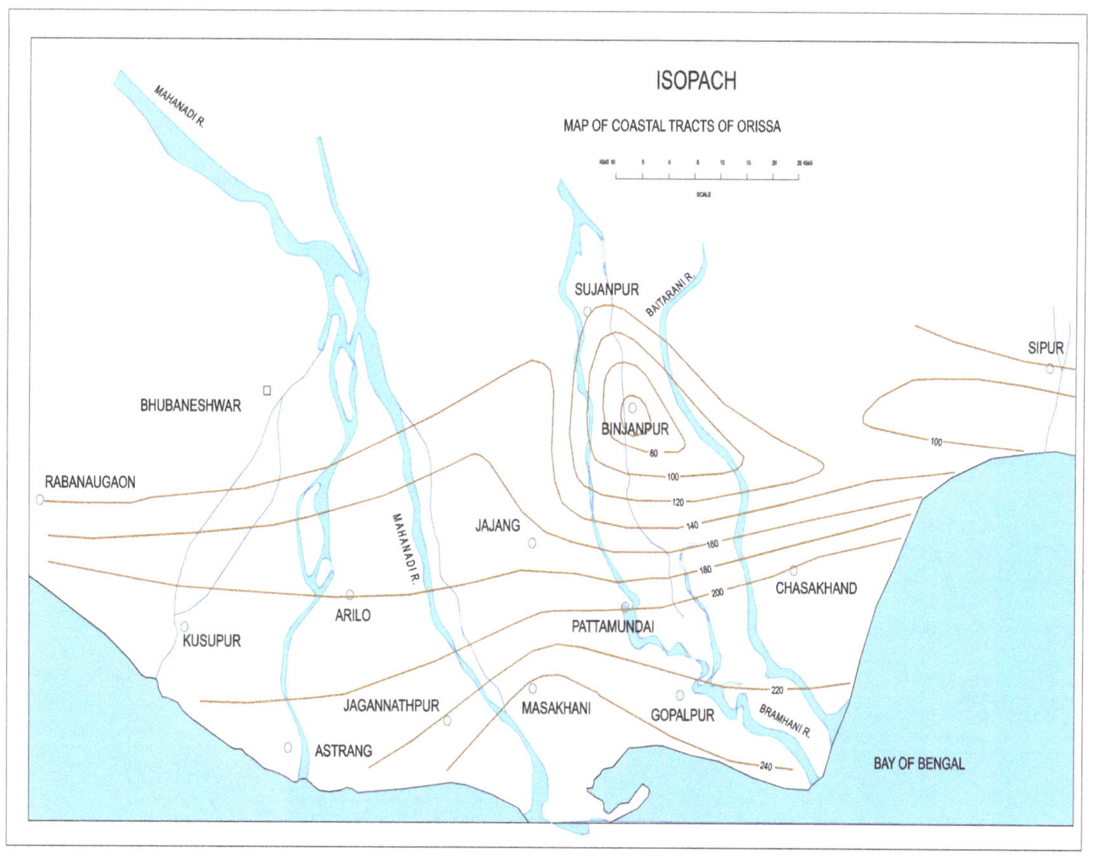

Fig. 4.62 Isopach distribution lines in Mahanadi-Brahmini-Baitarni delta complex (Orissa)

(Source: Radhakrishna, I. &Dutt, D.K.,1976)

The hydrogeological parameters for selected wells in the coastal Mahanadi-Brahmini-Baitarni basins are furnished in Table 4.26. with site locations of the sites presented in Fig.4.63. (Data source: Radharkishna, I. and Dutt, D.K., 1976; Bhatnagar, N.C., et al 1970).

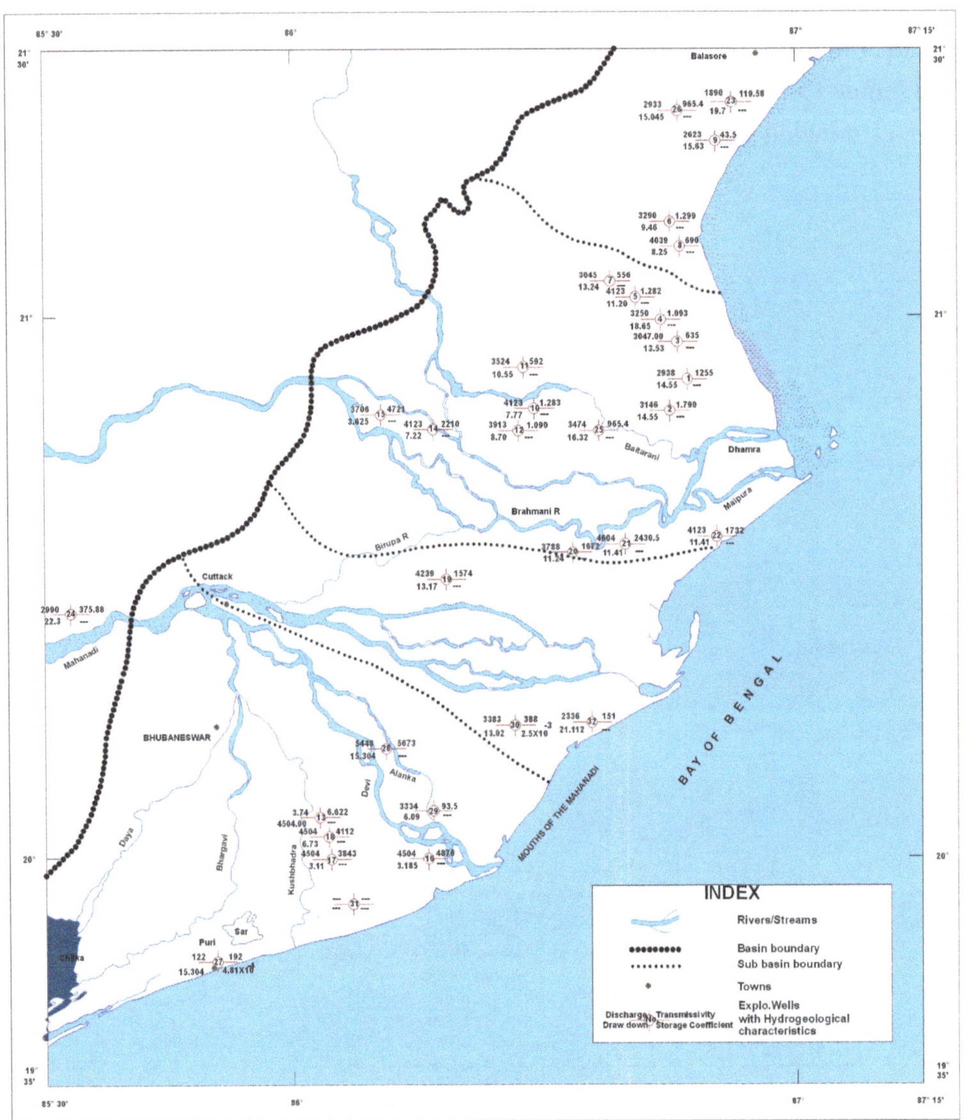

Fig. 4.63 Exploratory well site locations in Mahanadi-Brahmini-Baitarni delta complex (Orissa)

Table 4.26 Hydrogeological parameters in the Mahanadi-Brahmini-Baitarni delta basin

S.No	Location	Depth drilled (m)	Range of Aquifers tapped (mbgl)	Thickness of the zones (m)	SWL (mbgl)	Discharge (lpm) / Drawdown (m)	Specific Capacity (lpm/m)	T (m^2/day)	S	Remarks
1	2	3	4	5	6	7	8	9	10	11
1	Chasakhand	406.9	255.0-341.0	52	0.09(agl)	2938	201.9	1255	...	
						14.55				
2	Motto	316.09	188.0-270.0	53	0.5(agl)	3146	217	1,790	...	
						14.55				
3	Khraang	303.86	153.0-260.0	76	0.5(agl)	3047.00	225.2	635	...	
						13.53				
4	Charotharaph	300	176.0-264.0	74	0.30(agl)	3250	174.2	1,093	...	
						18.65				
5	Harishingpur	316.09	168.0-272.0	85	1.60(agl)	4123	368.1	1,282	...	
						11.20				
6	Daulatpur	313.62	108.0-193.0	59	0.45(agl)	3290	347.7	1,299	....	
						9.46				
7	Kolaha	309	96.0-180.0	42	0.06(agl)	3045	230	556	...	
						13.24				
8	Mandari	278	25.0-123.0	52	2.34	4039	489.6	690	...	
						8.25				
9	Gopalpur (churamari)	281	77.0-146.0	57	2.38	2623	167.8	435	...	
						15.63				
10	Dakshinabad	251.16	42.0-87.0	36	1.60(agl)	4123	530.7	1,283	...	
						7.77				
11	Dubal	346.1	112.0-229.0	68	2.54	3524	334	592	...	
						10.55				
12	Sarapur	215.18	51.0-163.0	68	1.95	3913	450	1,090	...	
						8.70				

Table 4.26 ***Contd...***

S.No	Location	Depth drilled (m)	Range of Aquifers tapped (mbgl)	Thickness of the zones (m)	SWL (mbgl)	Discharge (lpm) / Drawdown (m)	Specific Capacity (lpm/m)	T (m^2/day)	S	Remarks
1	2	3	4	5	6	7	8	9	10	11
13	Patalia	305.8	24.0-86.0	54	1.08	4504.00	1204	6,622	…	
						3.74				
14	Erbank	170	36.0-107.0	54	3.48	4123	571	2210	…	
						7.22				
15	Sujanpur	163.78	29.0-61.0	29	3.14	3706	1022	4721	…	
						3.625				
16	Juinti	305	28.0-68.0	32	1.55	4504	1414	4876	…	
						3.185				
17	Kundra	305.39	25.0-75.0	36	1.41	4504	1448	3843	…	
						3.11				
18	Kusupur	266.45	34.0-94.0	38	1.72	4504	669	4112	…	
						6.73				
19	Jajang	305.77	125.0-225.0	60	0.91(agl)	4239	321.9	1574	…	
						13.17				
20	Pattamundai	304	99.0-230.0	46	0.68	3788	337.1	1672	…	
						11.24				
21	Adjhori (sansarpur)	262	126.0-260.0	53	0.86	4604	403.6	2430.5	…	
						11.41				
22	Madanpur	307	130.0-200.0	50	0.80 (agl)	4123	361.3	1732	…	
						11.41				
23	Kasabajaypur	286	33.0-105.0	35	2.925	1890	95.94	119.58	…	
						19.7				
24	Nuapatna	191	140.0-185.0	33	0.4	2990	134.1	375.88	…	
						22.3				
25	Kumandihi	219.6	134.0-195.0	52	0.74	3474	210.3	965.4	…	
						16.52				

Table 4.26 ***Contd…***

S.No	Location	Depth drilled (m)	Range of Aquifers tapped (mbgl)	Thickness of the zones (m)	SWL (mbgl)	Discharge (lpm) / Drawdown (m)	Specific Capacity (lpm/m)	T (m^2/day)	S	Remarks
1	2	3	4	5	6	7	8	9	10	11
26	Shipura	330	20.0-168.0	71	3.9	2933	194.9	253.5	…	
						15.045				
27	Puri; Puri	602	149.0-228.1	119	1.991	122	7.97	192	4.81×10^{-4}	
						15.304				
28	Arilo; Cuttack	473.08	59.8-88.7	29	2.845	5448	355.9	5673	-	
						15.304				
29	Machgaon; Cuttack	612.41	30.8-50.5	20	1.545	3334	547.45	935	-	
						6.09				
30	Ersama; Cuttack	609.62	184.0-208.0	10	0.676 agl	3383	259.83	388	2.5×10^{-3}	
						13.02				
31	Konark; Puri	601	12.9-30.5	15	0.7	-		-	-	
32	Garjanga; Cuttack	612.6	320.0-350.0	20	0.815 agl	2336	110.6	151	--	
						21.112				

Regional groundwater flow net analysis

A groundwater flow trend map across different profiles for both near surface and deep aquifer systems are provided in the Figure 4.64. The groundwater flow across different lengths of different sub basins is provided in Table 4.27.

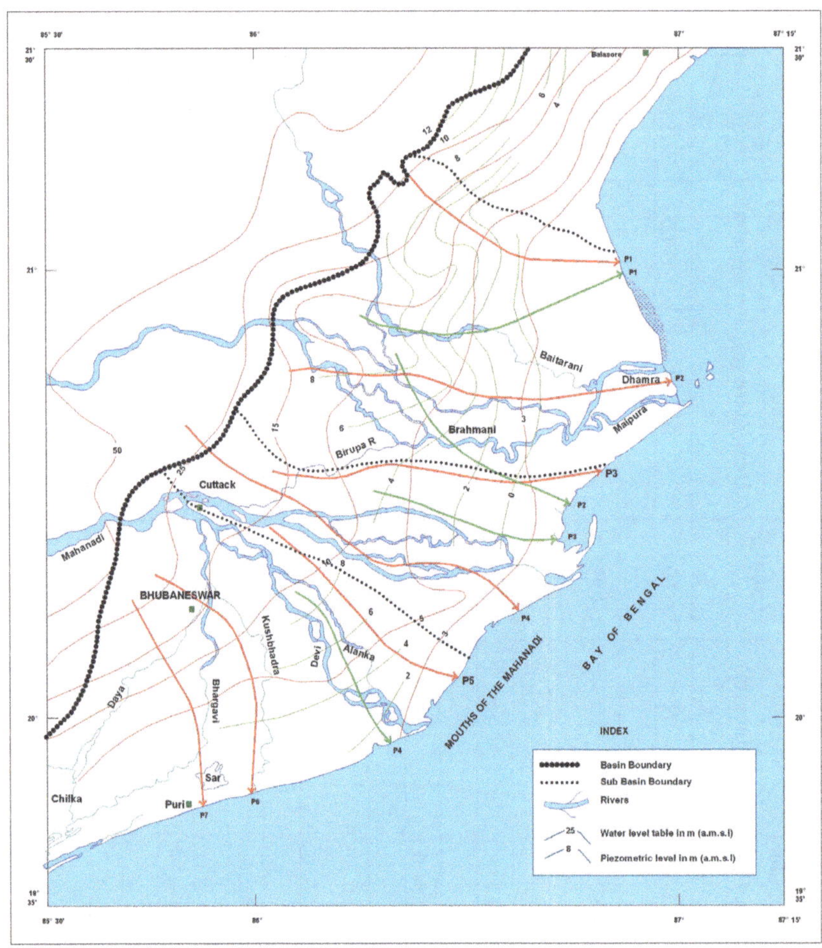

Fig. 4.64 Groundwater flow map for near surface and deep aquifer systems in Mahanadi-Brahmini-Baitarni delta complex (Orissa)

Table 4.27 Groundwater flow parameters across different profiles in Mahanadi-Brahmini-Baitarni delta basin Near surface Aquifers

S.No.	Basin	Profile	Avg.Transmissivity (T)(m^2/day)	Gradient (I) (m/kms)	Length of the Coast (L) (kms)	Total flow (Q)(MCM)
1.	Brahmani	P1	1600	0.57	21	6.99
		P2	1600	0.17	42	4.17
2.	Mahanadi	P3	2200	0.18	30	4.34
		P4	2200	0.19	42	6.41
3.	Daya	P5	3000	0.21	38	8.74
		P6	3000	0.47	30	15.44
		P7	3000	0.68	30	22.34

Table 4.27 ***Contd...***

S.No.	Basin	Profile	Avg.Transmissivity (T)(m^2/day)	Gradient (I) (m/kms)	Length of the Coast (L) (kms)	Total flow (Q)(MCM)
Deep system						
1.	Brahmani	P1	1000	0.23	47	3.95
		P2	1000	0.17	46	2.85
2.	Mahanadi	P3	800	0.12	47	1.65
3.	Daya	P4	3000	0.18	80	15.77

The total estimated groundwater flow across the whole coastal margins of Mahanadi-Brahmini-Baitarni basins is of the order of 92.65 MCM which comes to about 2.83 % of the dynamic potentials of the basin.

Seasonal Water Level Fluctuations: The seasonal water level fluctuations for a 5 year period (1975-80) at different site locations in different geological horizons are shown in the Table 4.28 and are reflected in the Fig. 4.65. The average fluctuation in the coastal alluvial/sedimentary tracks is about 2.71 m.

Table 4.28 Seasonal fluctuations in selected observation stations

S.No	Name of the Site	Geological Horizon	Avg. Fluctuation (m) for a 5 Year period (1975-80)
1.	Agarpara	Alluvium	2.69
2.	Banth	Laterite	2.12
3.	Bhadrak	Laterite	1.26
4.	Tihidi	Alluvium	2.57
5.	Sabrang	Laterite	2.32
6.	Basudevpur	Laterite	3.88
7.	Bhandaripokhri	Alluvium	2.34
8.	Dhamnagar	Alluvium	4.60
9.	Chandbali	Alluvium	2.90
10.	Panikoli	Laterite	5.29
11.	Sujanpur	Alluvium	2.71
12.	Pattamundai	Alluvium	1.60
13.	Mahakalpara	Alluvium	3.10
14.	Jagatpur	Alluvium	2.50
15.	Cuttack	Alluvium	1.38
16.	Kujang	Alluvium	1.92
17.	Jagatsingpur	Alluvium	2.83
18.	Machgon	Alluvium	2.86
19.	Gopalpur	Alluvium	1.14
20.	Kendrapara	Alluvium	1.76
21.	Pichikuli	Alluvium	9.40
22.	Tangi	Alluvium	6.00
23.	Puri	Alluvium	1.26
24.	Kakatpur	Alluvium	1.35
25.	Balianta	Alluvium	3.09

Table 4.28 ***Contd...***

S.No	Name of the Site	Geological Horizon	Avg. Fluctuation (m) for a 5 Year period (1975-80)
26.	Konark	Alluvium	1.22
27.	Gobakund	Alluvium	1.16
28.	Brahmagini	Alluvium	1.59
29.	Bhusandpur	Alluvium	4.42
30.	Sakhigopal	Alluvium	1.55
31.	Nimapara	Alluvium	1.28

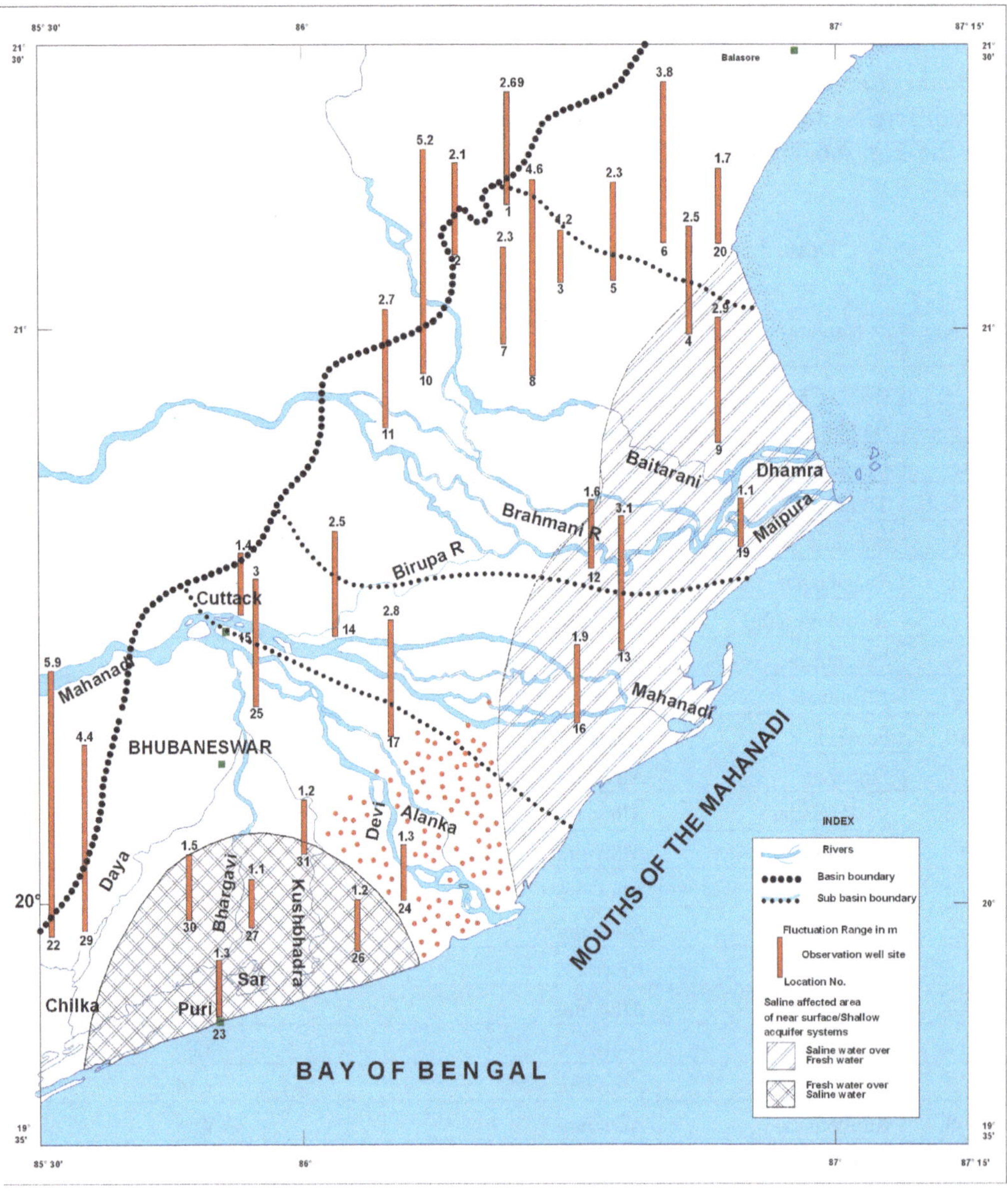

Fig. 4.65 Network observation well stations and their fluctuation trends in Mahanadi-Brahmini-Baitarni delta complex (Orissa)

Resource potentials in Brahmini-Baitarni sub basin

- Area of the Brahmini-Baitarni sub basins : 6544 sq.km
- Saline& brackish affected Area : 2762.7 sq.km
- Coverage of Alluvials + Tertiaries : 5417 sq.km
- Dynamic potentials for Alluvials/sedimentaries (excluding saline/brackish affected area) :

(i) Rainfall infiltration method : 870.58 MCM

(ii) Fluctuation method : 1355.36 MCM

(iii) Average dynamic potentials : 1113 MCM

- Resource potentials in deep confined aquifer systems (static) (Area × Thickness × S.Coefficient) :

(i) Alluviual area (Fresh) ($2654.3 \times 60 \times 2 \times 10^{-4}$) : 31.85 MCM

(ii) Alluvial area covered deep confined freshwater systems overlained by saline/brackish water ($2762.7 \times 100 \times 8 \times 10^{-4}$) : 221 MCM

- Total estimated resource potentials in deep confined aquifer systems : 252.85 MCM

Resource potentials in Mahanadi sub basin

- Area of the Mahanadi sub basins : 3490 sq.kms
- Saline& brackish affected Area : 1708.6 sq.kms
- Coverage of Alluvium+Tertiaries : 3418 sq.kms
- Coverage of Khondalites : 72 sq.kms
- Dynamic potentials for Alluvial sedimentaries (not affected by salinity & brackish):

(i) Rainfall infiltration method : 393.55 MCM

(ii) Fluctuation method : 528.3 MCM

- Dynamic potential for Khondalites by fluctuation method : 0.72 MCM
- Average dynamic potentials for Mahanadi sub basin : 461.64 MCM
- Resource potentials in deep confined aquifer systems (static) (Area x Thickness x S.Coefficient) :

(i) Alluvial area (fresh) ($1782 \times 60 \times 8 \times 10^{-4}$) : 85.536 MCM

(ii) Alluvial area covered deep confined freshwater systems overlained by saline/brackish water ($1708 \times 100 \times 4 \times 10^{-4}$) : 68.32 MCM

- Total estimated resource potentials in deep confined aquifer systems : 154 MCM

Resource potentials in Daya-Bhargavi-Kushbhadra sub basin

- Area of the Daya-Bhargavi-Kushbhadra sub basins : 5890 sq.km
- Coverage of Alluvium+Tertiaries : 3236 sq.km
- Dynamic potentials for Alluvial sedimentaries (not affected by salinity & brackish) :

(i) Rainfall infiltration method : 1356.1 MCM

(ii) Fluctuation method : 2025.26 MCM

(iii) Average dynamic potentials : 1690.73 MCM

- Resource potentials in deep confined aquifer systems (static) (Area × Thickness × S.Coefficient)

(i) Alluviual area (fresh) ($3236 \times 60 \times 8 \times 10^{-4}$) : 155.328 MCM

- Total estimated dynamic + static potentials in Mahanadi-Brahmini-Baitarni delta basin (3265.37 + 562.17) : 3827.54 MCM

4.4.8 Bhagiradhi-Hoogly delta region

Bhagiradhi-Hoogly delta region is a part of the major delta plains of 'Ganges delta' essentially stretched in the southern part of West Bengal as a distributary system of Ganges. Bhagiradhi river emerges from the '*Farrakka barrage*' initially as a downstream canal, joins the main Bhagiradhi river track (believed to be earlier course of Ganga river) in Murshidabad district and stretches over a distance of about 264 kms before joining Bay of Bengal. The river mainly runs through the district regions of Murshidabad, Baradhaman, Nadia, Hoogli, Haora, North and South 24 Paraganas and East Medinepur. Several tributaries from the western high lands join the main Bhagiradhi-Hoogli river, like Mayurakshi, Ajoy, Damodar, Rupnarayan and Kasai all along its course. All these rivers coupled with the main Bhagiradhi-Hoogly river make the southern part of West Bengal region perpetually a green belt with rich agricultural production base and with potential water resources. A characteristic ecological feature in this basin is the existence of Sunderbans with an aerial spread of the order of 4123 km^2 (as per latest govt. of India 1997 estimations). The tidal impacts are seen up to 150-170 km upstream. A drainage map of Bhagiradhi-Hoogly river delta and its tributaries coupled with isohyetels in South West Bengal region are shown in Fig. 4.66. The basin receives over 80% of the rainfall mainly from the southwest monsoon and the rest of the rainfall gets scattered during the remaining period of the year. The maximum rainfall of the order of 2500 mm occurs in the northwest region of the basin and the average minimum rainfall with an order of 1200 mm occurs in the southeastern parts of the basin. The overall average rainfall in the region is of the order of 1480 mm.

Several dams/barrages along these rivers have come up over the last several decades facilitating large surface water storages with good canal networks for irrigation. The Mayurakshi river waters are being used for irrigation purposes through its barrage near 'Suri' in Bhirbhum district in the north western upper delta region. The river Damodar in the middle of the delta terrain constitutes a rich irrigation belt through a number of barrages and dams supporting large irrigation potentials in the districts of Baradhaman, Bankura and Hugli. Similarly the Kasai dam in the south western region on the river Kasai serves as another major irrigation source covering the districts of Medinapur and Bankura. In addition to these surface water resources, and in view of large alluvial stretches covering these river basins, large networks of shallow and deep tube wells have come up supporting the irrigation potentials in a big way. The surface and Groundwater potentials of the districts, particularly covering the Bhagiradhi-Hugli river track are mentioned in the Table 4.29.

The total irrigation potential in the upper and lower delta regions covering all the southern districts of West Bengal under the hydrological regime of Bhagiradhi-Hugli river are of the order of 1,76,85,772 hectares. The irrigation potentials generated through Groundwater comes to the order of 26,00,008 hectares which comes to the order of 15% of the total irrigation potential, leaving a fairly big gap of irrigation potential yet to be developed in these tracks.

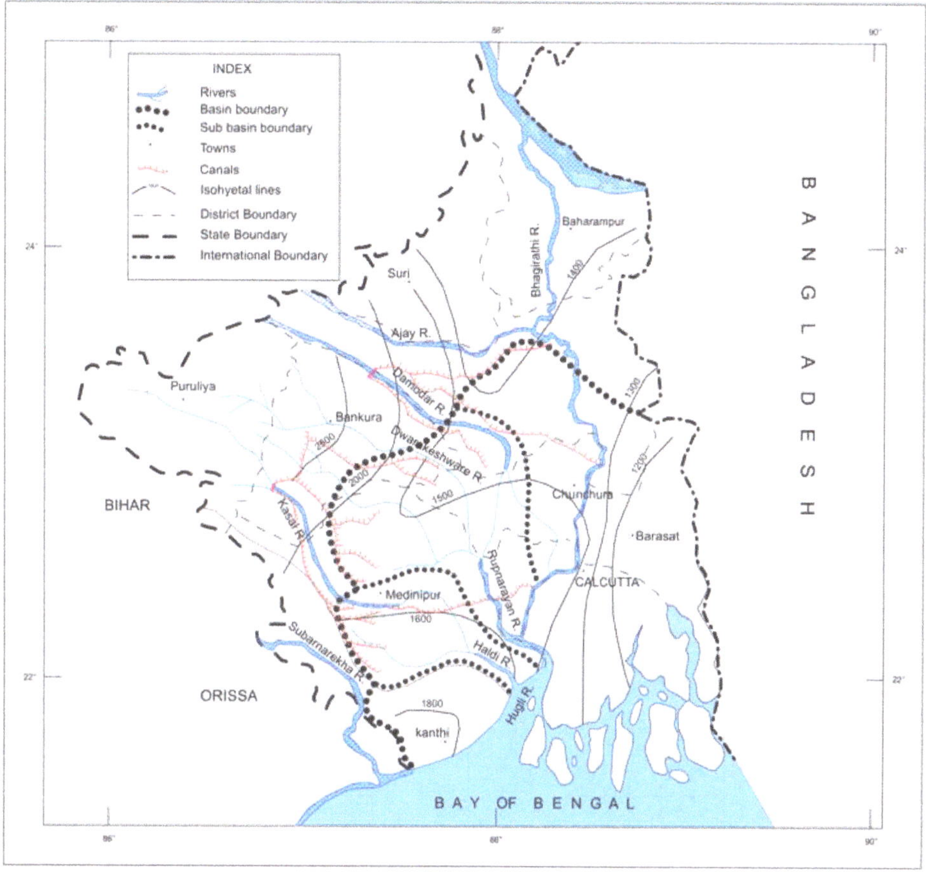

Fig. 4.66 Drainage and Isohyetal trends in Bhagiradhi-Hoogli delta basin (West Bengal)

Regional hydrogeological characteristics and system parameters

A regional geological map for the southern part of West Bengal (Bhagiradhi-Hoogli basin) is shown in Fig. 4.67. The entire basin could be broadly divided into three units based on geological and hydrogeological characteristics namely (i) The western high lands and marginal plains, (ii) The upper deltaic plains and (iii) The southern deltaic plains. The western highlands and marginal plains are characterized by rock types belonging to Archeans, Precambrians, Mesozoics, Tertiaries and stretches of older alluvium capped with laterites. These western highlands and marginal plains cover the districts of West Medinapur, Puruliya and western parts of Bankura and Birbhum districts. A major part of the extreme western terrain is considered to be dry and drought prone, though the region witnesses a heavy average rainfall of the order of 1975 mm. The Archean high land region essentially constitutes granitic and gneissic rock types with intrusions of dolerites. These are abundantly exposed in the Puruliya district and also in north western regions of Medinapur, Bankura Barddhaman and Birbhum. The Groundwater in these rock types occur in both unconfined and semi confined conditions in weathered and fractured strata down to depths varying from 10-60 m below ground level. The weathered portion however is restricted to a depth range between 15-30 m below ground level and fractured terrain up to 60 m below ground level. Groundwater development in this terrain largely takes place through dug wells and shallow tube wells with yields ranging from 100-150 lpm and with transmissivity values ranging from 5-10 m^2/day. The regional water levels in these tracks have a range varying from 8-16 m (post monsoon) and 5-14 m (pre monsoon).

Table 4.29 Irrigation potentials coverage of districts under Bhagiradhi-Hoogly delta basin

District	Surface Irrigation									Groundwater Irrigation							Gross Irrigation (ha)
	Tanks (ha)		Canals (ha)	Lift Irrigation (ha)		Surface Water Irrigation (ha)		Other sources (ha)	Total (ha)	Deep Tubewells (ha)		Shallow Tubewells (ha)		Dug Wells (ha)		Total (ha)	
	No.	Area Irrigated (ha)	Area Irrigated (ha)	No.	Area Irrigated (ha)	No.	Area Irrigated (ha)			No.	Area Irrigated (ha)	No.	Area Irrigated (ha)	No.	Area Irrigated (ha)		
NORTH																	
BARDDHAMAN	...		300940	...	31308	...	24355	...	356603	626	44584	51710	317063	88	140	361787	718390
NADIA	...			...	18552	...	245	...	18797	757	27580	94789	92193	...	...	119773	138570
MURSHIDABAD	...	10810	51190	...	14471	...	4193	...	80664	588	21211	83157	170556	...	...	191767	272431
SOUTH																	
HOOGLY	...	...	...	...	31900	...	1100		33000	509	21699	20292	85883	...	...	107582	140582
HOARA	...	...	...	...	4134000	...	3369121		7503121	183	512686	1116	353073	...	...	865759	8368880
N.24 PARAGANAS	...	...	...	...	1070	...	3790	43508	48368	314	14670	62142	149026	...	...	163696	212064
S.24 PARAGANAS	...	34712	52110	...	33913	...	29979.4		150714	28	343.4	9278	13631	...	...	13974.4	164688
EAST MEDINAPUR	...	22382	42699	...	28741	...	7219		101041	293	4587	15300	42559	...	...	47146	148187
WEST																	
PURILIYA	...	....	....	311	344400	19467	5942849		6287249			2	400	12796	402591	402991	6690240
WEST MEDINAPUR	...	....	....	....	34859	....	19229	1740	55828	499	19611	66221	198797	8981	9279	227688	283516
BANKURA	21152	32350	109800	....	7982	....	....	....	150132	315	2678	27754	47985	2167	2167	52830	202962
BIRBHUM	14681	25400	184020	....	17580	....	18296	54950	300246	7500	44645	....	....	448	370	45015	345261

(Source: District statistical handbook, Govt. of West Bengal, 2001-02)

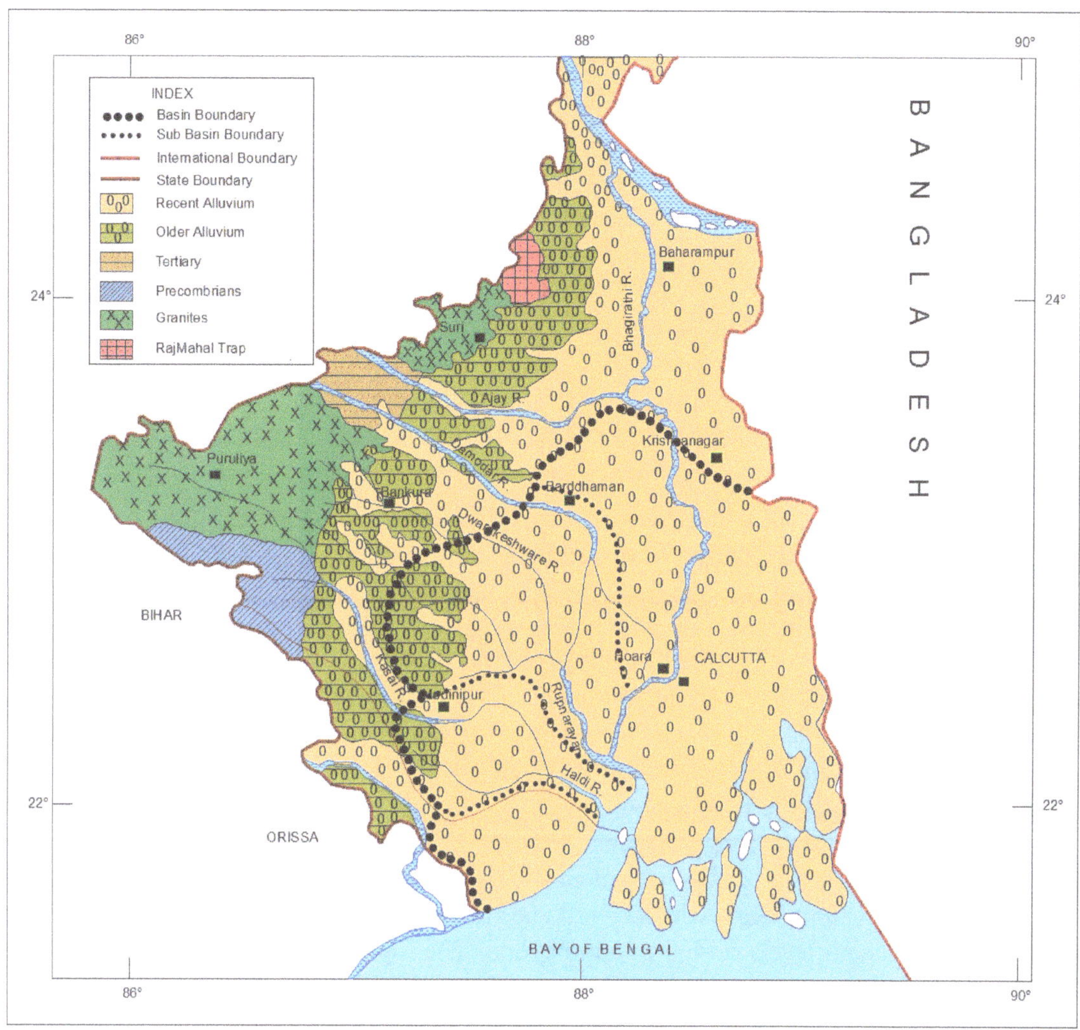

Fig. 4.67 Regional Geological map of Bhagiradhi-Hoogly delta basin (West Bengal)

Marginal upland regions: The marginal land forms sandwiched between high land areas and Recent alluvium are characterized by rock types belonging to Precambrians, Mesozoics, Tertiaries and Older alluvium. They are distributed all along the foot hill margins of high hill regions and are occasionally caped with Laterites. Groundwater in these tracks occurs in both unconfined and deep confined situations up to depths varying from 30-150 m below ground level. The yields of these aquifers range between 20-120 m^3/hour with transmissivity values ranging from 800 to 100 m^2/day. The regional water levels have a range between 5-10 m below ground level (pre monsoon) and 2-5 m below ground level (post monsoon).

Northern deltaic plains: The northern deltaic plains cover the districts of Murshidabad, Nadia, Eastern alluvial plains of Birbhum, Bankura and Barddhaman. These stretches lie in the upstream regions of Bhagiradhi and also in the down slope regions of Mayurakshi, Ajoy and the Damodar river tributaries and are filled with extensive alluvial plain deposits with thicknesses varying from 100-150 m below ground level. Below these alluvial tracks huge thicknesses of Older alluvium and Tertiaries exist up to depths varying between 500 to 1000 meters and beyond with thicknesses increasing towards central and eastern regions of the delta basin. The overall combined thicknesses of granular zones being tapped in both Recent alluvials Quaternaries and Tertiary sediments vary between 100-350 m. The yields of these aquifer systems vary from 100-200 m^3/hour at drawdowns varying from 8-12 m bgl and with transmissivities ranging from 800 to 5000 m^2/day. The aquifer thicknesses, yields and transmissivities systematically increase towards east. The central and eastern regions of this terrain is excellently suitable for heavy duty tubewells which can yield discharges of the order varying between 150 to 200m^3/hour and beyond. However the region east of Bhagiradhi in the districts of Murshidabad, Barddhaman and Nadia the arsenic levels at depths between 60 - 90 m below ground level are reported above the optimum ppm level of 0.05 mgL^{-1}. It is therefore necessary to avoid tapping these arsenic prevailing zones while designing the deep well structures. At the same time a level of caution is to be exercised when deep zones are being developed. There is again a threat that the arsenic contaminated water in the shallow aquifer systems could migrate vertically downward through the aquitards and pollute the deep waters. It is therefore necessary that a regional management plan with optimal level of developmental strategies are required to be implemented while developing the deeper systems.

Sothern/lower deltaic plains: The south/lower deltaic plains cover the districts of 24 Paraganas (North and South), Hoogly, Hoara, and East Medinapur. The East Medinapur, Hoara and South 24 Paraganas districts lie immediately close to the coast and bare the brunt of all coastal processes including saline water intrusion into the fresh water aquifers. The other two districts mainly Hoogly and North 24 Paraganas lie in the upper reaches of lower deltaic plains and escape the impacts of coastal processes and saline water intrusion to a large extent. The lower south coastal districts are intensely covered by marshy tracks and tidal channels. The world famous Sunderbans ecological terrain lies mostly in South 24 Paraganas. Quaternaries and Tertiaries are the two major geological units in the lower reaches of the basin. Quaternaries in general are represented by fine sands, silts and clays with cyclic changes in the sedimentation and have a depth range varying between 120 to 300 m below which Tertiaries are normally encountered. These blankets of clay deposits dominate in the lower belts up to depth levels varying between 100 to 180 m below which sand facies dominate but again with alternate stretches of clays and sands up to depth levels of 400 m and beyond. The top layers represented by fine silts and clays however in the upper reaches in the districts of Hoara and North 24 Paraganas are gradually transformed into fine and coarse sand facies facilitating development of groundwater potentials both in shallow levels and also in deep levels. The lower reaches of the delta in the districts of Medinapur and South 24 Paraganas are dominated by saline water intrusion up to depth levels varying between 90 to 170 m. As a consequence the development of groundwater potential is totally restricted up to depth levels of 90 to 170 m and fresh water aquifers could be tapped only at deeper levels beyond these depths. The Groundwater yields for these deeper aquifer horizons vary between 2500 to 3333 lpm at drawdowns varying between 5 to 16 m. The range of transmissivities for these aquifers varies between 1800 to 5500 m^2/day. In general the transmissivities increase from west to east. Similarly the Quaternaries and Tertiaries in the upper reaches of the lower deltaic plains in the districts of Hoogly and North 24 Paraganas and upper half of Medinapur district are free from salinity impacts and therefore shallow and deep wells could be designed from depths varying from 20 to 400 m below ground level and beyond. The shallow well systems constructed tapping the zones between 30 to 90 m have yielded discharges varying between 1666 to 2500 lpm at relatively less draw downs of 4 to 8 m. Similarly the deep systems tapping the aquifers at 90 to 350 m depth levels have also given rise to the yields varying from 2500 to 3333 lpm at drawdown varying between 6 to 15 m.

Lithofacies maps indicating the aquiclude and aquifer boundaries and their depth levels in the lower reaches in the districts of East Medinapur and South 24 Paraganas are given in Fig. 4.68.

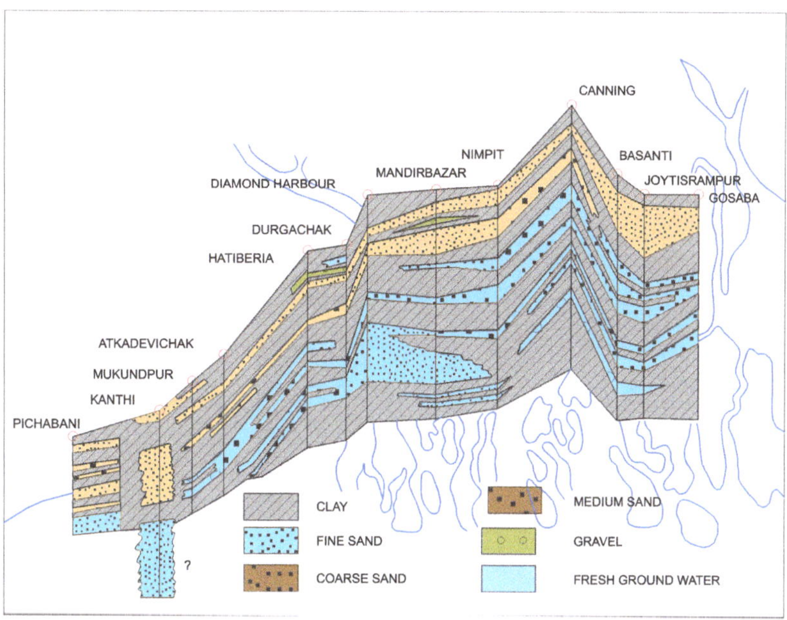

Fig. 4.68(a) A lateral lithological cross section parallel to the coast line in parts of Medinapur and South 24 Paraganas of Bhagiradhi and Hoogly delta complex (West Bengal) showing the depth variations of fresh-saline water boundaries

(Source: CGWB, 1988)

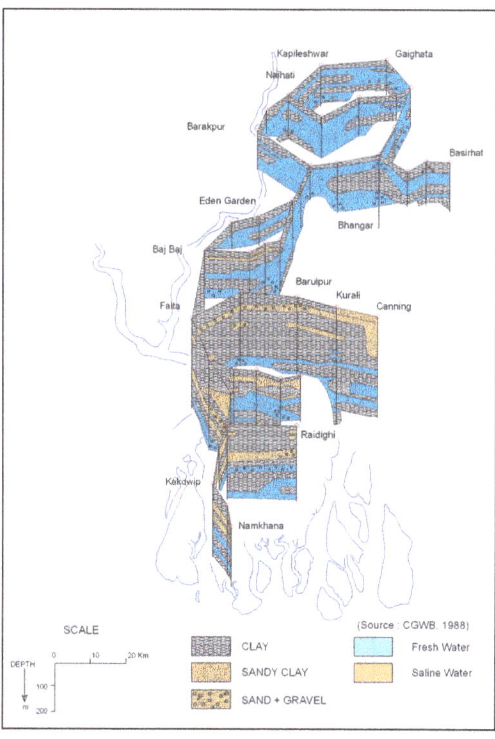

Fig. 4.68 (b) A lithological cross section in parts of South and North 24 paraganas of Bhagiradhi-Hoogly delta complex (West Bengal) showing the depth variations of fresh-saline water boundaries

(Source: CGWB, 1988)

Table 4.30 Regional hydrogeological parameters for selected exploratory wells in Bhagiradhi-Hoogly delta basin

S.No	Location	Depth drilled (m)	Range of Aquifers tapped (mbgl)	Thickness of the zones (m)	SWL (mbgl)	Discharge (lpm)	Specific Capacity (lpm/m)	T (m^2/day)	S	Remarks
		Constructed (mbgl)				Drawdown(m)				
1	2	3	4	5	6	7	8	9	10	11
Midinipur district										
1	Egra	83.2	…		3.4	4309.6	1346	…	…	
						3.2				
2	Salboni	149.6	13.0-185.0		12.72	681.8	69.6	1200		
						9.82				
3	Panchami	189.8	…		7	2500	416.6	…	…	
						6				
4	Maharajpur	102.8	…		14.9	1386	231	…	…	
						6				
5	Mamudpur	151.4	…		4.9	3233.5	538.9	…	…	
						6				
6	Lepsa	80.7	…		6	2434	405.7	…	…	
						6				
7	Brindabanpur	115	…		10.29	596	214.3	…	…	
						2.78				
8	Bengunbari	102	…		8.5	1664	277.3	…	…	
						6				
9	Radhaballavpur	128	…		6.4	1966	327.7	…	…	
						6				
10	Haldia	…	…		…	3333.33	…	1500	4.7 × 1-4	
						….				
11	Kanthi	…	120.0-185.0		…	1833	305.5	…	…	
						6				
12	Digha	…	104.0-172.5		…	750	375	…	…	
						2				
13	Tamluk	…	…		…	2500	500	…	…	
						5				
14	Gopiballabhpur	…	…		…		…	…	…	

Table 4.30 *Contd…*

S.No	Location	Depth drilled (m) Constructed (mbgl)	Range of Aquifers tapped (mbgl)	Thickness of the zones (m)	SWL (mbgl)	Discharge (lpm) Drawdown(m)	Specific Capacity (lpm/m)	T (m^2/day)	S	Remarks
14	Debra	…	21.0-60.0		…	366.67	…	…	…	
15	Medinipur	…	21.0-149.0		…	1916	239.5	1100		
						8				
16	Ghatal	…	…		…	2333	424.18	…	…	
						5.5				
17	Shyam Sundarpur	…	…		…	…	…	…	…	
	Hoogly									
18	Arambagh		45.0-98.0	53		3333.3	555.5	4700	2.5×10^{-3}	
						6				
19	Balagarh		91.0-114.0	23		3333.3	555.5	4700	2.5×10^{-3}	
						6				
20	Chanditala		145.0-162.0	17		3333.3	555.5	4700	2.5×10^{-3}	
						6				
21	Chinsurah-Mogra		79.0-154.0	75		3333.3	555.5	4700	2.5×10^{-3}	
						6				
22	Khanakul		87.0-137.0	50		3333.3	555.5	4700	2.5×10^{-3}	
						6				
23	Polba-Dadpur		98.0-131.0	33		3333.3	555.5	4700	2.5×10^{-3}	
						6				
24	Pursurah		96.0-121.0	25		3333.3	555.5	4700	2.5×10^{-3}	
						6				
25	Serampur-Uttarpara		113.0-280.0	167		3333.3	555.5	4700	2.5×10^{-3}	
						6				
26	Singur		105.0-113.0	8		3333.3	555.5	4700	2.5×10^{-3}	
						6				
	Howrah district									
27	Narsingh Dutta College, Howrah	204.82	51.81-204.90	126.88	-	1515.36			…	
						…				

Table 4.30 *Contd…*

S.No	Location	Depth drilled (m) Constructed (mbgl)	Range of Aquifers tapped (mbgl)	Thickness of the zones (m)	SWL (mbgl)	Discharge (lpm) Drawdown(m)	Specific Capacity (lpm/m)	T (m^2/day)	S	Remarks
28	B.E.College, shibpore	224.94	45.72-224.94	132.58	-	1,051	350.3		…	
						3				
29	G.T. Road, Park, Liluah	140.44	58.21-140.50	85.67	7.24	1674.46	1157.1		…	
						1.447				
	South 24 Parganas district									
30	Kurali	588	153.00-295.97	38.2		2158.8	472.38	3019	3×10^{-4}	
						4.57				
31	Goalberia	589.5	257.00-342.00	70		3468	227.40	1309	…	
						15.25				
32	Kashipur	607.5	150.00-275.00	52		3694.8	724.4	6514	….	
						5.1				
33	Gangasagar	350	207.00-319.00	64		1434	220.6	506	…	
						6.5				
34	Rudranagar	650	210.00-284.00	30		1435.8	121.78	397	…	
						11.79				
35	Kakdwip	442	203.47-215.30	11.83		2498.4	165.78	1446.4	…	
						15.07				
36	Jalerhat	280	236.00-402.00	42		1319.4	106.14	1725	$1.5 \times 1-^{3}$	
						12.43				
37	Jalaberia	350	271.00-332.00	27		375.6	31.9	464.6	…	
						11.75				
38	Anchana	395	261.00-371.00	53		1102	66.7	1106	2×10^{-4}	
						16.52				
39	Baruipur	400	210.00-280.00	52		1446	154.4	1727	9×10^{-4}	
						9.36				
40	Budge Budge	614.6	….			1036.8	452.7	1975.7	…	
						2.29				
41	Namkhana	…	…			2103.83	125.9	4643	1.0×10^{-3}	
						16.74				
	North 24 Parganas district									
42	Berachampa	334.64	31.82-170.80	46.13	2.98	3181.20	641.3	5219.4	1.10×10^{-3}	
						4.96				

Table 4.30 ***Contd…***

S.No	Location	Depth drilled (m)	Range of Aquifers tapped (mbgl)	Thickness of the zones (m)	SWL (mbgl)	Discharge (lpm)	Specific Capacity (lpm/m)	T (m^2/day)	S	Remarks
		Constructed (mbgl)				Drawdown(m)				
43	Khamarkhulta	330.38	65.61-203.41	98.43	3.97	3644.40	550	7035.5	-	
						6.62				
44	Dakshin Chatra	165.68	49.12-148.73	46.41	2.29	3772.80	729.7	7124.6	-	
						5.17				
45	Algaria	332.02	30.84-131.17	66.87	3.13	4164.60	840	5219.4	-	
						4.96				
46	Dholtukri	511.58	58.0-479.0	64		1469.40	202.3	…	…	
						7.26				
47	Adarshapalli(Khardha)	400.47	112.0-223.0	33		1723.8	363	6709	3.5×10^{-2}	
						4.748				

The hydrogeological parameters for selected wells in the coastal West Bengal basin are furnished below in table 4.30 with site locations shown in Fig. 4.69. (Data source: CGWB: 1982; 1994; 1995; 1999; Deshmukh, D.S., et al (1973) and Chakladar, S., et al., (1990).

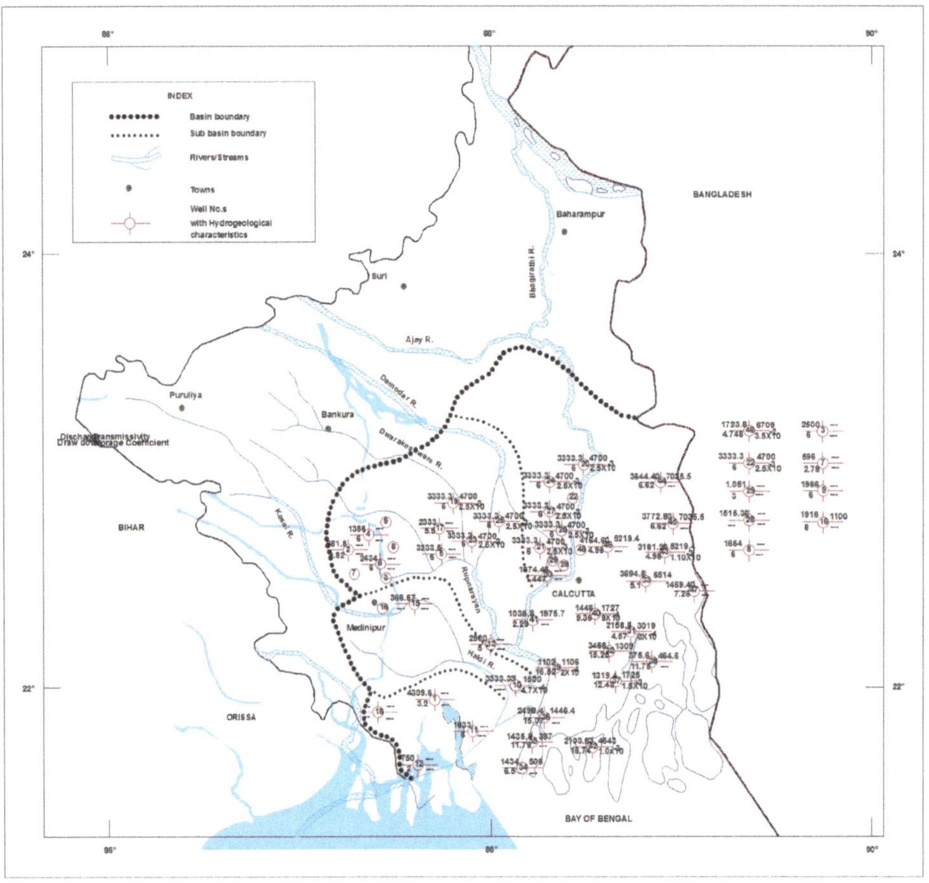

Fig. 4.69 Exploratory well site locations in Bhagiradhi-Hoogly delta complex (West Bengal)

Groundwater flow trends and analysis: A groundwater flow trend map across different profiles for both near surface and deep aquifer systems are provided in the Figure 4.70. The Groundwater flows across different lengths of different sub basins are provided in Table 4.31.

Table 4.31 Groundwater flow parameters across different profiles in Bhagiradhi-Hoogly delta basin

Near Surface Aquifers						
S.No	Basin	Profile	Avg.Transmissivity (T) (m^2/day)	Gradient (I) (m/km)	Length of the Coast (L) (Km)	Total flow (Q) (MCM)
1	Subharnarekha + Kasai	P1	800	0.66	100	19.27
2	Damodar + Rupnarayan	P2	800	0.50	20	2.92
3	Bhagiradhi-Hoogly	P3	1000	0.20	120	8.76

*Total flow for near surface system across coast line is 30.952 MCM

Table 4.31 ***Contd...***

Deep Aquifers						
S.No	**Basin**	**Profile**	**Avg.Transmissivity (T) (m^2/day)**	**Gradient (I)(m/km)**	**Length of the Coast (L) (Km)**	**Total flow (Q) (MCM)**
1	Subharnarekha + Kasai	P1	1700	0.15	100	9.30
2	Damodar + Rupnarayan	P2	2000	0.17	20	2.48
3	Bhagiradhi-Hoogly	P3	3200	0.11	60	7.7
		P4	3200	0.125	60	8.76

*Total flow for piezometric system across coast line is 28.24 MCM

The total estimated groundwater flow across the whole coastal margins of Bhagiradhi-Hoogly basins is of the order of 59.12 MCM which comes to about 0.945% of the dynamic potentials of the basin.

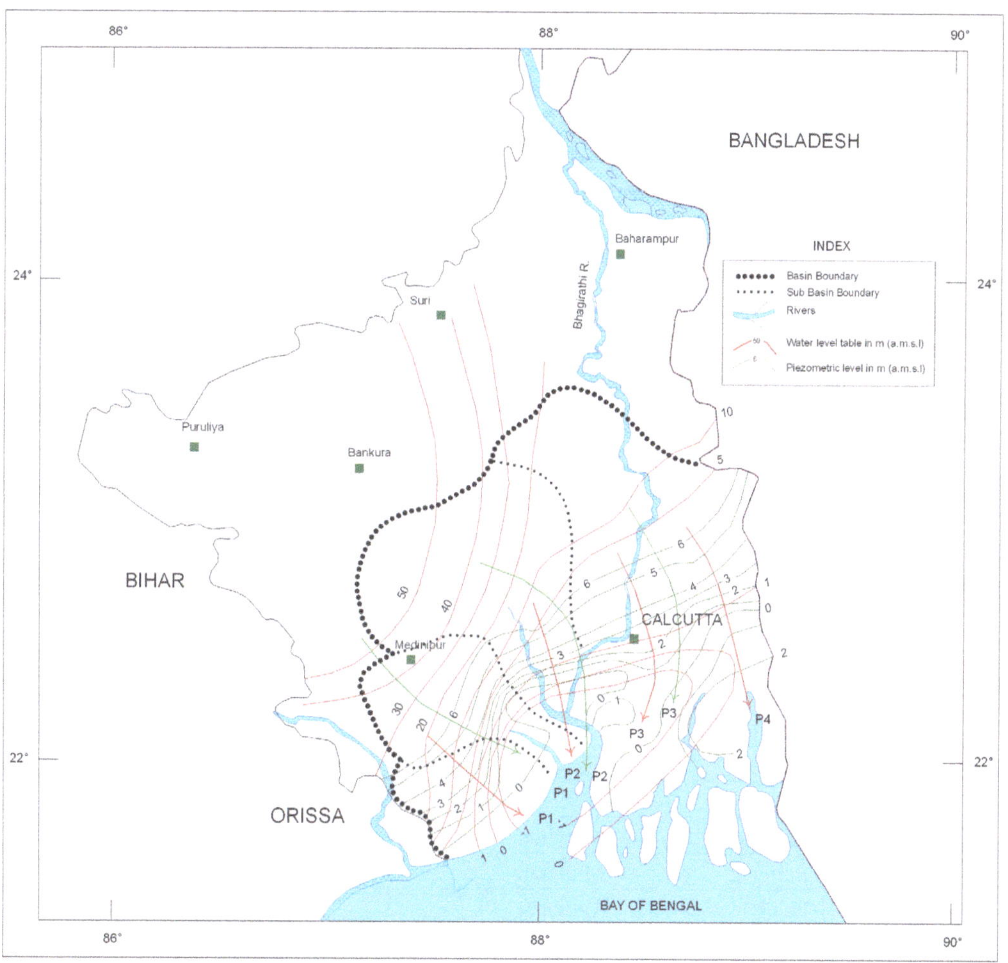

Fig. 4.70 Groundwater flow map for near surface and deep aquifer systems in Bhagiradhi-Hoogly delta complex (West Bengal)

Seasonal water level fluctuations: The seasonal water level fluctuations for a 5 year period at different site locations in different geological horizons are shown in the Table 4.32 and are reflected in the Figure 4.71. The average fluctuation in the coastal Alluvial/sedimentary tracks is about 2.73 m.

Table 4.32 Seasonal fluctuations in selected observation stations

S.No	Name of the Site	Geological Horizon	Avg. Fluctuation (m) for a 5 Year period (1975-80)
West Medinapur district			
1	Medinapur	Alluvium	2.55
2	Kharagpur	Alluvium	2.85
3	Keshiyari	Alluvium	2.00
4	Salbani	Alluvium	2.50
5	Garhbeta	Alluvium	2.29
6	Binpur	Alluvium	4.09
7	Jhargram	Alluvium	1.17
8	Chandrakona	Alluvium	2.29
9	Kespur	Alluvium	2.50
10	Nayagram	Alluvium	2.22
11	Debra	Alluvium	2.55
12	Dantan	Alluvium	2.00
13	Gopiballabhpur	Alluvium	2.49
North 24 Paraganas			
1	Bagdah	Alluvium	2.32
2	Barasat	Alluvium	2.31
3	Basirhat	Alluvium	1.29
4	Deganga	Alluvium	2.27
5	Dum Dum	Alluvium	2.24
6	Gobardanga	Alluvium	2.52
7	Gopalnagar	Alluvium	2.21
8	Ichapur	Alluvium	2.76
Hoara			
1	Amta	Alluvium	3.11
2	Domjur	Alluvium	2.00
3	Jagatballavpur	Alluvium	3.20
4	Panchla	Alluvium	2.00
5	Udaynarayanpur	Alluvium	2.00
6	Howrah	Alluvium	1.20
South 24 Paraganas			
1	Jadavpur	Alluvium	11.79
2	Bariupur	Alluvium	4.40
3	Raidighi	Alluvium	2.50
4	Canning	Alluvium	1.62
5	Jaynagar	Alluvium	3.05
6	Kakdwip	Alluvium	2.62
7	Diamondharbour	Alluvium	3.84

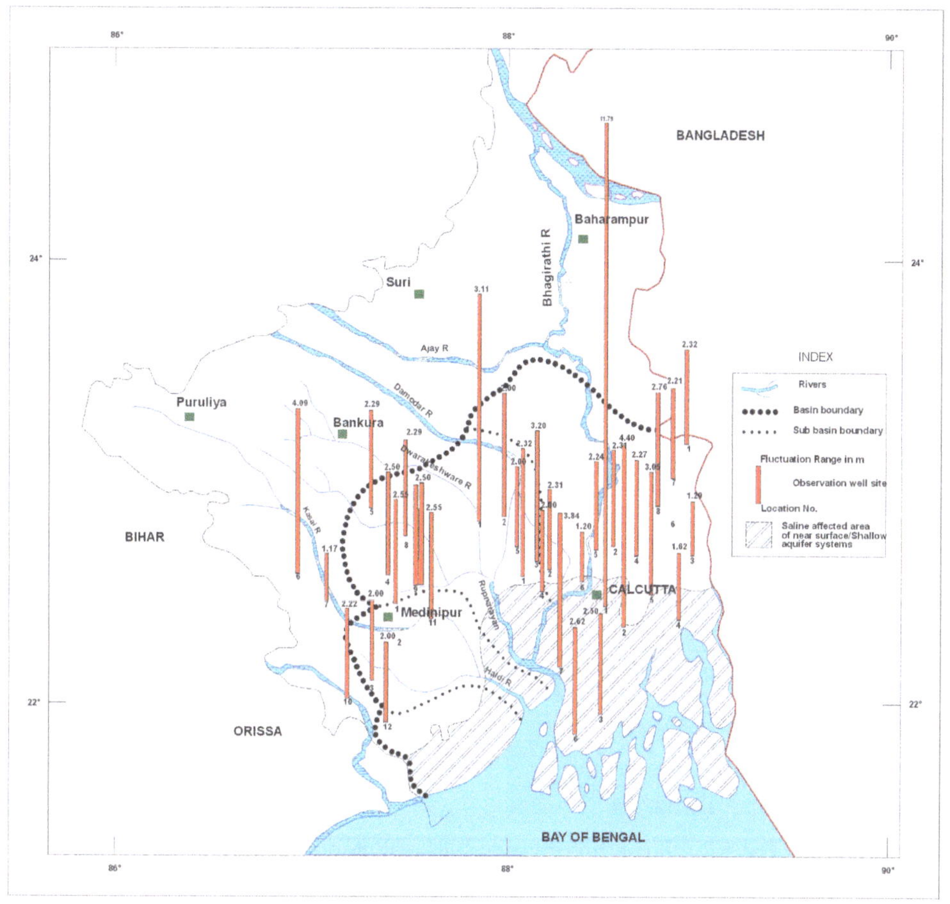

Fig. 4.71 Network of observation well stations and their fluctuation trends in Bhagiradhi – Hoogly delta complex (West Bengal)

Resource potentials in Subharnarekha + Kasai sub basin

- Area of the Subharnarekha + Kasai sub basin : 6440 sq.km
- Saline affected Area : 2530 sq.km
- Coverage of Alluvium + Tertiaries : 6440 sq.km
- Dynamic potentials for Alluvial sedimentaries (not affected by salinity):

(i) Rainfall infiltration method : 884.1 MCM

(ii) Fluctuation method : 1468 MCM

(iii) Total average dynamic potentials : 1176.5 MCM

- Resource potentials in deep confined aquifer systems (static) (Area × Thickness × S.Coefficient) :

(i) Alluviual area ($3910 \times 250 \times 1 \times 10^{-4}$) : 97.75 MCM

(ii) Alluvial area (deep fresh water systems overlained by saline aquifer systems) ($2530 \times 130 \times 6 \times 10^{-4}$) : 197.34 MCM

- Total estimated resource potentials in deep confined aquifer systems : 295 MCM

Resource potentials in Damodhar + Roopnarayan sub basin

- Area of the Damodhar + Roopnarayan sub basin : 8500 sq.km
- Saline effected Area : 1275 sq.km
- Coverage of Alluvium+Tertiaries : 8500 sq.km
- Dynamic potentials for Alluvial sedimentaries (not affected by salinity):

(i) Rainfall infiltration method : 1650.6 MCM

(ii) Fluctuation method : 2531.6 MCM

(iii) Average dynamic potentials : 2090 MCM

- Resource potentials in deep confined aquifer systems (static)
  (Area × Thickness × S.Coefficient) :

(i) Alluvial ($7225 \times 250 \times 1 \times 10^{-4}$) : 180.62 MCM

(ii) Alluvial area (deep fresh water systems overlained by saline aquifer systems)
($1275 \times 130 \times 6 \times 10^{-4}$) : 99.45 MCM

- Total estimated resource potentials in deep confined aquifer systems
  : 280 MCM

Resource potentials in Bhagiradhi-Hoogly sub basin

- Area of the Bhagiradhi-Hoogly sub basin : 19100 sq.km
- Saline affected Area : 11500 sq.km
- Coverage of Alluvium + Tertiaries : 19100 sq.km
- Dynamic potentials for Alluvial sedimentaries (not affected by salinity) :

(i) Rainfall infiltration method : 1736.3 MCM

(ii) Fluctuation method : 4240 MCM

(iii) Average dynamic potentials : 2988 MCM

- Resource potentials in deep confined aquifer systems (static)
  (Area × Thickness × S.Coefficient) :

(i) Alluvial ($7600 \times 250 \times 1 \times 10^{-4}$) : 190 MCM

(ii) Alluvial area (deep fresh water systems overlained by saline aquifer systems)
($11500 \times 100 \times 4 \times 10^{-4}$) : 460 MCM

- Total estimated resource potentials in deep confined aquifer systems
  : 650 MCM
- Total estimated (dynamic+static) potentials in Bhagiradhi-Hoogly basin
  : 7479.5 MCM

4.4.9 Narmada-Tapti delta region

Narmada and Tapti are the major river basin systems along the west coast with stretches of 1312 and 724 km with catchments of the order of 98,796 and 65,144 sq.kms respectively. These two river systems however have built up only small estuary delta systems of the order <200 sq.km with their apexes near Bharuch and Surat respectively. The tidal impacts are observed upstream up to 40-30 km from the coast line. The lower estuary deltaic regions are characterized by marshy areas and islands in between the estuary distributaries. The upper deltaic regions however have good wide spread alluvial terrain covering an area of the order of 2325 sq.km (Narmada outfall region) and about 1251 sq.km (Tapti outfall region) respectively with thicknesses varying from 20 to 120 m or some times even more. These stretches of alluvial sediments indicate probable paleo channel networks in both upper outfall regions of these two rivers. Both the river systems have vast networks of irrigation projects through several dam structures. The nearest dam structures in the vicinity of outfall regions are Sardar - Sarovar river valley project on Narmada and Ukai project on Tapti. These two dam structures have created huge irrigation potentials through several canal networks in the vicinity of the outfall areas and at the same time serving to enhance the recharge potentials of the sub surface Groundwater systems in the outfall deltaic tracks. A drainage map of Narmada and Tapti coastal outfall regions with different sub basin boundaries are presented in figure 4.72 mainly covering the districts of Bharuch (40.05%), Surat (33.45%) and Vadodara (5.23%). The annual average precipitation levels and their trends across these two basins are also shown in figure. The average annual rainfall in these two basins are of the order of 959 mm (Narmada basin) and 1205 mm (Tapti basin) respectively.

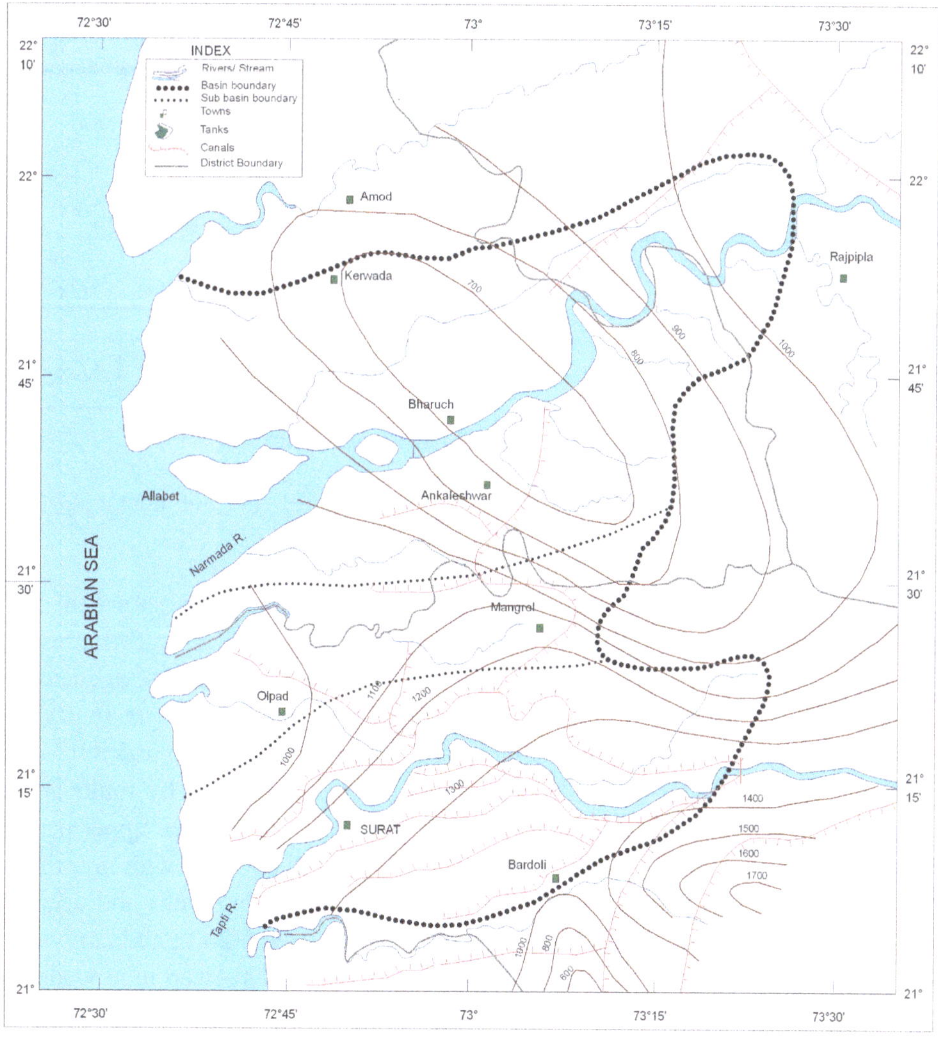

Fig. 4.72 Drainage and isohyetal trends in Narmada-Tapti delta basin (Gujarat)

The regional precipitation trends show a low closed loop line with 700-800 mm in the central part of the delta track gradually increasing towards outfall and upstream areas of Narmada. But in the case of Tapti the northwestern regions of Tapti basin show low level with gradually increasing towards southeastern side and also towards outfall region. Both the basins receive over 90% of the rainfall from the southwest monsoon spread over the period of July to September. The delta regions have characteristic tropical climate with mean summer temperatures at 38.5 °C (March to June) with mean winter temperature of 15 °C (November to February). The maximum (November to January) and minimum (May to June) humidity levels for Narmada are of the order of 71.4 % and 41.05%. Similarly, the maximum and minimum humidity levels for Tapti are of the order of 82% and 45%.The annual evapotranspiration is of the order of 1626 mm and 1636 mm respectively which are more than the average annual precipitation levels. The total gross irrigation potentials through tanks and canals is of the order of 83571 hectares. Similarly gross irrigation potential created through dug wells and tube wells is of the order of 65995 hectares. The percentage of irrigation through groundwater is about 44% of the total irrigation potentials. The basin/ district wise irrigation potentials are reflected below in Table 4.33.

Table 4.33 Surface and groundwater irrigation potentials in the delta region

Delta area	Surface Irrigation				Groundwater Irrigation			Gross Irrigation (ha)
	Tanks (ha)	**Canals (ha)**	**Other sources (ha)**	**Total (ha)**	**Tube wells (ha)**	**Dug wells (ha)**	**Total (ha)**	
Narmada delta: Bharuch/ Vadodora	680	22440	160	**23280**	3720	22840	**26560**	**49840**
Tapti delta: Surat	0	60192	99	**60291**	0	39435	**39435**	**99726**

(Source: Directorate of Agriculture, Gujarat, 2006-07)

The total irrigation potentials in both the basins are of the order of 1,49,566 hectares.

Regional hydrogeological characteristics and system parameters

A regional geological map within the framework of the basin boundary is shown in Fig. 4.73 The different geological formations outcropping in the basin include Deccan Traps, Tertiaries, Quaternaries and Recent. A number of exploratory wells were drilled by both central and state agencies up to depths ranging between 92 to 610 m (Narmada basin), the deepest one being at Vagra (610). Similarly the depth variations of exploratory wells in Tapti area have a range between 50 to 305 m (Tapti basin), the deepest one being at Kosmada (305 m). The Deccan Traps are represented by multi layered systems with each layer comprising two distinct units namely the vesicular basalt in the upper part and massive basalt in the lower part. The massive basalts in general are hard and compact with no primary porosity. The upper vesicular basalts however act as good aquifer zones. Both dug wells and deep wells appear feasible in these tracks. The depth of dug wells have a range between 5-20 m and can yield discharges varying between 200 to 1200 lpm at drawdowns between 4 to 7 m. The deep tube wells have a range between 100 to 150 m and have yields varying between 100-600 lpm at draw downs of 6-15 m, reflecting relatively poor yield characters at deeper levels.

Tertiaries, Quaternaries and Recent are widely spread out and constitute important aquifers in the basin. The Tertiaries though outcrop in the western part of the basin appears to underlie the Quaternaries and Recent in most parts of the Narmada and Tapti basin areas. The Tertiaries are represented by sandstones and some times by lime stones and yield significant quantities of Groundwater. The combined yield characteristics of the aquifer systems in these three units of Narmada basin have a range varying between 330-3366 lpm at draw downs varying from 2-8 m and with transmissivities varying from 148.8 to 6264 m^2/d. The general storage coefficients are of the order varying between 1×10^{-4} and 2×10^{-5}. Similarly the yield characteristics in respect of Tapti basin have a range varying between 2298 to 216 lpm and with transmissivities varying from 1302 to 20 m^2/d. The general storage coefficients are of the order varying between 5×10^{-4} and 8×10^{-5}.

Regional geological sections parallel to the Narmada & Tapti rivers in the outfall regions are presented in Figures 4.74. The depth of these bore holes in Narmada basin area have a range between 311-302 m and in case of Tapti basin range between 54-160 m. The major lithological units identified in these sections are Quaternaries and Tertiaries.

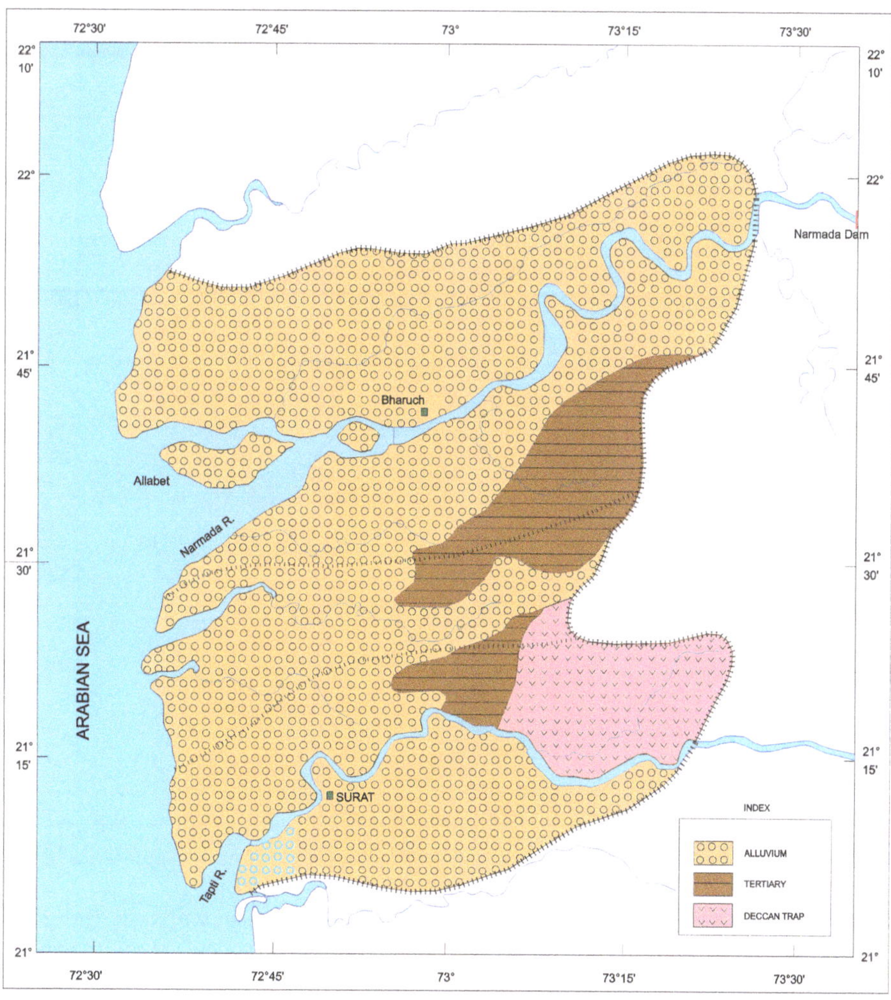

Fig. 4.73 Regional geological map of Narmada-Tapti delta basin (Gujarat)

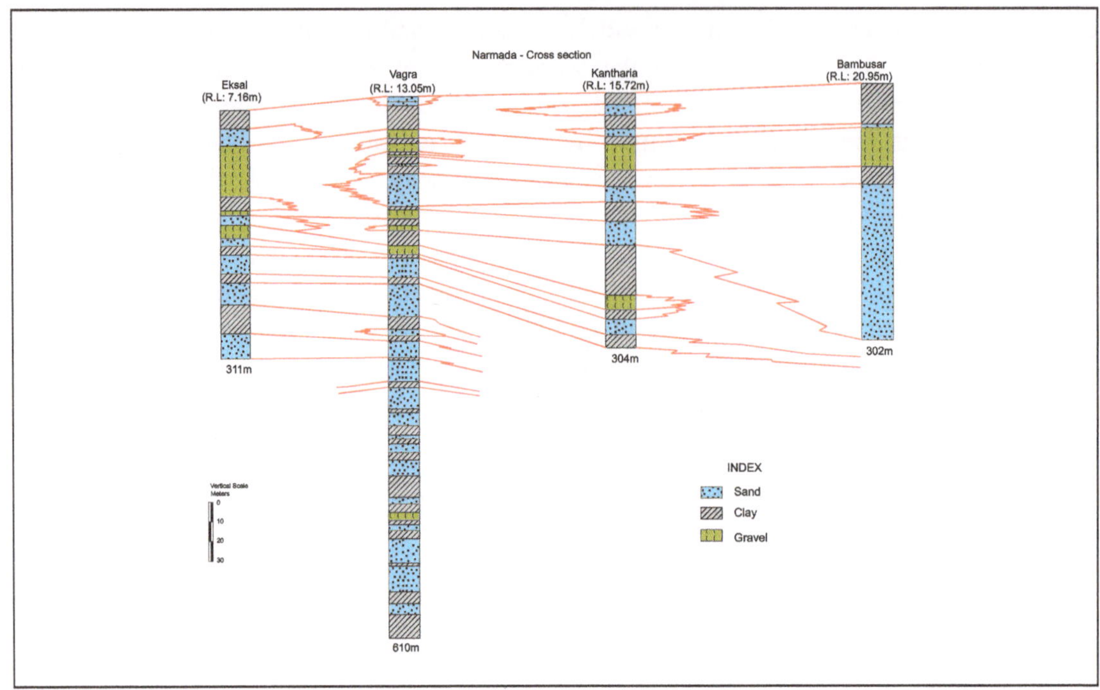

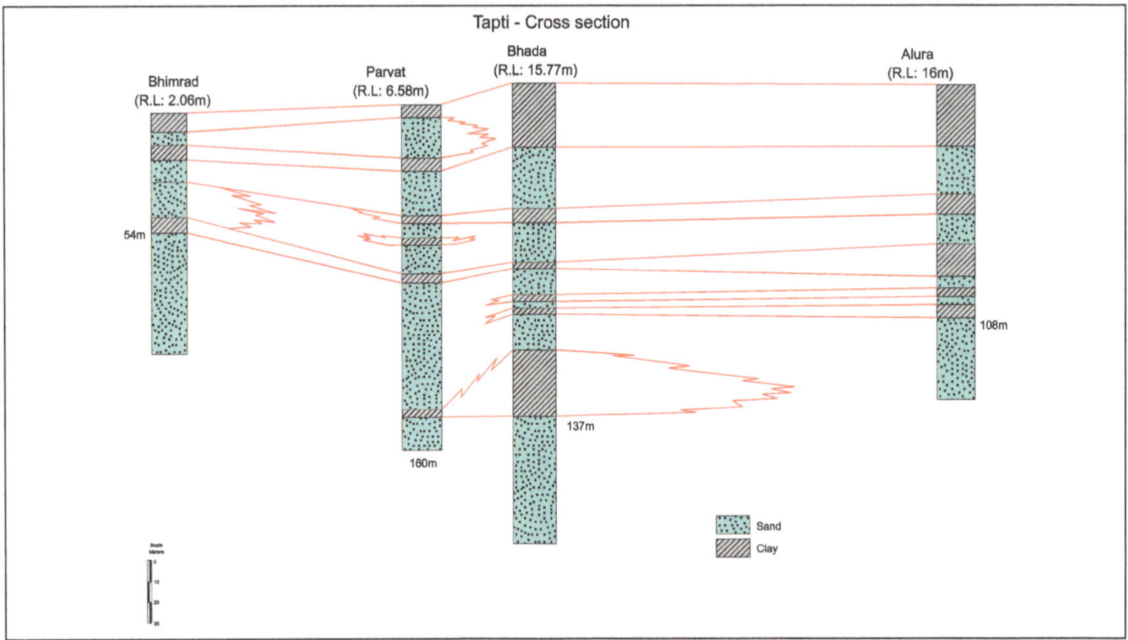

Fig. 4.74 (i) and (ii) Selected hydrogeological cross sections across Narmada-Tapti delta basin (Gujarat)

The relavant hydrogeological parameters for selected exploratory wells in these two basins are presented in Table 4.34 and the locations of the well sites are shown in Fig. 4.75. (Data source: Krupanidhi, K.V.J.R., 1986 and Ashok Kumar, 2011).

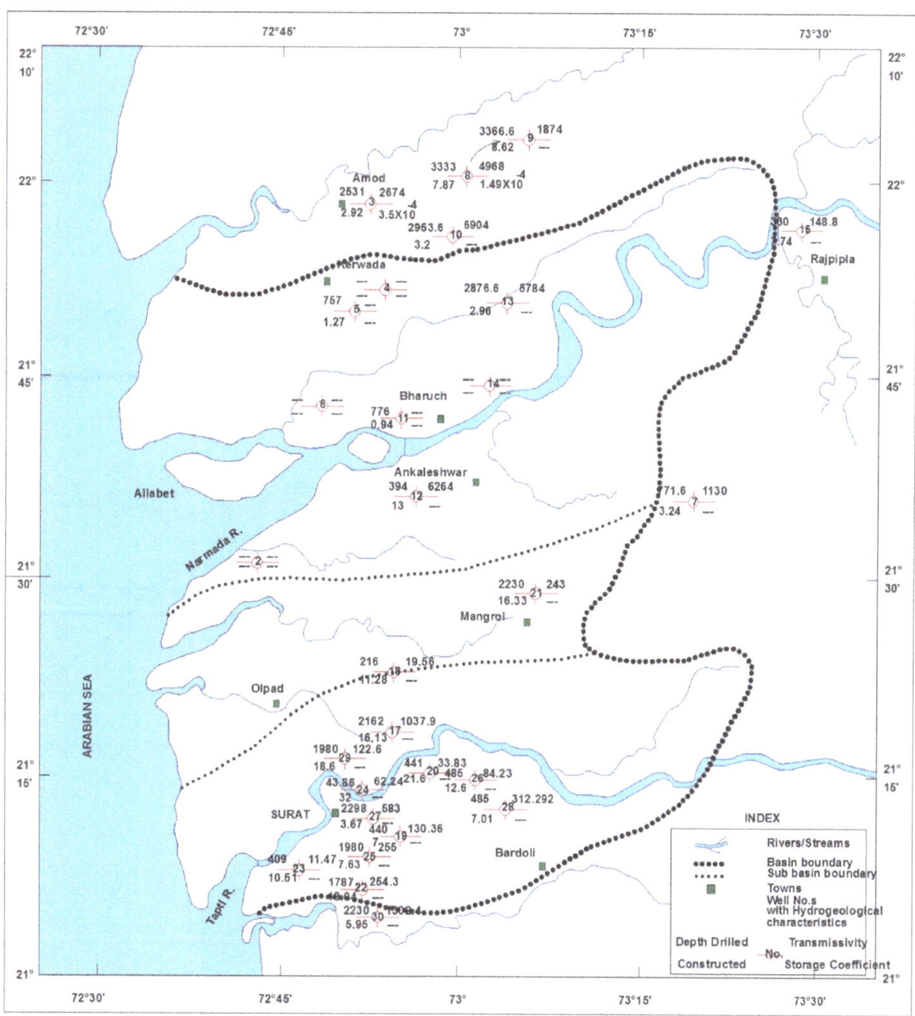

Fig. 4.75 Exploratory well site locations in Narmada-Tapti delta basin (Gujarat)

Table 4.34 Regional hydrogeological parameters for selected exploratory wells in Narmada-Tapti delta region

S.No	Location	Depth drilled (m)	Aquifer zones tapped (mbgl)	Thickness of the zones (m)	SWL (mbgl)	Discharge (lpm) / Drawdown(m)	Specific Capacity (lpm/m)	T (m^2/day)	S	Remarks
1	2	3	4	5	6	7	8	9	10	11
Narmda Delta region										
1	**Jantran**	206.3								Abandoned due to high salinity
2	**Katpur**	230.75								Abandoned due to high salinity
3	**Amod**	310.85	48.39-68.4	20	9.28	2531	866.7	2674	3.5×10^{-4}	
						2.92				
4	**Samni**	185.15	30.0-35	23	14.47					
			40-46							
			77-89.2							
5	**Vagra**	610.3	38-47	9	10.24	757	596	…		
						1.27				
6	**Eksal**	311.6								Abandoned due to high salinity
7	**Motia**	224.6	19.6-29.6	10	5.66	771.6	238.3	1130		
						3.24				
8	**Matar1**	457.81	34.06-45.69	57	19.75	3333	390.5	4968	1.49×10^{-4}	
			56.55-82.39			7.87				
			100-120							
9	**Matar2**	125.83			19.64	3366.6	423.5	1874		
						8.62				
10	**Kothi Vantarsa**	295.11	58.86-66.78	42	16.39	2953.6	923	5904		
			71.0-79.0			3.2				
			84.29-98.19							
			103.95-115.88							
11	**Kanthariya**	304.72	44.1-51.8	8	13.89	776	825.5	…..		
						0.94				

Table 4.34 ***Contd…***

S.No	Location	Depth drilled (m)	Aquifer zones tapped (mbgl)	Thickness of the zones (m)	SWL (mbgl)	Discharge (lpm) / Drawdown(m)	Specific Capacity (lpm/m)	T (m^2/day)	S	Remarks
12	**Umarwada**	200.25	45.75-64.55	19	2.24	394	30.3	6264		
						13				
13	**Bambusar**	302.46			16.6	2876.6	971.8	5784		
						2.96				
14	**Jhadeshwar**	262.28	62-74.2	12	17.08					
15	**Rasela**	92.05	49.0-59.0	16	25.29	330	42.63	148.8		
			62.5-64.5			7.74				
			68.5-72.5							
Tapti delta region										
16	**Kosad**	197	34-54	26.2	6.5	1485	67	103.84		
			58.39-64.59			22.15				
17	**Bhada**	137.6	34.2-55.4	40.5	10.1	2162	134	1037.9		
			61.9-81.2			16.13				
18	**Antroli (Kamrej)**	149	19.94-36.84	16.9	8.65	216	19	19.56		
						11.28				
19	**Antroli (Palsana)**	50	15.0-44.0	29	4.35	440	62.8	130.36		
						7				
20	**Vav**	52.13	10-15.2	28.2	4.88	441	20.4	33.83		
			23.1-46.0			21.6				
21	**Kosamba**	304.8	60-74	66.6	6	2230	136.5	243		
			78-92.1			16.33				
			114.1-135.1							
			151.8-169.3							
22	**Bhetsan**	168	59.9-83.8	38.5	7.2	1787	99	254.3		
			89.5-104.1			18.04				
23	**BhimRad**	54.87			2	409	38.9	11.47		
						10.51				

Table 4.34 ***Contd…***

S.No	Location	Depth drilled (m)	Aquifer zones tapped (mbgl)	Thickness of the zones (m)	SWL (mbgl)	Discharge (lpm) / Drawdown(m)	Specific Capacity (lpm/m)	T (m^2/day)	S	Remarks
24	**Uttran**	87.25	32-38	37.43	14.1	43.86	1.37	62.24		
			43.6-54.04			32				
			62.24-77.2							
			79.0-85.03							
25	**Saniya Kherada**	197	37.6-40.8	26.3	6.2	1980	259.5	255		
			43.8-49.0			7.63				
			53.0-70.9							
26	**Jokha**	52.52	12-24.1	21.2	1.19	485	38.4	84.23		
			36.3-45.4			12.6				
27	**Parvat**	160	33-52.9	33.7	6.5	2298	626	583		
			56.2-63.0			3.67				
			66.0-73							
28	**Alura**	108.2	28-48.5	45.9	3.42	485	69	312.292		
			57.9-73			7.01				
			88.7-93							
			96.3-102.3							
29	**Sandhier**	137	44-50	20	2.65	1980	106.4	122.6		
			56-70			18.6				
30	**Vanj**	190	40-74.1	93.42	6.28	2230	374.7	1302.4		
			80.7-96.0			5.95				
			100.02-123.0							
			131.44-141							
			159.7-187.8							

Groundwater flow trends and analysis: A regional groundwater flow trend map for Narmada and Tapti outfall basins for both near surface and deep aquifer systems are provided in the Figure 4.76. The groundwater flows across different lengths of different sub basins are provided in Table 4.35.

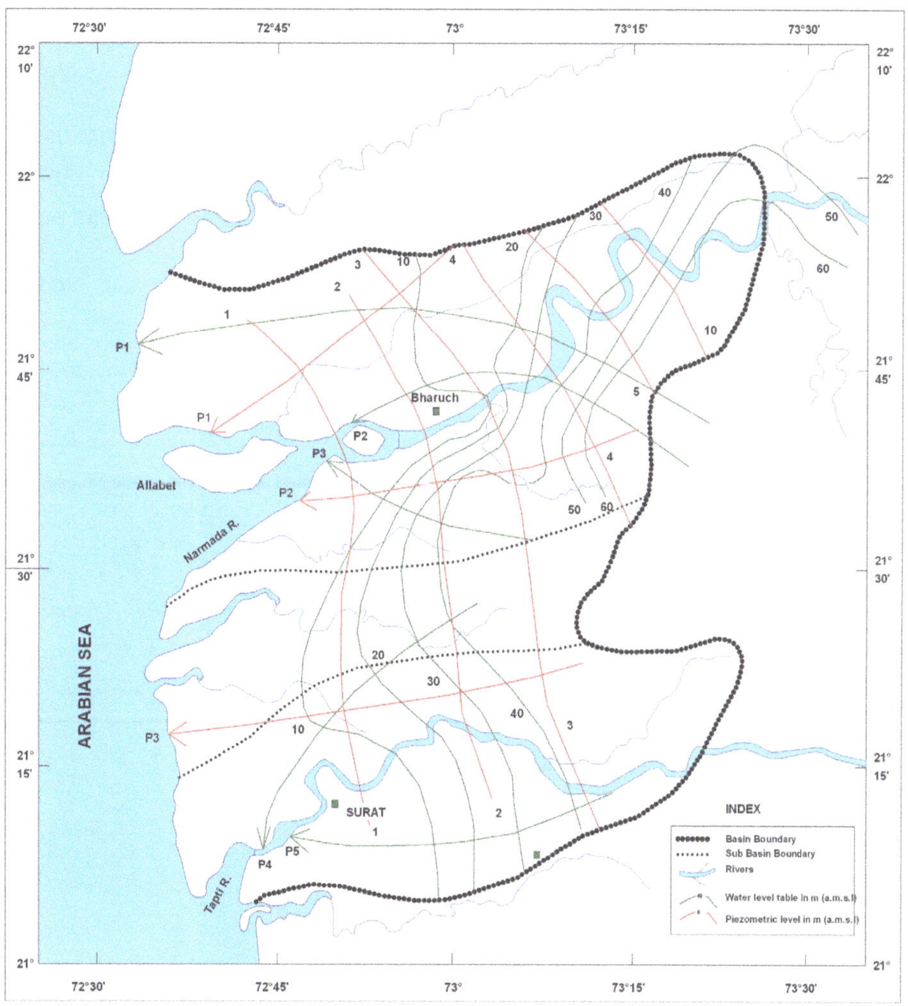

Fig. 4.76 Groundwater flow map for near surface and deep aquifer systems in parts of Narmada-Tapti delta basins (Gujarat)

Table 4.35 Groundwater flow across different profiles in Narmada-Tapti delta basin

S.No	Basin	Profile	Avg.Transmissivity (T) (m^2/day)	Gradient (I) (m/km)	Length of the Coast (L) (km)	Total flow (Q) (MCM)
Near surface aquifers						
1	Narmada	P1	40	1.67	20	0.49
		P2	40	2.17	17	0.54
		P3	40	3.38	19	0.94
2	Tapti	P4	30	1.36	19	0.28
		P5	30	1.43	9	0.14

Table 4.35 *Contd...*

S.No	Basin	Profile	Avg.Transmissivity (T) (m^2/day)	Gradient (I) (m/km)	Length of the Coast (L) (km)	Total flow (Q) (MCM)
Deep aquifers						
1	Narmada	P1	1500	0.11	28	1.69
		P2	1500	0.09	28	1.38
2	Tapti	P3	350	0.07	28	0.25

The total estimated groundwater flow across the whole coastal margins of Narmada and Tapti delta basins is of the order of 5.71 MCM which comes to about 0.6 % of the dynamic potentials of the basin.

Seasonal water level fluctuations: The seasonal water level fluctuations for 7 year period at selected observation sites are shown in Fig. 4.77 and are also reflected in Table 4.36. Majority of the observations are drilled in Quaternaries and Tertiaries. The average fluctuation in Narmada and Tapti are of the order of 3.44 m and 2.16 m respectively.

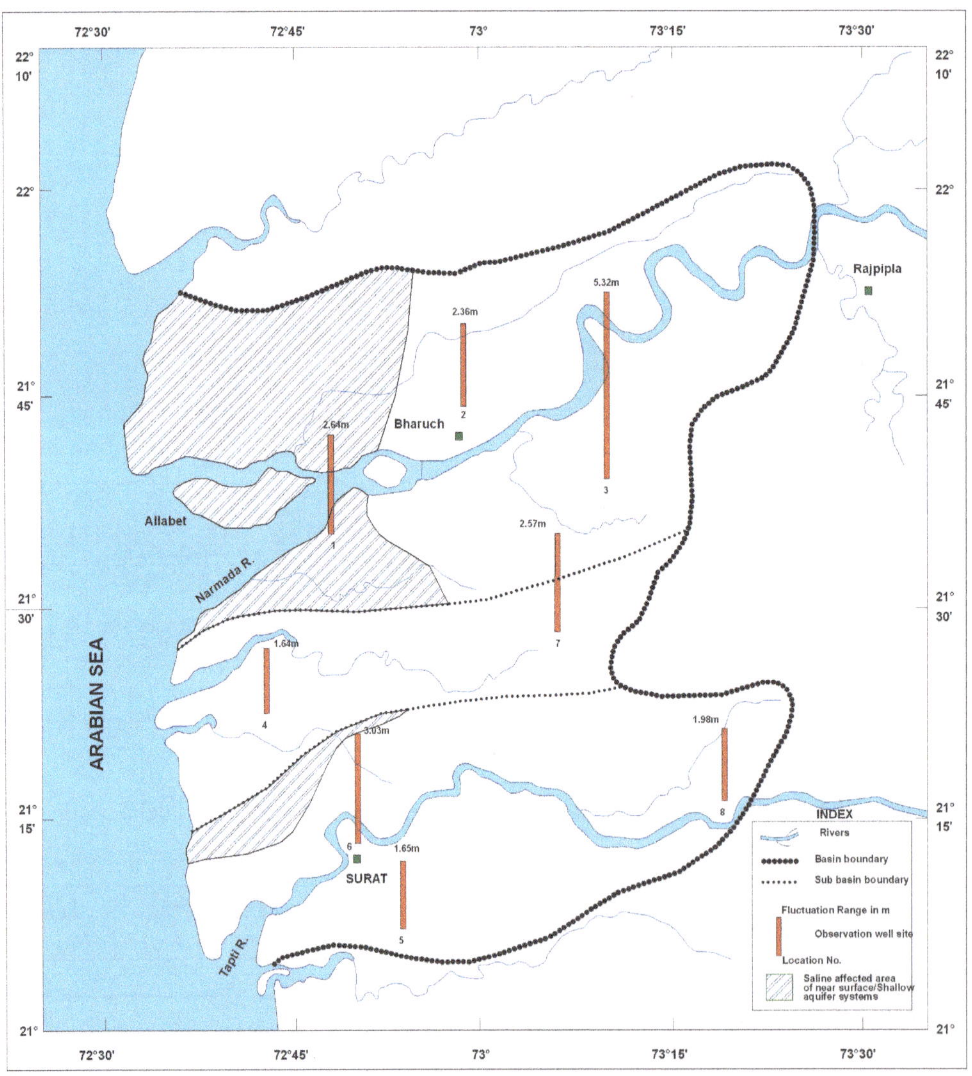

Fig. 4.77 Network of observation well stations and their fluctuation trends in Narmada-Tapti delta basins (Gujarat)

Table 4.36 Seasonal fluctuations in selected observation stations

S.No	Name of the Site	Geological Horizon	Avg. Fluctuation (m) for a 7 Year period (1978-84)
Narmda delta region			
1	Utraj	Alluvium	2.64
2	Bharuch	Alluvium	2.36
3	Valia	Sandstone	5.32
Tapti Delta region			
4	Olpad	Alluvium	1.64
5	Sachin Town	Quaternary Alluvium	1.65
6	Surat	Alluvium	3
7	Kosamba	Sandstone	2.57
8	Mandvi	Basalt	1.98

Resource potentials in Narmda sub basin

- Area of the Narmada sub basin : 2653 sq.km
- Saline affected Area : 957 sq.km
- Coverage of Alluvials not affected by salinity : 1696 sq.km
- Dynamic potentials for Alluvial sedimentaries (not affected by salinity):

(i) Rainfall infiltration method : 300 MCM

(ii) Fluctuation method : 774.5 MCM

(iii) Average Dynamic Potentials : 537.25 MCM

- Resource potentials in deep confined aquifer systems (static) (Area × Thickness × S.Coefficient)

(i) Alluvial area (fresh) ($1368.7 \times 40 \times 2.5 \times 10^{-4}$) : 13.7 MCM

(ii) Tertiary area (fresh) ($327.36 \times 120 \times 2.5 \times 10^{-4}$) : 9.82 MCM

- Total estimated resource potentials in deep confined aquifer systems : 23.5 MCM
- Total estimated (dynamic + static) potentials in Narmada delta basin : 560.75 MCM

Resource potentials in Tapti sub basin

- Area of the Tapti sub basin : 1830 sq.km
- Saline affected Area : 151 sq.km
- Coverage of Alluvials not affected by salinity : 1679 sq.km
- Dynamic potentials ofs Alluvial sedimentaries (not affected by salinity) :

(i) Rainfall infiltration method : 342 MCM

(ii) Fluctuation method : 528 MCM

(iii)	Average dynamic Potentials	:	435 MCM
•	Resource potentials in deep confined aquifer systems (static)		
	(Area × Thickness × S.Coefficient)	:	
(i)	Alluvial area (fresh) ($1100 \times 40 \times 7 \times 10^{-4}$)	:	30.8 MCM
(ii)	Tertiary (fresh) ($125.9 \times 120 \times 7 \times 10^{-4}$)	:	10.58 MCM
	• Total estimated resource potentials in deep confined aquifer systems	:	41.38 MCM
	• Total estimated (dynamic + static) potentials in Narmada + Tapti basin	:	476.37 MCM

CHAPTER 5

GROUNDWATER QUALITY CHARACTERISTICS AND TREND ANALYSIS IN COASTAL DELTA COMPLEXES

Evaluation of groundwater quality parameters forms an integral part of groundwater studies in any region or a basin and are considered more important in coastal deltaic/sedimentary systems in view of several natural and anthropogenic factors confronting the coastal systems. The chemical composition of groundwaters depends upon several factors like (i) climate and temperature, (ii) precipitation levels, (iii) catchment rocks and decomposition processes, (iv) biochemical, ecological, anthropogenic and environmental factors and (v) surface and sub-surface flow dynamics. The chemical composition of groundwaters in a region therefore gives a whole lot of information on several hydrological and hydrogeochemical processes such as rock water interactions, groundwater flow paths, recharge and discharge processes, environmental, anthropogenic and ecological impacts.

Rain water which is the primary source for both surface and groundwater in general contains only small amounts of dissolved mineral matter and is normally in equilibrium with atmospheric carbon dioxide ($1 \times 10^{-3.5}$ atmospheres). However when it reaches the earth's surface soil and rock layers, the carbon dioxide pressure levels get elevated due to decay of organic matter and root respiration. The quantity and mineral matter dissolved depend upon the chemical composition, texture and structure of the soil and rock material as well as the hydrogen ion concentration (pH) and redox potential (eH) of the water. The carbon dioxide in solution derived from atmosphere and through the organic processes of the soil increases its solvent capacity as it moves through different layers underground. Salts such as NaCl, KCl, $MgCl_2$, $MgSO_4$ and $NaHCO_3$ are easily dissolved through the water percolation and movement and consequently these ions get enriched in groundwater along their pathways and movement depending upon the solubility levels of various minerals. Groundwater passing through igneous rocks dissolves small quantities of mineral matter because of relative insolubility of rock matrix. Minerals in the sedimentary rocks relatively are more soluble and facilitate increasing levels of ionic constituents in groundwater. Bicarbonates, sulfates and chlorides as anions and calcium, magnesium, sodium, potassium as cations are normally found associated more in groundwater of sedimentary rocks when compared to groundwater in igneous rocks. Bicarbonates are found in abundance in the catchment/recharge areas and get reduced towards the outfall/discharge areas in most groundwater systems. Sulfates are also found abundantly in the catchment areas particularly in the areas containing sulphide ore bodies, sedimentary rocks containing rich sulphide and anhydrite materials and pyrite bearing beds. Chloride constituents however are found in greater amounts in the vicinity of sewage, connate water along tidal inlets of coastal regions and adjacent to coastal marshy areas and salt pans. Nitrates are found in increasing levels in the irrigation tracks. Iron sometimes is noticed in increasing levels where ferric rock materials are available in abundance particularly in the catchment areas and iron rich lateritic tracks. In general the following sequence of abundance of ions is noticed along the groundwater flow paths of saturated zones from the recharge areas to discharge areas in coastal sedimentary tracks.

$$HCO^-_3 \rightarrow (HCO^-_3 + SO^{2-}_4) \rightarrow (SO^{2-}_4 + Cl^-) \rightarrow Cl^-$$

Fresh water Brackish/Saline water

In general a comprehensive program of water quality studies should include (i) physical parameters like color, odour, turbidity and temperature, (ii) organic parameters such as BOD (Biochemical Oxygen Demand) and COD (Chemical Oxygen Demand) and total organic carbons and (iii) inorganic chemical quality parameters such as pH (hydrogen ion concentration); total cations (calcium, magnesium, sodium and potassium) and total anions (bicarbonates, carbonates, sulphates, chlorides, nitrates and fluorides). In addition there are several trace metals and non metals such as strontium, boron, beryllium, lithium, barium, titanium, vanadium, chromium, cobalt, nickel, molybdenum, copper, silver, lead, zinc, cadmium, mercury, arsenic, antimony, selenium, bromine and Iodine etc. In addition there are certain radioactive trace elements in waters such as uranium, thorium, radium and radioactive nuclides such as $lead_{210}$, tritium, $carbon_{14}$, $strontium_{90}$ etc. Of these tritium and $carbon_{14}$ are widely used to study hydrological processes and recharge estimations in different groundwater basins (Kauffman, S. and Libby, W.S., 1954; Clark, I.D. and Fritz, P., 1997; Kay, R.T., et al, 2002; Geyh, M., 2000). The present chapter however restricts itself to characterize the near surface and deep groundwater systems of major deltas along the east and west coast of India in terms of major ionic concentrations, their trends, analysis and their utility for agriculture, irrigation and drinking water purposes.

5.1 MAJOR CHEMICAL CONSTITUENTS AND THEIR CHEMISTRY

5.1.1 Alkalinity and hydrogen ion concentration (pH)

The alkalinity of a solution is defined as the capacity of its solutes to react and neutralize acid. The alkalinity in most natural waters is primarily due to the presence of dissolved carbon species, particularly bicarbonates and carbonates. Other constituents that contribute towards alkalinity are silicates, hydroxides, borates and certain organic compounds (Hem, J.D., 1989). Alkalinity is expressed as an equivalent concentration of dissolved calcite ($CaCO_3$). The pH value of a solution is expressed normally as negative base (log_{-10}) of the hydrogen ion activity in moles per liter. In pure natural waters, the dissolved molar concentration of H^+ (hydrogen) and OH^- (hydroxide) ions are equal (10^{-7} moles/liter), equivalent to pH 7. The pH value becomes less than 7 when H^+ ions exceed OH^- ions and becomes more than 7 if OH^- ions exceed H^+ ions. By definition when the number of hydrogen ions equals the number of hydroxyl ions the solution is characterized as '*neutral*' and the hydrogen ion activity pH 7. On the other hand when there are no hydrogen ions pH tends to become 14 and the solution is termed as '*alkaline*'. Again when there are no hydroxyl ions the solution becomes a pure acid and pH value equals to 1 and the solution is termed '*acidic*'. The hydrogen ion activity in a solution is often controlled by several interrelated chemical reactions and is also affected by temperature and pressure and is a useful index to estimate the state of equilibrium reactions involved. Dissolved carbon dioxide in water is one of the most important factor in establishing the pH value in water and is represented by the following reactions (Hem, J.D., 1989).

$$CO_2 + H_2O = H^+ + HCO^-_3 \qquad .....(5.1)$$

$$H_2CO_3 = H^+ + HCO_3^- \qquad .....(5.2)$$

$$HCO_3^- = H^+ + CO_3^{2-} \qquad .....(5.3)$$

In addition to the above facts the pH is also influenced by the presence of phosphates, silicates, borates, fluorides etc. Oxidation-reduction processes, decomposition of organic matter can substantially alter the pH values. The near surface fresh groundwater systems in general have pH values close to the neutral pH values, where as water samples derived from deep fresh water bodies, springs, mine waters etc. will have substantial impacts on pH values and tend to be '*acidic*' because of the temperature and pressure conditions, rock water interactions and slow velocities. The groundwater samples derived under the impacts of oceans and tidal areas will have their pH values increased towards alkalinity.

5.1.2 Specific electric conductance

Electrical conductance or conductivity is the ability of a substance to conduct electric current. Specific electrical conductance is the conductance of a body of unit length and unit cross section at a specified temperature. Electrical conductivity is also the reciprocal of resistivity and is normally expressed in μ mhos/cm. The relationship between conductivity and resistivity is expressed as below:

$$\text{Specific conductance } (\mu \text{ mhos/cm}) = \frac{10^4}{\text{Specific resistance (ohm/m)}} \quad .....(5.4)$$

It is also to be noted that the resistivity of a solute changes significantly with temperature. It is therefore required that all resistivity values are standardized to 25°C (Jones, P.H. and Buford, J.B., 1951). All natural waters contain some salt in solutions. A broad relationship between salts in solution and the conductance exists as indicated below:

$$\text{Sp. conductance in } \mu \text{ mhos/cm at } 25\ ^{\circ}\text{C} = \frac{\text{Total dissolved solids in ppm}}{\text{F}} \quad .....(5.5)$$

where 'F' factor has a wide range of values between 0.5 to 1. In most groundwaters the value of the factor F has a range between 0.6 and 0.7. The sp. Conductance of atmospheric waters in general has a value varying between 5 to 30 μ mhos/cm, the potable groundwater has a range between 30 to 2000 μ mhos/cm and the sea water has a range between 45,000 to 50,000 μ mhos/cm and the brine water has more than 100,000 μ mhos/cm (Davis, S.N. and DeWiest, J.M., 1991).

5.1.3 Total dissolved solids

All the mineral constituents dissolved in water constitute dissolved solids. These include bicarbonates, sulphates, chlorides, nitrates (anions) and calcium, magnesium, potassium and sodium (cations). In addition trace metals and radio active constituents occur as traces. The total dissolved solids of groundwater in the catchment areas may range from 20 ppm to 1000 ppm and could increase to a level of 10,000 ppm in the discharge areas and ultimately could become over 100,000 ppm in saline-brine areas/desert tracks.

5.1.4 Bicarbonates

The common sources of carbonates and bicarbonates in groundwater are: (i) dissolved carbon dioxide in the rain, (ii) carbon dioxide in the soil, (iii) dissolution of carbonate rocks and (iv) carbon dioxide generated through bacterial oxidation of organic matter. Water charged with carbon dioxide dissolves carbonate minerals on its transit through soils and rocks to give rise to bicarbonates. The pH of water reflects the form in which carbon dioxide is present in solution. The term alkalinity is often used as a synonymon to total contents of bicarbonates and carbonates in groundwater (Davis, S.N. and DeWiest, J.M., 1991). Under normal conditions the bicarbonate contents in groundwater varies from 100 to 800 ppm. Concentration above 1000 ppm can occur in certain igneous/metamorphic/tectonic processes such as active fault zones and in areas of metasomatic activities (Matthess, G., 1982;

Davis, S.N. and DeWiest, J.M., 1991). Some chemical reactions leading to the accumulation of bicarbonate ion in groundwater are mentioned below:

$$CO_2 + H_2O = H_2CO_3 \quad .....(5.6)$$

$$H_2CO_3 = H^+ + HCO^-_3 \quad .....(5.7)$$

$$CaCO_3 + CO_2 + H_2O = Ca^{++} + 2HCO^-_3 \quad .....(5.8)$$

$$2NaAlSiO_8 + 2CO_2 + 11H_2O = Al_2Si_2O_5\,(OH)_4 + 2HCO^-_3 + 2Na^+ + 4H_4SiO_4 \quad .....(5.9)$$

5.1.5 Chlorides

The chloride levels in rainwater are generally low, less than 10 ppm but occurs abundantly in oceanic waters to an order of about 19,000 ppm and could go up to 100,000 ppm in brine waters, evaporates and in paleo marine waters. Chloride levels in groundwater in general are less than 30 ppm in rain fed regions but increases rapidly to an order varying from 200 ppm to 600 ppm due to rock water interactions, anthropogenic impacts and saline/brackish water intrusions (Matthess, G., 1982). Most chlorides in groundwaters are either due to atmospheric sources, rock water interactions, sea water contaminations and some times due to connate water. Chlorides sometimes are found by way of replacement of hydroxide in biotite and hornblende bearing igneous and metamorphic rocks and sometimes as inclusions in glassy rocks. Chloride salts are relatively free from chemical reactions and do not enter into oxidation/reduction processes easily and are also not easily adsorbed on mineral surfaces. Chlorides therefore are used as a conservative tracer sometimes for these reasons (Hem, J.D., 1989). Most chlorides in groundwater are present as sodium chloride, but sometimes the chloride contents may exceed sodium due to *'base exchange'* phenomena. Sewage wastes salt ponds, evaporate deposits sometimes contribute to the increased levels of chlorides in groundwater.

5.1.6 Sulphates

Sulphate is an oxidised product of sulphur. The rain water contains rather low amounts of sulphate but increase substantially during contacts with sedimentary rocks and rocks containing sulphide minerals. Sulphur is a widely distributed element in reduced form in several igneous and sedimentary rocks, sulphides of heavy metals, volcanic eruptions and coal bearing rocks etc. Minerals like gypsum, pyrite and anhydrite can contribute heavily for the generation of sulphates in groundwater. The oxidation and hydrolysis processes of pyrite could produce soluble sulphates as given below.

$$2FeS_2 + 7H_2O + 15O \rightarrow 2Fe(OH)_3 + 4H_2SO_4 \quad .....(5.10)$$

$$FeS_2 + 7O + H_2O \rightarrow FeSO_4 + H_2SO_4 \quad .....(5.11)$$

$$2FeSO_4 + H_2SO_4 + O \rightarrow Fe_2(SO_4)_3 + H_2O \quad .....(5.12)$$

The sulphuric acid then could react with calcium carbonate found in several sedimentary rocks to give rise to calcium sulphate.

$$H_2SO_4 + CaCO_3 \rightarrow H_2HCO_3 + CaSO_4 \quad .....(5.13)$$

Sulphate sometimes is lost by reduction process caused by bacteria, CO_2 and release of hydrogen through decomposition of organic matter.

$$CaSO_4 + CH_4 = CaS + CO_2 + 2H_2O \quad .....(5.14)$$

$$CaS + 2CO_2 + 2H_2O = H_2S + Ca(HCO_3)_2 \quad .....(5.15)$$

$$CaSO_4 + 4H_2 = CaS + 4H_2O \quad .....(5.16)$$

Sulphate reduction bacteria is the main cause of hydrogen sulphide gas in groundwater and sometimes partially in coal bearing areas. Groundwater associated with oil and gas water, magmatic activity and sometimes in deep aquifers attains high concentrations of hydrogen sulphides in excess of 100 mg/l. The groundwater in igneous and sedimentary rocks in general contains sulphides less than 100 mg/l.

5.1.7 Nitrates

Nitrogen is present in substantial amounts in atmosphere and in minor amounts in crustal layers. The nitrogen occurs in water as nitrite (NO_2^-) and as nitrate (NO_3^-) and in cationic form as ammonium (NH_4^+) and sometimes at intermediate oxidation states as a part of organic solutes (Hem, J.D., 1989). The important sources of nitrate are (i) atmospheric precipitation, (ii) sewage and animal sources, (iii) agricultural belts induced with pesticides and fertilizers and (iv) geological strata. The contributions from atmospheric and geological sources for nitrate accumulation are rather small. The major sources are anthropogenic, sewage and animal wastes and agricultural practices with excessive use of fertilizers and irrigation return wastes. High levels of nitrates in general are noticed in shallow groundwaters of rich agricultural belts such as coastal deltas. Nitrogen can undergo numerous reactions in the subsurface layers, the main reactions are (i) immobilization and mineralization, (ii) nitrification, (iii) de nitrification and (iv) plant uptakes and recycling (Keeney, D., 1990). Immobilization is the biological assimilation of inorganic forms of nitrogen by plants and microorganisms to form organic compounds such as amino acids, sugars, proteins and nuclei acids. Mineralization is the inverse of immobilization. It is the formation of ammonia and ammonium ions during microbial digestion of organic nitrogen. Nitrification is the microbial oxidation process of ammonia and ammonium ion to nitrite and then to nitrate. Nitrification is a key reaction leading to the movement of nitrogen from land surface to water table. De nitrification is the biological process that utilizes the nitrate to oxidize organic matter into energy usable by microorganism. This process converts the nitrate to more reduced form yielding nitrogen gas. The uptake of nitrogen by plants also removes nitrogen from soil columns and gets released to the atmosphere when the plant dies (Rees, T.F., et al 1995).

5.1.8 Calcium

Calcium is a major constituent of many rock forming minerals like plagioclase, pyroxenes, and amphibolites. This is also an essential element for plant and animal life forms and occurs abundantly in solutes of natural waters. The most common form of calcium in sedimentary rocks are carbonates like calcite ($CaCO_3$) and dolomite ($CaMg(CO_3)_2$). Lime stones consist mainly calcite and admixtures of dolomites. Some other minerals of calcium in sedimentary rocks are gypsum ($CaSO_4.2H_2O$), anhydrite ($CaSO_4$) and fluoride (CaF_2). Sandstones and shales and other detrital deposits contain calcium carbonate as cementing matrix. Calcium is also present as adsorbed ion on negatively charged surfaces. Most carbonates and sulphates of calcium are easily soluble in water, where silicates are not easily soluble, but break down into clay minerals on weathering. Because of its solubility character, calcium is found abundantly in groundwater. Calcium carbonate is soluble on the impacts of carbon dioxide in a dissolved form in water. The reactions are as follows.

$$CaCO_3 + H_2O + CO_2 = Ca(HCO_3)_2 \quad .....(5.17)$$

The presence of sulphuric acid in water could cause dissolution of calcium carbonate due to oxidation and hydrolysis process.

$$H_2SO_4 + CaCO_3 \rightarrow H_2CO_3 + CaSO_4 \quad .....(5.18)$$

The range of calcium content in groundwater is largely dependent on the solubility of calcium carbonate, sulphate, and sometimes chloride and also varies with the partial pressure of carbon dioxide in air when in contact with water. Under conditions where partial pressure of carbon dioxide is low in groundwaters and water saturated with calcium carbonate, calcium gets precipitated as calcium carbonate in the form concretions. A part of the calcium sometimes get deflected due to ion exchange reactions.

$$2Na - clay + Ca^{++} = Ca^{++}\text{–}clay + 2Na^{+} \quad .....(5.19)$$

In most coastal waters of India, calcium is a predominant cation and gets reduced in its content where saline water intrusion is a dominant phenomenon.

5.1.9 Hardness

Hardness is a term related to the concentrations of certain metallic ions in water, particularly magnesium and calcium, and is usually expressed as an equivalent concentration of dissolved calcite ($CaCO_3$). In hard water, the metallic ions of concern may react with soap to produce an insoluble residue. These metallic ions may also react with negatively charged ions to produce a solid precipitate when hard water is heated (Freeze, R.A. and Cherry, J.A., 1979). Hard waters can thus consume excessive quantities of soap, and cause damaging scale in water heaters, boilers, pipes and turbines. Many of the problems associated with hard water, however, can be mitigated by using water-softening equipment.

5.1.10 Magnesium

Magnesium (Mg^{2+}) is an alkaline earth metal and is essential for plant and animal nutrition. Calcium and magnesium by and large in water chemistry hold similar effects, but the geochemical properties however are different. Magnesium ions are smaller when compared to other cations like sodium and calcium and have stronger charge, density and greater affinity to water molecules.

The magnesium occurs quite prominently in natural waters, though not as a major cation. It is known to occur widely in igneous and metamorphic and sedimentary rocks such as pyroxenites, amphibolites, dunites, schists, limestones and dolomites. Magnesium occurs in these igneous and metamorphic rocks as insoluble silicates but breaks down on weathering processes into more soluble carbonates and clay minerals. Magnesium carbonate in presence of carbonic acid gets converted into soluble bicarbonates. The levels of magnesium in fresh groundwaters are relatively less than that of calcium but their concentration in sea water appears quite high. It therefore helps to identify the salinity ingress sometimes in coastal region.

$$MgCO_3 + CO_2 + H_2O = Mg(HCO_3)_2 \quad .....(5.20)$$

The solubility of magnesium carbonate in water in the presence of carbon dioxide is nearly ten times that of calcium carbonate.

5.1.11 Sodium

Sodium is the most abundant element of alkali metal group. It occurs mostly in feldspars, pyroxenes and amphiboles of igneous rocks. Weathering processes of these minerals releases primarily soluble sodium products. The most important source of sodium are rich sodium salt deposits, evaporates in the desert tracks and sea water. Most sodium salts are readily soluble in water. Sodium in groundwater ranges from 1 ppm in hard rock catchment areas to 100,000 ppm in coastal brine water. A characteristic feature of sodium is its linear co variation with chloride. Sodium ions are not precipitated in groundwater like calcium, but under goes ion exchange reactions with other alkaline earths such as calcium and magnesium.

5.1.12 Potassium

Potassium in general is marginally less common in the minerals of igneous rocks, but appears present abundantly in several sedimentary rocks. However on the contrary its concentration in groundwater is far less than sodium including ocean waters. The probable reasons being its resistance for weathering and decomposition and its refixation in the weathered products and clay minerals. Unlike sodium, potassium plays an important role in the sustenance of plants and animals. The main sources of potassium are (i) rainwater, (ii) weathered products of silicate minerals, (iii) fertilizers, (iv) irrigation returns and (v) biological factors.

Potassium ions are larger than sodium and are absorbed less strongly than sodium anion exchange reactions, but enter into the clay mineral structures in a special way. In illite clay mineral potassium ions are incorporated in spaces between different crystal layers, and cannot be easily removed by ion exchange reactions. The potassium levels in natural fresh groundwaters range between less than 1 to 15 ppm and increases substantially in brine and ocean waters. Potassium concentrations in solutions are relatively high at places where fertilizers application to soils are high and sometimes controlled depending upon the nature of clay contents and organic matter.

5.1.13 Fluoride (Fluorine)

The fluorine compounds have low solubility in natural waters when compared to other compounds of halogens. Fluoride in natural waters gets released through weathering processes of alkali and siliceous igneous and sediment rocks. The fluoride concentration in most natural waters are small less than 1mg/l, but groundwater exceeding this limit are reported in several parts of India. The common fluoride bearing minerals are: (i) calcium fluoride (CaF_2), (ii) Flour apatite, (iii) amphiboles and (iv) volcanic and fumarolic gases.

5.1.14 Arsenic

Arsenic exists as a compound associated with other elements such as Oxygen, Chlorine and Sulphur and is commonly distributed / associated with minerals and ores containing copper and led. The stable forms of Arsenate (As^{5+}) are Arsenite (As^{3+}) oxy anions. Small amount of arsenic can be toxic to humans and is considered undesirable for human consumption. If the concentration limits go beyond 50 ppm level (U.S. Environmental protection agency 1976) arsenate or arsenite can be dominant inorganic form in groundwater. Monovalent arsenate (H_2AsO_4) could be expected between pH 3 and pH 7 and divalent species (HAs_4^{2-}) could occur between pH 7 and pH 11 (Hem, J.D., 1989; Dowdle, P.R., et al 1996; Harrington, J.M., et al 1998; Welch, A.H., et al 2000). Arsenic could also result from anthropogenic sources such as mining and smelting operations, agriculture applications and waste disposal sites. Arsenic contamination is a serious concern and remediation measures are still problematic. Arsenic contaminations in India are reported particularly in the deltaic alluvial tracks of West Bengal (Mandal, B.K., et al 1996).

5.1.15 Iron

Iron is one of the abundant elements and occurs widely in igneous, metamorphic and sedimentary rocks and is associated widely in the rock forming minerals such as pyroxenes, amphiboles, biotites, and magnetites etc. Iron occurs mainly in two states one in the ferrous form (Fe^{2+}) and Ferric form (Fe^{3+}) (Hem, J.D., 1989). These minerals are exposed to water precipitate as ferrous sulphides under reducing conditions and also could precipitate as ferric oxides under oxidizing conditions. The precipitates of iron are strongly controlled by pH and redox conditions where microorganisms play important role. The most common type of iron in solution is the ferrous iron (Fe^{2+}) and is widely reported in the groundwater where lateritic and iron rich formation in the catchment areas of the deltaic tracks in India are exposed.

5.2 METHODS, GRAPHICAL TECHNIQUES AND DATA PRESENTATIONS

The total dissolved contents in groundwaters in general are expressed by weight and volume or by chemical equivalents. A brief description of methodologies involved are briefed below.

5.2.1 Weight and weight volume method

Concentrations of ions and total dissolved salts are reported by weight per volume units or milligrams per liter. One part per million means one part of dissolved substance by weight is present in one million parts by weight of the solution. This is approximately equal to the weight of the dissolved matter in milligrams in one liter of water. This holds good for low concentration of water, but changes with specific gravity and increase in concentrations for which correction needs to be applied.

ppm (parts per million) = milligrams per kilogram = milligrams for liter of solution/Sp.gravity of the solution.

Sometimes the concentrations are expressed as 'parts per thousands' while representing composition of sea water (Rankama, Kalervo, and Sahama, T.G., 1950). Parts per billion or parts per trillion sometimes are used while reporting trace concentrations.

5.2.2 Chemical equivalence method

The weight and weight volume method may not be adequate while attempting to understand its significance in terms of several hydro geochemical and hydrological variations. The combination of dissociation of cations and anions are more governed by their equivalent weights rather than gravimetric weights. The equivalent weight of an ion equals to its atomic or molecular weight divided by its valancy. Therefore the ratio of ionic concentration in ppm to the equivalent weight of that ion gives rise to the concentration levels in equivalents per million (epm) or milli equivalents per liter (meq/l) depending upon the data expression. The conversion factors for converting milligrams per liter to milli equivalent per liter are given in several textbooks dealing with water chemistry (Hem, J.D., 1989).

5.2.3 Brief methodological estimations of major ions

The general practice in several field and laboratories involved in estimations of chemical analyses of waters are the pH and Specific Conductance are measured in the field immediately after collection of samples and then the samples are packed in sterilized polythene water samples, brought to the laboratory and conduct analysis for different anions and cations. These procedures are done as quickly as possible with minimum loss of time between the sample collection and analysis in the laboratory. Estimations of these ions are performed as per standard methods adopted by American public health association (1989) which are briefly described in Table 5.1.

Table 5.1 Standard methods for analyses of major ions

Nature of the ion	Method
Calcium(Ca) and Magnesium(Mg)	Titration using 0.05 NEDTA solution
Sodium(Na) and Potassium(K)	Flame photometer method
Bicarbonates and Carbonates	Titration method using 0.01N H_2SO_4 Solution
Chloride(Cl)	Titration method using 0.05N $AgNO_3$ Solution
Sulphate (SO_4)	Spectrophotometer method.

5.2.4 Graphical techniques for presentation of major ions

Several graphical techniques are in vogue to present and display the water quality data and are quite useful to estimate the changing patterns of water quality in a region, identify the mixing of groundwaters, estimate probable groundwater recharge and discharge areas and understand the salinity incursion problems. Most methods involving graphical analyses are designed to represent simultaneously the total solute concentration and the proportions estimated against each ionic species for one analyses or group of analyses. The units in which the graphical analyses expressed are milli equivalent per litre. Collins, W.D. (1923) presented a vertical bar graph with their heights proportional to the concentration of determined anions and cations. The bar is divided by a vertical line with the left half representing cations and the right half anions Figure 5.1. These are again divided by horizontal lines to show the concentration of the major ion. The height of the bar is proportional to the total epm values of cations and anions expressed in milliequivalents per litre. The height differences between the two bars should be same to both cations and anions, but for some minor errors of ionic estimation or other minor ions which are not included. The left half of the bars represents cations and the right half represents anions. This could be modified by plotting additional bars also as and when required depending upon other addition ions.

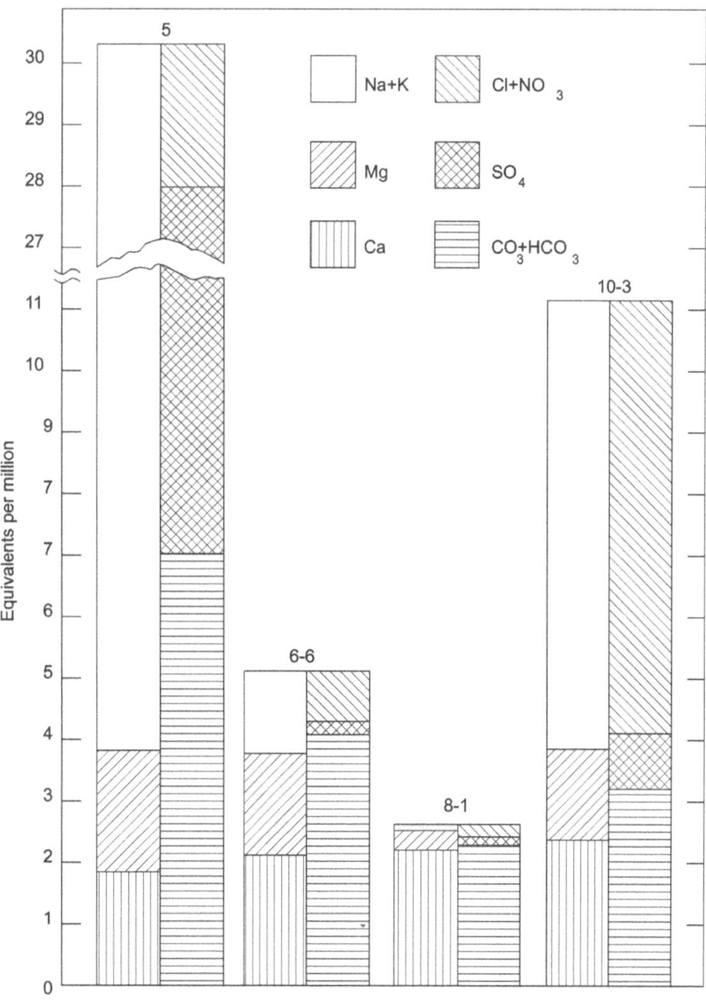

Fig. 5.1 Vertical bar graphs for representing analyses of groundwater quality (after Hem, J.D., 1989)

Another method of plotting chemical quality with radiation vectors proposed by Moucha (1949) is illustrated in Fig. 5.2. The distances from the centre of the vector diagram represents the concentration levels of the ions in terms of milliequivalents per liter. Though this is not widely used, it has the advantage of viewing the ionic differences at a glance quickly.

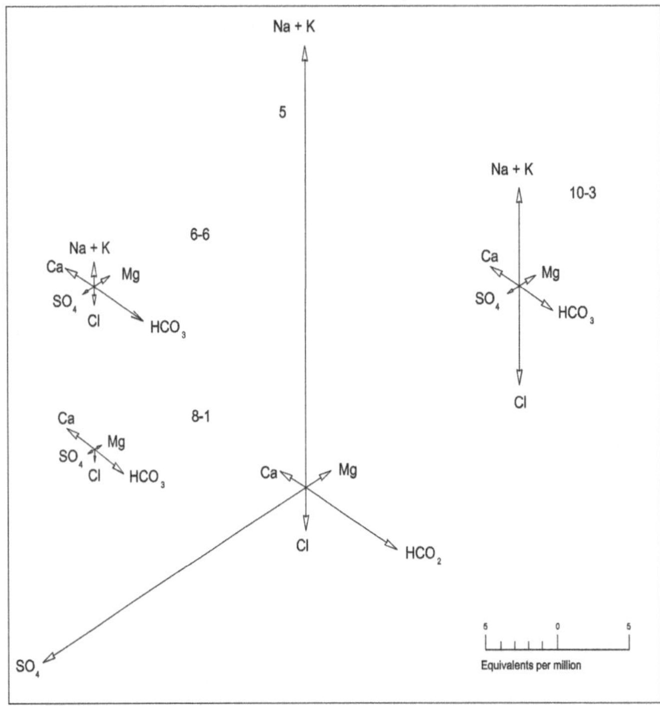

Fig. 5.2 Vector diagrams for representing analyses of groundwater quality (after Hem, J.D., 1989)

Another procedure of expression of analyses was suggested by Stiff, H.A. (1951). This diagram as shown in Fig. 5.3 used four parallel horizontal lines extending to left and right from the centre of the axis of the diagram. The cationic compositions are represented as the left side of the four parallel lines and like wise anions are represented along the four parallel lines in the right. All the ionic concentrations are expressed in the equivalents per litre. The resulting points are connected to give rise an irregular polygonal shape. The shape and size of the diagram indicates readily the total as well as individual ionic concentrations at a glance.

Similarly sometimes 'Pie diagrams' are also in use as shown in Fig. 5.4. In this various concentrations are represented in a circle with different components of ions proportional to the space in between each of the radii and the total circular space represents the total ionic concentrations. Expression of different ions in a circular pattern is easy to present and the variations at each site on a map with large sampling points could be assessed at a glance. It is therefore useful to study the aerial distribution of different point concentration levels.

Schoeller, H. (1935) developed another approach wherein different ions and their composition trends are shown as vertical axes, with each axes representing one anion or cation, all expressed in milliequivalents per litre and then combining these lines with straight lines. The ionic concentrations on each of the vertical scales are represented in logarithmic scales. Waters representing similar concentration trends appear parallel to each other. This type of presentation not only shows absolute values but also reflects on concentration differences among various ions.

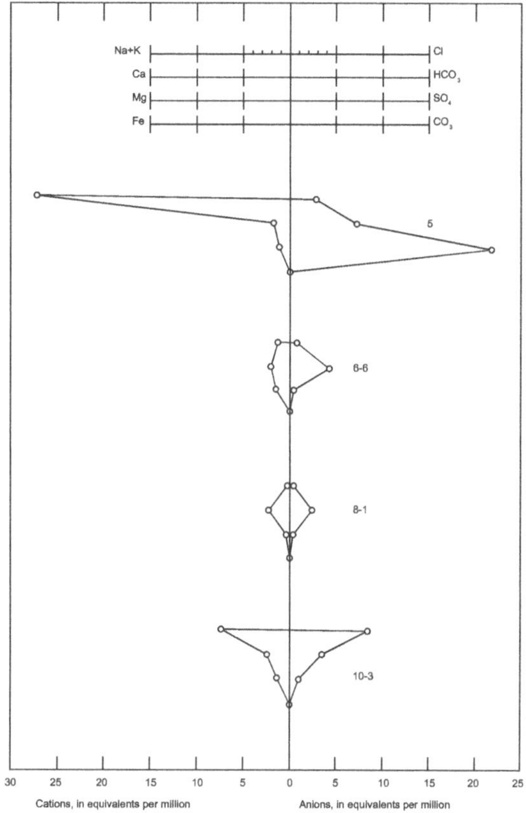

Fig. 5.3 Pattern diagrams for representing analyses of groundwater quality (after Hem, J.D., 1989)

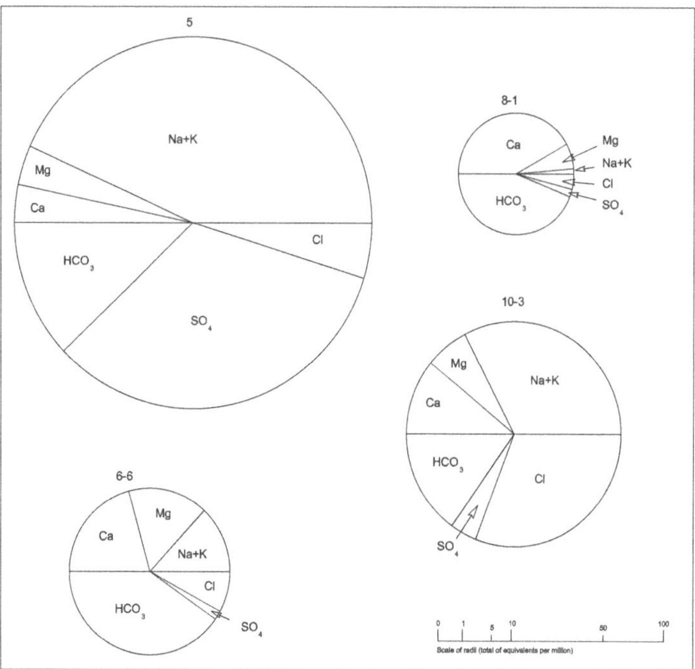

Fig. 5.4 Circular diagrams for representing analyses of groundwater quality (after Hem, J.D., 1989)

5.2.5 Spatial variations, ionic cross plots and ionic ratios

The distributional trends of EC, TDS and the other major ions like Ca, Mg, Na + K, HCO_3 + CO_3, SO_4 and Cl in different basins provide important inputs to understand and delineate the recharge and discharge areas, ion exchange processes and salinization processes etc. These hydrogeochemical base maps related to spatial distribution of various ions could be prepared by way of plotting values of ionic compounds in either ppm or epm at each well site and then drawing equal contour lines.

Cross plots between various cations and anions vs TDS would help to resolve to understand the contribution of source rocks, resident times and ionic exchange processes. The slope characteristics divergence of points with increasing levels of TDS would indicate the degree of involvement of different ions and their diagenetic changes, ionic exchange processes and salinization processes etc.

The ionic ratios and their inter plots against either TDS levels or chloride levels and their graphical trends work as an important indicators to delineate the recharge and discharge boundaries exchange well sites and boundaries related to pollution contaminations (anthropogenetic and natural) and help to resolve and quantify the impacts of saline/brackish water intrusion processes. The ratios such as Na/Cl, SO_4/Cl, Mg/Cl, Cl/HCO_3 + CO_3, Mg/Ca, Ca/HCO_3 + SO_4 are reported to help as an important indicator to resolve the boundaries between saline and brackish waters in different coastal systems. (Chen K.P. and Jiao, J.J., 2007; Jin-Yong Lee and Sung-Ho Song., 2007; Ahmed Zaharin Aris., et al, 2009; Ghabayen S.M.S., et al, 2006).

Based on several iterations of analytical data for different ionic ratios the following theoretical values to resolve the saline/brackish water boundaries have been adopted. (i) Na/Cl: saline water = 0.85, brackish water = 1.00; (ii) SO_4/Cl: saline water = 0.05, brackish water = 0.20; (iii) Mg/Cl: saline water = 0.15, brackish water = 0.35; (iv) Cl/ HCO_3 + CO_3: saline water = 6.6, brackish water = 0.50; (v) Mg/Ca: saline water = 3.5-5.00, brackish water = 1.00; (vi) Ca/ HCO_3 + SO_4: saline water = 1.00, brackish water = 0.50.

5.2.6 Water quality classification and hydrochemical facies

Classification and characterization of water quality is of paramount importance. A number of trilinear and rectangular diagrams are in vogue for this type of analysis and characterization. Palmer, C. (1911); Hill, R.A. (1940); Durov, S.A. (1948); Piper, A.M. (1944); Schoeller, H. (1962 & 1965) and Chadha, D.K. (1999) have been used for quite sometime by different workers in different basin areas. Of them Piper, A.M. (1953) has been widely used and help classifying in a major way and to understand how the different ions get combined and identify factors responsible to such mixing in different combinations. The piper trilinear diagram is shown in Fig. 5.5. Here the cations are expressed as percentages of total cations in milliequivalents per litre plot as a single point on the left triangle and similarly the anions are plotted as percentage of total anions on the right triangle. These two points are then projected into the top central diamond shaped area and represent the overall water compositional characteristic as a single point. This single point is uniquely related to the total ionic distribution. The point is normally represented by a circle with its area proportional to the total dissolved solids. This diagram has an added advantage of plotting number of samples of a well field and study distributional trends of different sites and their variations which in turn would help assessing the hydrochemical variations and factors responsible for such variations. Problems involving mixtures of natural waters, impacts of waters due to precipitation, base exchange relationships etc. can be solved through this diagram.

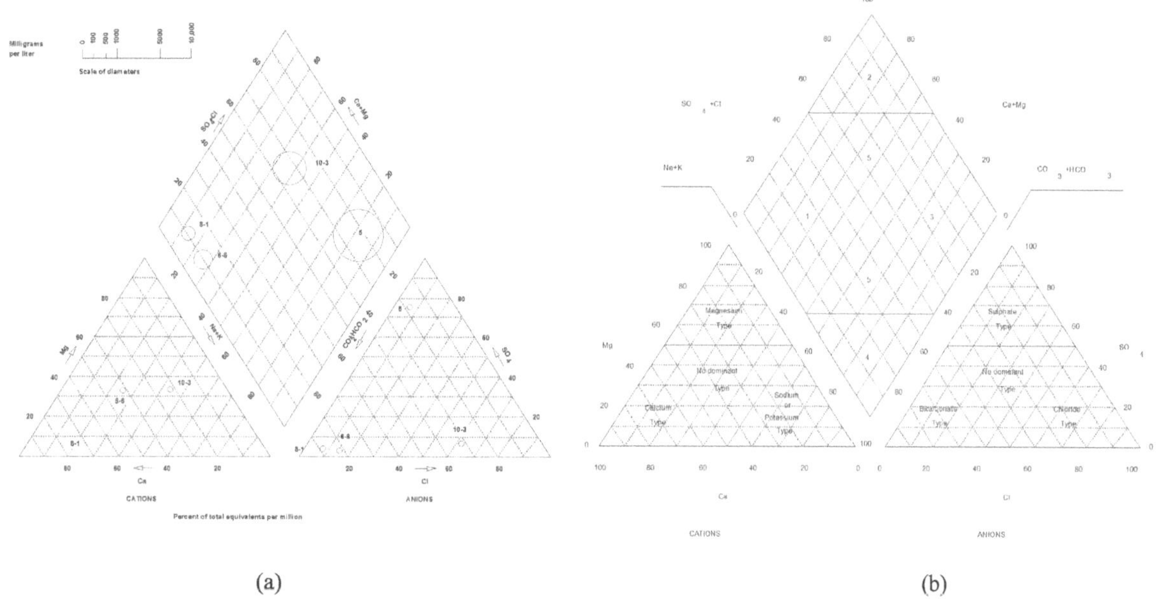

(a) (b)

Fig. 5.5 Trilinear diagram for representing analyses of groundwater quality (after Hem, J.D., 1989)

A simple and useful way of describing and identifying the water quality types and related hydro chemical facies was presented by Piper, A.M. (1953), the frame work of which was described earlier in methods and techniques. The above mentioned brief presents dominant cations and anions present in solution in a systematic triangular patterns and assimilates intimately representing the whole compositional pattern as a single point in the diamond shaped diagram as shown in Fig. 5.5(a). The dominant ions in triangular structure mean the ions whose concentrations exceed 50% of the total ions in solution. A diagram describing various types of dominant ions and the related water types and hydro chemical trends are shown in the Fig. 5.5 (b).

Apart from the various water types depicted in the base triangular diagrams on either side of the 'diamond shaped' segment of the plate, the following hydrogeochemical facies could be recognized.

Segment 1 - Represents combined concentrations of calcium, magnesium and bicarbonate ions exceeding 50% of the total dissolved constituents (characterized as $Ca^{2+} - Mg^{2+} - HCO_3$ type of waters)

Segment 2 - Represents combined concentration of alkali metals, sulphates and chlorides exceeding 50% of the total dissolved solids (characterized as $CA^{2+} - Mg^{2+} - Cl^{-} - SO_4^{2-}$ type of waters)

Segment 3 - Represents combined concentrations of alkali metals, sulphates and chlorides exceeding 50% of the total ions. ($Na^{+} - K^{+} - Cl^{-} - SO_4^{2-}$ type of waters)

Segment 4 - Represents combined Sodium, Potassium and Bi-carbonates exceeding 50% of the total ions ($Na^{+} - K^{+} - Cl^{-}$ - SO_4^{2-} type of waters)

Segment 5 - No specification – anions exceed 50% of the total ions.

5.2.7 Hydrogeochemical facies

The lateral variations in the hydro chemical characters of groundwaters in an aquifer or group of aquifers could be represented as maps by plotting individual site concentrates of any single ionic concentration or group of ionic concentrations or their ratios could be plotted on a map to study the regional water quality changes. Presentation of total dissolved solids and individual ions of different well sites in a map will help to understand the distributional trends in a more comprehensive way. This is done normally by drawing equal concentration lines between point to point to estimate its spatial variations. Similarly the ionic ratios and their trends such as HCO_3/Cl, Ca/Mg, SO_4/Cl and Na/Cl etc., could also be drawn to estimate relative dominance of each of the ion against others and synthesize the geochemical processes involved in such variations, while attempting such maps on a regional scale. It would be necessary to group them aquifer wise taking into consideration the homogeneity of the system and to obtain realistic chemical changes in simple aquifer systems.

5.2.8 Correlation matrix and factor analysis

Correlation between different cations and anions could be carried out using Pearson's correlation approach. A correlation matrix analysis is a bi variant method applied to describe the degree of relationship between different hydrochemical parameters. A high correlation coefficient (near 1 or –1) means a good relationship between two variables. A value near zero means no relationship between them at a significance level of less than 0.05. It could be said more precisely the parameters with values $r > 0.7$ are considered to be strongly correlated and values with r between 0.5 and 0.7 could be termed as "moderate correlation".

Factor analysis: Factor analysis is a statistical tool deployed widely for interpretation of hydrochemical data and the interrelationships between different parameters. The most important feature of factor analysis is its ability to reduce a large number of variables to a smaller number of factors. This facilitates to (i) to detect and identify groups of well correlated original variables, (ii) to reduce the number of variables under investigation, and (iii) to produce new combinations of original variables (groups) that can then be used as new variables in some further analysis (Davis, J.C., 1973). There are two types of factor analysis (i) R-mode factor analysis investigates the inter-relationships among the original variables and (ii) Q-mode factor analysis determines inter-relationships among the observations (samples) (Davis, J.C., 1973). Factor analysis takes data in a correlation matrix and rearranges in a manner that explains the structure of the chemical system. A correlation coefficient matrix measures the variances of each hydrochemical constituents and explains the relationships with each other. The variance/co variances in correlation coefficients of the variables are computed. The correlation coefficients are expressed by the equation,

$$r_{x,y} = \frac{\sum (x - x_m)(y - y_m)}{\sqrt{\left[\sum (x - x_m)\right]^2 \left[\sum (y - y_m)\right]^2}} \qquad \text{.....(5.21)}$$

In the above expression the correlation coefficients ($r_{x,y}$) is the sum of the products (samples) of the deviations of the x measurements and y measurements of each sample from the mean values of x and y respectively for a set of samples. Eigen values and Eigen vectors could be calculated from the co-variences matrix. Then the data can be transformed into factors. The eigen values and the percentages of variance associated with each factor could be summed up to express as a cumulative eigen value and percentage of variance. This criteria was proposed by Kaiser, H.F., 1958 and is widely used (Matalas, N.C. and Reicher, B.J., 1967; Davis, J.C., 1973; Reyment, R.A. and Joreskog, K.H., 1993).

In the present system analyses SPSS package described by Nie et al is adopted. The data was standardized according to the criteria proposed by Davis, J.C., 1973. This procedure facilitates to obtain a new rotated factor varimax where each factor is described in terms of different variables.

5.2.9 Water quality criteria

Groundwaters though free from suspended particles or turbidity, quality varies widely and therefore certain standards are prescribed for different uses such as drinking, industry and irrigation purposes.

Drinking water standards: A brief summary of different standards being adopted in India and the world over are summarized in Table 5.2. The drinking waters should be free from color, turbidity, odour and micro-bacterial hazards, and should contain low salinities, relatively soft and free from pollutants and poisonous materials. The world quality standards as well as Indian standards are presented in the Table 5.2.

Table 5.2 Standards for physical and chemical quality of drinking water

Quality	W.H.O International Standards, 1971		Indian Standards Institution, 1983	
	Highest desirable	Maximum permissible	Highest desirable	Maximum permissible
1	2	3	4	5
Physical				
Turbidity (JTU units)	5	25	10	25
Colour, Hazen-units (on platinum cobalt scale)	5	50	5	50
Taste and odour	Unobjectionable	-	-	Unobjectionable
Chemical				
pH	7.0-8.5	6.5-9.2	6.5-8.5	6.5-9.2
Total Dissolved Solids (mg/l)	500	1500	500	1500
Total hardness as $CaCO_3$ (mg/l)	100	500	300	600
Calcium (mg/l)	75	200	75	200
Iron (as Fe) (mg/l)	0.05	1.5	0.3	1
Manganese (as Mn)(mg/l)	0.1	1	0.1	0.5
Copper (as Cu) (mg/l)	0.05	1.5	0.05	1.5
Zinc (as Zn)(mg/l)	5	15	5	15
Chloride (mg/l)	200	600	250	1000
Sulphate (mg/l)	200	400	150	Up to 400 if Mg does not exceed 30 mg/l
Phenolic substances (as phenol)(mg/l)	0.001	0.002	0.001	0.002
Fluorides (mg/l)	0.6-0.9	0.8-1.78	0.6-1.2	1.5
#Nitrates (mg/l)	-	45#	45	No relaxation
Toxic constituents				
*Arsenic(mg/l)	-	0.05	0.05	do-
*Mercury(mg/l)	-	0.0001	0.001	do-
*Cadmium(mg/l)	-	0.01	0.01	do-
*Chromium(hexavalent)(mg/l)	-	-	0.05	do-
*Cyanide(as CN)(mg/l)	-	0.05	0.05	do-
*Lead(mg/l)	-	0.1	0.1	do-
*Selenium(mg/l)	-	0.01	0.01	do-
Radioactivity				
Gross alfa-emitters(Pci/l)	-	3	-	-
Gross beta emitters(Pci/l)	-	30	-	-

In areas where the nitrate concentration is known to be in excess of the stated concentration, the public should be warned of the potential danger of using the water for infant feeding.

* The concentrations listed under columns3 and 4, against toxic constituents, are mandatory upper limits and should not be exceeded.

5.3 METHODOLOGIES FOR ASSESSING SUITABILITY FOR IRRIGATION

The important characteristics/properties of water to be considered for irrigation in general are electrical conductivity, salinity, sodium percentage, residual sodium carbonate and sodium adsorption ratio.

5.3.1 Electrical Conductivity (EC)/Specific conductance and sodium percentage (Na%)

Electrical conductivity (EC) and sodium concentrations may be considered as important factors to classify irrigation water. The groundwater used for irrigation normally contains measurable quantities of dissolved substances called salts. They include relatively small but important amount of dissolved solids originating from weathering of rocks and soils and also from dissolution of lime, gypsum and other salt sources as water flows over these terrains and percolates as groundwaters. The salts present in the groundwater affects the growth of plants directly and also the soil structures. The sodium percentage (Na %) in groundwaters can be calculated by the following equation.

$$\%Na = \frac{(Na+K)100}{Ca+Mg+Na+K} \quad .....(5.22)$$

where all ionic concentrations are expressed in milli equivalents per litre

Wilcox, L.V. (1955) has come out with a classification of groundwater for irrigation purposes by way of plotting electrical conductivity against % sodium as presented in Figure 5.6 and classified into different categories of water based on sodium percentage.

(i) Excellent (< 20); (ii) Good (20-40); (iii) Permissible (40-60); (iv) Doubtful (60-80) and (v) Unsuitable (>80).

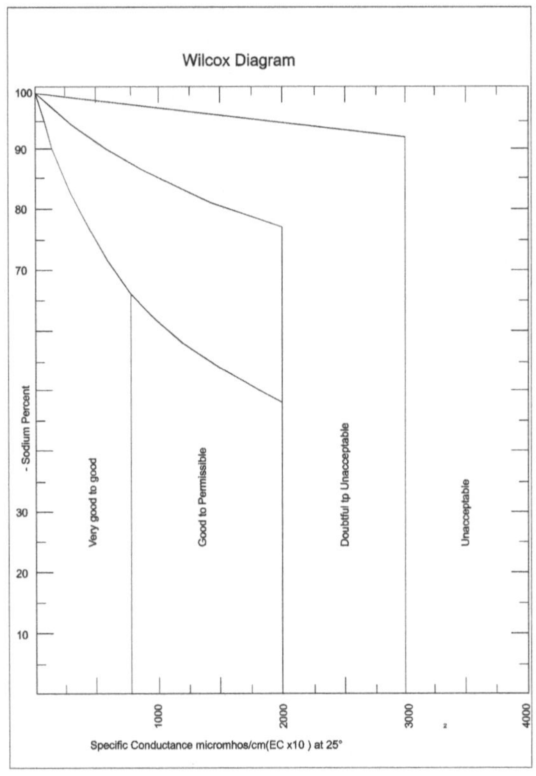

Fig. 5.6 Wilcox diagram

5.3.2 Estimations of Residual Sodium Carbonates (RSC) in groundwater

Precipitation of Ca and Mg in groundwater may occur when the sum of carbonates and bicarbonates are found in excess of calcium and magnesium. To quantify the effects of carbonate and bicarbonate, residual sodium carbonate (RSC) can be computed by the following equation.

$$RSC = (HCO_3 + CO_3) - (Ca + Mg) \text{ in milli equivalent per liter} \quad .....(5.23)$$

A high value of RSC in groundwater leads to an increase in the adsorption of sodium on soil (Eaton, 1950). Irrigation waters having RSC value > 2.5 meq l^{-1} are considered unsuitable for irrigation and values less than 1.25 meq l^{-1} are considered good for irrigation purposes. The values in between 1.25 and 2.5 are however considered as of doubtful nature.

5.3.3 Alkali and salinity hazard (SAR)

The US salinity laboratory of dept. of Agriculture (1954) recommends sodium adsorption ratio (SAR) as shown in figure and could be estimated as the following formula.

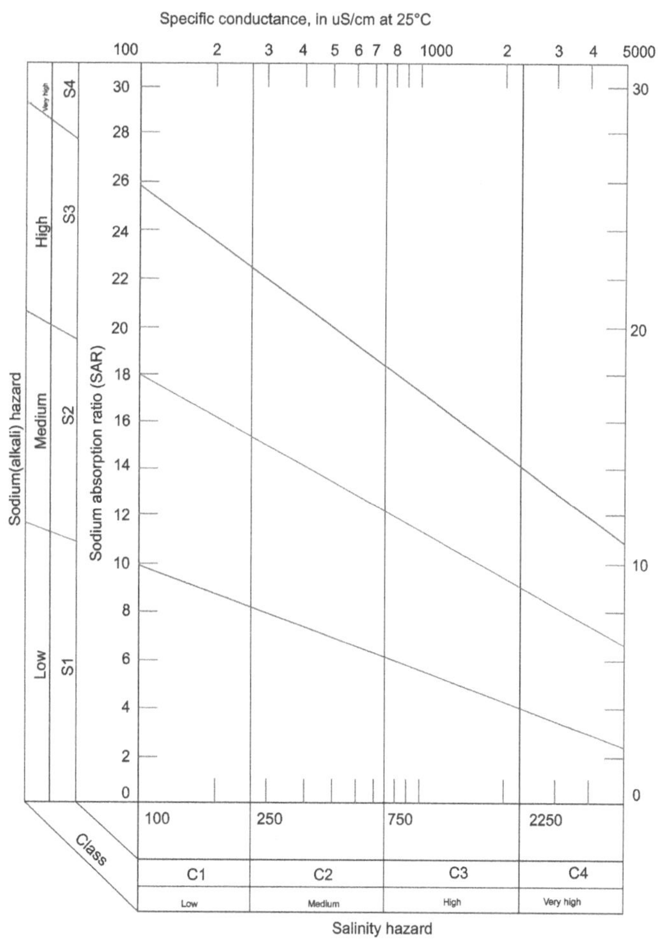

Fig. 5.7 Diagram for classification of irrigation water (after Richards)

$$SAR = \frac{Na}{\sqrt{(Ca+Mg)/2}} \quad .....(5.24)$$

where the concentrations are expressed in milliequivalents per liter.

The salinity or total concentrations of soluble salts can be expressed for the purpose of classification of irrigation waters as low (EC = < 250 μS cm^{-1}), medium (250-750 μS cm^{-1}), high (750-2250 μS cm^{-1}) and very high (2250-5000 μS cm^{-1}) while high salinity (EC) leads to formation of saline soils. A high sodium concentration leads to the formation of alkaline soils (Richards, 1954). The sodium or alkali in the irrigation waters is determined by absolute and relative concentration of cations expressed in the above formula of SAR. The plot values in the US salinity diagram where EC is taken as salinity hazard and SAR as alkalinity hazard demonstrates different categories of waters lying in the fields such as C_2S_1, C_3S_1, C_3S_2 and C_3S_3 indicating medium high salinity and low to high alkali waters as shown in Fig. 5.7 (Richards, L.A., 1954). High salinity waters can't be used as soils with poor drainage patterns and require special management practices such as salinity control and selection of salt tolerant plants. Low sodium waters (S_1) can be used for the irrigation on almost all soils with little danger of harmful affects. Medium sodium waters (S_2) will present an appreciable sodium hazard in fine textured soils and with high cation exchange capacity. However this water can be used on coarse textured soils with good permeability. High sodium water (S_3) may produce harmful effects of exchangeable sodium in most soils and require special management practices.

5.4 WATER CHEMISTRY, CHARACTERIZATION, HYDROGEO-CHEMICAL FACIES AND MULTIVARIATE TREND ANALYSIS OF GROUNDWATERS (NEAR SURFACE AND DEEP) IN PALAR, PONNAIYAR-VELLAR, CAUVERY AND VAIGAI DELTA SYSTEMS OF TAMIL NADU

The chemical composition of groundwaters in coastal deltas in principle is a function of precipitation, soil structure and characteristics. Infiltration and recharge, rock water interactions, groundwater flow and velocities, anthropogenic, natural and sea water intrusion impacts. The aquifer medium can play a major role in the chemical reactions such as dissolution, ion exchange, oxidation-reduction and adsorption processes. (Richter, B.C and Kreitlar, C.W., 1993; Wicks, C.M., et al 1995). In addition the groundwater draft and sea water intrusion processes in coastal systems could bring in a major change in the groundwater chemistry. Different approaches could be adopted to understand the hydrochemical processes such as (i) by way of drawing the aereal distribution of chemical parameters (ii) analyzing scatter plots between different parameters (iii) obtain hydrochemical facies changes through different graphical analyses such as Piper and Durov diagrams and (iv) through adoptation of multivariate statistical techniques/factor analysis.

Majority of the Indian studies revolve around pre and post monsoon seasonal variations and related hydrogeochemical processes, sporadically distributed and localized, in different parts of coastal regions. (Mondal N.C, et al; 2010; Raju, N.T, 2006; Rajmohan N.J, and Elongo, L (2004); Subbarao; N; (2006, Senthil Kumar G et al; (2008) and many others.

In the following paragraphs attempts have been made to apply the above methodologies for the near surface and deep systems of the coastal delta complexes of Tamil Nadu (Palar, Ponnaiyar-Vellar, Cauvery and Vaigai); Andhra Pradesh (Pennar, Krishna and Godavari); Orrissa (Mahanadi-Brahmini-Baitarni); West Bengal (Bhagiradhi-Hoogly) and Gujarat(Narmada- Tapti). Majority of the samples represented and analyzed (near Surface and deep systems) are picked up from the National hydrographic network stations and basic well data reports of CGWB, State Groundwater agencies of the respective regions and the results of the R&D programmes organized by NGRI. Majority of the data selected for analyses belong to the pre-monsoon periods. The chemical data trends for different coastal delta systems are described in terms of the following.

1. Spatial variations of specific conductance and statistical variations of different ions. Ionic cross plots in relation to total dissolved solids, selected ionic ratios and their cross plots in relation to chlorides

2. Trilinear plots, hydrochemical facies and regional trend analyses
3. Correlation matrices and factor analyses
4. Classification of groundwaters and irrigation suitabilities

5.4.1 Spatial variations of specific conductance, pH and statistical variations of different ions in Palar, Ponnaiyar-Vellar, Cauvery and Vaigai delta systems of Tamil Nadu

Palar delta region: The sample locations for both near surface and deep systems in respect of Palar delta region are shown in Figure 5.8(a).(Data source: CGWB: 1991; 1993; 2003; Sivashanmuganathan, K., 2008) The analytical results of specific conductance and pH in terms of max, min, mean and standard deviation values are presented in Table 5.3(a). The specific conductance values of near surface system of Palar delta region have a range varying from 4340 μ mhos/cm to 190 μ mhos/cm with an average value of 1255 μ mhos/cm and the deep systems have a range of values varying from 5000 μ mhos/cm to 422 μ mhos/cm with an average value of 1392 μ mhos/cm. The delta system is characterized by moderate to high levels of conductivity. A relatively moderate to high levels of

Fig. 5.8(a) Sample locations for both near surface and deep systems in Palar delta

conductivities are observed in the outfall regions of Araniyar (near surface > 3000 μ mhos/cm) and in the outfall regions of Cooum (deep systems > 2000 μ mhos/cm). The upstream areas are characterized by values < 500 μ mhos/cm (near surface) and < 1000 μ mhos/cm (deep systems). The values by and large reflect on the regions with high conductivity levels where salinisation processes are believed to be more intense such as the near surface systems in the outfall regions of Araniyar and the deep systems in the outfall regions of Cooum delta basin. The intensities of saline water intrusion processes in this region appear to be proportionally disposed in line with the groundwater developmental intensities. The variabilities/trends of specific conductance are shown in Fig. 5.8(b).

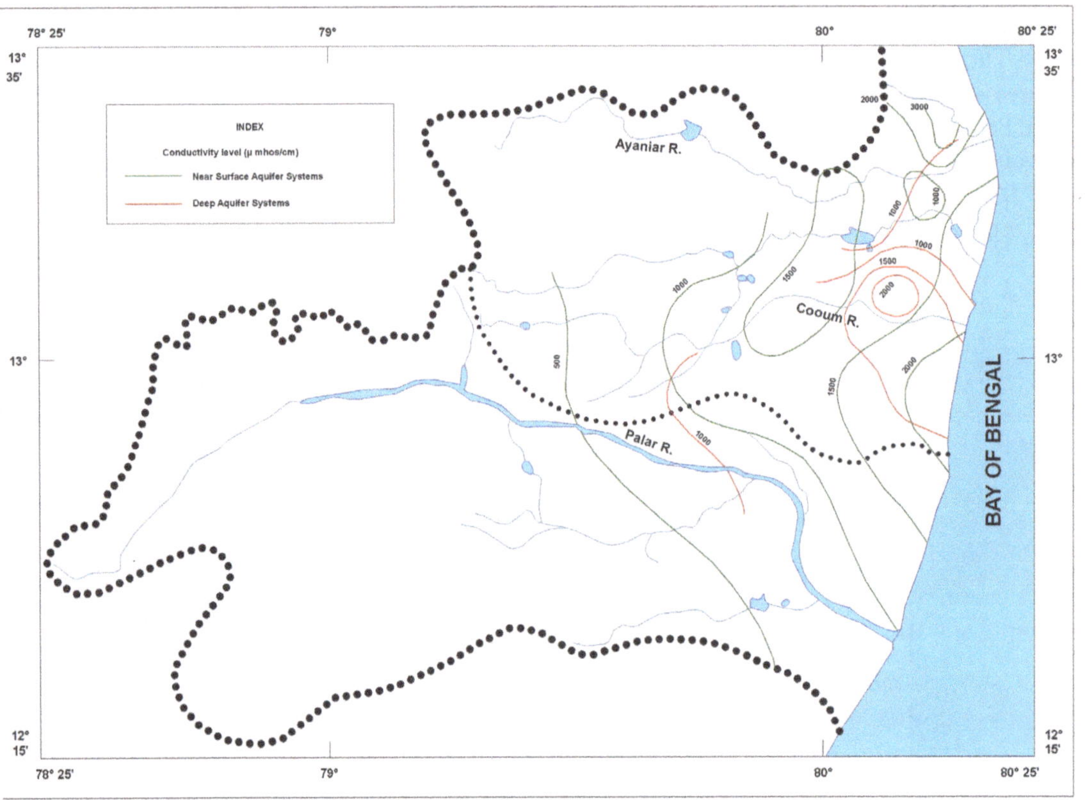

Fig. 5.8(b) Specific conductance trends in Palar delta

Table 5.3(a) The variations of specific conductance and pH along with different cations and anions of Palar delta region in Tamil Nadu

Parameter	Near Surface				Deep systems			
	Max.	**Min.**	**Mean**	**Std. deviation**	**Max.**	**Min.**	**Mean**	**Std. deviation**
pH	9.20	7.00	8.19	0.60	8.3	7.1	7.74	0.38
EC	3741.00	147.70	1184.30	875.40	5000	422	1392.6	996
Ca	216.00	5.00	53.60	44.40	154	4	48.4	40.9
Mg	151	1.2	32.8	32.4	106.00	1.20	26.13	26.16
Na	570.00	12.00	147.10	124.10	544.00	40.00	191.20	121.50
K	240.00	0.00	29.30	50.70	10.17	0.80	316.40	147.60
HCO_3	730	37	266.7	170.26	730.00	113.00	316.40	147.60
SO_4	419.8	0	76.3	92.27	250	0	79.18	73.73
Cl	1321	21	219	255.5	936.00	42.55	209.70	204.80

The pH values for near surface systems have a range varying between 9.2 to 7 with an average value of 8.19 and between 8.3 to 7.1 for deep systems with an average value of 7.74. The higher pH values for both near surface and deep systems are observed in the vicinity where intense salinisation processes are taking place particularly in the outfall regions of Araniyar and Cooum rivers.

Ponnaiyar-Vellar delta region: The sample locations for both near surface and deep systems in respect of Ponnaiyar-Vellar delta region are presented in Figure 5.9 (a). (Data source: CGWB, 1993; Chakrapani, R., 1989) The analytical results of specific conductance in terms of max, min, mean and standard deviations are presented in 5.3(b). The specific conductance values for near surface system have a range varying from 8240 μ mhos/cm to 100 μ mhos/cm with an average value of 1785 μ mhos/cm and deep systems have a range of values varying from 3370 μ mhos/cm to 107 μ mhos/cm with an average value of 826.7 μ mhos/cm. The specific conductance contour levels for near surface system indicate low levels all along the Ponnaiyar river up to the shore line boundary and increase moderately towards north and south in the vicinities of Gingee and Vellar basins and in their respective outfall areas. These observations reflect on the increasing levels of salinisation processes on either side of the Ponnaiyar delta basin. The deep freshwater bodies however appear to be relatively free from the salinisation processes but however isolated contour levels > 2000 μ mhos/cm indicate some amount of

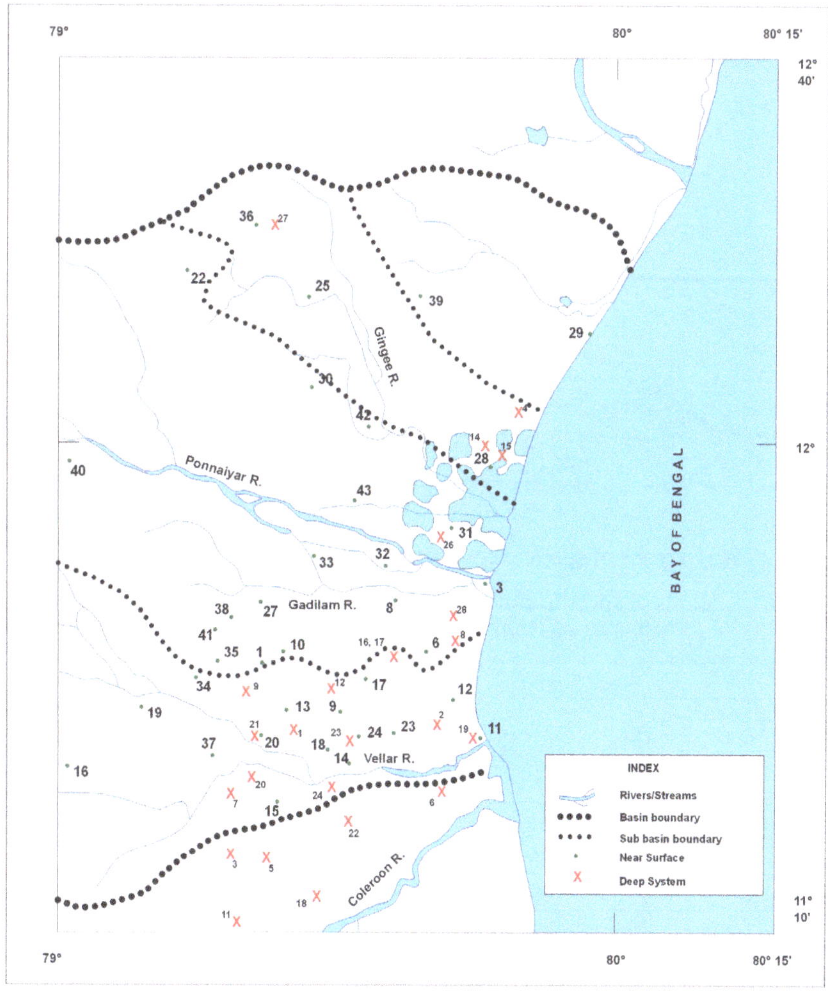

Fig. 5.9(a) Water quality sample locations in Ponnaiyar-Vellar delta region

saline water mixing near the coastal margins particularly in the Vellar outfall region. The groundwater developmental activities in the Vellar basin needs to be checked in this context. The variabilities/trends of specific conductance are shown in Fig. 5.9(b).

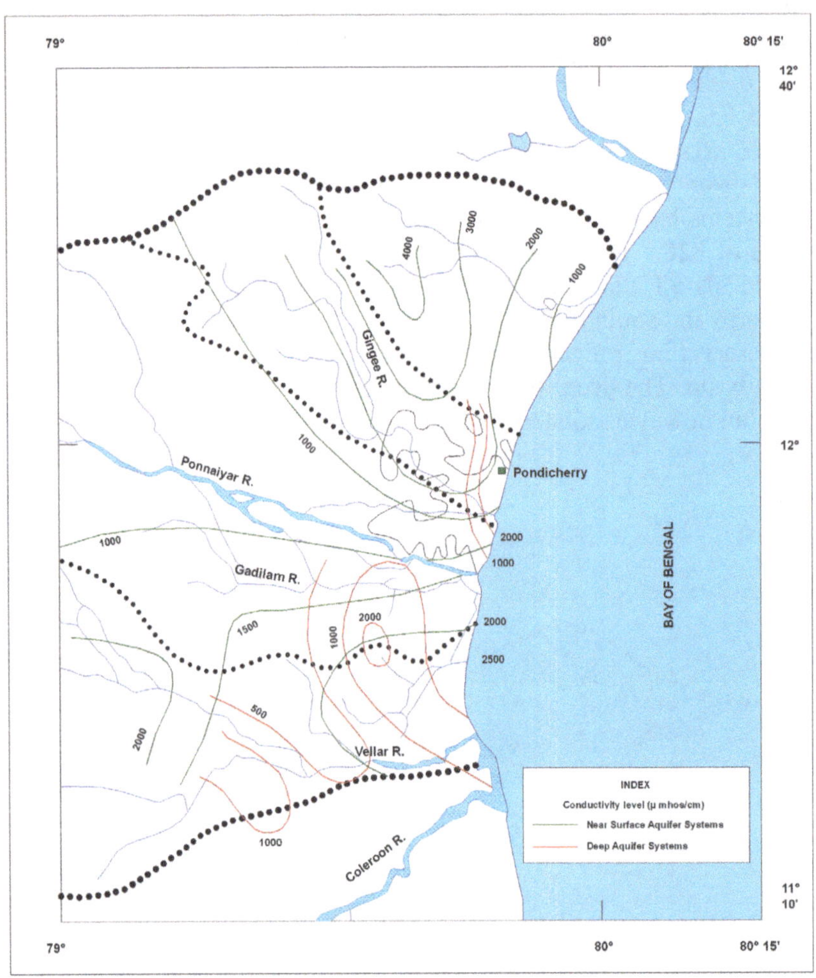

Fig. 5.9(b) Trends of specific conductance in Ponnaiyar-Vellar delta region

Table 5.3(b) The variations of specific conductance and pH along with different cations and anions of Ponnaiyar-Vellar delta region in Tamil Nadu

Parameter	Near Surface				Deep systems			
	Max.	Min.	Mean	Std. deviation	Max.	Min.	Mean	Std. deviation
pH	9.50	7.60	8.26	0.34	8.30	5.20	7.30	0.93
EC	8240.00	100.00	1654.40	1636.10	2447.69	107.00	738.86	643.45
Ca	280.00	8.00	80.70	66.23	229.40	6	44.8	43.6
Mg	236	3.6	41.2	45.38	35	2.00	13.60	8.77
Na	1357.00	8.70	210.30	258.60	450.00	5.00	95.30	121.70
K	172.00	0.00	18.80	32.37	20.00	0.00	5.60	3.96
HCO_3	540	14	220.3	133.2	474.00	19.00	203.30	130.10
SO_4	480	2.4	108	119.6	206.7	2.1	50	57
Cl	2425	18	387.8	487.1	843.00	8.40	123.30	190.10

The pH values for near surface systems have a range of values varying between 9.5 and 7.58 with an average of 8.25 indicating the tendency of water towards alkalinity. The deeper systems have a range of values varying between 8.3 to 5.2 with an average value of 7.9 reflecting their position in a neutral range. The above observations indicate that the intensity/mixing processes with saline water are more in the near surface systems than in the deeper systems of Ponnaiyar-Vellar delta regions.

Cauvery delta region: The sample locations for both near surface and deep systems in respect of Cauvery delta are presented in Figure 5.10(a). (Data source: CGWB: 1993; 1997; 2006). The analytical results of pH and conductivity in terms of max, min, mean and standard deviations are presented in Table 5.3(c). The specific conductance values for near surface systems have a range of values varying from 11,940 μ mhos/cm to 426 μ mhos/cm with an average value of 2399.5 μ mhos/cm and deep systems have a range of values varying from 4070 μ mhos/cm to 156 μ mhos/cm with an average value of 1096.5 μ mhos/cm. The trend variations of specific conductance for both near surface and deep systems by and large have low levels in the upstream areas (near surface 1000 μ mhos/cm and deep systems 500 μ mhos/cm) and increase gradually towards the outfall region (near surface 4000 μ mhos/cm and deep systems 4000 μ mhos/cm). The trends also indicate relatively high levels of conductivities and salinisation processes in the southeastern side of the basin, i.e. outfall regions of Vellar sub distributary and new delta system. The variabilities/trends of specific conductance are shown in Fig. 5.10(b).

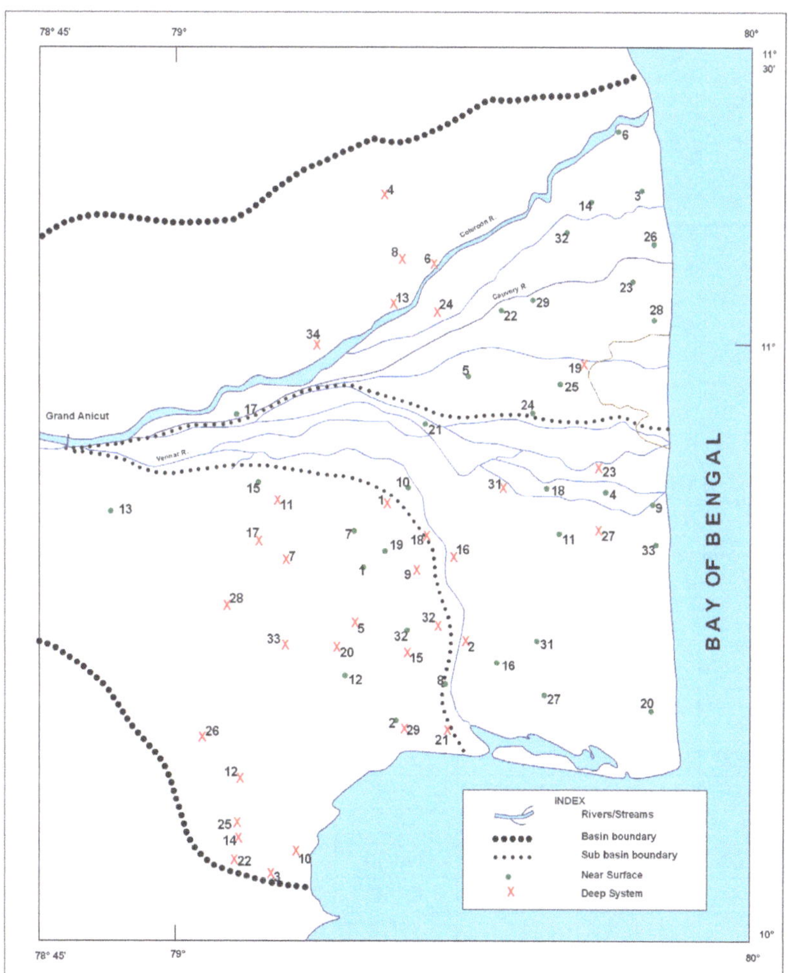

Fig. 5.10(a) Water quality sample locations in Cauvery delta region

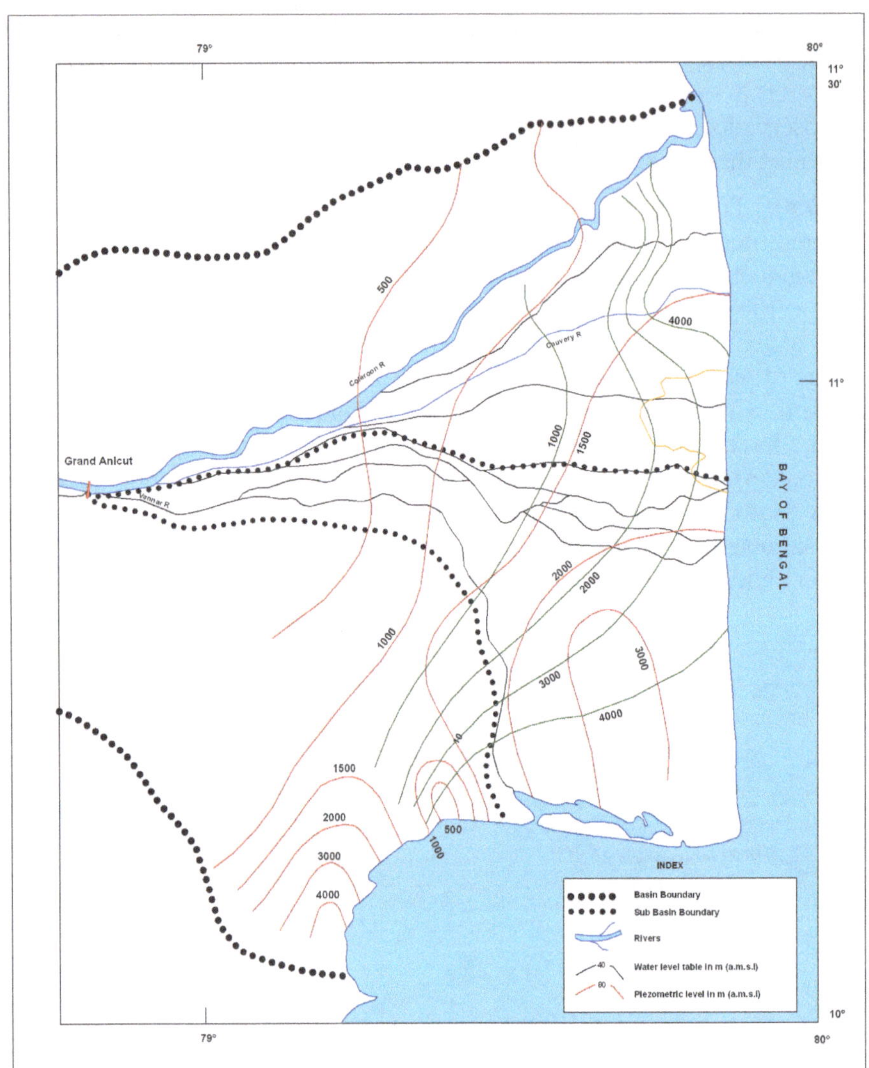

Fig. 5.10(b) Trends of specific conductance in Cauvery delta region

Table 5.3(c) The variations of specific conductance and pH along with different cations and anions of Cauvery delta region in Tamil Nadu

Parameter	Near Surface				Deep systems			
	Max.	Min.	Mean	Std. deviation	Max.	Min.	Mean	Std. deviation
pH	9.00	7.60	8.20	0.32	9.00	5.95	7.62	0.79
EC	11940.00	289.20	2247.50	2510.40	2770.77	298.80	994.38	613.02
Ca	304.00	8.00	70.50	66.20	166.00	8.00	52.6	31.9
Mg	352	4.9	50.3	69.2	64.8	3.4	20.10	13.50
Na	1725.00	20.00	320.50	390.50	447.00	14.00	130.30	128.70
K	469.00	0.78	43.40	85.50	115.00	0.00	12.60	19.80
HCO_3	1055	116	349.2	195.7	505.10	64.00	219.40	112.90
SO_4	720	0.05	76.2	159.38	201.6	3	57.67	53.47
Cl	3261	39	532.6	731	822.70	19.00	202.50	222.97

The pH values for near surface systems have a range of values varying between 9 and 7.6 with an average value of 8.27 and deep systems have a range of values varying between 9 and 5.95 with an average value of 7.61. The near surface aquifer systems tend more towards alkalinity indicating more dominant mixing of saline waters and the deep systems with values tending more towards neutrality.

Vaigai delta: The sample locations for both near surface and deep systems in respect of Vaigai delta are presented in Figure 5.11(a). (Data source: CGWB, 2002). The analytical results of pH and conductivity in terms of max, min, mean and standard deviations are presented in Table 5.3(d). The specific conductance values for both near surface and deep systems and their regional trends are presented in Fig. 5.11(b). The near surface systems have a range of values varying from 20,000 μ mhos/cm to 570 μ mhos/cm with an average value of 3401.7 μ mhos/cm. Similarly the deep systems have a range of values varying from 16050 μ mhos/cm to 925 μ mhos/cm with an average value of 3068.2 μ mhos/cm. The specific conductance contour levels for both near surface and deep systems exhibit rather moderate to high levels of salinity from 1000 to 4000 μ mhos/cm and 2000 to 10,000 μ mhos/cm respectively. The near surface systems exhibit low to moderate levels of conductance in the central part of the delta but increases gradually towards the outfall region while the deep systems exhibit low to moderate levels of conductance in the northeastern segments (Manimuttar delta region) and gradually increase towards the south of Vaigai river/delta system and also towards the outfall region of Gundar river.

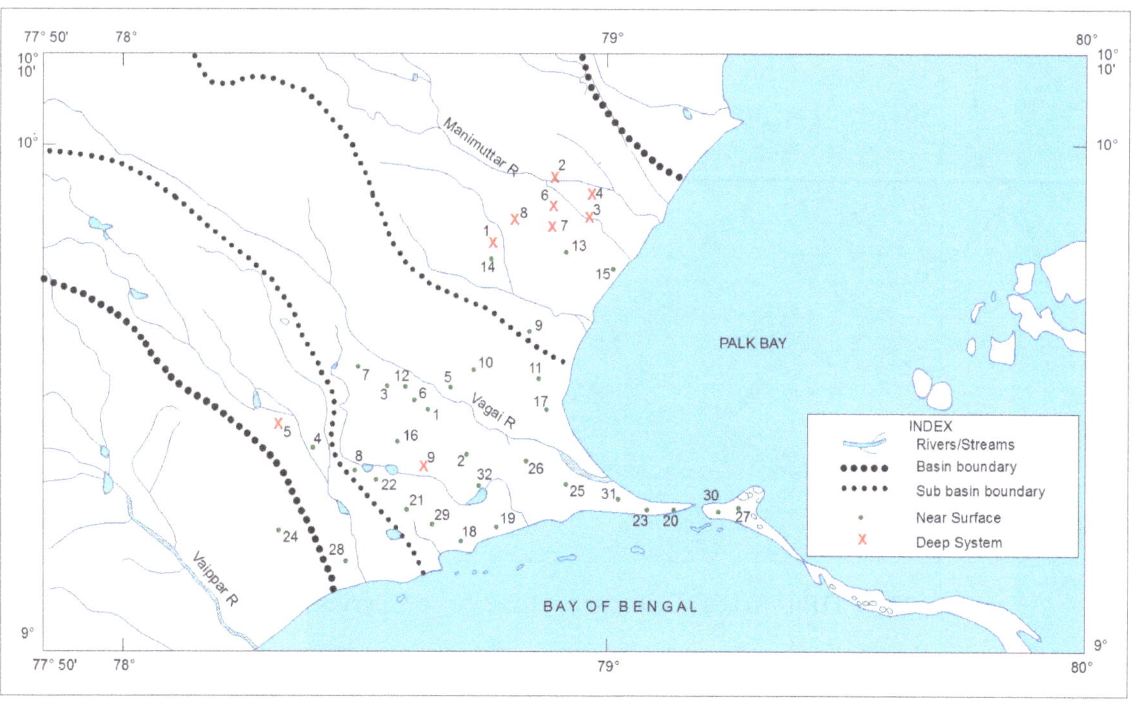

Fig. 5.11(a) Water quality sample locations in Vaigai delta region

The pH values for near surface have a range varying from 8.9 to 7.73 with an average value of 8.39 and deep systems have a range from 8.4 to 6.9 with an average value of 7.76. The data reflect on the near surface systems tending towards more alkalinity while the deep systems by and large maintain neutrality.

Table 5.3(d) The variations of specific conductance and pH along with different cations and anions of Vaigai delta region in Tamil Nadu

Parameter	Near Surface				Deep systems			
	Max.	Min.	Mean	Std. deviation	Max.	Min.	Mean	Std. deviation
pH	8.90	7.73	8.40	0.31	8.40	6.90	7.77	0.43
EC	20000	570.0	3401.0	4428.00	16050.0	925.0	3014.6	4896.72
Ca	858.00	14.00	96.60	164.80	808.4	10	135.7	253.3
Mg	525	11	78.09	102.04	102.1	6.08	28.24	30.49
Na	3105.0	46.00	532.10	749.80	85	2.7	13.70	26.80
K	168.00	1.00	30.01	42.27	85	2.7	13.70	26.80
HCO_3	835	79	327.2	165.2	341.00	237.9	294.40	35.40
SO_4	1200	9.6	152.21	220.22	860	40	175.4	258.7
Cl	6736	4	892.8	1534.9	5567.20	122.3	856.60	1768.00

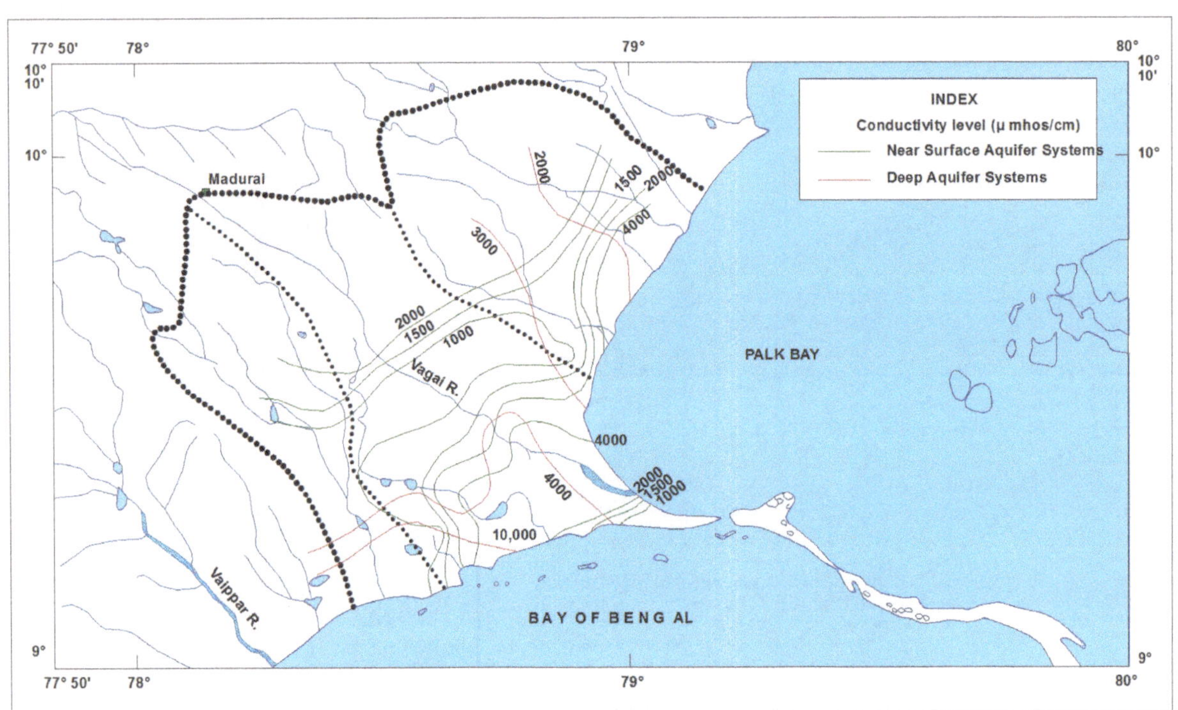

Fig. 5.11(b) Trends of specific conductance in Vaigai delta region

5.4.2 Cation-anion chemistry/Ionic ratios/Cross plots

The cationic and anionic abundances for near surface and deep systems in terms of max, min, mean and standard deviations for different delta systems of Tamil Nadu are presented in Tables. 5.3 (a),(b),(c) and (d). The sequential order of abundance of cations and anions (near surface and deep) in these four delta systems of Tamil Nadu are as follows.

Palar delta

(a) *Near surface systems:* $HCO_3 > Cl > SO_4$ and Na > Ca > Mg

(b) *Deep systems:* $HCO_3 > Cl > SO_4$ and Na > Ca > Mg

Cauvery delta

(a) *Near surface systems:* HCO_3 > Cl > SO_4 and Na>Ca>Mg

(b) *Deep systems:* HCO_3 > Cl > SO_4 and Na > Ca > Mg

Ponnaiyar-Vellar delta

(a) *Near surface systems:* HCO_3 > Cl > SO_4 and Na>Ca>Mg

(b) *Deep systems:* HCO_3 > Cl > SO_4 and Na > Ca > Mg

Vaigai delta

(a) *Near surface systems:* Cl > HCO_3 > SO_4 and Na > Ca > Mg

(b) *Deep systems:* Cl > HCO_3 > SO_4 and Na > Ca >Mg

The bicarbonate (HCO_3) and sodium (Na) are the dominant ions amongst the anions and cations respectively in all the delta systems excepting Vaigai where chloride (Cl) anion dominates in the anion sequence of the system.

Variations of cations and anions with respect to TDS: The maximum and minimum variations and the standard deviation in terms of different water quality parameters for both near surface and deep systems are presented earlier in Tables 5.3(a,b,c,d). The variabilities of cations and anions in relation to TDS (Total Dissolved Solids) for both near surface and deep systems for Palar, Ponnaiyar-Vellar, Cauvery and Vaigai deltas are presented below in Fig. 5.12(a,b,c,d).

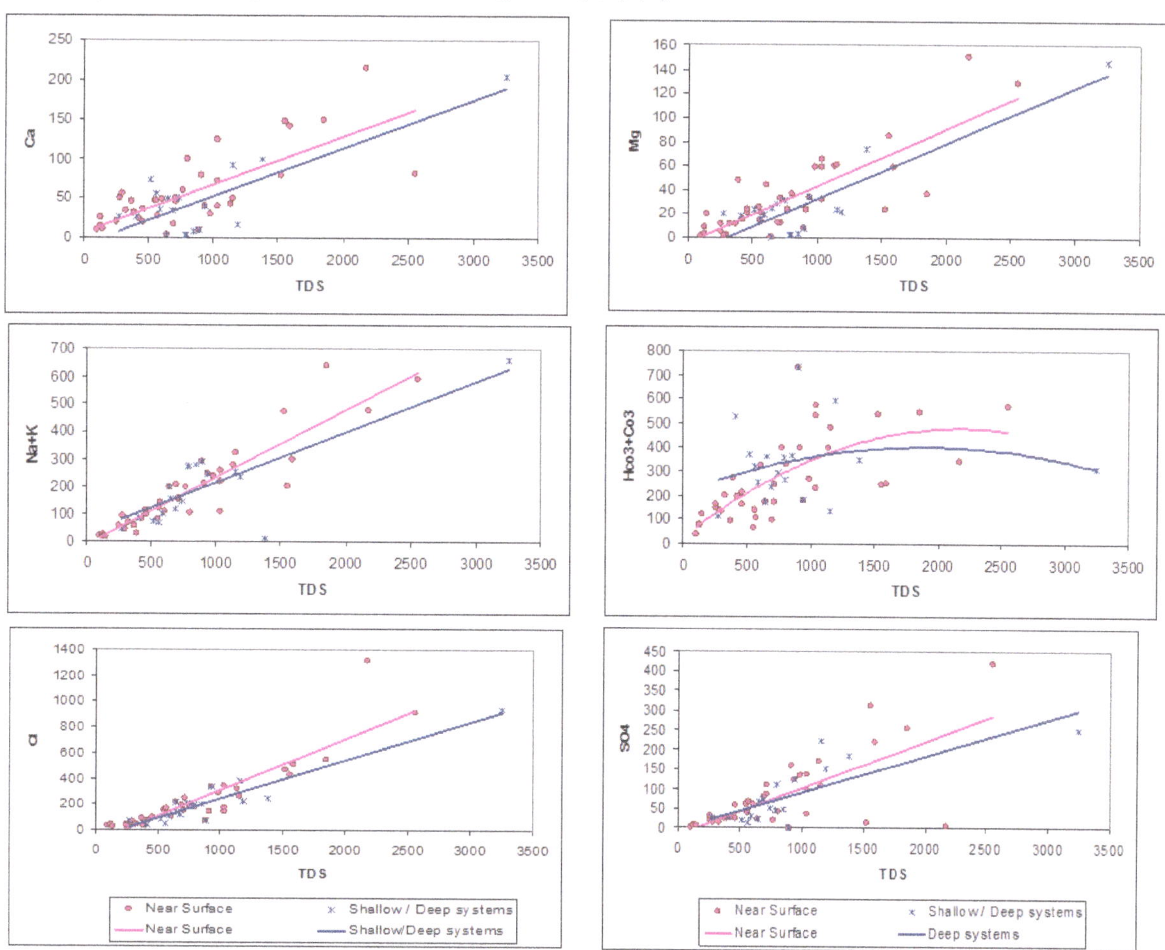

Fig. 5.12(a) Variations of cations and anions in relation to TDS in Palar delta

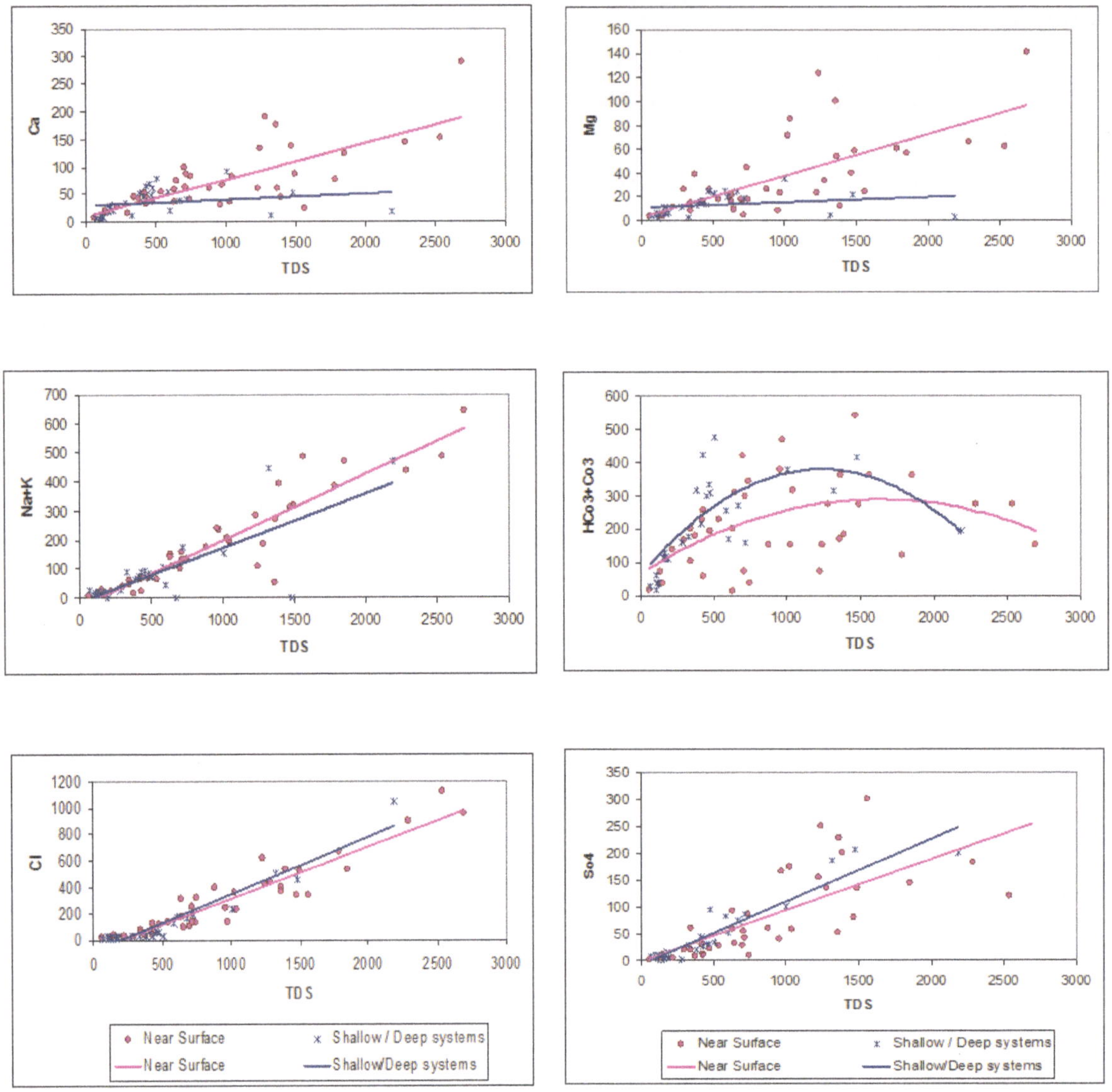

Fig. 5.12(b) Variations of cations and anions in relation to TDS in ponniyar

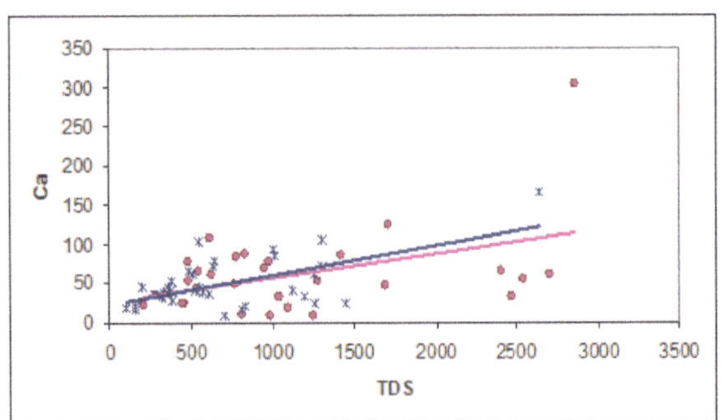

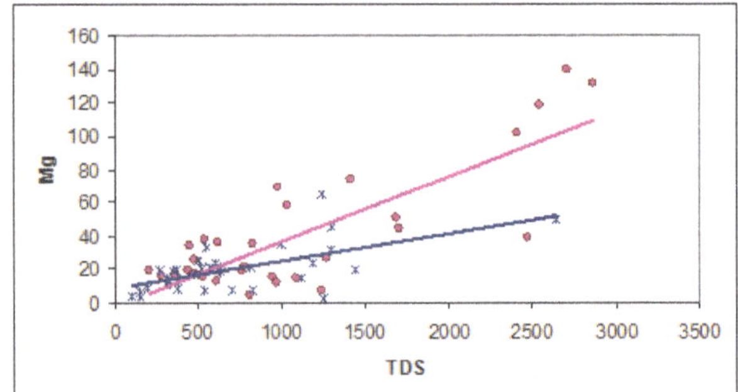

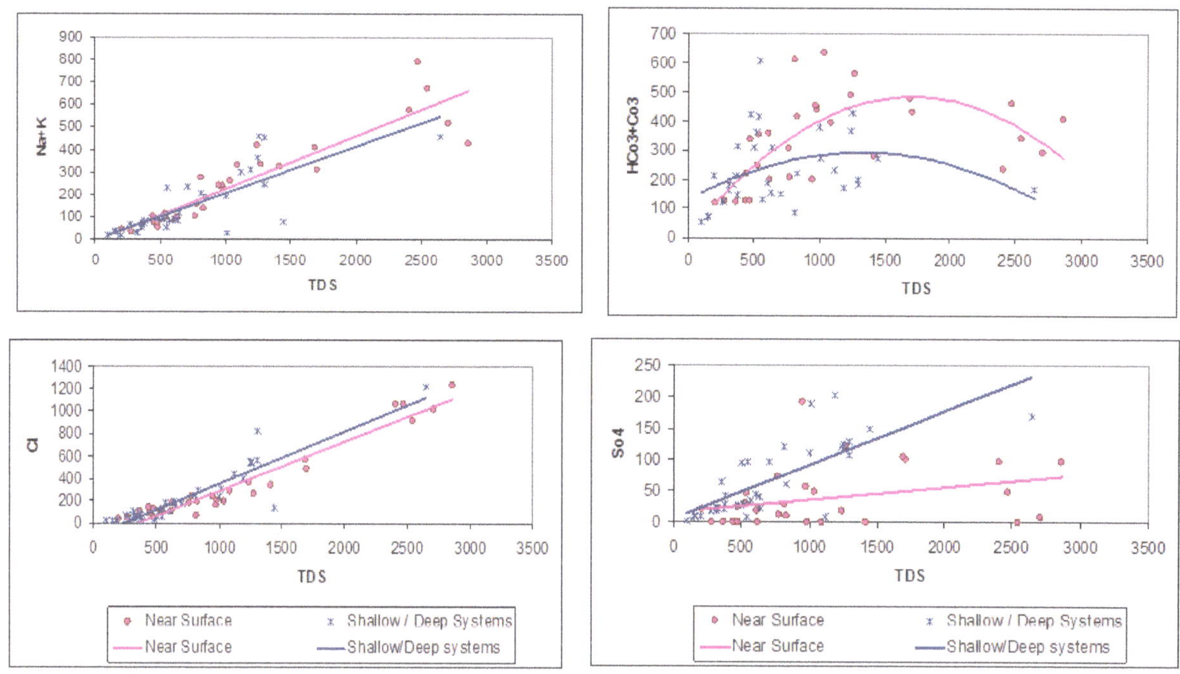

Fig. 5.12(c) Variations of cations and anions in relation to TDS in Cauvery

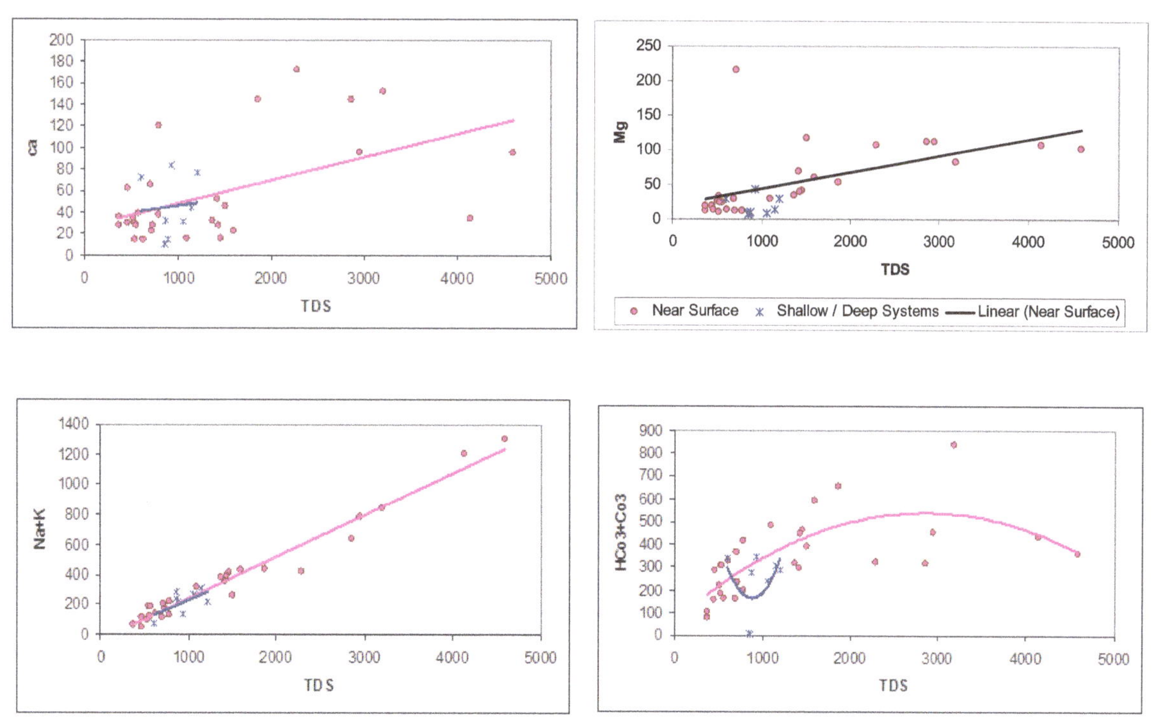

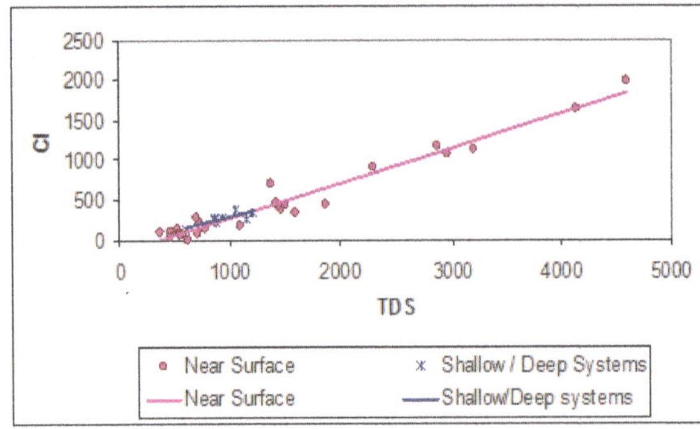

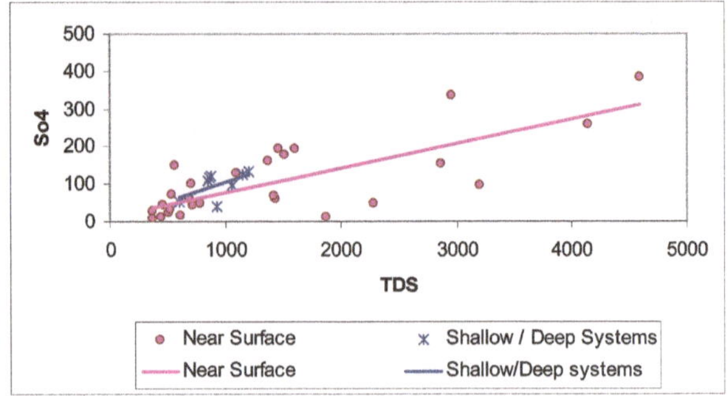

Fig. 5.12(d) Variations of cations and anions in relation to TDS in Vaigai delta

- The cross plots in general for all the delta systems (Palar, Ponnaiyar-Vellar, Cauvery and Vaigai) show an ascending linearity for both near surface and deep systems excepting in the case of HCO_3 vs. TDS where the trends are characterized by a curve linear relationship by and large in all the delta systems with marginal variations in shapes. The curve trends in general show an ascending order up to 1000-1500 ppm levels of TDS and then show declining levels gradually. The ascending levels indicate the region of fresh water non mixing area and the descending line reflects on the saline fresh water mixing process. All delta systems by and large exhibit non mixing trends up to 1000-1500 ppm TDS levels and there after show mixing trends for both near surface and deep systems.
- Trends of Ca and Mg in the case of Palar delta for both near surface and deep systems go parallel with each other reflecting similar involvement of hydrochemical processes in both near surface and deep systems with increasing levels of salinity. However in the case of Ponnaiyar-Vellar, Cauvery and Vaigai the trends for both near surface and deep systems initially (between 500-1000 ppm levels of TDS) gets converged along single line but gradually gets diverged with different slope characteristics with increasing levels of salinity.
- The Na + K trend values in Palar, Ponnaiyar-Vellar and Cauvery show convergence of points in respect of both near surface and deep systems upto 1000 ppm of TDS levels and then gets diverged with different slope characteristics with increasing levels of salinity. However in the case of Vaigai the mean trend values between Na + K vs TDS for both near surface and deep systems lie in a single line with increasing levels of salinity reflecting similar hydrochemical processes in both the systems.
- The SO_4 and Cl anions in both near surface and deep systems exhibit linear trend relationship with increasing levels of TDS and with different slope characteristics. The degree of gradient reflect on the intensity/involvement of respective anion.

Ionic ratios/Cross plots:

- The specific conductance / total dissolved solids / ionic ratios between selected cations and anions would help to estimate and assess the intensity of fresh and saline water mixing processes. The max , min and avg values of selected number of ionic ratios for the near surface and deep systems of Palar, Ponnaiyar-Vellar, Cauvery and Vaigai delta systems are furnished in the Table 5.4(a,b).

  The average Na/Cl ratios for near surface/deep systems in respect of Palar, Ponnaiyar, Cauvery and Vaigai are 1.40/1.80, 0.98/1.56, 1.22/1.25 and 2.96/1.16 respectively. The theoretical values for Na/Cl ratio for saline and brackish waters are of the order of 0.85 and 1.00 respectively. This means that majority of the near surface and deep systems in respect of Palar, Ponnaiyar, Cauvery and Vaigai lie in the field of fresh water but sometimes close to the brackish water boundary. However the average values in the case of near surface systems of Ponnaiyar lie in between the saline and brackish water zone. This gives an impression that majority of the near surface aquifers in the Ponnaiyar delta basin are in the threshold of mixing with brackish/saline water.

Table 5.4(a,b) Maximum, Minimum and Average values of Ionic ratios in different delta systems of Tamilnadu

(a) Near surface

Ionic Ratios	Palar			Ponnaiyar			Cauvery			Vaigai		
	Max	Min	Avg	Max	Min	Avg	Max	Min	Avg	Max	Min	Avg
Na / Cl	6.30	0.48	1.40	2.26	0.21	0.98	5.81	0.41	1.22	49.34	0.58	2.96
Mg / Cl	4.37	0.02	0.66	2.92	0.07	0.52	1.42	0.06	0.49	9.48	0.14	0.78
SO4 / Cl	0.82	0.00	0.29	0.89	0.02	0.24	3.67	0.01	0.38	3.14	0.02	0.37
$Cl/(HCO_3 + CO_3)$	3.83	0.20	1.42	33.81	0.30	4.20	7.76	0.20	2.35	90.56	0.02	6.57
$Ca/(HCO_3 + SO_4)$	1.89	0.08	0.53	5.07	0.1	0.87	1.74	0.05	0.54	2.28	0.06	0.51
Mg/Ca	3.24	0.07	1.09	3.60	0.09	0.92	14.21	0.20	1.71	12.71	0.16	2.08

(b) Deep systems

Ionic Ratios	Palar			Ponnaiyar			Cauvery			Vaigai		
	Max	Min	Avg	Max	Min	Avg	Ma x	Min	Avg	Max	Min	Avg
Na / Cl	6.30	0.06	1.80	4.87	0.00	1.56	4.11	0.07	1.25	1.72	0.72	1.16
Mg / Cl	1.25	0.02	0.49	2.57	0.01	0.82	2.18	0.01	0.61	0.71	0.05	0.22
SO_4 / Cl	0.56	0.05	0.28	1.17	0.05	0.42	1.17	0.01	0.33	0.40	0.10	0.26
$Cl/(HCO_3 + CO_3)$	5.18	0.22	1.51	11.61	0.09	1.23	8.40	0.12	1.80	37.85	0.62	5.62
$Ca/(HCO_3 + SO_4)$	0.75	0.03	0.37	1.13	0.07	0.52	1.44	0.09	0.60	1.83	0.08	0.54
Mg/Ca	2.01	0.40	1.00	1.52	0.07	0.64	4.50	0.11	0.86	1.00	0.21	0.67

- The average Mg/Cl ratios for near surface/deep systems of Palar, Ponnaiyar, Cauvery and Vaigai are 0.66/0.49, 0.52/0.82, 0.49/0.61 and 0.78/0.22 respectively. The theoretical values for Mg/Cl ratio for saline and brackish waters are of the order of 0.15 and 0.35 respectively. This shows that the majority of samples from near surface and deep systems lie in the field of fresh water but however, the average Mg/Cl values of near surface system in the case of Cauvery and deep system of Palar come close to the boundaries of brackish water zone signifying mixing tendencies of fresh and saline water. It is also signifying to note that the Mg/Cl ratio of the deep system in the case of Vaigai lies in the region of saline and brackish water zone signifying intense mixing of fresh and saline water.
- The average SO_4/Cl ratios for near surface/deep systems of Palar, Ponnaiyar, Cauvery and Vaigai are 0.29/0.28, 0.24/0.42, 0.38/0.33 and 0.37/0.26 respectively. The theoretical values for SO_4/Cl ratio for saline and brackish waters are of the order of 0.05 and 0.20 respectively. This signifies that the SO_4/Cl ratios of near surface system in the case of Palar and Ponnaiyar comes close to the threshold limit of the brackish water. Similarly the SO_4/Cl ratios for deep systems of Palar and Vaigai comes close to the threshold limits of the brackish water.

- The average $Cl/HCO_3 + CO_3$ ratios for near surface/deep systems of Palar, Ponnaiyar, Cauvery and Vaigai are 1.42/1.51, 4.20/1.23, 2.35/1.80 and 6.57/5.62 respectively. The theoretical values for $Cl/HCO_3 + CO_3$ ratio for saline and brackish waters are of the order of 6.6 and 0.50 respectively. This $Cl/HCO_3 + CO_3$ ratio characterizes that majority of samples in the case of Palar, Ponnaiyar and Cauvery comes in between the boundaries of saline water and brackish water but in the case of Vaigai both near surface and deep systems comes close to the saline water boundary signifying intense saline water mixing processes.
- The average $Ca/(HCO_3 + SO_4)$ ratios for near surface/deep systems of Palar, Ponnaiyar, Cauvery and Vaigai are 0.53/0.37, 0.87/0.52, 0.54/0.60 and 0.51/0.54 respectively. The theoretical values for $Ca/(HCO_3 + SO_4)$ ratio for saline and brackish waters are of the order of 1.00 and 0.50 respectively. This $Ca/(HCO_3 + SO_4)$ ratio indicates that majority of the samples of the near surface systems of Palar, Cauvery and Vaigai are close to the brackish water line where as Ponnaiyar is represented by values close to the saline water boundary. The deep systems excepting Palar lie within the fields of brackish saline water boundaries and therefore characterizes mixed trends by and large in all the four delta systems for both near surface and deep systems.
- The average Mg/Ca ratios for near surface/deep systems of Palar, Ponnaiyar, Cauvery and Vaigai are 1.09/1.00, 0.92/0.64, 1.71/0.86 and 2.08/0.67 respectively. The theoretical values for Mg/Ca ratio for saline and brackish waters are of the order of 3.5 and 1.00 respectively. The ratio data reflects that only near surface water of Ponnaiyar delta are free from saline water mixing and the rest of the near surface systems in all the three delta systems appears to carry mixed trends. In the case of deep systems all the delta systems are characterized by non mixing trends excepting Palar which shows mixing trends.

Ionic ratios Vs Cl cross plots: The ratios considered for cross plots to assess the intensity of mixing are: Na/Cl vs Cl, SO_4/Cl vs Cl, Mg/Cl vs Cl, Mg/Ca vs Cl, $Ca/(HCO_3+SO_4)$ vs Cl and $Cl/(HCO_3 + CO_3)$ vs Cl. The graphical plots of ionic ratios between Na/Cl vs. Cl, Mg/Cl vs. Cl, SO_4/Cl vs. Cl, $Cl/(HCO_3 + CO_3)$ vs Cl and $Ca / (HCO_3 + SO_4)$ vs Cl characterizing the probable saline fresh water mixing trends in both near surface and deep systems of Palar, Ponniyar-Vellar, Cauvery and Vaigai delta systems are presented in Fig. 5.13.

Palar

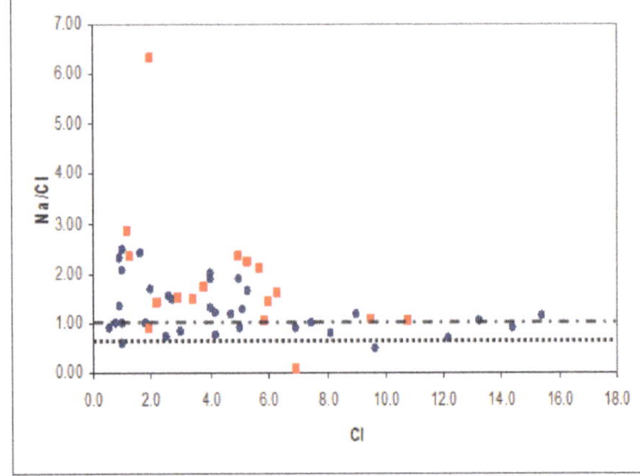

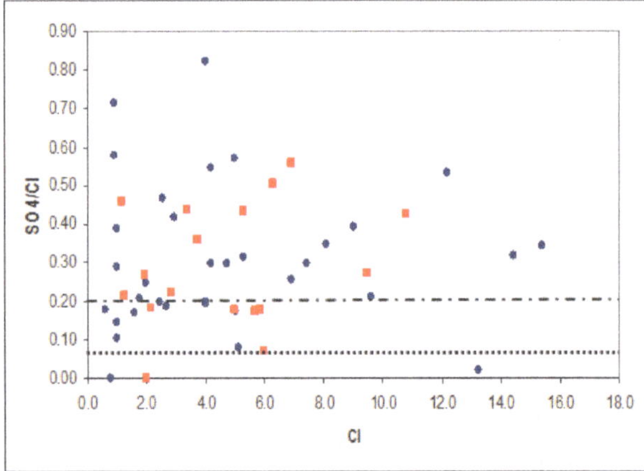

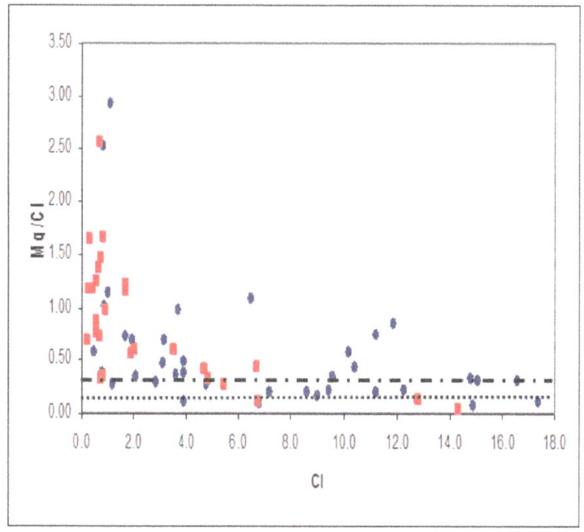
Mg/Cl
Cl

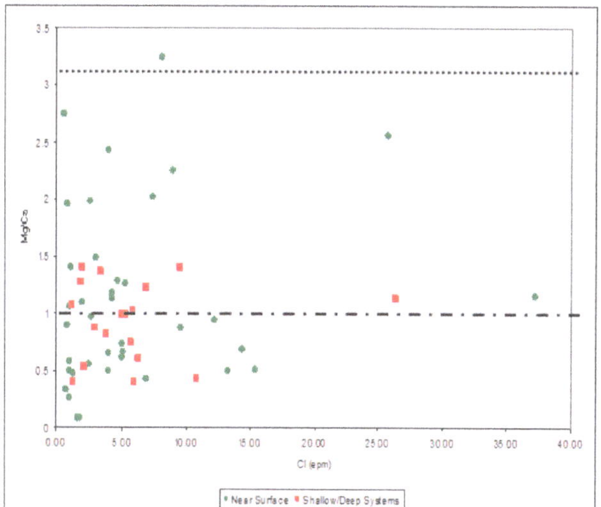
Mg/Ca
Cl (epm)
Near Surface
Shallow/Deep Systems

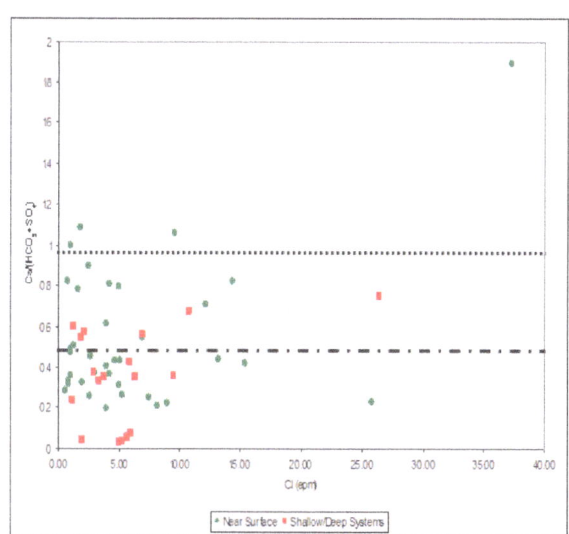
Cl (epm)
Near Surface
Shallow/Deep Systems

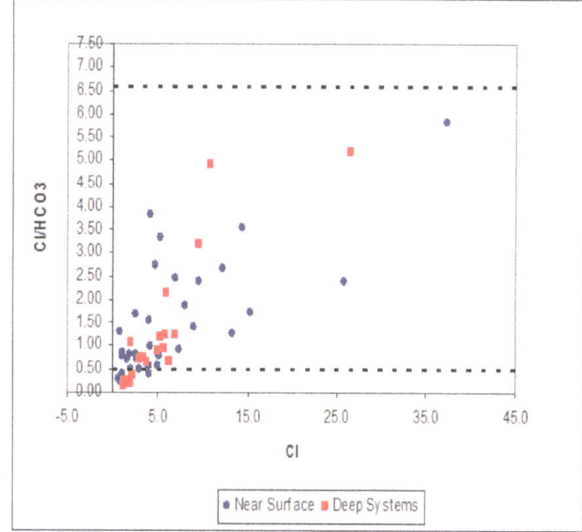
Cl/HCO3
Cl
Near Surface
Deep Systems

Ponnaiyar

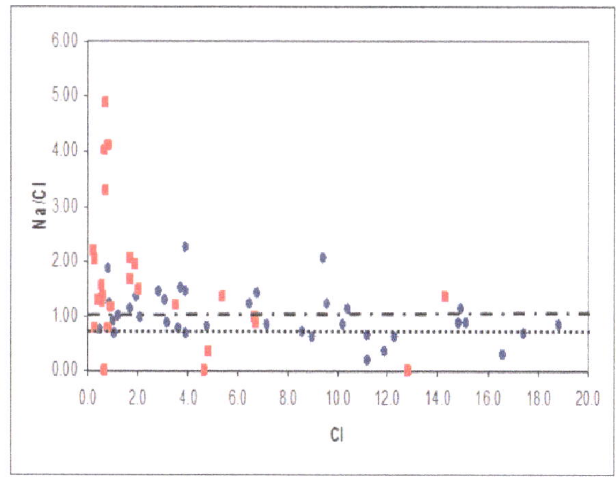
Na/Cl
Cl

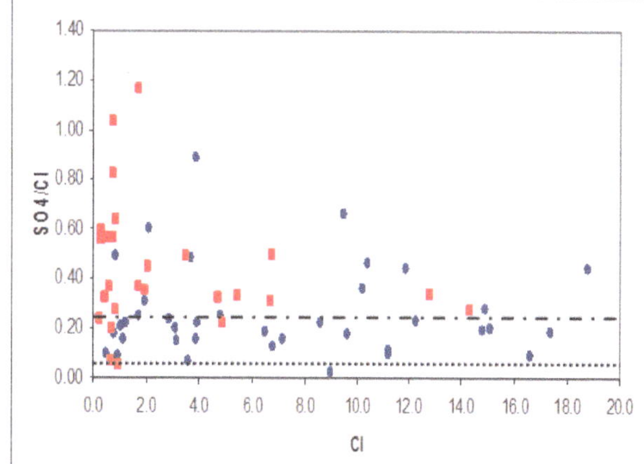
SO4/Cl
Cl

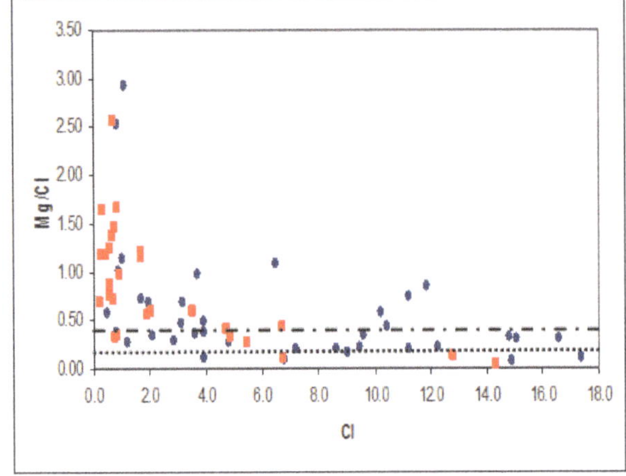

Mg/Cl
3.50
3.00
2.50
2.00
1.50
1.00
0.50
0.00
0.0 2.0 4.0 6.0 8.0 10.0 12.0 14.0 16.0 18.0
Cl

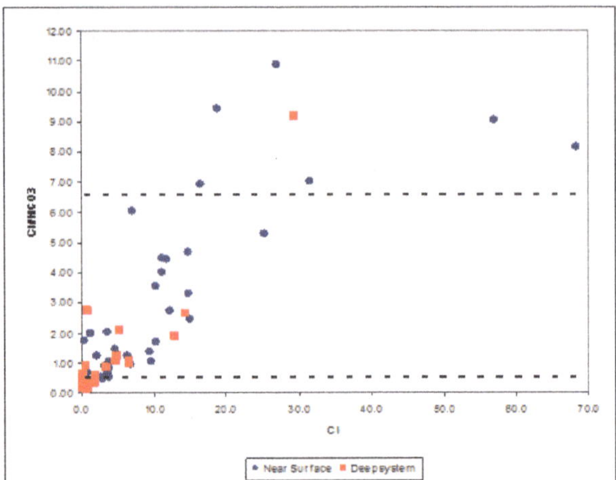

Cl/HCO3
12.00
11.00
10.00
9.00
8.00
7.00
6.00
5.00
4.00
3.00
2.00
1.00
0.00
0.0 10.0 20.0 30.0 40.0 50.0 60.0 70.0
Cl
Near Surface
Deepsystem

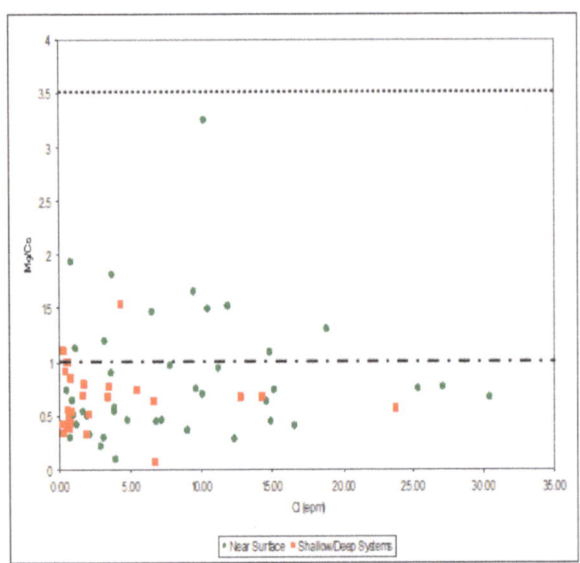

Mg/Ca
4
3.5
3
2.5
2
1.5
1
0.5
0
0.00 5.00 10.00 15.00 20.00 25.00 30.00 35.00
Cl (epm)
Near Surface
Shallow/Deep Systems

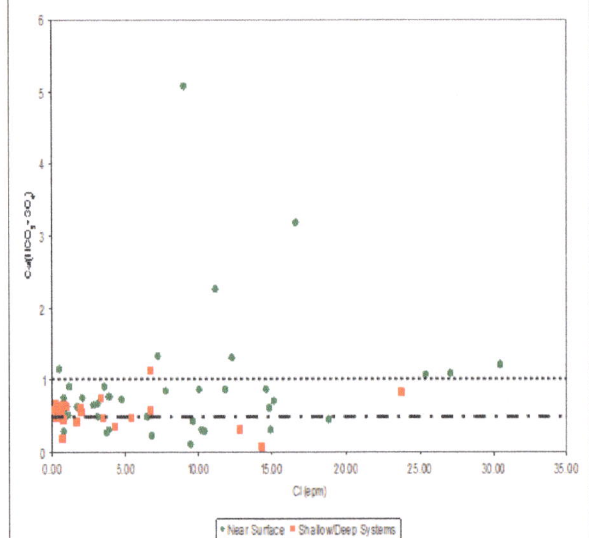

Ca/(HCO3+SO4)
6
5
4
3
2
1
0
0.00 5.00 10.00 15.00 20.00 25.00 30.00 35.00
Cl (epm)
Near Surface
Shallow/Deep Systems

Cauvery

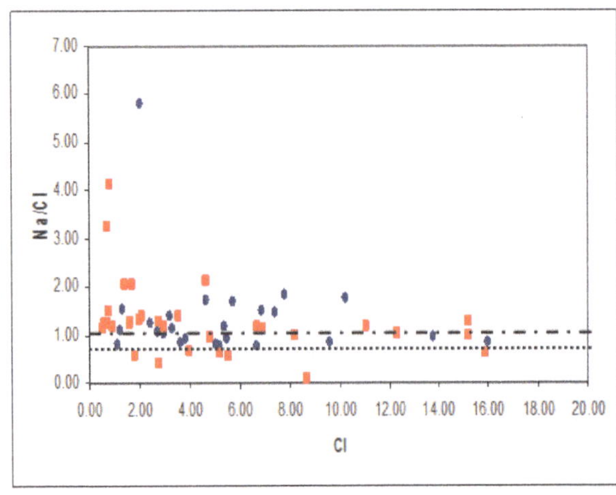

Na/Cl
7.00
6.00
5.00
4.00
3.00
2.00
1.00
0.00
0.00 2.00 4.00 6.00 8.00 10.00 12.00 14.00 16.00 18.00 20.00
Cl

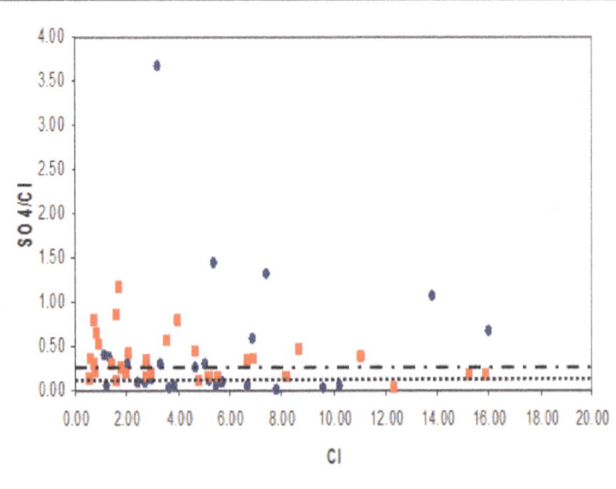

SO4/Cl
4.00
3.50
3.00
2.50
2.00
1.50
1.00
0.50
0.00
0.00 2.00 4.00 6.00 8.00 10.00 12.00 14.00 16.00 18.00 20.00
Cl

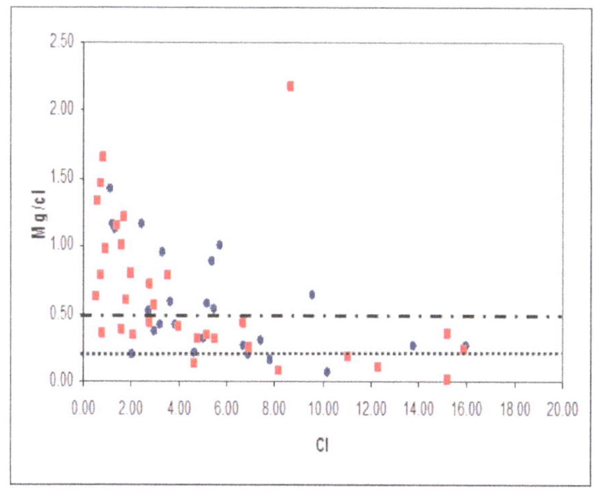
Mg/cl
Cl

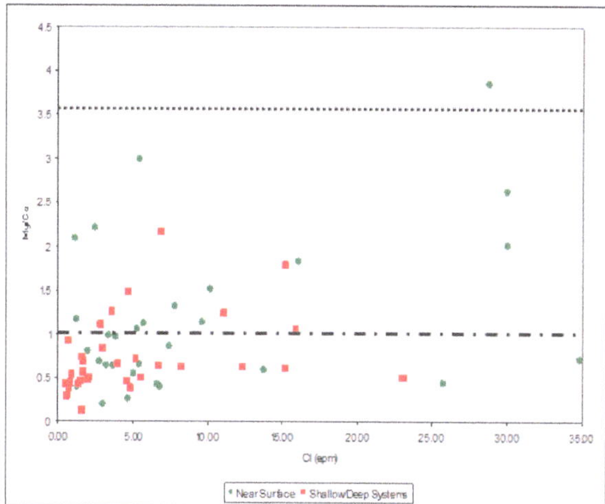
Cl (epm)
Near Surface Shallow/Deep Systems

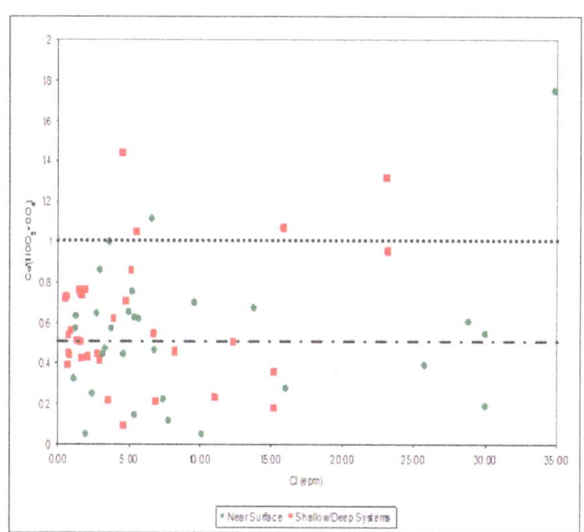
Cl (epm)
Near Surface Shallow/Deep Systems

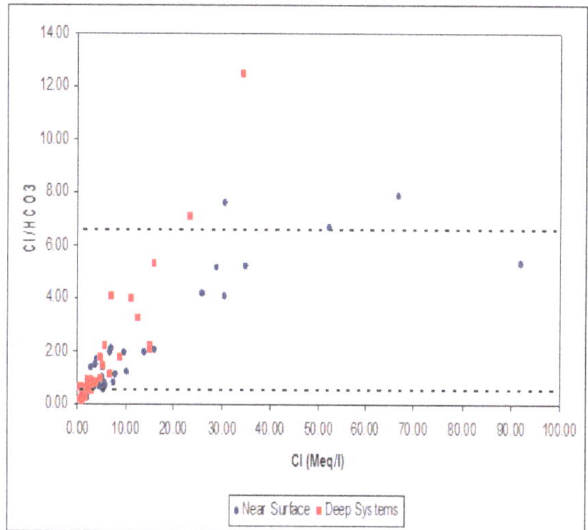
Cl/HCO3
Cl (Meq/l)
Near Surface Deep Systems

Vaigai

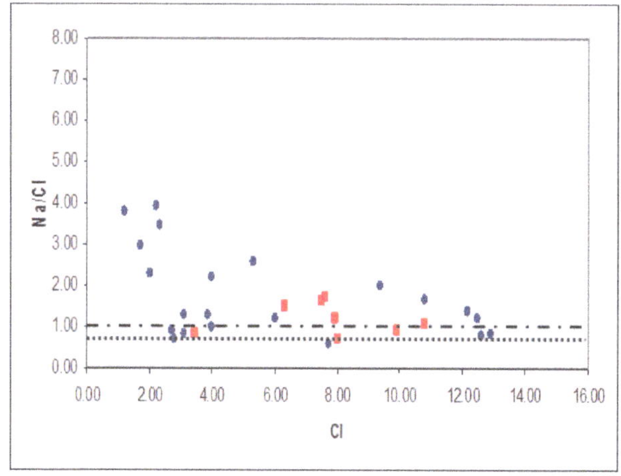
Na/Cl
Cl

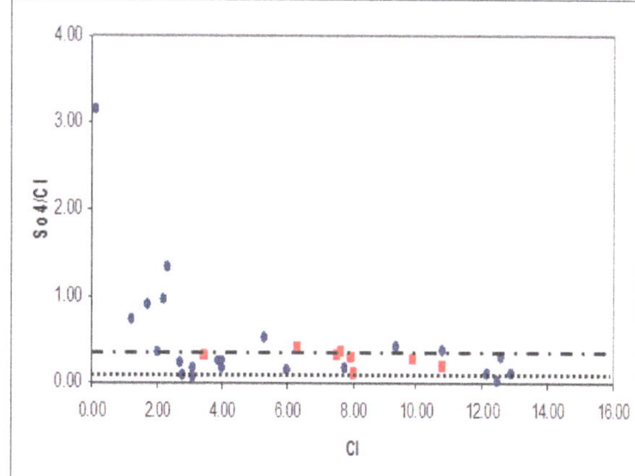
So4/Cl
Cl

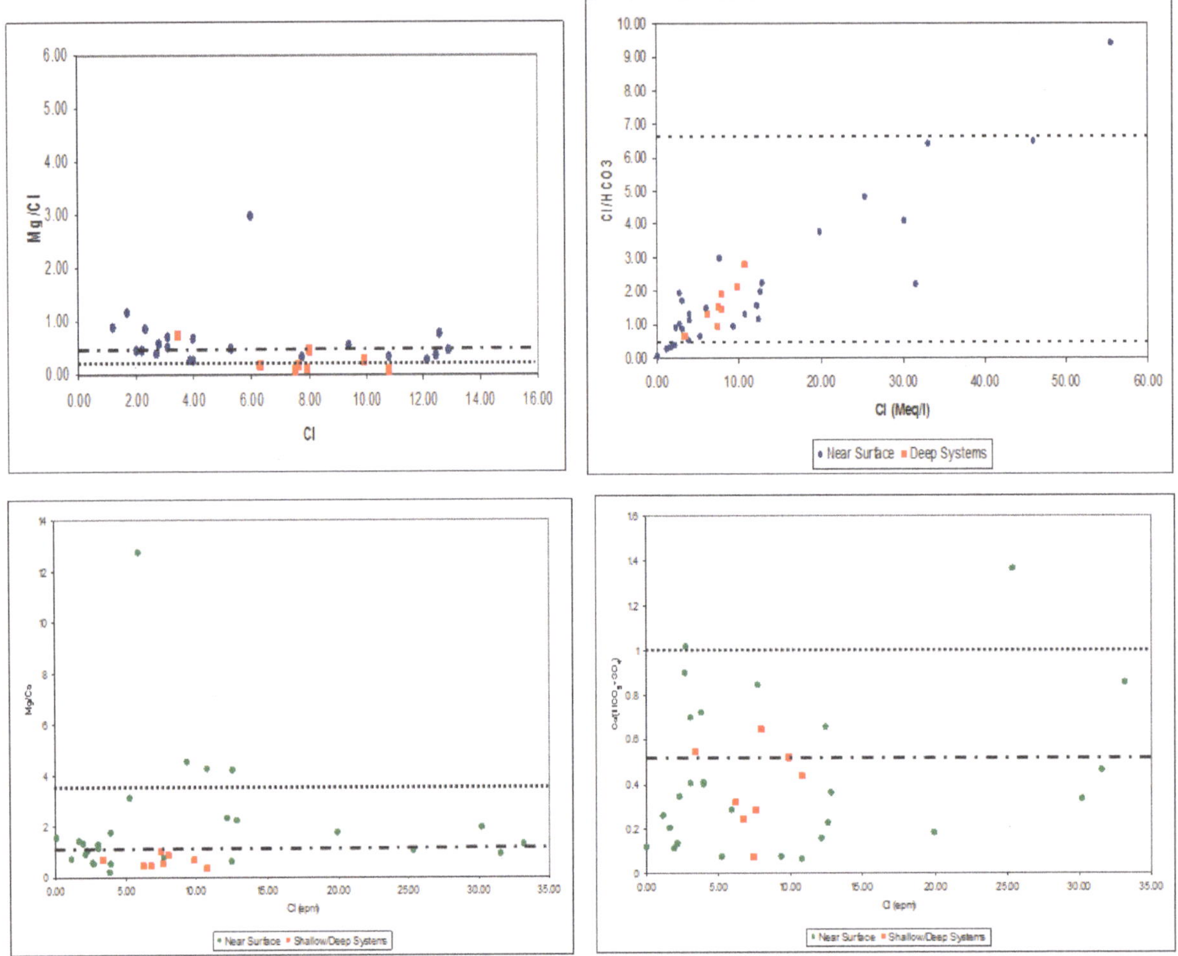

Fig. 5.13 Ionic ratios Vs Cl cross plots (near surface and deep systems) for Palar, Ponnaiyar- Vellar, Cauvery and Vaigai delta systems of Tamil Nadu

A brief analyses in respect of the above graphic plots discussing the significance of fresh saline water trends in different delta systems of Tamil Nadu are presented below in the following paragraphs.

Palar delta: *Na/Cl vs Cl* cross plot indicate the percentage of near surface and deep systems lying in the field of fresh water are respectively 53% and 88% and the rest 47% (near surface) and 12% (deep systems) lie in the brackish/saline water fields signifying that the deep systems are relatively free from saline water contamination when compared to the near surface systems. SO_4/Cl vs *Cl* cross plot indicate 66% (near surface) and 65% (deep) lie in the field of fresh water and the rest 34% (near surface) and 35% (deep) lie in the field of brackish/saline water signifying both near surface and deep systems are relatively less contaminated with brackish/saline water. *Mg/Cl vs Cl* cross plot reflect that 54% (near surface) and 83%(deep) lie in the field of fresh water and the rest 46%(near surface) and 17% (deep) lie in the field of brackish/saline water signifying greater mixing of saline water in the near surface systems when compared to deep systems. *Mg/Ca vs Cl* cross plot indicate 54% (near surface) and 50%(deep) lie in the field of fresh water and the rest 46%(near surface) and 50% (deep) lie in the field of brackish/saline water reflecting equal mixing of brackish/saline water in both near surface and deep systems. $Ca/(HCO_3 + SO_4)$ *vs Cl* cross plot indicate that 60% (near surface) and 68% (deep) lie in the field of fresh water and the rest 40% (near surface) and 32% (deep) lie in the field of brackish/saline water reflecting a majority of samples in both near surface and deep systems are free from salinity.

Cl/HCO_3 vs Cl cross plot indicate that about 17% (near surface) and 23% (deep) lie in the field of fresh water and the rest 83% (near surface) and 77% (deep) lie in the field of brackish/saline water reflecting dominant mixing of saline water in both near surface and deep systems.

Ponnaiyar delta: *Na/Cl vs Cl* cross plot trends indicate 40% of samples in near surface and 71% in deep systems lie in the fresh water field and the rest 60% (near surface) and 29% (deep) lie in the field of brackish/saline water indicating greater amount of mixing in the near surface systems when compared to deep systems. *SO_4/Cl vs Cl* cross plot trends indicate that 37% (near surface) and 80% (deep) lie in the fresh water field and the rest 63% (near surface) and 20% (deep) lie in the field of brackish/saline water reflecting a majority of samples in the near surface are contaminated with brackish/saline water when compared to the deep systems where majority of the samples in deep systems are free from contamination. *Mg/Cl vs Cl* cross plot trends indicate 45% (near surface) and 75% (deep) lie in the field of fresh water and the rest 55%(near surface) and 25% (deep) lie in the field of brackish/saline water indicating greater mixing with salinity in the near surface systems when compared to deep systems. *Mg/Ca vs Cl* cross plot trends indicate 70% (near surface) and 86%(deep) lie in the field of fresh water and the rest 30% (near surface) and 14% (deep) lie in the field of brackish/saline water indicating that majority of the samples in both near surface and deep systems are free from salinity. *$Ca/(HCO_3+SO_4)$ vs Cl* cross plot trends shows that 33% (near surface) and 56% (deep) lie in the field of fresh water and the rest 67% (near surface) and 44% (deep) lie in the field of brackish/saline water reflecting majority of the samples from near surface systems are contaminated with brackish/saline water but in the case of deep systems nearly half of the samples shows contamination with brackish/saline water. *Cl/HCO_3 vs Cl* cross plot trends reflect on 10% of the samples from near surface and 35% of the samples from the deep lie in the fresh water field and the rest 90% (near surface) and 65% (deep) lie in the brackish/saline water field reflecting majority of the samples in the near surface systems are dominantly contaminated with brackish/saline water. Similarly the deep systems also reflect to moderate to dominant contamination with brackish/saline waters.

Cauvery delta: *Na/Cl vs Cl* cross plot reflects on 58% (near surface) and 60% (deep) samples lie in the field of fresh water and the rest 42% (near surface) and 40% (deep) lie in the field of brackish/saline water signifying majority of the samples in both near surface and deep systems are free from saline water contamination. *SO_4/Cl vs Cl* cross plot reflect on 43% (near surface) and 52% (deep) lie in the field of fresh water and the rest 57% (near surface) and 48% (deep) lie in the field of brackish/saline water reflecting dominant mixing processes in the near surface system and nearly 50% contamination in the deep systems. *Mg/Cl vs Cl* cross plot reflect on 56% (near surface) and 37% (deep) lie in the fresh water field and the rest 44% (near surface) and 63% (deep) lie in the field of brackish/saline water reflecting dominant mixing processes in the deep systems and nearly 50% contamination in the near surface systems. *Mg/Ca vs Cl* cross plot reflect on 55% (near surface) and 78% (deep) lie in the field of fresh water and the rest 45% (near surface) and 22% (deep) lie in the field of brackish/saline water indicating equal proportions of mixing in the near surface and relatively less amount of samples are subjected to mixing in the deep systems. *$Ca/(HCO_3 + SO_4)$ vs Cl* cross plot indicate 46% (near surface) and 50% (deep) are free from salinity and the rest 54%(near surface) and 50% (deep) are in the field of brackish/saline water reflecting by and large equal proportions of mixing in both near surface and deep systems. *Cl/HCO_3 vs Cl* cross plot reflect on 40% (near surface) samples and 22% (deep) lie in the fresh water field and the rest 60% (near surface) and 78% (deep) lie in the field of brackish/saline water reflecting majority of the samples in both near surface and deep systems are subjected to mixing processes.

Vaigai delta: *Na/Cl vs Cl* cross plot reflect on 66% of near surface samples and 62% of the deep systems lie in the field of fresh water and rest 34% (near surface) and 38% (deep) lie in the field of brackish/saline water indicating majority of the samples in both near surface and deep systems are free from salinity. *SO_4/Cl vs Cl* cross plot reflect on 40% of near surface samples and 14% of the deep systems lie in the field of fresh water and the rest 60% (near surface) and 86% (deep) lie in the field of brackish/saline waters reflecting majority of the samples in both near surface and deep systems are subjected to mixing processes. *Mg/Cl Vs Cl* cross plot indicate 55% of the near surface samples and 14% of the deep sytems lie in the field of fresh water where as the rest 45% (near surface) and 86%(deep) lie in the field of brackish/saline water reflecting equal proportions of mixing in the case of mixing and dominant mixing in the case of deep systems. *Mg/Ca Vs Cl* cross plot indicate 39% of the near surface systems and 100% of the deep systems lie in the field of fresh water and the rest 61% (near surface) and 0% (deep) lie in the field of brackish/saline waters reflecting the greater amount of mixing taking place in the near surface systems and no mixing in deep systems. *Ca/*(HCO_3 + SO_4) *Vs Cl* cross plot reflect on 66% of near surface system and 75% of the deep systems lie in the field of fresh water and the rest 34% (near surface) and 25% (deep) lie in the field of brackish/saline water reflecting majority of samples from both near surface and deep systems are free from salinity. *Cl/*HCO_3 *vs Cl* cross plot ratio indicate 11% of the samples from the near surface systems and 0% from the deep lie in the field of fresh water and the rest 89%(near surface) and 100% (deep) lie in the field of brackish/saline waters reflecting a predominant mixing processes in both near surface and deep systems.

The *Na/Cl*, SO_4*/Cl* and *Mg/Cl vs. Cl* trends in all the delta systems show higher order in the initial levels and gradually decrease with increasing levels of chloride. The ratios characterize the saline brackish water mixing at all levels of salinity (chloride levels). However the fresh, brackish water mixing is more dominantly seen with increase of chloride levels in almost all delta systems.

5.4.3 Trilinear plots/Characterization of ionic trends

The dominant cations and anions present in solution of Palar, Ponnaiyar-Vellar, Cauvery and Vaigai delta regions are presented systematically in the Piper triangular patterns which assimilates and represents the whole compositional pattern as a single point in the diamond shaped diagram as shown in Figures 5.14 (a,b,c,d).

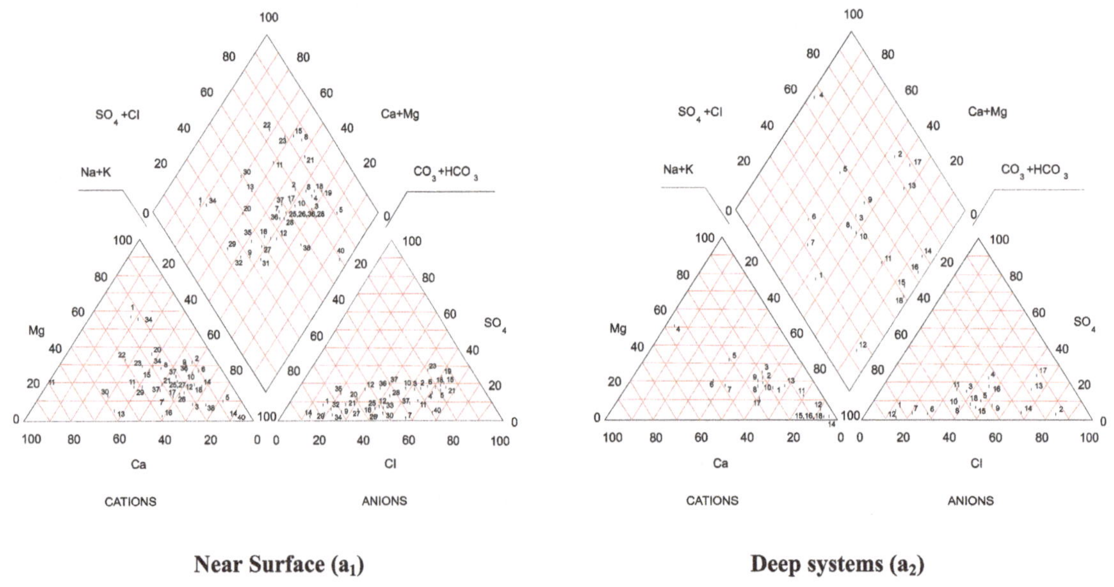

Near Surface (a_1) **Deep systems (a_2)**

Fig. 5.14(a) Trilinear plots for near surface (a_1) and deep systems (a_2) in palar delta

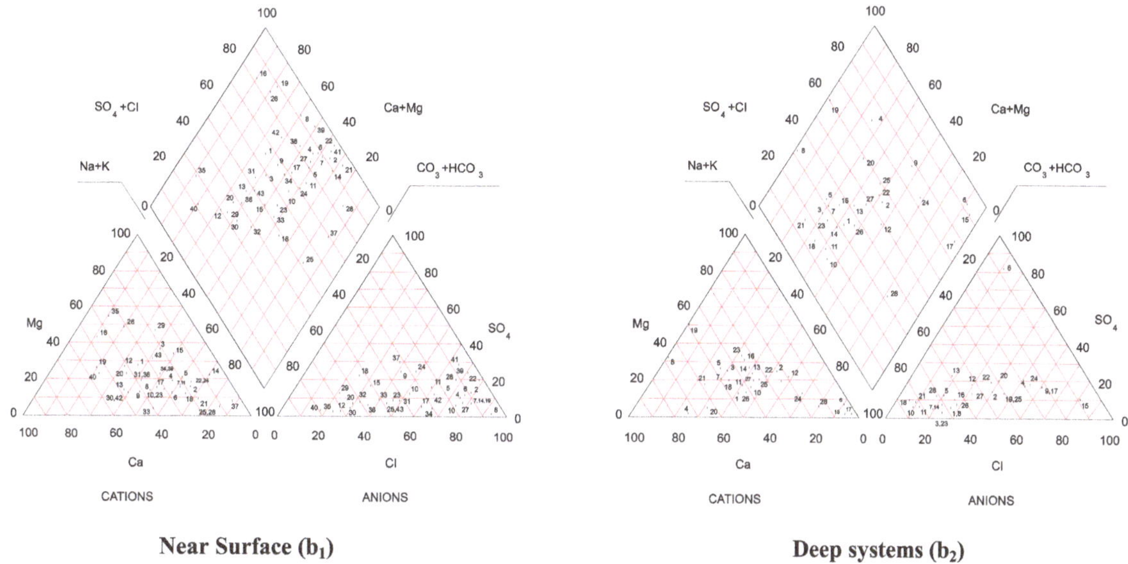

Fig. 5.14(b) Trilinear plots for near surface (b_1) and deep systems (b_2) in Ponnaiyar-Vellar delta

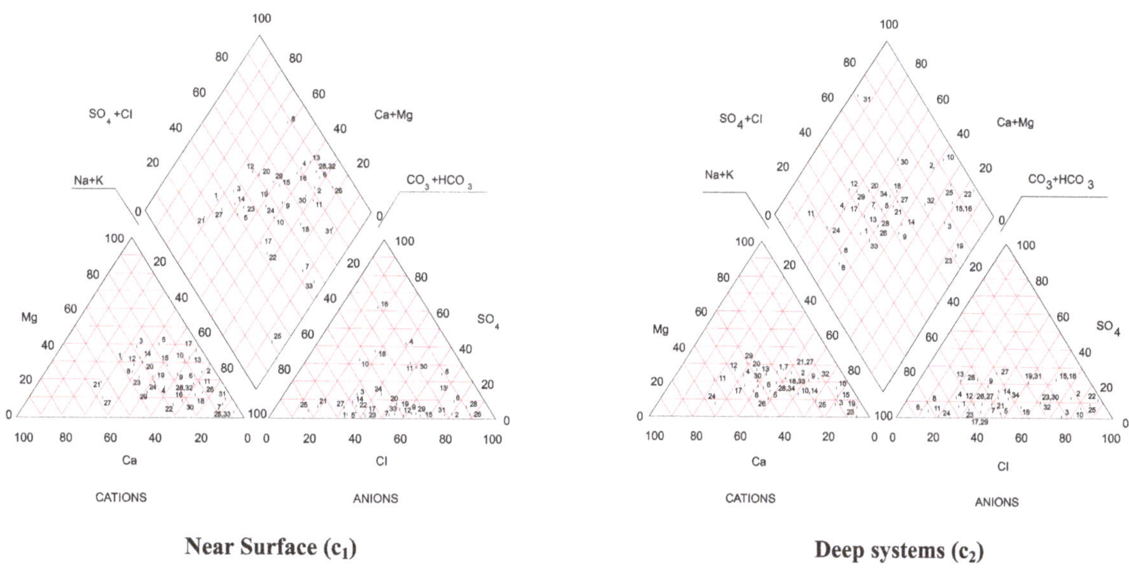

Fig. 5.14(c) Trilinear plots for near surface (c_1) and deep systems (c_2) in Cauvery delta

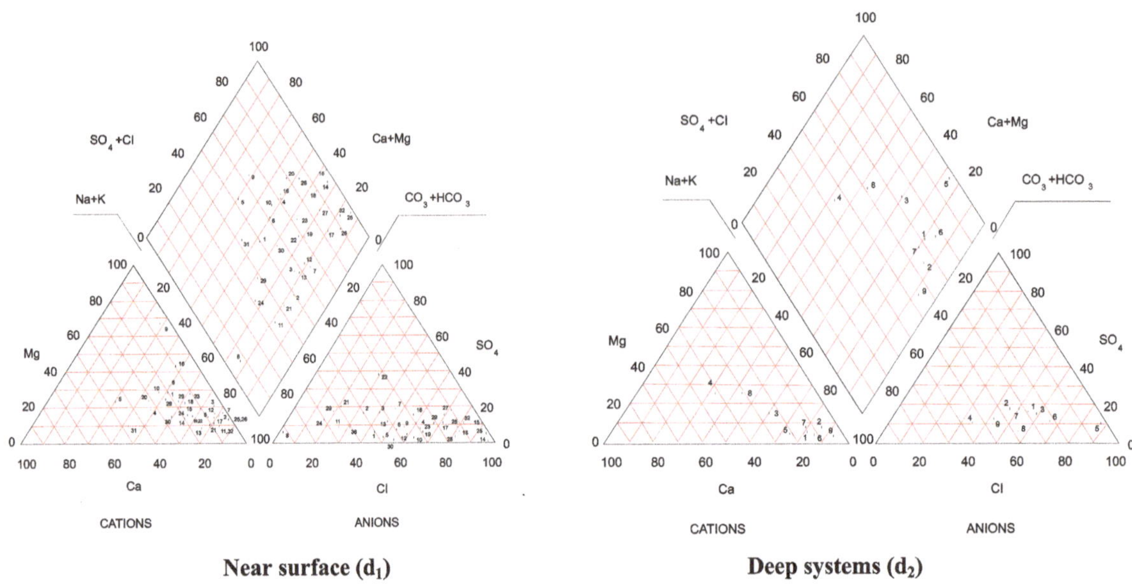

Near surface (d_1) **Deep systems (d_2)**

Fig. 5.14(d) Trilinear plots for near surface (d_1) and deep systems (d_2) in Vaigai delta

The hydrochemical characteristics as derived from piper trilinear diagrams for different delta systems of Tamil Nadu are presented in Tables 5.5, 5.6, 5.7 and 5.8 below.

Table 5.5 Characterization of ionic trends in Palar delta Number of Samples: (Near surface: 40; Deep systems: 18)

Subdivision of the diamond	Characteristics of corresponding subdivisions of diamond-shaped fields	Near Surface		Deep systems	
		No. of Samples	Percentage of samples in this category (%)	No. of Samples	Percentage of samples in this category (%)
1	Alakaline earth (Ca + Mg) Exceed alkalies (Na + K)	13	33	4	22
2	Alkalies exceeds alkaline earths	27	68	14	78
3	Weak acids (CO_3 + HCO_3) exceed strong acids (SO_4 + Cl)	17	43	7	39
4	Strong acids exceeds weak acids	23	58	11	61
5	Magnesium bicarbonate type	6	15	2	11
6	Clacium-Chloride type	0	0	1	6
7	Sodium-Chloride type	21	53	9	50
8	Sodium-Bicarbonate type	0	0	1	6
9	Mixed type (No cation-anion exceed 50%)	13	33	5	28

Table 5.6 Characterization of ionic trends in Ponnaiyar delta Number of Samples: (Near surface: 43; Deep systems: 28)

Subdivision of the diamond	Characteristics of corresponding subdivisions of diamond-shaped fields	Near Surface		Deep systems	
		No. of Samples	Percentage of samples in this category (%)	No. of Samples	Percentage of samples in this category (%)
1	Alakaline earth (Ca + Mg) Exceed alkalies (Na + K)	19	44	21	75
2	Alkalies exceeds alkaline earths	24	56	7	25
3	Weak acids (CO_3 + HCO_3) exceed strong acids (SO_4 + Cl)	11	26	17	61
4	Strong acids exceeds weak acids	32	74	11	39
5	Magnesium bicarbonate type	9	21	15	54
6	Calcium-Chloride type	4	9	1	4
7	Sodium-Chloride type	22	51	5	18
8	Sodium-Bicarbonate type	0	0	0	0
9	Mixed type (No cation-anion exceed 50%)	8	19	7	25

Table 5.7 Characterization of ionic trends in Cauvery delta Number of Samples : (Near surface: 33; Deep systems: 34)

Subdivision of the diamond	Characteristics of corresponding subdivisions of diamond-shaped fields	Near Surface		Deep Systems	
		No. of Samples	Percentage of samples in this category (%)	No. of Samples	Percentage of samples in this category (%)
1	Alakaline earth (Ca + Mg) Exceed alkalies (Na + K)	11	33	20	59
2	Alkalies exceeds alkaline earths	22	67	14	41
3	Weak acids (CO_3 + HCO_3) exceed strong acids (SO_4 + Cl)	10	30	15	44
4	Strong acids exceeds weak acids	23	70	19	56
5	Magnesium bicarbonate type	7	21	15	44
6	Calcium-Chloride type	0	0	1	3
7	Sodium-Chloride type	19	58	14	41
8	Sodium-Bicarbonate type	1	3	0	0
9	Mixed type (No cation-anion exceed 50%)	6	18	4	12

Table 5.8 Characterization of ionic trends in Vaigai delta Number of Samples: (Near surface: 32; Deep systems: 9)

Subdivision of the diamond	Characteristics of corresponding subdivisions of diamond-shaped fields	Near Surface		Deep Systems	
		No. of Samples	Percentage of samples in this category (%)	No. of Samples	Percentage of samples in this category (%)
1	Alakaline earth (Ca + Mg) Exceed alkalies (Na + K)	8	25	2	22
2	Alkalies exceeds alkaline earths	24	75	7	78
3	Weak acids (CO_3 + HCO_3) exceed strong acids (SO_4 + Cl)	8	25	1	11
4	Strong acids exceeds weak acids	24	75	8	89
5	Magnesium bicarbonate type	1	3	1	11
6	Clacium-Chloride type	0	0	0	0
7	Sodium-Chloride type	17	53	7	78
8	Sodium-Bicarbonate type	1	3	0	0
9	Mixed type (No cation-anion exceed 50%)	13	41	1	11

5.4.4 Hydrogeochemical facies

The derived spatial distribution of different hydrochemical facies identified in Palar, Ponnaiyar-Vellar, Cauvery and Vaigai delta systems for near surface and deep systems are presented and discussed below. The hydrogeochemical facies are the distinct cation and anion concentrations described within a defined composite category (Back, W., 1960 and 1966; Morgan, C.O. and Winner, M.D., 1962 and Seaber, P.R., 1962).

Hydrogeochemical facies in Palar delta basin

The different hydrogeochemical facies identified in Palar delta basin are presented below and depicted spatially in Fig. 5.15(a,b):

1. Ca-Mg-HCO_3 type 2. Na-K-HCO_3 type 3. Ca-Mg-Cl-SO_4 type 4. Na-K-Cl-SO_4 type

Facies no.1. Ca-Mg-HCO$_3$: The near surface and deep systems have an average total dissolved solids of the order of 251 ppm and 539 ppm respectively. The average Ca + Mg contents in the near surface and deep systems are 52 ppm and 90 ppm out of total average cation contents of 102 and 161 ppm respectively which comes to about 51.5% and 55.9% of total cations. The average bicarbonate contents in respect of near surface and deep systems are 190 ppm and 344 ppm out of the total anionic contents of 261 ppm and 422 ppm respectively which comes to about 73% and 81.5% of the total anions.

Facies no.2. Na-K-HCO$_3$: This zone is represented by average total dissolved solids of the order of 553 ppm and 828 ppm for near surface and deep systems respectively. The Na + K are the major cations in this zone and contain about 66.9% for near surface and 41.4% for deep systems of average total cations. This zone flanks zone no.1 in a major portion of the area but in a small area in the middle sector south of Palar river, where Na-K-Cl-SO_4 zone dominates. However in the case of deep systems this zone lies in an isolated patch with in the frame work of Na-K-Cl-SO_4 zone. The bicarbonate waters in this zone constitute about 69.7% for near surface system and 48.2% for deep systems respectively out of total anion contents.

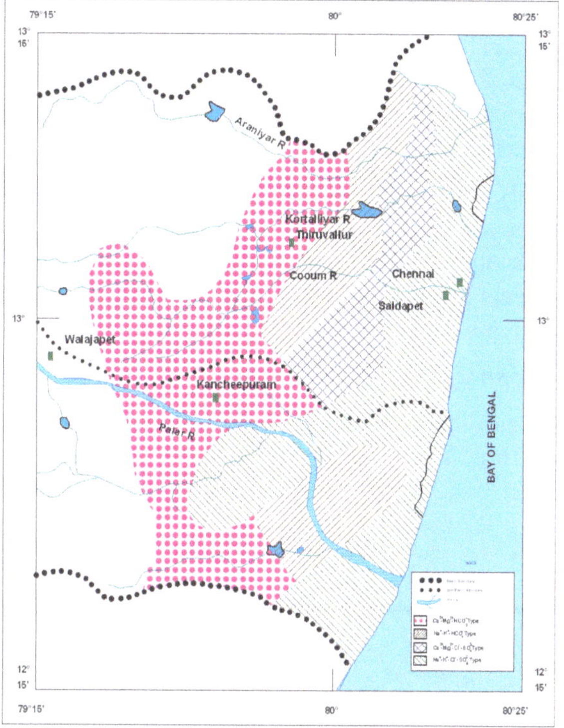

Near surface(a)

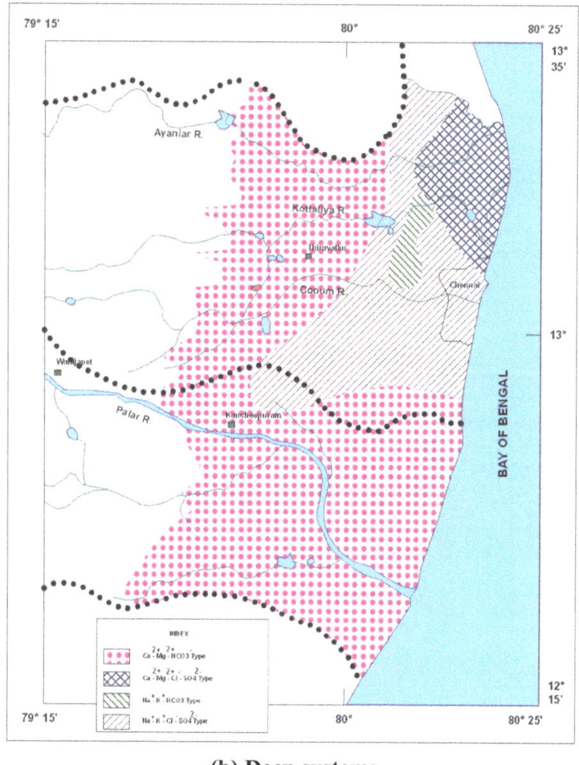

(b) Deep systems

Fig. 5.15(a)(b) Hydrogeochemical facies of near surface (a) and deep systems (b) in Palar delta

Facies no.3. Ca-Mg-Cl-SO_4: This zone is represented by total dissolved solids of the order of 684 ppm and 745 ppm for near surface and deep systems respectively. The Ca + Mg contents are the major cations in this zone and contain about 50.2% for near surface and 26.5% for deep systems of average total cations. The Cl-SO_4 contents are the major anions in this zone and contain about 70.5% for near surface and 27.5% for deep systems out of total anion contents.

Facies no.4. Na-K-Cl-SO_4: This zone is represented by total dissolved solids of the order of 966 ppm and 1093 ppm for near surface and deep systems respectively. The Na + K contents are the major cations in this zone and contain about 74% for near surface and 79% for deep systems out of average total cations. The Cl-SO_4 contents are the major anions in this zone and contain about 58.5% for near surface and 61.6% for deep systems out of total anion contents.

Hydrogeochemical facies in Ponnaiyar delta basin

The different hydrochemical facies identified for both near surface and deep systems in Ponnaiyar basin are spatially depicted in Fig. 5.16(a, b) and are described in the following paragraphs.

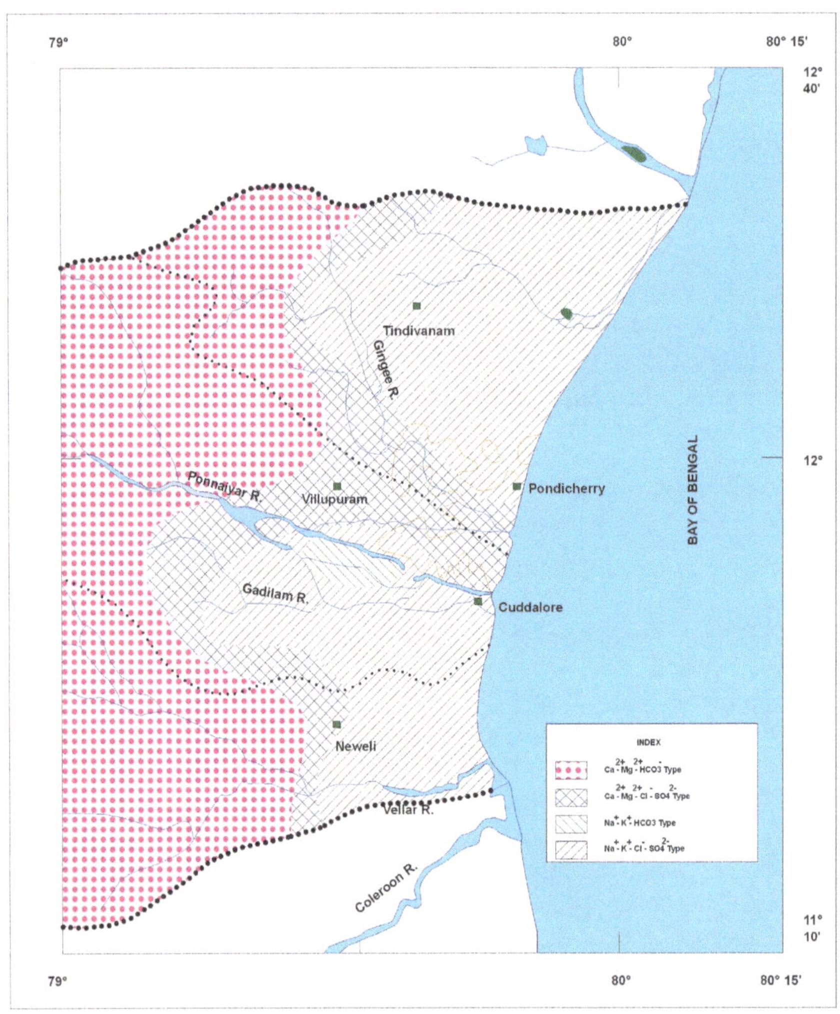

(a) Near surface

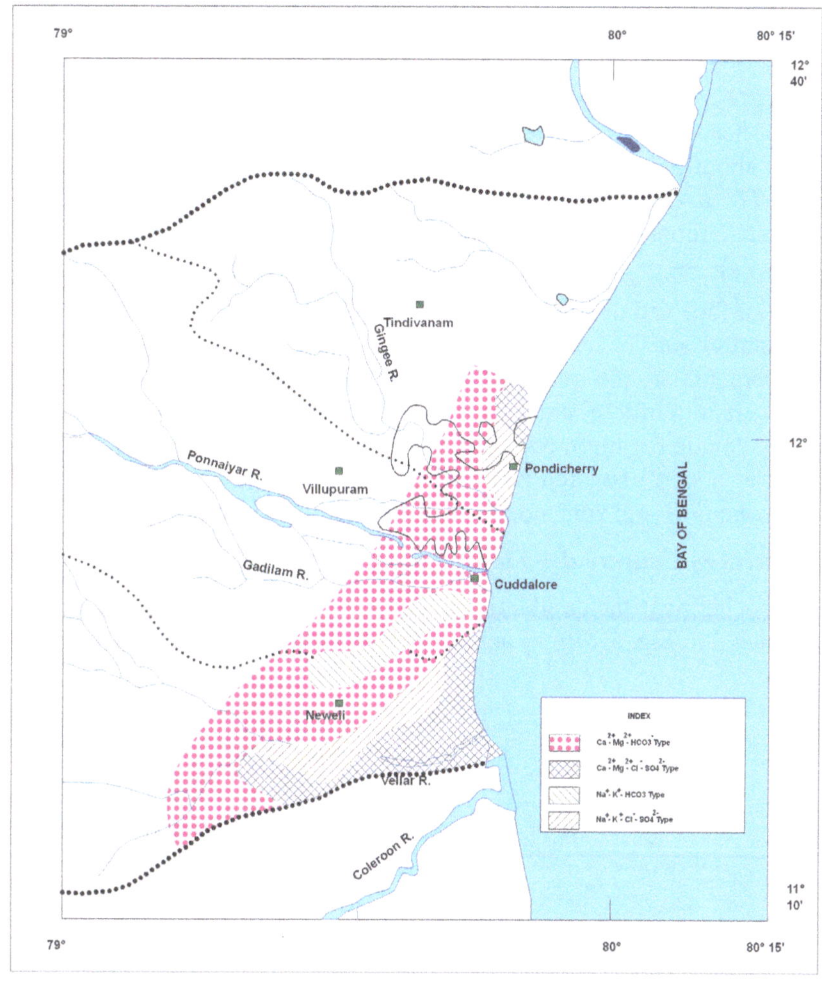

(b) Deep systems

Fig. 5.16 Hydrogeochemical facies of near surface (a) and deep systems (b) in Ponnaiyar delta

The different hydrochemical facies identified are:

1. Ca-Mg-HCO_3 type
2. Ca-Mg-Cl-SO_4 type
3. Na-K-HCO_3 type
4. Na-K-Cl-SO_4 type

Facies no.1. Ca-Mg-HCO$_3$: The near surface and deep systems have an average total dissolved solids of the order of 365 ppm and 296 ppm respectively. The Ca + Mg are the major cations in this zone and contain about 54% for near surface zone and 53% for deep systems out of the total average cation contents. The average bicarbonate contents in respect of near surface and deep systems are 72% and 79% of the total anionic contents. This zone probably represents the main recharge zone for the entire delta region.

Facies no.2. Ca-Mg-Cl-SO$_4$:This zone is represented by average total dissolved solids of the order of 824 ppm, 826 ppm for near surface and deep systems respectively. The Ca + Mg are the major cations in this zone and contain about 60% for near surface and 42% for deep systems out of the average total cationic composition. The anionic contents are represented by Cl and SO_4 and the total Cl + SO_4 contents contain about 59% for near surface and 57% for deep systems respectively. This zone flanks zone no.1 in case of near surface zone where as in the case of deep systems this zone flanks both zone no.1 and also the coastal boundaries.

Faceis no.3. Na-K-HCO_3: This zone is represented by average total dissolved solids of the order of 686 ppm, 203 ppm for near surface and deep systems respectively. The Na+K are the major cations in this zone and contain about 58% for near surface and 79% for deep systems out of the average total cationic composition. The average bicarbonate contents in respect of near surface and deep systems are 66% and 74% of the total anionic contents.

Facies no.4. Na-K-Cl-SO_4: This zone is represented by total dissolved solids of the order of 1516 ppm and 691 ppm for near surface and deep systems respectively. The Na + K contents are the major cations in this zone and contain about 73% for near surface and 84% for deep systems out of average total cations. The Cl-SO_4 contents are the major anions in this zone and contain about 75% for near surface and 73% for deep systems out of total anion contents. This zone occupies a major portion of the central and outfall parts of the deltaic region in respect of near surface zone. However in the case deep systems this zone is restricted to a small isolated part in the northeastern parts of the Pondicherry and occurs sandwiched between zone no.1 and zone no.2 in the southeastern outfall region of Vellar delta.

Hydrogeochemical facies in Cauvery delta basin

The different hydrogeochemical facies identified for both near surface and deep systems in Cauvery delta are spatially depicted in Fig. 5.17(a,b) and are described in following paragraph.

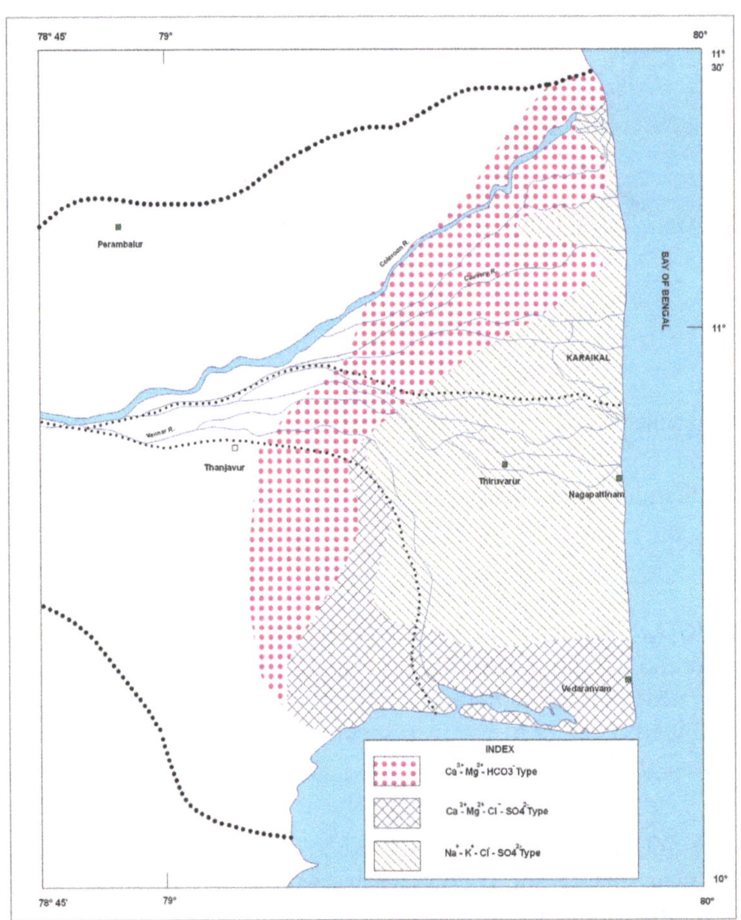

(a) Near surface

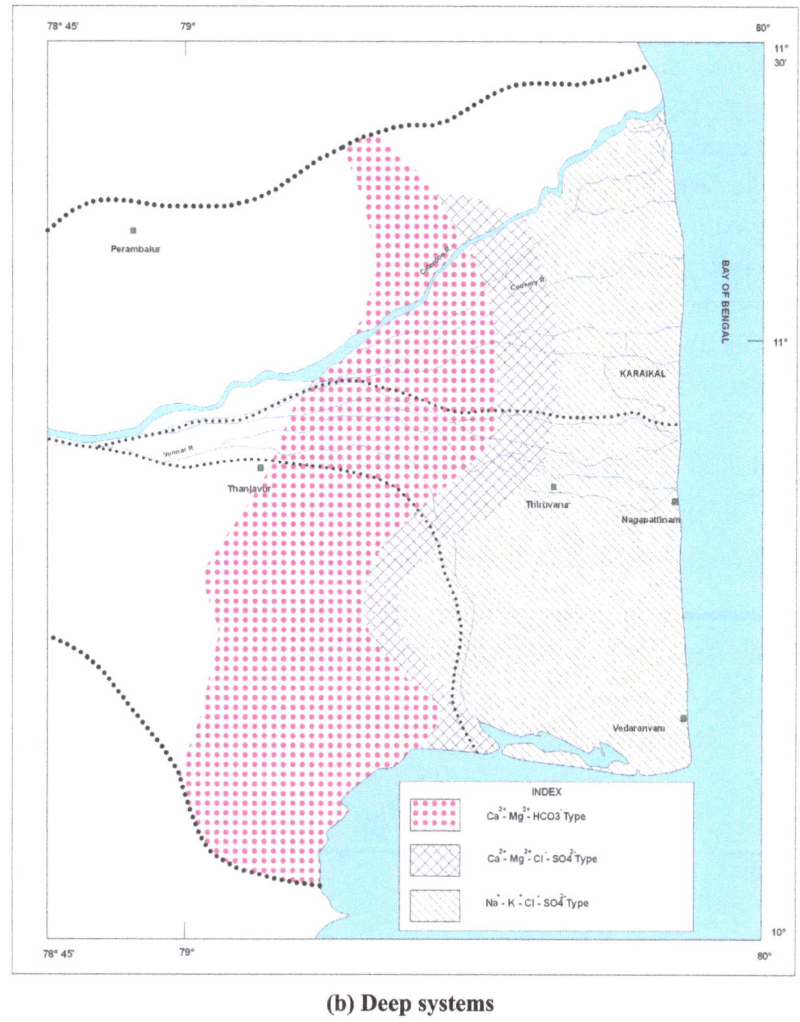

(b) Deep systems

Fig. 5.17 Hydrogeochemical facies of near surface (a) and deep systems (b) in Cauvery delta

1. Ca-Mg-HCO_3 type
2. Ca-Mg-Cl-SO_4 type
3. Na-K-Cl-SO_4 type

Faceis no.1. Ca-Mg-HCO_3: The near surface and deep systems in this zone have an average total dissolved solids of the order of 454 ppm and 316 ppm respectively. The Ca + Mg are the major cations in this zone and contain about 53% for near surface zone and 54% for deep systems out of the total average cations. The average bicarbonate contents in respect of near surface and deep systems are 71% and 74% of the total anionic contents. This facies zone represents the upstream portions of Cauvery delta region for both near surface and deep systems and probably act as the major recharge zone.

Facies no.2. Ca-Mg-Cl-SO_4: This zone is represented by average total dissolved solids of the order of 989 ppm, 546 ppm for near surface and deep systems respectively. The Ca+Mg are the major cations in this zone and contain about 48% for near surface and 52% for deep systems out of the average total cationic composition. The anionic contents are represented by Cl and SO_4 and the total Cl + SO_4

contents contain about 66% for near surface and 51% for deep systems respectively out of total anions. This zone flanks zone no.1 in the case of near surface system and also in the deep systems. However this zone is totally non represented in the northeastern portions of the near surface systems but extends predominantly in the southern and southeastern parts of the delta and is found extending up to the coastal boundary. In the case of deep systems however this zone flanks zone no.1 all through uniformly in the middle delta region.

Facies no.3. Na-K-Cl-SO_4 : This zone is represented by total dissolved solids of the order of 1955 ppm and 995 ppm for near surface and deep systems respectively. The Na + K contents are the major cations in this zone and contain about 80% for near surface and 76% for deep systems out of average total cations. The Cl + SO_4 contents are the major anions in this zone and contain about 68% for near surface and 67% for deep systems out of total average anion contents. This zone is represented dominantly in both near surface and deep systems, occupy a major portion in the central and outfall regions of the delta tracks and extend up to coastal boundaries.

Hydrogeochemical facies in Vaigai delta basin

The different hydrochemical facies identified for both near surface and deep systems for Vaigai delta are spatially depicted in Fig. 5.18(a,b) and are described in the following paragraphs.

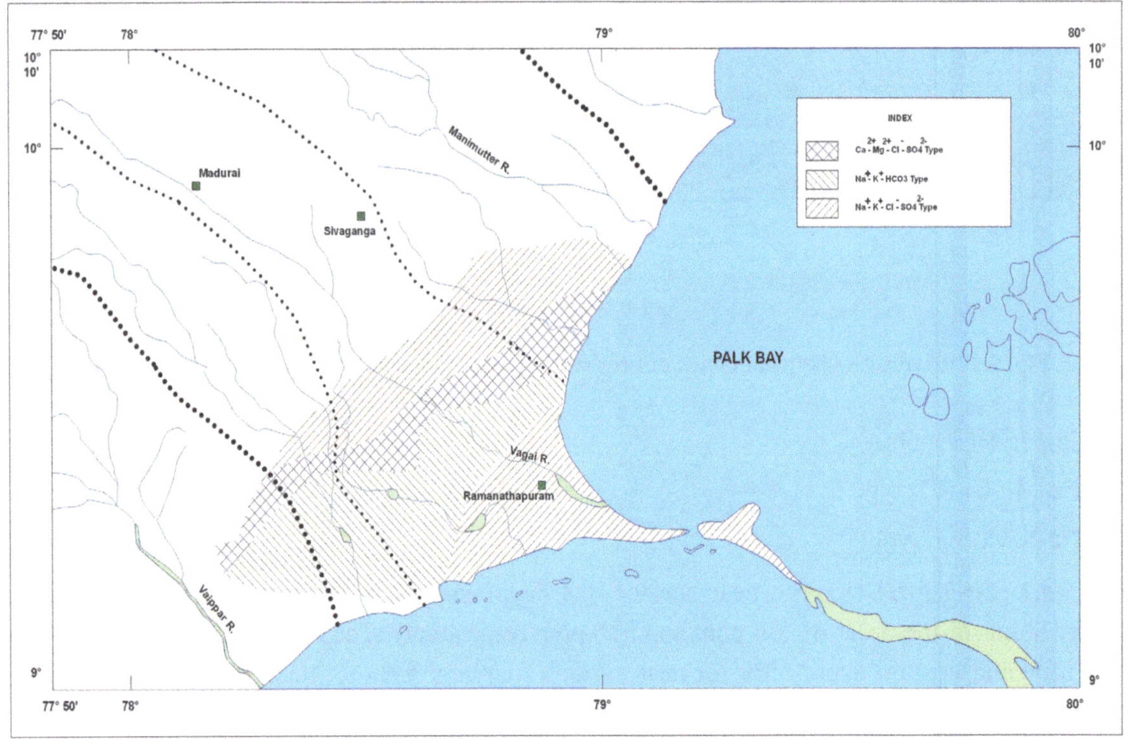

(a) Near surface

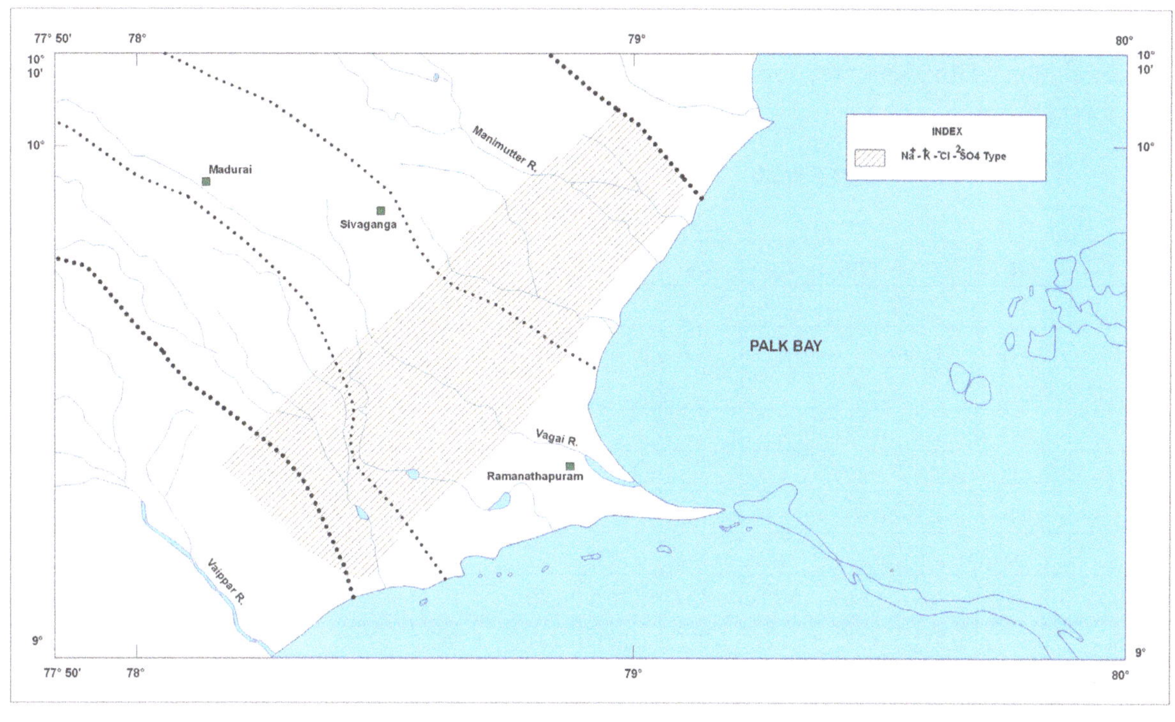

(b) Deep systems

Fig. 5.18 Hydrogeochemical facies of near surface (a) and deep systems (b) in Vaigai delta

1. Na-K-Cl-SO_4 type 2. Ca-Mg-Cl-SO_4 type 3. Na-K-HCO_3 type

Facies no.1. Na-K-Cl-SO_4: These facies zones is represented by both near surface system and deep system and have average total dissolved solids of the order of 3579 ppm and 2301 ppm respectively. The Na + K are the major cations in this zone and contain about 78% and 79% for near surface and deep systems respectively of the total average cationic contents. The anions are represented by Cl and SO_4 with over 80% for near surface and 82% for deep systems of the total average anions. This zone is found lying in both upstream portions of the delta for near surface system and in the case of deep systems this is the only facies represented in the middle portions of the deltaic tracks. These facies boundaries however could be improved if more data is available.

Facies no.2. Ca-Mg-Cl-SO_4: This zone is restricted only to the near surface system and not represented in the case of deep systems. The average total dissolved solid for near surface system is about 664 ppm for near surface. The Ca + Mg are the major cations of this zone and contain about 48% out of the average total cations. The anionic contents are represented by Cl and SO_4 and the total Cl + SO_4 contents contain about 52% of the total anions. This zone flanks the zone no.1 towards the downstream side.

Facies no.3. Na-K-HCO_3: This zone is also similarly not represented in the deep systems but observed in the only near surface systems. This zone with average total dissolved solids of the order of 641 ppm. The Na + K are the major cations in this zone and contain about 80% out of the average total cationic composition. The HCO_3 is the major anion in this zone and contain about 68% of the total anionic contents.

5.4.5 Correlation matrix and factor analysis

The correlation matrix analyses for Palar delta in near surface and deep systems are presented in Table 5.9(a, b) below.

Table 5.9(a,b) Correlation matrix analysis for Palar delta

(a) Near surface

	pH	EC	TDS	Ca	Mg	Na	K	HCO_3	SO_4	Cl	NO_3	F	TH
pH	1												
EC	0.379	1											
TDS	0.379	1	1										
Ca	0.091	0.684	0.684	1									
Mg	0.411	0.809	0.809	0.438	1								
Na	0.374	0.956	0.956	0.535	0.704	1							
K	0.289	0.479	0.479	0.446	0.196	0.371	1						
HCO_3	0.385	0.665	0.665	0.27	0.437	0.702	0.406	1					
SO_4	0.352	0.952	0.952	0.671	0.875	0.881	0.414	0.462	1				
Cl	0.321	0.943	0.943	0.651	0.838	0.896	0.363	0.407	0.979	1			
NO_3	–0.051	–0.234	–0.234	–0.077	–0.185	–0.226	–0.151	–0.192	–0.185	–0.222	1		
F	0.132	0.32	0.32	0.28	0.083	0.341	0.27	0.67	0.107	0.071	–0.482	1	
TH	0.336	0.886	0.886	0.802	0.88	0.74	0.413	0.437	0.922	0.887	–0.149	0.202	1

(b) Deep systems

	pH	EC	TDS	Ca	Mg	Na	K	HCO_3	SO_4	Cl	NO_3	F	TH
pH	1												
EC	–0.348	1											
TDS	–0.348	1	1										
Ca	–0.626	0.462	0.462	1									
Mg	–0.384	0.599	0.599	0.752	1								
Na	0.24	0.378	0.378	–0.495	–0.459	1							
K	0.374	0.338	0.338	–0.079	0.129	0.456	1						
HCO_3	0.054	0.121	0.121	–0.132	–0.125	0.245	–0.113	1					
SO_4	–0.279	0.791	0.791	0.549	0.501	0.25	0.393	–0.396	1				
Cl	–0.191	0.714	0.714	0.296	0.325	0.463	0.596	–0.454	0.887	1			
NO_3	–0.137	–0.328	–0.328	–0.159	–0.413	–0.059	–0.346	–0.099	–0.319	–0.254	1		
F	–0.129	–0.356	–0.356	0.11	–0.156	–0.375	–0.257	–0.018	–0.335	–0.411	0.807	1	
TH	–0.404	–0.011	–0.011	0.713	0.442	–0.524	0.057	–0.466	0.333	0.25	–0.192	0.1	1

Palar delta

Near surface system: A good correlation is observed between EC and Mg (r = 0.809), Na (r = 0.956), SO_4 (r = 0.952), Cl (r = 0.943) and TH (r = 0.886) and moderately with Ca (r = 0.684) HCO_3 (r = 0.665) indicating all major cations and anions are playing a reasonably good role in the evolution of hydrochemical processes of the delta system. Ca correlates moderately with Na (r = 0.535), SO_4 (r = 0.671), and Cl (r = 0.651). Mg strongly correlates with Na (r = 0.704), SO_4 (r = 0.875), Cl (r = 0.838) and with TH (r = 0.880). Na correlates well with HCO_3 (r = 0.702), SO_4 (r = 0.881) Cl (r = 0.896) and TH (r = 0.740). Similarly HCO_3 shows correlation with F (r = 0.670). SO_4 strong correlates with Cl (r = 0.979) and TH (r = 0.922).

Deep system: A correlation is observed between EC and Mg (r = 0.599), SO_4 (r = 0.791), and Cl (r = 0.714). Ca correlate positively with Mg(r = 0.752), and SO_4 (r = 0.549). Mg correlates with SO_4 (r = 0.501). Similarly K correlates with Cl (r = 0.596). SO_4 strong correlates with Cl (r = 0.887). NO_3 correlates with F (r = 0.807).

Factor analysis: Factor analysis has been applied for near surface and deep systems and the results are furnished below in Table 5.10(a,b).

Table 5.10(a) Factor analysis for near surface and deep systems of Palar delta

Near Surface					Deep systems				
Component	Eigen value	% of Variance	Cumulative Eigen value	Cumulative %	Component	Eigen value	% of Variance	Cumulative Eigen value	Cumulative %
1	7.725	59.421	7.725	59.421	1	5.031	38.701	5.031	38.701
2	1.675	12.882	9.4	72.304	2	3.082	23.705	8.113	62.407
3	1.033	7.949	10.433	80.253	3	1.626	12.511	9.739	74.918
4	0.942	7.246	11.375	87.499	4	1.492	11.473	11.231	86.391
5	0.727	5.589	12.102	93.088	5	0.808	6.216	12.039	92.607
6	0.492	3.783	12.594	96.871	6	0.531	4.086	12.57	96.693
7	0.287	2.207	12.881	99.078	7	0.266	2.047	12.836	98.74
8	0.093	0.716	12.974	99.794	8	0.081	0.62	12.917	99.36
9	0.016	0.119	12.99	99.914	9	0.069	0.529	12.986	99.889
10	0.009	0.068	12.999	99.982	10	0.012	0.096	12.998	99.985
11	0.002	0.012	13.001	99.994	11	0.002	0.015	13	100
12	0.001	0.006	13.002	100	12	2.8E-16	2.1E - 15	13	100
13	2.3E - 16	1.8E - 15	13.002	100	13	–1.7E-17	–1.3E-16	13	100

Table 5.10(b) Loading for varimax rotated factor matrix for Palar delta

Near surface systems				Deep systems				
Variable	Factor 1	Factor 2	Factor 3	Variable	Factor 1	Factor 2	Factor 3	Factor 4
pH	0.2	–0.003	0.86	pH	–0.333	–0.66	0.372	0.128
EC	0.906	0.287	0.278	EC	0.935	0.213	0.185	–0.166
TDS	0.906	0.287	0.278	TDS	0.935	0.213	0.185	–0.166
Ca	0.769	0.238	–0.207	Ca	0.326	0.892	–0.031	0.107
Mg	0.832	–0.014	0.3	Mg	0.358	0.769	0.31	0.022
Na	0.804	0.306	0.348	Na	0.57	–0.762	0.08	–0.114
K	0.361	0.383	0.218	K	0.423	–0.343	0.405	0.411
HCO_3	0.333	0.612	0.57	HCO_3	–0.008	–0.128	0.101	–0.901
SO_4	0.967	0.077	0.19	SO_4	0.823	0.247	0.168	0.385
Cl	0.964	0.062	0.143	Cl	0.834	–0.018	0.17	0.502
NO_3	–0.114	–0.685	0.151	NO_3	–0.132	–0.17	–0.953	0.025
F	0.026	0.92	0.151	F	–0.277	0.137	–0.824	0.037
TH	0.942	0.118	0.114	TH	–0.053	0.7	0.065	0.568

Table 5.10(a) shows the eigen values, % of variance, cumulative eigen value, cumulative percent of variance for the factor analysis of hydrochemical data of Palar delta basin. The first three factors for the near surface system and the first four factors for deep systems were selected to represent the hydrochemical processes. The first three factors and four factors by and large explains 80.2% (Near surface system) and 86.3% (deep systems) of the total variances.

Near surface systems

Factor 1 explains the largest proportion (59.42%) of the total variance and has high positive (0.80-0.96) loadings for EC, Cl, Ca, Mg, Na, SO_4 and TH and the rest of the parameters are weakly loaded and are insignificant. The strong positive loadings on Mg, SO_4 and Cl could be attributed to salinisation processes and Ca, Na could be related to weathering and ion exchanging processes in the aquifer material.

Factor 2 accounts for 12.882% of the total variance and shows high loadings on F (0.920) and moderately loading on HCO_3 (0.612). This high loading factor indicates recharge impacts and ion exchange processes in the upstream area.

Factor 3 accounts for 7.949 % of the total variance and exhibits high positive loading on pH (0.860) could be attributed to anthropogenic activity and industrial disposal sites.

Deep systems

Similarly, ***Factor1*** explains the largest proportion (38.7%) of the total variance and has high positive (0.82-0.93) loadings for EC, SO_4 and Cl and the rest of the parameters are low loadings (<0.5)(Table11(b)) and are insignificant. The factors with high loading values are attributed to the salinisation processes.

Factor 2 accounts for 23.7% of the total variance and shows high loadings (0.76-0.89) on Ca and Mg. This high loading factor indicates mixed hydro chemical processes related to weathering in the upstream areas and salinisation processes in the central and outfall regions of the coast.

Factor 3 accounts for 12.5% of the total variance and exhibits high negative loading on NO_3 (–0.95) and F (–0.82) and are insignificant.

Factor 4 accounts of 11.4% of the total variance and exhibits moderate loading on Cl (0.502) and TH (0.568).

Table 5.11(a,b) Correlation matrix analysis for Ponnaiyar

(a) Near surface

	pH	EC	TDS	Ca	Mg	Na	K	HCO_3	SO_4	Cl	NO_3	F	TH
pH	1												
EC	–0.227	1											
TDS	–0.227	1	1										
Ca	–0.362	0.643	0.643	1									
Mg	–0.153	0.942	0.942	0.541	1								
Na	–0.222	0.974	0.974	0.475	0.905	1							
K	0.395	0.329	0.329	0.319	0.19	0.26	1						
HCO_3	0.179	0.632	0.632	0.344	0.598	0.621	0.326	1					
SO_4	–0.146	0.92	0.92	0.618	0.79	0.901	0.411	0.662	1				
Cl	–0.304	0.985	0.985	0.613	0.923	0.973	0.277	0.533	0.877	1			
NO_3	0.26	0.019	0.019	0.269	0.136	–0.133	0.322	–0.014	–0.089	–0.07	1		
F	–0.222	0.324	0.324	0.252	0.278	0.311	–0.008	0.581	0.389	0.237	0.019	1	
TH	–0.324	0.898	0.898	0.854	0.864	0.793	0.322	0.515	0.779	0.894	0.146	0.233	1

(b) Deep systems

	pH	EC	TDS	Ca	Mg	Na	K	HCO_3	SO_4	Cl	TH
pH	1										
EC	0.469	1									
TDS	0.469	1	1								
Ca	0.284	0.551	0.551	1							
Mg	0.413	0.466	0.466	0.385	1						
Na	0.43	0.917	0.917	0.196	0.29	1					
K	0.305	0.685	0.685	0.314	0.505	0.658	1				
HCO_3	0.496	0.503	0.503	0.671	0.557	0.277	0.26	1			
SO_4	0.398	0.885	0.885	0.488	0.367	0.821	0.471	0.464	1		
Cl	0.354	0.907	0.907	0.249	0.327	0.951	0.727	0.141	0.739	1	
TH	0.271	0.435	0.435	0.912	0.553	0.071	0.203	0.766	0.467	0.069	1

Ponnaiyar delta

Near surface systems: A good correlation is observed between EC and Mg (r = 0.942), Na (r = 0.974), SO_4 (r = 0.92), Cl (r = 0.985) and TH (r = 0.898) and moderately with Ca (r = 0.643) and HCO_3 (r = 0.632).Ca moderately correlates with Mg (r =0.541), SO_4 (r =0.618) and Cl (r = 0.613). Mg strongly correlates with Na (r = 0.905), SO_4 (r = 0.79), Cl (r = 0.923) and with TH (r = 0.864) moderately with HCO_3 (r = 0.598). Na correlates well with SO_4 (r = 0.901), Cl (r = 0.973) and TH (r = 0.793). HCO_3 moderately correlates with SO_4 (r =0.662), Cl (r =0.533) and F (r = 0.581). Similarly SO_4 shows good correlation with Cl (r = 0.877) and TH (r = 0.799).

Deep systems: A strong correlation is observed between EC and Na (r = 0.917), SO_4 (r = 0.885), and Cl (r = 0.907) and moderately correlate with Ca(r = 0.551), K (r = .685) and HCO_3 (r = 0.503). Ca strongly correlate with TH (r = 0.912) and moderately with HCO_3 (r = 0.671). Mg moderately correlates with K (r = 0.505), HCO_3 (r = 0.557) and TH (r = 0.553). Na correlates strongly with SO_4 (r = 0.821), Cl (r = 0.951) and moderately with K (r = 0.658). K strongly correlate with Cl (r = 0.727). SO_4 strong correlates with Cl (r = 0.739).

Table 5.12(a) Factor analysis for near surface and deep systems of Ponnaiyar-Vellar delta

Near Surface					Deep systems				
Component	**Eigen value**	**% of Variance**	**Cumulative Eigen value**	**Cumulative %**	**Compo-nent**	**Eigen value**	**% of Variance**	**Cumulative Eigen value**	**Cumula-tive %**
1	7.663	58.945	7.663	58.945	1	6.367	57.886	6.367	57.886
2	1.734	13.336	9.397	72.281	2	2.182	19.837	8.549	77.723
3	1.254	9.65	10.651	81.931	3	0.877	7.976	9.426	85.698
4	1.066	8.2	11.717	90.131	4	0.735	6.682	10.161	92.38
5	0.636	4.89	12.353	95.021	5	0.398	3.62	10.559	96
6	0.345	2.655	12.698	97.675	6	0.27	2.459	10.829	98.459
7	0.161	1.242	12.859	98.917	7	0.144	1.31	10.973	99.768
8	0.098	0.757	12.957	99.674	8	0.017	0.154	10.99	99.922
9	0.034	0.263	12.991	99.937	9	0.006	0.055	10.996	99.978
10	0.008	0.061	12.999	99.997	10	0.002	0.022	10.998	100
11	0	0.002	12.999	100	11	1.3E-16	1.23E-15	1.10E+01	100
12	4.6E − 05	0	1.3E + 01	100					
13	4.2E − 16	3.2E-15	1.3E + 01	100					

Table 5.12(b) Loading for varimax rotated factor matrix for Ponnaiyar-Vellar delta

Near surface systems					Deep systems		
Variable	**Factor 1**	**Factor 2**	**Factor 3**	**Factor 4**	**Variable**	**Factor 1**	**Factor 2**
pH	–0.239	0.917	–0.023	–0.003	pH	0.406	0.409
EC	0.978	0.007	0.172	0.076	EC	0.905	0.396
TDS	0.978	0.007	0.172	0.076	TDS	0.905	0.396
Ca	0.623	–0.256	0.107	0.602	Ca	0.188	0.865
Mg	0.909	0.013	0.152	0.102	Mg	0.314	0.618
Na	0.96	0.023	0.17	–0.129	Na	0.982	0.046
K	0.329	0.665	–0.036	0.395	K	0.739	0.187
HCO_3	0.544	0.352	0.672	–0.045	HCO_3	0.178	0.87
SO_4	0.899	0.098	0.267	0.006	SO_4	0.785	0.394
Cl	0.988	–0.07	0.06	0.009	Cl	0.981	0.015
NO_3	–0.087	0.235	0.017	0.874	TH	0.04	0.964
F	0.154	–0.164	0.945	0.069			
TH	0.895	–0.142	0.072	0.345			

The first four factors (Near surface system) and two factors (deep systems) by and large explains 90.1% and 77.7% of the total variances respectively.

Near surface systems

Factor 1 explains the largest proportion (58.9%) of the total variance and has high positive (0.89-0.98) loadings for EC, Mg, Na, SO_4 and Cl and has moderate loadings on Ca (0.62), HCO_3 (0.54).

Factor 2 accounts for 13.3% of the total variance and has high loading on pH (0.917) and moderate loading on K (0.66).

Factor 3 accounts for 9.6% of the total variance and has high loading on F (0.945) and moderate loading on HCO_3 (0.672).

Factor 4 accounts for 8.2% of the total variance and has high loadings on NO_3 (0.874) and moderate loading on Ca (0.602).

Deep systems

Factor 1 explains the largest proportion 57.8% of the total variance and has high positive loadings (0.739-0.982) on EC, Na, K, SO_4 and Cl.

Factor 2 accounts for 19.8% of the total variance and has high loadings on Ca, HCO_3 and TH and moderate loading on Mg (0.618).

Table 5.13(a)(b) Correlation matrix analysis for Cauvery

(a) Near surface

	pH	**EC**	**TDS**	**Ca**	**Mg**	**Na**	**K**	**HCO_3**	**SO_4**	**Cl**	**NO_3**	**F**	**TH**
pH	1												
EC	-0.054	1											
TDS	–0.054	1	1										
Ca	–0.081	0.593	0.593	1									

Table 5.13 *Contd...*

	pH	EC	TDS	Ca	Mg	Na	K	HCO_3	SO_4	Cl	NO_3	F	TH
Mg	–0.207	0.898	0.898	0.712	1								
Na	0.039	0.951	0.951	0.341	0.752	1							
K	–0.318	0.363	0.363	0.333	0.414	0.225	1						
HCO_3	–0.371	0.284	0.284	-0.022	0.165	0.302	0.438	1					
SO_4	–0.22	0.596	0.596	0.335	0.654	0.566	–0.074	0.079	1				
Cl	0.01	0.984	0.984	0.595	0.879	0.936	0.387	0.188	0.488	1			
NO_3	0.134	–0.016	–0.016	0.114	–0.029	–0.053	–0.048	0.058	–0.17	–0.028	1		
F	0.053	0.375	0.375	–0.106	0.229	0.502	–0.137	0.216	0.164	0.373	0.432	1	
TH	–0.086	0.749	0.749	0.87	0.842	0.543	0.453	0.119	0.507	0.727	0.136	–0.011	1

(b) Deep systems

	pH	EC	TDS	Ca	Mg	Na	K	HCO_3	SO_4	Cl	TH
pH	1										
EC	0.125	1									
TDS	0.17	0.994	1								
Ca	0.075	0.492	0.527	1							
Mg	0.11	0.671	0.694	0.539	1						
Na	0.143	0.964	0.954	0.369	0.578	1					
K	0.127	0.285	0.29	0.098	0.343	0.208	1				
HCO_3	0.185	0.285	0.325	0.26	0.278	0.254	0.457	1			
SO_4	0.296	0.734	0.73	0.209	0.516	0.725	0.31	0.186	1		
Cl	0.084	0.937	0.933	0.552	0.681	0.938	0.203	0.12	0.599	1	
TH	–0.017	0.564	0.597	0.824	0.777	0.453	0.201	0.303	0.336	0.6	1

Cauvery delta

Near surface systems`

A good correlation is observed between EC and Mg ($r = 0.898$), Na ($r = 0.951$), Cl ($r = 0.984$) and TH ($r = 0.749$) and moderately with Ca ($r = 0.593$) and SO_4 ($r = 0.596$). Ca correlate strongly with Mg ($r = 0.712$) and TH ($r = 0.87$) and moderately with Cl ($r = 0.595$). Mg strongly correlates with Na ($r = 0.752$), Cl ($r = 0.879$) and with TH ($r = 0.842$) moderately with and SO_4 ($r = 0.654$). Na correlates well with Cl ($r = 0.936$) and moderately with SO_4 ($r = 0.566$) and F ($r = 0.502$). Similarly Cl shows correlation with TH ($r = 0.727$).

Deep systems

A strong correlation is observed between EC and Na ($r = 0.964$), SO_4 ($r = 0.734$), and Cl ($r = 0.937$) and moderately correlate with Mg ($r = 0.671$) and TH ($r = 0.564$). Ca correlate strongly with TH ($r = 0.824$) and moderately with Mg ($r = 0.539$) and Cl ($r = 0.552$). Mg moderately correlates with Na ($r = 0.578$), SO_4 ($r = 0.516$) and Cl ($r = 0.681$). Na correlates strongly with SO_4 ($r = 0.725$), and Cl ($r = 0.938$). SO_4 correlates moderately with Cl ($r = 0.599$).

Table 5.14(a) Factor analysis for near surface and deep systems of Cauvery delta

Near surface					Deep systems				
Comp-onent	**Eigen value**	**% of Variance**	**Cumulative Eigen value**	**Cumulative %**	**Comp-onent**	**Eigen value**	**% of Variance**	**Cumulative Eigen value**	**Cumulative %**
1	6.495	49.96	6.495	49.96	1	5.965	54.228	5.965	54.228
2	1.745	13.421	8.24	63.381	2	1.368	12.433	7.333	66.661
3	1.522	11.709	9.762	75.09	3	1.337	12.15	8.67	78.811
4	1.273	9.794	11.035	84.883	4	0.895	8.14	9.565	86.951
5	0.919	7.07	11.954	91.954	5	0.599	5.442	10.164	92.392
6	0.464	3.57	12.418	95.524	6	0.385	3.504	10.549	95.896
7	0.279	2.145	12.697	97.669	7	0.309	2.814	10.858	98.71
8	0.171	1.317	12.868	98.986	8	0.089	0.808	10.947	99.517
9	0.071	0.545	12.939	99.531	9	0.045	0.408	10.992	99.925
10	0.059	0.457	12.998	99.987	10	0.006	0.052	10.998	99.977
11	0.001	0.011	12.999	99.998	11	0.002	0.023	1.10E + 01	100
12	0	0.002	1.30E+01	100					
13	–2.6E–18	–2.0E-17	1.3E+01	100					

Table 5.14(b) Loading for varimax rotated factor matrix for Cauvery delta

Near surface systems					Deep systems			
Variable	**Factor 1**	**Factor 2**	**Factor 3**	**Factor 4**	**Variable**	**Factor 1**	**Factor 2**	**Factor 3**
pH	–0.004	–0.027	–0.718	0.29	pH	0.233	-0.256	0.505
EC	0.927	0.31	0.144	0.075	EC	0.91	0.337	0.145
TDS	0.927	0.31	0.144	0.075	TDS	0.894	0.369	0.182
Ca	0.373	0.853	–0.035	0.021	Ca	0.229	0.855	0.045
Mg	0.805	0.498	0.15	–0.046	Mg	0.518	0.63	0.233
Na	0.96	0.03	0.101	0.102	Na	0.943	0.204	0.099
K	0.082	0.523	0.688	0.041	K	0.125	0.127	0.775
HCO_3	0.204	–0.138	0.825	0.191	HCO_3	0.022	0.306	0.778
SO_4	0.728	0.045	–0.041	–0.342	SO_4	0.815	–0.011	0.285
Cl	0.894	0.343	0.087	0.1	Cl	0.873	0.41	–0.019
NO_3	–0.101	0.146	–0.065	0.892	TH	0.307	0.897	0.102
F	0.526	–0.431	0.029	0.634				
TH	0.554	0.782	0.075	0.047				

The first four factors (near surface systems) and three factors (deep systems) by and large explains 84.8 % and 78.8 % of the total variances respectively.

Near surface systems

Factor 1 explains the largest proportion (49.9%) of the total variance and has high positive loadings (0.72-0.92) for EC, Mg, SO_4 and Cl and has moderate loadings on F (0.52), TH (0.54).

Factor 2 accounts for 13.4% of the total variance and has high loading on Ca (0.85), TH (0.78) and moderate loading on K (0.52).

Factor 3 accounts for 11.7% of the total variance and has high loading on HCO_3 (0.825) and moderate loading on K (0.688).

Factor 4 accounts for 9.7% of the total variance and has high loadings on NO_3 (0.892) and moderate loading on F (0.634).

Deep systems

Factor 1 explains the largest proportion 54.2% of the total variance and has high positive loadings (0.81-0.94) on EC, Na, SO_4 and Cl and has moderate loading on Mg (0.518).

Table 5.15(a,b) Correlation matrix analysis for Vaigai

(a) Near surface

	pH	EC	TDS	Ca	Mg	Na	K	HCO_3	SO_4	Cl	NO3	F	TH
pH	1												
EC	0.435	1											
TDS	0.435	1	1										
Ca	–0.515	0.84	0.84	1									
Mg	–0.46	0.891	0.891	0.841	1								
Na	–0.372	0.983	0.983	0.741	0.853	1							
K	–0.372	.98	.98	0.274	0.311	0.325	1						
HCO_3	0.106	–0.044	–0.044	–0.117	–0.133	–0.053	0.362	1					
SO_4	–0.477	0.879	0.879	0.82	0.879	0.854	0.274	-0.1	1				
Cl	–0.42	0.995	0.995	0.837	0.894	0.982	0.326	-0.13	0.873	1			
NO_3	–0.314	0.081	0.081	–0.057	0.006	0.082	0.324	0.266	0.038	0.032	1		
F	0.257	–0.084	–0.084	–0.145	–0.013	–0.034	–0.205	0.08	0.15	–0.093	0.015	1	
TH	–0.513	0.925	0.925	0.968	0.92	0.854	0.324	-0.1	0.9	0.921	0.006	-0.097	1

(b) Deep systems

	pH	EC	TDS	Ca	Mg	Na	K	HCO_3	SO_4	Cl	TH
pH	1										
EC	–0.086	1									
TDS	–0.086	1	1								
Ca	–0.139	0.995	0.995	1							
Mg	–0.416	0.91	0.91	0.939	1						
Na	–0.038	0.998	0.998	0.989	0.884	1					
K	–0.076	0.997	0.997	0.993	0.898	0.997	1				
HCO_3	–0.576	–0.452	–0.452	–0.413	–0.185	–0.475	–0.438	1			
SO_4	–0.006	0.995	0.995	0.986	0.88	0.996	0.992	–0.504	1		
Cl	–0.075	0.999	0.999	0.995	0.907	0.998	0.997	–0.461	0.994	1	
TH	–0.468	0.894	0.894	0.924	0.997	0.865	0.882	–0.154	0.861	0.889	1

Factor 2 accounts for 12.4% of the total variance and has high loadings on Ca (0.85), TH (0.897) and moderate loading on Mg (0.63).

Factor 3 accounts for 12.1% of the total variance and has high loadings on K (0.775), HCO_3 (0.778) and has moderate loading on pH (0.505).

Vaigai delta

Near surface systems: A good correlation is observed between EC and Ca (r = 0.84), Mg (r = 0.891), Na (r = 0.983), K (r = 0.98), SO_4 (r = 0.879), Cl (r = 0.995) and TH (r = 0.925) indicating all major cations and anions are playing a reasonably good role in the evolution of hydrochemical processes of the delta system. Ca correlate strongly with Na (r = 0.741), Mg (r = 0.841), SO_4 (r = 0.82), Cl (r = 0.837) and TH (r = 0.963). Mg strongly correlates with Na (r = 0.853), SO_4 (r = 0.879), Cl (r = 0.894) and with TH (r = 0.92). Na correlate well with SO_4 (r = 0.854), Cl (r = 0.982) and with TH (r = 0.854). Similarly SO_4 correlates strongly with Cl (r = 0.873) and with TH (r = 0.9).

Deep systems: A strong correlation is observed between EC and Ca (r = 0.995), Mg (r = 0.91), Na (r = 0.998), K (r = 0.997), SO_4 (r = 0.995), Cl (r = 0.999) and TH (r = 0.894). Ca correlates strongly with Mg(r = 0.939), Na(r = 0.989), K(r = 0.993), SO_4(r = 0.986), Cl(r = 0.995), TH(r = 0.924). Mg correlates strongly with Na(r = 0.884), K (r =0.898), SO_4(r = 0.88), Cl(r =0.907) and TH(r = 0.997). Na correlates strongly with K(r = 0.997), SO_4(r = 0.996), Cl(r = 0.998) and TH(r = 0.865). K strongly correlates with SO_4(r = 0.992), Cl(r = 0.997) and TH(r = 0.882). Similarly SO_4 strongly correlates with Cl(r = 0.994) and TH(r = 0.861).

Table 5.16(a) Factor analysis for near surface and deep systems of Vaigai delta

Near surface					Deep systems				
Component	Eigen value	% of Variance	Cumulative Eigen value	Cumulative %	Component	Eigen value	% of Variance	Cumulative Eigen value	Cumulative %
1	7.689	59.148	7.689	59.148	1	8.854	80.493	8.854	80.493
2	1.631	12.548	9.32	71.696	2	1.78	16.181	10.634	96.673
3	1.217	9.36	10.537	81.056	3	0.283	2.575	10.917	99.248
4	0.979	7.534	11.516	88.59	4	0.072	0.65	10.989	99.898
5	0.584	4.494	12.1	93.084	5	0.006	0.054	10.995	99.952
6	0.443	3.405	12.543	96.489	6	0.004	0.034	10.999	99.987
7	0.23	1.77	12.773	98.259	7	0.001	0.007	11	99.994
8	0.125	0.963	12.898	99.222	8	0.001	0.006	11.001	100
9	0.09	0.689	12.988	99.91	9	6.9E-16	6.2E-15	11.001	100
10	0.008	0.058	12.996	99.968	10	–3.2E-17	–2.9E-16	11.001	100
11	0.003	0.021	12.999	99.989	11	–4.5E-16	–4.1E-15	1.10E + 01	100
12	0.001	0.011	1.30E + 01	100					
13	1.5E-16	1.15E-15	1.30E + 01	100					

Table 5.16(b) Loading for varimax rotated factor matrix for Vaigai delta

Near surface systems				Deep systems		
Variable	Factor 1	Factor 2	Factor 3	Variable	Factor 1	Factor 2
pH	–0.528	–0.114	0.54	pH	–0.137	–0.943
EC	0.981	0.019	0.062	EC	0.997	–0.046
TDS	0.981	0.019	0.062	TDS	0.997	–0.046
Ca	0.903	–0.106	–0.097	Ca	0.999	0.012
Mg	0.936	–0.084	0.059	Mg	0.936	0.309
Na	0.942	0.005	0.125	Na	0.991	–0.092
K	0.372	0.698	0.023	K	0.992	–0.049
HCO_3	–0.089	0.739	0.344	HCO_3	–0.435	0.797
SO_4	0.922	–0.082	0.183	SO_4	0.989	–0.123
Cl	0.977	–0.056	0.042	Cl	0.996	–0.057
NO_3	0.078	0.729	–0.109	TH	0.921	0.354
F	–0.09	–0.143	0.85			
TH	0.971	–0.063	–0.023			

The first three factors (Near surface system) and two factors (deep systems) by and large explains 81% and 96.6% of the total variances respectively.

Near surface systems

Factor 1 explains the largest proportion (59.1%) of the total variance and has high positive loadings (0.88-0.973) for EC, Ca, Mg, Na, SO_4, Cl and TH.

Factor 2 accounts for 12.5% of the total variance and has high loading (0.711-0.769) on K, HCO_3 and NO_3.

Factor 3 accounts for 9.3% of the total variance and has high loading on F (0.805) and moderate loading on pH (0.626).

Deep systems

Factor 1 explains the largest proportion 80.5% of the total variance and has high positive loadings (0.94-0.99) except HCO_3.

Factor 2 accounts for 16.1% of the total variance and has high loadings on pH (0.932). Rest of the parameters are weakly loaded.

5.4.6 Classification and irrigation suitability of groundwaters based on sodium percentages, sodium adsorption ratios and residual sodium carbonate

Sodium percentage: The classification of groundwater with respect of percentage of sodium in Palar, Ponnaiyar-Vellar, Cauvery and Vaigai for the near surface and deep systems are discussed in the following paragraph and are presented below in Tables 5.17(a,b,c,d).

Palar delta basin: The sodium % in near surface system of Palar delta has a range of values between 18.3 - 95.6 and for deep system between 30 – 96.4. About 82.5% of the samples of the near surface system fall in the range of excellent to permissible limits and the rest 17.5% of the samples lie in the range of doubtful to unsuitable categories. In the case of deep systems about 61% of the samples lie in the range of good to permissible limits and 39% of the samples lie in the range of doubtful to unsuitable categories. As per the Bureau of Indian Standards (BIS 1991) the maximum sodium percentage of 60% is recommended for irrigation purposes.

Table 5.17(a) Sodium percentage and water classes of Palar delta basin

Sodium (%)	Water Class	No. of Samples	
		Near Surface system	Deep system
< 20	Excellent	1 (18.2)	0
20 - 40	Good	10 (2.33-38.8)	4 (30.07-38.9)
40 - 60	Permissible	22 (40.5-57.1)	7 (52.7-58.7)
60 - 80	Doubtful	6 (60.4-77.7)	2 (62.3-66.8)
> 80	Unsuitable	1 (95.68)	5 (90-96.4)

Ponnaiyar delta basin: The sodium percentages in the case of Ponnaiyar delta have a range of values between 11.9% to 83.7% and for deep systems between 22.9% to 94.6%. About 7% of the samples from the near surface systems lie in the range of excellent to permissible limits and the rest 23% lie in the doubtful to unsuitable categories. In the case of deep systems 82 % of the sample lie in the range of excellent to permissible limits while 18% of the samples lie in the range of doubtful to unsuitable categories.

Table 5.17(b) Sodium percentage and water classes of Ponnaiyar delta basin

Sodium (%)	Water Class	No. of Samples	
		Near Surface system	Deep system
< 20	Excellent	3 (11.9-19.4)	0
20 - 40	Good	13 (20.9-39.9)	14 (22.9-39.4)
40 - 60	Permissible	17 (42-59.2)	9 (40.1-58.2)
60 - 80	Doubtful	8 (60.4-78.2)	4 (66.7-77.5)
> 80	Unsuitable	2 (82-83.7)	1 (94.6)

Cauvery delta basin : The sodium percentages in the case of near surface systems of Cauvery delta have a range of values between 23.5% to 94.1% and for deep systems between 9 to 86.6%. About 67% of the samples from the near surface systems fall in the range of excellent to permissible limits and the rest 33% of the samples lie in the doubtful to unsuitable categories. In the case of deep systems 76% of the samples lie in the range of excellent to permissible limits and the rest 24% lie in the range of doubtful to unsuitable categories.

Table 5.17(c) Sodium percentage and water classes of Cauvery delta basin

Sodium (%)	Water Class	No. of Samples	
		Near Surface system	Deep system
< 20	Excellent	0	2 (9-17.4)
20 - 40	Good	7 (23.5-38.7)	12 (22-39)
40 - 60	Permissible	15 (40.3-59.8)	12 (40.1-59.6)
60 - 80	Doubtful	7 (60.8-79.5)	6 (63.3-78.4)
> 80	Unsuitable	4 (86.3-94.1)	2 (86.5-86.6)

Vaigai delta basin: The sodium percentages in near surface systems of Vaigai delta have a range of values between 27.4 % to 82.8 % and for deep systems between 32.9% to 91.8 %. About 50% of the samples from the near surface systems lie in the range of excellent to permissible limits and the rest 50 % lie in the doubtful to unsuitable categories. In the case of deep systems 33% of the samples lie in the range of excellent to permissible limits and the rest 67% of the samples lie in the categories doubtful to unsuitable.

Table 5.17(d) Sodium percentage and water classes of Vaigai delta basin

Sodium (%)	Water Class	No. of Samples	
		Near Surface system	Deep system
< 20	Excellent	0	0
20 – 40	Good	2 (27.4-29.9)	1 (32.1)
40 – 60	Permissible	14 (40.3-59.7)	2 (42.6-58.9)
60 - 80	Doubtful	11 (60.6-79.7)	5 (73.2-79.3)
> 80	Unsuitable	5 (80.3-82.8)	1 (91.8)

- The above data structure indicates that the majority of the samples of both near surface and deep systems of Palar, Ponnaiyar and Cauvery delta fall in the range of excellent to permissible ranges and only minor amount of samples lie in the range of doubtful to unsuitable categories. However in the case of Vaigai delta 50% of the near surface system fall in excellent to permissible range and the rest 50% fall in doubtful to unsuitable range. However in the case of deep systems of Vaigai delta only 33% of the samples lie in the excellent to permissible limits while 67% fall in the doubtful to unsuitable limits indicating threshold limits of contamination for the near surface system and heavy contamination for deep system.

- Groundwaters containing excessive sodium contents can damage the soil structure as sodium replaces Ca and Mg and also affects soil permeability and hardening of soils.

Sodium Adsorption Ratios (SAR)

The classification of groundwater samples based on SAR (Sodium Adsorption Ratios) for Palar, Ponnaiyar-Vellar, Cauvery, Vaigai are described below and shown graphically in Figures 5.19(a, b, c and d).

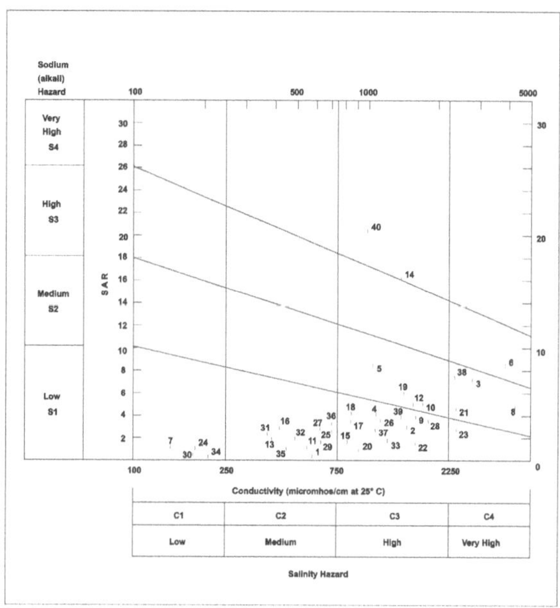

Near surface (a_1)

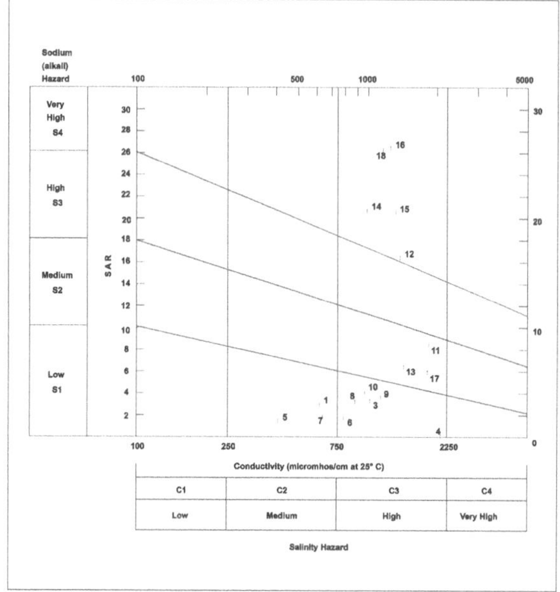

Deep systems (a_2)

Fig. 5.19(a) Sodium adsorption ratios for near surface (a_1) and deep systems (a_2) in Palar delta system

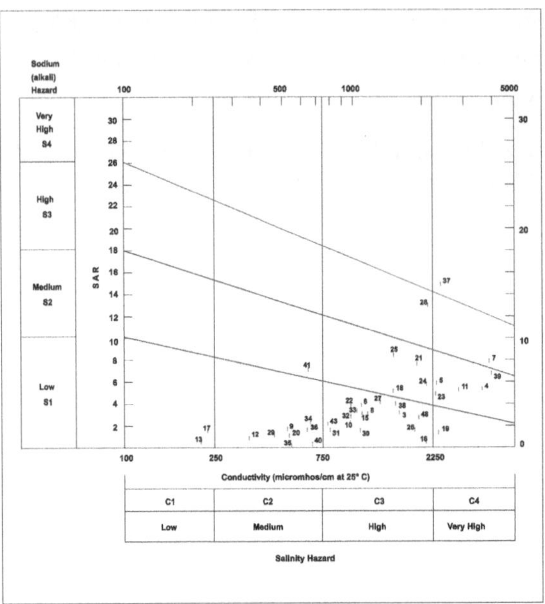

Near surface (b_1)

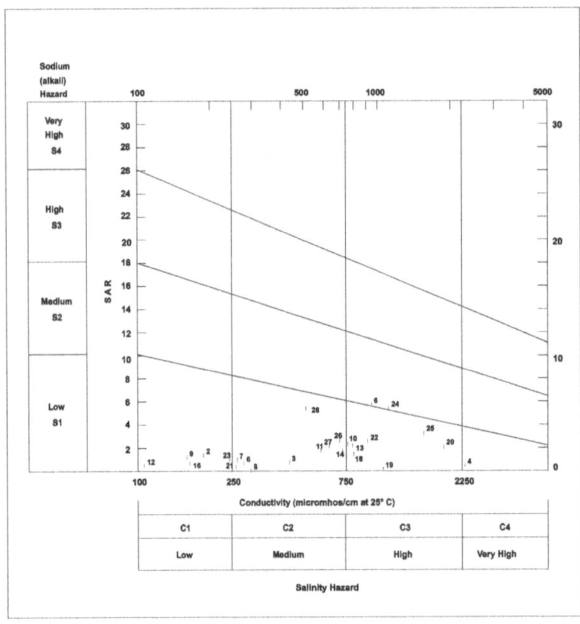

Deep systems (b_2)

Fig. 5.19(b) Sodium adsorption ratios for near surface (b_1) and deep systems (b_2) in Ponnaiyar delta system

Palar delta

Near surface: About 38% of the samples from the near surface system of Palar delta lie in the fields of C_1S_1 and C_2S_1 (low salinity/low alkalinity to medium salinity/low alkalinity). About 36% of the samples lie in the fields of C_3S_1, C_4S_1 (high salinity/low alkalinity to very high salinity/ low alkalinity). The rest of the samples about 20% of the samples are spread in the fields of C_3S_2, C_4S_2 (high salinity/medium alkalinity to very high salinity/medium alkalinity). About 8% of the samples lie in the fields of C_3S_4, C_4S_3 (high salinity/very high alkalinity to very high salinity/high alkalinity).

In the case of deep systems 50% of the samples lie in the fields of C_2S_1, C_3S_1 (medium salinity/low alkalinity to high salinity/low alkalinity). About 45% of the samples lie in the fields of C_3S_2 and C_3S_4 (high salinity/medium alkalinity to high salinity/very high alkalinity).

Ponnaiyar delta

Near surface: About 24% of the samples of the near surface system lie in the fields of C_1S_1, C_2S_1 (low salinity/low alkalinity to medium salinity/low alkalinity). About 39% of the samples lie in the fields of C_3S_1, C_4S_1 (high salinity/low alkalinity to very high salinity/low alkalinity). In addition about 21% of the samples lie in the fields of C_3S_2 to C_4S_2 (high salinity/medium alkalinity to very high salinity/medium alkalinity). The rest of the samples running to 6% lie in the fields of C_3S_3, C_4S_3 and C_4S_4 (high salinity/high alkalinity, very high salinity/high alkalinity and very high salinity/very high alkalinity).

In the case of deep systems about 54% of the samples lie in the fields of C_1S_1 and C_2S_1 (low salinity/low alkalinity to medium salinity/low alkalinity). About 29% of the samples lie in the fields of C_3S_1, C_4S_1 (high salinity/low alkalinity to very high salinity/low alkalinity) and 7% of the samples lie in the field C_3S_2 (high salinity/medium alkalinity).

Cauvery delta

Near surface: About 18% of the samples lie in the field of C_2S_1 (medium salinity/low alkalinity). About 33% of the samples lie in the field of C_3S_1 (high salinity/low alkalinity). About 18% of the samples lie in the fields of C_3S_2, C_3S_3 and C_3S_4 (high salinity/medium alkalinity, high salinity/high alkalinity and high salinity/very high alkalinity). In addition about 21% of the samples lie in the field of C_4S_2, C_4S_3 and C_4S_4 (very high salinity/medium alkalinity, very high salinity/high alkalinity and very high salinity/very high alkalinity).

Deep systems: About 35% of the samples lie in the field of C_1S_1, C_2S_1 (low salinity/low alkalinity and medium salinity/low alkalinity). About 47% of the samples lie in the fields of C_3S_1, C_3S_2 (high salinity/low alkalinity to high salinity/medium alkalinity). In addition about 15% of the samples lie in the fields of C_3S_3, C_3S_4 and C_4S_3 (high salinity/high alkalinity, high salinity/very high alkalinity and very high salinity/high alkalinity).

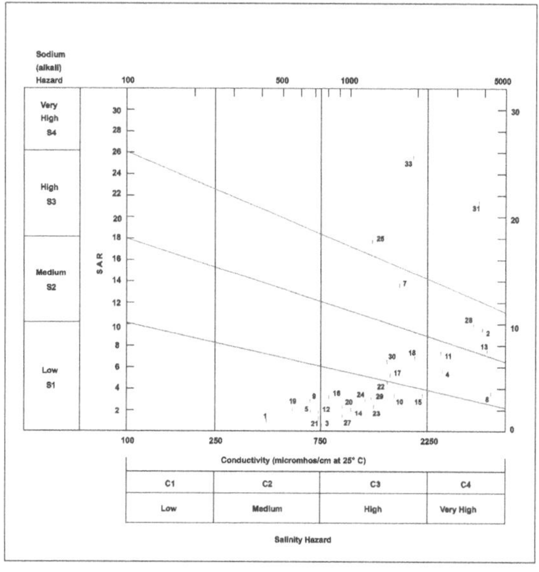

Near surface (c_1)

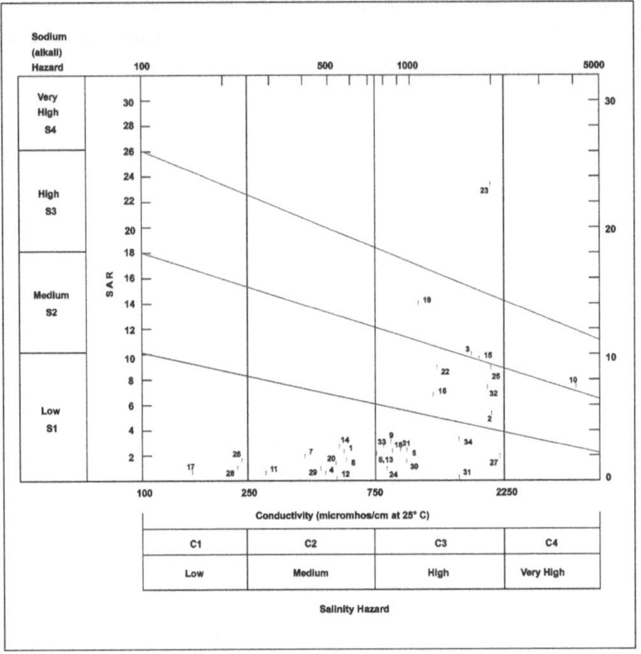

Deep systems (c_2)

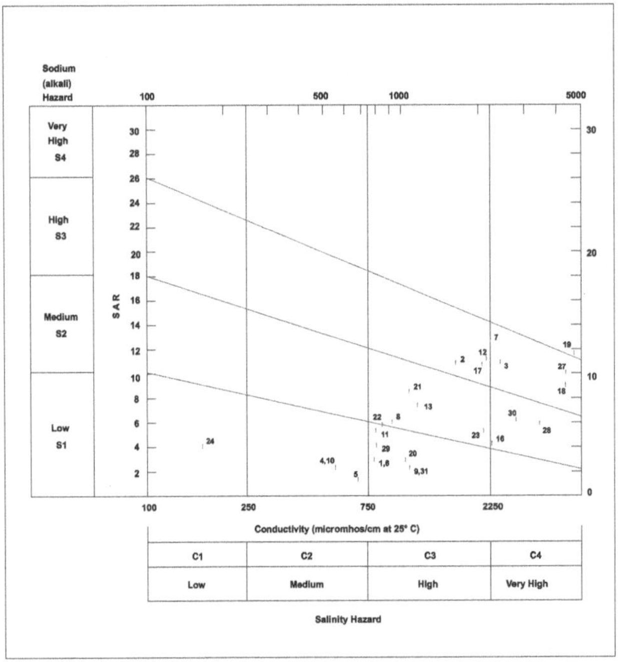

Near surface (d_1)

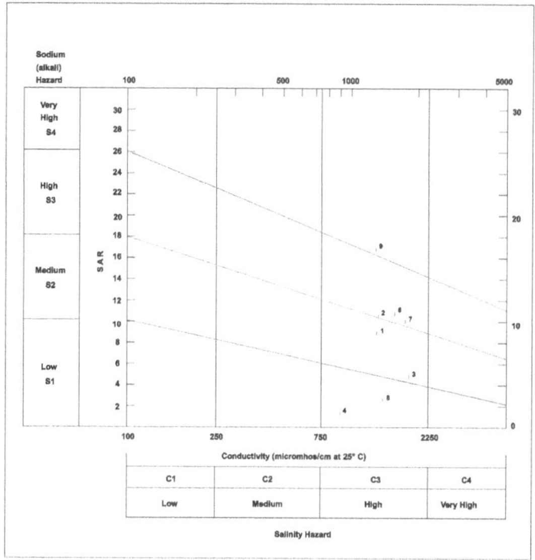

Deep systems (d_2)

Fig. 5.19(d) Sodium adsorption ratios for near surface (d_1) and deep systems (d_2) in Vaigai delta systems

Vaigai delta

Near surface: About 12% of the samples lie in the fields of C_1S_1, C_2S_1 (low salinity/low alkalinity to medium salinity/low alkalinity). About 33% of the samples lie in the fields of C_3S_1, C_3S_2 (high salinity/low alkalinity to high salinity/medium alkalinity). About 18% of the samples lie in the fields of C_3S_3 to C_4S_2 (high salinity/high alkalinity to very high salinity/medium alkalinity). In addition about 16% of the samples lie in the field of C_4S_3 and C_4S_4 (very high salinity/high alkalinity to very high salinity/very high alkalinity).

Deep systems: There are nil number of samples in the field of C_1S_1, C_2S_1 (low salinity/low alkalinity to medium salinity/low alkalinity). About 44% of the samples lie in the fields of C_3S_1 and C_3S_2 (high salinity/low alkalinity to high salinity/medium alkalinity). In addition about 44% of the samples lie in the field of C_3S_3 and C_3S_4 (high salinity/high alkalinity to high salinity/very high alkalinity).

Residual Sodium Carbonate (RSC)

The RSC values for the near surface and deep systems for Palar, Ponnaiyar, Cauvery and Vaigai delta system are presented in Tables 5.18(a,b,c,d). The RSC with values less than 1.25 is considered good and values more than 2.5 are considered unsuitable. The values between 1.25 and 2.5 are of doubtful category.

Palar delta: Majority of the near surface (87.5%) and deep system (78%) have values less than 1.25 and the rest of the samples fall in the categories of doubtful and unsuitable nature.

Table 5.18(a) Residual sodium carbonate values in Palar delta

RSC (epm)	Remark on Quality	No. of Samples	
		Near Surface system	Deep system
< 1.25	Good	35 (-17.6-1.13)	10 (-11.3-0.8)
1.25 - 2.5	Doubtful	3 (1.5-2.48)	4 (1.3-2.48)
> 2.5	Unsuitable	2 (2.5-2.8)	4 (3.9-10.7)

Ponnaiyar delta: About 93% of the near surface and 82% of the deep systems have values less than 1.25 and are considered as good quality. Rest of the sample percentages goes in to doubtful and unsuitable categories.

Table 5.18(b) Residual sodium carbonate values in Ponnaiyar delta

RSC (epm)	Remark on Quality	No. of Samples	
		Near Surface system	Deep system
< 1.25	Good	40 (-20.3-1.17)	23 (-5.4-1.04)
1.25 - 2.5	Doubtful	0	4 (1.5-2.04)
> 2.5	Unsuitable	3 (2.5-3.58)	1 (2.3-4.18)

Cauvery delta: About 76% of the near surface system and 85% of the deep system have values less than 1.25 and fall in the category of good quality and the rest of the samples percentages fall in the doubtful and unsuitable categories.

Table 5.18(c) Residual sodium carbonate values in Cauvery delta

RSC (epm)	Remark on Quality	No. of Samples	
		Near Surface system	Deep system
< 1.25	Good	25 (-23.6-0.7)	29 (-9.6-1.01)
1.25 - 2.5	Doubtful	1 (2.4)	3 (1.49-2.3)
> 2.5	Unsuitable	7 (2.6-9)	2 (3.3-4.2)

Vaigai delta: About 72% of the near surface systems and 56% of the deep systems have values less than 1.25 and are considered good quality for irrigation the rest of the percentage of samples fall in the doubtful and unsuitable categories.

Table 5.18(d) Residual sodium carbonate values in Vaigai delta

RSC (epm)	Remark on Quality	No. of Samples	
		Near Surface system	Deep system
< 1.25	Good	23 (-83.8-0.9)	5 (-44.5-0.53)
1.25 - 2.5	Doubtful	2 (2.0-2.3)	2 (1.3-1.75)
> 2.5	Unsuitable	7 (3.1-5.0)	2 (2.7-3.9)

5.5 WATER CHEMISTRY, CHARACTERIZATION, HYDROGEO-CHEMICAL FACIES AND MULTIVARIATE TREND ANALYSIS OF GROUNDWATERS (NEAR SURFACE AND DEEP) IN KRISHNA, GODAVARI AND PENNAR DELTA SYSTEMS OF ANDHRA PRADESH

5.5.1 Spatial variations of specific conductance, pH and statistical variations of different ions in Krishna, Godavari and Pennar delta systems of Andhra Pradesh

Krishna delta region: The sample locations of both near surface and deep systems in respect of Krishna delta region and the trends of specific conductance values are shown in Fig. 5.20(a) and (b). (Data source: Radhakrishna, I., 2001). The analytical results of specific conductance and pH in terms of max, min, mean and standard deviations values are presented in Table 5.19(a).The specific conductance values for near surface systems of Krishna delta have a range of values varying from 40000 μ mhos/cm to 1176 μ mhos/cm with an average value of 5,401 μ mhos/cm and deep systems have a range of values varying from 32,610 μ mhos/cm to 561 μ mhos/cm with an average of value 8282 μ mhos/cm. A major portion of the Krishna delta region is characterized by relatively/moderately high levels of conductivity except in the regions closed to the flanks of mid west region of Krishna delta and also all along the eastern flanks of delta region. These are the areas where paleochannel network systems are predominantly observed. The upstream areas are characterized by values less than 2000 μ mhos/cm (near surface) and less than 1000 μ mhos/cm (deep systems). The middle and outfall regions have a characteristic range of values varying from 4000 to 20,000 μ mhos/cm and sometimes beyond. These values indicate that the major portion of the Krishna delta has become brackish to saline reflecting the intensity of salinization processes.

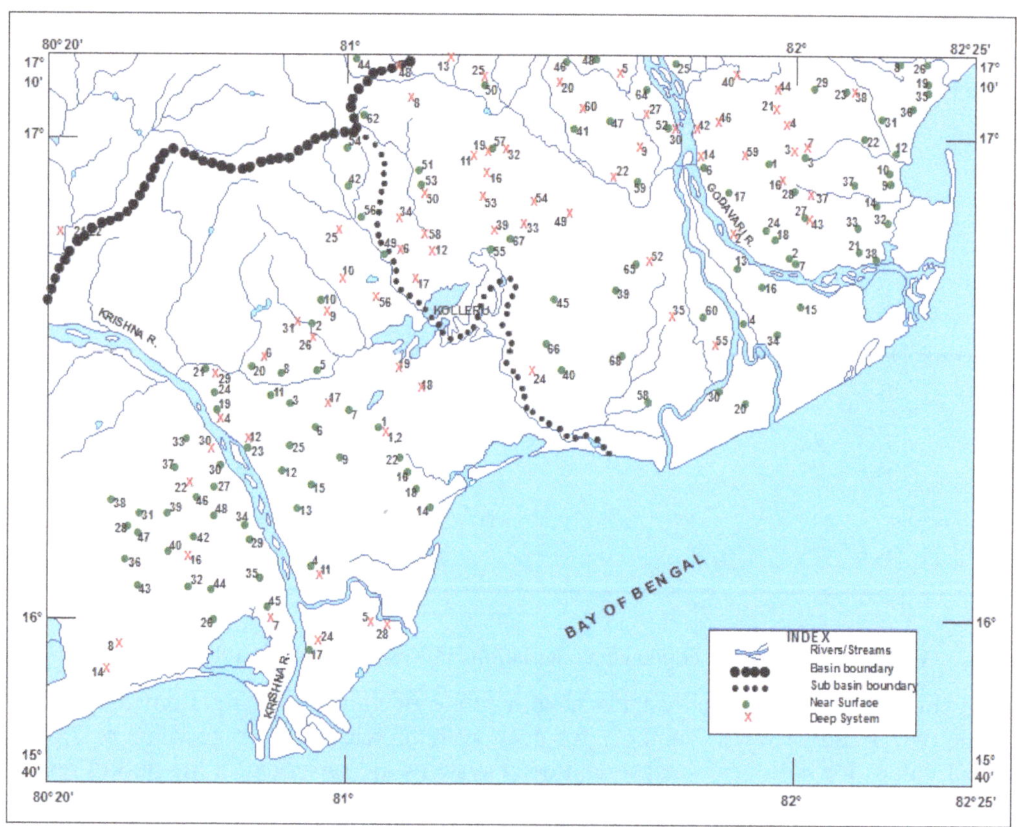

Fig. 5.20(a) Sample locations for near surface and deep systems of Krishna and Godavari deltas

Table 5.19(a) The variations of specific conductance and pH along with different cations and anions of Krishna delta region in Andhra Pradesh

Parameter	Near Surface				Deep systems			
	Max.	**Min.**	**Mean**	**Std.dev**	**Max.**	**Min.**	**Mean**	**Std.dev**
pH	8.50	7.40	7.70	0.27	9	6.7	7.6	0.42
EC	40000.00	1176.10	5401.60	6309.00	32610	561	8282.6	9179.8
Ca	165.00	30.00	92.60	33.70	1192	9	261.7	316
Mg	780	24	133.38	147.8	927.00	4.20	204.70	258.50
Na	8575.00	75.00	913.40	1326.10	5851.00	65.00	1344.10	1641.40
K	210.00	8.00	39.60	35.70	180.00	1.00	27.90	34.90
HCO_3	490	4	229	154.7	532.00	104.00	284.90	103.00
SO	1260	40	293	235.4	2000	20	579.2	652.2
Cl	14600	150	1630.8	2306.1	11210.00	19.50	2624.40	3199.70

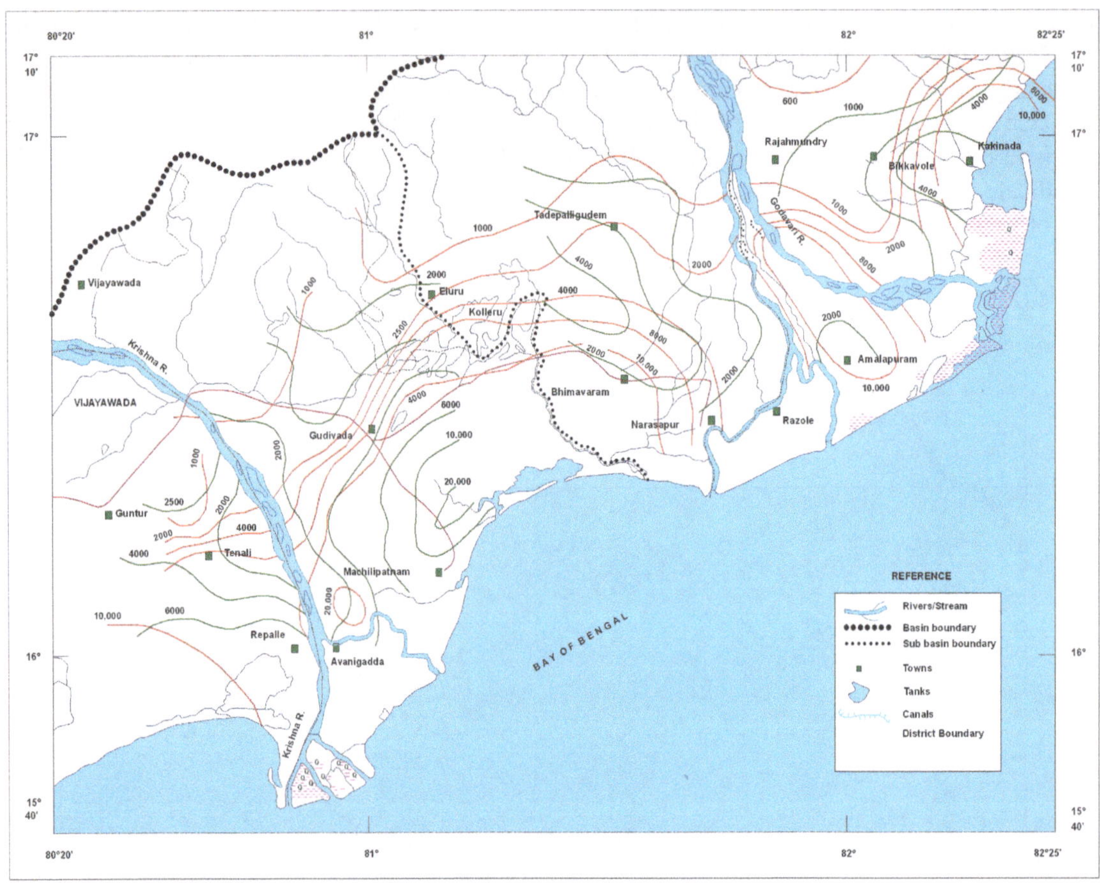

Fig. 5.20(b) Specific conductance variations in Krishna, Godavari delta regions

The pH values for near surface systems have a range of values varying from 8.5 to 7.4 with an average value of 7.7 and between 9 and 6.7 for deep systems with an average value of 7.6. In general the higher pH values for both near surface and deep systems are observed in the outfall regions where the salinsation processes are most intense.

Godavari delta region : The sample locations for both near surface and deep systems are presented in Fig. 5.20(a) and specific conductance trends are reflected in Fig. 5.20(b). (Data source: CGWB: 1995; 1999). The analytical results of specific conductance and pH in terms of max, min, mean and standard deviation values are presented in Table 5.19(b). The specific conductance values for near surface system of Godavari delta have a range of values varying from 15380 to 394 μ mhos/cm with an average value of 2288 μ mhos/cm and deep systems have a range of values varying from 20880 to 208 μ mhos/cm with an average value of 1438 μ mhos/cm. The regions flanking the eastern and western banks of the Godavari river have values less than 1000 μ mhos/cm and gradually increase to a level of 2000 μ mhos/cm indicating a large portions of the near surface system of the delta are suitable for irrigation. Here also it is observed that a good network of paleo channels are the controlling factors for existence of fresh water storage potentials. The freshwater of deep systems however with values less than 1000 μ mhos/cm and some times less than 2000 μ mhos/cm are observed only in the upstream portions of the regions and the deep systems in the entire delta region have values more than 2000 μ mhos/cm and go upto the levels of 10000 μ mhos/cm. The trend values of specific conductance of deep systems however show that they don't have any control on the river systems or paleo channel network in the region unlike the near surface systems where the river and the paleo channel network appears to have a close control on the trend variations.

Table 5.19(b) The variations of specific conductance and pH along with different cations and anions of Godavari delta region in Andhra Pradesh

Parameter	Near Surface				Deep systems			
	Max.	Min.	Mean	Std.deviation	Max.	Min.	Mean	Std.deviation
pH	9.00	7.30	7.90	0.40	8.8	6.4	7.5	0.55
EC	15380.00	394.00	2288.00	2327.10	20880	208.4	1438.2	2928.1
Ca	286.00	14.00	94.50	62.40	800	3.6	75.38	102.21
Mg	359	4.9	65.33	63.44	1121.90	3.60	50.90	143.40
Na	1340.00	23.00	238.40	273.20	3193.20	2.00	145.80	434.20
K	522	1.2	53.6	91.9	81.00	0.00	17.30	19.90
HCO_3	1820	92	451	285.6	3551.20	73.00	292.40	432.00
SO_4	916	5	149.9	199.6	714.1	5.95	39.7	92
Cl	3040	30	416.9	565.1	7161.00	15.00	321.20	973.80

The pH values for near surface systems of Godavari delta systems have a range of values varying between 9 and 7.3 with an average value of 7.9 and between 8.8` and 6.4 for deep systems with an average value of 7.5. The higher pH values for both near surface and deep systems are seen where the specific conductance levels are increasingly of higher order.

Pennar delta region: The sample locations and specific conductance values and their trend variations are reflected in Fig. 5.21(a,b). (Data source: Radhakrishna, I. and Krishna, G.S.R., 1990). The analytical

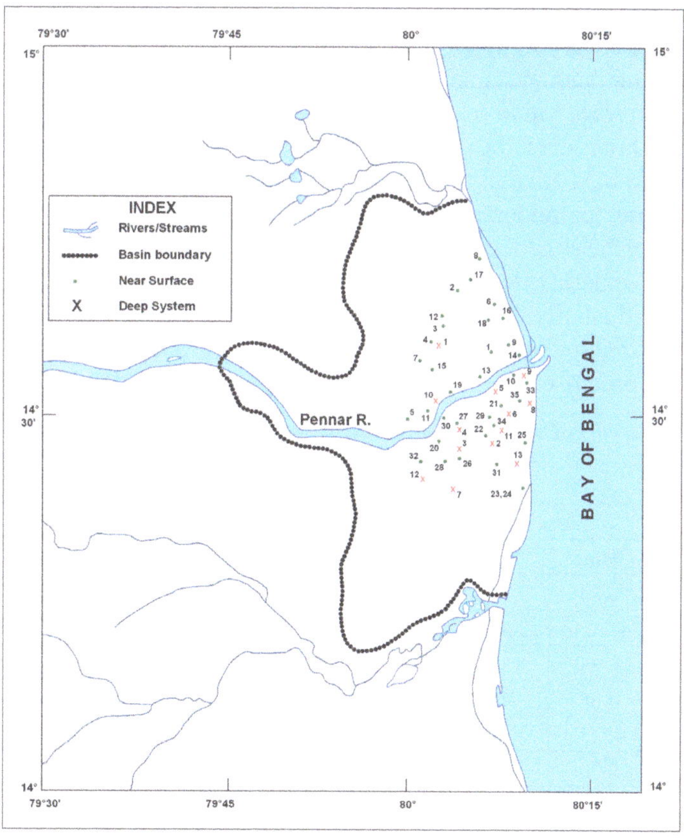

Fig. 5.21(a) Sample locations for near surface and deep systems in Pennar delta

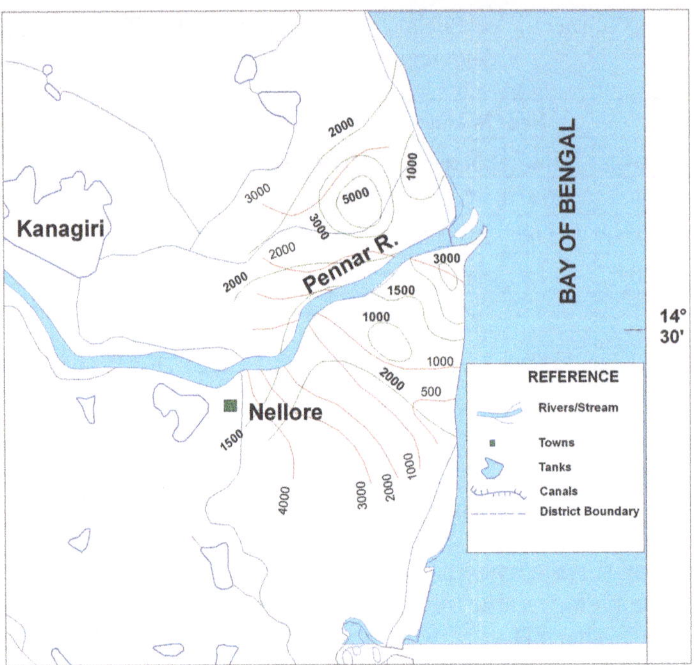

Fig. 5.21(b) Specific conductance variations in Pennar delta regions

results of specific conductance and pH in terms of max, min, mean and standard deviation values are presented in Table 5.19(c). The specific conductance values of near surface system of Pennar delta region have a range of values varying from 13000 μ mhos/cm to 410 μ mhos/cm with an average value of 2496 μ mhos/cm and the deep systems have a range of values varying from 13000 μ mhos/cm to 460 μ mhos/cm with an average value of 2475.5 μ mhos/cm. The northern part of the delta system in general have moderate to high levels of conductivity both in respect of near surface and deep systems excepting the areas close to the banks of the Pennar river. The southern part of the delta particularly the mid delta region have low to moderate conductivity values signifying the areas with fresh water storages both in respect of near surface and deep systems. This is the region where relatively good network of paleochannels exist as described earlier in chapter 3. The upstream areas however have characteristically high conductivity levels excepting a small stretch in the southeastern sector of the delta. The regions in the vicinity of estuary of the delta have moderate to high levels of conductivity signifying the salinization processes.

Table 5.19(c) The variations of specific conductance and pH along with different cations and anions of Pennar delta region in Andhra Pradesh

Parameter	Near Surface				Deep systems			
	Max.	**Min.**	**Mean**	**Std.deviation**	**Max.**	**Min.**	**Mean**	**Std.deviation**
pH	8.30	6.50	7.32	0.39	8	7	7.5	0.24
EC	13000.00	410.70	2496.90	2535.70	13000	460	2475.5	3383.8
Ca	1024.00	24.00	133.40	212.40	340	12	89.8	104.5
Mg	224	4.88	40.9	49.8	122.00	4.87	31.90	33.95
Na	1620.00	47.00	441.20	388.20	1600.00	75.00	371.50	449.20
K	460.00	0.00	54.30	119.50	44.00	1.00	11.46	12.50
HCO_3	1230	70	454.8	239.4	720.00	220.00	380.00	161.90
SO_4	1900	20	319	400.3	850	10	213.8	238.7
Cl	4432	57	584.2	909.9	2694.00	32.00	494.90	766.60

The pH values for near surface systems have a range of values varying between 8.3 and 6.5 with an average value of 7.32 and for deep systems between 8 and 7 with an average value of 7.5. The locations with high pH values probably indicate the intensity of salinisation processes taking place in this part of the delta.

5.5.2 Cation-anion chemistry/Ionic ratios/Cross plots

The sequential order of abundance of cations and anions (near surface and deep) in these three delta systems of Andhra Pradesh are as follows.

Krishna delta region

(a) *Near surface:* Cl > SO_4 > HCO_3 and Na > Mg > Ca > K

(b) *Deep:* Cl > SO_4 > HCO_3 and Na > Ca > Mg > K

Godavari delta region

(a) *Near surface:* HCO_3 > Cl > SO_4 and Na > Ca > Mg > K

(b) *Deep:* Cl > HCO_3 > SO_4 and Na > Ca > Mg > K

Pennar delta region

(a) *Near surface:* Cl > HCO_3 > SO_4 and Na > Ca > K > Mg

(b) *Deep:* Cl > HCO_3 > SO_4 and Na > Ca > Mg > K

The chloride (Cl) and sodium (Na) are the dominant ions amongst the anions and cations respectively in all the delta systems excepting the near surface systems of Godavari delta region where the bicarbonate (HCO_3) anion marginally dominates in the anion sequence of the systems.

Variations of cations and anions with respect to TDS

The maximum and minimum variations and the standard deviation in terms of different water quality parameter for both near surface and deep systems are presented earlier in Tables 5.22(a,b,c). The variations of cations and anions in relation to TDS for both near surface and deep systems for Krishna, Godavari and Pennar are presented in Fig. 5.22(a,b,c).

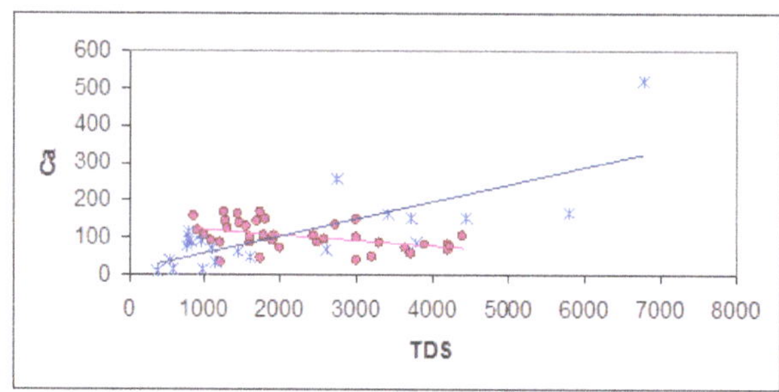

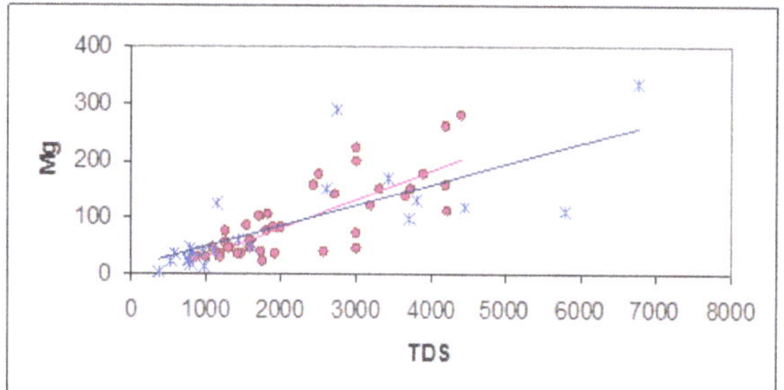

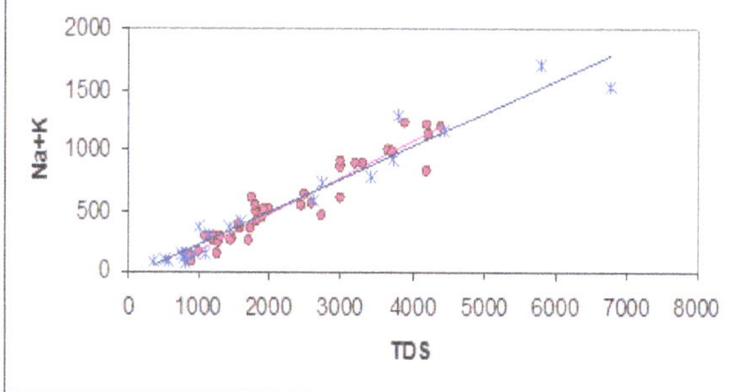

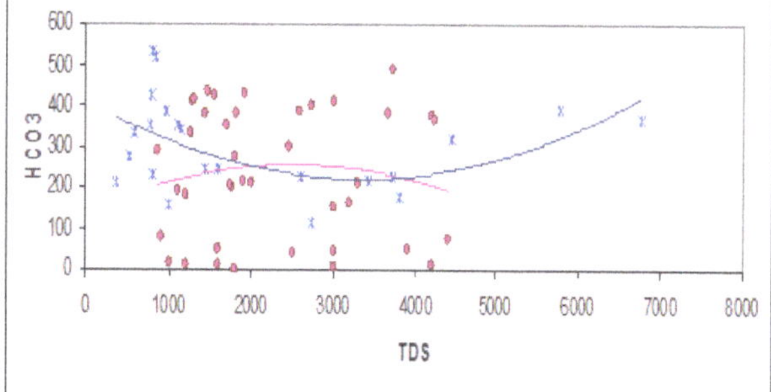

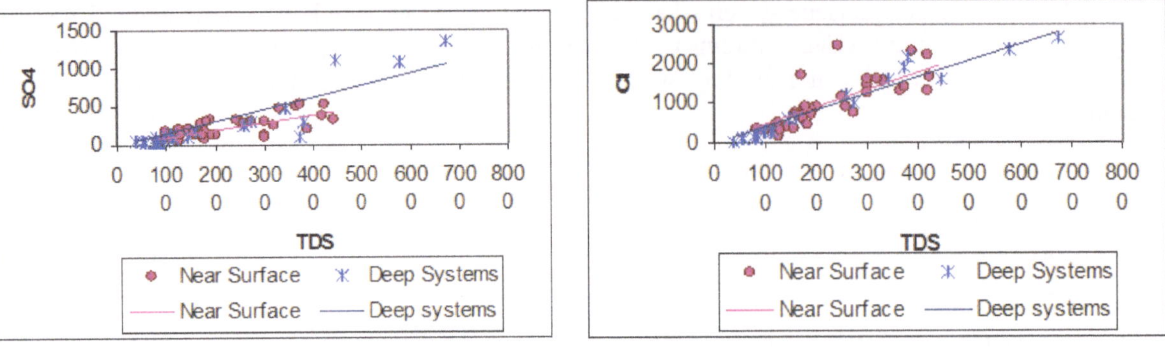

Fig. 5.22(a) Variations of cations and anions in relation to TDS in Krishna delta region

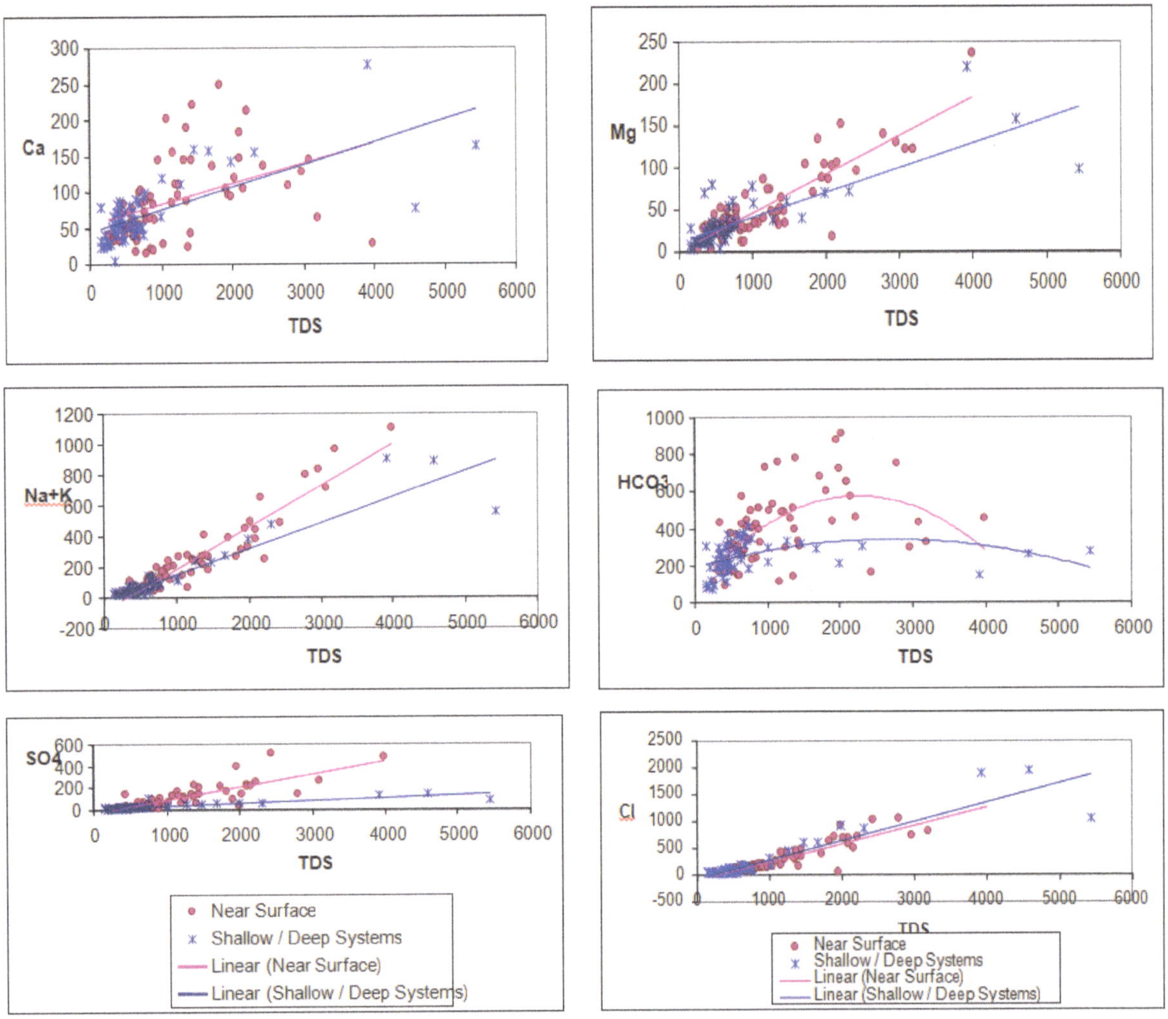

Fig. 5.22(b) Variations of cations and anions in relation to TDS in Godavari delta region

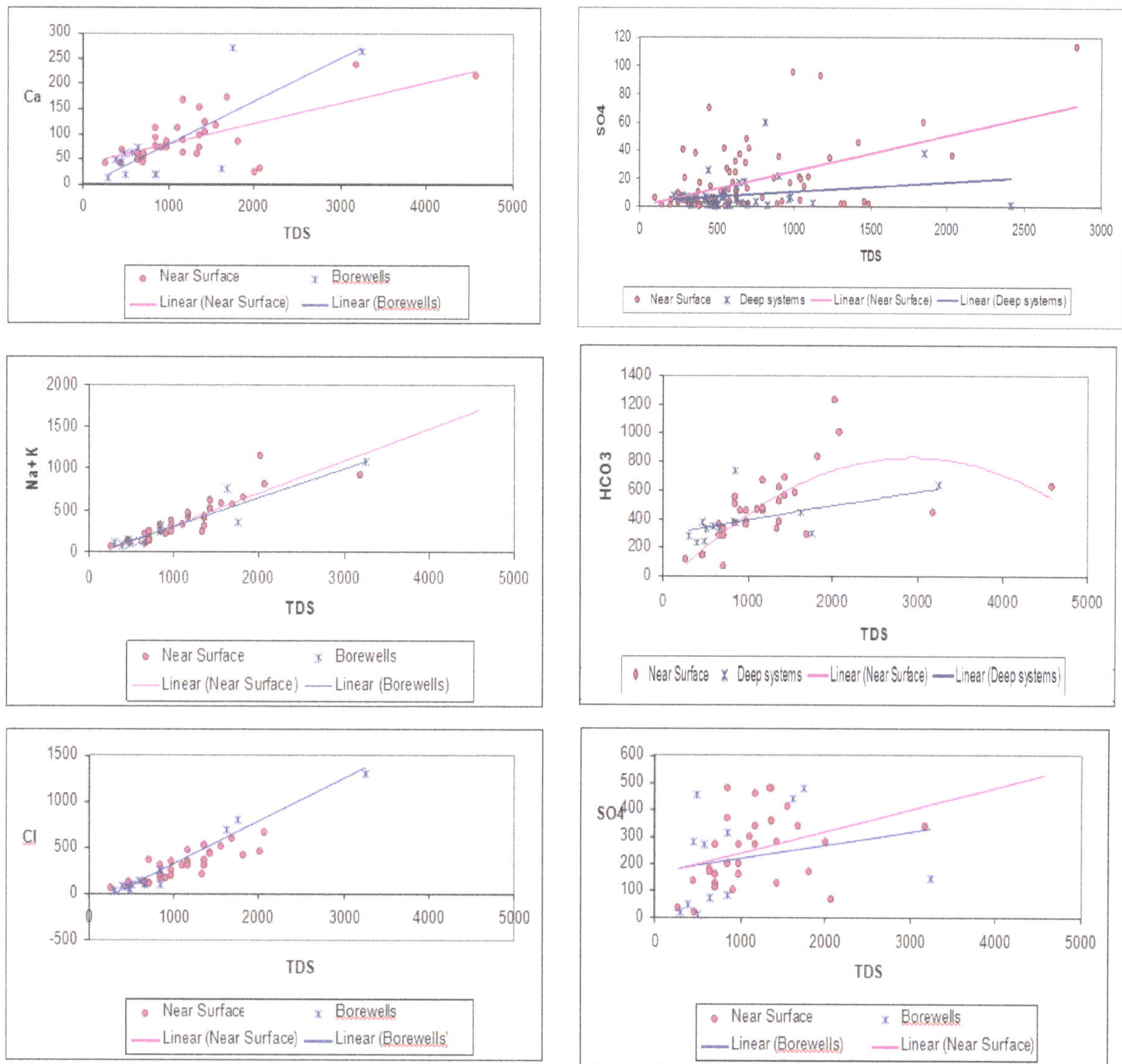

Fig. 5.22(c) Variations of cations and anions in relation to TDS in Pennar delta region

- The cross plots for all the delta systems (Krishna, Godavari and Pennar) show an ascending linearity for both near surface and deep systems excepting in the case of HCO_3 vs. TDS where the trends are characterized by a curve linear relationship by and large in all the delta systems with different geometrical shapes. The curve trends in respect of Godavari, both near surface and deep show an ascending order up to 1000 -1500 ppm levels of TDS and then show declining trends characterizing saline water mixing processes. The HCO_3 vs. TDS trends in respect of Krishna delta show different geometrical shapes. The near surface systems show an ascending order up to a level of 2000-3000 ppm level of TDS and then show declining trend as it happened normally with other delta systems characterizing mixing of saline water after 3000 ppm levels of TDS. The deep systems in contrast show declining levels up to 3500 ppm levels of TDS and than gradually show rising trends with increasing levels of TDS reflecting probable mixing of fresh water at deeper levels.
- The trends of Ca and Mg in the case of Godavari and Pennar show an ascending order of linearity with increasing levels of TDS values for both near surface and deep systems with different slope

characteristics. However the slope characteristics of Ca levels of near surface and deep systems in Godavari delta though initially differ marginally gets converged into a single line after 3000 ppm level of TDS. Similarly the Mg levels of both near surface and deep systems in case of Pennar delta emerges as a single line with same slope characteristics. The Ca trends of near surface systems of Krishna delta take a linearly declining trend with the increasing levels of TDS where as the Ca values in the deep systems show an ascending order of linearity with increasing levels of TDS. The Mg trends for both near surface and deep systems in Krishna delta are both disposed as single line up to 3000 ppm levels of TDS and then get diverged with different slope characteristics with increasing levels of salinity.

- The Na + K trend in the case of Krishna delta region for both near surface and deep systems are disposed linearly as a single line with increasing levels of salinity signifying similar hydrochemical involvements of Na + K with increasing levels of salinity. However the Na + K trends in the case of Godavari and Pennar for both near surface and deep systems show ascending and linearly disposed trends with increasing levels of salinity and with different slope characteristics. In both delta systems the near surface system show high gradient levels when compared to the deep systems reflecting more Na + K contributions in the near surface systems relatively to the deep systems.
- The SO_4 and Cl anions in both near surface and deep systems of all the deltas exhibit linear trend relationships with increasing levels of TDS values and with different slope characteristics. The degree of gradient reflects on the intensity/involvement of respective anion.

Ionic ratios/Cross plots

- The specific conductance/ total dissolved solids/ionic ratios between selected cations and anions would help to estimate and assess the intensity of fresh and saline water mixing processes. The max, min and avg values of selected number of ionic ratios for the near surface and deep systems of Krishna, Godavari and Pennar delta systems are furnished below in Table 5.20(a,b).

Table 5.20(a,b) Maximum, Minimum and Average values of ionic ratios in different delta systems of Andhra Pradesh

(a) Near surface

S.No	Ionic ratios	Krishna			Godavari			Pennar		
		Max	Min	Avg	Max	Min	Avg	Max	Min	Avg
1.	Na/Cl	2.36	0.22	0.90	13.30	0.24	1.30	2.56	0.56	1.63
2.	Mg/Cl	1.07	0.09	0.30	5.35	0.10	0.85	0.72	0.12	0.26
3.	SO_4/Cl	1.03	0.05	0.21	6.01	0.03	0.38	2.21	0.04	0.68
4.	Cl/(HCO_3 + CO_3)	696.93	0.63	72.09	10.65	0.09	1.62	24.60	0.50	2.59
5.	Ca/(HCO_3 + SO_4)	2.34	0.09	0.69	1.65	0.08	0.52	5.29	0.05	0.55
6.	Mg/Ca	12.85	0.27	2.76	13.89	0.12	1.47	1.51	0.18	0.55

(b) Deep systems

S.No	Ionic ratios	Krishna			Godavari			Pennar		
		Max	Min	Avg	Max	Min	Avg	Max	Min	Avg
1.	Na/Cl	6.27	0.41	1.12	3.47	0.01	0.88	5.30	0.61	2.11
2.	Mg/Cl	1.13	0.04	0.39	7.32	0.22	1.01	1.39	0.08	0.42
3.	SO_4/Cl	1.51	0.04	0.25	0.67	0.05	0.21	7.30	0.07	1.29
4.	Cl/(HCO_3 + CO_3)	116.54	0.16	21.19	10.64	0.09	1.62	6.44	0.20	1.69
5.	Ca/(HCO_3+SO_4)	3.05	0.11	0.64	1.65	0.07	0.51	0.96	0.05	0.36
6.	Mg/Ca	5.24	0.31	1.55	32.18	0.16	1.34	1.59	0.40	0.72

- The average Na/Cl ratios for near surface/deep systems in respect of Krishna, Godavari and Pennar are 0.90/1.12, 1.30/0.88 and 0.63/2.11 respectively. The theoretical values for Na/Cl ratio for saline and brackish waters are of the order of 0.85 and 1.00 respectively. The average values in respect of the near surface systems and deep systems of Krishna and Godavari lie either in the field of brackish/saline water zones or come close to the brackish water boundary. The average values of the rest of the systems in Godavari and Pennar lie in the field of fresh water.
- The average Mg/Cl ratios for near surface/deep systems in respect of Krishna, Godavari and Pennar are 0.30/0.39, 0.85/1.01 and 0.26/0.42 respectively. The theoretical values for Mg/Cl ratio for saline and brackish waters are of the order of 0.15 and 0.35 respectively. This indicates that the near surface systems of Krishna and Pennar and the deep systems of Krishna come close to the brackish/saline water boundaries and the rest of the systems appear to be free from mixing processes.
- The average SO_4/Cl ratios for near surface/deep systems in respect of Krishna, Godavari and Pennar are 0.21/0.25, 0.38/0.21 and 0.68/1.29 respectively. The theoretical values for SO_4/Cl ratio for saline and brackish waters are of the order of 0.005 and 0.20 respectively. The above ratio reflects that the average values of near surface of Krishna and the deep systems of Krishna and Godavari come close to the brackish/saline water boundaries and the rest of the systems appear to be free from brackish/saline water mixing processes.
- The average Cl/(HCO_3 + CO_3) ratios for near surface/deep systems in respect of Krishna, Godavari and Pennar are 72.09/21.19, 1.62/1.62 and 2.59/1.69 respectively. The theoretical values for Cl/(HCO_3+CO_3) ratio for saline and brackish waters are of the order of 6.60 and 0.50 respectively. The ratio data reflects on intense salinity processes in near surface and deep systems of Krishna delta and the rest of the near surface and deep systems of Godavari and Pennar lie within the boundaries of brackish/saline water fields.
- The average Ca/(HCO_3 + SO_4) ratios for near surface/deep systems in respect of Krishna, Godavari and Pennar are 0.69/0.64, 0.52/0.51 and 0.55/0.36 respectively. The theoretical values for Ca/(HCO_3 + SO_4) ratio for saline and brackish waters are of the order of 1.00 and 0.50 respectively. The average values of the near surface and deep systems of Krishna comes in the brackish/saline water boundaries and in the case of Godavari and Pennar both near surface and deep systems comes close to the brackish water boundary with exception to the deep system of Pennar where the waters appear to be relatively free from salinity.
- The average Mg/Ca ratios for near surface/deep systems in respect of Krishna, Godavari and Pennar are 2.76/1.55, 1.47/1.34 and 0.55/0.72 respectively. The theoretical values for Mg/Ca ratio for saline and brackish waters are of the order of 3.5 and 1.00 respectively. The average values shows that the near surface and deep systems of Krishna and Godavari lie in the fields of brackish/saline water zones where as the near surface and deep systems of Pennar reflects on non-mixing processes with saline water.

Ionic ratios vs Cl cross plots: The ratios considered for cross plots to assess the intensity of mixing are: Na/Cl vs. Cl, SO_4/Cl vs.Cl, Mg/Cl vs. Cl, Cl/(HCO_3 + CO_3) vs Cl, Mg/Ca vs Cl and Ca/(HCO_3 + SO_4) vs Cl. The graphical plots of ionic ratios between Na/Cl vs. Cl, SO_4/Cl vs.Cl, Mg/Cl vs Cl, Cl/(HCO_3 + CO_3) vs Cl, Mg/Ca vs Cl and Ca/(HCO_3 + SO_4) vs Cl characterizing the probable saline–fresh water mixing trends in both near surface and deep systems of Krishna, Godavari and Pennar delta systems are presented in Fig. 5.23.

Krishna delta

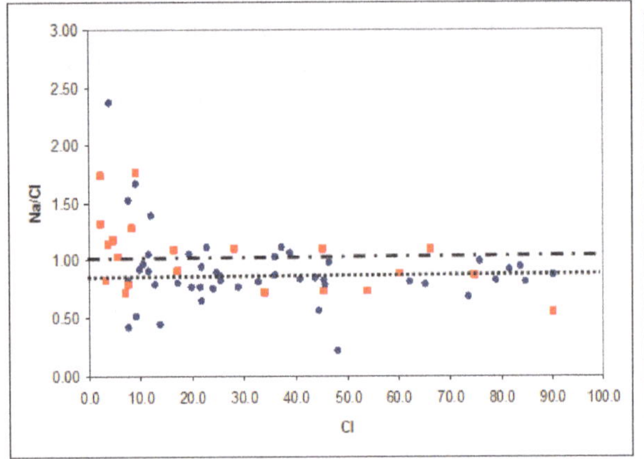

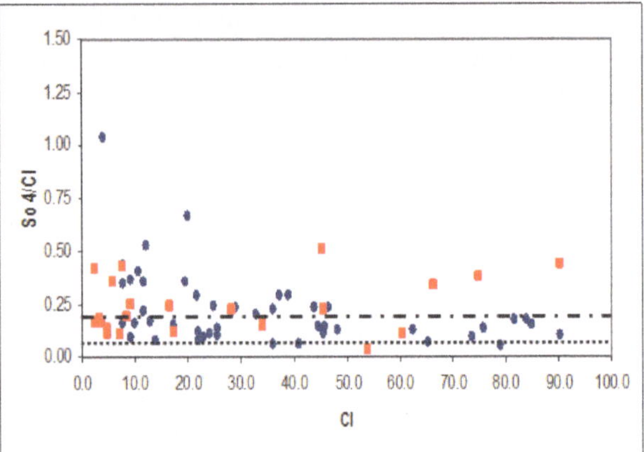

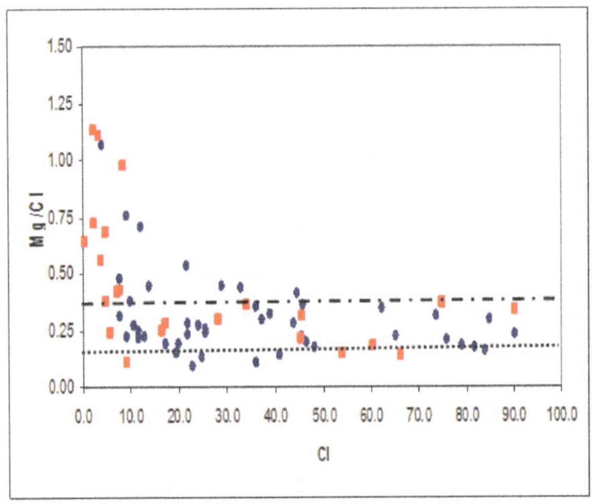

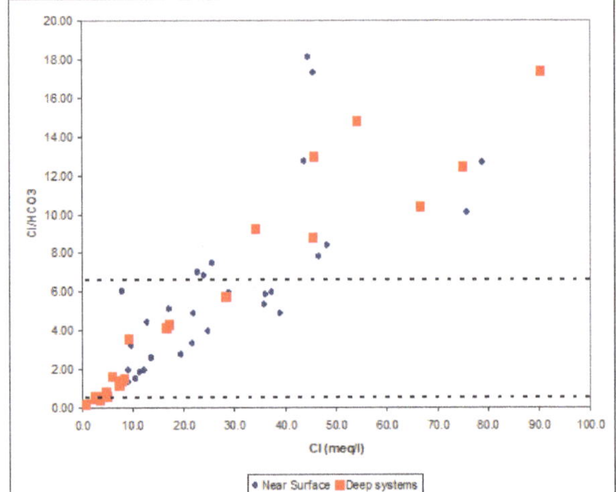

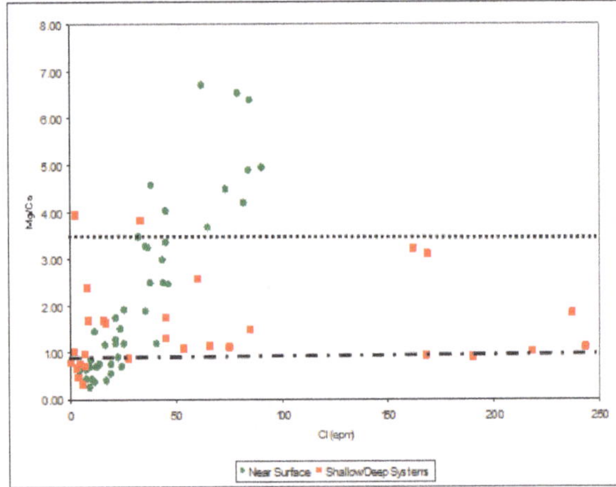

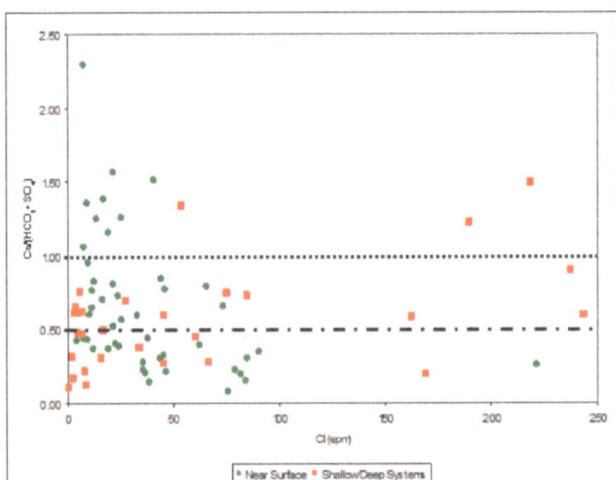

Godavari delta

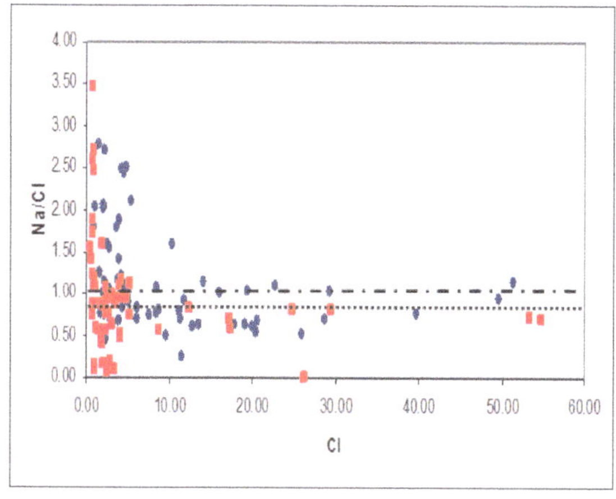

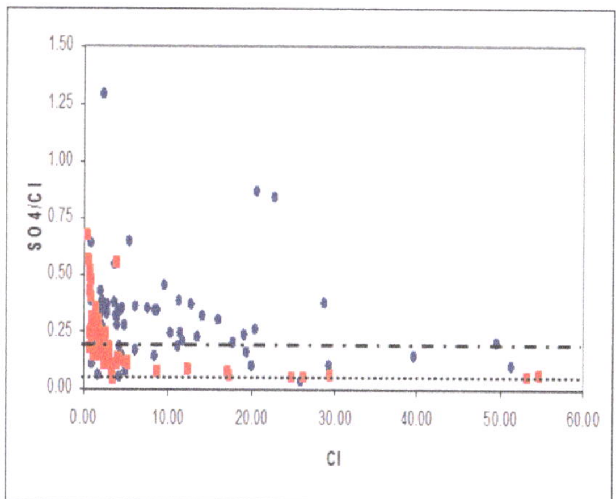

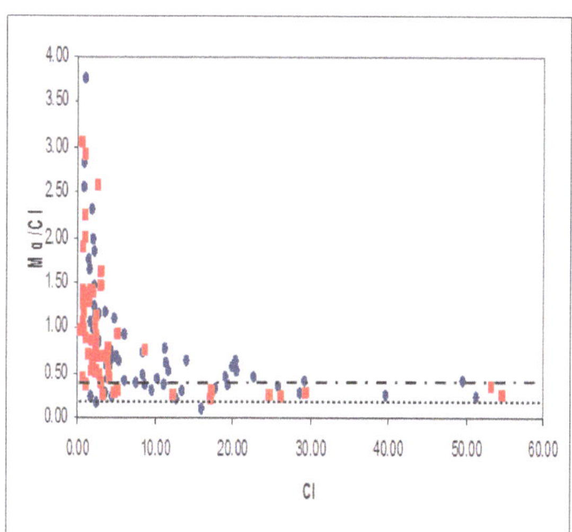

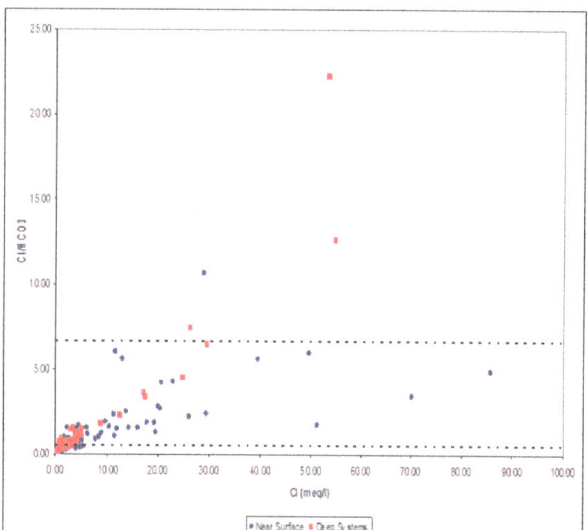

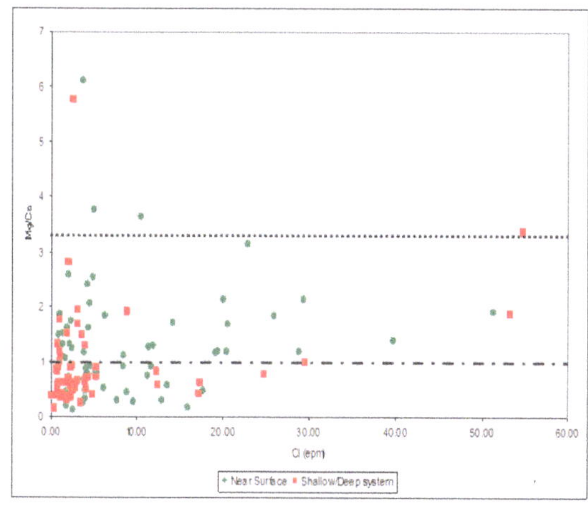

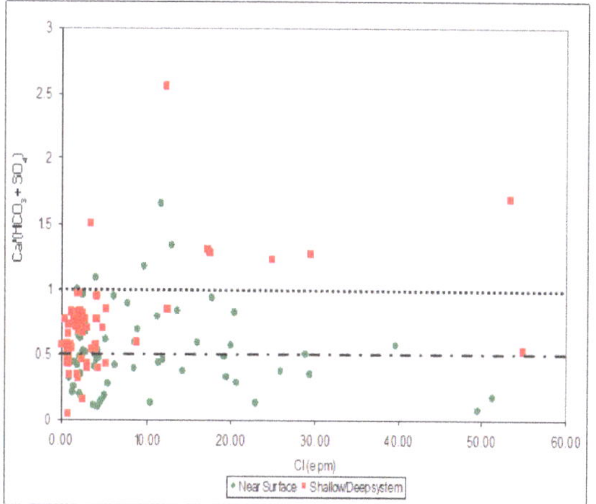

Pennar delta

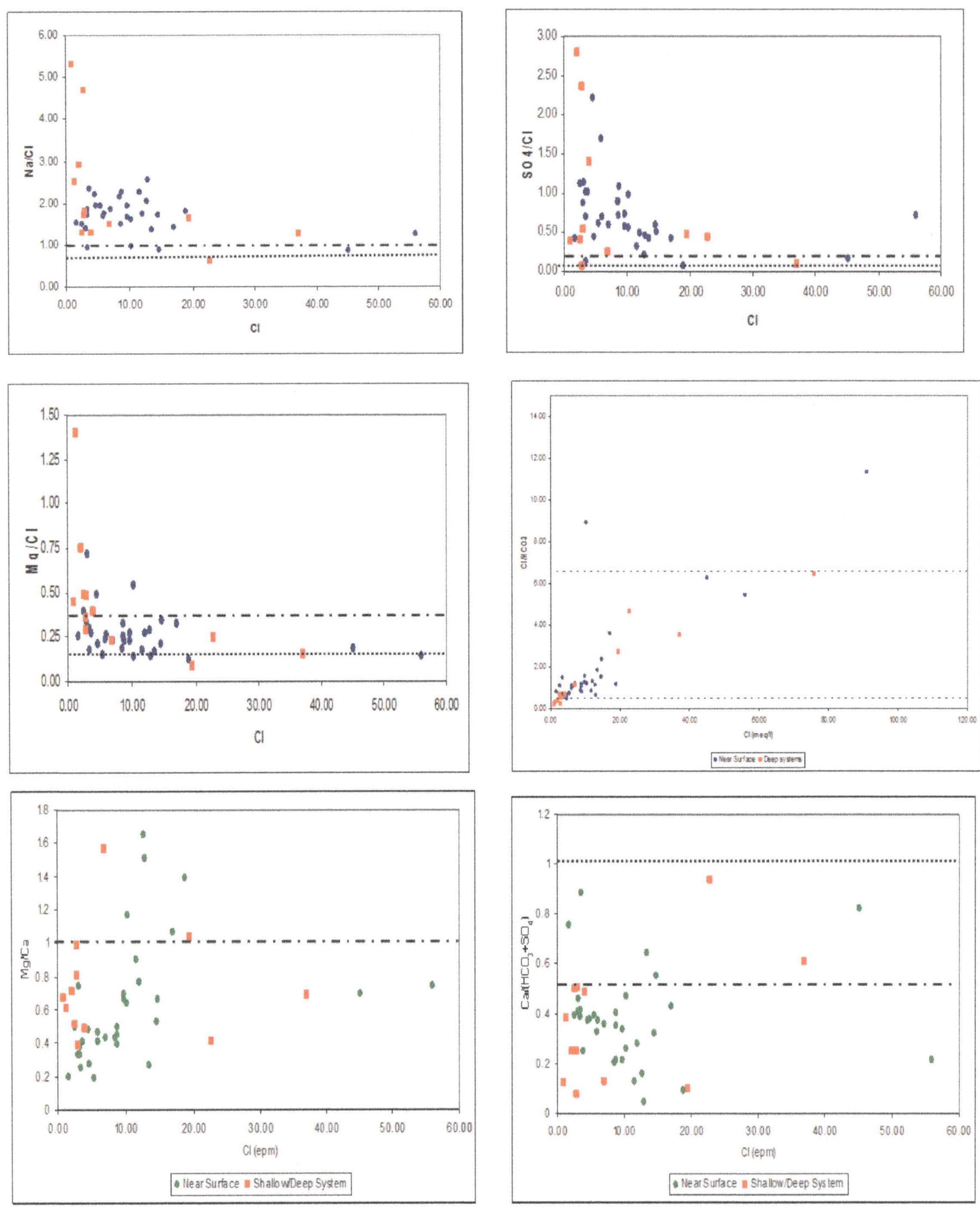

Fig. 5.23 Ionic ratios/Graphical plots in Krishna, Godavary and Pennar delta region

A brief analyses in respect of the above graphic plots discussing the significance of fresh saline water mixing trends in different delta systems of Andhra Pradesh are presented below in the following paragraphs.

Krishna delta: *Na/Cl vs Cl* cross plot indicate the percentage of near surface and deep systems lying in the field of fresh water are respectively 21% and 52% and the rest 79% (near surface) and 48% (deep systems) lie in the brackish/saline water fields signifying that the near surface systems are dominantly saline and the deep systems by and large are disposed in equal proportions. *SO_4/Cl vs Cl* cross plot indicate 38% (near surface) and 55% (deep systems) lie in the field of fresh water and the rest 62% (near surface) and 45% (deep systems) lie in the field of brackish/saline water signifying dominant mixing of saline water in the near surface zone and near to equal proportions of mixing in the deep systems. *Mg/Cl vs Cl* cross plot reflect on 23% (near surface) and 43% (deep systems) lie in the field of fresh water and the rest 77% (near surface) and 57% (deep systems) lie in the field of brackish/saline water indicating dominant mixing of brackish/saline water in the both near surface and deep systems. *Cl/HCO_3 vs Cl* cross plot indicate the percentage of near surface and deep systems lying in the field of fresh water are respectively 0% and 22% and the rest 100% (near surface) and 78% (deep systems) lie in the brackish/saline water fields signifying high intensities of brackish/saline water mixing in the both near surface and deep systems. *Mg/Ca Vs Cl* cross plot indicate the percentage of near surface and deep systems lying in the field of fresh water are respectively 33% and 31% and the rest 67% (near surface) and 69%(deep systems) lie in the brackish/saline water fields indicating moderately dominant mixing of brackish/saline water in both near surface and deep systems. *Ca/(HCO_3 + SO_4) vs Cl* cross plot indicate the percentage of near surface and deep systems lying in the field of fresh water are respectively 51% and 48% and the rest 49% (near surface) and 52% (deep systems) lie in the brackish/saline water fields reflecting by and large equal proportions of brackish/saline water mixing in both near surface and deep systems.

Godavari delta: *Na/Cl vs Cl* cross plot shows the 43% of near surface and 37% of deep systems lie in the field of fresh water and the rest 57% (near surface) and 63% (deep systems) lie in the field of brackish/saline water indicating moderate mixing of brackish/saline water in both near surface and deep systems. *SO_4/Cl vs Cl* cross plot reflect on 65% of near surface systems and 43% of deep systems lie in the field of fresh the rest 35% (near surface) and 57% (deep systems) are disposed in the field of brackish/saline water reflecting only relatively small amounts of mixing of brackish/saline waters in the near surface system and nearly equal proportions of mixing in the deeper systems. *Mg/Cl vs Cl* cross plot indicate the percentage of near surface and deep systems lying in the field of fresh water are respectively 73% and 71% and the rest 27% (near surface) and 29% (deep systems) lie in the brackish/saline water fields reflecting that relatively small amounts of mixing of brackish/saline water in both near surface and deep systems. *Cl/HCO_3 vs Cl* cross plot indicate the percentage of near surface and deep systems lying in the field of fresh water are respectively 13% and 28% and the rest 87% (near surface) and 72% (deep systems) lie in the brackish/saline water fields reflecting high intensities of brackish/saline water mixing in both near surface and deep systems. *Mg/Ca vs Cl* cross plot shows that 36% of the near surface systems and 60% of the deep systems lie in the field of fresh water and the rest 64% (near surface) and 40%(deep systems) are disposed in the field of brackish/saline water indicating moderately high intensities of mixing of brackish/saline water in the near surface zone and relatively small amounts of mixing in the deep systems. *Ca/(HCO_3 + SO_4) vs Cl* cross plot shows that 58% of the near surface systems and 22% of deep systems lie in the field of fresh water and the rest 42% (near surface) and 78% (deep systems) lie in the brackish/saline water fields reflecting nearly equal proportions of mixing in the near surface systems and dominant mixing of brackish/saline water in the deep systems.

Pennar delta: *Na/Cl vs Cl* cross plot ratios indicate that 87% of the near surface systems and 92% of the deep systems lie in the field of fresh water and the rest 13% (near surface) and 8% (deep systems) lie in the field of brackish/saline water reflecting insignificant mixing in both near surface and deep systems. *SO_4/Cl vs Cl* cross plot shows that 87% of the near surface systems and 82% of the deep systems lie in the field of fresh water and only 13% (near surface) and 18% (deep) lie in the field of brackish/saline water reflecting again insignificant amount of mixing in both near surface and deep systems. *Mg/Cl vs Cl* cross plot shows that 13% of the near surface and 50% of the deep systems lie in the field of fresh water and the rest 87% (near surface) and 50% (deep systems) lie in the brackish/saline water fields reflecting high intensity of mixing in the near surface and equal proportions of mixing in the deep systems. *Cl/HCO_3 vs Cl* cross plot indicate 0% of the near surface systems and 27% of the deep systems lie in the field of fresh water and the rest 100% (near surface) and 73% (deep systems) lie in the brackish/saline water indicating high intensities of saline water mixing in the near surface and moderately high intensities of mixing in the deep systems also. *Mg/Ca vs Cl* cross plot reflect on 84% of the near surface and 83% of the deep systems lie in the field of fresh water and the rest 16% (near surface) and 17% (deep systems) lie in the field of brackish/saline water indicating relatively insignificant mixing. *Ca/(HCO_3 + SO_4) vs Cl* cross plot indicate 85% of the near surface systems and 83% of the deep systems lie in the field of fresh water and the rest 15% (near surface) and 17% (deep systems) are disposed in the field of brackish/saline water reflecting insignificant mixing in both near surface and deep systems.

5.5.3 Trilinear plots/Characterization of ionic trends

The dominant cations and anions present in solution of Krishna, Godavari and Pennar delta regions are presented systematically in the Piper triangular patterns shaped diagram are shown in Figures 5.24(a, b and c).

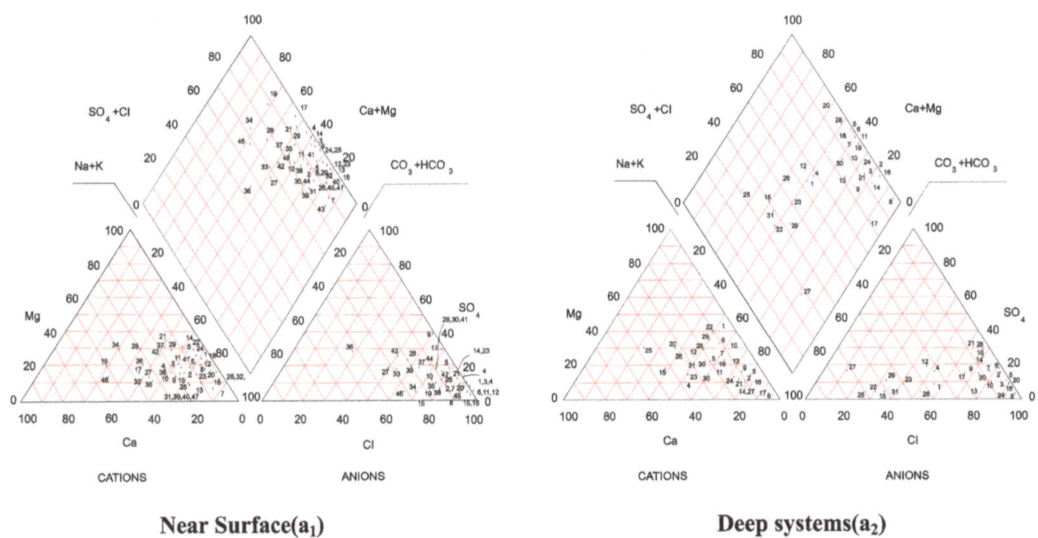

Fig. 5.24(a) Trilinear plots for near surface (a_1) and deep systems (a_2) in Krishna delta

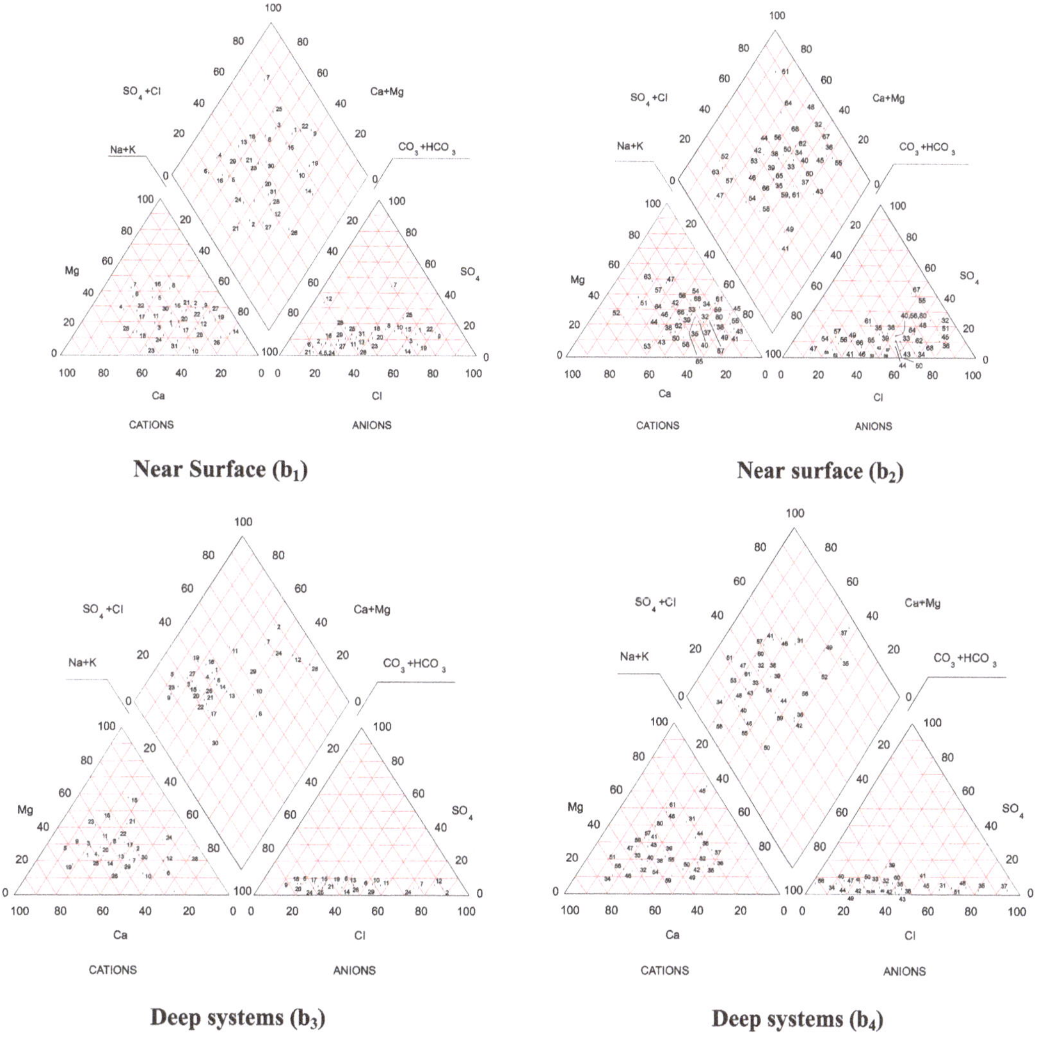

Near Surface (b$_1$) **Near surface (b$_2$)**

Deep systems (b$_3$) **Deep systems (b$_4$)**

Fig. 5.24(b) Trilinear plots for near surface (b$_1$ and b$_2$) and deep systems (b$_3$ and b$_4$) in Godavari delta

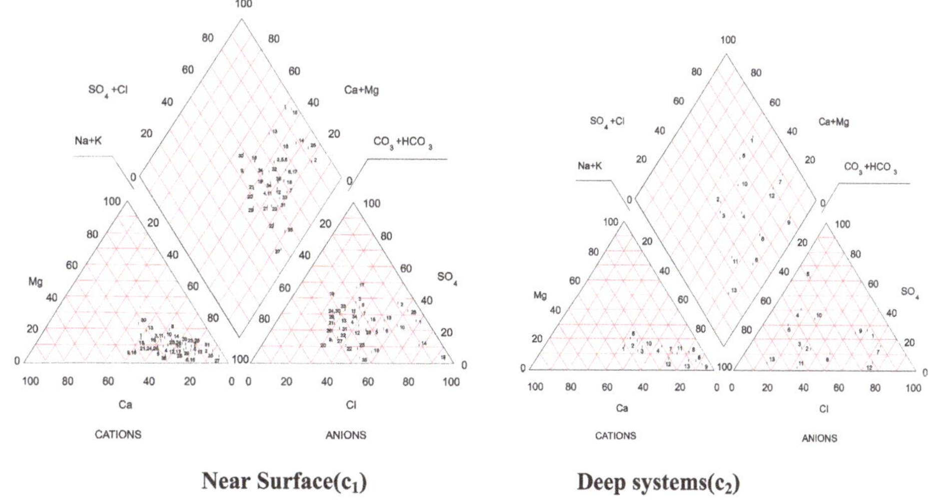

Near Surface(c$_1$) **Deep systems(c$_2$)**

Fig. 5.24(c) Trilinear plots for near surface (c$_1$) and deep systems (c$_2$) in Pennar delta

Hydro chemical characteristics as derived from piper trilinear diagram for Krishna, Godavari and Pennar delta systems are presented in table below.

Table 5.21 Characterization of ionic trends in Krishna delta

Number of Samples: (Near surface: 48, Deep systems: 31)

Subdivision of the diamond	Characteristics of corresponding subdivisions of diamond-shaped fields	Near surface		Deep systems	
		No. of Samples	Percentage of samples in this category (%)	No. of Samples	Percentage of samples in this category (%)
1	Alakaline earth (Ca+Mg) Exceed alkalies (Na + K)	6	13	7	23
2	Alkalies exceeds alkaline earths	42	88	24	77
3	Weak acids (CO_3 + HCO_3) exceed strong acids (SO_4 + Cl)	0	0	7	23
4	Strong acids exceeds weak acids	48	100	24	77
5	Magnesium bicarbonate type	0	0	4	13
6	Clacium-Chloride type	2	4	1	3
7	Sodium-Chloride type	41	85	21	68
8	Sodium-Bicarbonate type	0	0	1	3
9	Mixed type (No cation-anion exceed 50%)	5	10	4	13

Table 5.22 Characterization of ionic trends in Godavari delta

Number of Samples: (Near surface: 68, Deep systems:61)

Subdivision of the diamond	Characteristics of corresponding subdivisions of diamond-shaped fields	Near surface		Deep systems	
		No. of Samples	Percentage of samples in this category (%)	No. of Samples	Percentage of samples in this category (%)
1	Alakaline earth (Ca + Mg) Exceed alkalies (Na + K)	35	51	48	79
2	Alkalies exceeds alkaline earths	33	49	13	21
3	Weak acids (CO_3 + HCO_3) exceed strong acids (SO_4 + Cl)	28	41	44	72
4	Strong acids exceeds weak acids	40	59	17	28
5	Magnesium bicarbonate type	15	22	42	69
6	Clacium-Chloride type	2	3	0	0
7	Sodium-Chloride type	21	31	10	16
8	Sodium-Bicarbonate type	0	0	0	0
9	Mixed type (No cation-anion exceed 50%)	30	44	9	15

Table 5.23 Characterization of ionic trends in Pennar delta

Number of Samples : (Near surface:35, Deep systems:13)

Subdivision of the diamond	Characteristics of corresponding subdivisions of diamond-shaped fields	Near surface		Deep systems	
		No. of Samples	Percentage of samples in this category (%)	No. of Samples	Percentage of samples in this category (%)
1	Alakaline earth (Ca+Mg) Exceed alkalies (Na + K)	1	3	3	23
2	Alkalies exceeds alkaline earths	34	97	10	77
3	Weak acids (CO_3 + HCO_3) exceed strong acids (SO_4 + Cl)	2	6	5	38
4	Strong acids exceeds weak acids	33	94	8	62
5	Magnesium bicarbonate type	0	0	1	8
6	Clacium-Chloride type	0	0	0	0
7	Sodium-Chloride type	32	91	6	46
8	Sodium-Bicarbonate type	0	0	1	8
9	Mixed type (No cation-anion exceed 50%)	3	9	5	38

5.5.4 Hydrogeochemical facies

Hydrogeochemical facies in Krishna delta region

The different hydrochemical facies identified for both near surface and deep systems for Krishna delta are spatially depicted in the Fig. 5.25(a, b) and are described in the following paragraphs.

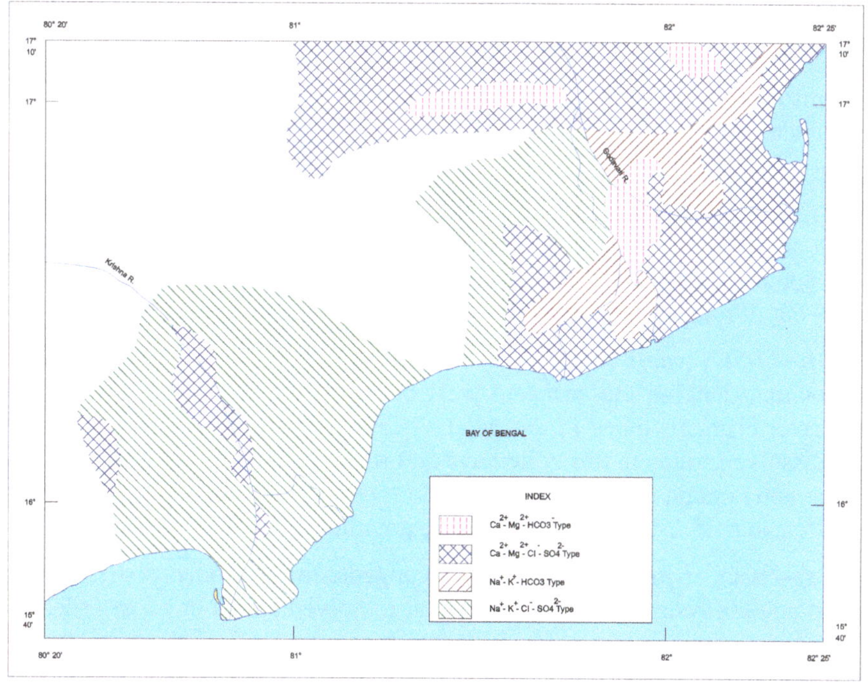

(a) Near surface

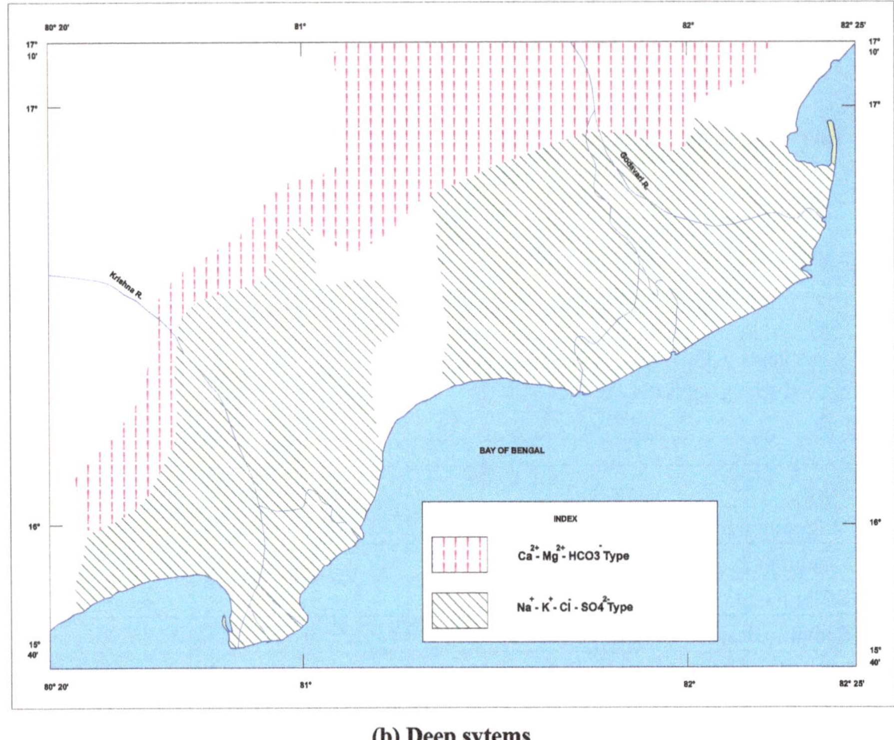

(b) Deep sytems

Fig. 5.25 Hydrogeochemical facies of near surface (a) and deep systems (b) in Krishna & Godavari delta systems

The different hydrochemical facies identified for both near surface and deep systems in the Krishna delta system are as follows

1. Na-K-Cl-SO_4 type
2. Ca-Mg-Cl-SO_4 type

The deep systems however appears to have been characterized by Ca -Mg-HCO_3 type and the entire delta mosaic is characterized by Na-K-Cl-SO_4 type.

Facies no.1 (Na-K-Cl-SO_4 type): The near surface and deep systems in this delta have an average total dissolved solids of the order of 3728.4 ppm and 7021.6 ppm respectively. The Na-K are the major cations in the delta and contains about 82% (near surface) and 77.2% of the total average cations. The average Cl + SO_4 contents in respect of both near surface and deep are about 88% (near surface) and 94% (deep system) out of total average anionic contents.

Facies no.2 (Ca-Mg-Cl-SO_4 type): This facies is represented only in the near surface systems and are restricted to the regions flanking the east and west banks of the river and to a certain extent in the portions of paleo river channels/lobes in the delta system and sometimes flanking the canal systems. The average total dissolved solids in this zone have an order of 1045.4 ppm. The Ca + Mg are the major cations in this zone and contain about 51% out of the total average cationic composition. The average Cl + SO_4 content are about 63.2% of the total average anionic contents.

Facies no.3 (Ca-Mg-HCO_3 type): This facies zone appears to have been restricted only in the upper regions of the deep aquifer systems. The average total dissolved solids have an order of 698 ppm. The Ca + Mg are the major cations in this zone and contain about 53.4% of the total average cations. The average bicarbonate contents in this zone are of 72.7% of total average anionic contents.

Hydrogeochemical facies in Godavari delta region

The different hydrochemical facies identified for both near surface and deep systems for Godavari delta are spatially depicted in the above Figs 5.24(a,b) and are described in the following paragraphs.

The different hydrochemical facies identified in Godavari delta for both near surface and deep systems are as follows

1. Ca-Mg-HCO_3 type
2. Ca-Mg-Cl-SO_4 type
3. Na-K-HCO_3 type
4. Na-K-Cl-SO_4 type

Facies no.1 (Ca-Mg-HCO_3 type): The near surface and deep systems in this delta basin have an average total dissolved solids 608 ppm and 386 ppm respectively. Ca+Mg are the major cations in this zone and contain about 59% (near surface) and 60% (deep systems) of the total average cations. The average bicarbonate contents in respect of both near surface and deep systems are 73.9% and 74% of the total anionic contents. This facies zone is represented as an isolated pocket in the near surface systems regionally surrounded by Ca-Mg-Cl-SO_4 in the upstream portions and in the mid delta region this flank by different types of chemical facies zones namely Na-K-Cl-SO_4 type Na-K-HC_3 type and Ca-Mg-Cl-SO_4 type. This facies zone is regionally wide spread in the upstream portions of the deep systems.

Facies no.2 (Ca-Mg-Cl-SO_4 type): This facies is represented only in the near surface system and significantly absent in the deep systems of the delta region. The average total dissolved solids is of the order of 1214 ppm. The Ca + Mg are the major cations in this zone contain about 50.4% of the total average cations. The average contents of Cl-SO_4 are 55.9% of the total average anionic contents. This facies zone is widely represented in both upstream portions of the delta as well as the outfall regions. The reasons for regional domination of this facies in the upstream portions could be both anthropogenic and salinisation processes.

Facies no.3 (Na-K-HCO_3 type): This zone is also restricted to only near surface systems and significantly found absent in the deep systems. This zone is sandwiched in the eastern portions between Ca-Mg-Cl-SO_4 type and Na-K-Cl-SO_4 type but in the western portions this zone is mostly surrounded by Na-K-Cl-SO_4 type, Ca-Mg- Cl-SO_4 and Ca-Mg-HCO_3 type. The average total dissolved solids in this zone is of the order of 943 ppm. The Na+K are the major cations of this zone and contain about 71% of the total average cationic contents. The average HCO_3 contents of this zone are about 70.8% of the total average anioic contents.

Facies no.4 (Na-K-Cl-SO_4 type): This zone is widely represented in both near surface and deep systems of the delta. Infact a major portions of the upstream and downstream regions of the deep systems is covered by this facies zone only. In the case of near surface system in the western portions of the delta this zone lies adjoinig Ca-Mg-Cl-SO_4, Na-K-HCO_3 type and Ca-Mg-HCO_3 type. However in the eastern portions this zone lies adjacent to Na-K-HCO_3 type, Ca-Mg-Cl-SO_4 type and the shore line boundary. The average total dissolved solids in both near surface and deep systems of the delta have an order of 2513 ppm and 3424 ppm respectively. The Na+K are the major cations in this zone and contain about 73% (near surface) and 66% (deep systems) of the total cationic contents. The average Cl + SO_4 contents in respect of both near surface and deep systems are 66.8% and 72.9% of the total anionic contents respectively.

Hydrogeochemical facies in Pennar delta region

The different hydrochemical facies identified for both near surface and deep systems for Pennar delta are spatially depicted in Fig. 5.26(a,b) and are described in the following paragraphs.

The different hydrochemical facies identified for both near surface and deep systems in the Pennar delta are as follows:

1. Na-K-HCO_3 type
2. Na-K-Cl-SO_4 type

Facies no.1 (Na-K-HCO_3 type): The near surface and deep systems in this delta region have an average total dissolved solids of the order of about 1678 ppm and 642.5 ppm respectively. The Na + K are the major cations in this zone and contain about 90.6% (near surface) and 79.7% (deep systems) of the total average cationic contents. The average bicarbonate contents in this zone in respect of both near surface and deep systems are 62.5% (near surface) and 67.9% (deep systems) of the total average anionic contents. This facies are restricted to the central portions of the south delta system (near surface and deep systems) where paleo channel network is predominantly observed (refer chapter-III).

Facies no.2 (Na–K–Cl–SO_4 type): This facies is represented in the bulk portion of the delta in respect of both near surface and deep systems and have an average total dissolved solids of the order of 1604 ppm and 2508.8 ppm respectively. The Na+K are the major cations of this zone and contain about 73% (near surface) and 80.3% (deep systems) of the total average cations. The average Cl + SO_4 in respect of both near surface and deep systems are 68.3% and 71% of the total anionic contents respectively.

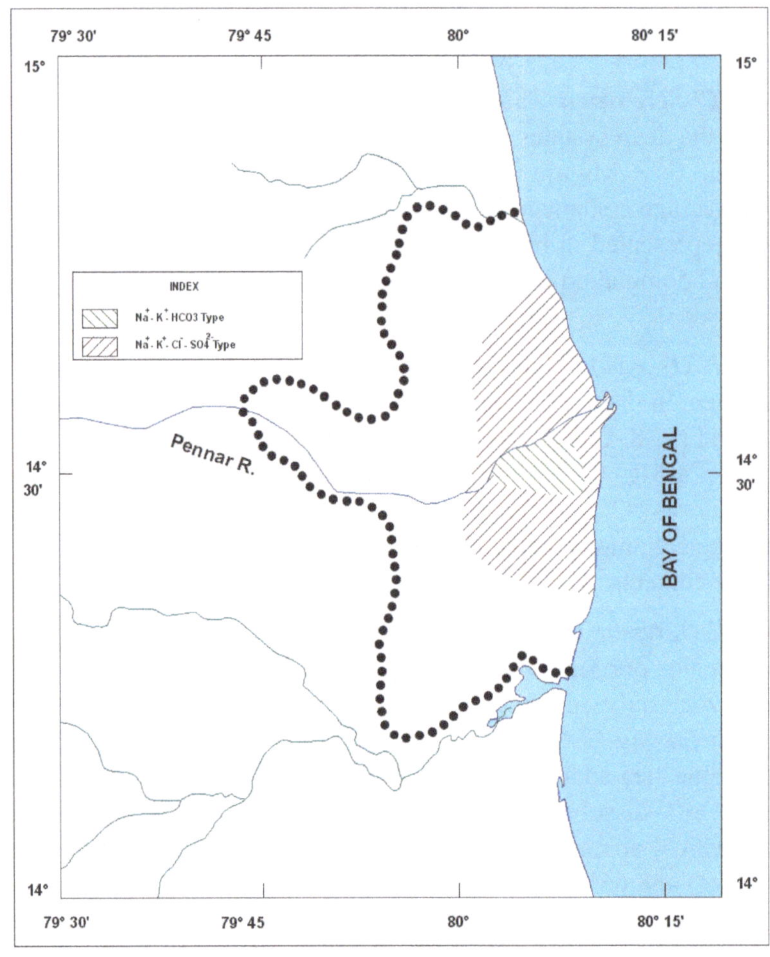

(a) Near surface

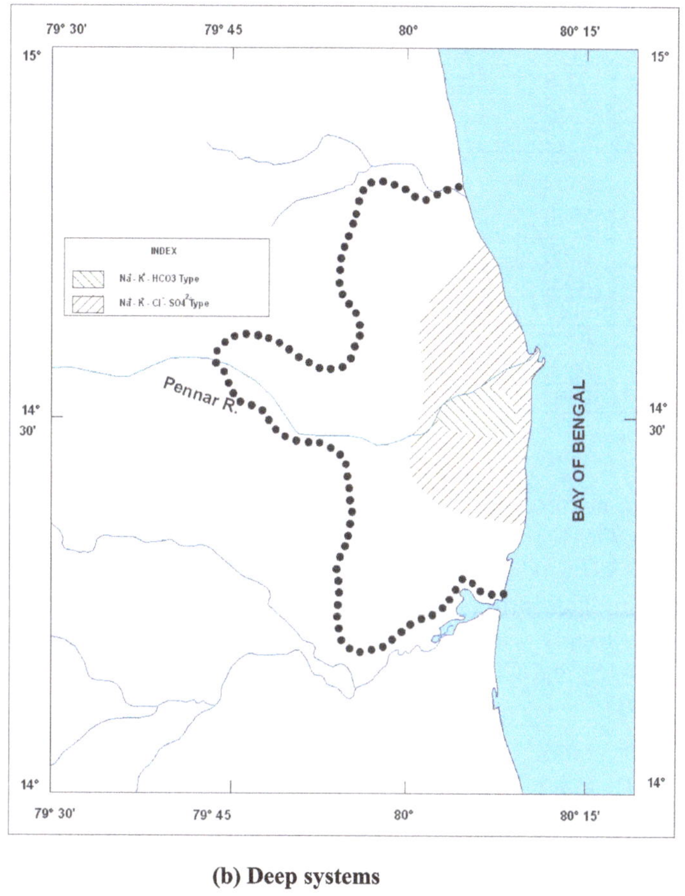

(b) Deep systems

Fig. 5.26 Hydrogeochemical facies of near surface (a) and deep systems (b) in Pennar delta system

5.5.5 Correlation matrix and factor analysis

The correlation matrix analyses for Krishna, Godavari and Pennar delta's near surface and deep systems are presented in table below:

Table 5.24(a,b) Correlation matrix analysis for Krishna delta

(a) Near surface

	pH	**EC**	**TDS**	**Ca**	**Mg**	**Na**	**K**	**HCO_3**	**SO_4**	**Cl**	**NO_3**	**TH**
pH	1											
EC	0.732	1										
TDS	0.724	0.999	1									
Ca	–0.415	–0.139	–0.134	1								
Mg	0.698	0.927	0.933	–0.064	1							
Na	0.705	0.991	0.992	–0.158	0.905	1						
K	0.704	0.882	0.875	–0.135	0.763	0.855	1					
HCO_3	–0.091	–0.183	–0.198	0.059	–0.313	–0.219	–0.101	1				
SO_4	0.77	0.865	0.859	–0.2	0.828	0.844	0.815	–0.068	1			
Cl	0.708	0.991	0.993	–0.119	0.934	0.995	0.846	–0.266	0.833	1		
NO_3	0.082	0.093	0.102	–0.124	0.214	0.093	0.064	–0.288	0.165	0.118	1	
TH	–0.359	–0.32	–0.335	0.333	–0.381	–0.355	–0.277	0.729	–0.271	–0.387	–0.282	1

(b) Deep systems

	pH	EC	TDS	Ca	Mg	Na	K	HCO_3	SO_4	Cl	TH
pH	1										
EC	–0.215	1									
TDS	–0.215	0.999	1								
Ca	–0.274	0.765	0.783	1							
Mg	–0.02	0.776	0.795	0.854	1						
Na	–0.176	0.973	0.967	0.61	0.665	1					
K	–0.261	0.801	0.783	0.492	0.47	0.789	1				
HCO_3	–0.317	–0.205	–0.197	–0.122	–0.218	–0.184	–0.243	1			
SO_4	–0.173	0.855	0.872	0.751	0.838	0.806	0.583	–0.073	1		
Cl	–0.225	0.995	0.993	0.782	0.765	0.958	0.809	–0.242	0.818	1	
TH	–0.162	0.92	0.911	0.771	0.79	0.856	0.795	–0.292	0.729	0.921	1

Krishna delta

Near surface systems: A good correlation is observed between pH and EC (r = 0.732), TDS (r = 0.724), Na (r = 0.705), SO_4 (r = 0.77), K (r = 0.704), Cl (r = 0.708) and moderately with Mg (r = 0.698). EC correlates strongly with TDS (r = 0.999), Mg (r = 0.927), Na (r = 0.991), K (r = 0.882), SO4(r = 0.865) and Cl (r = 0.991). Mg correlates strongly with Na (r = 0.905), SO_4 (r = 0.828), K (r =0.763) and Cl (r = 0.934). Na correlates strongly with K (r = 0.855), SO4 (r = 0.844) and Cl (r = 0.995). K strongly correlate with SO4 (r = 0.815) and Cl (r = 0.846). HCO_3 strongly with TH (r = 0.729). SO4 strongly correlates with Cl (r = 0.833).

Deep systems: A strong correlation is observed between EC and TDS (r = 0.999), Ca (r = 0.765), Mg (r = 0.776), Na (r = 0.973), K (r = 0.801), SO4 (r = 0.855), Cl (r = 0.995) and TH (r = 0.92). Ca correlates strongly with Mg (r = 0.854), SO_4 (r = 0.751), Cl (r = 0.782), TH (r = 0.771) and moderately with Na (r = 0.61). A good correlation is observed between Mg and SO_4 (r = 0.838), Cl (r = 0.958) and TH (r = 0.79) and moderate correlation with Na (r = 0.665). Na correlates strongly with K (r = 0.789), SO_4 (r = 0.806), Cl (r = 0.958) and TH (r = 0.856). K strongly correlate with Cl (r = 0.809), TH (r = 0.795) and moderately correlate with SO_4 (r = 0.583). SO_4 strongly correlate with Cl (r = 0.818) and TH (r = 0.729). Cl correlates strongly with TH (r = 0.921).

Factor analysis: Factor analysis has been applied for near surface and deep systems and the results are furnished below in Table 5.25(a,b).

Table 5.25(a) Factor analysis for near surface and deep systems of Krishna delta

Near Surface					Deep systems				
Compo-nent	Eigen value	% of Variance	Cumulative Eigen value	Cumulative %	Compo-nent	Eigen value	% of Variance	Cumulative Eigen value	Cumulative %
1	7.298	60.815	7.298	60.815	1	7.595	69.045	7.595	69.045
2	1.768	14.732	9.066	75.547	2	1.355	12.314	8.95	81.358
3	1.116	9.297	10.182	84.844	3	0.852	7.748	9.802	89.107
4	0.84	6.997	11.022	91.842	4	0.59	5.368	10.392	94.475
5	0.323	2.691	11.345	94.533	5	0.268	2.433	10.66	96.908
6	0.241	2.008	11.586	96.541	6	0.181	1.644	10.841	98.552
7	0.187	1.558	11.773	98.099	7	0.107	0.972	10.948	99.524
8	0.154	1.284	11.927	99.383	8	0.047	0.428	10.995	99.951
9	0.068	0.566	11.995	99.949	9	0.005	0.045	11	99.996
10	0.005	0.044	12	99.993	10	0	0.003	11	99.999
11	0.001	0.005	12.001	99.998					
12	0	0.002	1.20E + 01	100					

Table 5.25(b) Loading varimax rotated factor for Krishna delta

Near surface system				Deep systems		
Variable	Factor 1	Factor 2	Factor 3	Variable	Factor 1	Factor 2
pH	0.751	–0.061	–0.478	pH	–0.231	0.817
EC	0.986	–0.106	–0.041	EC	0.988	–0.008
TDS	0.983	–0.125	–0.032	TDS	0.99	–0.013
Ca	–0.061	0.142	0.952	Ca	0.828	–0.098
Mg	0.918	–0.258	0.041	Mg	0.841	0.133
Na	0.968	–0.143	–0.047	Na	0.933	0.003
K	0.901	–0.026	–0.093	K	0.8	–0.014
HCO_3	–0.091	0.91	–0.116	HCO_3	–0.231	–0.804
SO_4	0.896	–0.037	–0.185	SO_4	0.88	–0.068
Cl	0.967	–0.193	–0.006	Cl	0.984	0.008
NO_3	0.038	–0.561	–0.077	TH	0.94	0.09
TH	–0.232	0.833	0.254			

The first three factors (near surface system) and two factors (deep systems) by and large explains 84.8% and 81.3% of the total variances respectively.

Near surface systems

Factor 1 explains the largest proportion (60.8%) of the total variance and has high positive loadings (0.75-0.98) for pH, EC, TDS, Mg, Na, K, SO4 and Cl.

Factor 2 accounts for 14.7% of the total variance and has high loading on HCO_3 (0.91) and TH (0.833).

Factor 3 accounts for 9.2% of the total variance and has high loading on Ca (0.952).

Deep systems

Factor 1 explains the largest proportion 69% of the total variance and has high positive loadings (0.8-0.99) on EC, TDS, Ca, Mg, Na, K, SO_4, Cl and TH.

Factor 2 accounts for 12.3% of the total variance and has high loadings on pH (0.81).

Table 5.26(a,b) Correlation matrix analysis for Godavari delta

(a) Near surface

	pH	EC	TDS	Ca	Mg	Na	K	HCO_3	SO_4	Cl	NO	F	TH
pH	1.000												
EC	–0.259	1.000											
TDS	–0.246	0.977	1.000										
Ca	–0.544	0.615	0.615	1.000									
Mg	–0.152	0.928	0.932	0.528	1.000								
Na	–0.098	0.894	0.925	0.399	0.831	1.000							
K	–0.29	0.486	0.515	0.246	0.396	0.427	1.000						
HCO_3	–0.311	0.653	0.684	0.408	0.565	0.725	0.295	1.00					
SO_4	–0.254	0.788	0.809	0.497	0.752	0.675	0.704	0.314	1.000				
Cl	–0.197	0.962	0.97	0.599	0.928	0.91	0.387	0.606	0.724	1.000			
NO_3	0.281	–0.167	–0.156	–0.098	–0.142	–0.173	0.14	–0.311	–0.1	–0.172	1.000		
F	0.127	–0.163	–0.163	–0.342	–0.177	–0.034	–0.164	0.431	–0.249	–0.248	–0.297	1.000	
TH	–0.35	0.921	0.924	0.792	0.934	0.76	0.389	0.578	0.743	0.915	–0.148	–0.268	1.000

(b) Deep system

	pH	EC	TDS	Ca	Mg	Na	K	HCO_3	SO_4	Cl	TH
pH	1										
EC	–0.028	1									
TDS	–0.028	1	1								
Ca	–0.077	0.947	0.947	1							
Mg	0.008	0.934	0.934	0.882	1						
Na	–0.019	0.983	0.983	0.924	0.97	1					
K	–0.034	0.363	0.363	0.421	0.412	0.345	1				
HCO_3	–0.095	0.542	0.542	0.593	0.684	0.564	0.379	1			
SO_4	0.054	0.883	0.883	0.81	0.861	0.909	0.31	0.25	1		
Cl	–0.037	0.984	0.984	0.971	0.926	0.978	0.376	0.544	0.878	1	
TH	–0.027	0.959	0.959	0.923	0.985	0.983	0.398	0.631	0.884	0.959	1

Godavari delta

Near surface systems: A good correlation is observed between EC and TDS (r = 0.977), Mg (r = 0.928), Na (r = 0.894), SO_4 (r = 0.788), Cl (r = 0.962) and TH (r = 0.921) and moderately correlated with Ca (r = 0.615) and HCO_3 (r = 0.65). Ca correlate strongly with TH (r = 0.792) and moderately with Mg (r = 0.528) and Cl (r = 0.599). Mg strongly correlates with Na (r = 0.831), SO_4 (r = 0.752), Cl (r = 0.928), TH (r = 0.934) and moderately with HCO_3 (r = 0.565). Na correlate well with HCO_3 (r = 0.725), Cl (r = 0.91), TH (r = 0.76) and moderately with SO_4 (r = 0.675). K strongly correlates with SO_4 (r = 0.704). HCO_3 moderately correlate with Cl (r = 0.606) and TH (r = 0.578). SO_4 strongly correlate with Cl (r = 0.724). Cl strongly correlate with TH (r = 0.915).

Deep systems: A strong correlation is observed between EC and Ca (r = 0.947), Mg (r = 0.934), Na (r = 0.983), SO_4 (r = 0.883), Cl (r = 0.984) and TH (r = 0.959) and moderately with HCO_3 (r = 0.542). Ca correlates strongly with Mg (r = 0.882), Na (r = 0.924), SO_4 (r = 0.81), Cl (r = 0.971) and TH (r = 0.923) and correlate moderately with HCO_3 (r = 0.593). Mg correlates strongly with Na (r = 0.97) SO_4 (r = 0.861), Cl (r = 0.926) and TH (r = 0.985) and moderately correlate with HCO_3 (r = 0.684). Na correlates strongly with SO_4 (r = 0.90), Cl (r = 0.978), TH (r = 0.983) and moderately with HCO_3 (r = 0.564). HCO_3 moderately correlate with Cl (r = 0.544) and TH (r = 0.631), SO_4 strongly correlate with Cl (r = 0.878) and TH (r = 0.884). Cl strongly correlate with TH (r = 0.959).

Table 5.27(a) Factor analysis for near surface and deep systems of Godavari delta

Near Surface					Deep systems				
Component	Eigen value	% of Variance	Cumulative Eigen value	Cumulative %	Component	Eigen value	% of Variance	Cumulative Eigen value	Cumulative %
1	7.502	57.705	7.502	57.705	1	8.083	73.484	8.083	73.484
2	1.73	13.306	9.232	71.01	2	1.077	9.795	9.16	83.279
3	1.323	10.177	10.555	81.188	3	0.906	8.232	10.066	91.511
4	0.975	7.497	11.53	88.684	4	0.659	5.994	10.725	97.505
5	0.689	5.298	12.219	93.982	5	0.178	1.618	10.903	99.123
6	0.33	2.536	12.549	96.518	6	0.057	0.516	10.96	99.639
7	0.216	1.662	12.765	98.18	7	0.021	0.188	10.981	99.827
8	0.134	1.029	12.899	99.209	8	0.012	0.106	10.993	99.932
9	0.051	0.393	12.95	99.602	9	0.005	0.046	10.998	99.978
10	0.028	0.215	12.978	99.818	10	0.002	0.022	11	100
11	0.019	0.144	12.997	99.962	11	–8.9E-16	–8.16E-15	1.10E + 01	100
12	0.003	0.022	1.30E + 01	99.984					
13	0.002	0.016	1.30E + 01	100					

Table 5.27(b) Loading varimax rotated factor for Godavari delta

Near surface system				Deep systems		
Variable	**Factor 1**	**Factor 2**	**Factor 3**	**Variable**	**Factor 1**	**Factor 2**
pH	–0.08	–0.906	–0.105	pH	0.153	–0.807
EC	0.959	0.193	0.077	EC	0.972	0.16
TDS	0.977	0.176	0.078	TDS	0.972	0.16
Ca	0.512	0.704	–0.106	Ca	0.922	0.26
Mg	0.94	0.101	0.04	Mg	0.949	0.215
Na	0.928	–0.033	0.211	Na	0.98	0.151
K	0.544	0.134	–0.245	K	0.348	0.459
HCO_3	0.618	0.151	0.647	HCO_3	0.511	0.55
SO_4	0.813	0.206	–0.182	SO_4	0.924	–0.062
Cl	0.95	0.161	0.043	Cl	0.968	0.176
NO_3	–0.04	–0.363	–0.681	TH	0.965	0.21
F	–0.167	–0.291	0.834			
TH	0.884	0.375	–0.011			

The first three factors (near surface systems) and two factors (deep systems) by and large explains 81.8% and 83.2% of the total variances respectively.

Near surface systems

Factor 1 explains the largest proportion (57.7%) of the total variance and has high positive loadings (0.81-0.97) for EC, TDS, Mg, Na, SO_4, Cl and TH and has moderate loadings on K (0.544) and HCO_3 (0.618).

Factor 2 accounts for 13.3 % of the total variance and has high loading on Ca (0.704).

Factor 3 accounts for 10.1 % of the total variance and has high loading on F (0.834) and moderate loading on HCO3 (0.647).

Deep systems

Factor 1 explains the largest proportion 73.4% of the total variance and has high positive loadings (0.92-0.98) on EC, TDS, Ca, Mg, Na, SO_4, Cl and TH and has moderate loading on HCO_3 (0.511).

Factor 2 accounts for 9.7% of the total variance and has moderate loading on HCO_3 (0.55) and has high negative loading on pH (-0.807).

Table 5.28(a,b) Correlation matrix analysis for Pennar delta

(a) Near surface

	pH	**EC**	**TDS**	**Ca**	**Mg**	**Na**	**K**	**HCO_3**	**SO_4**	**Cl**	**F**	**TH**
pH	1											
EC	–0.049	1										
TDS	–0.053	0.998	1									
Ca	–0.027	0.895	0.894	1								
Mg	–0.021	0.924	0.928	0.942	1							
Na	–0.063	0.939	0.945	0.735	0.829	1						
K	0.048	0.833	0.849	0.77	0.772	0.772	1					
HCO_3	–0.046	0.219	0.236	–0.061	0.044	0.398	0.322	1				
SO_4	–0.054	0.565	0.593	0.493	0.633	0.697	0.607	0.189	1			
Cl	–0.028	0.982	0.98	0.95	0.954	0.889	0.806	0.068	0.559	1		
F	–0.076	0.097	0.1	0.119	0.205	0.189	0.045	–0.015	0.463	0.118	1	
TH	–0.036	0.915	0.916	0.991	0.973	0.774	0.776	–0.026	0.534	0.962	0.127	1

(b) Deep systems

	pH	EC	TDS	Ca	Mg	Na	K	HCO_3	SO_4	Cl	F	TH
pH	1											
EC	–0.026	1										
TDS	–0.032	0.999	1									
Ca	–0.235	0.841	0.849	1								
Mg	–0.176	0.933	0.938	0.963	1							
Na	0.078	0.953	0.944	0.728	0.864	1						
K	–0.128	0.783	0.783	0.794	0.78	0.638	1					
HCO_3	0.154	0.78	0.776	0.538	0.701	0.877	0.427	1				
SO_4	0.005	0.881	0.871	0.806	0.83	0.834	0.909	0.62	1			
Cl	–0.045	0.983	0.981	0.864	0.957	0.967	0.719	0.797	0.851	1		
F	0.352	0.234	0.219	–0.136	–0.001	0.309	0.126	0.227	0.212	0.202	1	
TH	–0.209	0.861	0.867	0.976	0.983	0.794	0.711	0.646	0.761	0.907	–0.13	1

Pennar delta

Near surface systems: EC correlates strongly with TDS (r = 0.998), Ca (r = 0.895), Mg (r = 0.924), Na (r = 0.939), K (r = 0.833), Cl (r = 0.982), TH (r = 0.915) and moderately with SO4 (r = 0.565). Ca correlates strongly with Mg (r = 0.942), Na (r = 0.735), K (r = 0.77), Cl (r = 0.95) and TH (r = 0.991). Mg strongly correlates with Na (r = 0.829), K (r = 0.77), Cl (r = 0.954), TH (r = 0.973) and moderately correlate with SO_4 (r = 0.633). Na strongly correlates with K (r = 0.772), Cl (r = 0.889), TH (r = 0.774) and moderately with SO_4 (r = 0.697). K strongly correlate with Cl (r = 0.806), TH (r = 0.776) and moderately with SO_4 (r = 0.607). SO4 moderately correlate with Cl (r = 0.559) and TH (r = 0.534). Cl correlate strongly with TH (r = 0.962).

Deep systems: A strong correlation is observed between EC and TDS (r = 0.99), Ca (r = 0.841), Mg (r = 0.933), Na (r = 0.953), K (r = 0.783), HCO_3 (r = 0.78), SO_4 (r = 0.881), Cl (r = 0.983), TH (r = 0.861). Ca correlates strongly with Mg (r = 0.963), Na (r = 0.728), K (r = 0.794), SO4 (r = 0.806), Cl (r = 0.864) and TH (r = 0.976) and moderately with HCO_3 (r = 0.538). Mg strongly correlates with Na (r = 0.864), K (r = 0.78), HCO_3 (r = 0.701), SO_4 (r = 0.83), Cl (r = 0.957) and TH (r = 0.983). Na strongly correlates with HCO_3 (r = 0.877), SO4 (r = 0.834), Cl (r = 0.967), TH (r = 0.794) and moderately with K (r = 0.638). K strongly correlate with SO_4 (r = 0.909), Cl (r = 0.719) and TH (r = 0.711). HCO_3 strongly correlates with Cl (r = 0.797) and moderately with SO_4 (r = 0.62) and TH (r = 0.646). SO_4 strongly correlate with Cl (r = 0.851) and TH (r = 0.761). Cl strongly correlate with TH (r = 0.907).

Table 5.29(a) Factor analysis for near surface and deep systems of Pennar delta

Near Surface					Deep systems				
Component	**Eigen value**	**% of Variance**	**Cumulative Eigen value**	**Cumulative %**	**Component**	**Eigen value**	**% of Variance**	**Cumulative Eigen value**	**Cumulative %**
1	7.711	64.257	7.711	64.257	1	8.523	71.025	8.523	71.025
2	1.287	10.723	8.998	74.98	2	1.683	14.022	10.206	85.047
3	1.201	10.004	10.199	84.984	3	0.735	6.126	10.941	91.173
4	0.983	8.189	11.182	93.173	4	0.605	5.039	11.546	96.212
5	0.388	3.237	11.57	96.409	5	0.25	2.084	11.796	98.297
6	0.249	2.074	11.819	98.483	6	0.108	0.902	11.904	99.198
7	0.139	1.159	11.958	99.642	7	0.077	0.641	11.981	99.839
8	0.034	0.283	11.992	99.925	8	0.018	0.15	11.999	99.989
9	0.006	0.047	11.998	99.972	9	0.001	0.009	12	99.999
10	0.002	0.019	12	99.991	10	0	0.001	12	100
11	0.001	0.005	12.001	99.996	11	1.E-05	8.9E-05	12.0000108	100
12	0	0.004	1.20E+01	100	12	8.3E-07	6.9E-06	12.0000116	100

Table 5.29(b) Loading varimax rotated factor for Pennar delta

Near surface system				Deep systems		
Variable	**Factor 1**	**Factor 2**	**Factor 3**	**Variable**	**Factor 1**	**Factor 2**
pH	0.02	–0.345	–0.061	pH	–0.095	0.774
EC	0.969	0.05	0.168	EC	0.979	0.126
TDS	0.971	0.064	0.186	TDS	0.979	0.11
Ca	0.963	0.025	–0.162	Ca	0.917	–0.308
Mg	0.964	0.151	–0.049	Mg	0.978	–0.147
Na	0.861	0.212	0.374	Na	0.929	0.276
K	0.844	0.021	0.307	K	0.822	–0.085
HCO_3	0.061	0.057	0.968	HCO_3	0.766	0.332
SO_4	0.57	0.612	0.195	SO_4	0.906	0.093
Cl	0.991	0.051	0.005	Cl	0.981	0.093
F	0.064	0.908	–0.131	F	0.126	0.818
TH	0.976	0.05	–0.124	TH	0.932	–0.258

The first four factors (near surface systems) and three factors (deep systems) by and large explained 84.9% and 85% of the total variances respectively.

Near surface systems

***Factor* 1** explains the largest proportion (64.2%) of the total variance and has high positive loadings (0.84-0.99) for EC, TDS, Ca, Mg, Na, K, Cl and TH.

***Factor* 2** accounts for 10.7% of the total variance and has high loading on F (0.908) and moderate loading on SO_4 (0.612).

***Factor* 3** accounts for 10 % of the total variance and has high loading on HCO_3 (0.968).

Deep systems

***Factor* 1** explains the largest proportion 71% of the total variance and has high positive loadings (0.76-0.98) on EC, TDS, Ca, Mg, Na, K, HCO_3, SO_4, Cl and TH.

***Factor* 2** accounts for 14% of the total variance and has high loadings on pH (0.774) and F (0.818).

5.5.6 Classification and irrigation suitability of groundwaters based on sodium percentages, sodium adsorption ratios and residual sodium carbonate

Sodium percentage

The classification of groundwaters with respect of percentage of sodium in Krishna, Godavari and Pennar delta systems are presented below in Table 5.30(a,b,c).

Table 5.30(a) Krishna delta

Sodium (%)	**Water Class**	**No. of Samples**	
		Near Surface system	**Deep system**
< 20	Excellent	0	0
20 - 40	Good	3 (27.4-32.5)	4 (22.7-39.6)
40 - 60	Permissible	13 (40-59.2)	12 (41.5-59.6)
60 - 80	Doubtful	27 (60.8-79.2)	10 (60.2-78.7)
> 80	Unsuitable	5 (81-84.4)	5 (80.4-94.4)

Table 5.30(b) Godavari delta

Sodium (%)	Water Class	No. of Samples	
		Near Surface system	Deep system
< 20	Excellent	6 (13.7-18.4)	21 (1.1-19.7)
20 - 40	Good	25 (23.3-39)	21 (21.9-39)
40 - 60	Permissible	29 (40-58.6)	17 (41.8-59.1)
60 - 80	Doubtful	8 (60.9-75.1)	2 (60.2-69.4)
> 80	Unsuitable	0	0

Table 5.30(c) Pennar delta

Sodium (%)	Water Class	No. of Samples	
		Near Surface systems	Deep system
< 20	Excellent	0	0
20 - 40	Good	0	1 (38.0)
40 - 60	Permissible	11 (43-58.9)	4 (41.1-51.6)
60 - 80	Doubtful	22 (60.3-78.3)	5 (60.4-78.7)
> 80	Unsuitable	2 (83.2-88.3)	3 (82.3-90.1)

Krishna delta: The sodium percentages in the near surface system of Krishna delta have a range of values between 27.4 and 84.4 and the deep systems between 22.7 and 94.4. About 33.3% of the samples from near surface fall in the range of excellent to permissible limits and the rest 66.7% of the samples lie in the range of doubtful to unsuitable categories. In the case of deep systems about 51.6% of the samples lie in the range of good to permissible limits and rest 48.4% of the samples lie in the range of doubtful to unsuitable categories.

Godavari delta: The sodium percentages in the near surface system of Godavari delta have a range of values between 13.7 to 75.10 and the deep systems between 1.1 to 69.4. About 89.2% of the samples from near surface fall in the range of excellent to permissible limits and the rest 10.8% of the samples fall in the doubtful to unsuitable categories. In the case of deep systems about 85.7% of the samples lie in the range of excellent to permissible limits and the rest 14.3% of the samples lie in the range of doubtful to unsuitable categories.

Pennar delta: The sodium percentages in the near surface system of Pennar delta have a range of values between 43.00 to 88.3 and deep systems between 38.00 to 90.10. About 31.4% of the samples from near surface fall in the range of excellent to permissible limits and the rest 68.6% fall in the range of doubtful to unsuitable categories. In the case of deep systems about 38.4% of the samples lie in the range of good to permissible limits and the rest 61.6% lie in the range of doubtful to unsuitable categories.

Sodium Adsorption Ratios (SAR): The classification of groundwater samples based on SAR (Sodium Adsorption Ratio) for all the delta systems, both near surface and deep are presented below in Figures 5.27(a, b and c).

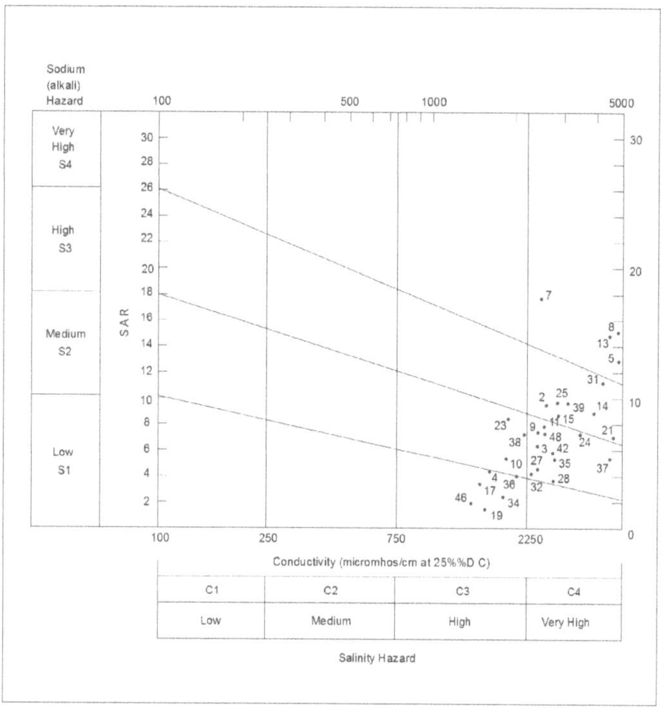

Near surface (a_1)

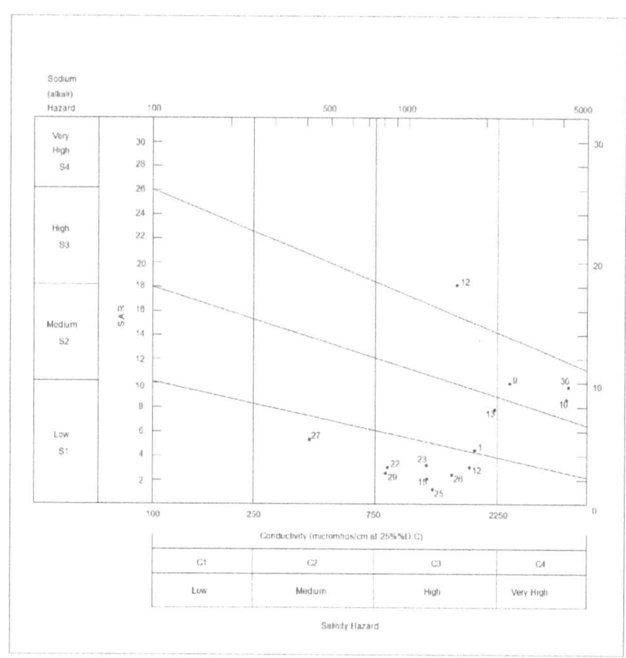

Deep systems (a_2)

Fig. 5.27(a) Sodium adsorption ratios for near surface (a_1) and deep systems (a_2) in Krishna delta

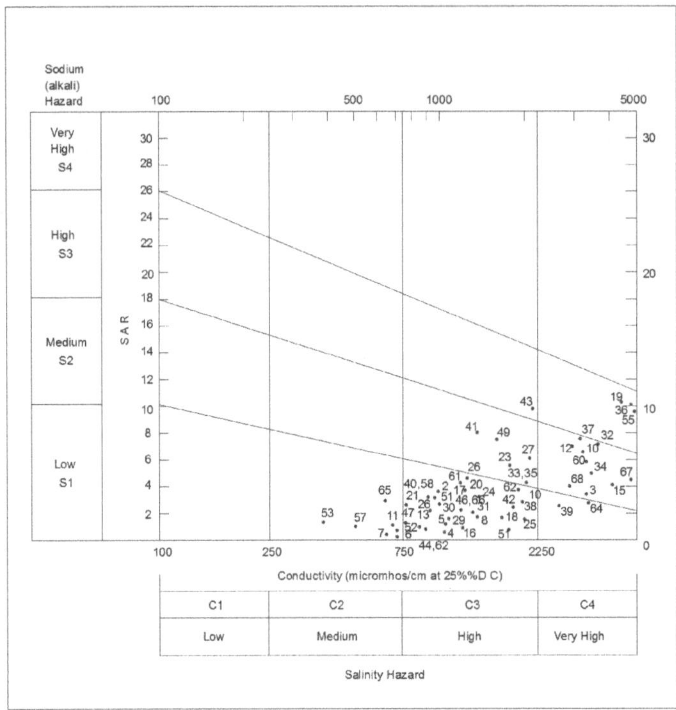

Near Surface (b_1)

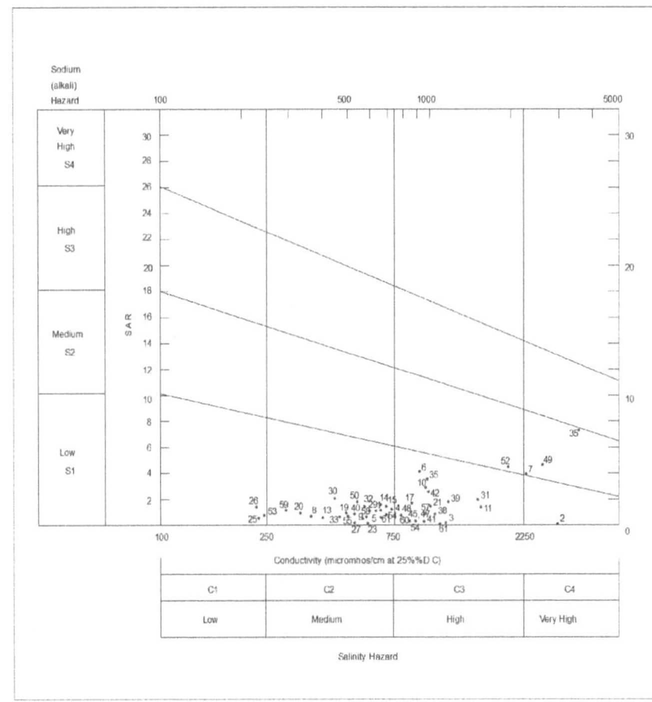

Deep systems (b_2)

Fig. 5.27(b) Sodium adsorption ratios for near surface (b_1) and deep systems (b_2) in Godavari delta

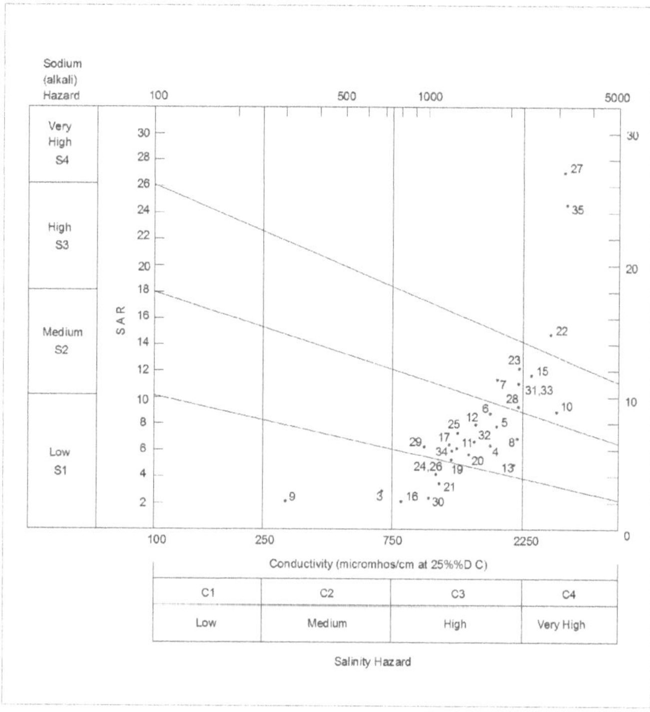

Near Surface (c_1)

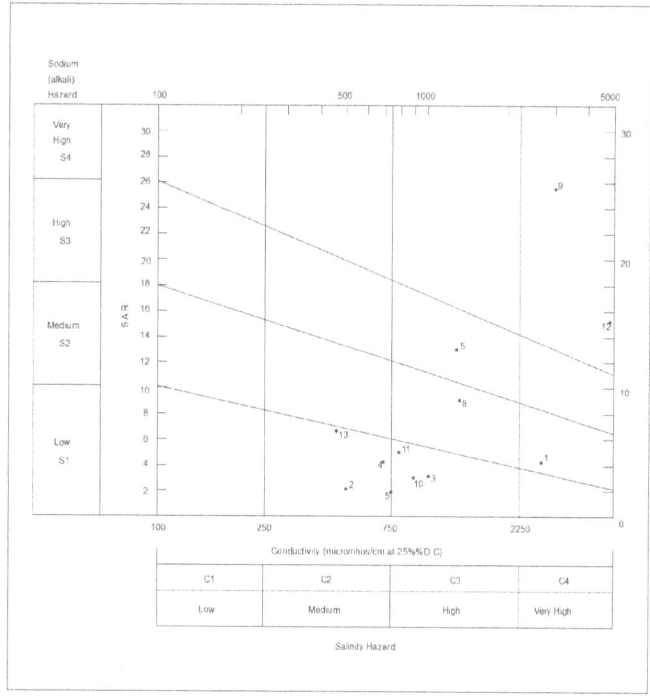

Deep systems (c_2)

Fig. 5.27(c) Sodium adsorption ratios for near surface (c_1) and deep systems (c_2) in Pennar delta

Krishna delta: The near surface system of Krishna delta have nil numbers in the field of C_1S_1 and C_2S_1 (low salinity and med. Salinity/low alkalinity fields). About 29.03% of the samples lie in the fields of C_3S_1 and C_3S_2 (High salinity and med. alkalinities). About 70.97% of the samples lie in C_4S_2, C_4S_3 and C_4S_4 fields. (Very high salinities and high alkalinities). Similarly the deep systems have only 7.14% of samples represented in the fields of C_1S_1 and C_2S_2 (Low salinities and low/med, in alkalinities). About 71.42% of the samples lie in the fields of C_3S_1, C_3S_2, C_3S_3 and C_3S_4 (High salinities and low to high alkalinities).

Godavari delta: The near surface systems of Godavari delta have only 8.62% of reorientation in the fields of C_1S_1 and C_2S_2 (low salinity and low to med. alkalinities). About 65.51% of the samples lie in the fields of C_3S_1, C_3S_2 and C_3S_3 (High salinities and low to high alkalinities). The rest 25.86% lie in the fields of C_4S_1, C_4S_2 and C_4S_3 (very high salinities and low to high alkalinities). Similarly the deep systems are represented by 50% of the samples in the fields of C_1S_1 and C_2S_2 (low salinities and low to med. alkalinities). About 41.30% of the samples lie in the fields of C_3S_1 and C_3S_2 (High salinities and low to med. alkalinities). And about 8.69% of the samples lie in the fields of C_4S_1 and C_4S_2 (very high salinities and low to med. alkalinities).

Pennar delta: The near surface systems of Pennar delta have again only about 6.45% of the samples in the fields of C_1S_1 and C_2S_1 (low salinities and low to med. alkalinities), about 70.96% of the samples lie in the fields of C_3S_1, C_3S_2 and C_3S_3 (High alkalinities and low to high alkalinities levels), and the rest 22.58% of the samples lie in the fields of C_4S_1, C_4S_3, and C_4S_4 (Very high salinities and low to very high alkalinities). The deep systems are represented by about 33.33% in the fields of C_1S_1 and C_2S_2 (low salinities and low to med. alkalinities) about 41.66% are represented in the fields of C_3S_1, C_3S_2, C_3S_3 (High salinities and low to high alkalinities), and the rest 25% of the samples lie in the fields of C_4S_2, C_4S_4 and C_4S_4 (very high salinities and med. to high alkalinities).

Residual Sodium Carbonate (RSC)

The RSC values for the near surface and deep systems for Krishna, Godavari and Pennar delta systems are presented in Table 5.31(a, b and c) as shown below.

Table 5.31(a) Krishna delta

RSC (epm)	Remark on Quality	No. of Samples	
		Near Surface systems	Deep systems
< 1.25	Good	48 (-68.1-1.1)	30 (-111.3-1.0)
1.25 - 2.5	Doubtful	0	0
> 2.5	Unsuitable	0	1 (2.7)

(b) Godavari delta

RSC (epm)	Remark on Quality	No. of Samples	
		Near Surface systems	Deep systems
< 1.25	Good	53 (-26.1-1.0)	56 (-74.1-1.2)
1.25 - 2.5	Doubtful	7 (1.3-2.4)	5 (1.3-1.5)
> 2.5	Unsuitable	8 (2.8-11.7)	0

(c) Pennar delta

RSC (epm)	Remark on Quality	No. of Samples	
		Near Surface systems	Deep systems
< 1.25	Good	22 (-63.5-1.0)	7 (-15.2-0.6)
1.25 - 2.5	Doubtful	5 (1.4-2.4)	2 (2.0-2.5)
> 2.5	Unsuitable	8 (2.5-17.2)	4 (3.6-6.7)

Krishna delta: Nearly 100% of the samples from near surface systems and 96.77% of the samples from deep systems have value less than 1.25 and are considered to be good of quality. All the rest of the samples are in doubtful to unsuitable categories.

Godavari delta: About 77.94% of the samples from near surface and 91.80% of deep systems have value less than 1.25 and are considered to be of good quality. All of the rest of the samples fall under the categories of doubtful and unsuitable.

Pennar delta: About 62.85% of the near surface and 53.84% of the deep systems have values less than 1.25 and are considered to be good quality. All of the rest of the samples fall under the categories of doubtful to unsuitable.

5.6 WATER CHEMISTRY, CHARACTERIZATION, HYDROGEO-CHEMICAL FACIES AND MULTIVARIATE TREND ANALYSIS OF GROUNDWATERS (NEAR SURFACE AND DEEP) IN MAHANADI-BRAHMINI-BAITARNI DELTA SYSTEMS OF ORISSA

5.6.1 Spatial variations of specific conductance, pH and statistical variations of different ions in Mahanadi-Brahmini-Baitarni delta systems of Orissa

The sample locations for both near surface and deep systems in respect of Mahanadi delta systems are presented in Fig. 5.28(a,b). (Data source: Radharkishna, I. and Dutt, D.K., 1976). The analytical results of major ions in terms of max, min, mean and standard deviation are presented in Table 5.32.

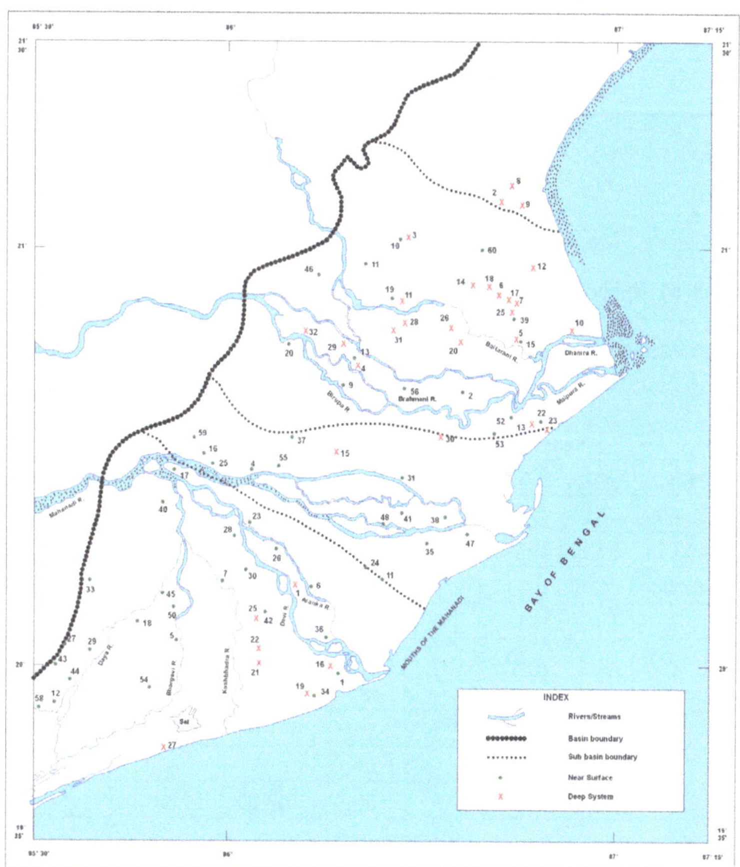

Fig. 5.28(a) Sample locations of near surface and deep systems in Mahanadi- Brahmini-Baitarni

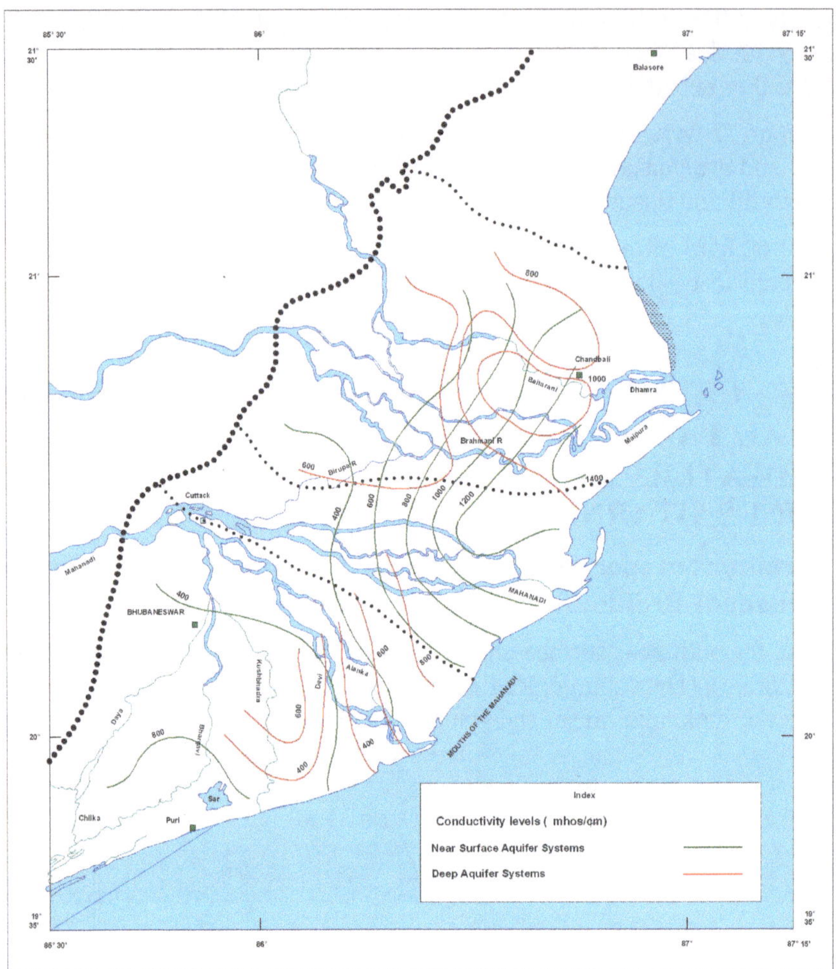

Fig. 5.28(b) Specific conductance variations in Mahanadi-Brahmini-Baitarni

Table 5.32 The variations of specific conductance and pH along with different cations and anions of Mahanadi-Brahmini-Baitarni in Orissa

Parameter	Near Surface				Deep systems			
	Max.	Min.	Mean	Std.deviation	Max.	Min.	Mean	Std.deviation
pH	10.60	7.16	8.34	0.73	8.6	7.2	7.93	0.48
EC	3400.00	120.00	620.55	540.73	1695.4	209.23	718.09	268.78
Ca	202.00	6.00	37.15	30.71	56	10	28.44	12.18
Mg	111	1.2	13.34	15.94	62.00	2.00	19.96	13.10
Na	357.00	6.40	64.96	70.71	295.00	22.00	99.15	54.03
K	203.00	0.00	19.15	37.38	10.50	2.00	5.36	1.80
HCO_3	871	12	164.9	138	421.00	42.00	214.0	87.90
SO_4	384	0	25.87	52.58	195	0	49.19	45.24
Cl	681	11	89.63	110.08	509.00	11.00	113.2	88.33

The specific conductance values of the near surface systems have a range varying from 120 to 3400 μ mhos/cm with an average value of 620.55 μ mhos/cm and deep systems values have a range varying from 209.23 to 1695.38 μ mhos/cm with an average value of 718.09 μ mhos/cm. The contour trends for both near surface and deep systems are presented in Fig. 5.28(a,b).

The near surface system of Mahanadi delta is characterized by low levels in the upstream portions and also all along the central region of the Mahanadi delta upto the shore area. The maximum levels go up to 800 μ mhos/cm in the south western part of the delta and 1200 μ mhos/cm in the north eastern part of the delta. Similarly the near surface systems in the upstream portions of the Brahmini-Baitarni delta regions have values in between 400 to 600 μ mhos/cm and increase gradually to a maximum level of 1400 μ mhos/cm towards the downstream/shore line areas. The deep systems of Mahanadi delta show a trend less than 600 μ mhos/cm in the upstream areas and increase marginally to 800 μ mhos/cm in the outfall regions towards the shore line. The same trend is reflected in the Brahmini-Baitarni region but with a marginal increase up to 1000 μ mhos/cm and beyond in the outfall regions.

The pH for near surface system have a range of values varying between 7.23 and 10.6 with a mean value of 8.3 reflecting the tendency of water towards alkalinity. The deeper systems have a range varying between 7.2 and 8.6 with an average value of 7.93 reflecting the tendency towards neutral range. These observations indicate a more dominant mixing process of saline water intrusion processes in the near surface systems than in the deeper systems.

5.6.2 Cation-anion chemistry/Ionic ratios/Cross plots

The sequential order of abundance of cations and anions (near surface and deep) in the composite Mahanadi-Brahmini - Baitarni are as follows:

Mahanadi-Brahmini – Baitarni delta region

(a) *Near surface:* HCO_3 > Cl > SO_4 and Na > Ca > K > Mg

(b) *Deep systems:* HCO_3 > Cl > SO_4 and Na > Ca > Mg > K

The bicarbonate (HCO_3) and sodium (Na) are the dominant ions amongst the anions and cations in both near surface and deep systems of the composite Mahanadi-Brahmini-Baitarni delta systems of Orissa.

Variations of cations and anions with respect to TDS: The maximum and minimum variations and the standard deviations for different water quality parameters for both near surface and deep systems of the composite Mahanadi delta regions are presented in Table 5.32 earlier. The variations of cations and anions in relation to TDS for both near surface and deep systems of the composite Mahanadi delta are presented in Fig. 5.29.

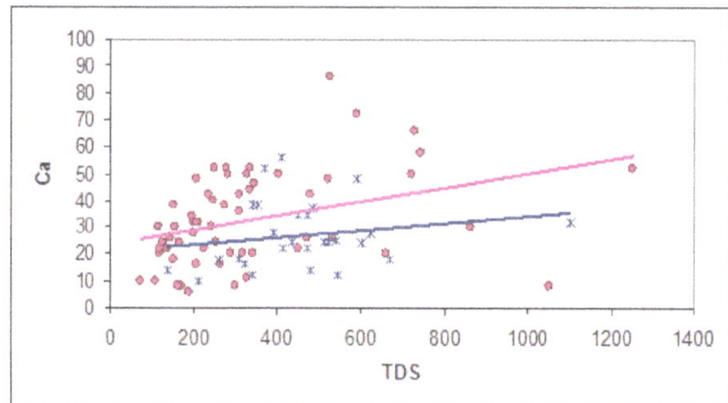

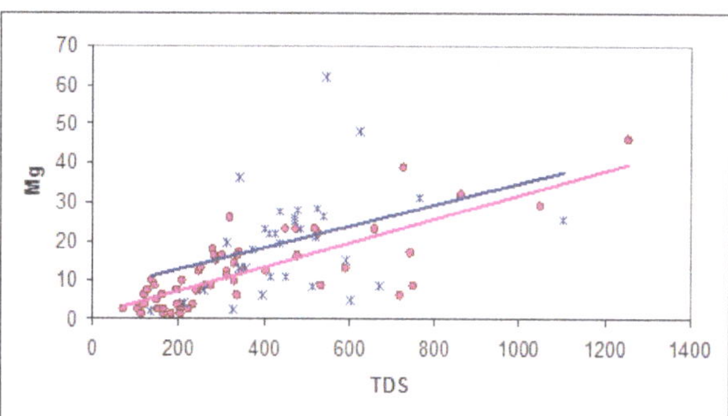

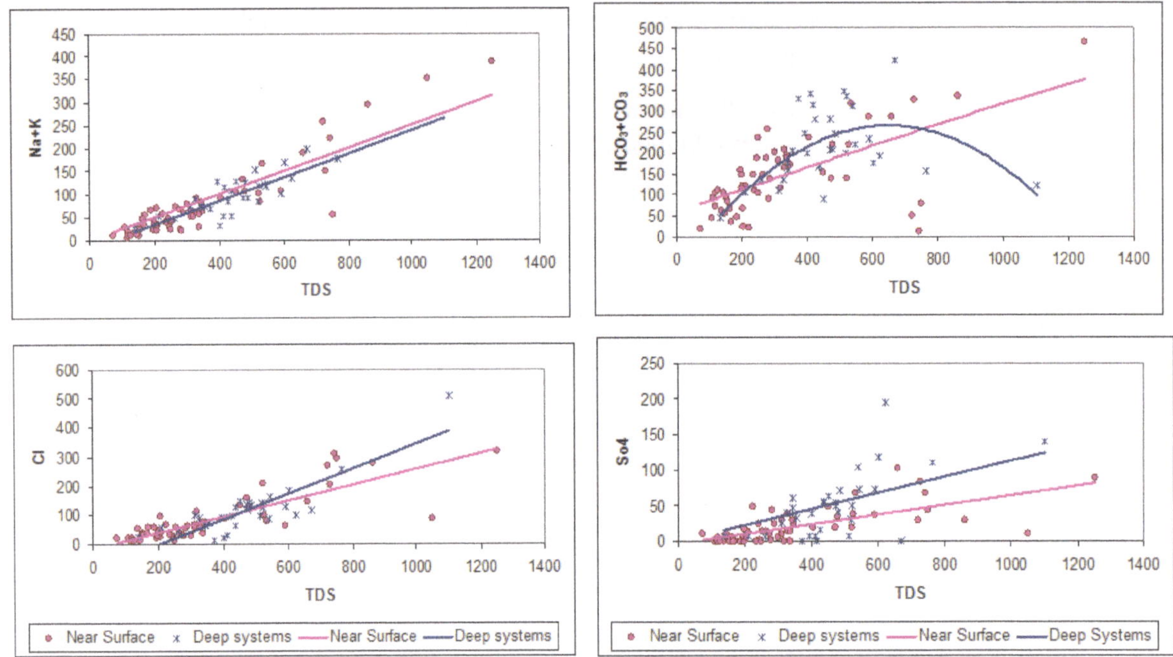

Fig. 5.29 Variations of cations and anions in relation to TDS in Mahanadi-Brahmini-Baitarni

- The cross plots in general for the composite Mahanadi-Brahmini-Baitarni shows an ascending order of linearity for both near surface and deep systems excepting the case HCO_3 vs. TDS where the deep systems trends are characterized by a curve linear relationship. However the near surface system show a near linearity with the increasing levels of TDS values. The curve linearity in respect of the deep systems show an ascending order up to 1000 – 1500 ppm levels of TDS and then show declining levels gradually characterizing fresh saline water mixing processes. In contrast, near surface systems show by and large an ascending linearity with increasing levels of TDS indicating dominant recharge processes. The trends indicate reflecting saline water mixing processes for the deep systems and non mixing trends of the near surface systems.
- The trends of Ca and Mg show increasing levels of linearity with moderately varying slope characteristics. The Ca levels of both near surface and deep systems appear to get converged at the initial levels of TDS but gradually get diverged with increasing levels of TDS. The Ca vs. TDS slope characteristics also reflect on the lesser involvement of Ca ions in the deeper systems when compared to the near surface systems where the gradient levels are relatively high. The trends for both near surface and deep systems go parallel reflecting equal involvement of Mg in both the systems.
- The Na + K vs. TDS trend for both near surface and deep systems show by enlarge close parallel linearities with parallel slope characteristics reflecting equal contribution of Na + K with increasing levels of TDS in both near surface and deep systems.
- The SO_4 and Cl anionic trends show by and large similar linearity. The Cl values for both near surface and deep systems show convergence up to the levels of 600 TDS and then gradually gets diverged with increasing levels of TDS. The graph shows relatively less slope characteristics in the near surface when compared to the deep systems reflecting a more dominant chloride contribution for the deep systems. The SO_4 vs. TDS graphical trends also reflect by enlarge same processes in the near surface and deep systems.

Ionic ratios/Cross plots

- The ionic ratios between selected cations and anions would serve as important indicators to assess and estimate the intensity of fresh and saline water mixing processes in coastal delta systems. The max, min and avg values of selected number of ionic ratios for near surface and deep systems of Mahanadhi-Brahmini-Baitarni delta systems are furnished in the Table 5.33(a,b) below.

Table 5.33(a,b) Maximum, Minimum and Average values of ionic ratios in different delta systems of Orissa

(a) Near surface

S.No	Ionic ratios	Mahanadi -Brahmini-Baitarni delta system		
		Max	Min	Avg
1.	Na/Cl	6.35	0.28	1.32
2.	Mg/Cl	2.50	0.04	0.61
3.	SO_4/Cl	0.64	0.00	0.19
4.	Cl/(HCO_3 + CO_3)	44.74	0.10	1.78
5.	Ca/(HCO_3 + SO_4)	4.60	0.03	0.71
6.	Mg/Ca	5.97	0.07	0.31

(b) Deep systems

S.No	Ionic ratios	Mahanadi -Brahmini-Baitarni delta system		
		Max	Min	Avg
1.	Na/Cl	8.69	0.88	1.71
2.	Mg/Cl	4.78	0.08	0.80
3.	SO_4/Cl	1.44	0.00	0.35
4.	Cl/(HCO_3 + CO_3)	9.35	0.06	1.12
5.	Ca/(HCO_3 + SO_4)	0.88	0.01	0.27
6.	Mg/Ca	8.33	0.21	0.74

- The average Na/Cl ratios for near surface/deep systems in respect of Mahanadi-Brahmini-Baitarni are 1.32/1.71 respectively. The theoretical values for Na/Cl ratio for saline and brackish waters are of the order of 0.85 and 1.00 respectively. This means that majority of the near surface and deep systems of Mahanadi-Brahmini-Baitarni delta complex systems lie in the field of fresh water and mixing with saline water appears to be insignificant.
- The average Mg/Cl ratios for near surface/deep systems of Mahanadi-Brahmini-Baitarni are 0.61/0.80 respectively. The theoretical values for Mg/Cl ratio for saline and brackish waters are of the order of 0.15 and 0.35 respectively. It indicates that majority of the near surface and the deep systems lie in the field of fresh water and no mixing appears to be taking place.
- The average SO_4/Cl ratios for near surface/deep systems of Mahanadi-Brahmini-Baitarni are 0.19/0.35 respectively. The theoretical values for SO_4/Cl ratio for saline and brackish waters are of the order of 0.005 and 0.20 respectively. This means that the near surface systems are close to the brackish water boundary and the deep systems are disposed in the fresh water field reflecting mixing processes in the near surface systems only.
- The average Cl/(HCO_3 + CO_3) ratios for near surface/deep systems of Mahanadi-Brahmini-Baitarni are 1.78/1.12 respectively. The theoretical values for Cl/(HCO_3 + CO_3) ratio for saline and brackish waters are of the order of 6.6 and 0.50 respectively. This reflects that by and large all the near surface and deep systems lie in the field of brackish to saline water indicating moderate mixing in both near surface and deep systems.

- The average $Ca/(HCO_3 + SO_4)$ ratios for near surface/deep systems of Mahanadi-Brahmini-Baitarni are 0.71/0.27 respectively. The theoretical values for $Ca/(HCO_3 + SO_4)$ ratio for saline and brackish waters are of the order of 1.00 and 0.50 respectively. This indicates that the near surface systems having moderate mixing lie marginally in the field of brackish water and the deep systems are totally free from saline water mixing.
- The average Mg/Ca ratios for near surface/deep systems of Mahanadi-Brahmini-Baitarni are 0.31/0.74 respectively. The theoretical values for Mg/Ca ratio for saline and brackish waters are of the order of 3.5 and 1.00 respectively. This indicates that both near surface and deep systems are in the field of fresh water.

Ionic ratios vs Cl cross plots: The ratios considered for cross plots to assess the intensity of mixing are: Na/Cl vs Cl, SO_4/Cl vs Cl, Mg/Cl vs Cl, Mg/Ca vs Cl, $Ca/(HCO_3 + SO_4)$ vs Cl and $Cl/(HCO_3 + CO_3)$ vs Cl. The graphical plots of ionic ratios between Na/Cl vs Cl, SO_4/Cl vs Cl, Mg/Cl vs Cl, Mg/Ca vs Cl, $Ca/(HCO_3 + SO_4)$ vs Cl and $Cl/(HCO_3 + CO_3)$ vs Cl characterizing the probable saline fresh water mixing trends in both near surface and deep systems of Mahanadi-Brahmini-Baitarni delta systems are presented in Fig. 5.30.

Mahanadi -Brahmini-Baitarni delta system

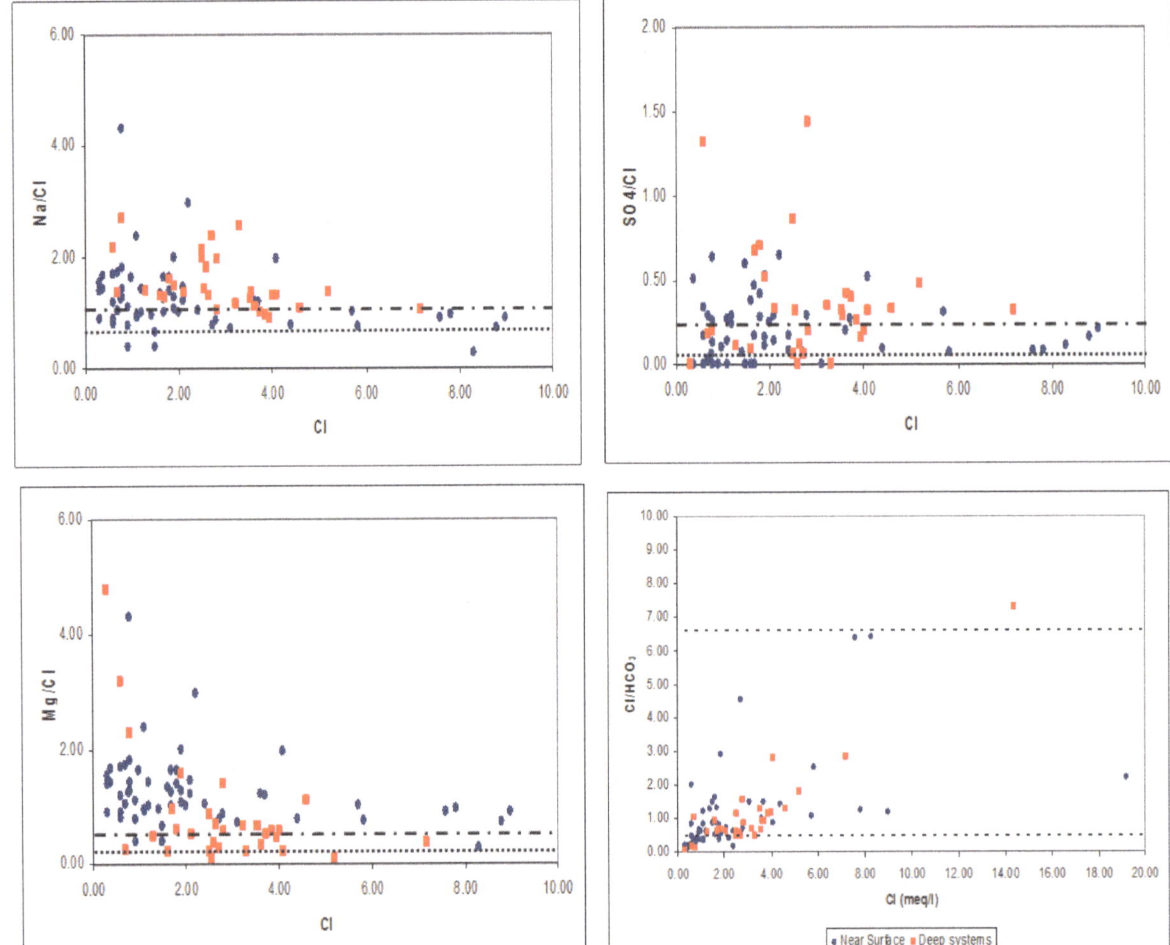

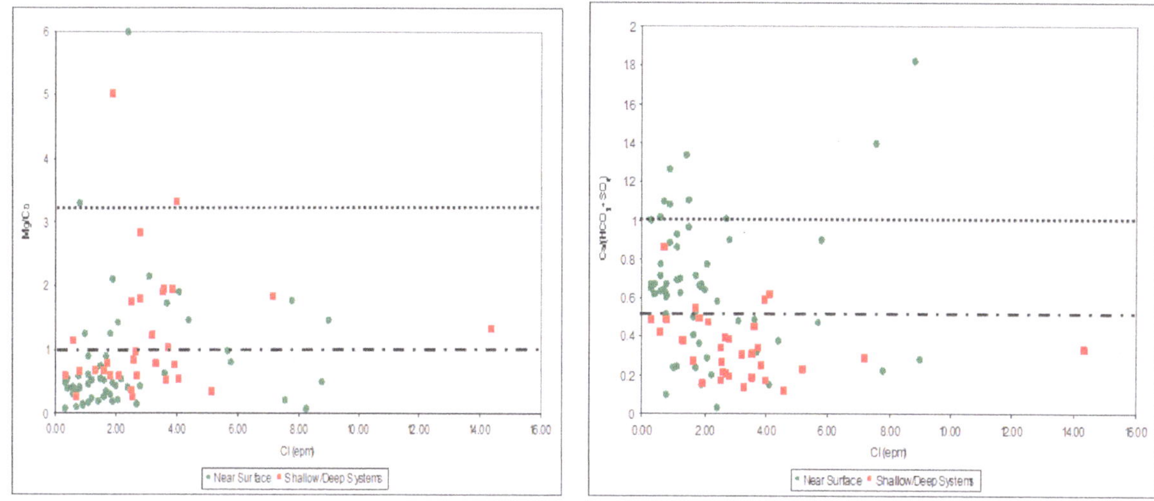

Fig. 5.30 Ionic ratios/Graphical plots in Mahanadi-Brahmini-Baitarni delta region

A brief analyses in respect of the above graphic plots discussing the significance of fresh saline water trends in Mahanadi-Brahmini-Baitarni delta system of Orissa are presented below in the following paragraphs.

Mahanadi-Brahmini-Baitarni delta: *Na/Cl vs Cl* cross plot indicate that 51% of near surface systems and 90% of the deep systems lie in the field of fresh water and the rest 49% (near surface) and 10% (deep) lie in the brackish/saline water fields reflecting by and large equal proportion of mixing in the near surface and nearly insignificant mixing in the deep systems. SO_4/Cl *vs Cl* cross plot indicate 40% of the near surface systems and 58% of the deep systems lie in the field of fresh water and the rest 60% (near surface) and 42% (deep) lie in the field of brackish/saline water reflecting a dominant mixing with saline water in the near surface systems and relatively a minor amount of mixing in the deep systems. *Mg/Cl vs Cl* cross plot indicate that 94% of near surface systems and 57% of the deep systems lie in the field of fresh water and only 6% (near surface) and 43% (deep) lie in the field of brackish/saline water reflecting insignificant mixing in the near surface systems and moderate mixing in the deep systems. Cl/HCO_3 *vs Cl* cross plot indicates that 28% of near surface systems and 11% of the deep systems lie in the field of fresh water and the rest 72% (near surface) and 89% (deep) lie in the field of brackish/saline water reflecting dominant mixing processes in both near surface and deep systems. *Mg/Ca vs Cl* cross plot indicate that 76% of the near surface systems and 60% of the deep systems lie in the field of fresh water and the rest 24% (near surface) and 40%(deep) lie in the field of brackish/saline water reflecting relatively small amounts of mixing in both near surface and deep systems. $Ca/(HCO_3 + SO_4)$ cross plot indicate 37% of the near surface and 87% of the deep systems lie in the field of fresh water and the rest 63% (near surface) and 13% (deep) lie in the brackish/saline water field indicating dominant mixing processes in the near surface zone and relatively insignificant mixing in the deep systems.

5.6.3 Trilinear plots/Characterization of ionic trends

The dominant cations and anions present in solution are systematically represented in the Piper Triangular patterns and are in Figure 5.31(a, b and c).

Hydrochemical characteristics as derived from piper trilinear diagram for Mahanadi-Brahmin-Baitarni delta system are presented in Table 5.34 below.

Deep systems (c)

Fig. 5.31 Trilinear plots for near surface (a and b) and deep systems (c) for Mahanadi delta system

Table 5.34 Characterization of ionic trends in Mahanadi-Brahmini-Baitarni

Number of Samples : (Near surface: 60, Deep systems: 32)

Subdivision of the diamond	Characteristics of corresponding subdivisions of diamond-shaped fields	Near surface		Deep systems	
		No. of Samples	Percentage of samples in this category (%)	No. of Samples	Percentage of samples in this category (%)
1	Alakaline earth (Ca + Mg) Exceed alkalies (Na + K)	36	60	7	22
2	Alkalies exceeds alkaline earths	24	40	25	78

Table 5.34 *Contd...*

Subdivision of the diamond	Characteristics of corresponding subdivisions of diamond-shaped fields	Near surface		Deep systems	
		No. of Samples	Percentage of samples in this category (%)	No. of Samples	Percentage of samples in this category (%)
3	Weak acids (CO_3 + HCO_3) exceed strong acids (SO_4 + Cl)	37	62	19	59
4	Strong acids exceeds weak acids	23	38	13	41
5	Magnesium bicarbonate type	28	47	4	13
6	Clacium-Chloride type	1	2	0	0
7	Sodium-Chloride type	15	25	10	31
8	Sodium-Bicarbonate type	1	2	2	6
9	Mixed type (No cation-anion exceed 50%)	15	25	16	50

5.6.4 Hydrogeochemical facies

Hyderogeochemical facies of Mahanadi delta

The hydrogeochemical facies and trends of Mahanadi-Brahmini-Baitarni delta basin are depicted in Figure 5.32(a,b).

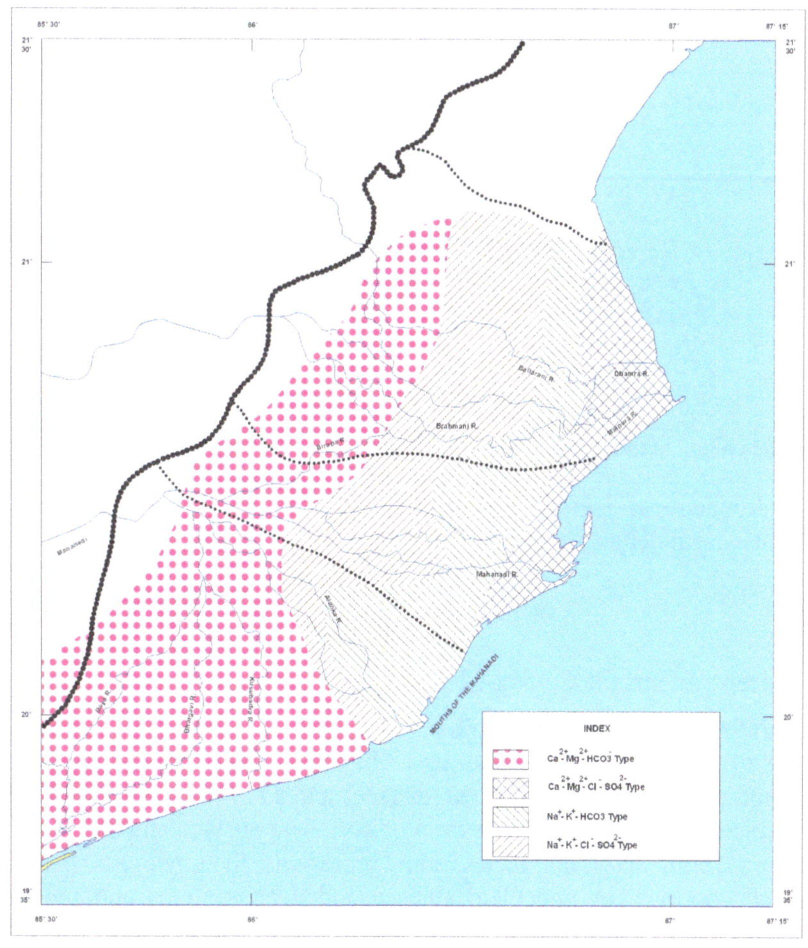

(a) Near surface

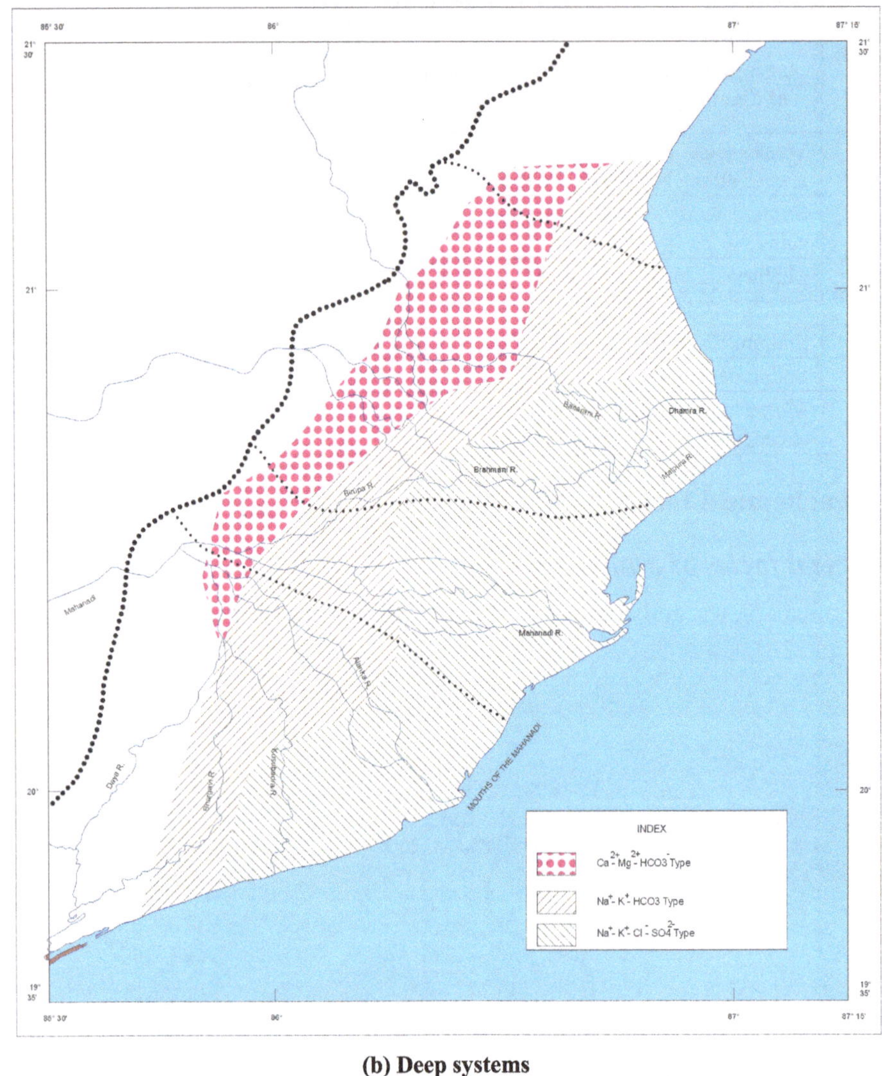

(b) Deep systems

Fig. 5.32 Hydrogeochemical facies for near surface (a) and deep systems (b) for Mahanadi delta system

The different hydrochemical facies identified for both near surface and deep systems in the Mahanadi- Brahmini-Baitarni delta system are as follows:

1. Ca-Mg-HCO_3 type
2. Na-K-HCO_3 type
3. Na-K-Cl-SO_4 type
4. Ca-Mg-Cl-SO_4 type

Note: In the case of deep systems the Ca-Mg-Cl-SO4 type is found absent.

Facies no.1(Ca-Mg-HCO$_3$ type): The near surface and deep systems in this delta basin have an average total dissolved solids of the order of 260 ppm and 411 ppm respectively. The Ca+Mg are the major cations in this zone and contain about 54% (near surface) and 51% (deep systems) of the total average cations. The average bicarbonate contents in respect of both near surface and deep systems are 71% and 80% of the total anion contents. This facies zone represents the upstream portion of Mahanadi-Brahmini-Baitarni delta regions for both near surface and deep systems and probably act as major recharge zone. It is also to be noticed in the southwestern delta regions this zone extends up to the coastal boundary.

Facies no.2 (Na-K-HCO$_3$ type): This zone is represented by average total dissolved solids of the order of 365 ppm and 431 ppm for near surface and deep systems respectively. The Na + K are the major cations in this zone and contain about 80% for near surface and 70% for deep system out of the total cationic composition. The anionic contents are represented by HCO_3 containing about 76% in near surface and 64% for deep systems respectively out of the total anionic contents. This zone is sandwiched between zone no.1 (Ca-Mg-HCO$_3$ type) and zone no.3 (Na-K-Cl-SO$_4$) in the case of both near surface systems and deep systems.

Facies no.3 (Na-K-Cl-SO$_4$): This zone is represented by total dissolved solids of the order of 618 ppm 541 ppm for near surface and deep systems respectively. The Na-K contents are the major cations in this zone and contain about 74% (near surface) and 75% (deep systems) out of total cations. The Cl + SO_4 contents are the major anions in this zone and contain about 57% (near surface) and 59% (deep system) out of total average anionic contents. This zone is sandwiched between zone no.2 (Na-K-HCO$_3$ type) and zone no.4 (Ca-Mg-Cl-SO$_4$) in the north eastern central part of the delta. But however this zone gets flanked against the coast in the central southwest part of the coast. But in the case of deep systems this zone flanks by and large against the cost of the entire delta mosaic.

Facies no.4: (Ca-Mg-Cl-SO$_4$ type): This zone is present in the near surface system of the delta and significantly found absent in the deep system and is represented by average total dissolved solids of the order of 330 ppm for near surface. Ca+Mg are the major cations in this system and contain about 59% out of the total average cationic composition. The anionic contents of this zone is represented by Cl and SO4 and the total Cl and SO_4 contents contain about 48% of the total average anions. This zone is significantly found to exist along the narrow coastal belt of Mahanadi-Brahmini-Baitarni track only.

5.6.5 Correlation matrix and factor analysis

The correlation matrix analysis for Mahanadi-Brahmini-Baitarni delta near surface and deep systems are presented in Table 5.35 below.

Table 5.35 Correlation matrix analysis for Mahanadi-Brahmini-Baitarni

Near surface

	Ph	EC	TDS	Ca	Mg	Na	K	HCO_3	SO_4	Cl	NO_3	F	Th
Ph	1												
EC	–0.185	1											
TDS	–0.185	1	1										
Ca	–0.332	0.551	0.552	1									
Mg	–0.205	0.891	0.887	0.429	1								
Na	–0.033	0.9	0.9	0.227	0.759	1							
K	–0.08	0.742	0.743	0.295	0.62	0.574	1						
HCO_3	–0.232	0.709	0.71	0.183	0.645	0.78	0.373	1					
SO_4	–0.171	0.832	0.832	0.511	0.874	0.699	0.562	0.438	1				
Cl	–0.127	0.914	0.912	0.603	0.827	0.753	0.748	0.382	0.838	1			
NO_3	0.004	0.221	0.228	0.568	–0.066	0.077	0.234	–0.178	0.063	0.319	1		
F	–0.105	0.134	0.134	0.037	0.024	0.167	0.143	0.198	–0.019	0.045	0.092	1	
THs	–0.324	0.835	0.834	0.872	0.817	0.558	0.525	0.469	0.801	0.834	0.326	0.037	1

Deep systems

	pH	EC	TDS	Ca	Mg	Na	K	HCO_3	SO_4	Cl	TH
pH	1										
EC	–0.162	1									
TDS	–0.162	1	1								
Ca	–0.428	0.167	0.167	1							
Mg	–0.254	0.369	0.369	–0.034	1						
Na	0.023	0.892	0.892	–0.114	0.14	1					
K	–0.007	0.618	0.618	–0.165	0.143	0.677	1				
HCO_3	–0.428	0.207	0.207	0.285	0.103	0.162	–0.109	1			
SO_4	0.103	0.644	0.644	0.029	0.516	0.498	0.592	–0.24	1		
Cl	0.008	0.861	0.861	–0.056	0.239	0.859	0.684	–0.178	0.565	1	
TH	–0.344	0.275	0.275	0.383	0.589	0.028	0.106	0.247	0.311	0.059	1

Near surface systems

A good correlation is observed between EC and Mg (r = 0.891), Na (r = 0.9), K (r = 0.74), HCO_3 (r = 0.709), SO_4 (r = 0.832), Cl (r = 0.914) and TH (r = 0.835) and moderately correlated with Ca (r = 0.551) indicating all major cations and anions are playing a reasonably good role in the evolution of hydrochemical processes of the delta system. Ca correlate strongly with TH (r = 0.872) and moderately with SO_4 (r = 0.51), Cl (r = 0.60), NO_3 (r = 0.568). Mg strongly correlates with Na (r = 0.759), SO_4 (r = 0.874), Cl (r = 0.827), TH (r = 0.81) and moderately with K (r = 0.6) and HCO_3 (r = 0.64). Na correlate well with HCO_3 (r = 0.78), Cl (r = 0.75) and moderately with K (r = 0.57), SO_4 (r = 0.699) and TH (r = 0.558). K strongly correlate with Cl (r = 0.748) and moderately with SO_4 (r = 0.562) and Similarly SO_4 correlates strongly with Cl (r = 0.838) and with TH (r = 0.8).

Deep systems

A strong correlation is observed between EC and Na (r = 0.89), Cl (r = 0.86) and moderately with K (r = 0.61) and SO_4 (r = 0.64). Mg correlates moderately with SO_4 (r = 0.51) and TH (r = 0.58). Na correlates strongly with Cl (r = 0.85) and moderately with K (r = 0.67). K moderately correlates with Cl (r = 0.68) and SO_4 (r = 0.592).SO_4 moderately correlates with Cl (r =0.565).

Table 5.36 Factor analysis for near surface and deep systems of Mahanadi-Brahmini-Baitarni

Near Surface					Deep systems				
Component	Eigen value	% of Variance	Cumulative Eigen value	Cumulative %	Component	Eigen value	% of Variance	Cumulative Eigen value	Cumulative %
1	7.417	57.053	7.417	57.053	1	4.904	44.585	4.904	44.585
2	1.703	13.097	9.12	70.149	2	2.265	20.588	7.169	65.172
3	1.155	8.884	10.275	79.034	3	1.344	12.219	8.513	77.392
4	1.106	8.51	11.381	87.544	4	0.85	7.724	9.363	85.116
5	0.592	4.557	11.973	92.101	5	0.615	5.589	9.978	90.705
6	0.532	4.089	12.505	96.19	6	0.469	4.266	10.447	94.97
7	0.288	2.218	12.793	98.408	7	0.341	3.099	10.788	98.07
8	0.111	0.854	12.904	99.262	8	0.14	1.271	10.928	99.341
9	0.091	0.703	12.995	99.965	9	0.042	0.383	10.97	99.723
10	0.003	0.026	12.998	99.991	10	0.03	0.277	11	100
11	0.001	0.007	12.999	99.998	11	2.1E – 16	1.9E – 15	1.1E + 01	100
12	0	0.002	1.3E + 01	100					
13	1.2E – 05	9.1E – 05	1.3E + 01	100					

Table 5.37 Loading varimax rotated factor matrix for Mahanadi-Brahmini-Baitarni

Near surface system					Deep systems			
Variable	**Factor 1**	**Factor 2**	**Factor 3**	**Factor 4**	**Variable**	**Factor 1**	**Factor 2**	**Factor 3**
Ph	–0.056	–0.01	–0.913	–0.099	pH	–0.002	–0.749	–0.205
EC	0.968	0.176	0.122	0.091	EC	0.935	0.22	0.214
TDS	0.967	0.18	0.121	0.094	TDS	0.935	0.22	0.214
Ca	0.39	0.717	0.467	–0.105	Ca	–0.043	0.687	0.155
Mg	0.928	–0.041	0.21	–0.107	Mg	0.171	0.064	0.862
Na	0.912	–0.088	–0.067	0.205	Na	0.963	0.04	–0.075
K	0.734	0.237	–0.135	0.147	K	0.771	–0.186	0.121
HCO_3	0.713	–0.354	0.252	0.297	HCO_3	0.059	0.811	–0.108
SO_4	0.859	0.13	0.171	–0.203	SO_4	0.609	–0.264	0.598
Cl	0.881	0.376	0.039	–0.073	Cl	0.923	–0.12	0.093
NO_3	0.04	0.912	–0.089	0.14	TH	0.035	0.397	0.786
F	0.054	0.076	0.076	0.929				
Th	0.753	0.436	0.412	–0.124				

The first four factors (Near surface system) and three factors (deep systems) by and large explains 87.5% and 77.3% of the total variances respectively.

Near surface systems

Factor 1 explains the largest proportion (57%) of the total variance and has high positive loadings (0.75-0.96) for EC, TDS, Mg, Na, K, HCO_3, SO_4, Cl and TH.

Factor 2 accounts for 13% of the total variance and has high loading on Ca (0.717) and NO_3 (0.912).

Factor 3 accounts for 8.8% of the total variance and has high negative loading on pH (–0.913).

Factor 4 accounts for 8.5% of the total variance and has high loadings on F (0.929).

Deep systems

Factor 1 explains the largest proportion 44.5% of the total variance and has high positive loadings (0.7-0.96) on EC, TDS, Na, K and Cl and moderately loading on SO_4 (0.609).

Factor 2 accounts for 20.5% of the total variance and has high loadings on HCO_3 (0.81) and moderate loading on Ca (0.687).

Factor 3 accounts for 12.2% of the total variance and has high loadings on Mg (0.862), TH (0.786) and moderate loading on SO_4 (0.598).

5.6.6 Classification and irrigation suitability of groundwaters based on sodium percentages, sodium adsorption ratios and residual sodium carbonate

Sodium percentage

The classification of groundwaters with respect of percentage of sodium in Mahanadi delta system is presented below in Table 5.38.

Table 5.38 Sodium percentage and water classes of Mahanadi-Brahmini-Baitarni

Sodium (%)	Water Class	No. of Samples	
		Near Surface system	Deep system
< 20	Excellent	6 (14 – 19.5)	0
20 – 40	Good	22 (20 – 38.2)	4 (26.3 – 39)
40 – 60	Permissible	24 (40.3 – 59.2)	17 (43 – 58.3)
60 – 80	Doubtful	6 (63.7 – 76.4)	10 (62.4 – 79.7)
> 80	Unsuitable	2 (84.3 – 86.2)	1 (82.8)

The sodium percentages in near surface systems of Mahanadi-Brahmini-Baitarni delta have a range of values between 14 and 86.2 and for deep system between 26.3 and 82.8. About 87% of the samples of the near surface systems fall in the range of excellent to permissible limits and the rest 13% of the samples lie in the range of doubtful to unsuitable categories. In the case of deep systems about 66% of the samples lie in the range of good to permissible limits and 34% of the samples lie in the range of doubtful to unsuitable categories.

Sodium Adsorption Ratio (SAR)

The classification of groundwater samples based on SAR (Sodium Adsorption Ratio) is presented in Figures 5.33(a,b,c) below.

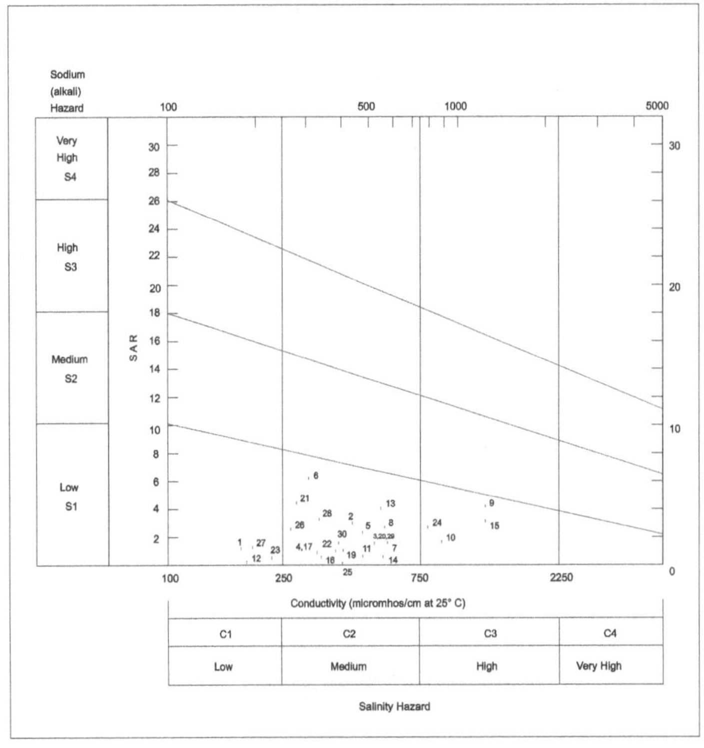

(a) Near surface

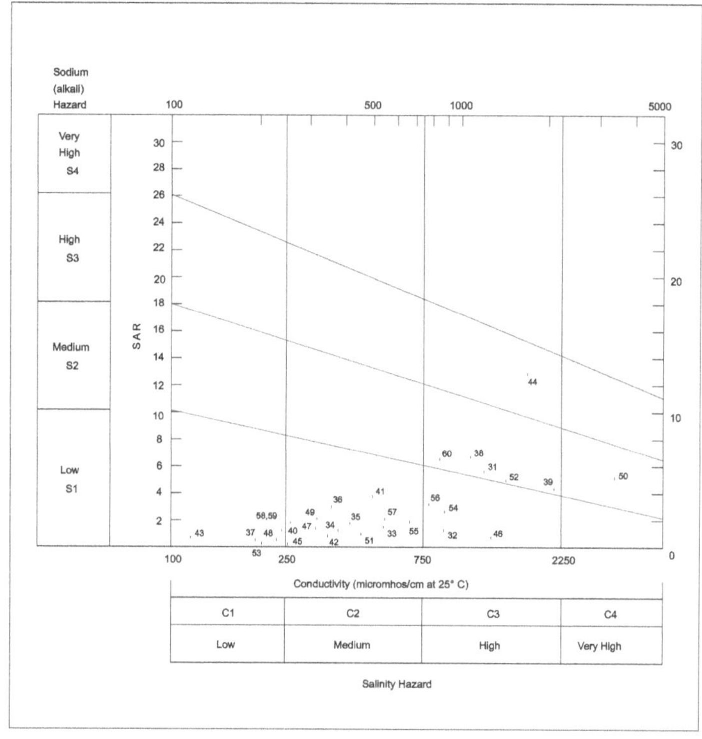

(b) Near surface

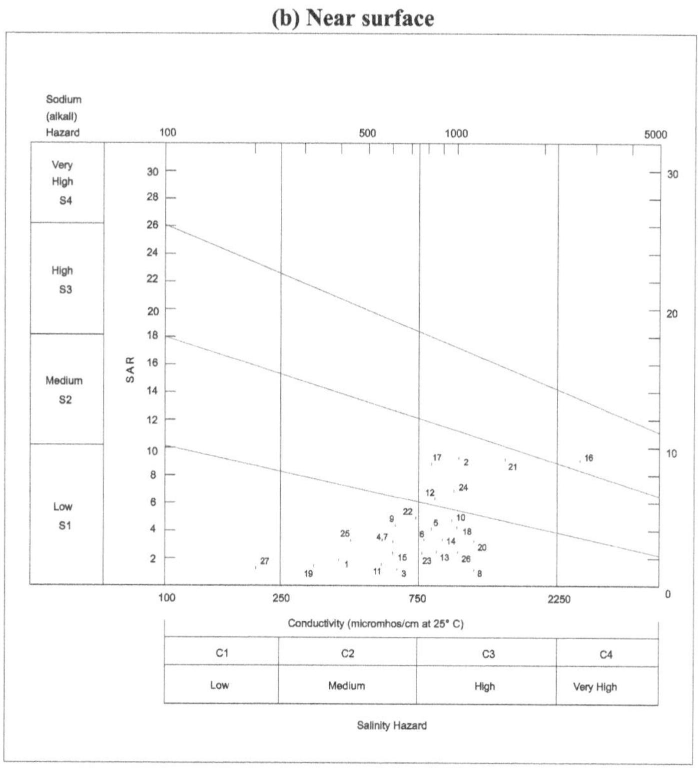

(c) Deep systems

Fig. 5.33 Sodium adsorption ratios for near surface (a and b) and deep systems (c) for Mahanadi delta system

About 69% of the samples from the near surface system of Mahanadi-Baitarni-Brahmini delta lie in the fields of C_1S_1 and C_2S_1 (low salinity/low alkalinity to medium salinity/low alkalinity). About 13% of the samples lie in the fields of C_3S_1 (high salinity/low alkalinity). The rest of the samples about 9% of the samples are spread in the fields of C_3S_2, C_4S_2 (high salinity/medium alkalinity to very high salinity/medium alkalinity).

In the case of deep systems 31% of the samples from the near surface lie in the fields of C_1S_1 and C_2S_1 (low salinity/low alkalinity to medium salinity/low alkalinity). About 31% of the samples lie in the fields of C_3S_1 (high salinity/low alkalinity). The rest of the samples about 15% of the samples are spread in the fields of C_3S_2 (high salinity/medium alkalinity).

Residual sodium carbonate (RSC)

The RSC values for the near surface and deep systems for Mahanadi-Brahmin-Baitarni delta system are presented in Table 5.39 below.

Table 5.39 Residual sodium carbonate values in Mahanadi-Brahmin-Baitarni

RSC (epm)	Remark on Quality	No. of Samples	
		Near Surface system	Deep system
< 1.25	Good	53 (– 9.5 – 1.2)	24 (–2.2 – 1.2)
1.25 – 2.5	Doubtful	5 (1.5 – 2.4)	5 (1.3 – 2.1)
> 2.5	Unsuitable	2 (3.2 – 12.4)	3 (3.2 – 5.3)

About 88% of the samples from near surface and 80% of the samples from deep system have values less than 1.25 and are considered as good quality. Rest of the percentage samples fall in doubtful and unsuitable categories.

5.7 WATER CHEMISTRY, CHARACTERIZATION, HYDROGEO-CHEMICAL FACIES AND MULTIVARIATE TREND ANALYSIS OF GROUNDWATERS (NEAR SURFACE AND DEEP) IN BHAGIRADHI-HOOGLY DELTA SYSTEMS OF WEST BENGAL

5.7.1 Spatial variations of specific conductance, pH and statistical variations of different ions in Bhagiradhi-Hoogly delta systems of West Bengal

Bhagiradhi - Hoogly delta system: The sample locations for both near surface and deep systems in respect of Bhagiradhi - Hoogly delta region is presented in Fig 5.34 and the spatial variations of specific conductance are shown in Fig. 5.35. (Data source: CGWB, 1998). The analytical results of specific conductance pH along with other cations and anions in terms of Max, Min, Mean and Std. deviation are presented in table 5.40. The specific conductance of near surface systems in the outfall region of Bhagiradhi - Hoogly delta have a range of values between 4365 μ mohs/cm and 150 μ mohs/cm with an average value of 1162 μ mohs/cm. The deep systems have a range of values from 3714 μ mohs/cm to 172 μ mohs/cm with an average value of 959 μ mohs/cm. The upstream portion of the outfall regions of the delta in general have a range of values less than 1000 μ mhos/cm and gradually increase towards the vicinity of Hoogly eastury. There appears to be a moderately high values in the Roopnarayan and Hugli interstream areas. Similarly the deep systems have also values less than 1000 μ mhos/cm in the upstream region and gradually increases towards the outfall region with values more than 2000 μ mhos/cm. The specific conductance values in respect of both near surface and deep systems in general increase from the upstream areas towards outfall regions. The intensity of salinisation processes are not so much intense when compared to the other regions both in respect of near surface and deep systems, probably because of intense network of freshwater drainage channels.

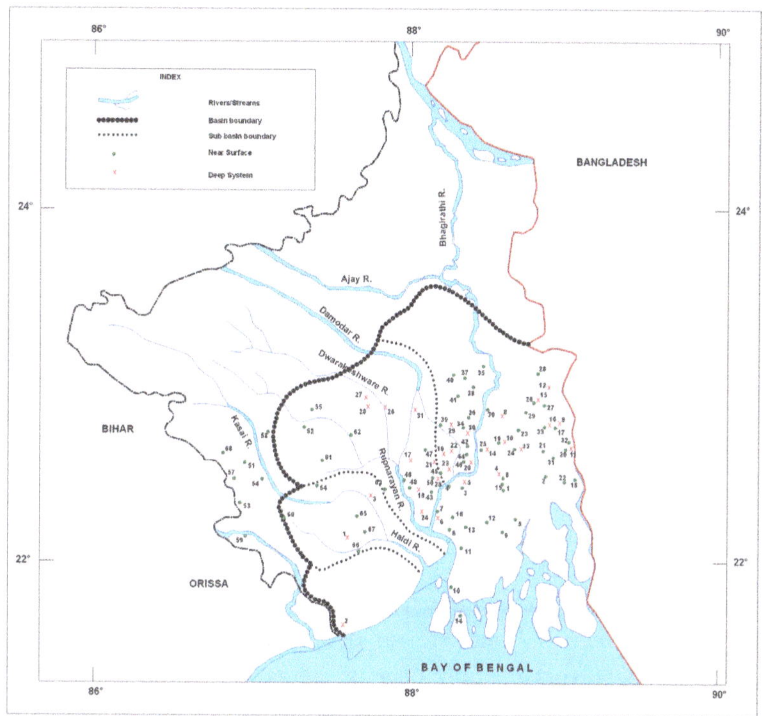

Fig. 5.34 Sample locations for near surface and deep systems in Bhagiradhi-Hoogly delta

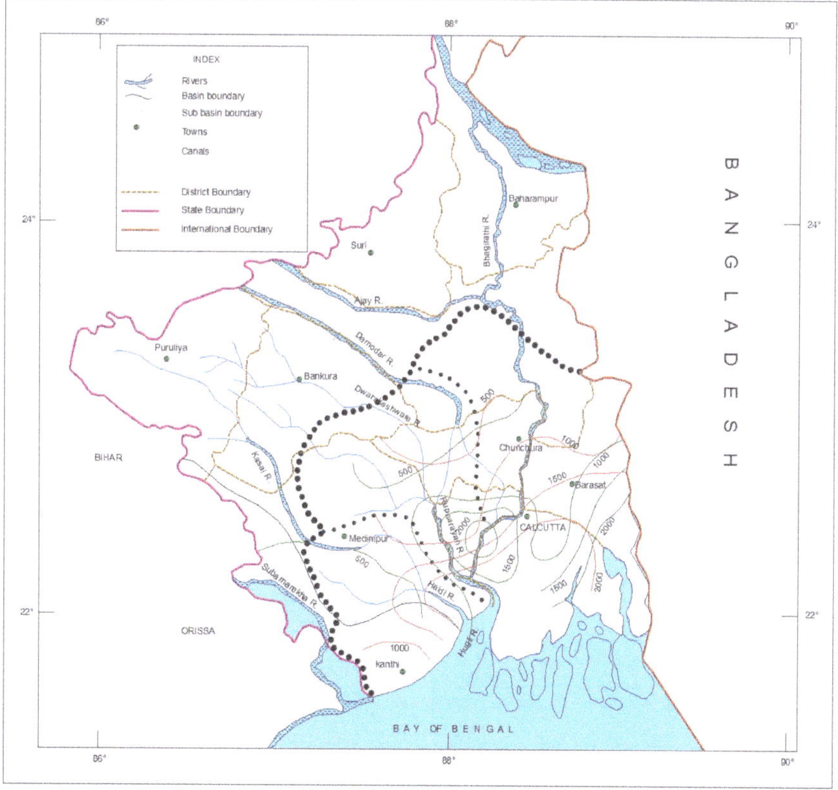

Fig. 5.35 Specific conductance variations in Bhagiradhi-Hoogly delta system

Table 5.40 The variations of specific conductance and pH along with different cations and anions of Bhagiradhi-Hoogly delta system

Parameter	Near Surface				Deep systems			
	Max.	Min.	Mean	Std.deviation	Max.	Min.	Mean	Std.deviation
pH	8.10	6.90	7.58	0.31	8.2	7.2	7.7	0.34
EC	4369.20	150.00	1162.00	813.00	3714	172.3	959.1	707.6
Ca	142.00	10.00	56.00	32.60	166	10	61.8	37.8
Mg	88	4.8	24.7	15.1	73.00	9.70	24.22	15.90
Na	911.00	5.00	136.60	156.60	690.00	5.10	105.70	140.40
K	101.00	0.97	13.00	20.80	37.00	1.00	6.13	9.13
HCO_3	1238	43	349.6	202.1	647.00	47.00	338.80	159.40
SO_4	250	1	24.5	44.9	60	1	7.9	12.7
Cl	1563	3	176.6	239.7	1275.00	11.00	140.10	245.60

The pH values for near surface systems have a range of values varying between 8.1 and 6.9 with an average value of 7.58 and between 8.2 and 7.2 for deep systems with an average value of 7.7. The higher pH values for both near surface and deep systems are observed in the areas where high conductivity levels are seen particularly in the regions where salinisation processes are more intense.

5.7.2 Cation-anion chemistry/Ionic ratios/Cross plots

The general sequential order of cations and anions in Bhagiradhi-Hoogly delta system of West Bengal is as follows:

Bhagiradhi-Hoogly

(a) *Near surface:* $HCO_3 > Cl > SO_4$ and Na > Ca > Mg

(b) *Deep systems:* $HCO_3 > Cl > SO_4$ and Na > Ca > Mg

The bicarbonate (HCO_3) and sodium (Na) are the dominant ions amongst the cations and anions in the Bhagiradhi-Hoogly delta system.

Variations of cations and anions with respect to TDS: The maximum and minimum variations and the std. deviation in terms of different water quality parameters for both near surface and deep systems are presented in Table 5.40. The variations of cations and anions in relation to TDS for both near surface and deep systems in Bhagiradhi-Hoogly delta system are presented below in Fig. 5.36.

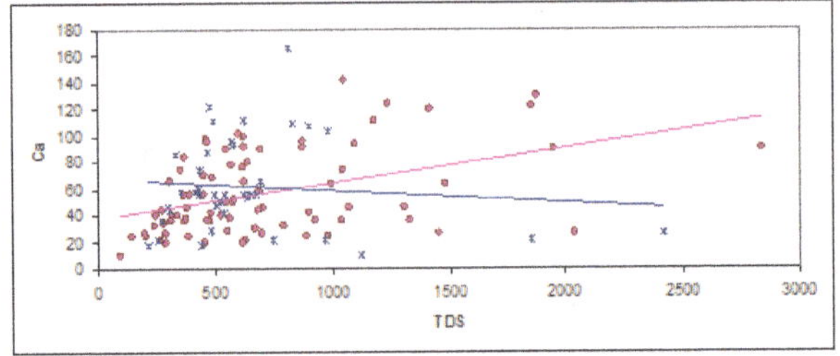

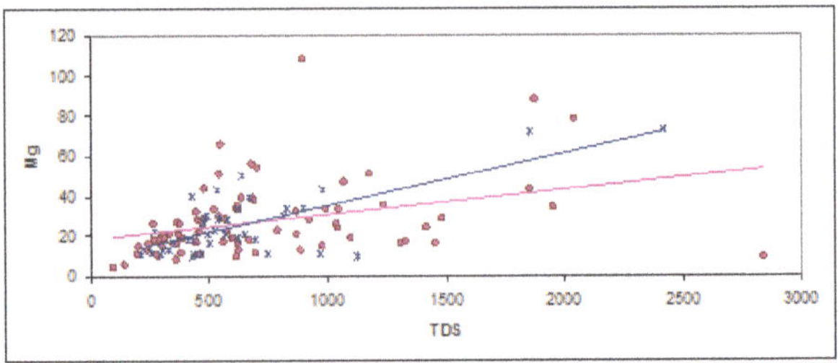

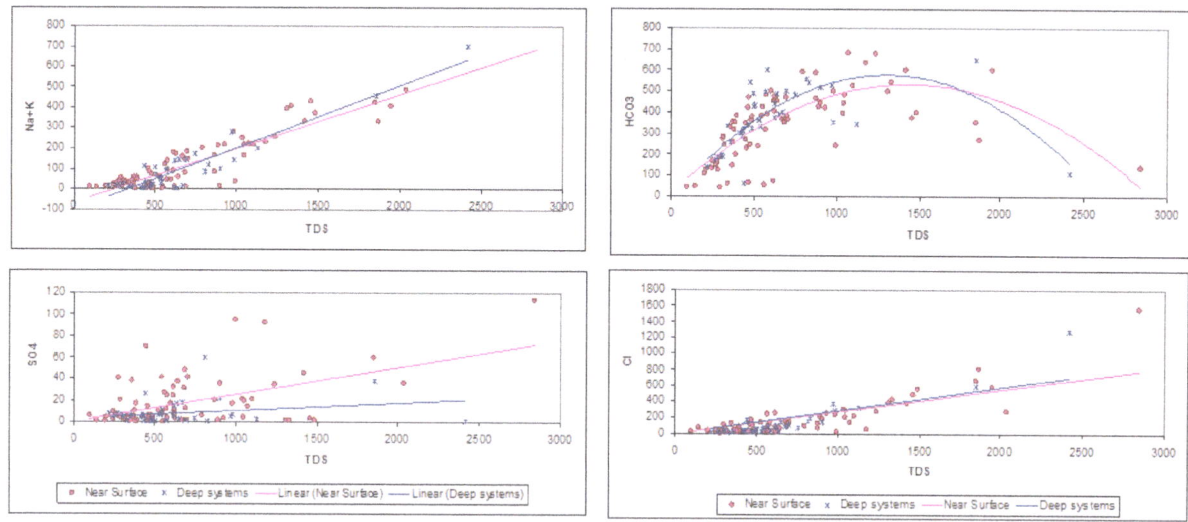

Fig. 5.36 Variations of cations and anions in relation to TDS in Bhagiradhi-Hoogly delta system

- The cross plots in general shows an ascending linearity for both near surface and deep systems except in the case of Ca vs. TDS of the deep systems where the trends show a descending linearity. In the case of HCO_3 vs. TDS the trends for both near surface and deep systems show characteristic curve linear relationship with increasing levels of TDS. The trends further show an ascending order up to 1000 ppm levels of TDS an then takes a gradually descending order with increasing level of salinity. The ascending linearity field characterizes the region of non-mixing and the descending field characterizes the mixing processes.
- The Ca levels in the case of near surface shows a low gradient ascending linearity while the deep systems show a descending linearity with increasing level of TDS reflecting cation exchange processes. The Mg levels however for both near surface and deep systems show an ascending linearity with different slope characteristics. The initial levels up to 1000 ppm levels of TDS are closely clustered and gets scattered widely after 1000 ppm TDS levels.
- The Na + K trend values of both near surface and deep systems show a characteristic linearity with increasing levels of TDS values both in the fresh water and brackish water region reflecting similar involvement of Na + K in both near surface and deep systems with increasing levels of salinity.
- The SO_4 and Cl anions for both near surface and deep systems exhibit ascending linearity but with different slope characteristics. The deep systems in the case of SO_4 vs. TDS show a very low gradient or nearly horizontal reflecting insignificant role of SO_4 whereas the near surface systems shows low gradient with increasing levels of salinity. The Cl vs. TDS graph shows that in both near surface and deep systems the gradient is reflected as a single line indicating involvement of Cl anions in equal proportions.

Ionic ratios/Cross plots

The specific conductance/total dissolved solids ionic ratios between selected cations and anions would help to estimate and assess the intensity of fresh and saline water mixing processes. The max, min and avg values of selected number of ionic ratios for the near surface and deep systems of Bhagiradhi-Hoogly delta system are presented in Table 5.41(a,b) below.

Table 5.41(a,b) Maximum, Minimum and Average values of ionic ratios in different delta systems of West Bengal

(a) Near surface

S.No	Ionic ratios	Bhagiradhi-Hoogly delta system		
		Max	Min	Avg
1.	Na/Cl	6.61	0.10	1.53
2.	Mg/Cl	17.50	0.02	1.37
3.	SO_4/Cl	2.81	0.00	0.22
4.	Cl/(HCO_3 + CO_3)	19.21	0.03	1.32
5.	Ca/(HCO_3 + SO_4)	1.78	0.06	0.54
6.	Mg/Ca	4.94	0.16	0.88

(b) Deep systems

S.No	Ionic ratios	Bhagiradhi-Hoogly delta system		
		Max	Min	Avg
1.	Na/Cl	3.96	0.68	1.60
2.	Mg/Cl	7.16	0.09	1.69
3.	SO_4/Cl	0.49	0.00	0.11
4.	Cl/(HCO_3 + CO_3)	19.95	0.04	1.17
5.	Ca/(HCO_3 + SO_4)	0.87	0.09	0.51
6.	Mg/Ca	5.39	0.21	0.94

- The average Na/Cl ratios for near surface/deep systems in respect of Bhagiradhi-Hoogly are 1.53/1.60 respectively. The theoretical values for Na/Cl ratio for saline and brackish waters are of the order of 0.85 and 1.00 respectively. This means that by and large all the samples in the near surface and deep systems lie in the field of fresh water.
- The average Mg/Cl ratios for near surface/deep systems of Bhagiradhi-Hoogly are 1.37/1.69 respectively. The theoretical values for Mg/Cl ratio for saline and brackish waters are of the order of 0.15 and 0.35 respectively. This means that majority of the samples of the near surface and deep systems lie in the field of fresh water.
- The average SO_4/Cl ratios for near surface/deep systems of Bhagiradhi-Hoogly are 0.22/0.11 respectively. The theoretical values for SO_4/Cl ratio for saline and brackish waters are of the order of 0.005 and 0.20 respectively. This indicates that near surface systems are close to the brackish water boundary and the deep system lie within the brackish/saline water zone reflecting moderate mixing processes in both the systems.
- The average Cl/(HCO_3 + CO_3) ratios for near surface/deep systems of Bhagiradhi-Hoogly are 1.32/1.17 respectively. The theoretical values for Cl/(HCO_3 + CO_3) ratio for saline and brackish waters are of the order of 6.6 and 0.50 respectively. This reflects that by and large all the samples both in the near surface and deep systems lie within brackish/saline water zone indicating moderate mixing processes in both the systems.
- The average Ca/(HCO_3 + SO_4) ratios for near surface/deep systems of Bhagiradhi-Hoogly are 0.54/0.51 respectively. The theoretical values for Ca/(HCO_3 + SO_4) ratio for saline and brackish waters are of the order of 1.00 and 0.50 respectively. This reflects that both near surface and deep systems are close to the brackish water boundary indicating moderate mixing processes.

- The average Mg/Ca ratios for near surface/deep systems of Bhagiradhi-Hoogly are 0.88/0.94 respectively. The theoretical values for Mg/Ca ratio for saline and brackish waters are of the order of 3.5 and 1.00 respectively. These ratios reflect that near surface and deep systems are in the fresh water zone signifying no mixing processes.

Ionic ratios vs Cl cross plots

The ratios considered for cross plots to assess the intensity of mixing are: Na/Cl vs Cl, SO_4/Cl vs Cl, Mg/Cl vs Cl, Cl/(HCO_3 + CO_3) vs Cl, Mg/Ca vs Cl and Ca/(HCO_3 + SO_4) vs Cl. The graphical plots of ionic ratios between Na/Cl vs Cl, SO_4/Cl vs Cl, Mg/Cl vs Cl, Cl/(HCO_3 + CO_3) vs Cl, Mg/Ca vs Cl and Ca/(HCO_3 + SO_4) vs Cl characterizing the probable saline fresh water mixing trends in both near surface and deep systems of Bhagiradhi-Hoogly delta system are presented in Fig.5.37.

Bhagiradhi-Hoogly delta

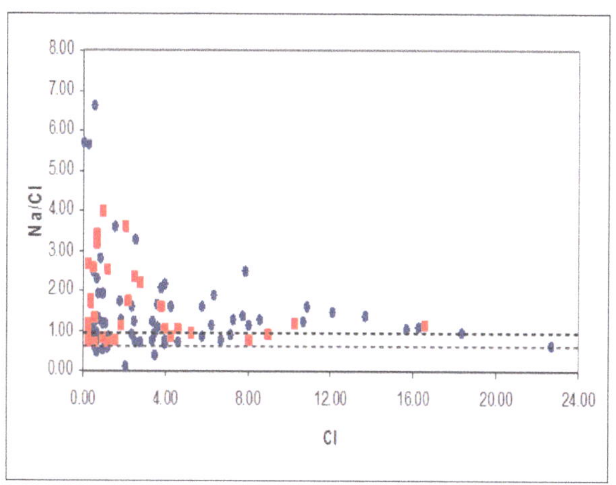

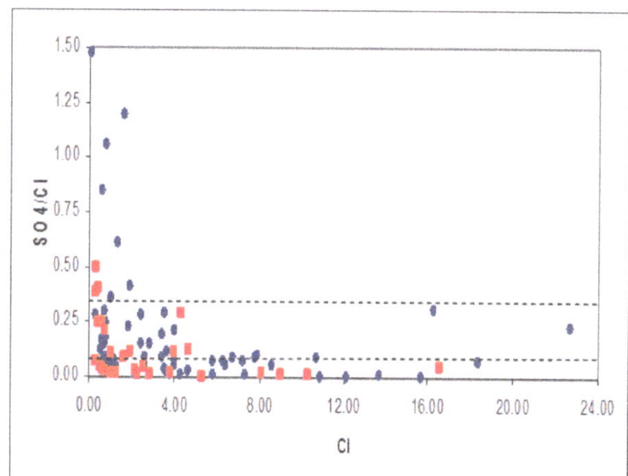

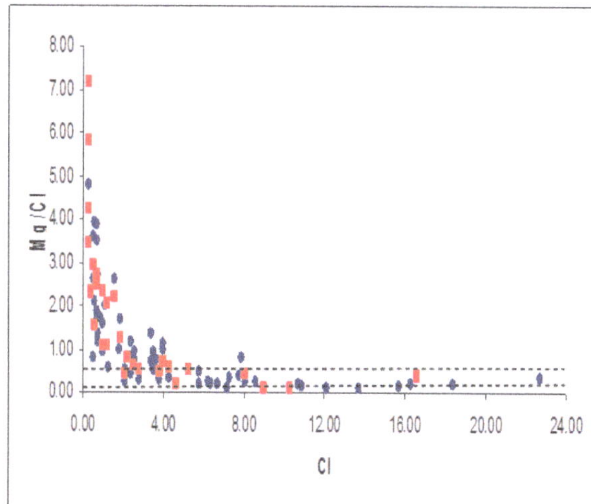

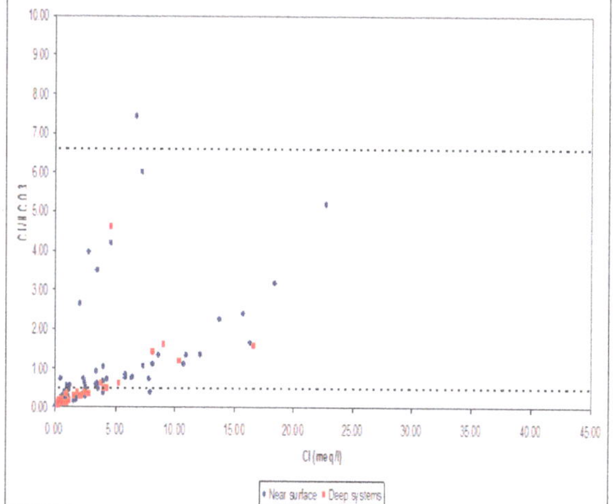

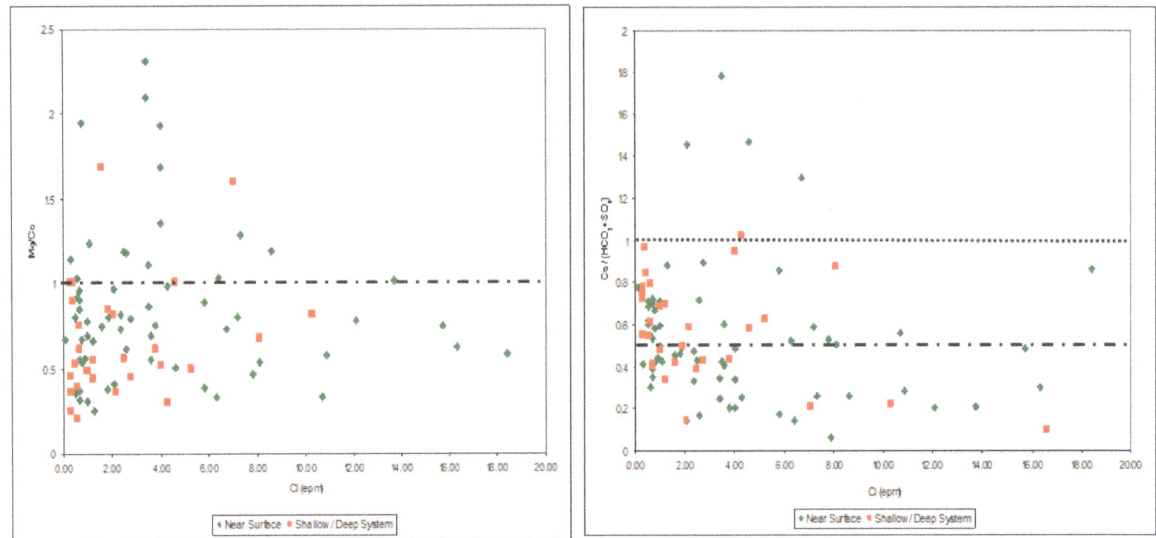

Fig. 5.37 Ionic ratios/Graphical plots in Bhagiradhi-Hoogly delta region

A brief analyses in respect of the above graphic plots discussing the significance of fresh saline water trends in Bhagiradhi-Hoogly delta system of West Bengal are presented below in the following paragraphs.

Bhagiradhi-Hoogly delta: *Na/Cl vs Cl* cross plot ratio indicate that 64% of the near surface systems and 68% of the deep systems lie in the field of fresh water and the rest 36% (near surface) and 32% (deep systems) lie in the field of brackish/saline water reflecting small amounts of mixing in both near surface and deep systems. *SO_4/Cl vs Cl* cross plot indicate that 14% of the near surface systems and 11% of the deep systems lie in the field of fresh water and the rest 86% (near surface) and 89% (deep) lie in the field of brackish/saline water reflecting intense salinization processes. *Mg/Cl vs Cl* cross plot indicate 58% of the near surface and 75% of the deep systems lie in the field of fresh water and the rest 42% (near surface) and 25% (deep) lie in the field of brackish/saline water reflecting moderate mixing in the near surface and nearly insignificant mixing in the deep systems. *$Cl/(HCO_3 + CO_3)$ vs Cl* cross plot shows that 20% of the near surface and 55% of the deep systems lie in the field of fresh water and the rest 80% (near surface) and 45% (deep) lie in the brackish/saline water zone reflecting intense salinisation in the near surface and moderate mixing in the deep systems. *Mg/Ca vs Cl* cross plot indicate that 75% of the near surface and 85% of the deep systems lie in the fresh water zone and the rest 25% (near surface) and 15% (deep) lie in the brackish/saline water zone reflecting minor amounts of mixing in both the systems. *$Ca/(HCO_3 + SO_4)$ vs Cl* cross plot indicate that 53% of the near surface systems and 43% of the deep lie in the field of fresh water and the rest 47% (near surface) and 57% (deep) lie in the field of brackish/saline water zone reflecting nearly equal proportions of mixing in both near surface and deep systems.

5.7.3 Trilinear plots/Characterization of ionic trends

The dominant cations and anions present in solution are presented systematically in the Piper Triangular patterns and are shown in Figure 5.38 (a, b and c) below.

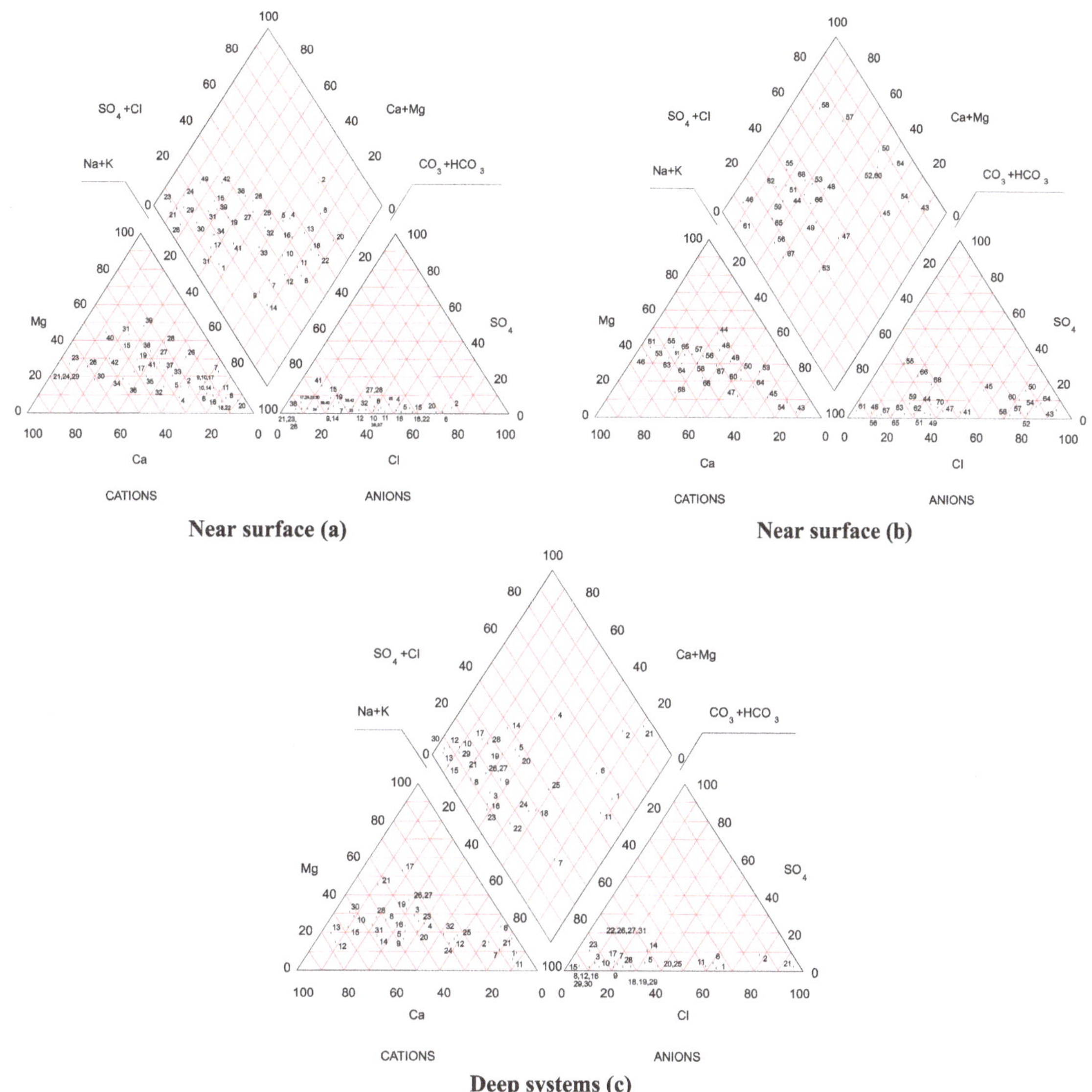

Fig. 5.38 Trilinear plots for near surface (a and b) and deep systems (c) for Bhagiradhi-Hoogly delta system

Hydrochemical characteristics as derived from piper trilinear diagram for Bhagiradhi-Hoogly delta system are presented in table below.

Table 5.42 Characterization of ionic trends in Bhagiradhi-Hoogly delta

Number of Samples : (Near surface: 68, Deep systems:31)

Subdivision of the diamond	Characteristics of corresponding subdivisions of diamond-shaped fields	Near surface		Deep systems	
		No. of Samples	Percentage of samples in this category (%)	No. of Samples	Percentage of samples in this category (%)
1	Alakaline earth (Ca + Mg) Exceed alkalies (Na + K)	38	56	21	68
2	Alkalies exceeds alkaline earths	30	44	10	32

Table 5.42 *Contd...*

Subdivision of the diamond	Characteristics of corresponding subdivisions of diamond-shaped fields	Near surface		Deep systems	
		No. of Samples	Percentage of samples in this category (%)	No. of Samples	Percentage of samples in this category (%)
3	Weak acids (CO_3 + HCO_3) exceed strong acids (SO_4 + Cl)	48	71	25	81
4	Strong acids exceeds weak acids	20	29	6	19
5	Magnesium bicarbonate type	36	53	20	65
6	Clacium-Chloride type	2	3	0	0
7	Sodium-Chloride type	17	25	5	16
8	Sodium-Bicarbonate type	1	1	1	3
9	Mixed type (No cation-anion exceed 50%)	12	18	5	16

5.7.4 Hydrogeochemical facies

Hydrogeochemical facies for Bhagiradhi-Hoogly delta

The hydrochemical facies and trends of Bhagiradhi-Hoogly delta basin are depicted in Figure 5.39(a,b) below.

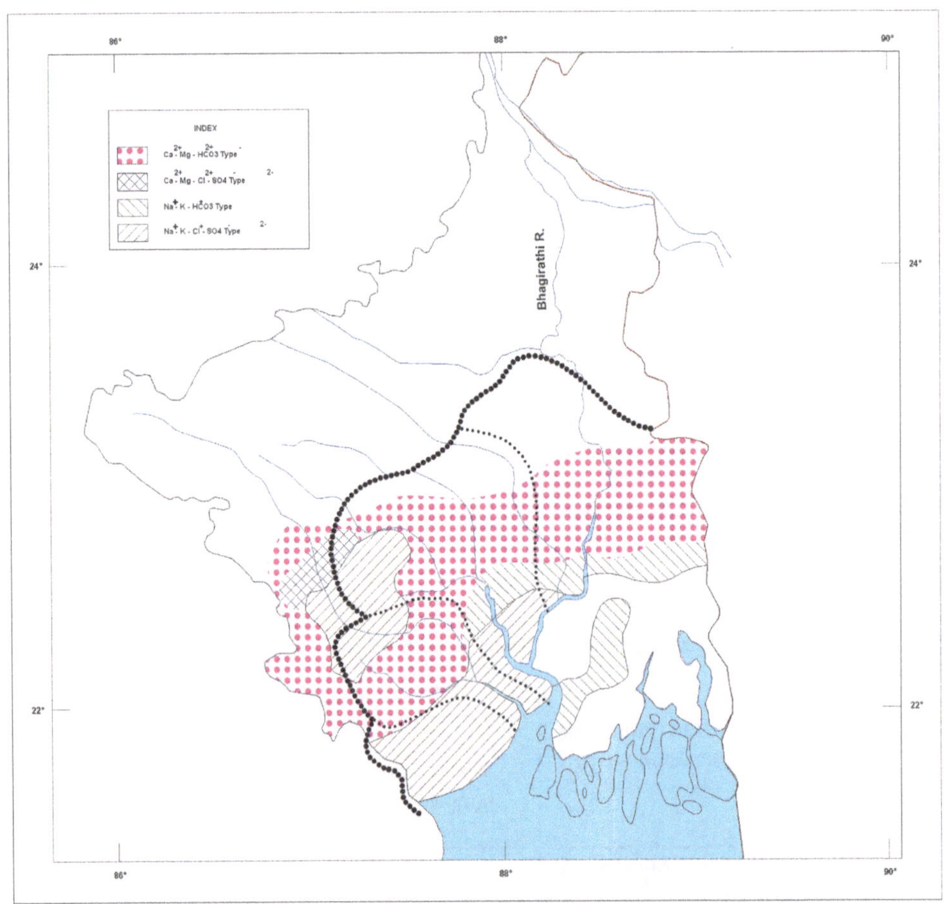

(a) Near surface

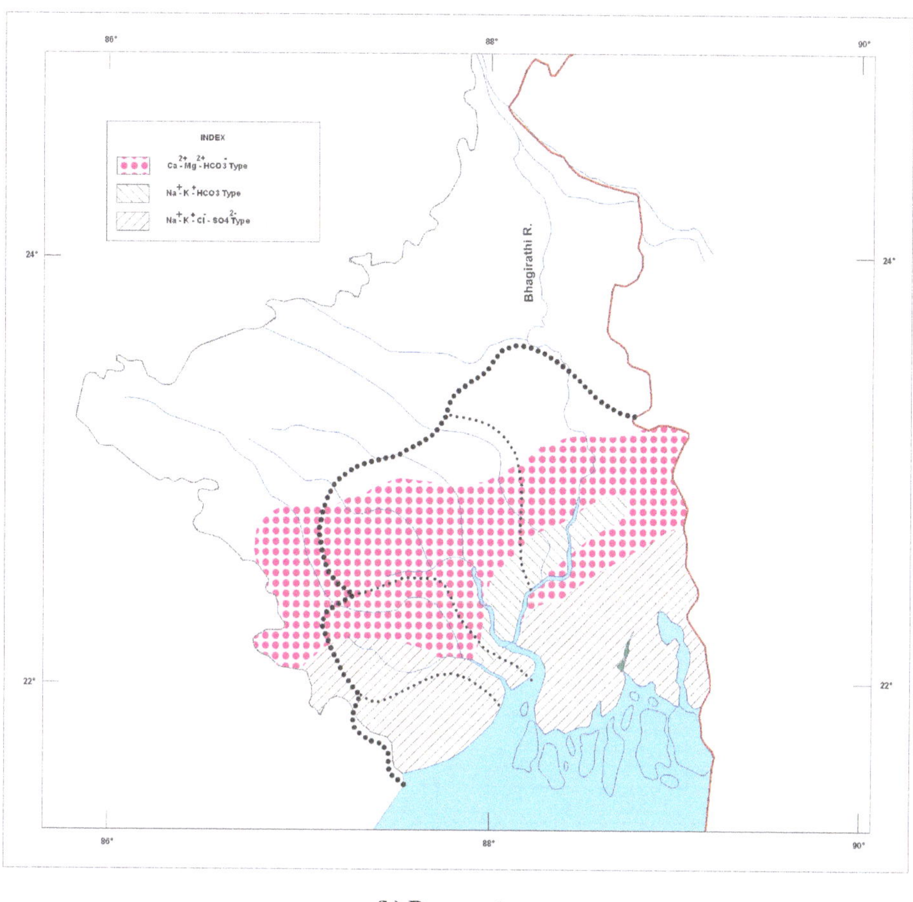

(b) Deep systems

Fig. 5.39 Hydrogeochemical facies for near surface (a) and deep systems (b) for Bhagiradhi-Hoogly delta system

The different hydrochemical facies identified for both near surface and deep systems in the Bhagiradhi-Hoogly delta system are as follows:

1. Ca-Mg-HCO_3 type
2. Na-K-HCO_3 type
3. Na-K-Cl-SO_4 type
4. Ca-Mg-Cl-SO_4 type (near surface only)

Facies no.1 (Ca-Mg-HCO_3 type): The near surface and deep systems of this zone have an average total dissolved solids of the order of 494 ppm and 435.8 ppm respectively. The Ca+Mg are the major cations of this zone and contain about 58.8% (near surface) and 68.6% (deep systems) of the total average cations. The average bicarbonate contents in respect of near surface and deep systems are 80.2% and 84.8% out of the total anionic contents. This facies zone is regionally wide spread in the upstream portions of the middle and outfall delta regions in respect of both near surface and deep systems.

Facies no.2 (Na-K-HCO_3 type): This zone in general is found sandwiched between Ca-Mg-HCO_3 type and Na-K-Cl-SO_4 type in the near surface zone but is represented as an isolated area within the Ca-Mg-HCO_3 type waters in the deep systems. It is also represented as isolated zone with in the Na-K-Cl-SO_4 type waters in the outfall regions of near surface system. This zone has an average total dissolved solids of the order of 930.1 ppm and 621.5 ppm for near surface and deep systems respectively. Na + K are the major cations of this and contain about 75.6% (near surface) and 67.9% (deep systems) of the total average cationic contents. The average HCO_3 content in respect of near surface and deep systems of this zone are 74.3% and 82.6% of the total anionic contents respectively.

Facies no.3 (Na-K-Cl-SO_4 type): This zone is regionally wide spread for both near surface and deep systems in the outfall regions of the delta and have total dissolved solids of the order of 1265 ppm and 1230 ppm for near surface and deep systems respectively. The Na+K contents are the major cations of this zone and contain about 78.3% (near surface) and 86.4% (deep systems) of the total cationic contents respectively. The Cl and SO4 are the major anions of this zone and contain about 59.6% for near surface and 67.5% for deep systems of the total anionic contents respectivley.

Facies no.4 (Ca-Mg-Cl-SO_4 type): This faceis zone is represented only in the near surface aquifer systems and is restricted to the north western portions (middle upstream portions of Kasai river) and is found sandwiched between Ca-Mg-HCO_3 type of waters and Na-K-Cl-SO_4 type. The average total dissolved solids in this zone is of the order of 337 ppm. The Ca + Mg are the major cations in this zone and contain about 60.9% of the total average cations. The Cl + SO_4 are the major anions in this zone and contain about 48.1 % of the total average anionic contents.

5.7.5 Correlation matrix and factor analysis

The correlation matrix analyses for Bhagiradhi-Hoogly delta near surface and deep systems are presented in Table 5.43(a,b) below.

Table 5.43(a,b) Correlation matrix analysis for Bhagiradhi-Hoogly

(a) Near surface

	pH	EC	TDS	Ca	Mg	Na	K	HCO3	SO4	Cl	NO3	F	TH
pH	1.000												
EC	–0.066	1.000											
TDS	–0.066	1.000	1.000										
Ca	0.109	0.452	0.452	1.000									
Mg	0.047	0.457	0.457	0.394	1.000								
Na	–0.089	0.923	0.923	0.193	0.195	1.000							
K	0.047	0.271	0.271	0.372	0.433	0.067	1.000						
HCO_3	–0.19	0.534	0.534	0.371	0.523	0.385	0.308	1.00					
SO_4	0.237	0.586	0.586	0.44	0.523	0.405	0.218	0.114	1.000				
Cl	0.008	0.869	0.869	0.303	0.194	0.919	0.053	0.105	0.539	1.000			
NO_3	0.549	–0.05	–0.05	0.113	–0.021	–0.091	0.324	–0.271	0.339	0.06	1.000		
F	0.215	–0.161	–0.161	–0.075	–0.108	–0.118	–0.178	–0.001	–0.062	–0.19	0.171	1.000	
TH	0.106	0.54	0.54	0.879	0.781	0.228	0.475	0.522	0.569	0.303	0.068	–0.115	1.000

(b) Deep systems

	pH	EC	TDS	Ca	Mg	Na	K	HCO_3	SO_4	Cl	NO_3	TH
pH	1											
EC	–0.023	1										
TDS	–0.035	0.998	1									
Ca	0.11	–0.029	–0.039	1								
Mg	0.022	0.967	0.962	0.093	1							
Na	0.003	0.987	0.986	–0.129	0.953	1						
K	–0.145	0.644	0.655	–0.192	0.614	0.618	1					
HCO_3	0.067	–0.451	–0.466	0.227	–0.413	–0.497	–0.273	1				
SO_4	0.06	0.901	0.899	0.12	0.931	0.883	0.625	–0.464	1			
Cl	–0.002	0.988	0.988	–0.056	0.963	0.992	0.616	–0.556	0.894	1		
NO_3	–0.058	0.255	0.278	0.037	0.251	0.243	0.361	–0.616	0.282	0.31	1	
TH	–0.25	0.046	0.039	0.624	0.069	–0.036	0.039	0.113	–0.003	0.018	–0.011	1

Near surface system: A good correlation is observed between EC and Na (r = 0.923), Cl (r = 0.869) and moderately with HCO_3 (r = 0.534), SO_4 (r = 0.586) and TH (r = 0.54). Ca correlates strongly with TH (r = 0.879). Mg strongly correlates with TH (r = 0.781) and moderately with HCO_3 (r = 0.523) and SO_4 (r =0.523). Na correlate strongly with Cl (r = 0.919). HCO_3 moderately correlates with TH (r = 0.522). SO_4 moderately correlates with Cl (r = 0.539) and with TH (r = 0.569).

Table 5.44 Factor analysis for near surface and deep systems of Bhagiradhi-Hoogly

Near Surface					Deep systems				
Component	**Eigen value**	**% of Variance**	**Cumulative Eigen value**	**Cumulative %**	**Component**	**Eigen value**	**% of Variance**	**Cumulative Eigen value**	**Cumulative %**
1	5.437	41.826	5.437	41.826	1	6.613	55.107	6.613	55.107
2	2.175	16.732	7.612	58.558	2	1.727	14.393	8.34	69.5
3	1.797	13.821	9.409	72.378	3	1.292	10.77	9.632	80.27
4	1.063	8.18	10.472	80.558	4	1.088	9.064	10.72	89.334
5	0.801	6.165	11.273	86.723	5	0.605	5.041	11.325	94.375
6	0.65	4.997	11.923	91.72	6	0.368	3.065	11.693	97.441
7	0.541	4.165	12.464	95.885	7	0.213	1.774	11.906	99.214
8	0.304	2.34	12.768	98.225	8	0.056	0.468	11.962	99.682
9	0.201	1.548	12.969	99.773	9	0.029	0.243	11.991	99.925
10	0.023	0.179	12.992	99.953	10	0.007	0.057	11.998	99.982
11	0.005	0.037	12.997	99.989	11	0.002	0.013	12	99.995
12	0.001	0.011	1.30E + 01	100	12	0.001	0.005	12.001	100
13	1.2E – 16	9.6E –16	1.30E + 01	100					

Table 5.45 Loading varimax rotated factor matrix for Bhagiradhi-Hoogly

Near surface system					Deep systems				
Variable	**Factor 1**	**Factor 2**	**Factor 3**	**Factor 4**	**Variable**	**Factor 1**	**Factor 2**	**Factor 3**	**Factor 4**
pH	–0.031	0.065	0.774	0.25	pH	0.01	–0.044	–0.039	0.95
EC	0.91	0.386	–0.086	–0.02	EC	0.984	0.007	0.123	–0.022
TDS	0.91	0.386	–0.086	–0.02	TDS	0.98	–0.002	0.146	–0.034
Ca	0.213	0.749	0.135	–0.015	Ca	–0.021	0.93	–0.027	0.226
Mg	0.172	0.811	–0.04	0.001	Mg	0.973	0.098	0.113	0.049
Na	0.973	0.048	–0.122	0.015	Na	0.974	–0.095	0.136	0.002
K	–0.023	0.648	0.189	–0.335	K	0.669	–0.105	0.194	–0.3
HCO_3	0.248	0.642	–0.484	0.259	HCO_3	–0.361	0.184	–0.792	0.027
SO_4	0.532	0.435	0.445	–0.054	SO_4	0.919	0.081	0.182	0.107
Cl	0.961	0.034	0.112	–0.133	Cl	0.965	–0.022	0.216	0.007
NO_3	–0.024	0.073	0.882	–0.021	NO_3	0.136	0.069	0.926	–0.035
F	–0.114	–0.077	0.178	0.912	TH	0.034	0.862	–0.048	–0.321
TH	0.231	0.927	0.074	–0.015					

Deep system : A strong correlation is observed between EC and TDS (r = 0.998), Mg (r = 0.967), Na (r = 0.987), SO_4 (r = 0.901), Cl (r = 0.988) and moderately with K (r = 0.644). Ca moderately correlates with TH (r = 0.624). Mg correlates strongly with Na (r = 0.953), SO_4 (r = 0.931), Cl (r = 0.963) and moderately with K (r = 0.614). Na correlates strongly with SO_4 (r =0.883), Cl (r = 0.992) and moderately with K (r = 0.618). K moderately correlates with SO_4 (r = 0.625), Cl (r = 0.616). SO_4 Strongly correlates with Cl (r = 0.894).

The first four factors (Near surface system) and three factors (deep systems) by and large explained 80.5% and 89.3% of the total variances respectively.

Near surface system

Factor 1 explains the largest proportion (41.8%) of the total variance and has high positive loadings (0.91-0.97) for EC, TDS, Na, Cl and moderate loading on SO_4 (0.532).

Factor 2 accounts for 16.7% of the total variance and has high loading on Ca (0.749), Mg (0.811), TH (0.927) and moderate loading on K (0.648), HCO_3 (0.642).

Factor 3 accounts for 13.8% of the total variance and has high loading on pH (0.774) and NO_3 (0.882).

Factor 4 accounts for 8.1% of the total variance and has high loading on F (0.912).

Deep systems

Factor 1 explains the largest proportion 55.1% of the total variance and has high positive loadings (0.96-0.98) on EC, TDS, Mg, Na, SO_4 and Cl and moderate loading on K (0.669).

Factor 2 accounts for 14.3% of the total variance and has high loadings on Ca (0.93) and TH (0.862).

Factor 3 accounts for 10.7% of the total variance and has high loading on NO_3 (0.926).

Factor 4 accounts for 9% of the total variance and has high loading on pH (0.95).

5.7.6 Classification and irrigation suitability of groundwaters based on sodium percentages, sodium adsorption ratios and residual sodium carbonate

Sodium percentage: The classification of groundwaters with respect to the percentage of sodium in Bhagiradhi-Hoogly delta system is presented below in Table.

Table 5.46 Sodium percentage and water classes of Bhagiradhi-Hoogly delta

Sodium (%)	Water Class	No. of Samples	
		Near Surface system	Deep system
< 20	Excellent	13 (4.5 – 18.4)	9 (3.9 – 19.5)
20 – 40	Good	23 (21.6 – 39.4)	12 (23.1 – 39.5)
40 – 60	Permissible	13 (43.1 – 59.9)	4 (40.3 – 56.5)
60 – 80	Doubtful	14 (61.4 – 79.4)	4 (69.5 – 79.9)
> 80	Unsuitable	5 (82.1 – 87.6)	2 (81.9 – 85.3)

The sodium percentages for the near surface systems of Bhagiradhi – Hoogly delta have a range of values from 4.5 to 87.6 and the deep systems between 9 and 85.3. About 72% of the samples from near surface systems fall in the range of excellent to permissible limits and the rest 28% of the sample lie in the categories of doubtful to unsuitable. In the case of deep systems 83% of the samples lie in the range of excellent to permissible limits and the rest 17% lie in the range of doubtful to unsuitable categories.

Sodium Adsorption Ratio (SAR): The classification of groundwater samples based on SAR (Sodium Adsorption Ratio) is presented in Fig. 5.40 (a, b and c) below.

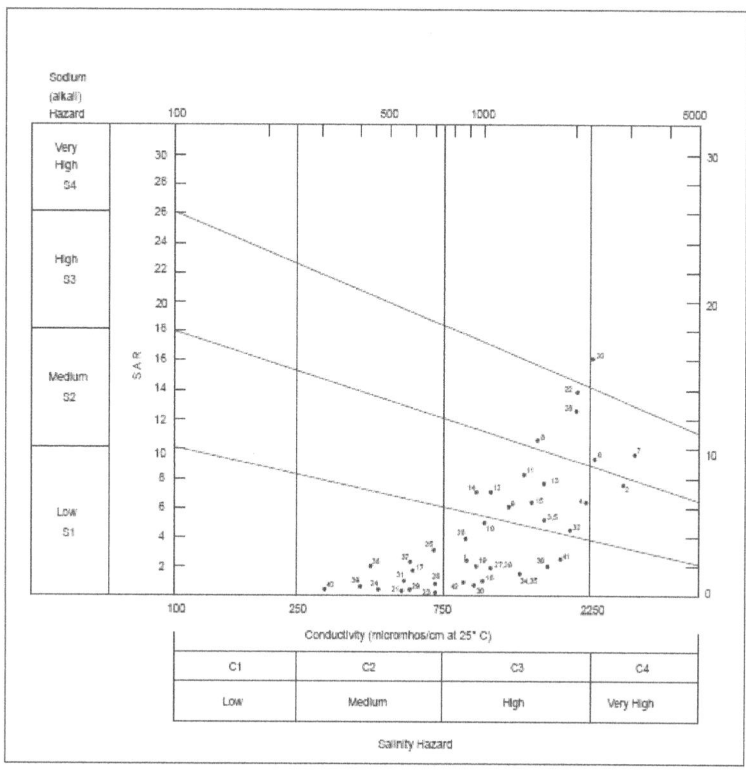

Near surface (a)

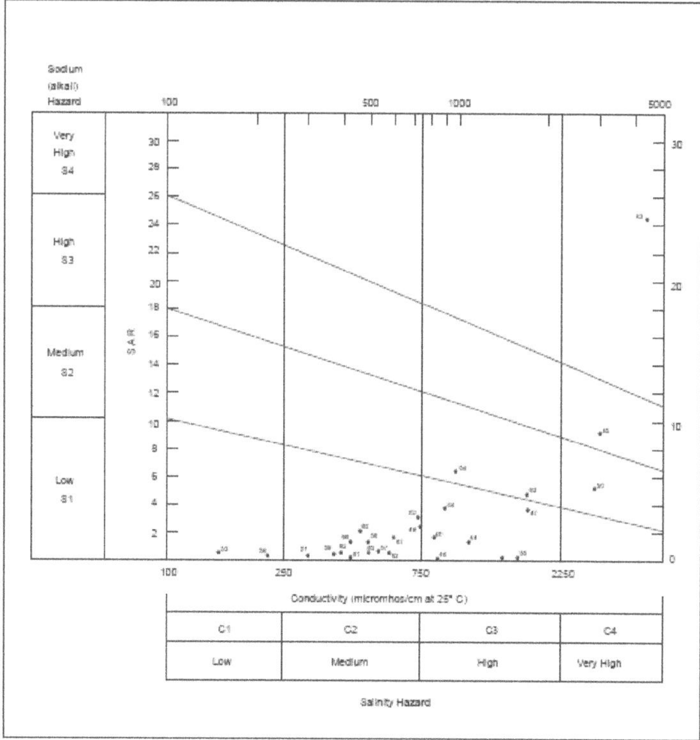

Near surface (b)

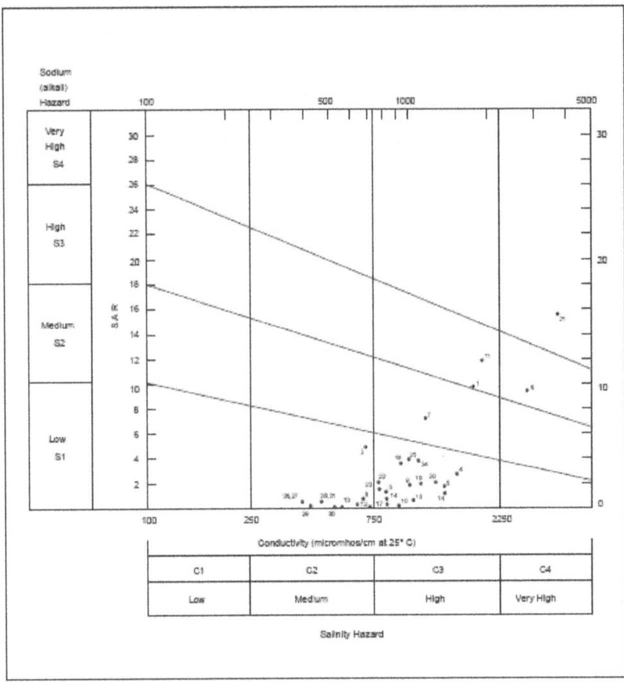

Deep systems (c)

Fig. 5.40 Sodium adsorption ratios for near surface (a and b) and deep systems (c) for Bhagiradhi-Hoogly delta system

About 41% of the samples from near surface systems of the Bhagiradhi-Hoogly delta fall in the fields of C_1S_1 and C_2S_1 (low salinity/low alkalinity to medium salinity/low alkalinity). About 45% of the samples lie in the fields of C_3S_1 and C_3S_2 (High salinities/low to medium alkalinities). All the rest of the samples lie in the fields of C_4S_2 and C_4S_3 (High salinities and low to medium alkalinities). About 32% of the samples from deep systems lie in the fields of C_1S_1 and C_2S_1 (Low salinity/low alkalinity to medium salinity/low alkalinity). About 50% of the sample lie in the fields of C_3S_1 (High salinity/low alkalinity), and the rest of the samples lie in the fields of high salinity/high alkalinities.

Residual sodium carbonate (RSC)

The RSC values for the near surface and deep systems for Bhagiradhi-Hoogly delta system are presented in Table 5.47 below.

Table 5.47 Residual sodium carbonate values in Bhagiradhi-Hoogly delta

RSC (epm)	Remark on Quality	No. of Samples	
		Near Surface system	Deep system
< 1.25	Good	43 (– 9.3 – 1.1)	20 (– 5.5 – 1.0)
1.25 – 2.5	Doubtful	9 (1.5 – 2.4)	4 (1.5 - 2.5)
> 2.5	Unsuitable	16 (2.5 – 12.6)	7 (2.8-5.9)

About 63% of the samples from near surface and 64% of the samples from deep systems have values less than 1.25 and are considered as good quality. All the rest of the samples fall in the doubtful to unsuitable categories.

5.8 WATER CHEMISTRY, CHARACTERIZATION, HYDROGEO-CHEMICAL FACIES AND MULTIVARIATE TREND ANALYSIS OF GROUNDWATERS (NEAR SURFACE AND DEEP) IN NARMADA-TAPTI DELTA SYSTEMS OF GUJARAT

5.8.1 Spatial variations of specific conductance, pH and statistical variations of different ions in Narmada-Tapti delta systems of Gujarat

The sample locations of both near surface and deep systems in respect of Narmada-Tapti delta system are presented in Fig. 5.41 and the specific conductance trends values are reflected in Fig. 5.41.(Data source: Krupanidhi, K.V.J.R., 1986; Ashok Kumar, 2011). The regional analytical results interms of max, min, mean and standard deviation values are presented in table 5.48. The specific conductance values for near surface systems of Tapti-Narmada delta region have a range of values varying from 3475 μ mhos/cm to 556 μ mhos/cm with an average value of 1451 μ mhos/cm and the deep systems have a range of values varying from 3881 μ mhos/cm to 1126 μ mhos/cm with an average value of 2698 μ mhos/cm reflecting dominant saline water mixing in the deeper aquifers when compared to the near surface systems. The northern part of outfall region of the Narmada delta is characterized by relatively high specific conductance value and gradually gets reduced towards the central part of the

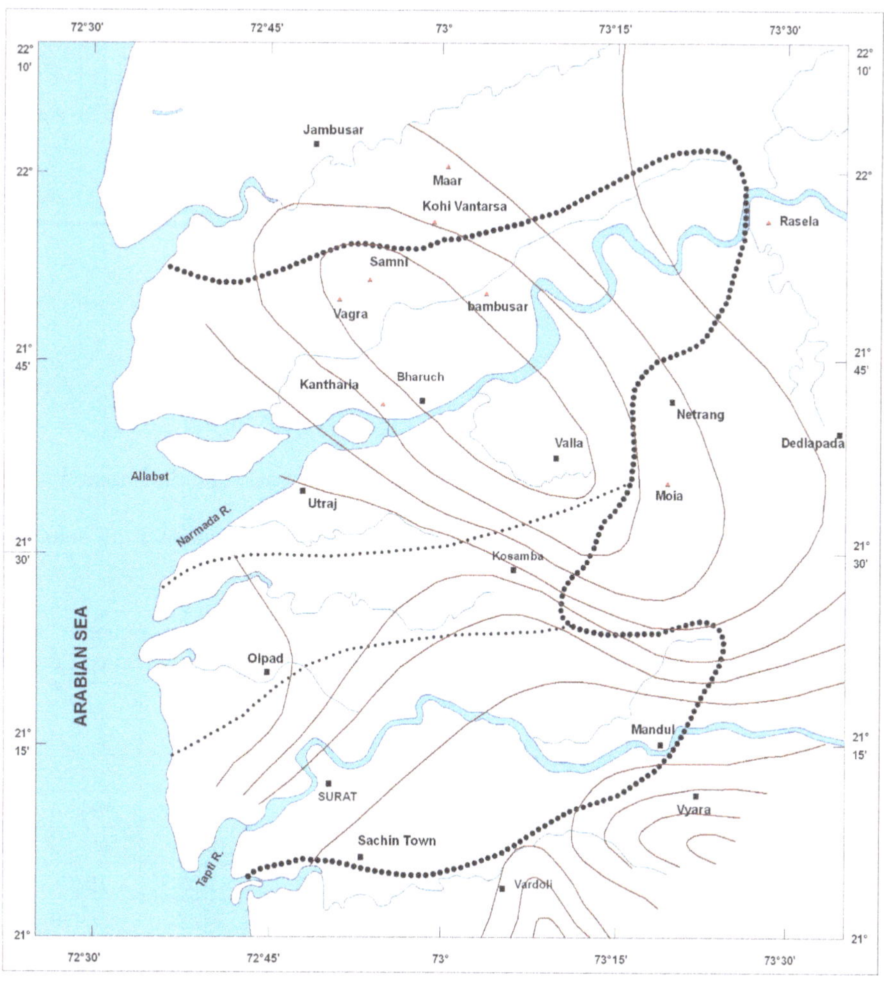

Fig. 5.41 Sample location sites for near surface and deep systems of Narmada-Tapti delta

outfall region and also towards the interstream area of Narmada and Tapti delta region. These values again gradually increase towards the southern side of the Tapti delta region. The data related to the deep aquifer systems is inadequate but appears to increase from northern side of the Narmada towards the central delta region.

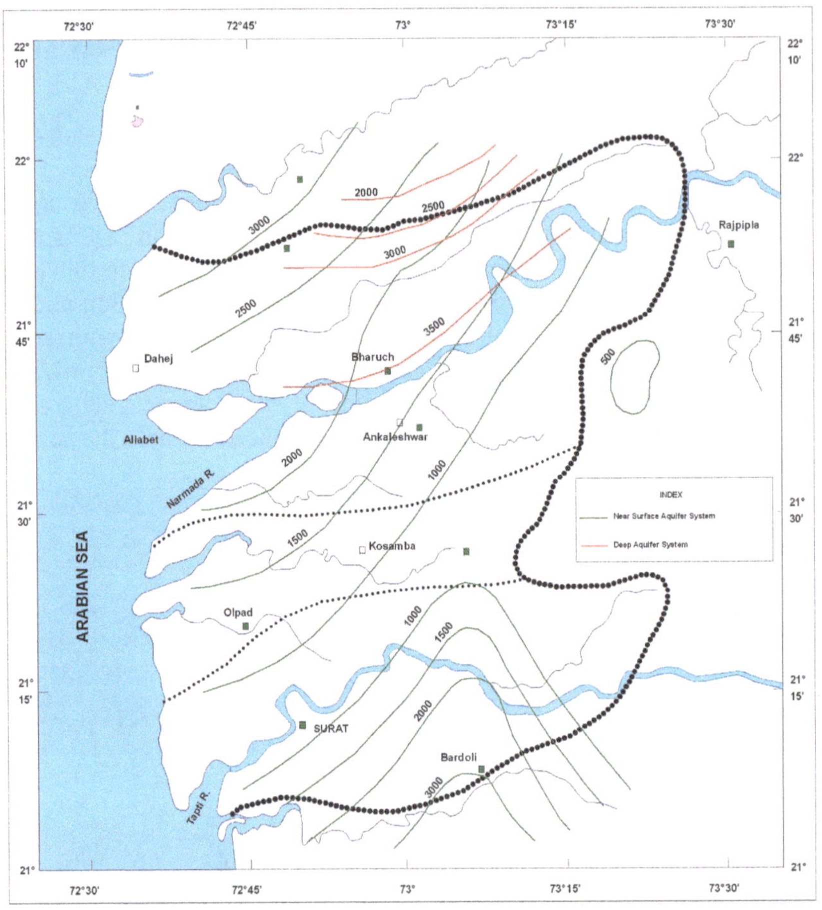

Fig. 5.42 Specific conductance variations in Narmada-Tapti delta region

Table 5.48 Water quality data for both Near surface and Deep systems of Narmada-Tapti delta region in terms of Max, Min, Mean and Standard deviations

Parameter	Near Surface				Deep systems			
	Max.	**Min.**	**Mean**	**Std.dev**	**Max.**	**Min.**	**Mean**	**Std.dev**
pH	8.65	7.65	8.20	0.27	8.4	5.8	7.7	0.8
EC	3475.38	556.92	1451.18	1026.65	3881.54	1126.15	2692.15	1061.03
Ca	332.00	7.40	70.42	83.02	330	26	92.4	87.4
Mg	112	11	42.54	29.25	216.00	22.00	100.30	82.63
Na	506.00	12.00	156.92	168.17	600.00	68.00	347.80	195.43
K	18.00	0.00	4.46	5.81	16.00	5.00	10.30	4.62
HCO_3	811	24	345.46	222.36	732.00	28.00	273.50	209.70
SO_4	244	5	53.46	70.56	210	8	99.3	80.7
Cl	1058	21	254.31	326.46	1570.00	200.00	811.70	449.96

The pH values for near surface systems have a range of values varying from 8.652 to 7.65 with a mean value of 8.20 and from 8.4 to 5.8 with a mean value of 7.7 for the deep systems. The sites of high pH values probably indicate the intensity of salinisation process in this part of the delta.

5.8.2 Cation-anion chemistry/Ionic ratios/Cross plots

The sequential orders of abundance of cations and anions for both near surface and deep systems in the Narmada-Tapti delta system are as follows:

Narmada-Tapti delta region

(a) *Near surface:* HCO_3 > CL > SO_4 and Na > Ca > Mg > K

(b) *Deep systems:* CL > HCO_3 > SO_4 and Na > Mg > Ca > K

The bicarbonate and sodium are the dominant ions of the near surface system amongst the anions and cations. The Chloride and the Sodium are the dominant ions amongst the anions and cations of the deep system indicating relatively high saline water mixing processes. This again confirms the earlier observation that deep systems are subjected to saline water mixing more dominantly than the near surface systems.

Variations of Cations and anions with respect to TDS: The max and min variations and the standard deviations in terms of different water quality parameters for both near surface and deep systems are presented in the Table 5.48. The variations of cations and anions in relation to TDS for both near surface and deep systems for Narmada-Tapti delta are presented in Fig. 5.43.

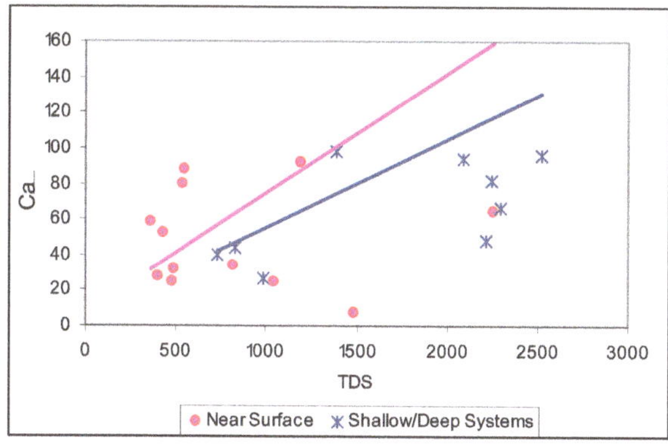

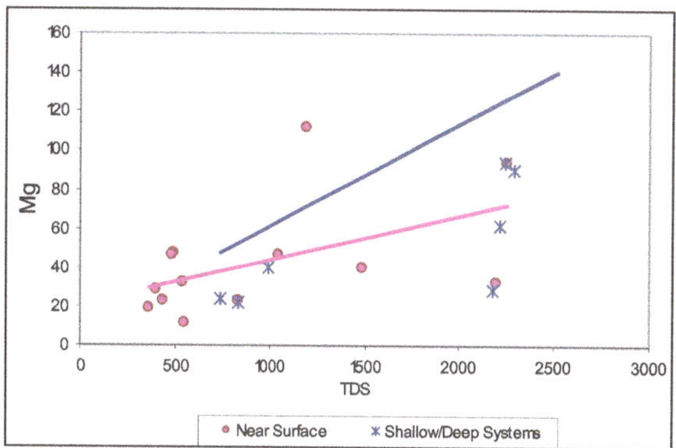

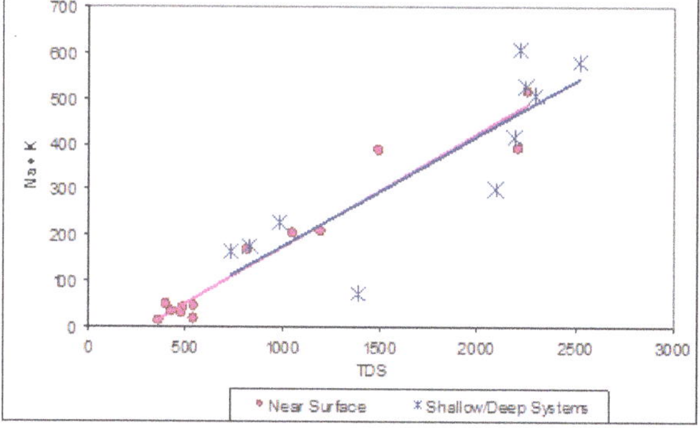

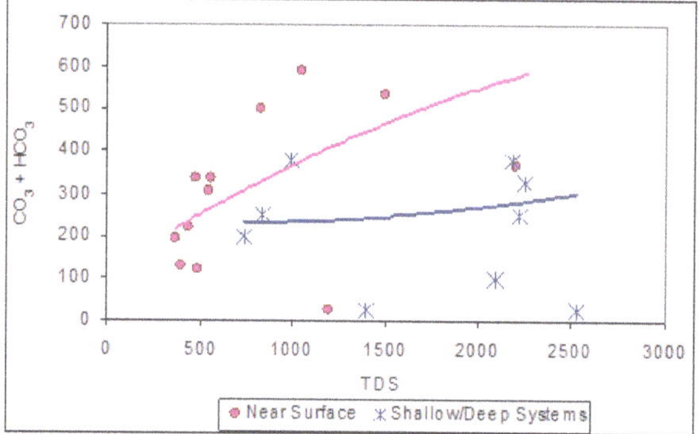

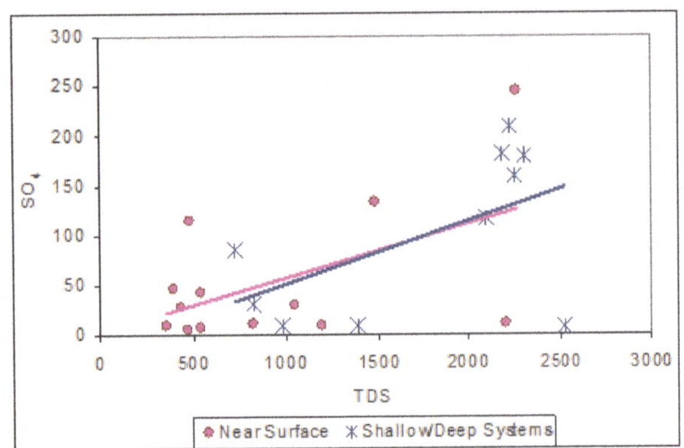

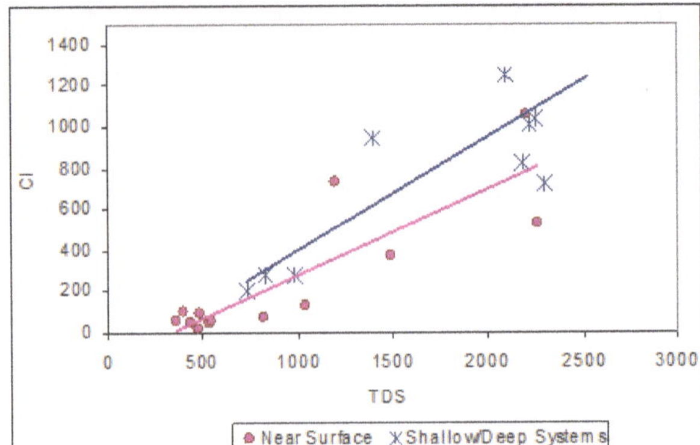

Fig. 5.43 Variations of cations and anions in relation to TDS in Narmada-Tapti delta region

- The HCO_3 vs TDS cross plots for both near surface and deep systems show partial curve linearity with increasing levels of TDS. The near surface systems show an increasing level up to 2000 TDS level reflecting dominance of recharge processes with little or insignificant mixing of salinity. The deep systems however show a near horizontality with increasing levels of TDS reflecting neither any significant recharge contribution nor saline water mixing.
- The Ca vs TDS graph shows culmination of both near surface and deep systems at initial levels of TDS and gradually divert with increasing levels of TDS and with different slope characteristics. The near surface system is characterized by high gradient when compared to the deep system reflecting the involvement of Ca activity more in the near surface systems. The Mg levels of both near surface and deep systems also show similar trends with culmination points at the low levels of TDS but gradually diverge with different slope characteristics. The gradient levels however in the case of Mg vs TDS graph show a different trend. The Mg levels of the near surface have a low gradient characteristic when compared to the deep systems where the gradient levels are relatively high reflecting higher involvement of Mg ions in the deep systems when compared to the near surface system. This also reflects on the probable higher intensities of salinization processes in the deep Systems.
- The Na + K trend values for both near surface and deep systems by and large align with similar slope trend characteristics reflecting similar intensities of Na + K involvement, with increasing levels of TDS in both near surface and deep systems.
- The SO_4 and Cl anions for both near surface and deep systems exhibit ascending linearity with different slope characteristics. The SO_4 vs TDS for both near surface and deep systems shows similar slope characteristics with increasing levels of TDS, reflecting similar contribution of SO_4. The slope gradients in the case of Cl vs TDS graph however differ with higher gradients for deeper systems and lower gradient for near surface reflecting relatively higher contributions of Cl in the deep systems.

Ionic ratios/Cross plots : The specific conductance/total dissolved solids/ionic ratios between selected cations and anions would help to estimate and assess the intensity of fresh and saline water mixing processes. The max, min and average values of selected number of ionic ratios for the near surface and deep systems of Narmada and Tapti delta systems are presented in Table 5.49(a,b).

Table 5.49(a,b) Maximum, Minimum and Average values of ionic ratios in Narmada and Tapti delta systems of Gujarat

(a) Near surface

S.No	Ionic ratios	Narmada and Tapti delta systems		
		Max	Min	Avg
1.	Na/Cl	3.08	0.32	1.24
2.	Mg/Cl	6.40	0.08	1.29
3.	SO4/Cl	0.95	0.00	0.27
4.	$Cl/(HCO_3 + CO_3)f$	28.52	0.10	3.08
5.	$Ca/(HCO_3 + SO_4)$	7.65	0.11	1.14
6.	Mg/Ca	8.89	0.15	1.73

(b) Deep systems

S.No	Ionic ratios	Narmada and Tapti delta systems		
		Max	Min	Avg
1.	Na/Cl	1.21	0.11	0.77
2.	Mg/Cl	0.67	0.09	0.34
3.	SO_4/Cl	0.31	0.00	0.10
4.	$Cl/(HCO_3 + CO_3)$	71.43	1.21	15.02
5.	$Ca/(HCO_3 + SO_4)$	8.55	0.20	2.12
6.	Mg/Ca	3.73	0.13	2.17

- The average Na/Cl ratios for near surface/deep systems in respect of Narmada-Tapti are 1.24/0.77 respectively. The theoretical values for Na/Cl ratio for saline and brackish waters are of the order of 0.85 and 1.00 respectively. This reflects that the near surface system is free from saline water mixing and the deep systems shows intense saline water mixing.
- The average Mg/Cl ratios for near surface/deep systems of Narmada-Tapti are 1.29/0.34 respectively. The theoretical values for Mg/Cl ratio for saline and brackish waters are of the order of 0.15 and 0.35 respectively. This again shows that the near surface system is free from saline water mixing and the deep systems are close to the brackish water boundary.
- The average SO_4/Cl ratios for near surface/deep systems of Narmada-Tapti are 0.27/0.10 respectively. The theoretical values for SO_4/Cl ratio for saline and brackish waters are of the order of 0.005 and 0.20 respectively. This indicates that the near surface system though free from mixing comes close to the brackish water boundary and the deep systems lie within brackish/saline water fields.
- The average $Cl/(HCO_3 + CO_3)$ ratios for near surface/deep systems of Narmada-Tapti are 3.08/15.02 respectively. The theoretical values for $Cl/(HCO_3 + CO_3)$ ratio for saline and brackish waters are of the order of 6.6 and 0.50 respectively. This indicates that near surface system shows moderate saline water mixing and the deep system shows intense saline water mixing.

- The average $Ca/(HCO_3 + SO_4)$ ratios for near surface/deep systems of Narmada-Tapti are 1.14/2.12 respectively. The theoretical values for $Ca/(HCO_3 + SO_4)$ ratio for saline and brackish waters are of the order of 1.00 and 0.50 respectively. This reflects that both near surface and deep systems are undergoing saline water mixing processes.
- The average Mg/Ca ratios for near surface/deep systems of Narmada-Tapti are 1.73/2.17 respectively. The theoretical values for Mg/Ca ratio for saline and brackish waters are of the order of 3.5 and 1.00 respectively. This also reflects that both near surface and deep systems are undergoing saline water mixing processes.

Ionic ratios vs Cl cross plots: The ratios considered for cross plots to assess the intensity of mixing are: Na/Cl vs Cl, SO_4/Cl vs Cl, Mg/Cl vs Cl, $Cl/(HCO_3 + CO_3)$ vs Cl, Mg/Ca vs Cl and $Ca/(HCO_3 + SO_4)$ vs Cl. The graphical plots of ionic ratios between Na/Cl vs Cl, SO_4/Cl vs Cl, Mg/Cl vs Cl, $Cl/(HCO_3 + CO_3)$ vs Cl, Mg/Ca vs Cl and $Ca/(HCO_3 + SO_4)$ vs Cl characterizing the probable saline fresh water mixing trends in both near surface and deep systems of Narmada-Tapti delta system are presented in Fig. 5.44.

Narmada-Tapti

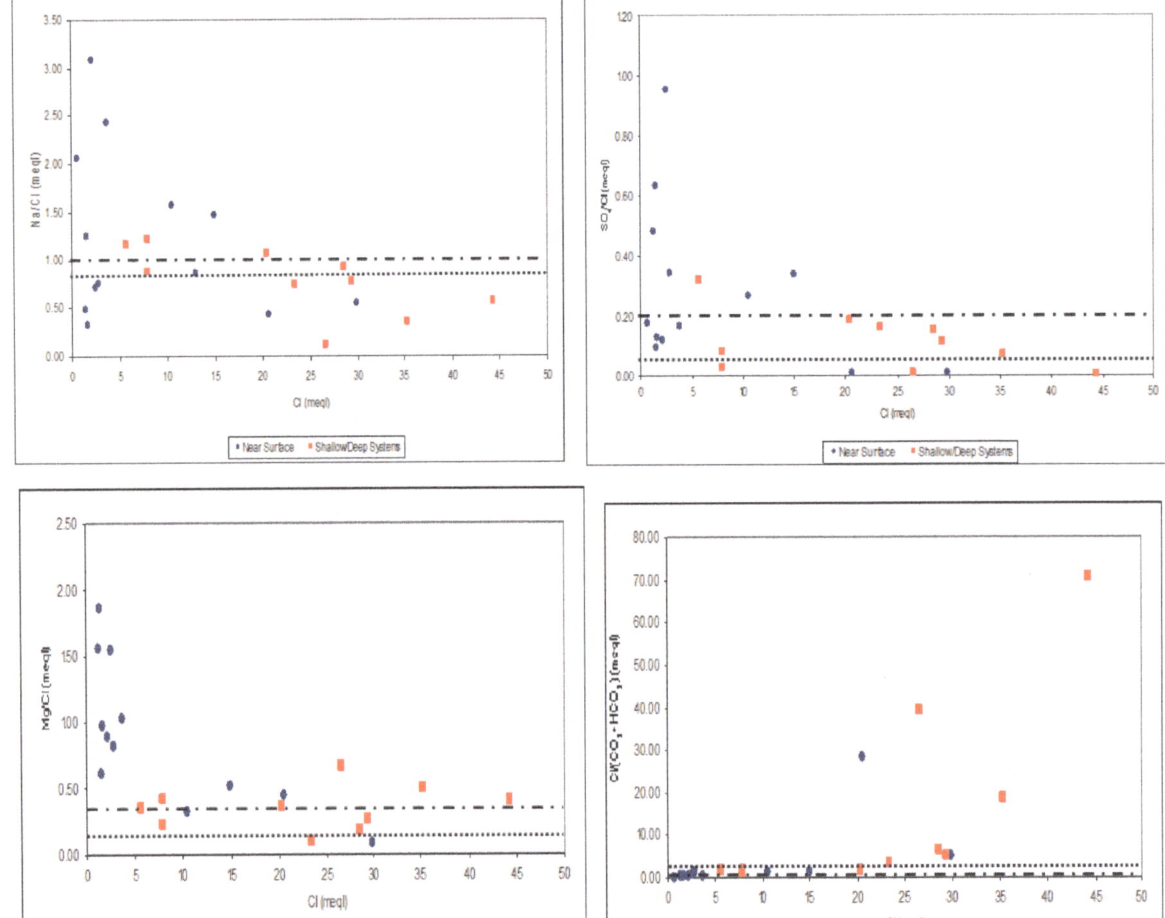

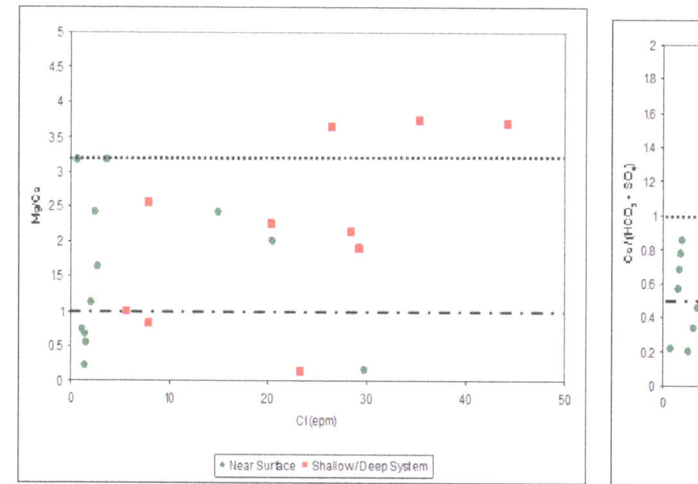

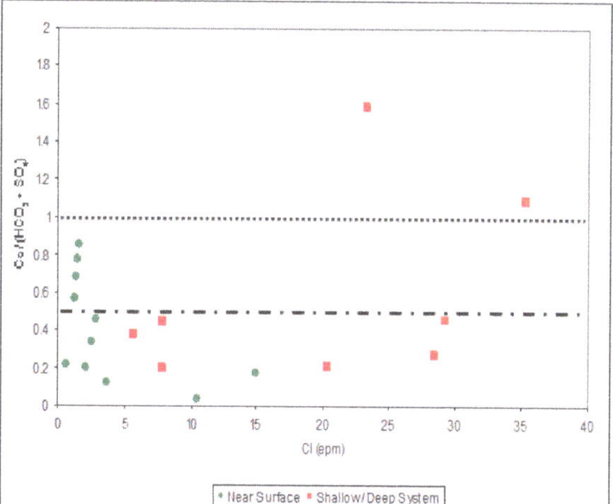

Fig. 5.44 Ionic ratios cross plots

A brief analyses in respect of the above graphic plots discussing the significance of fresh saline water trends in Narmada-Tapti delta system of Gujarat are presented below in the following paragraphs.

Narmada-Tapti: *Na/Cl vs Cl* cross plot indicates that 46% of the near surface and 30% of the deep system lie in the field of fresh water and the rest 54% (near surface) and 70% (deep) lie in the field of brackish/saline water reflecting that majority of the samples in both near surface and deep systems are getting mixed up with saline water. SO_4/Cl *vs Cl* cross plot indicate that 46% of the near surface and 10% of the deep system lie in the field of fresh water and the rest 54% (near surface) and 90% (deep) lie in the field of brackish/saline water. This reflects on moderate mixing in the near surface and intense mixing in the deep systems with saline water. *Mg/Cl vs Cl* cross plot indicates 83% of the near surface and 60% of the deep systems lie in the field of fresh water and the rest 17% (near surface) and 40% (deep) lie in the field of brackish/saline water. This reflects on insignificant mixing of the near surface systems and moderate mixing in the deep systems. $Cl/(HCO_3 + CO_3)$ *vs Cl* cross plot shows 54% of the near surface and 0% of the deep systems lie in the field of fresh water and the rest 46% (near surface) and 100% (deep) lie in the field of brackish/saline water reflecting moderate mixing processes in the near surface and intense mixing in the deep systems. *Mg/Ca vs Cl* cross plot shows 42% of the near surface and 30% of the deep systems lie in the field of fresh water and the rest 58%(near surface) and 70% (deep) lie in the brackish/saline water fields reflecting moderate mixing in the near surface systems and intense mixing in the deep systems. $Ca/(HCO_3 + SO_4)$ *vs Cl* cross plot indicates 64% of the near surface and 75% of the deep systems lie in the field of fresh water and the rest 36% (near surface) and 25% (deep) lie in the field of brackish/saline water reflecting moderate levels of mixing in both near surface and deep systems.

5.8.3 Trilinear plots/Characterization of ionic trends

The dominant cations and anions present in solution are presented systematically in the Piper Triangular patterns and are shown in Figure below 5.45(a,b).

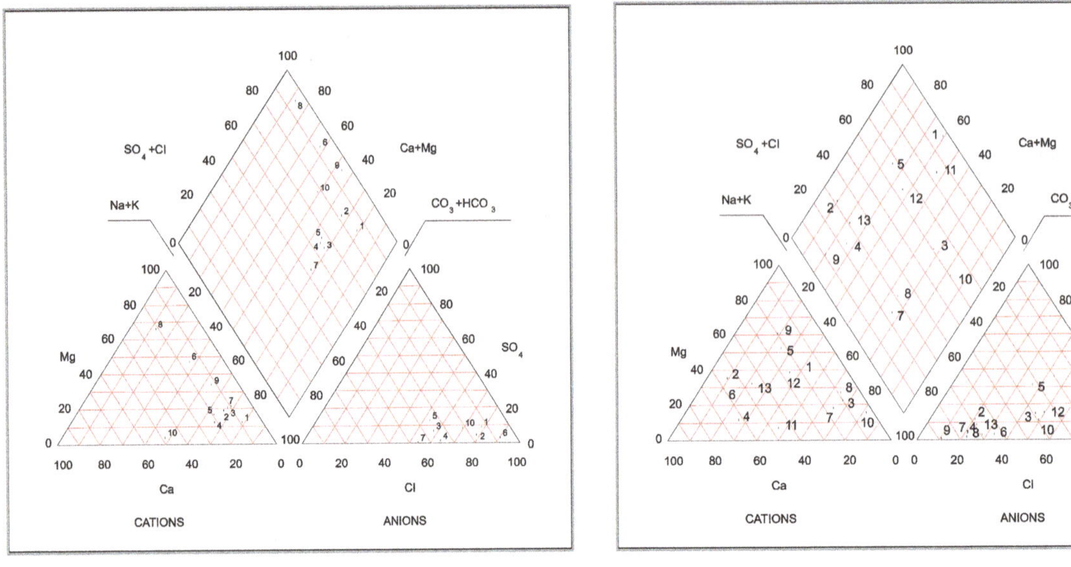

Near surface (a) **Deep systems (b)**

Fig. 5.45 Trilinear plots for near surface (a) and deep systems (b) for Narmada-Tapti delta system

Hydrochemical characteristics as derived from piper trilinear diagram for Narmada-Tapti delta systems are presented in Table 5.50 below.

Table 5.50 Characterization of ionic trends in Narmada-Tapti River

Number of Samples: (Near surface: 13, Deep systems: 10)

Subdivision of the diamond	Characteristics of corresponding subdivisions of diamond-shaped fields	Near surface		Deep systems	
		No. of Samples	Percentage of samples in this category (%)	No. of Samples	Percentage of samples in this category (%)
1	Alakaline earth (Ca + Mg) Exceed alkalies (Na + K)	9	69	3	33.3
2	Alkalies exceeds alkaline earths	4	31	7	70
3	Weak acids ($CO_3 + HCO_3$) exceed strong acids ($SO_4 + Cl$)	7	54	0	0
4	Strong acids exceeds weak acids	6	46	10	100
5	Magnesium bicarbonate type	5	38	0	0
6	Clacium-Chloride type	1	8	2	20
7	Sodium-Chloride type	2	15	7	70
8	Sodium-Bicarbonate type	0	0	0	0
9	Mixed type (No cation-anion exceed 50%)	7	54	1	10

5.8.4 Hydrogeochemical facies

Hydrochemical facies of Narmada and Tapti delta

The hydrochemical facies and trends of Narmada and Tapti delta basins are depicted in Figure 5.46(a,b).

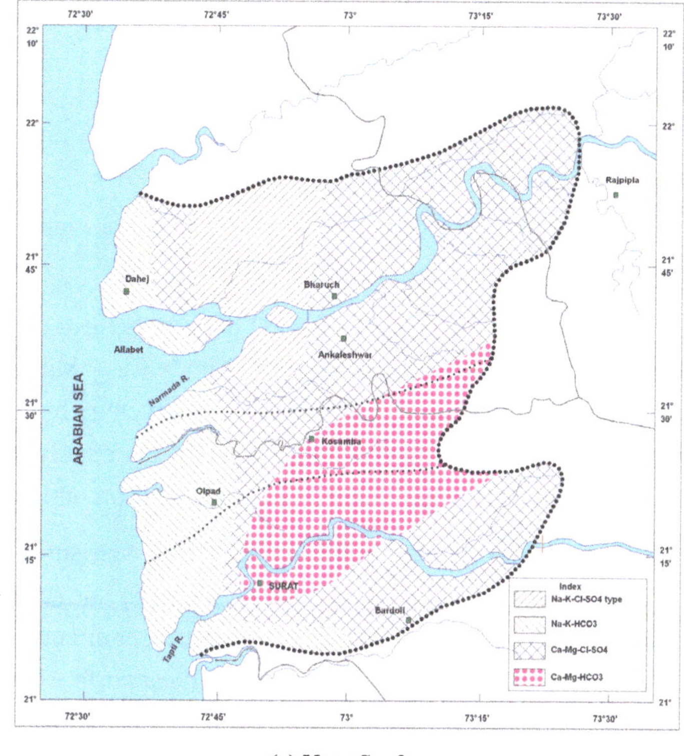

(a) Near Surface

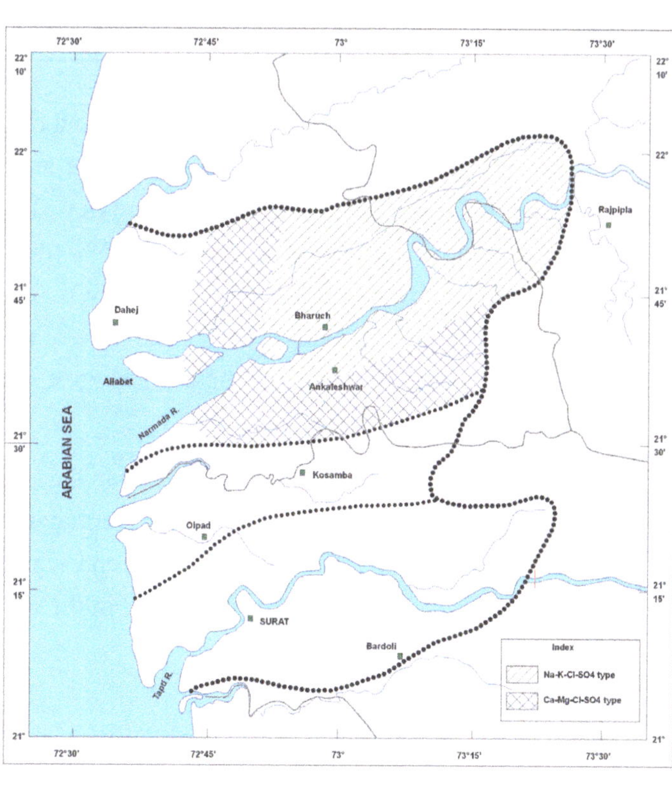

(b) Deep systems

Fig. 5.46 Hydrogeochemical facies for near surface (a) and deep systems (b) for Narmada-Tapti delta systems

The different hydrochemical facies identified for both near surface and deep systems in the Narmada and Tapti delta systems are as follows:

1. Ca-Mg-HCO_3 type
2. Ca-Mg-Cl-SO_4 type
3. Na-K-Cl-SO_4 type
4. Na-K-HCO_3 type

In the case of deep systems the Ca-Mg-HCO_3 and Na-K-HCO_3 are not identified probably because of inadequacy of the data.

Facies no.1 (Ca-Mg-HCO_3 type): This facies is represented only in the near surface of the delta basin and have an average total dissolved solids of the order of 472 ppm. Ca + Mg are the major cations in this zone and contain about 82% of the total average cations. The average bicarbonate contains in the system is 74% of the total anionic contents. This facies zone represents mainly the interstream portions of Narmada and Tapti basin area and partly in the middle portions of Tapti outfall delta region.

Facies no.2 (Ca-Mg-Cl-SO_4 type): This facies is represented both in near surface and deep systems of the delta but more dominantly in the the near surface systems. The near surface and deep systems have an average total dissolved solids of the order varying between 1071 ppm and 1889 ppm respectively. Ca+Mg are the major cations in this zone and contain about 59% (near surface) and 65% (deep systems) of the total average cations. The anionic contents are represented by Cl and SO_4. The average Cl + SO_4 in respect of near surface and deep systems are 85% and 91% of the total anionic contents. This facies dominantly represent in the upstream portions of the near surface systems and in the outfall regions of the deep systems.

Facies no.3 (Na-K-Cl-SO_4): This is also represented in both near surface and deep systems of the delta. The near surface and deep systems of this facies in the delta regions have an average total dissolved solids of the order varying between 1873 ppm and 1690 ppm respectivley. The Na+K are the major cations in this facies zone and contain about 73% (near surface) and 65% (deep systems) of the total average cations. The cations in this facies are represented by Cl and SO_4. The average Cl and SO_4 contents in respect of both near surface and deep systems are 60% and 81% of the total anionic contents respectively. This zone lies with in the frame work of Ca-Mg-Cl-SO_4 zone. But in the case of deep systems it stretches in the upstream portions of the Narmada delta.

Facies no.4 (Na-K-HCO_3): This zone is represented only in the near surface systems and has an average total dissolved solids of the order of 935 ppm. The Na+K are the major cations in this zone and contain about 65% out of the total cationic contents. The anions are represented by HCO_3 containing about 73% of the total anionic contents. This is dominantly represented in the outfall region close to the coastal boundary of the near surface zone.

5.8.5 Correlation matrix and factor analysis

The correlation matrix analyses for Narmada and Tapti delta's near surface and deep systems are presented in Table 5.51 below.

Table 5.51(a,b) Correlation matrix analysis for Narmada-Tapti delta

(a) Near surface

	pH	EC	TDS	Ca	Mg	Na	K	HCO_3	SO_4	Cl	TH
pH	1										
EC	–0.084	1									
TDS	–0.084	1	1								
Ca	0.022	0.542	0.542	1							
Mg	0.265	0.503	0.503	–0.02	1						
Na	–0.132	0.969	0.969	0.34	0.514	1					
K	–0.019	0.473	0.473	0.333	–0.022	0.461	1				
HCO_3	–0.335	0.604	0.604	–0.056	0.1	0.68	0.348	1			
SO_4	–0.057	0.517	0.517	–0.22	0.447	0.621	0.046	0.558	1		
Cl	0.093	0.849	0.849	0.763	0.518	0.749	0.394	0.11	0.164	1	
TH	0.201	0.744	0.744	0.778	0.576	0.607	0.281	–0.028	0.061	0.963	1

(b) Deep systems

	pH	EC	TDS	Ca	Mg	Na	K	HCO_3	SO_4	Cl
pH	1									
EC	–0.564	1								
TDS	–0.564	1	1							
Ca	–0.098	0.393	0.393	1						
Mg	0	0.435	0.435	–0.032	1					
Na	–0.517	0.856	0.856	0.146	0.054	1				
K	0.496	–0.408	–0.408	0.182	–0.385	–0.393	1			
HCO_3	–0.725	0.13	0.13	0.065	–0.569	0.298	–0.118	1		
SO_4	–0.65	0.548	0.548	0.333	–0.307	0.597	0.082	0.543	1	
Cl	–0.181	0.851	0.851	0.261	0.782	0.599	–0.388	–0.386	0.148	1

Near surface systems

A good correlation is observed between EC and Na (r = 0.969), Cl (r = 0.849), TH (r = 0.744) and EC moderately correlates with HCO_3 (r = 0.604), Ca (0.542), SO_4 (r = 0.517) and Mg (r = 0.503). Ca strongly correlates with TH (r = 0.778) and Cl (r = 0.763). Mg correlates moderately with Na (r = 0.514), TH (r = 0.576) and Cl (r = 0.518). Na correlates strongly with Cl (r = 0.749) and moderately correlates with HCO_3 (r = 0.68), SO_4 (r = 0.621) and TH (r = 0.607). Moderate correlation is observed between HCO_3 and SO_4 (r = 0.558). Cl correlates strongly with TH (r = 0.963).

Table 5.52 Factor Analysis for near surface and deep systems of Narmada-Tapti

Near Surface					Deep systems				
Component	Eigen value	% of Variance	Cumulative Eigen value	Cumulative %	Component	Eigen value	% of Variance	Cumulative Eigen value	Cumulative %
1	5.109	51.088	5.109	51.088	1	4.704	47.044	4.704	47.044
2	1.811	18.11	6.92	69.197	2	2.591	25.912	7.295	72.956
3	1.45	14.497	8.37	83.694	3	1.399	13.985	8.694	86.941
4	0.803	8.03	9.173	91.724	4	0.649	6.486	9.343	93.428
5	0.44	4.4	9.613	96.125	5	0.432	4.318	9.775	97.745
6	0.289	2.889	9.902	99.014	6	0.194	1.937	9.969	99.682
7	0.098	0.983	10	99.997	7	0.028	0.282	9.997	99.964
8	0	0.003	10	100	8	0.004	0.036	10.001	100
9	3.22E – 05	0	10.000032	100	9	1.38E-05	0	10.0010138	100
10	1.09E – 16	1.09E – 15	10.000032	100	10	–3.49E – 17	–3.49E – 16	10.0010138	100

Deep systems: A strong correlation is observed between EC and Na (r = 0.856), Cl (r = 0.851) and moderate correlation with SO_4 (r = 0.548). Mg correlates strongly with Cl (r = 0.782). Na moderately correlates with Cl (r = 0.599) and SO_4 (r = 0.597). HCO_3 moderately correlates with SO_4 (r = 0.543).

Table 5.53 Loading varimax rotated factor matrix for Narmada-Tapti

Near surface system				Deep systems			
Variable	Factor 1	Factor 2	Factor 3	Variable	Factor 1	Factor 2	Factor 3
pH	0	–0.111	0.772	pH	–0.445	–0.765	0.273
EC	0.752	0.653	–0.007	EC	0.969	0.22	0.068
TDS	0.752	0.653	–0.007	TDS	0.969	0.22	0.068
Ca	0.906	–0.238	0.047	Ca	0.36	0.077	0.735
Mg	0.154	0.619	0.644	Mg	0.616	–0.621	–0.311
Na	0.613	0.771	–0.064	Na	0.761	0.439	–0.016
K	0.606	0.105	–0.321	K	–0.468	–0.122	0.759
HCO_3	0.148	0.731	–0.537	HCO_3	–0.089	0.945	–0.068
SO_4	–0.085	0.899	0.016	SO_4	0.387	0.734	0.377
Cl	0.895	0.266	0.297	Cl	0.946	–0.312	–0.017

The first three factors in both near surface system and deep systems by and large explained 83.7% and 86.9% of the total variances respectively.

Near surface systems

***Factor* 1** explains the largest proportion (51.08 %) of the total variance and has high positive loadings (0.75-0.90) for EC, TDS, Ca and Cl.

***Factor* 2** accounts for 18.11% of the total variance and has high loading on Na (0.77), HCO_3 (0.73) and

SO_4 (0.89).

***Factor* 3** accounts for 14.49% of the total variance and has high loading on pH (0.772).

Deep systems

***Factor* 1** explains the largest proportion 47.07 % of the total variance and has high positive loadings (0.76-0.96) on EC, TDS, Na and Cl.

***Factor* 2** accounts for 25.9% of the total variance and has high loadings on HCO_3 (0.94) and SO_4 (0.73).

***Factor* 3** accounts for 13.9% of the total variance and has high loading on Ca (0.73) and K (0.75).

5.8.6 Classification and irrigation suitability of groundwaters based on sodium percentages, sodium adsorption ratios and residual sodium carbonate

Sodium percentage: The classification of groundwaters with respect of percentage of sodium in Narmada and Tapti delta systems are presented below in Table 5.54.

Table 5.54 Sodium percentage and water classes of Narmada-Tapti delta

Sodium (%)	Water Class	No. of Samples	
		Near Surface system	Deep system
< 20	Excellent	4 (9.52 – 19.61)	1 (11.49)
20 - 40	Good	4 (24.55 – 38.95)	2 (33.80 – 35.40)
40 - 60	Permissible	1 (45.86)	3 (47.63 – 59.85)
60 - 80	Doubtful	3 (6172 – 66.47)	4 (61.11 – 66.64)
> 80	Unsuitable	1 (81.14)	Nil

The sodium percentages for the near surface system of Narmada-Tapti delta regions have a range of values from 9.52 to 81.14 and the deep systems between 4.49 to 66.64. About 69.2% of the sample of near surface systems fall in the range of excellent to permissible limits, and the rest 30.8% of the samples lie in the categories of doubtful to unsuitable. In the case of deep systems about 50% of the samples lie in the categories of excellent to permissible limits and the rest 50% of the samples lie in the doubtful to unsuitable categories.

Sodium Adsorption Ratio (SAR): The classification of groundwater samples based on SAR (Sodium Adsorption Ratio) is presented in Figures 5.47(a, b) below.

About 39.45% of the samples from near surface systems of the Narmada-Tapti deltas fall in the fields of C_1S_1 and C_2S_1 (low salinity/low alkalinity to medium salinity/low alkalinity). Again about 39.45% of the samples lie in the fields of C_3S_1 and C_3S_2 (high salinity/low to medium alkalinity). The rest of the samples lie in the fields of C_4S_2 and C_4S_3 (high salinity and medium to high alkalinities). Significantly in the deep systems none of the samples lie in the fields of C_1S_1 and C_2S_1 (low salinity/low alkalinity to medium salinity/low alkalinity). About 40% of the samples lie in the fields of C_3S_1 and

C_3S_2 (high salinities and low to medium alkalinities). The rest 60% of the samples lie in the fields of C_4S_2, C_4S_3 and C_4S_4 (high salinities and medium to high alkalinities).

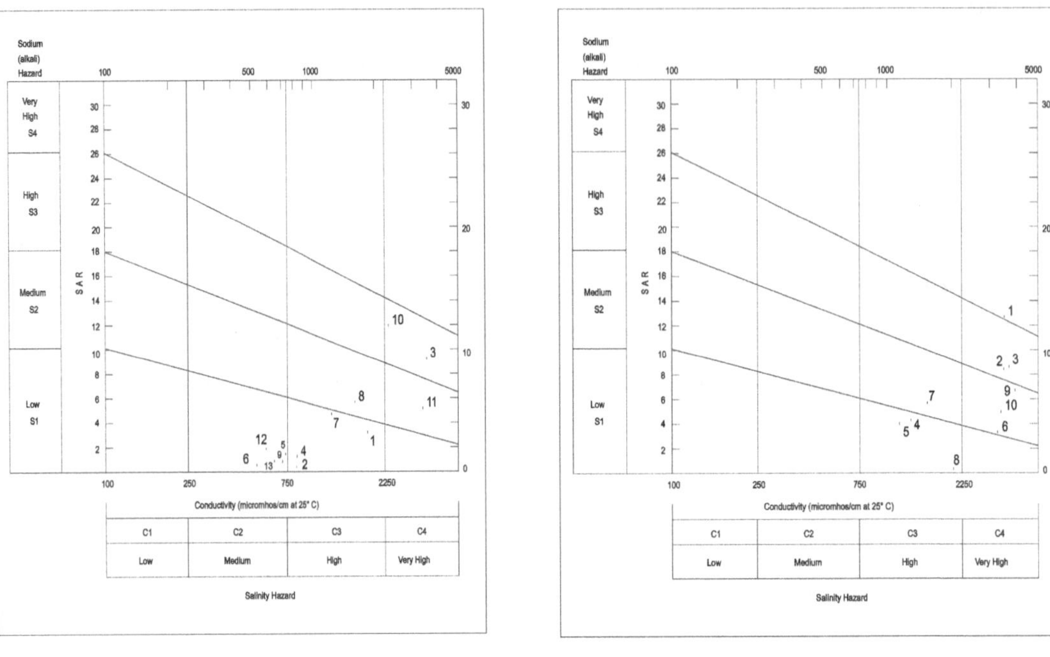

(a) Near surface (b) Deep systems

Fig. 5.47 Sodium Adsorption ratios for near surface (a) and deep systems (b) for Narmada-Tapti delta systems

Residual sodium carbonate (RSC)

The RSC values for the near surface and deep systems for Narmada and Tapti delta systems are presented in Table 5.55 below.

Table 5.55 Residual sodium carbonate values in Narmada-Tapti delta

RSC (epm)	Remark on Quality	No. of Samples	
		Near Surface system	Deep system
< 1.25	Good	9 (–13.2 to 0.53)	9 (–22.00 to 1.24)
1.25 - 2.5	Doubtful	1 (2.18)	1 (1.91)
> 2.5	Unsuitable	3 (4.61 to 5.09)	0

About 69.2% of the near surface and 90% of the deep systems have values less than 1.25 and are of good quality. All the rest of samples fall in the fields of doubtful to unsuitable.

CHAPTER 6

SALINE FRESH WATER INTERRELATIONSHIPS, ESTIMATIONS OF INTERFACE STRUCTURES AND MANAGEMENT/MITIGATION STRATEGIES

Coastal delta groundwater systems along the Indian coast constitute important resource belts for agriculture, irrigation and drinking water needs and play a vital role for the growth and development of the rural base. These systems often are interfaced with oceanic basins on one side and upland regions with major river valley projects on the other end and become targets for subsurface hydrodynamic and hydrogeochemical changes on either side causing saline water ingression inland into the fresh water systems (surface and sub surface). Physico chemical processes accompanying such ingression, aquifer reservoir characteristics and paleo hydrogeological processes in different sub systems of the deltas are the major elements required to be studied and quantified. These would help to develop appropriate optimization schemes, models with predictive capabilities on physico-chemical changes and space time variations and evolve required protective measures for sustained development of fresh water resources. The fresh groundwater resources in the coastal deltas and the inter delta regions over the last two or three decades have been subjected to large scale stresses in the form of groundwater withdrawal resulting in reduction of water levels and fresh water storage potentials causing saline water intrusion and mixing. This situation is being further threatened by global warming processes and sea level rise events, coupled with increasing anthropogenic levels which could trigger the sea water intrusion and contamination processes damaging the situation further. It is therefore vitally necessary to obtain the first hand information on the existing fresh Groundwater systems, their aerial spreads, subsurface boundary conditions and their relationships with adjoining saline water bodies. These are site specific problems and are required to be studied basin wise all along the coast in different delta systems at National level as a part of the National hydrological or Integrated Coastal Zone Management (ICZM) programs.

A major portion of our country's agricultural output is coming from the coastal deltaic regions. The deltaic tracks of the coastal regions mainly receive waters from surface sources (precipitation, rivers and canals) and the non-deltaic areas mainly depend upon Groundwater sources. But of late with the design and construction of major river valley projects at the upstream areas of the delta, there has been substantial reduction in the surface water inputs in the deltaic systems resulting in large withdrawal of Groundwater resources within the delta frame particularly in the tail end areas of the delta there by causing increasing levels of saline water mix up with the fresh water. A quantitative understanding of the patterns of movement and mixing between fresh and saline water and factors that influence this phenomenon are necessary to develop appropriate optimization schemes and at the same time protect this valuable fresh water resources. This calls for a careful and systematic analysis of our coastal systems.

A major factor that determines the movement of fresh water and salt water is the density. The density of typical saline water is of the order of 1.025 g/cm^3 and that of freshwater is 1.00 g/cm^3. Krieger, R.A., et al, (1957) gave the following classification to characterize the fresh-brackish-saline waters.

Fresh water	:	<100 parts per thousands
Slightly saline	:	1-3 parts per thousands
Moderately saline	:	3.-10 parts per thousands
Very saline	:	10-35 parts per thousands
Brine	:	> 35 parts per thousands

The average concentrations of major dissolved constituents of sea water are given in the following Table 6.1.

Table 6.1 Average concentrations of major dissolved constituents of sea water (Hem, J.D., 1989)

Constituent	Concentration (milligrams/liter)
Chloride	19,000
Sodium	10,500
Sulphate	2,700
Magnesium	1,350
Calcium	410
Potassium	390
Bicarbonate	142
Bromide	67
Strontium	8
Silica	6.4
Boron	4.5
Fluoride	1.3

A number of workers have made significant contributions towards understanding the saline-fresh water interrelationships in coastal aquifers since the beginning of 19th century. Ghyben, B.W., (1889) and Herzberg, A., (1901); Muskat, M., (1937); Hubbert, M.K., (1940); Glover, R.E., (1959); Henry, H.R., (1964); Cooper, H.H., et al, (1964); Kohout, F.A., (1964) and Custodio, E., et al, (1987). Some of the recent advances made through numerical analysis (two dimensional and three dimensional) include Shamir, U. and Dagan, G., (1971); Kashef, A.-A.I., (1975 (a,b)); Vappicha, V.N. and Nagaraja, S.H. (1976); Bear, J. and Kapuler, I., (1981); Pinder, G.F. and Cooper, H.H., (1970); Konikow, L.F. and Bredehoeft, J.D., (1978) and Voss, C.I., (1984). Majority of these techniques developed are two dimensional. Infact saline-water intrusion is a three dimensional phenomena. Recently three dimensional density dependent solute transport codes have been developed. Huyakorn, P.S., et al, (1987); Harbaugh, A.W. and McDonald M.G., (1996)., Harbagh, A.W., et al, (2000) and Guo, W., and Langevin, C.D., (2002), which would play a major role in solving problems related to saline water intrusion in coastal systems.

6.1 SALINE–FRESH WATER INTERACTIONS: PRINCIPLES AND PROCESSES

The saline-fresh water interface boundary is a complex and dynamic three dimensional phenomenon and could occur in a variety of combinations and shapes such as (i) Freshwater aquifers either overlying or underlying the saline water aquifers, (ii) Fresh water aquifers sandwitched between two saline water aquifers or vice versa, (iii) Alternate systems of fresh and saline water aquifers and (iv) Fresh water aquifers wedging towards the catchment or upstream areas. All these complex combinations of saline-fresh water aquifers make the well field development strategies rather difficult and demands a very accurate data structure and analyses related to (i) Thicknesses of fresh/saline aquifers and their interface boundaries, (ii) Nature of interface/diffusion boundary, (iii) Wedging of saline/fresh water aquifers and their geometrical variations, (iv) Estimations of depth levels for concrete plugging in the gravel packs in the deep well systems to avoid vertical saline water inflows and mixing, (v) Designing appropriate monitoring networks particularly near the wedge shaped aquiclude boundaries where saline and fresh water systems come into juxta positions and vi) Assessment of possible vertical leakages through aquitards from saline water aquifers to fresh water aquifers and vice versa. Several hydrogeological, surface and subsurface geophysical and geochemical tools are available to study and assess the interface configurations and their dynamics (Jones, P.H., 1951; Radhakrishna, I., et al 1990; Radhakrishna, I., 2001). Paul M. Barlow, (2003) reviewed some of the basic scientific principles and management issues related to the flow of fresh water and salt water in groundwater systems along the Atlantic coastal zone. Some of the basic principles, concepts and processes confronting saline-fresh water interactions in coastal delta systems are described below in the following paragraphs.

6.1.1 Ghyben - Herzberg concept

The coexistence of fresh and saline water was first reported by Ghyben, B.W., (1889) and Herzberg, A., (1901) while working in the German coastal regions. They estimated that the thicknesses of the fresh water columns near the coastal margins will be nearly forty times the height of water table from the mean sea level and they have attributed this phenomenon to the density differences between fresh and saline water and gave the following formula based on density fractionation near the shoreline boundaries for isotropic and steady state aquifer situations as shown in Fig. 6.1 and the related expressions are given in equations 6.1 and 6.2.

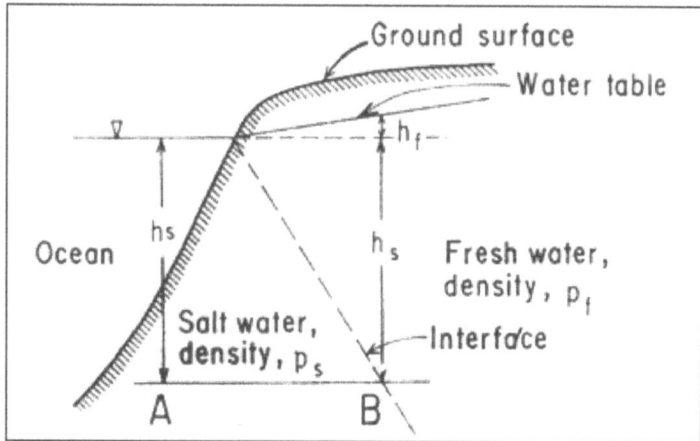

Fig. 6.1 Fresh and Groundwater relationship in isotropic unconfined coastal aquifer

(Source: Todd, D.K., 1980)

The total hydrostatic pressures at A and B respectively are:

$$P_A = \rho_s g h_s \text{ and } P_B = \rho_f + g h_f + \rho_f g h_s \qquad .....(6.1)$$

$$h_s = \frac{\rho_s}{\rho_s - \rho_f} h_f \qquad .....(6.2)$$

where h_s = depth below mean sea level where interface occurs

ρ_f = density of fresh water

ρ_s = density of saline water

h_f = elevation of the Groundwater table above mean sea level

g = acceleration due to gravity

This expression is popularly known as Ghyben-Herzberg relation. In typical sea water conditions, $\rho_S = 1.025\ g/cm^3$ and $\rho_f = 1.00\ g/cm^3$, so that $h_s = 40\ h_f$.

There could be some serious limitations to the Ghyben - Herzberg concept (1901). (i) If the static conditions were to exist in a natural domain, the slope of the water table would be 'zero' and the fresh water will rest upon salt water because of the density fractionation. (ii) The fresh water will be certainly in a state of dynamic motion with thicknesses varying and some amount of fresh water discharging into ocean. (iii) The saline-fresh water interface represents a state of dynamic equilibrium, with these two fluids mixing with a zone of dispersion, but not as a sharp boundary.

6.1.2 Cooper's concept

Cooper, H.H., et al (1964) improved the above concept and considered that in a relatively homogeneous porous media, the denser saline water tends to remain separated from the overlying fresh water with a *"zone of diffusion" or "dispersion"* as reflected in Fig. 6.2. This zone of diffusion was found to vary in thickness probably depending upon the porosity of the material and gradients of water levels. Cooper also observed through several field investigations that the zone of diffusion is associated with circulation of sea water and subsequently confirmed by Kohout, F.A., (1965).

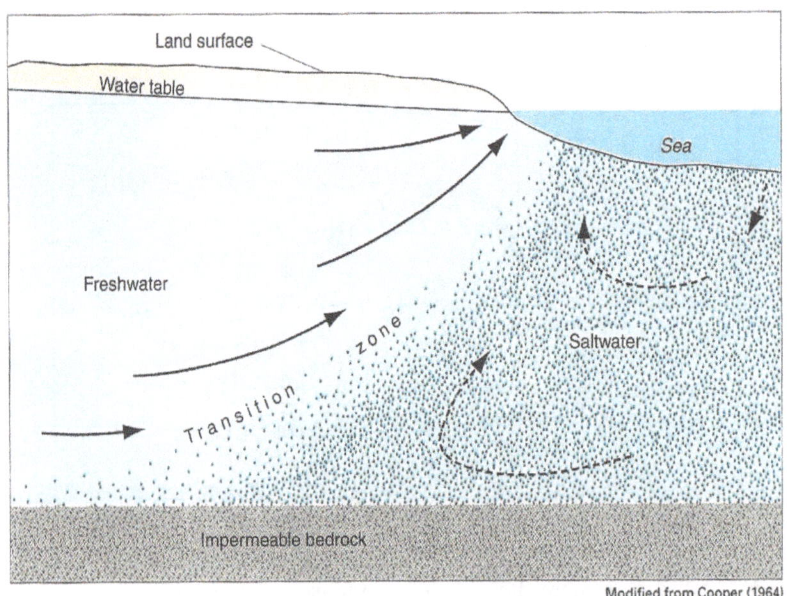

Fig. 6.2 Saline-fresh water transition zone in an idealized coastal aquifer showing the flow patterns

The transition is characterized by dissolved solids varying between 1000 mg/l (lower concentration limit) to 35,000 mg/l (upper concentration limit) and of chloride levels varying between 250 mg/l (lower limit) and 19,000 mg/l (upper limit). The trends of these chloride levels are well studied in a classical coastal aquifer system mainly Biscayne aquifer system near Miami in Florida (U.S.A) which is shown in Fig. 6.3.

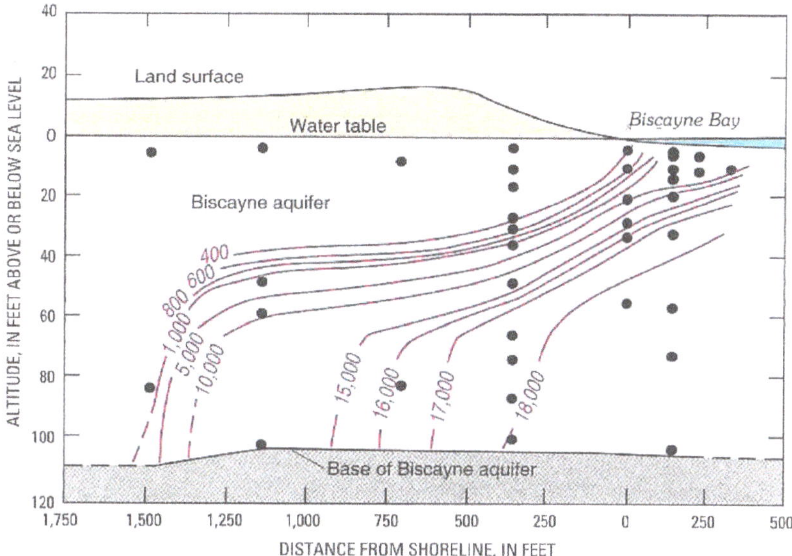

Fig. 6.3 Fresh-saline water transition zone in the Biscayne aquifer near Miami, florida (U.S.A) with chloride concentration levels

(Source: kohout,F.A.., 1964)

The fresh water flowing towards ocean side normally gets mixed up with the saline water by a process of dispersion and molecular diffusion (Todd, D.K and Huisman, L., (1959); De Josselin de Jong, G., (1959); Bear, J and Todd, D.K., (1960) and Scheidegger, A.E., (1960)). Bear, J. and Dagan, G. (1962, 1963, 1964 a,b), gave insight into sharp interface/dispersion and its dynamics. Henry, H.R. (1959, 1962 and 1964) developed solutions related to sharp interface under various conditions and quantified the effects of dispersion and density dependent fluid flow and saline water encroachment in coastal systems, using the advection-diffusion equation. This zone of dispersion will have large variabilities depending upon the hydraulic properties of the aquifers involved near the coastal boundaries, the flow path gradients, tidal ranges and wave action intensities and other related dynamic forces. These dynamic forces cause the fresh water and salt water zones to move seaward and landward in different times. The thicknesses of these transition zones could vary from a few meters to an order of about 300 meters depending upon the aquifer thicknesses, their depth variations and their connectivities to the ocean boundaries.

The same phenomena could exist in several coastal regions with multi layered systems as shown in Fig. 6.4. Each aquifer may have its own interface depending upon the aquifer gradients, vertical flows through aquitards and other hydrodynamic situations. The shallow coastal systems in general will have its interface more towards the land area and the deeper systems will have their interfaces relatively away from the shore line. The layers separating these different aquifers could be aquicludes or aquitards. This characteristic situation is noticed in multi aquifer systems in most parts of the deltaic regions of India. The sediments of the coastal margins could have been deposited either in fluviatile or in marine situations. The sea level fluctuations during Pleistocene times could have inundated in the past with saline water which are now stand exposed or underneath in the continental side. Majority of these aquifers slope towards the sea and could have been subjected to flushing due to several recharge cycles over a period of time and have eventually become repositories for fresh water. This is how fresh water is found beneath the saline water zones in many deltaic regions of Indian coast.

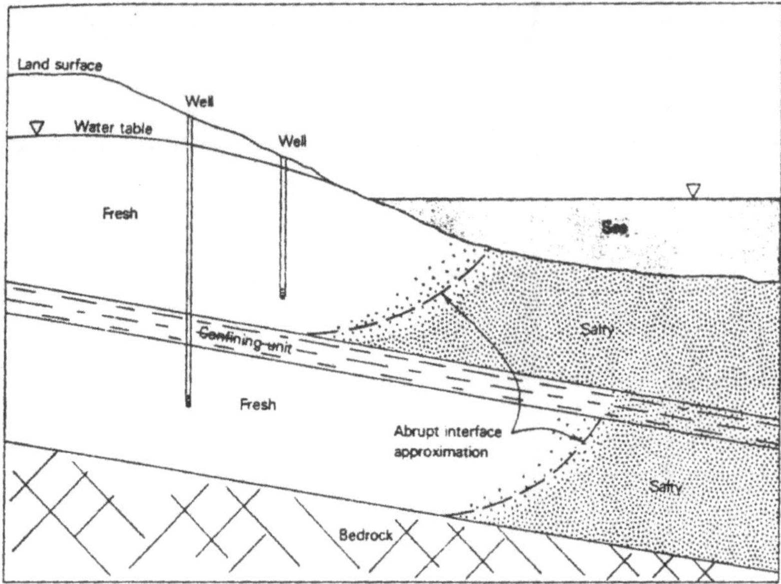

Fig. 6.4 Diffusion/dispersion zone in multi aquifer systems

6.1.3 Hubbert's concept

Hubbert, M.K. (1940) further improved upon the G-H relation by taking into consideration the pressure heads in each of the fluid systems. The governing equation could be derived by defining the potential heads in each of the fluids.

$$h_f = \frac{p}{\rho_f g} + Z \quad \text{and } h_s = \frac{p}{\rho_s g} + Z \qquad .....(6.3)$$

where Z is the elevation above datum; P is fluid pressure and g is the acceleration due to gravity.

Equating these two expressions gives:

$$Z = \frac{\rho_f}{(\rho_f - \rho_s)} h_f - \frac{\rho_s}{(\rho_f - \rho_s)} h_s \qquad .....(6.4)$$

where Z is the vertical location above datum of the point on the interface

ρ_s, ρ_f are densities of saline and fresh water and

h_f *and* h_s are fresh and saline water head levels.

The above formula defines the position of the interface under equilibrium conditions as reflected in Fig. 6.5 (A &B). The left one represents salt water as stationary and the right one represents both fluids in motion. The above formula is considered to be basic and at the same time a step forward from G-H relationship. Muskat, M., (1937) and Hubbert, M.K., (1940) had accounted for the dynamics of the fluid systems and considered that the movement and potential distribution of fluids play an important role in determining the interface geometry.

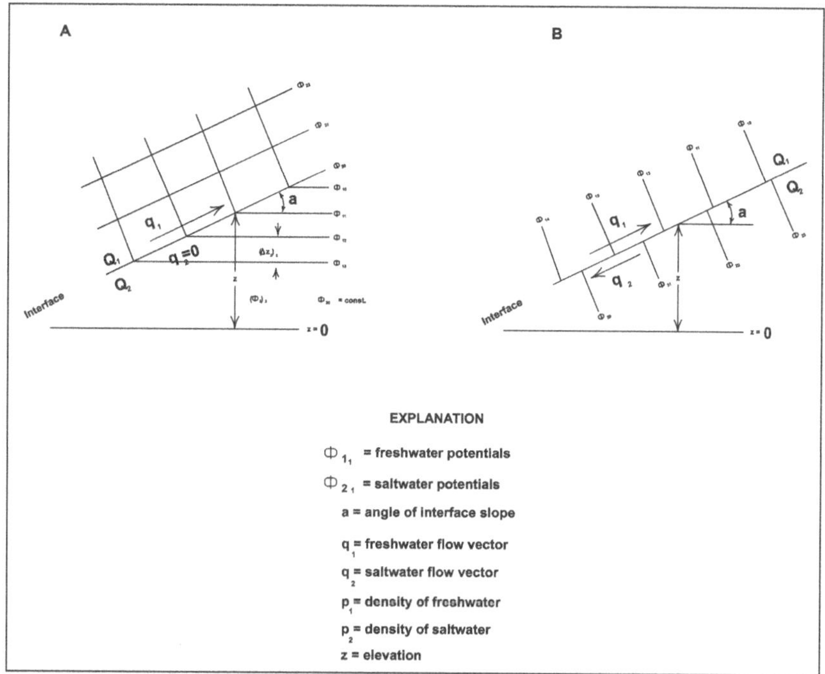

Fig. 6.5 Equipotential surfaces of both fluids and fluid interfaces (modified from Hubert, 1940) (A) When one fluid flows while the other is static and (B) when both fluids are moving

Source: T.E. Reilly and A.S. Goodman (1985)

6.1.4 Glover's concept

Glover (1959) developed a formula to describe the salt water-fresh water interface in a coastal aquifer that accounts for the movement and discharge of fresh water as reflected in Fig. 6.6.

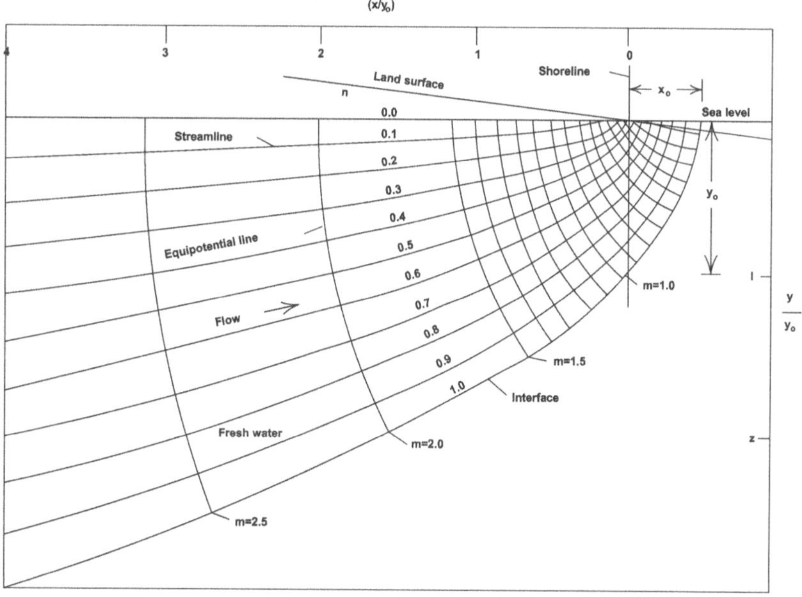

Fig. 6.6 Flow patterns near a coastal zone (Glover, R,E., 1959)

The interface between the fresh and saline water could be obtained from the following equation:

$$\gamma^2 - \frac{2Q}{\gamma k} X - \frac{Q^2}{\gamma^2 k^2} = 0 \qquad .....(6.5)$$

where Q is the fresh water flow per unit length of shore (L^2T^{-1})

K is the hydraulic conductivity of the medium (LT^{-1})

$\gamma = (\rho_s - \rho_f)/\rho_f$ (dimensionless)

ρ_s is the density of salt water (ML^{-3})

ρ_f is the density of fresh water (ML^{-3})

X is the distance from the shore line (L)

This differs from the G-H relation from the fact that it allows the discharge of the shore and balances pressures across the interface.

6.1.5 Tidal impacts near coastal margins

The coastal margins in general contains several tidal inlets and estuaries. The salt water advances upward in the mouth region during the periods of high tides, and injects relatively high density saline ions underground, replaces the fresh water gradually and builds up a saline water mound at the base of the aquifer. These processes over a period of time displaces all the fresh water and builds up a salt water mound. This phenomenon is illustrated below in Fig. 6.7 as demonstrated from the Miami coastal region of Florida. (The Year Book of Agriculture, U.S. Dept of Agriculture, 1955).

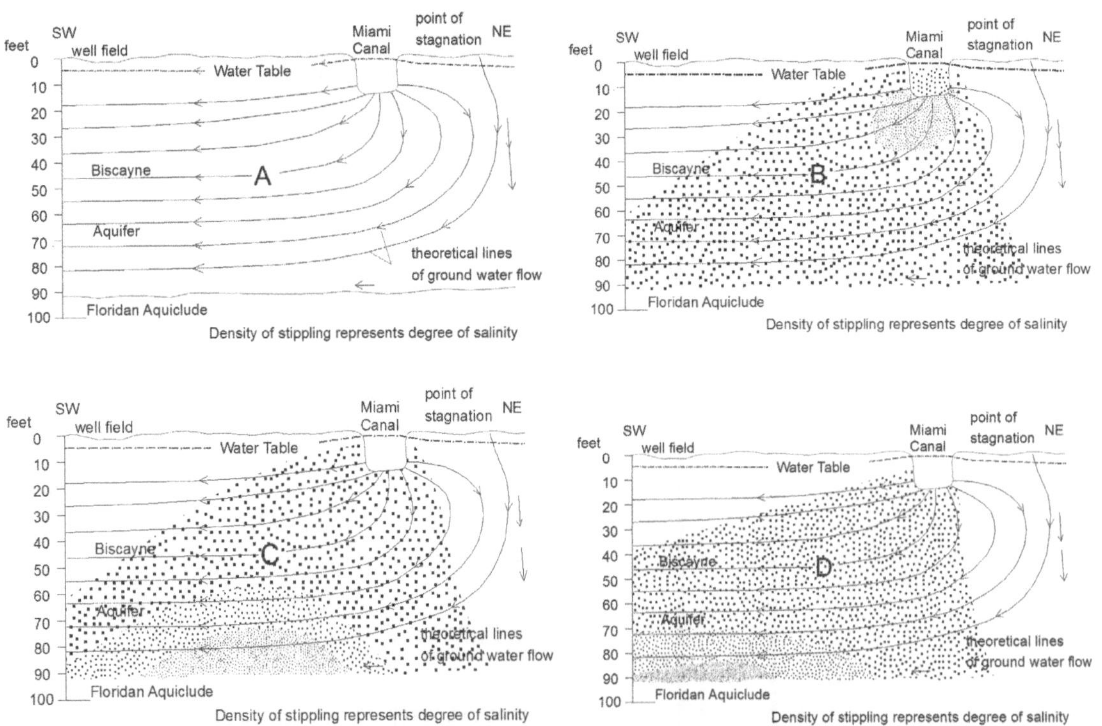

Fig. 6.7 Tidal impacts near coastal margins

(Source: The Yearbook of Agriculture, U.S. Dept. of Agriculture, 1955)

6.2 SALINE-FRESH WATER INTERFACE DYNAMICS – SOME THEORETICAL CONSIDERATIONS

On the basis of Ghyben-Herzberg relation a number of approximate formulae for the computation of the position of interface under a given equilibrium conditions have been developed (Santing, G., 1963). The approximate formulae are based on the following assumptions: (a) the aquifer is homogenous; (b) sharp interface; (c) the flow is one dimensional (without considering the vertical flow component).

6.2.1 Phreatic groundwater replenished by rain

The groundwater flow recharged uniformly by rainfall as shown in the Fig. 6.8 is expressed by the following equation:

$$q_0 + Wx = -K\left(h + \frac{\rho_f}{\rho_s - \rho_f}h\right)\frac{dh}{dx} \quad .....(6.6)$$

where q_0 is the flow at x = 0; W = recharge by rainfall.

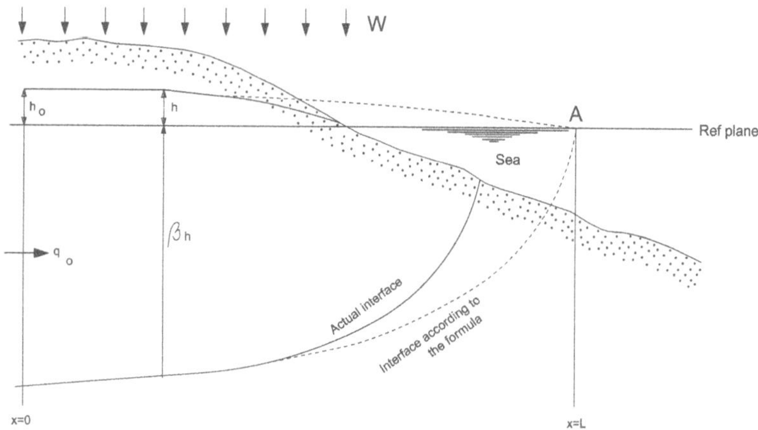

Fig. 6.8 Phreatic groundwater replenished by rain

By integrating equation (6.6) with the boundary conditions x = 0, q = q_0 and h = h_0 we get

$$h_0^2 - h^2 = \frac{2q_0x + Wx^2}{K\left(1 + \frac{\rho_f}{\rho_s - \rho_f}\right)} \quad .....(6.7)$$

which gives the relation between h and x.

For x = L, the theoretical length of the fresh water body, h = 0. Therefore the equation (6.7) gets reduced to

$$h_0^2\left(1 + \frac{\rho_f}{\rho_s - \rho_f}\right) = \frac{2q_0L}{K} + \frac{WL^2}{K} \quad .....(6.8)$$

If W and h_0 are known, and by taking L the distance from the point x = 0 to the coast, the discharge q_0 can be computed.

6.2.2 Oceanic islands

The depth of the interface in the oceanic islands could be determined by the shape and size of the island, permeability of the formations and the recharge rate as reflected in the Fig. 6.9.

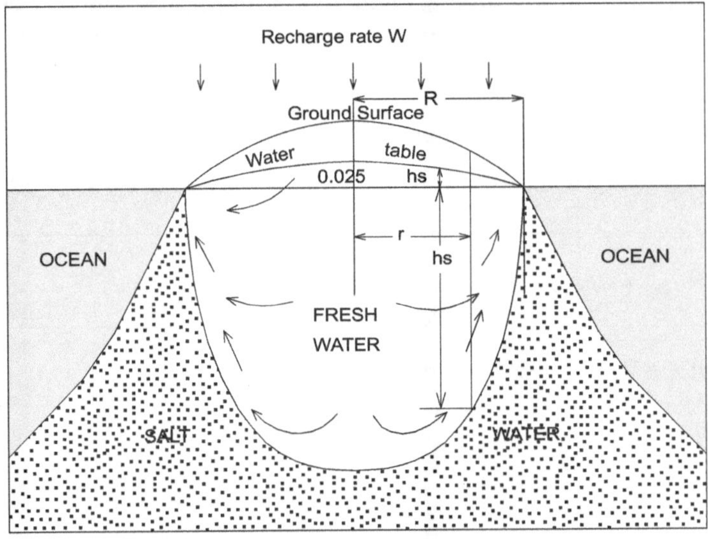

Fig. 6.9 Fresh water lens in an oceanic island under natural conditions

Based on the Dupuit assumption and the Ghyben-Herzberg relation an approximate interface depth can be determined for boundary conditions $h_s = 0$ when $r = R$

$$h_s^2 = \frac{W}{2K\left(\frac{\rho_s - \rho_f}{\rho_f}\right)\left(\frac{\rho_s}{\rho_f}\right)}\left(R^2 - r^2\right) \qquad \text{.....(6.9)}$$

$$= \frac{\rho_f^2 W}{2K\rho_s\left(\rho_s - \rho_f\right)}\left(R^2 - r^2\right) \qquad \text{.....(6.10)}$$

where h_s= depth of the interface below mean sea level, at radius r

R = radius of circular island

r = radius at a point where the depth of the interface is h_s

W = effective rate of recharge from rainfall

For normal densities of fresh and sea water (ρ_f=1.0 and ρ_s=1.025 g/cm^3), this reduces to

$$h_s^2 = \frac{W}{0.0512K}\left(R^2 - r^2\right) \qquad \text{.....(6.11)}$$

6.2.3 Semi confined aquifer system replenished by rain

The flow through a semi confined aquifer overlained by a poor conductivity layer as shown in the Fig. 6.10 could be expressed by the Darcy's equation as below.

$$q_0 + W_x = -Kb\frac{dh}{dx} = -\frac{Kb}{\rho_f/\left(\rho_s - \rho_f\right)}\cdot\frac{db}{dx} \qquad \text{.....(6.12)}$$

where W is the replenishment of groundwater by rain

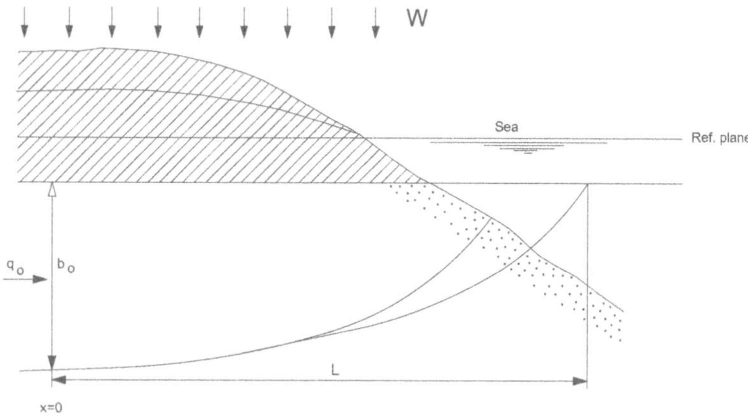

Fig. 6.10 Semi confined aquifer system replenished by rain

By integration and substitution of the boundary condition (x = 0, b = b_0) we get

$$\frac{1}{\rho_f/(\rho_s-\rho_f)}(b_0^{\ 2}-b^2)=\frac{2q_0x}{K}+\frac{2Wx^2}{K} \qquad .....(6.13)$$

For x = L, b = 0, we get

$$\frac{1}{\rho_f/(\rho_s-\rho_f)}b_0^{\ 2}=\frac{2q_0L}{K}+\frac{2WL^2}{K} \qquad .....(6.14)$$

The formula reduces to

$$\frac{1}{\rho_f/(\rho_s-\rho_f)}b_0^{\ 2}=\frac{2q_0L}{K} \qquad .....(6.15)$$

6.2.4 Confined aquifers

The flow through the confined aquifers as shown in the Fig. 6.11 could be expressed using the Darcy's equation as shown below:

$$q_0=-Kb\frac{dh}{dx}=-\frac{Kb}{\rho_f/(\rho_s-\rho_f)}.\frac{db}{dx} \qquad .....(6.16)$$

By integrating and substitution of the boundary conditions (for x= 0, b = b_0); we get

$$\frac{b_0^{\ 2}-b^2}{\rho_f/(\rho_s-\rho_f)}=\frac{2q_0x}{K} \qquad .....(6.17)$$

For x = L, the theoretical length of fresh water body, h = 0; hence

$$\frac{1}{\rho_f/(\rho_s-\rho_f)}b_0^{\ 2}=\frac{2q_0L}{K} \qquad .....(6.18)$$

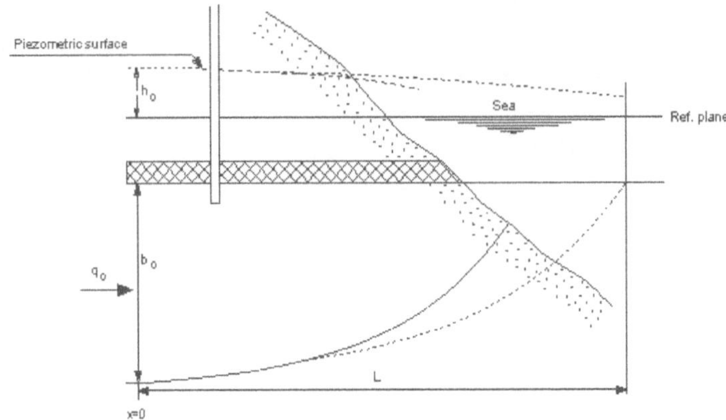

Fig. 6.11 Confined aquifers

6.3 GROUNDWATER ABSTRACTION SCENARIOS IN PHREATIC AND CONFINED AQUIFERS AND THEIR IMPACTS ON INTERFACE

Groundwater abstraction often changes the balance of equilibrium between the fresh water flow and the interface disposition. The position of interface can be computed based on the outflow.

6.3.1 Phreatic aquifer

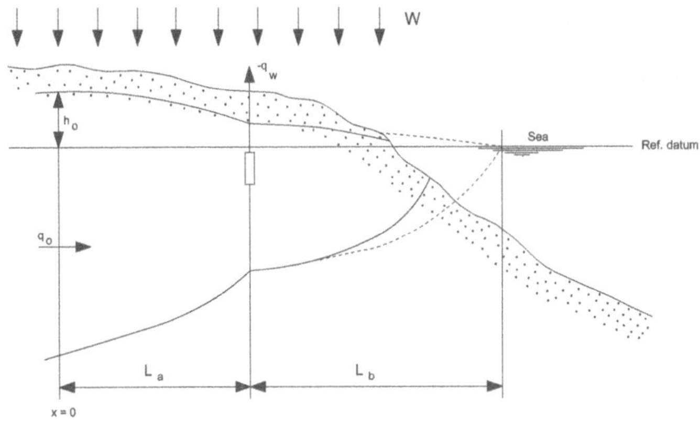

Fig. 6.12 Groundwater abstraction from a phreatic aquifer

By applying formula (6.7) to the zone La and formula (6.8) to the zone Lb, the following relation can be obtained:

$$\left(1+\frac{\rho_f}{\rho_s-\rho_f}\right)h_0^{\ 2}=\frac{2q_s(L_a+L_b)}{K}+\frac{W(L_a+L_b)^2}{K}+\frac{2q_wL_b}{K} \quad .....(6.19)$$

where q_w is the yield of the gallery or the series of wells per unit length of the aquifer

6.3.2 Confined aquifers

The abstraction situation in a confined aquifer is depicted below in Fig. 6.13.

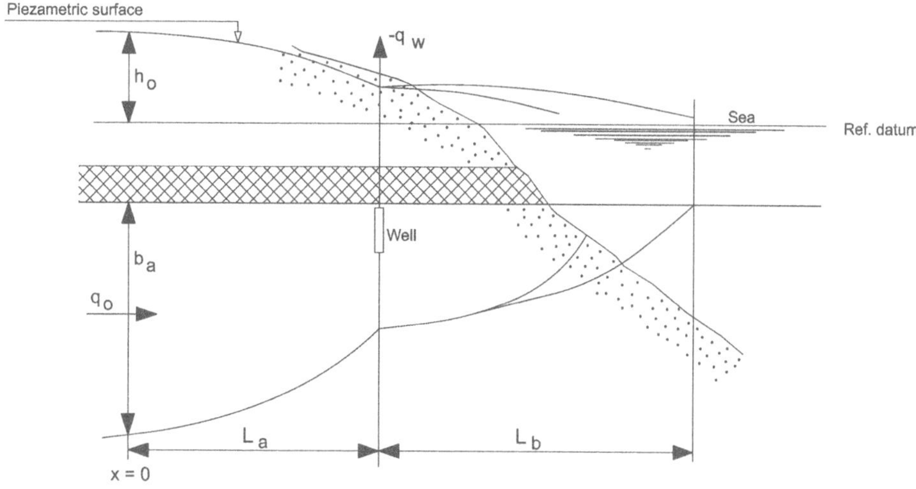

Fig. 6.13 Groundwater abstraction from a confined aquifer

By applying equation (6.17) to the La and equation (6.18) to the zone Lb the following relation can be obtained.

$$\frac{1}{\rho_f/(\rho_s-\rho_f)}=\frac{2q_0(L_a+L_b)}{K}+\frac{2q_wL_b}{K} \quad .....(6.20)$$

where q_0 is the yield of the gallery and the series of wells per unit length.

6.3.3 Interface upconing

Several fresh water well fields in coastal regions often overlie saline water aquifers. The fresh water pumped in such cases generate a local rise of interface beneath the well structure and this phenomenon is called "*upconing*" and is shown in Fig. 6.14.

The interface in such situation is considered to be horizontal at the start of pumping when $t = t_o$. The interface after certain amount of pumping time rises gradually and eventually will reach the bottom of the well and subsequently mixes with the fresh water being pumped in the well causing damage to the agricultural fields. This situation needs to be checked and optimization of pumping is required to be implemented without causing saline water mixing. This phenomenon originally was recognized by Muskat, M. and Wyckoff, R.D., (1935). Upconing is a complex phenomenon and is related to the type of well structure, pumping rate, nature of interface/diffusion boundaries and the aquifer properties such as porosity and permeability etc. Most investigators (Muskat, M and Wyckoff, R.D., (1935); Dagan, G. and Bear, J., (1968) and Haubold, R.G., (1975)) assume an abrupt interface between these two fluids and neglects the reality of the situation that the interface occurs as a transition zone.

An approximate solution for the upconing phenomena based on the Dupuit assumption and the Ghyben – Herzberg relation is given by Schmorak, S. and Mercado, A., (1969).

$$z=\frac{Q}{2\pi dk(\Delta p/\rho f)} \quad .....(6.21)$$

where $\Delta p = \rho_s - \rho_f$ and k is the hydraulic conductivity

Q is the pumping rate

z is the critical rise of saline water.

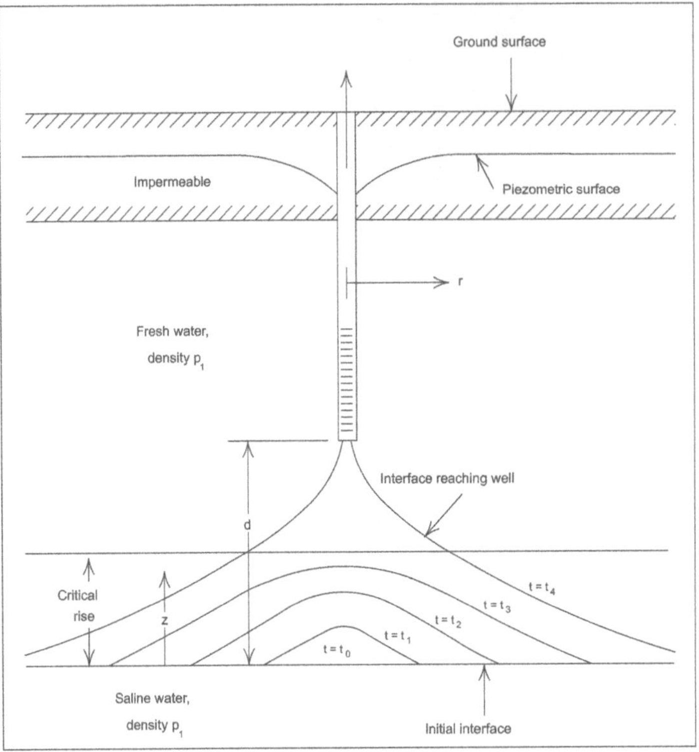

Fig. 6.14 Upconing phenomenon in coastal aquifers

(Source: Schmorak, S. and Mercado, A., 1969)

The above equation gives an ultimate rise of interface to a new equilibrium position i.e. directly proportional to the pumping rate Q. If the upconing exceeds a critical rise, the saline water accelerates towards the well and mixing with the fresh water takes place. The critical rise has been estimated as z/d = 0.3 to 0.5 Assuming an upper limit of z/d = 0.5, it follows a maximum pumping rate without saline water entering into the well as:

$$Q_{max} \leq \pi d^2 k(\Delta\rho / \rho_f) \qquad .....(6.22)$$

The saline/fresh water interface migrations inland and fresh water discharge into oceanic basin and their consequences are some of the critical factors need to be ascertained and quantified while formulating policies and programmes in coastal areas.

6.4 TECHNIQUES/METHODOLOGIES FOR IDENTIFICATION OF INTERFACE DEPTH PROFILES IN COASTAL SYSTEMS

Most coastal systems have multilayered aquifer configurations with variable interface depths and with different water quality configurations (Radhakrishna, I., et al, 1990). Though the interface depths for near surface(unconfined systems) could be theoretically estimated/predicted using G-H relation, the interface depth estimations for deep systems are not exactly possible because of changing hydrodynamics and anisotropic geological configurations and one has to adopt several hydrogeological, hydrochemical and geophysical approaches including shallow and deep drilling and borehole geophysical approaches. A brief description of the hydrogeological and geophysical approaches which are in vogue to understand and estimate the water yields/qualities/interface depths etc are briefed below.

6.4.1 Hydrogeological and hydrogeochemical approaches

The basic hydrogeological and hydrogeochemical data in a basin could be obtained through rigorous well inventory methods and when the data obtained are projected on a topographical/geological map would serve as an important base map indicating the nature of aquifers, the distributional patterns, yield and water quality characters. The well inventory includes (i) terrain mapping using normal geological techniques coupled with remote sensing techniques such as satellite images and high resolution satellite pictures, (ii) data related to different well structures like dug wells, dug com borewells, filter points, shallow and deep tubewells, (iii) depth variations of these well structures, water levels in each of the well structures and estimating the nature of aquifers such as unconfined, semiconfined and confined etc, (iv) yield and water quality characteristics in each of the well structures. Water quality characteristics include pH, conductivity levels, cations (Ca, Mg, Na, K) and anions (HCO_3, CO_3, Cl, SO_4). These parameters wholly or partially could be carried out in the field itself by setting up temporary mobile field laboratory. Similarly yield tests are done in the field either in the dugwells or in the borewells/tubewells. In general yield tests conducted in dugwells are incorrect due to storage interferences. However one can establish the appropriate yields through the recovery test procedures. Yield tests conducted in borewells/tubewells would yield by and large correct aquifer discharges against specific draw downs and also help to estimate the parameters like Storativity (S) and Transmissivity (T) and (v) reduction of water levels to mean sea levels. All these features when incorporated on a base topo map would serve as an important hydrogeological and hydrogeochemical base map reflecting the regional aquifer distribution trends, flow and water quality situations, recharge and discharge boundaries etc. At times when required data is sufficiently not authentic and the data gaps are sufficiently large it becomes necessary to go for additional wells either shallow or deep to obtain authentic information on the aquifer systems, distributional patterns and yield and water quality characteristics. Such wells could be categorized as (i) dugwells/dug com borewells, (ii) borewells or augur wells, (iii) driven wells (iv) jetted wells (v) shallow and deep drilled wells.

- ***Dugwells:*** Dugwells are large diameter wells varying in dia from 1m to 10m upto depths varying from 5m to 20m or sometimes beyond depending upon the depth of occurrence of the first unconfined or semi confined aquifers. These wells in alluvial coastal systems are normally lined with either bricks and cement or with concrete rings, and are exposed or unlined at the saturated regions of the aquifers. These wells facilitate to supply large amounts of water because of the storage space for water available in the well structure. These wells are normally available in plenty in the alluvial coastal regions or in the hard rock regions in the vicinity of the coast, and provide first hand information on the quantity and quality of the near surface aquifer systems.
- ***Borewells/Auger wells:*** Borewells/Auger wells are normally constructed in the coastal alluvial regions. These are small dia holes operated through hand augers and could be drilled upto depths of 30m or 60m. These wells are done in the regions where the aquifer depths are approximately known and could be fitted with slot and blank pipes depending upon the depth of aquifers encountered. Sometimes a cavity is allowed to be formed if the aquifer material is loose and the sand is bailed out through a compressor. These wells are also available in plenty and provide useful information related to quantity and quality of the near surface systems. At times fresh bore holes could be planned to acquire more authentic information on water levels, yields, water qualities etc and these are cost effective and relatively cheap to operate.
- ***Driven wells:*** Driven wells are done by way of driving a sharp pointed strainer and sequence of pipes attached to it successively as the drilling proceeds and stopped when the strainer encounters the required aquifer materials. These are also done upto depths varying from 20m to 40m. This type of wells also are available in the coastal alluvial regions and provide useful information

related to quantity/quality of the near surface systems. These are also cost effective methods and could be planned similarly as those of borewells and augerwells and gives accurate information.

- ***Jetted wells:*** A water jet sometimes through a drill bit is used to drill small dia holes upto depths varying from 20m to 40m. Sometimes percussion and rotary action could be imparted while drilling. An outer casing pipe is normally provided to avoid collapse of the bore hole and to facilitate drilled material to come out. The tools could be removed subsequently and an appropriate blank/slotted pipes could be installed depending upon the aquifer material encountered or sometimes a cavity may also be made at the bottom. A number of water supply systems could be constructed this way in the alluvial coastal regions and serve to obtain important input information related to quantity and quality of the near surface aquifer systems or sometimes moderately deep systems also. Designing this type of structures are bit costly and should be done sparingly where ever required depending upon the needs.
- ***Drilled wells:*** Moderate/deep drilling could be made through either percussion (cable tool) or rotary drilling methods. These are all widely used techniques where moderate to deep wells are required to be constructed for heavy duty operations. Depths ranging between 100m to 400m could be done easily through these techniques and a wealth of hydrogeological information could be obtained related to the availability of aquifers, yield and water quality characters at different depth levels. *Percussion* drilling is normally done by raising and dropping a sharp edged drilling bit attached to a string of tools suspended from the cable. By means of percussion action the loose or semi hard material could be crushed and broken and brought back to the surface either through a bailer or by using a sand pump. To prevent caving the borewell while drilling should be protected by casing pipes which could be increased stage by stage while drilling progresses. This method of drilling is very fast and widely used in loose or semi consolidated formations and reliable information related to water table depths, permeability of the formation yields and water qualities etc could be obtained at different depth levels. *The hydraulic rotary* drilling is widely used technique in several parts of the country for exploration and developmental purposes. The excavation of material is normally done through rotary action of the drill bit connected to a string of drill rods. A mud pump is connected through a kelley which facilitates mud circulation and removal of the drilled material on to the surface. Bentonite mud is normally used for this purpose which acts as good adhesive and protects the wall from caving etc. It would be necessary to monitor the density of the mud particularly while drilling through high pressured artesian aquifer systems and in the saline affected coastal regions and adopt necessary changes required in the mud density to keep the bore hole free from collapse. Monitoring mud resistivity while drilling help to detect changes in water qualities of aquifers. Normally mud resistivity > 6.00 ohm-meter is an indication that the aquifers are piercing through fresh water aquifer systems. The mud resistivity in the range of 3 to 6 ohm-meters indicate occurrences of brackish water zones and less than 3 ohm-meters reflect on the saline water zones. A mud resistivity profile while drilling help to assess the nature of the quality of the aquifer systems and build up approximate interface profile. Designing these structures are costly and are normally done through governmental agencies. However the data obtained through this method is highly reliable and accurate.
- ***Packer yield test (PYT):*** Packer yield tests for specific zones at different depths could be performed particularly in the coastal alluvial regions where saline-fresh water aquifers occur in complex combinations. These tests are feasible through direct rotary drilling – making a pilot bore hole to the required depth and subsequently by way of reaming the pilot bore holes at different stages of depth and fixing packers against the aquicludes adjoining the aquifers to be tested. A compressor test run could be made in each of the aquifer zones isolated through packers

to assess the approximate yields and also determine the aquifer properties of different aquifers. This basic data would facilitate to obtain reliable yield/water quality profiles for different aquifer zones, though the operational costs will be bit high.

- ***Reverse rotary drilling:*** Sometimes reverse rotary drilling is adopted where water is used instead of mud for circulation and to bring the material on to the surface. The water is sent underground through the annual space between drill rods and casing and the material is drawn back through the drill bit/rods on to the surface. Both direct rotary and reverse rotary techniques are widely used for exploration and developmental purposes and facilitates to obtain a wealth of information related to water levels, pressure heads of the individual aquifers, yield and water quality characters for different layers etc and help to prepare a detailed three dimensional hydrogeological, hydrochemical and saline –fresh water interface depth configuration maps.

6.4.2 Surface geophysical approaches

Geophysical methods/approaches measure the physical properties of the earth materials and can help in a great way to delineate the aquifer/aquiclude distribution systems and sometimes reflect on the aquifer qualities also in different hydrogeological basins. Although there are several variety of geophysical techniques available like electrical, electromagnetic, seismic and gravimetric etc, the electrical, electromagnetic and sometimes seismic methods are particularly useful in coastal systems (Murali Sabnavis, and Patangay, N.S., 2006). A brief description of the techniques with some case studies related to its ability to resolve saline-fresh water boundaries in coastal alluvial systems are outlined below.

Electical resistivity method: The method in brief involves generation of current of low frequency (less than one cycle per second) introduced into the ground through two metal electrodes and voltage measured through other two potential electrodes. By measuring the current flowing through the ground and voltage potentials obtained through other two electrodes and displacing the electrodes with different configurations, it is possible to compute the resistivity of the subsurface material (Zohdy, A.A.R., et al, 1974). The electrical resistivity of the material 'R' is expressed as:

$$R = \frac{A}{L} \times \frac{\Delta V}{I} \qquad \text{.....(6.23)}$$

where A is the cross sectional area of the current flow

L is the length of the flow path

V is the voltage drop

I is the electrical current and R is the electrical resistivity.

If A and B are the positive and negative current electrodes and M and N are the potential electrodes, the resistivity (apparent) of the given material is expressed as:

$$R = \left(\frac{2\pi}{\frac{1}{AM} - \frac{1}{BM} - \frac{1}{AN} - \frac{1}{BN}} \right) \frac{\Delta V}{I} \qquad \text{.....(6.24)}$$

There are several electrode configurations and the commonly deployed are (i) Wenner array (ii) Schlumberger array and (iii) the Dipole-Dipole array .

(i) The Wenner array configuration consists of four electrodes spaced at equal distances oriented in a straight line. $\overline{AM} = \overline{MN} = \overline{NB} = a$ (inter electrode separation) as depicted in the following Fig. 6.15.

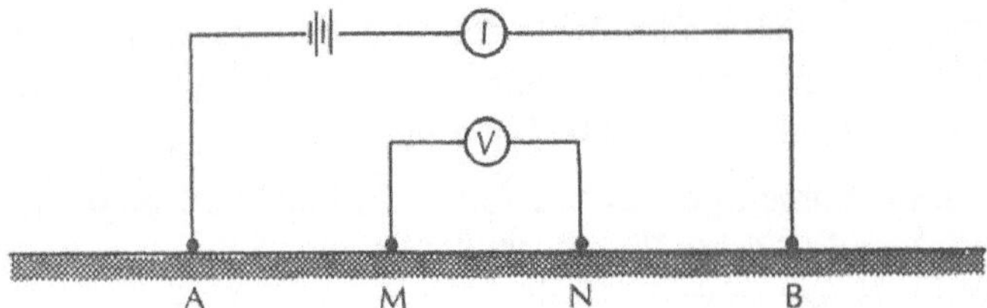

Fig. 6.15 Wenner configuration of electrodes

The apparent resistivity in the wenner configuration is expressed as:

$$\overline{R} = 2\pi a \frac{\Delta V}{I} \qquad .....(6.25)$$

Which could be solved from equation (6.24).

(ii) The Schlumberger array is also a linear array but with potential electrodes placed closely to the sounding point, as shown in Fig. 6.16.

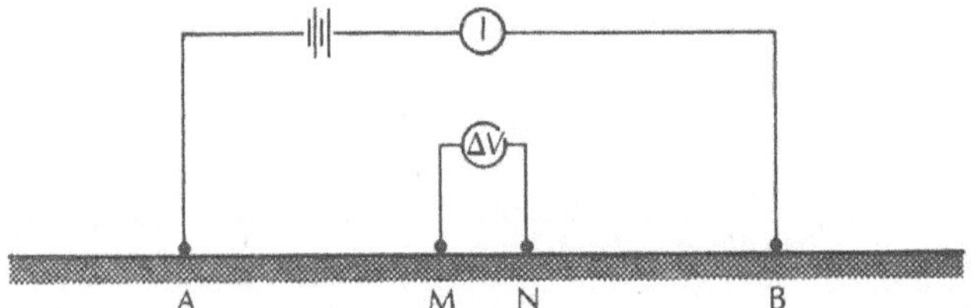

Fig. 6.16 Schlumberger configuration of electrodes

The electric current is sent through A and B electrodes and the potentials are measured through inner electrodes M and N. The distance between A and B is set to equal or greater than five times the value of $\overline{MN}$. The apparent resistivity is given by:

$$\overline{R} = \pi \frac{\left(\frac{\overline{AB}}{2}\right)^2 - \left(\frac{\overline{MN}}{2}\right)^2}{\overline{MN}} \frac{\Delta V}{I} \qquad .....(6.26)$$

(iii) The Dipole-Dipole array configuration has the current electrode pair separated from the potential electrode pair as shown in Fig. 6.17 below.

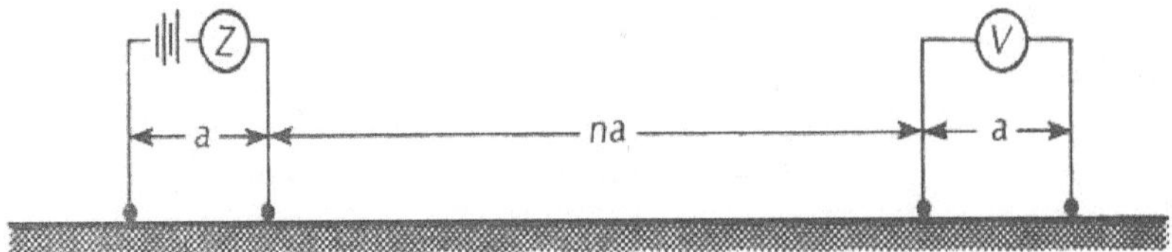

Fig. 6.17 Dipole-Dipole configuration of electrodes

The space between the current electrodes is set to be equal to the potential electrodes and the distance between electrode pairs 'na' which is a multiple of n and a. The apparent resistivity of the dipole-dipole array is given by:

$$\overline{R} = n(n+1)(n+2)a\frac{\Delta V}{I} \qquad .....(6.27)$$

The equipment required in the field to carry out the resistivity surveys are (i) a resistivity meter (ii) DC current source (iii) insulated cables (iv) pair of current electrodes(iron stakes) and (v) a pair of potential electrodes(porous pots with saturated copper sulphate solution and copper rods).

The resistivity surveys in general are made in two ways (i) Vertical Electrical Sounding (VES) where the apparent resistivity values with depth are obtained (ii) Horizontal profiling where the lateral variations of resistivities are measured.

A Vertical Electrical Sounding (VES) involves gradual shifting of electrode spacings. As and when the electrode spacing increases the distance between potential electrodes and the current electrodes also increases and the current will travel progressively to greater depths resulting in obtaining apparent resistivity values for materials at greater depths. Both Wenner and Schlumberger arrays give good results, but the Schlumberger is preferred because for each incremental measurements only the outer electrodes need to be moved every time whereas in Wenner all the four electrodes are required to be moved every time. The apparent resistivities obtained for each configuration are plotted on a logarithmic paper as a function of electrode spacing. The resulting sounding curves could be of various types depending upon the geological, hydrogeological and hydrochemical situations of the subsurface materials and also the electrode spread.

The shape and type of the curve depends upon electrode configuration, resistivities and thicknesses of the layers under consideration. A single homogenous layer will be represented by a straight line for all electrode spacings. A two layered case with an upper homogenous layer of thickness h, and resistivity ρ_1 underlained by infinite thickness ($h_2 = \infty$) and resistivity ρ_2. The initial section of the curve resembles that of a single layered curve but the later section departs either ascending or descending depending upon their relative resistivities and electrode configuration. The ascending curve represents a shallow zone with high resistivity. Ex: gravels, pebbles and cobles or weathered rock complex followed by basement rocks. The descending type curves reflects on the presence of low shallow resistivity layers with clays and zones of moderate to high salinity levels. Descending curves are normally observed in saline affected regions of coastal systems. A three layered case of thicknesses h_1, h_2 and h_3 with resistivities ρ_1, ρ_2 and ρ_3 will result in four different shapes of curves classified as given under and are shown in Fig. 6.18.

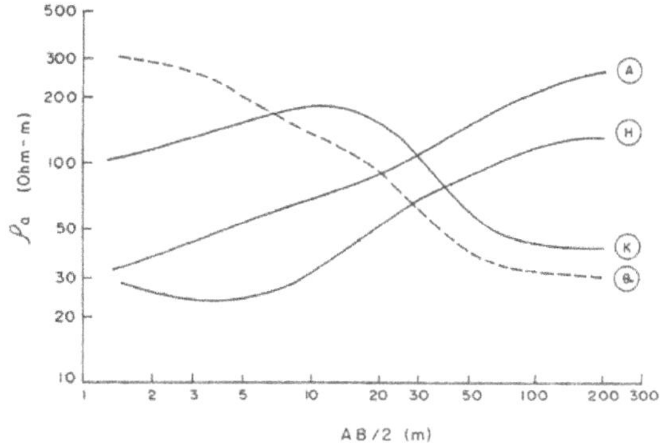

Fig. 6.18 Generalized sounding curves in different hydrogeological situations

(i) $\rho_1 > \rho_2 < \rho_3$ = H-type curve

(ii) $\rho_1 < \rho_2 < \rho_3$ = A-type curve

(iii) $\rho_1 < \rho_2 > \rho_3$ = K-type curve

(iv) $\rho_1 > \rho_2 > \rho_3$ = Q-type curve

In the case of four layered structure there could be several combination of curves such as HK, HA, KH, KQ, AA, AK, QQ, QH and several more complicated curves could be obtained in multilayered systems. The interpretation of all these curves obtained through VES data are done in different ways such as (i) curve matching (Mooney, H.M. and Wetchel, W.W., 1956; Orellana, E. and Mooney, H.M., 1966; Zohdy, A.A.R., et al 1974 and Ghosh, D.P., 1971); (ii) computer aided inverse techniques (Vozoff, K. and Jupp, D.L.B., 1975; Jupp, D.L.B and Vozoff, K., 1975); (iii) An inverse slope technique (Sankaranarayana, P.K. and Ramanujachary, K.R., 1967). The curve matching techniques in brief involves plotting of apparent resistivity values against AB/2 (half electrode separation) on log-log paper with the same moduli as those of master curves and read the curve in terms of two layer, three layer or four layer etc as the case may be. The best fit portion could be noted initially and the first layer resistivity and thickness could be obtained. The thicknesses and resistivities for 2nd, 3rd and 4th layers etc could be determined from $\frac{\rho 2}{\rho 1}, \frac{\rho 3}{\rho 1}, \frac{\rho 4}{\rho 1}$ and $\frac{h2}{h1}, \frac{h3}{h1}, \frac{h4}{h1}$ on the type curves. Sometimes the computer aided inversion data could be followed by way of generating a theoretical curve using filter coefficients and comparing with the field curves through a number of iterations till such time an exact matching is obtained (Koefed, O., 1979). An inverse slope method proposed by Sankaranarayana, P.K and Ramanujachary, K.R.(1967) is also used as an effective technique particularly in the hard rock regions where the I/R are plotted against the Wenner electrode spacing 'a'. A number of straight line segments with different resistivity boundaries could be obtained which gives directly the absolute resistivities of the corresponding layers.

Case study: A characteristic analysis of VES data obtained from the coastal Mahanadi delta region of Orissa (Satpathy, B.N., et al, 1976) is presented below. The locations of five sites from where VES data curves obtained are presented in Fig. 6.19. A set of VES data curves obtained in the coastal strip of puri district in Orissa are shown in Fig. 6.20(a) and the interpreted section is presented in Fig. 6.20(b). The general hydrogeological features along this track indicate presence of thick sands and gravels on the top below the soil cover with thicknesses varying from 40 meters to 100 meters from shoreline to upstream saturated with fresh water followed by alternate layers of sands and clays upto a depth of 300 meters and beyond (Radhakrishna, I. and Dutt, D.K., 1976).

The VES data and the interpretation made invariably doesn't exactly match with the field hydrogeological situations due to several limitations particularly in the coastal regions. The ambiguities contained in the VES data in terms of interpretation of saline fresh water boundaries have also been explained by Hodlur, G.K., et al (2006) by way of correlating the VES soundings with the borehole geophysical data (Radhakrishna, I., et al 1990; Radhakrishna, I., 2001) and found the error of deviation from the borehole data to the interpreted VES data is of the order varying between 30-40% or even more. It therefore becomes necessary that the VES data obtained invariably should be standardized as far as possible with the subsurface lithological and geophysical depth profiles to assess realistic situations.

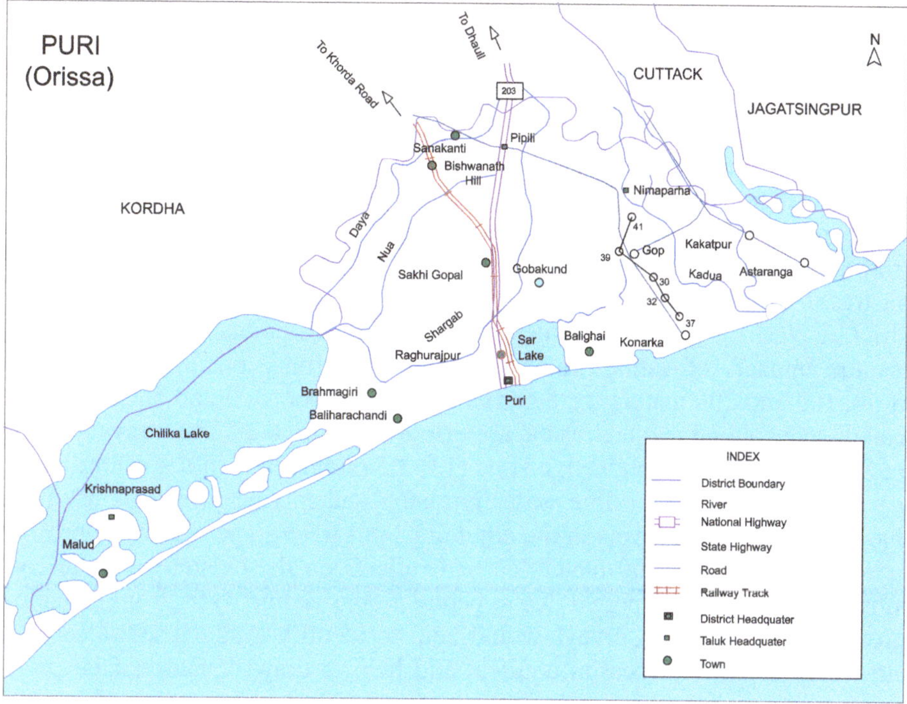

Fig. 6.19 Site locations for VES in puri district Mahanadi delta region

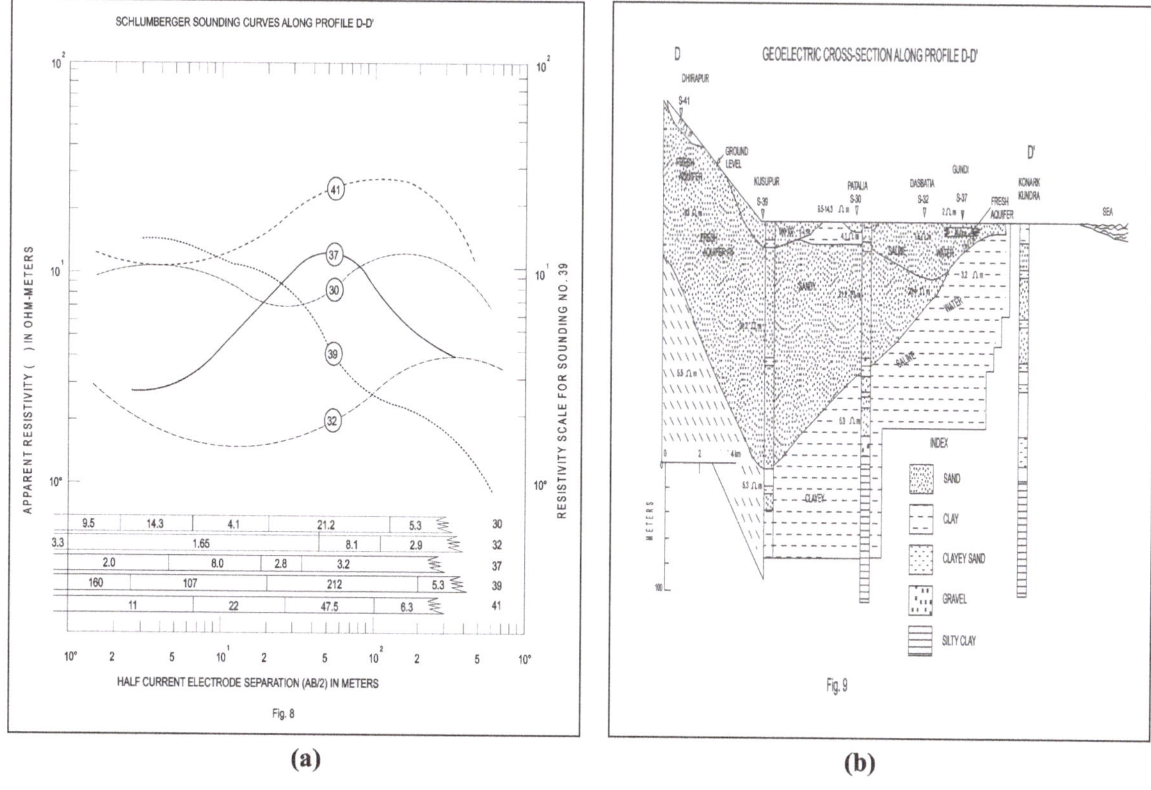

(a) (b)

Fig. 6.20 (a,b) VES data sounding curves and the interpreted hydrogeological section along with lithological logs.

Electrical Resistivity Profiling: Electrical resistivity profiling is a useful technique to study the lateral variations in terms of lithological and water quality variations in different sub surface strata. In horizontal profiling the electrode spacing is kept constant and the electrodes are moved in different profiles either along one direction or in different directions. This set up would yield valuable information related to apparent resistivity values laterally in different directions with different electrode spacings. The greater the electrode spacing the greater will be the depth of probing. When all apparent resistivity values are obtained either over a profile or in different directions, one could draw equi resistivity contour maps for a particular electrode spacing which gives highs and lows which in turn could be interpreted in terms of lithological and water quality changes. Choudhury, K. and Saha, D.K., (2004) demonstrated the advantages of electrical resistivity profiling near Digha coastal area in Midnapur district of West Bengal. The profiles were taken at two points PR1 and PR2 as shown in Fig. 6.21 to assess the impacts of saline water intrusion. Resistivity profiling at PR1 with electrode separations of 50, 100 and 200 meters were carried out near a borewell and another profile PR2 near a saline water canal were studied and the results are displayed in fig.6.22 The arrays near PR1 reflects on the existence of a relatively high conductive zone (1.0-3.5 ohm/meter) in the central part of the profile due to saline water intrusion. The saline water intrusion and mixing processes are believed to have taken place due to high pumping levels in the adjoining well fields. The resistivity profile PR2 with an electrode spacing of 50,100 and 200 meters carried reflects on the existence of the conductive zone (1.3-3.00 ohm/meter) which is interpreted due to percolation of saline water from the adjoining canal in to the underground aquifers. Such surveys would help to identify systems saturated either with fresh water or saline water in different coastal aquifers and help to draw developmental programs without failures.

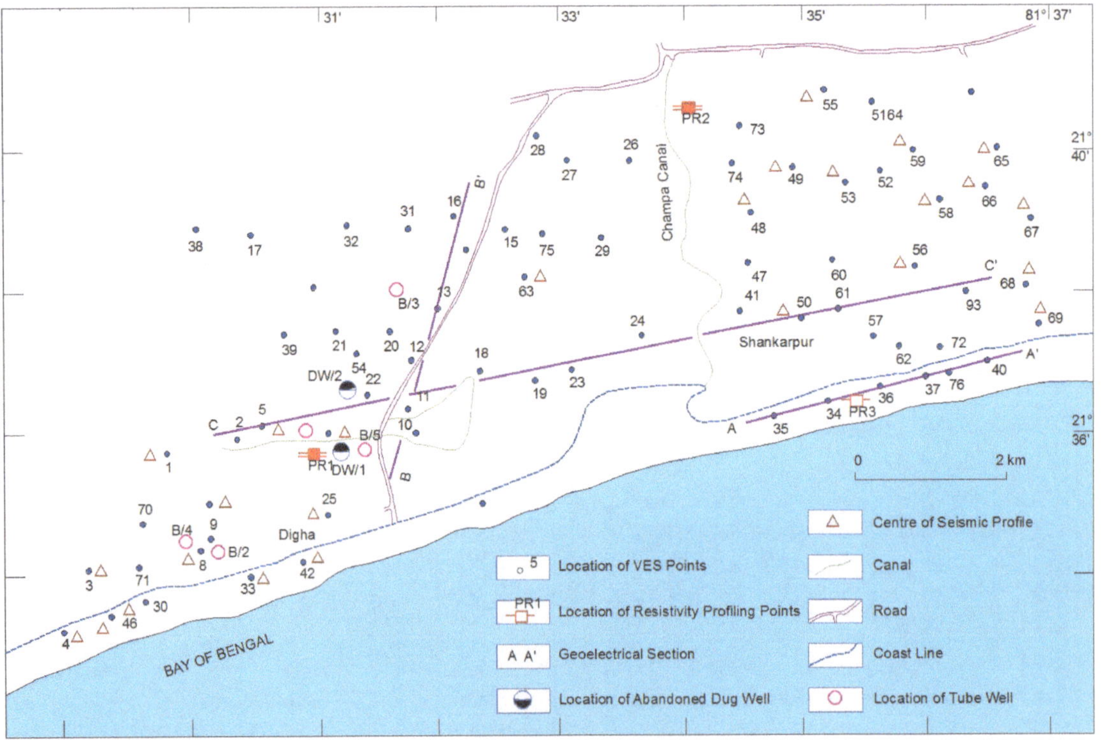

Fig. 6.21 Site locations for electrical resistivity profiling digha coastal area West Bengal

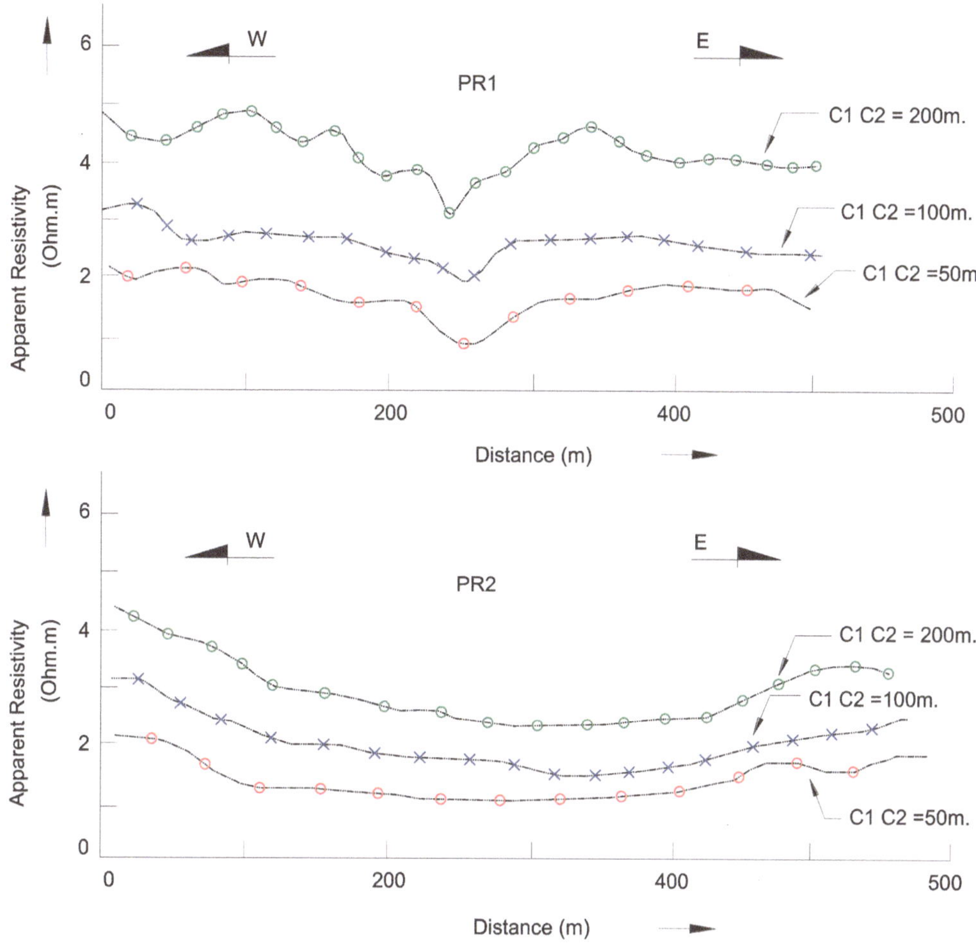

Fig. 6.22 Results of electrical resistivity profiling

Electrical Resistivity Tomography: Electrical Resistivity Tomography is a non destructive geoelectrical prospective tool that analyses subsurface materials in terms of the electrical behavior in the porous materials. The method is based on positioning array of electrodes along a transverse section, each electrode separated by a particular distance according to the required degree of resolution and the depth of investigation. All the electrodes are connected to the measuring equipment and using a specific sequential program created for each objective. High resolution electrical tomography 2-D resistivity profile could be obtained (Loke, M.H., et al, 2010) indicating information about the subsurface materials (granular, non granular zones and water qualities) and could be of great assistance in delineating areas affected by saline water intrusion processes. The data obtained can be presented in the form of subsurface images (Loke, M.H. and Barker, R.D., 1995;1996).

Satish, S., et al (2011) provided a case study near Chennai coast where a multi electrode resistivity imaging system IRISSYSCAL PRO-96 was used. Resistivity profiling was carried out in five profiles near Chennai in between Adayar river mouth and Muthukadu back waters. The images obtained across different profiles are reproduced below along with the interpreted results in relation to lithological frame work and saline water mixing areas.

Some of the characteristics in each of the profiles which are taken perpendicular to the coast are mentioned below.

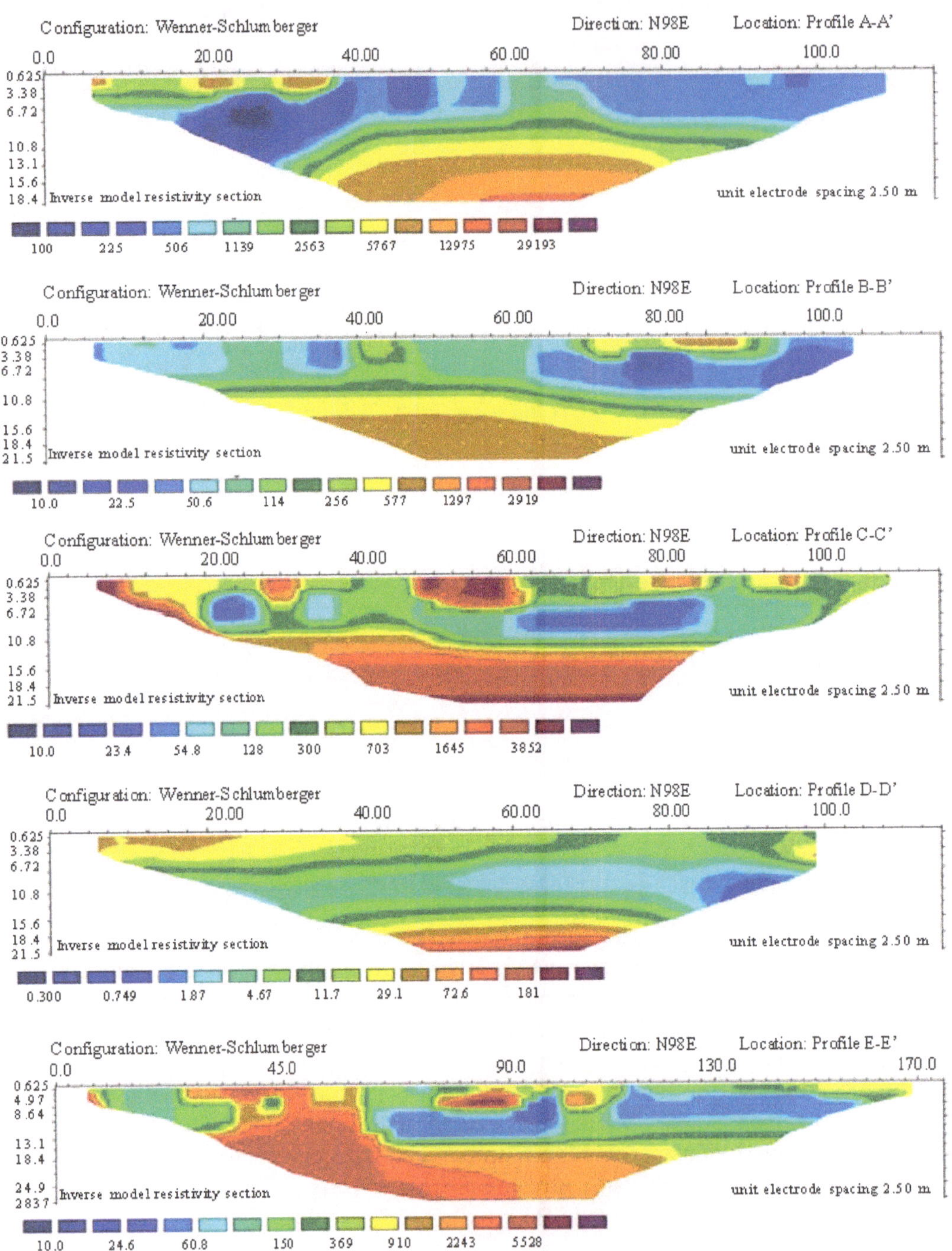

Fig. 6.23(a) High resolution electrical resistivity images across different profiles

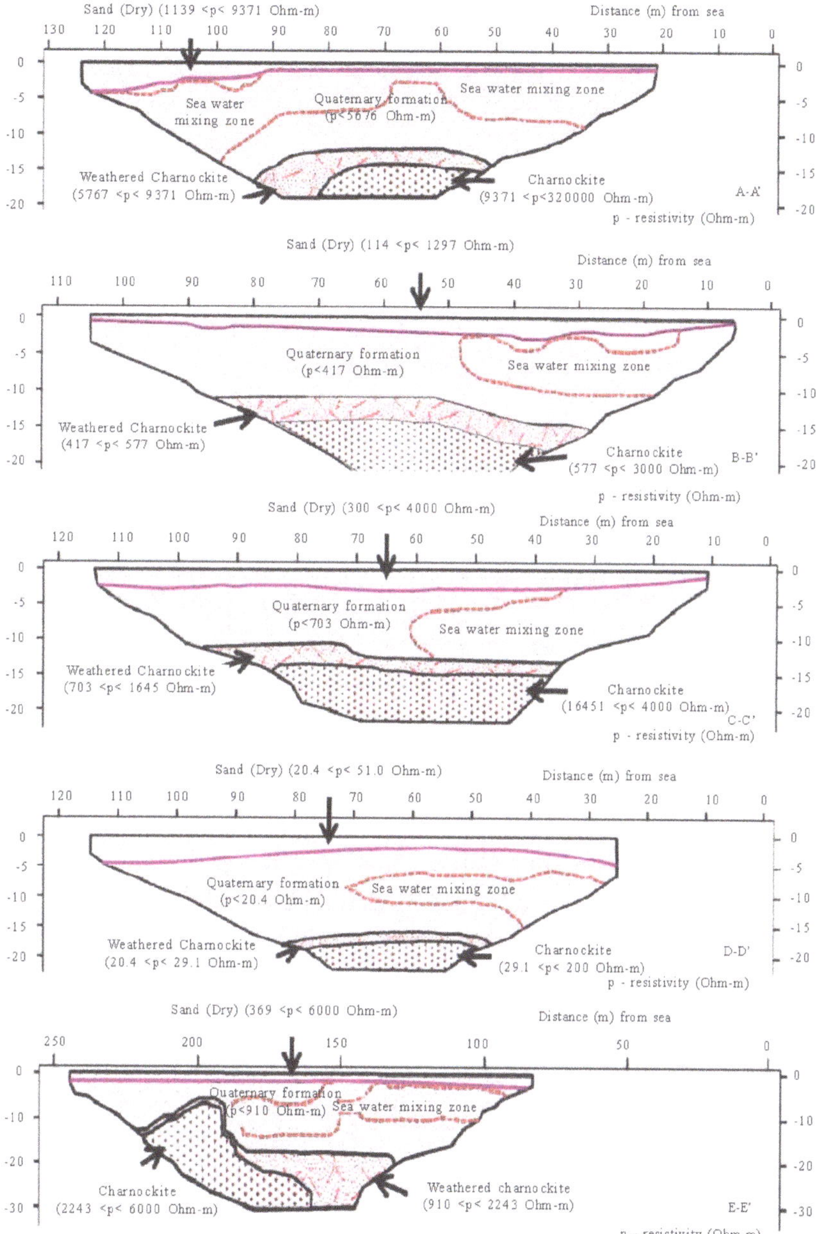

Fig. 6.23(b) Interpretations for different resistivity images across different profiles

(Source: Satish, S., et al 2011)

Profile 1 (A-A'): The profile A-A' is carried out over a length of 120 m at a distance of 10 m from the coast line. The maximum resistivity values recorded in this profile are of the order of 32,000 ohm-meter which may be related to the basement rock. The depth to the water table interpreted varies from 1 m to 4 m below ground level. The zone above the water table as the resistivity of the order of 9370 ohm-meter. The Quaternary formation extends upto a depth of 12.5 m below ground level. The low resistivity zones indicate the reflectance of the influence of the sea water mixing.

Profile 2 (B-B'): The profile B-B' is carried out over a distance of 100 m at a distance of 5 m from the coast line. A high apparent resistivity of the order of 3000 ohm-meter is recorded in this profile probably represents the weathered charnockite layer occurring below Quaternaries. The Quaternaries extend upto a depth of 11.5 m below ground level. The depth of the water table varies from 1-3 m below ground level. The unsaturated zone is characterized by a resistivity of the order of 1297 ohm-meter. The low resistivity zone is found to extend upto a distance of 40 m which represents the zone of saline water mixing.

Profile 3 (C-C'): The profile C-C' was carried out over a length of 100 m and at a distance of 10 m from the coast line. The maximum resistivity values recorded are of the order of 4000 ohm-meter which represents the weathered charnockite layer occurring below Quaternaries. The Quaternaries are found to extend upto a depth of 11.5 m below ground level. The water table depth varies from 1-3.5 m below ground level. The low resistivity zone represent the zone of saline water mixing which extends upto a distance of 50 m from the coast line.

Profile 4 (D-D'): The profile D-D' was carried out over a lenth of 100 m and at a distance of 5 m from the coast line. The maximum resistivity obtained over this profile is of the order of 2000 ohm-meter which probably represents weathered charnockite layer. The Quaternaries appear to extend upto a depth of 15.5 m below ground level. The water table depth varies from 1-4 m below ground level. The sea water influence/mixing is seen extending upto 47.5 m from the coastline as reflected by low resistivity zones.

Profile 5 (E-E'): The profile E-E' is obtained over a lenth of 170 m and at a distance of 80 m from the coast line. A high resistivity layer of the order of 6000 ohm-meter which probably represents the basement charnockite rock. The Quarternary sediments extends upto a depth of 11-20 m below ground level. The water table varies from 1-4 m below ground level. The low resistivity zone which extends upto a distance of 120 m represents the zone of saline water mixing.

Note: Majority of the interpreted geophysical profiles by and large match with the drilled borehole lithological logs and hydrological characteristics in the area. The techniques and methods applied are cost effective and useful to delineate the aquifer materials underneath the soil cover and also help to identify the zones affected by saline water intrusion.

Electromagnetic methods: Electromagnetic methods were originally developed for Mineral exploration specifically to detect massive metallic conductors in resistive host rock. Over the last several years many ground based and airborne electromagnetic methods have been developed and applied for mineral and Groundwater exploration. The use of electromagnetic techniques to Groundwater studies is relatively new. A complete description of these techniques are available in the paper published by different workers. Keller, G.V. and Frischknecht, F.C., (1966); Nabighian, M.N., (1988, 1991); Stewart, M.T., (1982); Fitterman, D.V. and Stewart, M.T., (1986) and McNeill, J.D., (1990). These techniques of late are found to contain potential application in Groundwater exploration particularly in coastal systems where resolution of saline-fresh water interrelationships need to be worked out.

The EM method basically measures the conductivity (inverse of resistivity) and is an inherent property of the rock material. The electrical current passing through the rock material is transmitted more by fluids in the pore space and the surface area of the pores than through the mineral grains. Pore fluids and the pore surface area are important factors in determining the bulk conductivity of the rock matrix. The electromagnetic methods use an electromagnetic field generated by a transmitter coil through which an alternating current is passed. This generates a magnetic field around a transmitter coil. This magnetic field induces an electrical field on the earth when the transmitter is held near the earth. The electrical field will travel through the ground with different strengths depending upon the conductivity levels of the ground. The strength is measured through a passive receiver coil. Changes in the primary field (Phase, amplitude and orientation) with time or distance can be related to electrical properties of the earth in the vicinity of the transmitter and receiver.

The depth of investigation in EM service can be varied by manipulating the transmitter frequency or inter coiling spacing (frequency domain) or by measuring the secondary field strength at a discrete time intervals after an abrupt shutoff of the primary field (time domain or transient method). The time domain electromagnetic methods (TDEM) are mainly used for depth soundings and the frequency domain electromagnetic methods (FDEM) surveys are used for mapping lateral changes of conductivity. In both these electromagnetic methods no direct contact is made by electrodes with the ground and therefore surveying rate can be very rapid than other electrical techniques. Reports containing the use of FDEM profiling are given by several workers (McNeill, J.D., 1980(a,b) and 1986). In recent years FDEMs are being integrated with GPS technologies to improve and increase the speed of work and for larger coverage of area. The results of FDEM surveys are presented as contour maps reflecting the variations in conductivity values. Variations of conductivity levels often are also associated with differences in lithological units, faulted boundaries and mineralized zones etc. The electromagnetic techniques are also used for rapid reconnaissance of Groundwater mapping (Richards, R.T., et al, 1995). Specific field examples characterizing the usefulness of FDEM techniques are given by Godio, A., et al, (1998) and Van Lissa, R.V., et al, (1987). To achieve better accuracy of the results the FDEMs can be integrated with resistivity profiling and resistivity vertical soundings. The time domain electromagnetic service can produce one dimensional and two dimensional geo electric cross sections. Kauffman, A.A. and Keller, G.V., (1983); McNeill, J.D., (1990) have discussed extensively on the subject. Several regional case histories are available to demonstrate the utility of TDEMs for Groundwater exploration and more specifically to map the salt water intrusion (Stewart, M.T., 1982; Stewart, M.T. and Gay, M.C., 1986; Hoekstra, P. and Blohm, M., 1990). Airborne electromagnetic systems are also being used for rapid acquisition of data over large and inaccessible terrains (Jackson, P., 1993; Paine, J.C., et al, 2000). Sometimes Helicopter-based electromagnetic services are also being conducted to map and identify and assess the saline water intrusion process (Fitterman, D.V., et al, 1999).

6.4.3 Sub surface geophysical approaches

Subsurface borehole Geophysics provides us with wealth of important insitu information related to the physical, chemical and structural properties of the formations penetrated in the drilled boreholes. Borehole geophysical logs are essentially developed for petroleum industry, but of late are being used widely for water industry particularly in the deep exploratory drilling operation in coastal sedimentaries where resolving bed boundaries and fresh-saline water variations is a problem. The borehole logs can also indicate high porosity and permeability zones as well as zones with fresh water and saline water. These logs can also be used for stratigraphic and Lithological correlations more precisely and sometimes even that parameters could be qualitatively predicted. A large number of techniques have come into operation such as (i) Resistivity logs, (ii) Self potential logs (iii) Temperature logs, (iv) Caliper logs and (v) Radiation logs such as Gamma radiation, Neutron logging and Gamma-Gamma logs etc. A number of workers contributed immensely towards the application and usefulness of the logs in to water well industry. (Keys, W.S. and Mac Cary, L.M., (1971); Baldwin, A.D., and Miller, J., (1979); Norris, S.E., (1972); Keys, W.S. and Brown, R.F., (1978); Mac Cary, L.M., (1983); Keys, W.S., (1986); Kwader, T., (1986); Radhakrishna, I. and Gangadhararao, T., (1990); Jones, P.H. and Buford, T.B., (1951)). Though a large number of Borehole techniques have come into existence, the commonly deployed techniques in water industry are (i) Resistivity and Self potential logs (ii) temperature logs (iii) Caliper logs and (iv) Natural Gamma logs. Resistivity and Self potential logs are Resistivity logs widely used in the Groundwater and the rest of the logs are used in a limited way depending upon the need. The usefulness of these techniques and specific purpose they serve are illustrated by Keys, W.S and Mac Cary, L.M., (1971); are presented below in Table 6.2.

Table 6.2 A summary of log applications and their utilities for identification of different geophysical characteristics/parameters

Required information on the properties of Rocks, Fluid, Wells, or the Ground-Water System	Widely Available Logging Techniques that might be utilized
Lithology and stratigraphic correlation of aquifers and associated rocks	Electric, sonic, or caliper logs made in open holes; nuclear logs made in open or cased holes
Total porosity or bulk density	Calibrated sonic logs in open holes, calibrated neutron or gamma-gamma logs in open or cased holes
Effective porosity or true resistivity.	Calibrated long-normal resistivity logs
Clay or shale content.	Gamma logs
Permeability.	No direct measurement by logging. May be related to porosity, injectivity, sonic amplitude
Secondary permeability-fractures, solution	Caliper, sonic, or borehole televiewer or television logs
Openings	Calibrated neutron logs
Specific yield of unconfined aquifers	Possible relation to formation factor derived from electric logs
Grain size	
Location of water level or saturated zones	Electric, temperature, or fluid conductivity in open hole or inside casing, neutron or gamma-gamma logs in open hole or outside casing
Moisture content.	Calibrated neutron logs
Infiltration.	Time-interval neutron logs under special circumstances or radioactive tracers
Direction, velocity, and path of ground-water flow	Single-well tracer techniques-point dilution and single-well pulse; multiwell tracer techniques
Dispersion, dilution, and movement of waste	Fluid conductivity and temperature logs, gamma logs for some radioactive wastes, fluid sampler
Source and movement of water in a well	Injectivity profile; flowmeter or tracer logging during pumping or injection; temperature logs
Chemical and physical characteristics of water, including salinity, temperature, density and viscosity	Calibrated fluid conductivity and temperature in the well; neutron chloride logging outside casing; multi-electrode resistivity
Determining construction of existing wells, diameter and position of casing, perforations, screen	Gamma-gamma, caliper, collar, and perforation locator; borehole television
Guide to screen setting.	All logs providing data on the lithology, water-bearing characteristics, and correlation and thickness of aquifers
Cementing.	Caliper, temperature, gamma-gamma; acoustic for cement bond
Casing corrosion.	Under some condition, caliper or collar locator
Casign leaks and (or) plugged screen	Tracer and flowmeter

(Source: Keys, W.S. and Mac Cary, L.M., 1971)

Resistivity log: The resistivity is measured by lowering current and potential electrodes into the borehole and measuring the electrical resistivity of the surrounding medium and its variation with depth. These measurements are done normally in uncased boreholes. The resistivity at any point is a function of fluid content in well, well diameter, rock formation and the pore waters in the rock. The resistivity of a clean granular aquifer is expressed as:

$$R_t = F \times R_w \quad .....(6.28)$$

where R_w is the water resistivity and F is a constant that represents the effect of pore space (Archie, G.E., 1942). This constant is called formation resistivity factor and is given by the formula:

$$F = \frac{A}{\phi^m} \quad .....(6.29)$$

where ϕ is the effective porosity (fraction of total volume) and A and m are the non dimensional numbers that represent the effect of porosity distribution and continuity, which vary from rock to rock and the values could be determined from laboratory measurement.

Further it is to be noted that the resistivity of water R_w decreases when salinity increases. At a given temperature the water resistivity is related to the total dissolved solids (expressed in PPM) and by the expression as under:

$$R_w = K/PPM \quad ......(6.30)$$

where K is a factor that is constant for a given salt.

A clean granular rock having a porosity more than 10% of the rock, the A and m have the following average values.

(i) Rocks with little or no cement matrix: A = 0.62 and m = 2.15.

(ii) Rocks with more cement matrix: A = 1 and m = 2.

Combining all the equations (6.28, 6.29 and 6.30) and using A and m values as given above, the resistivity of a clean granular aquifer is given by the following equation:

$$R_t = \frac{0.62K}{\phi^{2.15} PPM} \quad .....(6.31)$$

A number of different electrode configurations are used. These include point resistance log, short normal, long normal and lateral configurations as shown in Fig. 6.24.

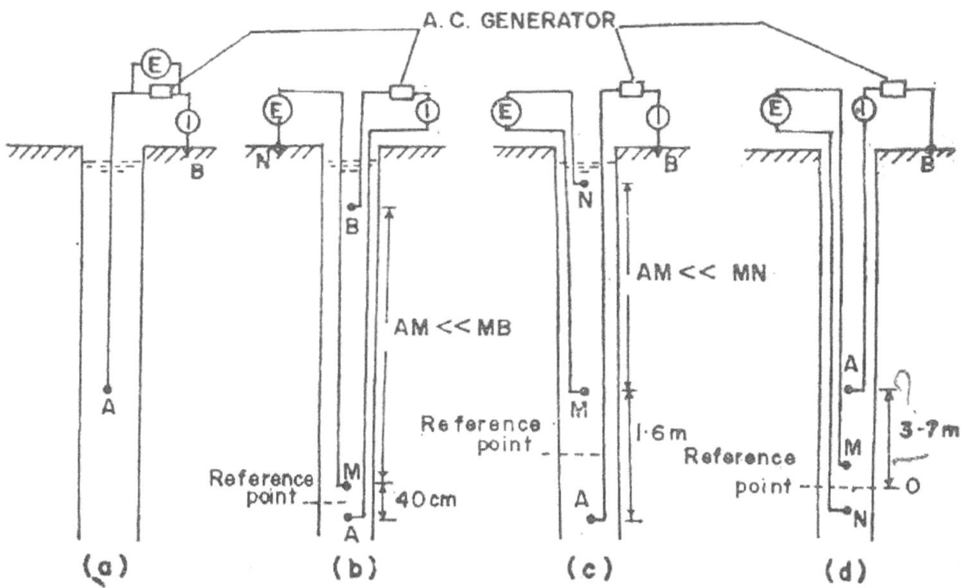

Fig. 6.24 Electrode arrangements for (a) single point (b and c) 40 cm and 1.6 m normals and (d) lateral devices

A point resistance or single electrode log is done by lowering one electrode into the borehole and one at the surface and the resistance is recorded. Such logs are useful in making qualitative boundaries and the interstitial water qualities of the surrounding rocks in the borehole. The second and third categories of logs namely short normal and long normal are used for semi qualitative or quantitative purposes and have greater radius of influence surrounding the borehole. The short and long normal devices contain two current electrodes A and B and two potential electrodes M and N. The spacing between A and M in the case of short normal is 40 cms and 160 cms in the case of long normal. The logs obtained this way are to be interpreted using log charts for accurate estimations of true resistivities (Schlumberger well surveying company, 1978). A fairly accurate estimations of resistivities could be obtained from normal logs if the aquifer thicknesses are large and uniform and contain relatively high porosities, less mud invasions and with fresh water interstitial waters. The short normal arrangement gives rise to the resistivity values of the mud invaded zone and the long normal records the values beyond the mud invaded zone. The lateral device has three electrodes the electrode spacing is measured from the current electrode 'A' to the mid point 'O' between electrodes M and N and is standardized at 5.7 m. This device often gives the best results when bed thicknesses exceed 2AO (about 12m).

Spontaneous potential logs: Spontaneous potential logs (SP) record the natural potentials which occur in a borehole caused by electrochemical reactions between mud, formation waters and clay. The SP log response in general is affected by several factors such as borehole diameter, bed thicknesses, shaliness of the formation, quality of the formation water, mud filtrate invasion, and temperature. The electrochemical potential generated in such logs is given by the following empirical formula.

$$SP = -K\log\frac{R_{mf}}{R_w} \qquad \text{.....(6.32)}$$

where SP= log deflection (mv) against a zone; K = a factor generally taken to be equal to 70, but deviates marginally with temperature, 60 + 0.133T where T = borehole temperature in (F°); R_{mf} = resistivity of the mud infiltrate ($\Omega\, m^{-1}$); R_w = resistivity of the formation water ($\Omega\, m^{-1}$).

The above equation is essentially based on the following assumptions: (i) Both the borehole fluid and formation waters are sodium chloride solutions (ii) The shale formations are ideal ion selective membranes and the sand formations have no ion sieving properties. (iii) The borehole fluid has a much greater resistivity than the combined resistivity of the sand and shale (Keys, W.S. and Mc Cary, L.M., 1971). The above relationship is used effectively in the oil industry to estimate 'R' through SP logs since NaCl is the dominant salt in solution and the inter relationships between concentration, activity and resistivity are well established. This cannot be applied directly to the water industry because the medium generally is not dominated by NaCl waters and the activity of several divalent cations will significantly affect SP response. Self potential is really a function of activity rather than the resistivity ratio of the borehole fluid and the formation water. The self potential measures static SP (SSP) for clean thick sand and is related to the activity by the following equation (Gondouin, M., et al, 1957).

$$SSP = -K\log\left[(a)_w/(a)_{mf}\right] \qquad \text{.....(6.33)}$$

$$SSP = -K\log\left[(a_{Na} + a_{Ca} + a_{Mg})W/(a_{Na})mf\right] \qquad \text{.....(6.34)}$$

where a_{Na}, a_{Ca}, a_{Mg} are the solution activities of sodium, calcium and magnesium ions.

In equation 6.32 above the mud filtrate is considered as NaCl solution and the relationship between activity and resistivity is shown in Fig. 6.25(a) below.

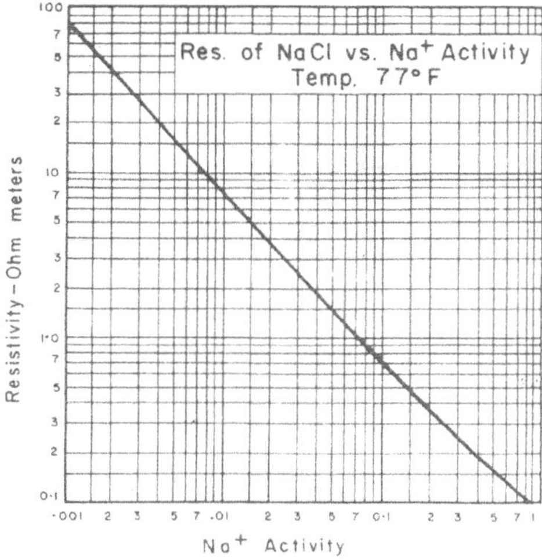

Fig.6.25(a) Relation between sodium ion and resistivity for sodium chloride solution at 77°F activity

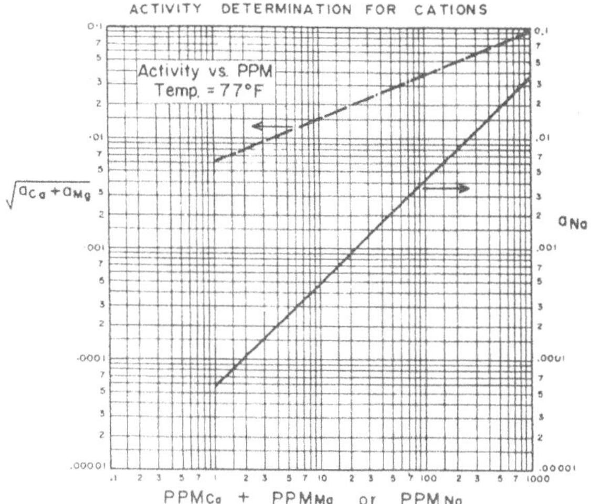

Fig. 6.25(b) Cation concentration versus activity for dilute solution at 77 °F

This facilitates derivation of activity of mud filtrate when R_{mf} is known. Similarly the relation between cation concentration and activity for dilute solutions is given in Fig. 6.25(b). The solid line represents the relationship for Na ions and the dashed line for Ca and Mg ions. The total activity required in equation 6.33 can be computed, which can be used to check SSP values for different zones if complete chemical analyses are available.

In addition to the cations, the anions also have a role to play in determining the resistivity of solutions. In most fresh groundwater systems, bicarbonate (HCO_3) is the dominant anion. The resistivities of NaCl and $NaHCO_3$ solutions (for the same Na concentrations) are different. The HCO_3 ion contributes only 27% as much conductivity as an equal weight of Cl ion. In other words, R_w of $NaHCO_3$ solution will be 1.75 times greater than R_w of an NaCl solution with the same Na concentration (Alger, R.P., 1966).

The foregoing discussions indicate clearly that SP response is a parameter related to the type of ions present in the formation water and the resistivity thus obtained can be termed "equivalent formation water resistivity" (R_{we}) instead of R_w of equation 6.31. One could therefore, build up graphical interrelationships between R_w and R_{we} if sufficient analytical data are available representing the groundwater regime. If the aquifers are clayey, the clay-affected SP should be corrected by finding V_{cl} (fractional volume of clay) using the following equation 6.34.

$$SSP = PSP/(1 - V_{cl}) \qquad .....(6.35)$$

V_{cl} can be estimated from the natural gamma log as and when required.

SP baseline shift is a common phenomenon in coastal aquifers. The shift in SP baseline takes place wherever there is a change in the quality regime. The SP deflections are measured from the shifted baseline positions and corrections could be made accordingly. In general though there are some constraints involved in the interpretation of SP logs, one could definitely use these logs in a system where extensive granular zones are available and the chemical regime is gradual. A detailed case study of interpretations of SP logs for coastal regions is presented ahead in the same chapter.

Nuclear logging: Nuclear logging techniques could be broadly categorized into two: (i) that measures the natural radioactivity of the rock and fluids (ii) that measures the radioactivity induced into the rock formations through an artificial source. These techniques have the advantage in that these could be operated in the cased boreholes also and obtain the required formational details Meyers, G.D., (1992) reviewed all nuclear methodologies in detail. The nuclear logs that are commonly deployed in the groundwater industry are (a) natural gamma (b) gamma-gamma and (c) neutron.

Natural gamma logging: The natural gamma logging method facilitates measurement of natural gamma radiation that originates from unstable isotopes of potassium, uranium and thorium. The shales and clays contain high gamma activity and less activity in sands making it possible to differentiate between sands, clays and shales, etc (Norris, S.E., 1972). Gamma ray logging unit has basically a gamma ray detector mounted in a scintillation counter in the form of a sonde which is moved along the walls of the borehole which records the intensities of gamma rays all along the borehole. The intensity is expressed in terms of counts per minute. The natural gamma ray log is a potential unit which can be used for lithological and stratigraphic correlations based on the intensities of radiation in different log profiles.

Gamma-Gamma logging: The gamma radiation originating from a source probe and recording after it backscatters and attenuates within the borehole and the surrounding formations is termed as gamma-gamma log. A source could be either Cobalt-60 or Cesium-137. Gamma-gamma logging could be used effectively to identify lithology, bulk rock density and porosities of the rocks. The porosities are determined by the equation given below.

$$\alpha = \frac{\rho_G - \rho_B}{\rho_G - \rho_F} \qquad \text{.....(6.36)}$$

where ρ_G is the grain density, ρ_B is the bulk density and ρ_F is the fluid density.

Grain density could be measured from the drill cuttings and the fluid density is assumed as 1.00 gm/cm^3. These logs could be used to approximately assess the specific yield of the rock aquifer from the differences of bulk densities measured above and below the water table and also could be used for estimations of casing lengths and cement ceiling positions/grouting, etc.

Neutron logging: Neutron logging like other tools is useful in groundwater investigations in several ways, it has a neutron source and a detector mounted in a sonde which produces anomalies related to hydrogen contents in the borehole rocks. Several neutron probe sources are available such as Beryllium combined with Radium-226, Plutonium-239 and Americium-241. The neutrons emitted are slowed and scattered by collisions with nuclei of hydrogen atoms. The hydrogen atom has approximately the same mass of that of a neutron. The hydrogen content is directly proportional to the water contents in pore space. Neutron logs are therefore useful to measure the moisture levels above the water table and the porosities of rocks below water table (Peterson, F.L., 1974). By measuring the moisture levels above and below the water table it is possible to estimate this specific yields of the rock formations.

6.5 ESTIMATIONS OF INTERFACE STRUCTURES IN COASTAL DELTA COMPLEXES

Self potential and electrical resistivity logs have been widely used in a qualitative way to define the bed boundaries, their thicknesses, lithological correlations, aquifers reservoir boundaries and also to define saline fresh water boundaries which would serve as an important database for well field development and estimation of mechanism of saline water ingression processes etc. Very limited efforts so far have been made to quantify the water quality parameters in terms of different ionic components in multi aquifer coastal systems. Bore hole geophysical logs involving both electrical resistivity and self potential logs play a vital role in this regard. Estimation of true resistivity of formational waters in both logging procedures is very crucial. Jones, P.H. and Bufford, J.B., (1951) attempted quantitative interpretation of resistivity data and obtained reasonably good results. Resistivity of formation water (R_w) has a definite relation to dissolved solids and one can estimate R_w values by taking into consideration the formation resistivity factor. Kwader, T., (1986) estimated R_w values for Tertiary carbonate and granular formations by cross plotting saturated formation resistivity (R_0) against bulk porosity. These data, when plotted on Hingle porosity cross plots yielded graphical solutions for estimations of R_w values. Alger, R.P. and Harrison, C.W., (1989), using true formation resistivity (R_{mf}) and spontaneous potentials presented solutions to quantifying formation water quality, permeability (hydraulic conductivity) and effective porosity (storativity). Spontaneous potential logs can also provide comparable evaluations of hydrogeochemical parameters for different zones, however, the number of data and case studies available is rather limited. Alger, R.P., (1966) demonstrated the usefulness of Self Potential (SP) logs for Gulf Coast aquifers and obtained good results. Vonhoff, J.A. (1966) found a workable empirical relationship between spontaneous potential deflection on the electric log and water quality in glacial aquifers. Dyck, J.H., et al, (1972) also stressed the importance of SP logs based on their studies on Kipling sand aquifers. Hearst, J.R. and Nelson, P.H., (1985) made critical analyses and reviewed major approaches towards solving many groundwater problems including interpretation of SP curves. In view of the non-availability of good case studies on spontaneous potential log interpretation for water quality problems an attempt is made to evaluate the water qualities in the coastal groundwater systems of Mahanadi delta in Orissa.

6.5.1 Application

The following method is adopted to evaluate chemical parameters for unknown zones through SP curves.

1. A number of sites are selected for which water analyses and electrical log data are available. A linear relationship between R_w and total dissolved solids (TDS) for the groundwater system under consideration is obtained.
2. The R_w-R_{we} interrelationship for NaCl waters and $NaHCO_3$ waters are plotted on double log scale and the factor of deviation is obtained.
3. Based on analytical results selected for a set of samples, scatter trend diagrams between TDS and different anions and cations are prepared. These graphical trends serve as standard diagrams for the area under consideration.
4. R_w values for unknown zones are calculated and TDS and other ions are interpreted with the help of graphical trends.

6.5.2 Derivation of hypothetical chemical analyses

The vertical and lateral changes in a groundwater system help to build up the interrelationship between R_w and TDS. This is shown in Fig.6.26 for the coastal groundwater system of Orissa in India (Radhakrishna, I, and Gangadhara Rao, T.; 1990), which is based on analyses of groundwater samples from selected wells and their (R_w) values.

The R_w-R_{we} relationship for NaCl waters is given by a diagonal line in Fig. 6.26(b). The R_{we} values using chemical analyses in the coastal groundwater system are plotted against each other and a straight line is drawn by least square fitting, which is displaced upwards such that for every value of R_{we}, the R_w value is 1.80 times greater than for NaCl solutions. The predominant cations and anions in solution are sodium and bicarbonate. The activity of the solution is primarily dependent on cation concentration and the R_w values of NaCl and $NaHCO_3$ values are different. The derived R_w values for the groundwater of different zones at each of the well sites are given in Table 6.3 below.

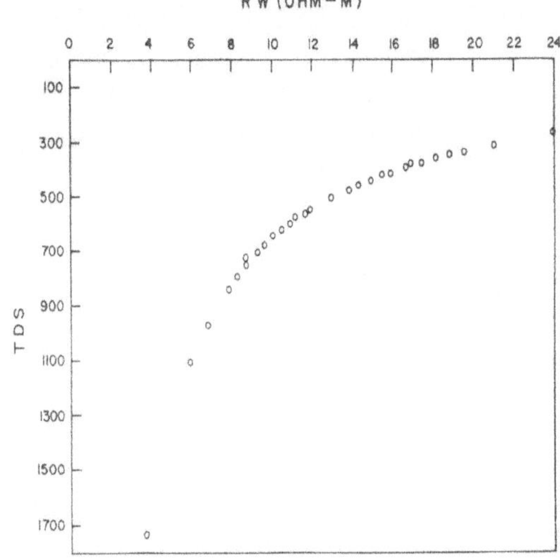

Fig.6.26(a) Relationship between formation water resistivity (R_w) and total dissolved solids (TDS) (p.p.m.) for coastal groundwaters of Orissa, India

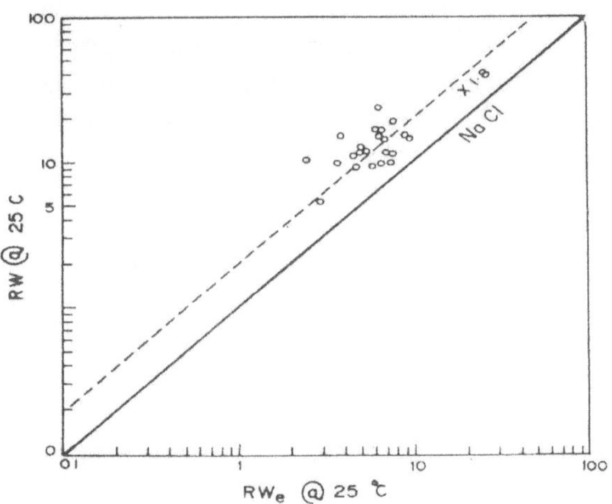

Fig.6.26(b) Relationship between formation water resistivity (R_w) and equivalent formation water Resistivity (R_{we}), (both in Ω m) for coastal groundwaters of Orissa, India

Table 6.3 Results obtained from Self Potential log analyses

Site and location	Zone depth(m)	SP	R_{mf}	R_{we}	$\frac{R_{mf}}{R_{we}}$	R_w
SP log data for shallow aquifer systems						
Arilo	0-60	+22.5	6.50	13.63	0.48	24.525
(Long.86°13'38'' Lat.20° 12'15'')	60-96	+12.5	6.50	9.81	0.66	17.650
	96-107	–15.0	6.50	3.97	1.64	7.143
Machagaon						
(Long.86°16'0'' Lat.20° 4'0'')	0-50	+22.5	5.00	10.48	0.48	18.865
	50-100	–25.0	5.00	2.20	2.27	3.954
Patalia	20-40	+10.0	6.00	8.34	0.72	15.006
((Long.86°2'35'' Lat.19° 57'25'')	40-80	0.0	6.00	6.00	1.00	8.430
	80-110	–7.5	6.00	4.69	1.28	8.430
SP log data for deeper aquifer systems						
Jajang	85-105	–47.5	5.90	1.24	4.76	2.226
(Long.86°25'52'' Lat.19° 31'24'')	105-250	–15.0	5.59	3.60	1.64	6.483
Kherang						
(Long.86°45' Lat.20° 59'30'')	120-140	–50.0	9.50	1.83	5.19	3.301
	150-165	–25.0	9.50	4.17	2.28	7.513
	180-190	–20.0	9.50	4.92	1.93	8.356
	190-210	–20.0	9.50	4.96	1.91	8.900
	220-240	–30.0	9.50	3.54	2.68	6.374
	240-265	–40.0	9.50	2.55	3.73	4.587
Adjhori	65-95	+10.00	6.00	8.34	0.72	15.006
(Long.86°45' Lat.20° 34')	96-140	+14.6	6.00	9.63	0.62	17.330
	140-155	+29.0	6.00	15.37	0.39	27.760
	165-220	–7.5	6.00	4.69	1.28	8.430

The trends of the scatter plots between TDS and various cations and anions are shown in Fig. 6.27 and 6.28. Almost all the ions are represented by linear trends with different slopes except bicarbonate value which are represented by a field with certain boundary values.

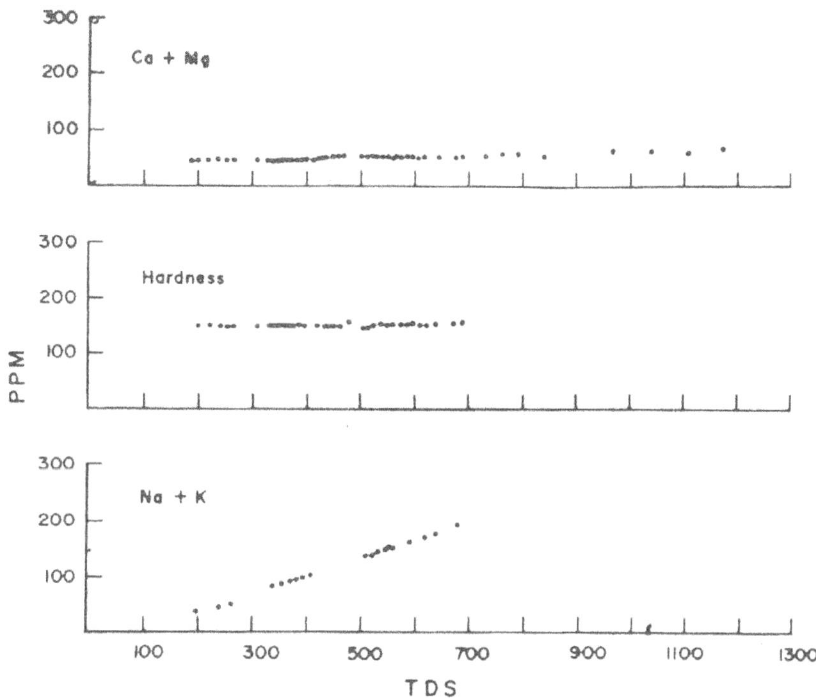

Fig. 6.27 Relationship between total dissolved solids (TDS) and cations (both in p.p.m.) for coastal groundwaters of Orissa, India

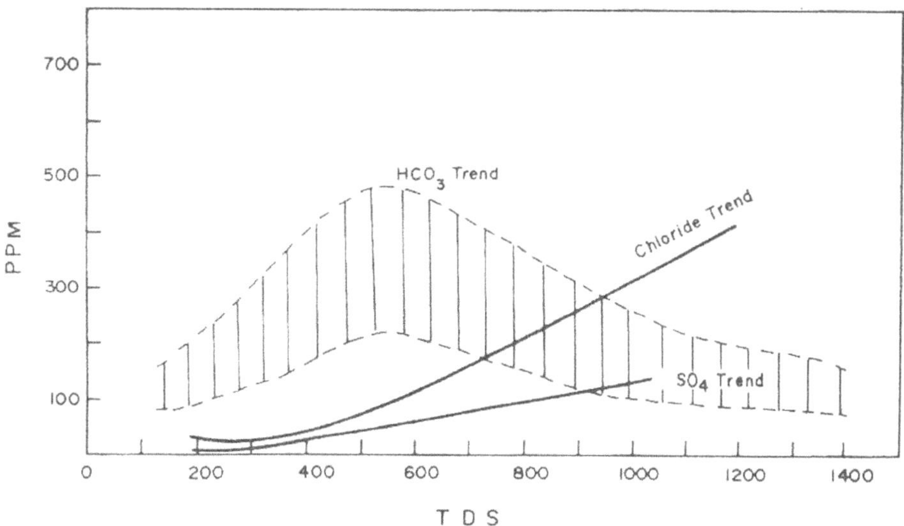

Fig. 6.28 Relationship between total dissolved solids (TDS) and anions (both in p.p.m.) for coastal groundwaters of Orissa, India

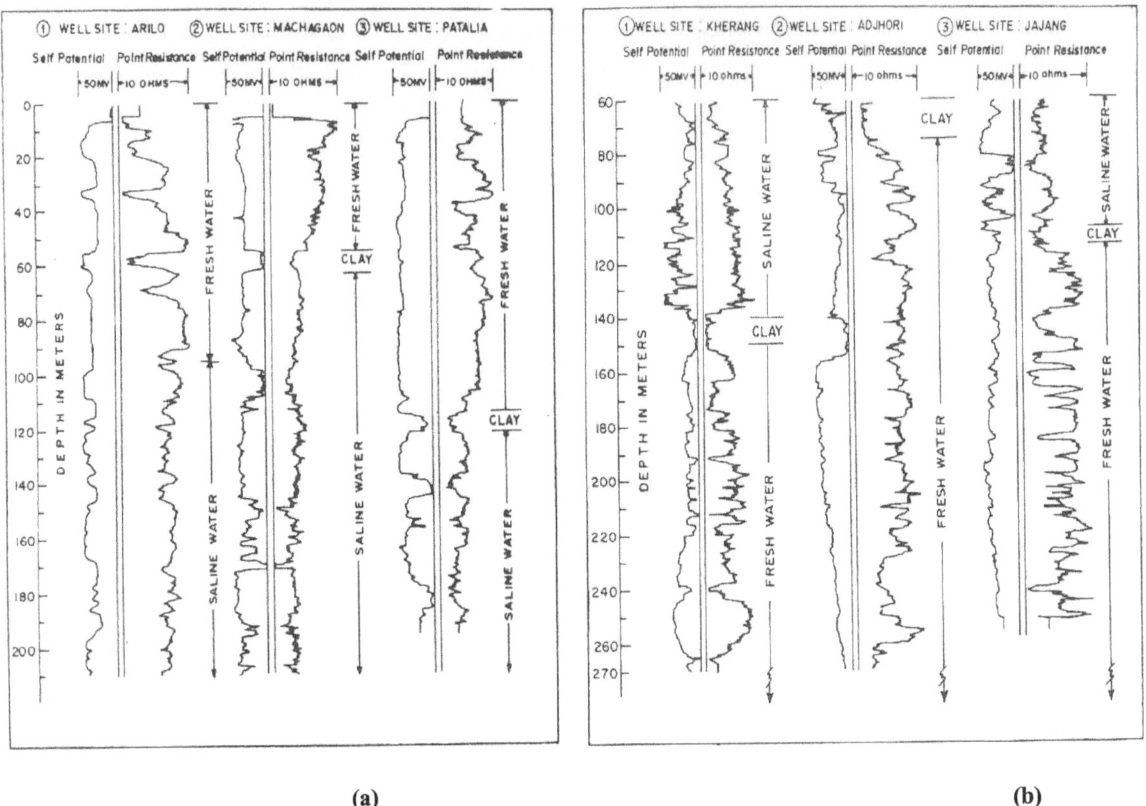

(a) (b)

Fig. 6.29(a,b) Electrical log charactertistics of shallow and deeper fresh aquifer system for coastal groundwaters of Orissa, India

6.5.3 Analyses and interpretation

Electrical logs of six wells representing both shallow and deep fresh groundwater systems were selected for analysis of individual zones of a combination of different zones exhibiting similar shapes and amplitudes Fig. 6.29 (a,b). The shallow fresh groundwater system is characterized by low positive SP anomalies (Arilo, Machagaon and Patalia) and the deeper fresh groundwater system is characterized by low negative SP anomalies (Kherang, Adjhori and Jajang) with exception of the zone between depth levels 95 and 160 meters at Adjhori where low positive anomalies were recorded. The resistivities for the fresh water aquifers are relatively higher when compared to those of saline water aquifers. Invariably, the boundaries of these two fluid systems are separated by aquicludes, but locally the shallower fresh water regime is in direct contact with saline water without any significant separation.

Hydrogeochemical parameters interpreted from SP logs for selected specific zones with the help of equation 6.31 and Fig.6.26(b) are given in Table 6.4. The agreement is fairly good for both shallow and deep groundwater systems. This is because of the quality of logs obtained and the presence of regionally existence thick aquifer zones. The approach could be adopted in most alluvial and sedimentary areas where multi-aquifer systems exist.

Table 6.4 Hydrogeochemical parameters (laboratory and interpreted values)

Values based on laboratory		Values interpreted from SP logs (p.p.m.)								Analysis of water samples (p.p.m.)						
Well site and zone depth (m)	**SP. cond. (µmhos cm^{-1})**	**TDS**	**Cl**	**SO_4**	**HCO_3**	**Ca + Mg**	**Na + K**	**TH[1]**	**SP. cond. (µmhos cm^{-1})**	**TDS**	**Cl**	**SO**	**HCO**	**Ca + Mg**	**Na + K**	**TH**
Hydrogeochemical parameters for shallower groundwater systems at selected wells/zones																
Arilo (60-90)	725	471	44	-	232	45	-	135	560	365	38	20	150-372	50	70	153
Patalia (80-110)	1041	677	142	-	329	16	-	145	1200	780	200	95	60-387	60	220	160
Machagaon (0-50)	747	485	47	-	232	34	-	105	525	340	30	165	140-350	48	83	173
Hydrogeochemical parameters for deeper groundwater systems at selected wells/zones																
Jajang (115-200)	1648	707	199	-	115	28	-	135	1570	1020	323	140	100-253	60	300	160
Kherang (190-210)	830	540	115	50	161	17	135	141	1015	660	150	77	195-450	52	190	160
Adjhori (165-220)	983	639	180	-	165	31	-	205	1200	780	200	95	60-387	69	220	160

[1]TH= total hardness

The above methodology would help to define and characterize the water quality parameters for both shallow and deeper aquifer systems. It would also be possible to prepare regional interface configurational maps (depth wise) in different delta systems and account for the mechanisms of saline water intrusion processes and adopt appropriate well field structure designs for development of fresh water resources.

6.5.4 Interface structure for Mahanadi-Brahmini-Baitarni delta region

The saline-fresh water interface boundaries, interface depth profiles and the nature of interface etc are presented in Table 6.5 and the interface structure map for the entire delta region is presented in Fig. 6.30(b) based on the interpretation of selected Electrical logs in the delta region. The map indicates that most of the outer and mid delta region of Mahanadi-Brahmini and Baitarni is affected by saline water intrusion processes. The saline water intrusion has taken place basically in the top layers of the delta with depth levels varying from 60m to 250m below ground level, below which fresh water aquifer systems occur. Maximum depth levels of interface are seen in the middle and outer Mahanadi delta. The interface depth levels narrows down to both towards north, south and also towards the inland side of the delta. The interface separating upper saline water aquifers and below fresh water systems are aquicludes and aquitards which pinch down inland and increases in thicknesses towards the coast. The interface structure however in the south Mahanadi delta the situation becomes reverse with fresh water sitting over saline waters, with interface depth levels varying from 20m towards the coast to an order of about 120m upland. The interface boundary however is diffusion boundary. A diffusion boundary line is drawn where the interface emerges close to near surface and comes in juxta position with adjoining fresh water zones. Based on the interface structural map, the entire delta region could be divided into four zones namely A,B,C and D.

Zone A represents upper delta region, the region between the apex point and the diffusion boundary. This region is by and large free from saline water intrusion hazards and contain aquifers derived from Recent alluvials, Quaternaries and Tertiaries. A large number of dug wells, shallow and moderately deep systems exist in this zone with moderate to high discharge well fields constructed upto depth levels varying between 80m to 150m below ground level. However these fresh water well fields are required to be optimized and maintain the balance between recharge and discharge potentials in this zone, failing which the diffusion zone could migrate inland and contaminate the fresh water well fields.

Zone B represents the region between diffusion boundary and the shoreline covering parts of central and outfall regions of Mahanadi, Brahmini and Baitarni deltas. This region is characterized by saline brackish waters sitting over fresh water aquifers, with aquicludes and aquitards acting as interfaces. These aquicludes and aquitards pinch out towards inland regions and widens towards the coastal boundaries. Majority of the top saline water aquifer zones, probably belong to Recent alluvials and Quaternaries, and the fresh water bearing aquifers belong to Tertiaries. These fresh water aquifers below the interfaces occur as deep confined aquifers with artesian and semi artesian pressure heads, and are suitable for design and construction of high discharge well fields. Every care should be taken to see that all the boreholes intended for developmental purposes should be electrically logged, identify the interface boundary, and plug the annular space in the gravel packs to avoid vertical migration of saline water to avoid consequent contamination of fresh water aquifers while pumping.

Zone C represents the south Mahanadi delta region where saline-fresh water systems have taken a reversal, with fresh water aquifers sitting over saline water aquifers with a diffusion boundary as interface. The interface depths vary from 20m (towards the coast) to an order of about 120m (away from the coast). Well fields with moderate to heavy discharges could be designed in these tracks while taking care to see that the upconing of saline water below may not take place.

Zone D represent the zone where saline fresh water mixing takes place. The dimensions and geometry of this zone depends upon the dynamics involved in both the systems.

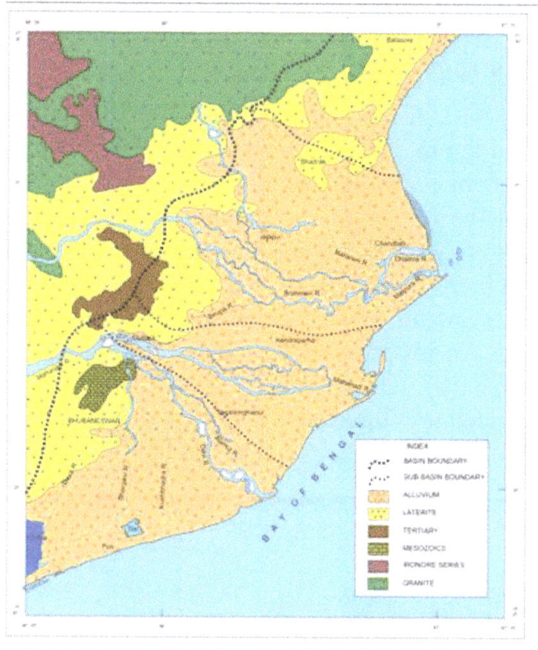

Fig. 6.30(a) Geological map for Mahanadi delta

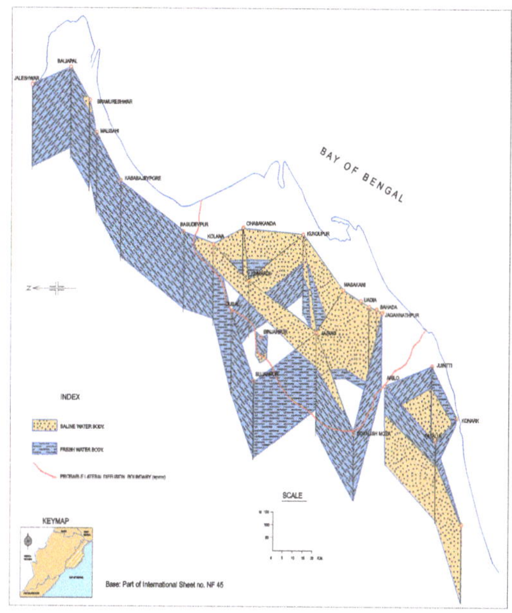

Fig. 6.30(b) Interface structure map of Mahanadi delta

(Source: Radhakrishna, I., 1981)

Table 6.5 Saline-freshwater interface depths and characteristics

S.no.	Delta Region	Well site	Interface depth (m)	Nature of interface disposition
1	South Mahanadi Delta	Puri Konark Patalia Arilo	28 30 100 100	Diffusion zone with freshwater overlying brackish water.
2	North Mahanadi Delta	Jajang Masakani Dubakona Madanpur	120 290 50 60	Aquiclude zone with brackish water overlying freshwater
3	Baitarni-Brahmini delta	Musanga Kherang Dakshinabad Mandari	60 120 Nil Nil	Aquiclude zone with brackish water overlying freshwater. No interface- all through freshwater

Inline with the above data analyses, attempts are made to build the interface structure maps for other delta regions of east coast, where the geophysical and hydrogeological data are sufficiently available. As and when the data base becomes large with more data network, the maps accordingly could be modified in future.

6.5.5 Interface structure for Ponnaiyar-Vellar delta region

The saline-fresh water interface boundaries, their depth levels and the nature of the interface etc for Ponnaiyar-Vellar delta region are presented in Table 6.6 and the interface structure is projected in Fig. 6.31(b) based on interpretations made through Electrical log data available at fifteen well sites. The map indicates that the saline water intrusion basically exists in the top layers of sedimentaries and runs parallel to the coast. The interface thicknesses vary from 40m to 60m in the north to depth levels of the order of 110-120m below ground level. The interface depths increases gradually from north to the south parallel to the coastal boundary with occasional migrations inland depending upon the intensity of Groundwater withdrawals and their impact on the interface. An imaginary line separating the regions affected by saline water and fresh water is shown in Fig. 6.31(b) which represents zone of diffusion. Based on the interface structure the delta region could be divided into three zones namely A, B and C.

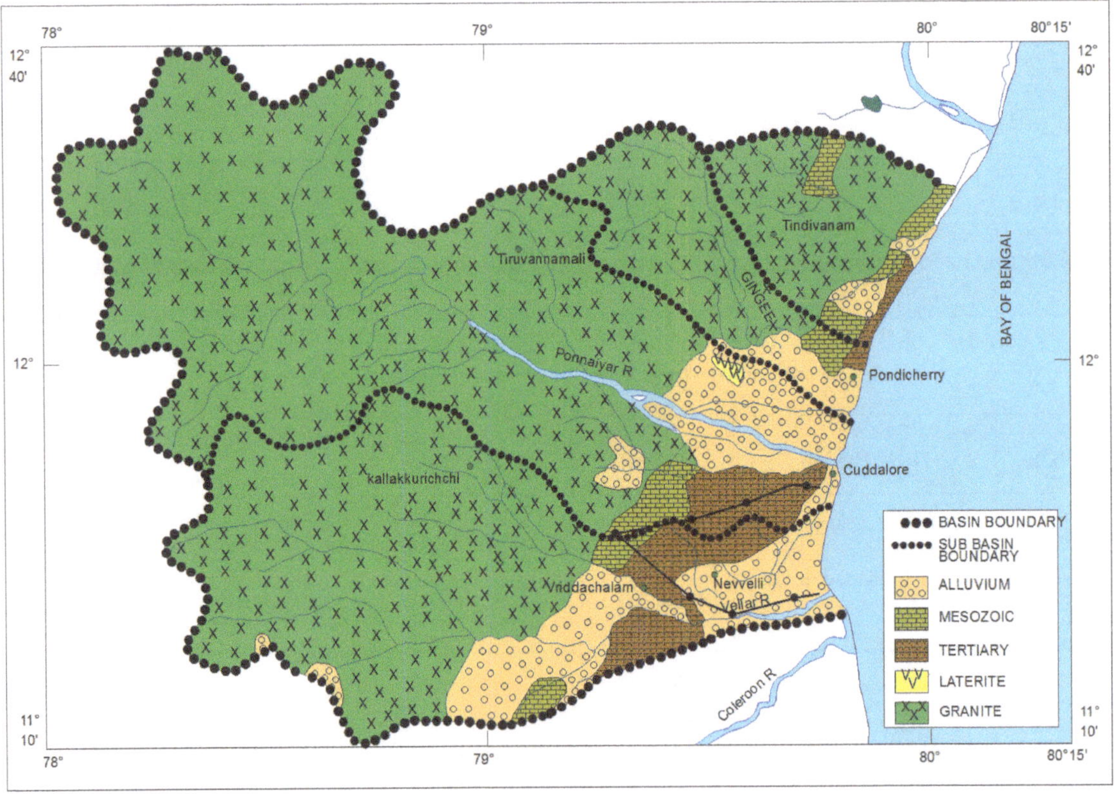

Fig. 6.31(a) Geological map of Ponnaiyar delta

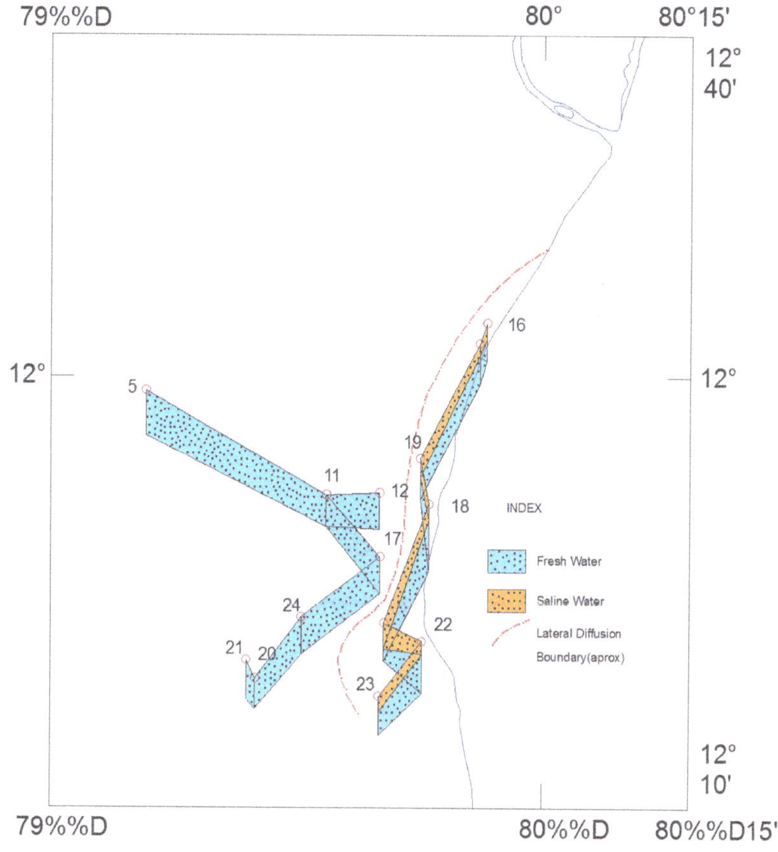

Fig. 6.31(b) Interface structure of Ponnaiyar delta

Zone A represents the region with fresh water, bounded by hard rocks on upstream side and the diffusion boundary. This region by and large is free from salinity hazards and contain aquifer zones belonging to Recent sediments, Quaternaries and Tertiaries and facilitate design and construction of moderate to high yielding well fields. A rigorous monitoring practices are required to be exercised and see that the recharge-discharge balances are maintained and avoid further migration of salinity inland into fresh water aquifer systems.

Zone B represents the region between diffusion boundary and the shoreline and contain mainly rock types belonging to Recent, Quaternaries, Tertiaries and Mesozoics (Cretaceous). Majority of the top layers with varying depth profiles are affected by saline water intrusion processes and the depth of the variations for selected well sites are given in Table 6.6. The aquifer zone below the interface depths by and large contain fresh water upto depth levels varying from 300 to 350 m. Moderate to high discharge well fields could be designed but care should be taken to carryout the electrical logs in every borehole drilled in these tracks and delineate the interface boundary and also adopt cement/concrete plugging in the gravel packs so that vertical migration of saline water may not take place while developing the fresh water well fields.

Zone C represents an imaginary line separating the fresh and saline water affected area where probable mixing takes place. The geometry and dimensions of this zone need to be assessed in the field through rigorous hydrogeological and hydrogeochemical inventories.

Table 6.6 Saline-freshwater interface depths and characteristics

S.no.	Delta regions	Well sites	Interface depth (m)	Nature of interface disposition
1.	Ponnaiyar delta region	* Manjuguppam	0-130 brackish 130-350 fresh	Brackish water zone overlies fresh water with an aquiclude boundary.
		* Sangolikuppam	0-80 saline 100-180 brackish 180-300 fresh	Saline/brackish water overlies fresh water with an aquiclude boundary.
		Koonimedu(16)	0-145 saline 145-300 fresh	Saline water overlies fresh water with an aquiclude boundary.
2.	Interstream area	Kuchipalayam (5)	0-350 fresh	All fresh water (free from salinity hazards)
		* Sangolikuppam	0-80 saline 100-180 brackish 180-300 fresh	Saline/brackish water overlies fresh water with an aquiclude boundary.
		Panrutti (11)	0-300 fresh	No saline water hazard.
		Nellikuppam (12)	0-300 fresh	No saline water hazard.
		Koonimedu (16)	0-145 saline 145-300 fresh	Saline water overlies fresh water with an aquiclude boundary.
		Kattusagai (17)	0-300 fresh (but close to the diffusion boundary)	No immediate saline water hazard.
		Chavadi (18)	0-80 near saline/brackish 80-400 fresh	Saline/brackish water overlies fresh water with an aquiclude boundary.
		Keelalinjipattu (19)	0-110 saline 110-300 fresh	Saline water zone overlies fresh water with an aquiclude boundary.
3.	Vellar delta region	Srimushnam (20)	0-300 all fresh	No saline water hazard.
		Valayamdevi (21)	0-300 all fresh	No saline water hazard.
		Portonova (22)	0-100 saline 100-230 marginally brackish 230-450 fresh	Saline/brackish water overlies fresh water with an aquiclude boundary.
		Chidambaram (23)	0-110 saline 110-300 fresh	Saline water zone overlies fresh water with an aquiclude boundary.
		Neyveli (24)	0-300 all fresh	No immediate saline water hazard.
* could not be displayed in the map for some technical difficulties				

6.5.6 Interface structure for Cauvery delta region

The saline-fresh water interface boundaries, the interface depth levels and the nature of interface etc for Cauvery delta region are presented in Table 6.7 and the interface structure map is presented in Fig. 6.32(b) based on the interpretation and analyses of several electrical logs available in the region. The map indicates that most of the outfall region of the delta is affected by saline water intrusion processes. The saline water intrusion basically has taken place in the top layers of the aquifer systems. The interface thicknesses have a range varying from 80 m to 228 m below ground level below which fresh

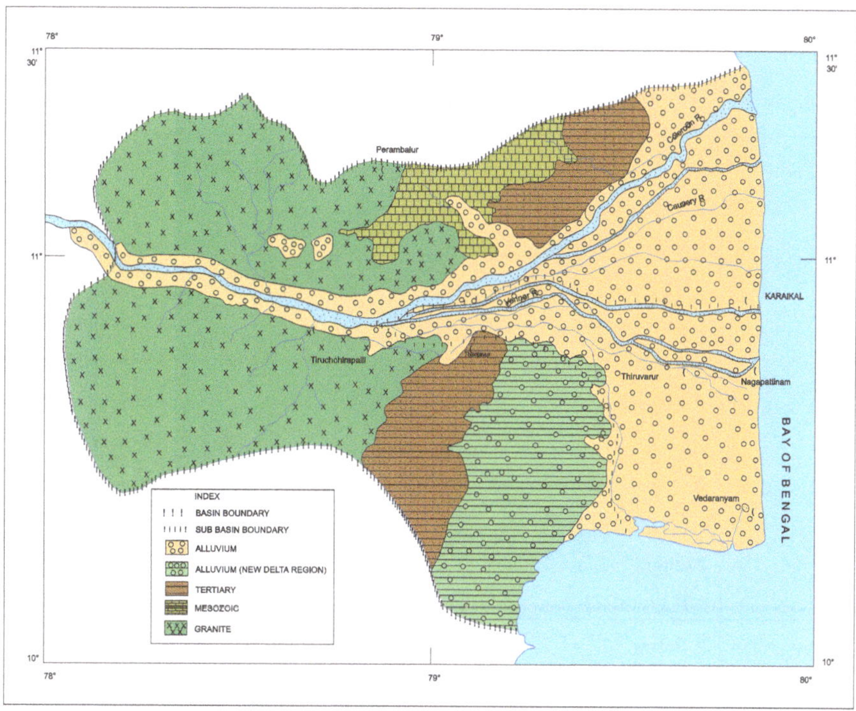

Fig. 6.32(a) Geological map of Cauvery delta

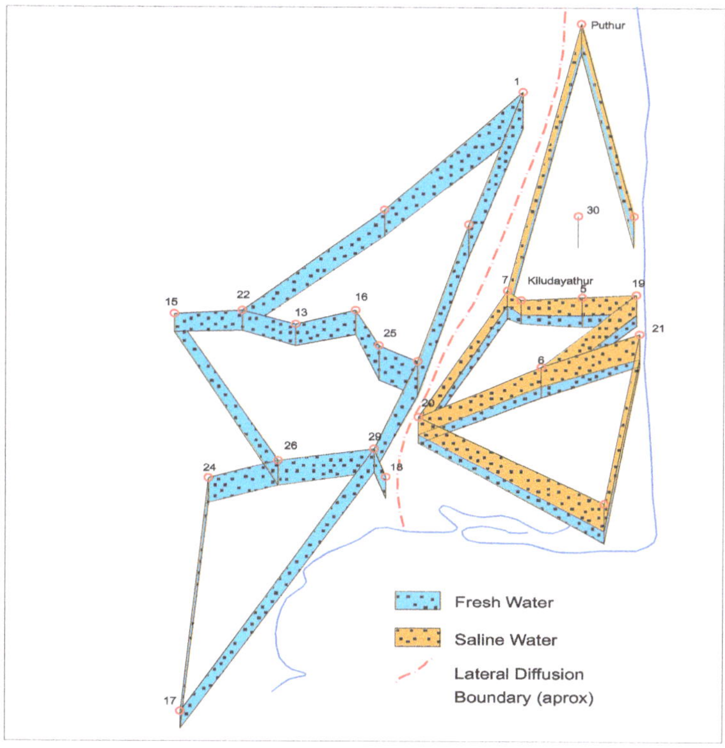

Fig. 6.32(b) Interface structure of Cauvery delta

Table 6.7 Saline-freshwater interface depths and characteristics

S no.	Delta Regions	Well sites	Interface depth (m)	Nature of interface disposition
1.	Upper delta region	Aruvappadi (1)	-	All fresh water (0-300m)
		*Marudandanallur	-	All fresh water (0-300m)
		Kattuthottam (22)	-	All fresh water (0-300m)
		Vallam (15)	-	All fresh water (0-300m)
		Sadayarkoil(13)	-	All fresh water (0-300m)
		Muvanallur (16)	-	All fresh water (0-300m)
		Adanuratti (25)	-	All fresh water (0-300m)
		Muttakurchchi (29)	-	All fresh water (0-300m)
		Nambivoyal (26)	-	All fresh water (0-300m)
		NeyveliVadapadi (24)	-	All fresh water (0-300m)
		Melpattu (17)	-	All fresh water (0-300m)
		* Kattur (27)	-	All fresh water (0-300m)
2.	Lower delta region	* Puttur	0-80 saline 80-250 fresh	Aquiclude zone with brackish water overlying fresh water
		Papancheri (19)	0-228 saline 228-320 fresh	Aquiclude zone with brackish water overlying fresh water
		Kivalur (5)	0-168 saline 168-395 fresh	Aquiclude zone with brackish water overlying fresh water
		Vilamal (7)	0-198 brackish/saline 198-426 fresh	Aquiclude zone with brackish water overlying fresh water
		Prataparamapuram(21)	0-167 saline 167-334 fresh	Aquiclude zone with brackish water overlying fresh water
		Kachchanam (6)	0-243 saline/ brakish 243-395 fresh	Aquiclude zone with brackish water overlying fresh water
		Tirumakottai (20)	0-213 brackish 213-304 fresh	Aquiclude zone with brackish water overlying fresh water
		*Vedarnyam	0-350 saline(?) 350-450 fresh (?)	Aquiclude zone with brackish water overlying fresh water

* could not be displayed in the map for some technical difficulties

water aquifers are encountered. The interface depth narrows down towards north (80 m below ground level) and gradually deepens (250 m below ground level) towards south. Similarly the interface narrows towards west, inland of the delta region. The region affected by the saline water intrusion also narrows down toward north and gradually widens towards the southern parts of the delta. An imaginary line is drawn (diffusion zone) separating the fresh and saline water affected regions, which probably represents a zone of mixing. This may be a dynamic zone which could migrate either landward or seaward depending upon the dynamics involved in the adjoining systems. Based on the interface structure prepared, the delta region could be divided into three zones namely A, B and C.

Zone A represents the upper delta region from apex point, adjoining the hard rock regions upto the diffusion line. This region is by and large free from saline water hazards and contain aquifers from Recent alluvials, Quaternaries, Tertiaries and Mesozoics. A large number of dug wells with high discharges exist with Recent alluvials/Quarternary systems. Deep well structures/well fields with moderate to high discharges can be designed and constructed upto depths varying between 250 to 300 m or even beyond without immediate intervention of salinity hazards. However the well field systems and the withdrawal rates are required to be optimized in such a way that the withdrawal rates balances with the recharge potentials, failing which the diffusion zone could migrate landward further and contaminate the fresh water well fields.

Zone B represents the area between diffusion boundary and the coastline and contain the aquifer systems derived from Recent alluvials, Quaternaries, Tertiaries and Mesozoics. Majority of the top layers with depths varying from 80m to 250 m are affected by saline water intrusion. The depth variations for selected well sites are given in the table 6.7. The aquitard zones below these interface depths upto depth levels varying from 350 to 450 m contain fresh water aquifers. Moderate to high discharge well fields could be designed tapping these zones, but care should be exercised to carry out electrical logs at every site, identify the interface boundaries and adopt cement/concrete plugging in gravel packs so that the vertical saline water flow could be checked while developing the fresh water well fields.

Zone C represents an imaginary line separating fresh and saline water affected regions in the delta where probable mixing takes place. The dimensions and the geometry of the zone need to be assessed in the field through rigorous hydrogeological and hydrogeochemical mapping and monitoring.

6.5.7 Interface structure for Vaigai delta region

The saline-fresh water interface boundaries, the interface depth levels and the nature of interface boundary etc for Vaigai delta region are presented in Table 6.8 and the interface structure map is presented in Fig. 6.33 (b) based on selected electrical log profiles available in this region, particularly in the middle and eastern parts of the Vaigai and outfall region of Manimuttar river. The interface structure for the western region of Vaigai is not clear where sufficient data is not available. The middle and outfall regions of the delta is affected by saline water intrusion processes, where the top and deep layers upto depths varying between 100 m to 280 m contain saline aquifers below which fresh water zones are encountered, probably belonging to Tertiaries and Mesozoics. The interface depths narrow down towards north and also towards western upstream side, and increases towards eastern flanks of Vaigai delta. An imaginary diffusion line is drawn between saline affected regions and fresh water zones in the top layers. The diffusion zoned goes by and large parallel to the coast and pinches down towards the coast line in the western flanks of Vaigai. Based on the data available the delta region could be divided into three zones namely A, B and C.

Zone A represents the upper delta region where the near surface zone and the shallow systems are fresh. A number of dug wells and shallow filter points tapping the zones between 10 to 30m are available. Below this fresh water alluvial zone, Quaternaries, Tertiaries and Mesozoics are available, but sufficient exploratory data is not available whether these aquifers contain fresh water or saline water. However one could confess that this region acts as a recharge zone to several fresh water aquifers in the lower delta zone, should contain fresh water. However the exploratory data available is not sufficient to draw any specific conclusions.

Zone B represents the area between the diffusion line and the shoreline boundary and contain the aquifers derived from Recent alluvials, Quaternaries, Tertiaries and Mesozoics. The top aquifer layers upto depths varying between 100m to 250m below ground level are contaminated with saline water intrusion process, below which fresh water aquifers with moderate yields are available upto depth levels varying between 350 to 400 m below ground level, and one can design and construct moderate well fields in these tracks. It is necessary that every borehole is logged electrically to define and identify the interface boundaries and plug the annular space in the gravel pack by cement grouting to avoid vertical saline/brackish water and contamination of fresh water well fields.

Zone C represents an imaginary diffusion zone and its existence is required to be established through several hydrogeological and hydrogeochemical inventories.

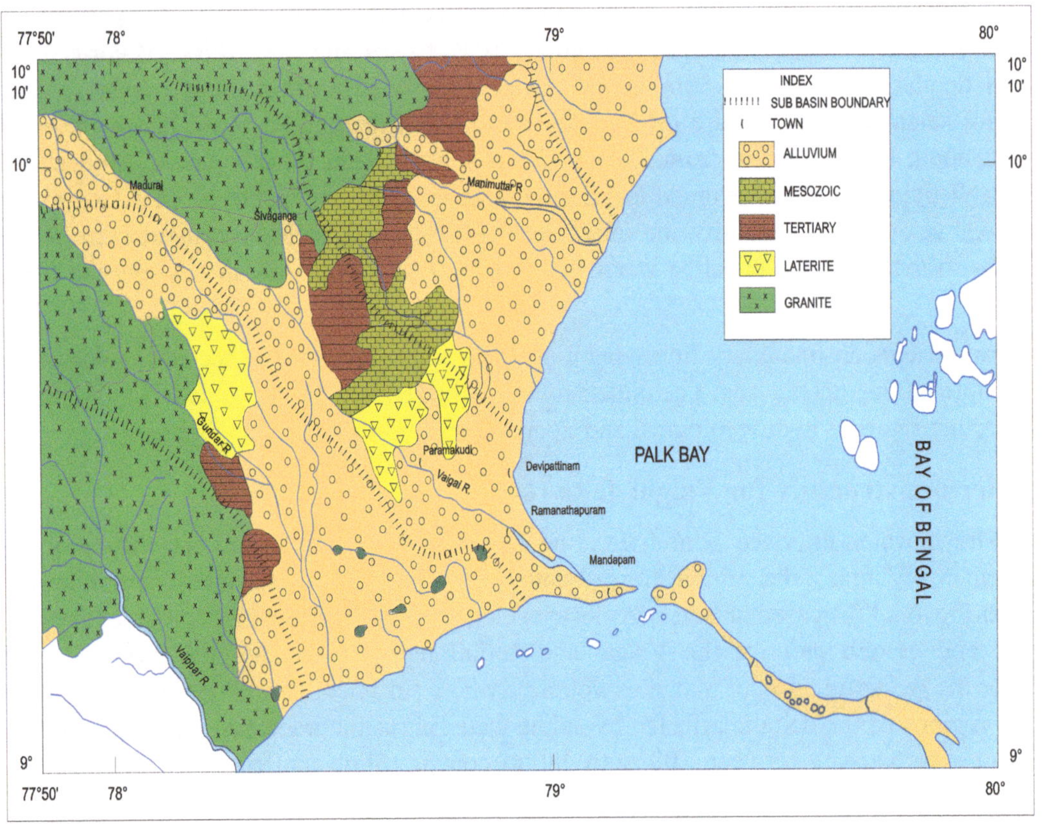

Fig. 6.33(a) Geological map of Vaigai delta

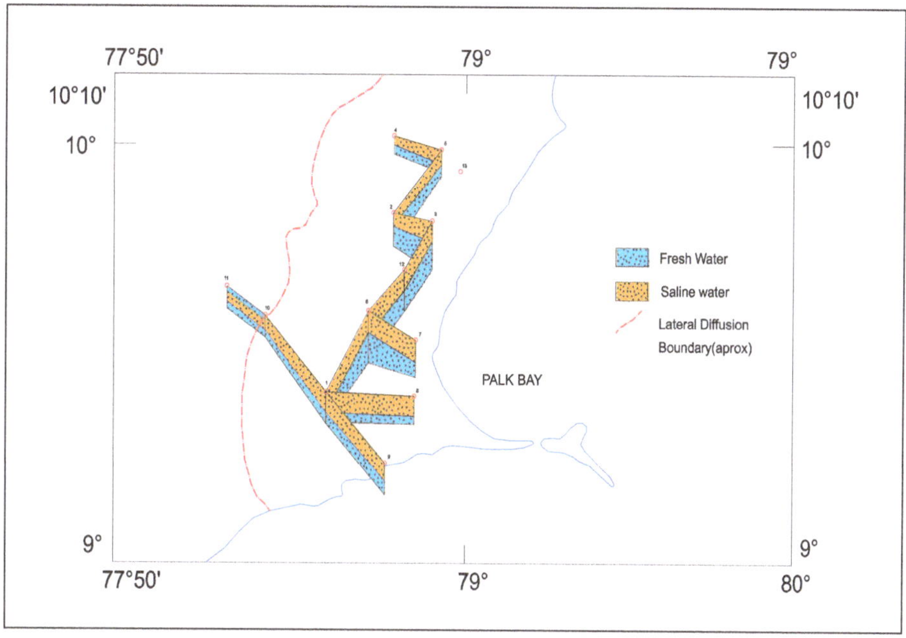

Fig. 6.33(b) Interface structure of Vaigai delta

Table 6.8 Saline-freshwater interface depths and characteristics

S.No	Delta regions	Well sites	Interface depth (m)	Nature of interface disposition
1.	Upper delta region	Kamudakudi (10)	0-30 fresh Below 30 saline(?)	Aquiclude zone with brackish water overlying freshwater
		Melperungarai (7)	0-50 fresh Below 50 saline(?)	Aquiclude zone with brackish water overlying freshwater
2.	Lower delta region	Palangulam (1)	0-254 saline 254-390 fresh	Aquiclude zone with brackish water overlying freshwater
		Tiruvandanai (2)	0-100 saline 100-390 fresh	Aquiclude zone with brackish water overlying freshwater
		Melpanaiyur (3)	0-260 saline 260-669 fresh	Aquiclude zone with brackish water overlying freshwater
		Devakkottai (4)	0-100 saline 100-180 fresh	Aquiclude zone with brackish water overlying freshwater
		Kannangudi (5)	0-150 saline 150-263 fresh	Aquiclude zone with brackish water overlying freshwater
		Tharikomban (6)	0-280 saline Below 280 fresh	Aquiclude zone with brackish water overlying freshwater
		Naranamangalam (7) Pullangudi (8)	0-280 saline 280-304 fresh	Aquiclude zone with brackish water overlying freshwater
			0-280 saline 280-369 fresh	Aquiclude zone with brackish water overlying freshwater

Note: The interface depths are based on preliminary interpretations of available electrical logs.

6.5.8 Interface structure for Bhagiradhi-Hoogly delta region

The saline-fresh water interface boundaries, interface depth profiles and nature of interface etc are presented in Table 6.9 and the interface structure map for the outfall delta region is presented in Fig.6.34 (b), based on the data available and some selected electrical logs (Sinha Ray, S.P. and Das, C.R., 2002). The map indicates that the saline water intrusion processes are prevalent in the lower delta region covering the districts of south 24 paraganas, parts of east Medinapur and Hoara, affecting the top layers and with depth of variations ranging between 130 to 250m below ground level. Maximum interface depths are observed in the south 24 paraganas in the extreme southern and south eastern part of the delta and the depth levels reduce towards west (east Medinapur) and north (Hoara). An imaginary line is drawn in between regions separating the regions of saline affected area and the fresh water area. This probably represents the area where mixing processes are taking place. Based on the interface structure map, the delta region could be divided into three zones A, B and C.

Zone A covers large tracks of delta (middle and upper parts) not affected by salinity. The aquifers in this zone belong to Recent, Quaternaries and Tertiaries and constitute potential fresh Groundwater systems, where moderate to high well discharge fields could be designed and constructed upto depth levels varying from 300 to 400 m below ground level. At the same time optimization of fresh water resources, monitoring of water levels and water qualities are important and see that recharge-discharge balances are maintained and the diffusion line may not migrate inland and contaminate further fresh water well fields. There is another important constraint in this zone namely 'Arsenic contamination'. The arsenic contaminated aquifers appears to be present all along the Bhagirathi-Hoogly river track and the adjoining marginal areas at depths varying between 40 to 80 m below ground level. It would become necessary therefore to go for tapping the deeper systems while sealing/plugging arsenic contaminated aquifers to avoid vertical migration of these arsenic contaminated waters into deeper fresh water zones. It is also required to implement scientifically planned optimisation schemes while developing there deeper systems.

Zone B represents the area between the diffusion boundary and the coastline where the saline water aquifer zones occupy the top layers. The saline water impacts in these tracks are seen upto depth levels varying from 60 to 200 m below ground level. Maximum interface depth levels are seen upto 250 m in the southern and south eastern sectors of the delta (South 24 Paraganas) and reduce gradually towards western (east Medinapur) and northern sides (Hoara). Below these interface depth levels, potential fresh water confined aquifers occur with artesian and semi artesian pressure heads and with moderate to high yields and transmissivities which probably belong to the Tertiary group of sediments. Here also while designing the well fields, and tapping the deep confined systems below the interface levels concrete plugging is required to be implemented in gravel packs to avoid vertical migration of saline water. At the same time care should be exercised to see that the developmental potential created should not exceed the recharge potentials failing which vertical migration takes place through aquitards into the depressurized fresh water aquifers. Monitoring of water levels and water qualities and building up appropriate mathematical models therefore are of paramount importance.

Zone C represents the diffusion zone which is an imaginary trend line separating region affected by salinity and the region free from salinity. It represents the probable zone of mixing and the geometry and dimensions of this zone keeps changing depending upon the fluid dynamics in this part.

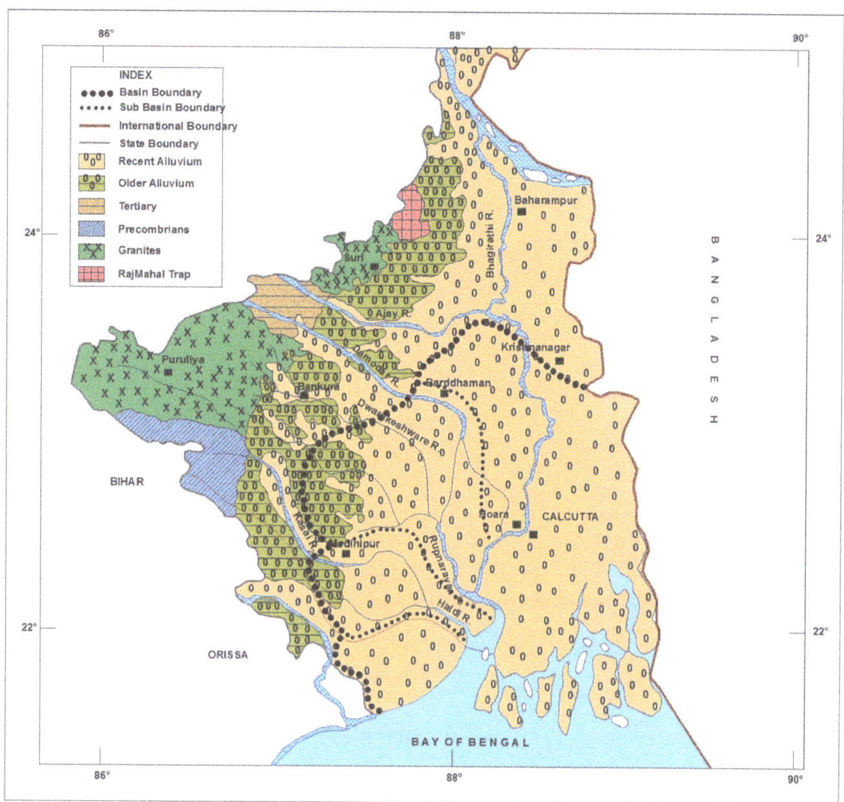

Fig. 6.34(a) Geological map of Bhagiradhi-Hoogly delta

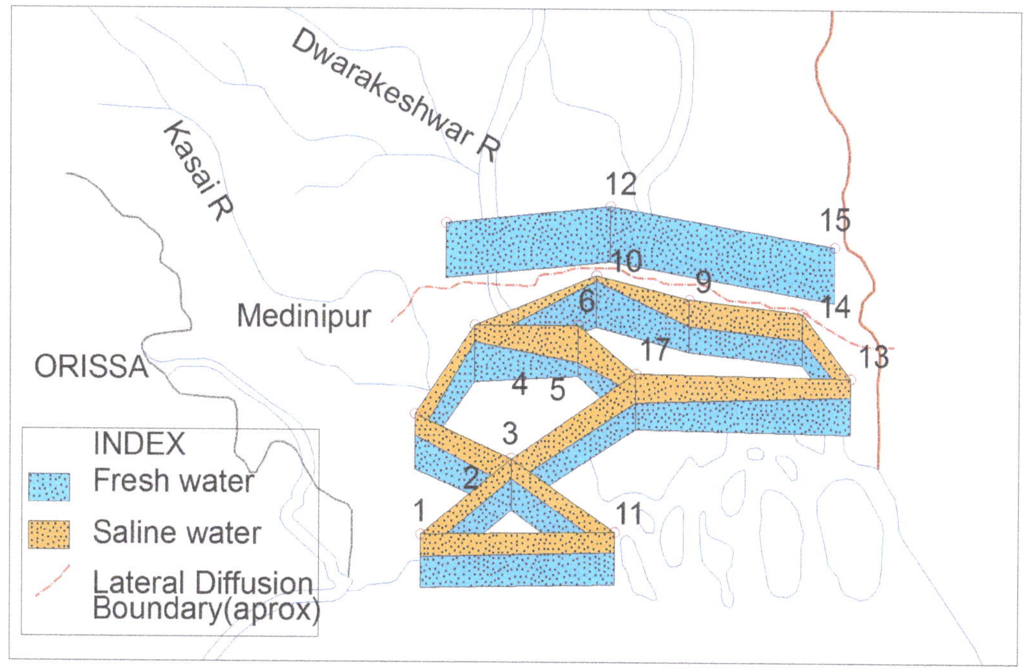

Fig. 6.34(b) Interface structure of Bhagiradhi- Hoogly delta

Table 6.9 Saline-freshwater interface depths and characteristics

S.no	Delta region	Well sites	Interface depth (m)	Nature of disposition
1.	Bhagirathi-Hoogly delta region	Pichabani (1)	0-180 saline Below 180 fresh	Aquiclude zone with brackish water overlying freshwater
		Kakadevichak (3)	0-190 saline Below 190 fresh	Aquiclude zone with brackish water overlying freshwater
		Falta (6)	0-180 saline Below 180 fresh	Aquiclude zone with brackish water overlying freshwater
		Baruipur (9)	0-200 saline Below 200 fresh	Aquiclude zone with brackish water overlying freshwater
		Budj Budj (10)	0-130 saline Below 130 fresh	Aquiclude zone with brackish water overlying freshwater
		Namkhana (11)	0-200 fresh Below 200 fresh	Aquiclude zone with brackish water overlying freshwater
		Shibpore (12)	0-300 fresh	All fresh water
		Gosaba (13)	0-200 saline Below 200 fresh	Aquiclude zone with brackish water overlying freshwater
		Canning (14)	0-250 saline Below 250 fresh	Aquiclude zone with brackish water overlying freshwater
		Dholtukri (15)	0-300 fresh	All fresh water
		Mandir Bazar (17)	0-200 saline Below 200 fresh	Aquiclude zone with brackish water overlying freshwater

6.6 ESTIMATIONS OF SUBMARINE GROUNDWATER DISCHARGE (SGD) IN THE COASTAL DELTA SYSTEMS

The discharge of groundwater into coastal zone has received increased attention over the last one decade or so. It is now well recognized that the processes/fluxes may represent an important path way for material transport (chemical constituents and nutrients) from terrestrial waters to the marine environment and sometimes vice-versa. These fluxes could act sometimes as beneficiaries to the ecosystems wherever required. Sometimes these fluxes also could be damaging wherever contaminated sub-surface waters flow from the upstream areas into the marine ecosystems. Quantification of these fluxes/processes though very complex is an important part of the hydrological cycle and help to assist the safe withdrawal rates of fresh water on the coastal terrestrial watersheds. Johannes, R.E. (1980) was one of the first to discuss the groundwater as the potential path way for nutrients to the coastal estuaries. Burnett, W.C., et al (2003) defines SGD as the discharge of groundwater across the sea floor and into the ocean. Zektser, I.S., et al (1983) defines SGD as net groundwater discharge into the ocean. Church, T.M. (1996) defines SGD as the direct groundwater outflow across the ocean-land interface into the ocean which would also include recirculated sea water. Zektser, I.S. and Dzhamalov, R.G. (2007) made a comprehensive review on the subject groundwater / sea water interactions in a new book entitled Submarine Groundwater. The groundwater discharge near the shorelines is a result of changing hydraulic gradients from the land towards the sea. Submarine groundwater discharge in general takes place in most continental ocean boundaries but the quantity and quality of seepage waters vary from place to place. In regions where large surface discharges occur (large deltaic tracts) the subsurface discharge could also be large. Similarly in the bay areas where high porous material exist in the shallow areas the discharge could be sufficiently large. The discharge could be very low in the interstream areas where more significant porous material exist. The subsurface discharge could also take place through deep multilayered confined aquifer systems. The amount of discharge from such zones depends upon the gradient and hydraulic permeability of the aquifers. The deep drilling operations and the geophysical studies conducted over the east coastal delta regions of India indicate fairly thick freshwater aquifer material exist at deeper levels with extensions into the coastal boundaries and probably into the ocean side. Estimations of SGD in the deep confined coastal systems are rather difficult. However some preliminary estimations are projected in the following paragraphs.

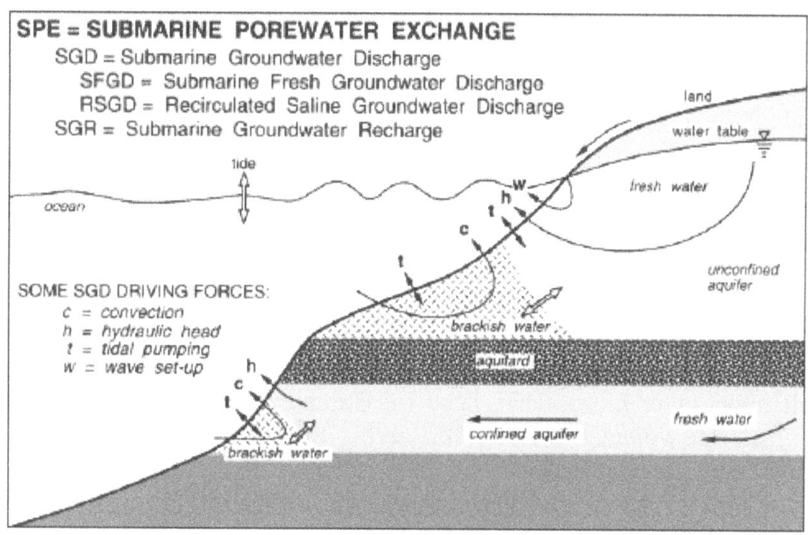

Fig. 6.35 Schematic depiction of processes associated with SGD

(Source: Taniguchi, M., et al, 2002)

The near surface and shallower systems near the coastal margins could contribute large SGD's into the ocean. These include fresh waters and also recirculated brackish waters. The global estimates of SGD's could vary between 0.2 to 10% of the river flows (Garrels, R.M. and MacKenzie, F.T., 1971) at the same time high SGD's have been reported in parts of Atlantic bight to the extent of 40% of the river water flux (Moore, W.S., 1996). Similarly large fluxes of SGD's are reported in the mouth regions of Ganges-Brahmaputra delta region (Moore, W.S., 1997). Some of the experiments conducted in the coastal zone of Osaka bay Japan have also indicated high discharge flux of the order of 36% of the river discharge. A conceptual drawing indicating the processes and distribution of fresh and saline water are shown below in the Fig. 6.35.

6.6.1 Driving forces

A variety of driving forces are involved in controlling the SGD's such as (i) Terrestrial hydraulic gradient, permeability and thicknesses of the aquifer material, sea bottom sediments and their spatial variabilities. (ii) Tidal and wind driven changes in sea level that influences the hydraulic gradient and also wave induced tidal pumping at near shore sea bottom sediments. (iii) density differences between fresh and saline Groundwaters, hyper saline surface water in salt marshes and tidal flats can result in buoyant forces and density driven convection (Simmons, C.T., et al 1991). (iv) Seasonal inflow and outflow that results as a consequence of recharge. (v) geothermal heating. SGD therefore is a complex phenomena that contains mixtures of fluids and various driving forces which become difficult to quantify.

6.6.2 Measurements/approaches

Several methods are in vogue to measure and quantify the subsurface seepages into oceanic bottom sediments. These include (i) seepage meters (ii) hydraulic gradient method (Darcy's law) (iii) tracer studies (iv) numerical models.

Seepage meter: A seepage meter provides simple inexpensive method to measure the subsurface flow from coastal terrestrial aquifers into oceanic sediments. Lee, D.R. (1977) designed and developed a simple equipment consisting of 55 gallons steel drum with one end fitted with a sample port and a plastic bag. The volume of water that enters the bag over a known time and area will give the seepage rate, these meters are labour intensive and appear to under-estimate discharge due to the increased hydraulic frictions associated with the meter. Of late these seepage meters are being automated by use of thermal (Taniguchi, M. and Fukuo, Y., 1993) and ultrasonic techniques for measuring flow rates (Paulsen, R.J., et al 2001).

Hydraulic gradient method (Darcy's law): Groundwater seepage rates could also be measured through design and construction of several piezometers and piezometric nests (Freeze, R.A. and Cherry, J.A., 1979). The Darcy's law/expression, $Q = KIA$ has been widely used in several hydrological studies. The volumetric flow rate of water Q is directly proportional to the hydraulic gradient I, the aquifer hydraulic conductivity K and the cross sectional area of the aquifer A. A number of piezometers designed and constructed along with the geophysical log profiles could help to estimate not only the fresh, brackish and saline water aquifer boundaries but also help to obtain the flow rates into the oceanic bottom sediments through multi layer piezometric head measurements in different seasons. This approach would help to estimate the volumetric flow rates across the coastal boundaries and into the oceanic sediments, if the hydraulic gradient maps, hydraulic conductivities aquifer wise are precisely estimated. Some preliminary estimates for selected Indian coastal deltas are furnished below based on the available piezometric and geophysical data.

Tracer studies: Tracers such as Radium, Radon and Methane provide an integrated signal allowing quantification of SGD's throughout the shallow coastal waters (Rama and Moore, W.S., 1996; Krest, J.M., et al 1999; Charette, M.A., et al 2003; and Breier, J.A., et al 2004). The ideal tracer substance for SGD measurements is that which can dissolve in fluids so that its movement could be measured in space and time. The tracer substance should exhibit substantial enrichment in groundwater relative to other potential end members and also should behave in a conservative manner within the coastal zone. Radon and Radium isotopes appear to be best in this regard for SGD measurements. Methane is a product of anaerobic decay and is found in high concentrations with sufficient organic matter but is subject to microbial uptake and its production is not strictly conservative and so is not a dependable parameter for SGD measurements (Bugna, G.C., et al 1996). In recent years several studies are conducted using radium isotopes and ^{222}Rn to assess the SGD's into ocean. Moore, W.S. (1996) reported large enrichments of ^{222}Rn in coastal waters of south Atlantic bight and found that the SGD's are of the order of 40% of the river water flux.

Numerical Techniques: Variable density groundwater flow models could also be used to assess the groundwater discharge if sufficient accurate hydrogeological data such as recharge inputs, hydraulic conductivity levels along with geophysical log data are available. Langevin, C.D. (2001) used USGS

SEAWAT variable density groundwater transport code to model SGD's in Biscayne bay, Florida with mixed results. More rigorous work analysis probably are required to establish the usefulness of the numerical modeling techniques.

As far as the Indian coastal systems are concerned no significant SGD measurements so far are quantified excepting the work of Moore, W.S. (1997) in the Gangetic-Brahmaputra complex where the SGD's are reported to be of high order. In absence of such data in other coastal delta systems of India a preliminary estimation of SGD flows along the major coastal delta systems based on piezometric gradient analyses are presented below.

6.6.3 Preliminary estimations of SGD flows along the major coastal delta basins

Table 6.10 Cauvery delta basin

S.No	Basin	Profile	Avg. Transmissivity (m^2/day)(T)	Gradient (m/Km) (I)	Length of the Coast (km)(L)	Total flow (Q) (MCM)
Near Surface Aquifers						
1	Cauvery Delta	P1	2500	0.63	40	22.81
		P2	2500	0.58	23	12.2
		P3	2500	0.43	17	6.6
		P4	2500	0.54	13	6.44
		P5	2500	0.60	25	13.68
		P6	2500	0.54	35	17.35
		P7	2500	0.69	44	27.68
Deep Aquifers						
1	Cauvery Delta	P1	1500	0.61	42	13.93
		P2	1500	0.45	17	4.23
		P3	1500	0.69	38	14.4
		P4	1500	0.80	33	14.45
		P5	1500	1.32	38	27.37

- Total aerial extensions of the basin considered : 21,000 sq.kms
- The estimated recharge potentials in the basin area : 4436.89 MCM
- Estimated subsurface groundwater flow (SGD) across the coastal margin of Cauvery delta is of the order of 181.14 MCM which is 4.08% of the total recharge potentials in the basin.

Table 6.11 Mahanadi-Brahmini-Baitarni delta basin

Near Surface Aquifers

S.No	Basin	Profile	Avg. Transmissivity (m^2/day) (T)	Gradient (m/Km)(I)	Length of the Coast (km)(L)	Total flow (Q)
1	Brahmani	P1	1600	0.57	21	6.99
		P2	1600	0.17	42	4.17
2	Mahanadi	P3	2200	0.18	30	4.34
		P4	2200	0.19	42	6.41
3	Daya	P5	3000	0.21	38	8.74
		P6	3000	0.47	30	15.44
		P7	3000	0.68	30	22.34

Table 6.11 *Contd...*

S.No	Basin	Profile	Avg. Transmissivity (m^2/day) (T)	Gradient (m/Km)(I)	Length of the Coast (km)(L)	Total flow (Q)
Deep system						
1	Brahmani	P1	1000	0.23	47	3.95
		P2	1000	0.17	46	2.85
2	Mahanadi	P3	800	0.12	47	1.65
3	Daya	P4	3000	0.18	80	15.77

- Total aerial extensions of the basin considered : 15,924 sq.kms
- The estimated recharge potentials in the basin area : 3265.37 MCM
- Estimated subsurface groundwater flow (SGD) across the coastal margin of Mahanadi-Brahmini-Baitarni delta is of the order of 92.65 MCM which is 2.83% of the total recharge potentials in the basin.

Table 6.12 Bhagiradhi-Hoogly delta basin (Ganges Delta)

S.No	Basin	Profile	Avg.Transmissivity (m^2/day) (T)	Gradient (m/Km) (I)	Length of the Coast (Km)(L)	Total flow (Q) in MCM
Near Surface Aquifers						
1	Subharnarekha+Kasai	P1	800	0.66	100	19.27
2	Damodar+Rupnarayan	P2	800	0.50	20	2.92
3	Bhagirathi-Hoogly	P3	1000	0.20	120	8.76
Deep Aquifers						
1	Subharnarekha+Kasai	P1	1700	0.15	100	9.30
2	Damodar+Rupnarayan	P2	2000	0.17	20	2.48
3	Bhagirathi-Hoogly	P3	3200	0.11	60	7.7
		P4	3200	0.125	60	8.76

- Total aerial extensions of the basin considered : 34,040 sq.kms
- The estimated recharge potentials in the basin area : 6254.5 MCM
- Estimated subsurface groundwater flow (SGD) across the coastal margin of Bhagirathi-Hoogly delta is of the order of 59.12 MCM which is 0.945% of the total recharge potentials in the basin.

6.7 MANAGEMENT/MITIGATION STRATEGIES

Salt water intrusion is a major threat to fresh water resources in the coastal delta complexes of India and elsewhere in the world. Fresh water aquifers in the coastal delta regions are the most productive groundwater systems supplementing irrigation and agricultural needs to the tune of about 40-50%. The seaward flow of fresh groundwater under natural conditions prevents salt water from entering inland into the fresh water systems. The saline water intrusion could take place through several regional and localized hydrological processes such as: (i) over exploitation of fresh groundwater resources which are hydraulically connected to either laterally with the sea or estuaries or with paleo marine saline water aquifers underlying the fresh water zones. (ii) development of river valley projects in the upstream areas of the deltas intercepting large quantities of normal flows and creating imbalances in the hydrogeological environment in the downstream sides of the delta. (iii) reduction of recharge and fresh water storage potentials in both near surface and shallow/deep systems of the deltas. (iv) the tidal impacts and their migrations inland and seaward. The saline waters because of their heavier densities could migrate to the bottom of the aquifer systems during different tidal cycles and displace fresh waters underground. (v) the climate change and sea level rise events. The expected rise of sea level with in the next 50-150 years is of the order varying between 0.5 to 1.5 m. This could inundate large areas along the coast and could trigger saline water intrusion into the fresh water systems.

The fresh water systems in the Indian coastal deltas occur as unconfined aquifers in the near surface/shallow regions and also as deep confined systems at different levels. These two systems invariably are separated by brackish/saline water zones. The thicknesses of these zones could vary between 20 m to 150 m. The near surface/shallow systems are largely restricted to paleochannels, levees and beach sands etc with varying thicknesses and geometries. Large number of well fields in the form of filter points, shallow tubewells, dugwells and dug cum borewells have been operationalized along different coastal delta systems withdrawing heavy groundwater. These morphostructures could either be linked hydraulically with river/canal systems or remain isolated. These systems which are hydraulically linked with perennial river/canal systems give rise to sustainable yields for long periods of time. The systems which are isolated and developed could either get desaturated or converted into brackish water systems through the process of upconing from the brackish/saline water aquifers underneath the fresh water zones. The second group of fresh water aquifer systems are deep confined aquifers which normally underlie the brackish/saline water zones at different depth levels. The deep fresh water systems underlying the saline water bodies are multilayered and have their hydraulic interconnections with the upstream recharge areas and the associated surface water bodies like rivers, canals and tanks. These deep confined fresh water aquifers though deposited under marine conditions probably represent saline water flushed out systems, filled subsequently with fresh water through several cycles of recharge processes. These systems probably could extend beyond the shore line boundary and could get discharged into the ocean bottom through different leaching processes and also as springs depending upon the nature of hydrogeological framework at the ocean bottom. Protection and management of these fresh water aquifers either at near surface or at deeper levels is of paramount importance. The fresh water bodies in the coastal delta complexes when subjected to heavy draft situations sometimes overtaking the recharge potentials could lead to saline water migration inland either laterally or through vertical migration from the underneath saline water zones. These processes could damage or contaminate the fresh water well fields. Similarly creation of heavy draft in the deep confined bodies beyond the optimum limits could destabilize the fresh water pressure heads and induce vertical downward migration through aquitards or through lateral migration. All these complicated processes could be solved if one understands and optimizes the system before hand prior to the developmental strategies are introduced and evolve a reliable optimization scheme through numerical models.

6.7.1 Numerical models

Aquifer simulation models help to predict the behavior of subsurface flow and transport processes. These involve solutions of partial differential equations governing the flow and solute transport to obtain solute concentrations in space and time. The coupled partial differential equations could be solved in a conceptual framework using two approaches. The first approach known as sharp interface approach, which considers fresh water and salt water as two immiscible fluids of different densities separated by a sharp interface boundary. The second one is the miscible interface with diffusion/dispersion processes with varying thicknesses and boundaries. The partial differential equations describing the groundwater flow and solute transport could be solved mathematically using either analytical solutions or numerical solutions. Majority of the analytical solutions have already been described in the earlier paragraphs. In the numerical solutions continuous variables are replaced with discrete variables that are defined at the nodal points of the grids. The continuous differential equation which defines hydraulic heads or solute concentrations everywhere in the system is replaced by a finite number of algebraic equations that define the hydraulic head or concentrations at specific points. This system of algebraic equations are solved using matrix techniques. The non-linear solute transport equation is generally more difficult to solve numerically than the groundwater flow equation, because of the mathematical complexities in the transport equation. In view of these complexities, majority of the numerical models or simulation models are likely to face several errors which could probably be minimized if one uses finer discretization in space and time and could probably reduce numerical errors.

Majority of the models concerned with saline water intrusion programmes deal with groundwater flow models and solute transport models either independently or combinedly. Groundwater flow models help basically to evolve the regional steady state flow in the aquifer systems, changes in the hydraulic heads caused by changes in discharge or recharge, changes in the head near well fields, dewatering well systems, injection wells and surface and groundwater interactions. A solute transport model basically helps to understand salt water intrusion processes near the coastal boundaries, leachate movements from land fills, contamination plumes from seepage ponds, movement of pesticides from agricultural fields and other types of groundwater pollution.

In order to apply the above models to different field situations one needs to understand the aquifer /aquiclude configurations of the system. This includes aerial extensions and thicknesses of all aquifers and aquicludes, the dispositions of surface water bodies like tanks, canals and rivers etc. Hydraulic properties such as transmissivities/permeabilities and storativities etc are required to be established. It is also important to establish the hydraulic connectivities between surface water bodies and the aquifers. Estimations of natural recharge and discharge and their boundaries through continuous monitoring of water table and potentiometric mapping is also important. These parameters if obtained with reasonable accuracy will help to obtain a good flow dynamic models. Similarly a solute transport model needs information like effective porosities, aquifer dispersivities, fluid density variations and total solute concentrations in the basin. These input parameters in the solute transport model will help to derive the solute movement in relation to the stresses involved in the system.

A numerical model initially is required to be calibrated by taking the initial estimates of model parameters and solving the model to check how exactly it reproduces some known conditions of the aquifer. The models in general are initially calibrated against the steady state groundwater heads (water table/ potentiometric surface). In the event of unmatched situations the aquifer parameter values and recharge potentials are varied until the model closely matches with the known water table/potentiometric surface conditions. This process is termed as model calibration. In the event of a transient situation when the water levels /potentiometric surfaces are dropped, appropriate stresses in the model system could be given in line with the existing well field draft situations. The model parameters like S and T etc could be varied so as to match with the transient dynamic situations. Such model calibrations if done could be useful to develop predictive analysis of the system.

A number of models have been developed over the last forty years to solve the problems related to saline water intrusion. They range from simple one dimensional analytical solutions based on Ghyben-Herzberg principle to two and three dimensional numerical computer models. Ghyben-Herzberg principle is based on the hydrostatic equilibrium between fresh and saline water and the algebraic expression provides a first approximation of the extent of salt water intrusion. After the development of Ghyben-Herzberg principle several analytical solutions have been published describing different boundary conditions and cross sections (Glover, R.E., 1959; Hubert, M.K., 1940; Van Der Veer, P., 1977; Bear, J. and Dagan, G., 1964a; Bear, J. and Maulem, Y., 1974). The analytical models can be solved rapidly with a programmable calculator. The use of two dimensional vertical plane models based upon mathematical equations of fluid flow and the transport of salt concentrations represent next level of development in the numerical model analyses. The three dimensional models offer the most complete mathematical representation of salt water intrusion problems. These models are effective in taking into account the irregularities of the coastline and aquifer, aquiclude boundaries, anthropogenic impacts, heterogeneities in aquifer systems and spatial variations of salt concentrations. These models provide a greater insight and yield accurate results and also act as powerful tools` to predict the future hydrogeological and hydrogeochemical regime and in the environmental management programmes.

The most commonly used numerical algorithms include the finite difference and finite element methods. Both methods solve complicated boundary conditions by dividing the entire study region into grids and meshes. The governing equation of the hydrogeological problem either in steady state (Poisson equation or Laplace equation) and transient (diffusion and convection) equation can be solved for each grid or mesh and ultimately the problems could be solved by extending the solution to the entire study area. Some of the significant contributions related to the salt water intrusion problems (analytical, sharp interface and miscible interface models) are described in the following paragraphs.

6.7.2 Analytical models

The flow and transport processes in coastal aquifer systems are quite complex and are highly non linear. A transition zone of varying densities and thicknesses exist between salt water and fresh water. Several workers earlier have ignored this transition zone and assumed as a sharp interface. The analytical solutions for sharp interface conditions in a coastal system was presented by Henry, H.R. (1959); Bear, J. and Dagan, G. (1964(a,b)); Hantush, M.S (1968); and Strack, O.D.L. (1976). Subsequently several advances have taken place in terms of sharp interface and miscible interface models. A chronological order of these developments were given by Reilly, T.E and Goodman, A.S. (1985); Bobba, A.G. (1993).

6.7.3 Sharp interface models

The sharp interface models which solve the coupled fresh water and salt water flow equations have been developed with different numerical techniques (Shamir, U. and Dagan, G. (1971), Vappicha, V.N. and Nagaraja, S.H. (1976), Pinder, G.F. and Page, R.H. (1977) and Mercer, J.W., et al (1980). Wilson, J.L. and Sa Da Costa A. (1982) developed a 2-D sharp interface model that stimulates fresh water and salt water as two immiscible fluids coupled with an interface boundary.

The vertically integrated equations are solved using finite element approach. Polo, J.F. and Ramis, F.R. (1983) developed a similar model using finite difference approach which was unconditionally convergent and stable. Alternate Direction Implicit (ADI) was used for faster convergence. The model was validated using analytical solutions for steady state and non steady state cases.

Essaid, H.I. (1986, 1990a) developed a Quasi 3-D finite difference model that simulates the fresh and salt water flow separated by a sharp interface in a multilayered coastal aquifer system including the effects of salt water dynamics on fresh water. The flow in the system is assumed to be horizontal and could be integrated in vertical direction also. The coupled partial differential equation were solved for fresh water, salt water and interface elevations using strongly implicit procedure. The flow equations in each of the layers were solved and were coupled to the adjacent layers through vertical leakage parameters. The results were verified through analytical solutions and were applied to field problems.

Huyakorn, P.S., et al (1996) developed a sharp interface numerical model to simulate salt water intrusion. The model employs a mixed formulation of one fluid potential and pseudo-saturation as dual dependent variables. The numerical solution incorporates upstream weighting and non linear algorithms with several enhanced features like aquitard leakages, well hydraulic parameters and a robust Newton-Raphson procedure with automatic time stepping. A more realistic approach for modeling sea water intrusion will be three dimensional flow and transport processes with a mixed zone. As the saline fresh water mixes the density of the mixed fluid varies spatially. Huyakorn, P.S., et al (1987); Galeati, G., et al., (1992); Das, A. and Datta, B. (2000) and Putti, M. and Paniconi, C. (1995).

6.7.4 Miscible interface models

The Miscible flow models consider interface as a diffusion boundary with various thicknesses and densities and the saline-fresh water systems are assumed to be in dynamic equilibrium with each other. Several computer codes in recent years have been developed to simulate flow and solute transport problems. Some of the popularly used numerical codes dealing with saline water intrusion problems are presented below along with some selected case studies.

Method of Characteristics: Pinder, G.F. and Cooper, H. H. (1970) were one of the first to develop transient two dimensional numerical solution using method of characteristics. Konikow, L.F. and Bredehoeft, J.D. (1978) also developed and demonstrated the applicability of method of characteristics by way of coupling the groundwater flow equation with the solute transport equation. This is useful method for simulating mass transport in Groundwater. It is usually used in conjunction with a finite-difference model that simulates the flow of the groundwater. The finite-difference grid then becomes the coordinate system for the movement of the fluid and the contained solute. The method of characteristics follows the fluid and computes the changing concentrations of the solute. In the two-dimensional problem there are three equations: one each for the x-velocity, the y-velocity and the concentration. The solution of each yields what is known as characteristics curve. Points, representing solutes, are placed in various cells of the finite-difference grid. The computer code then moves each of them a distance proportional to the length of the time increment and the groundwater velocity at that point. This simulates convective transport and the remaining factors of the mass-transport equation are solved with finite-difference techniques. This has been used to estimate the chloride migration rates of Pennar delta region under different states (Radhakrishna, I. and Chowdary, M.V.R., 1998). An improved version of MOC known as MOCDENSE was published by Sanford, W.E. and Konikow, L.F. (1985). A case study is given for MOC in later paragraphs.

Sutra Model: Voss, C.I., (1984) developed SUTRA model (Saturated, Unsaturated, Transport). This is a finite-element groundwater flow and energy (heat) and solute (contaminant) transport code which was developed by U.S. Geological Survey. The model can be applied aerially or in a cross-section. It uses a two-dimensional, combined finite element and integrated finite difference method to approximate the equations that describe the two interdependent processes being simulated. When used to simulate salt-water movement in aquifers in cross-section, the two interdependent processes are the density-dependent saturated groundwater flow and the transport of dissolved solids in the groundwater. Either local or regional-scale sections having dispersed or relatively sharp transition zones between salt water and fresh water may be simulated. The results of a SUTRA simulation of salt-water movement show distributions of fluid pressures, dissolved solids concentrations, and the magnitude and direction of fluid velocities as they vary with time. Almost all aquifer properties that are entered into the model may vary in value throughout the simulated section. Sources and boundary conditions may vary with time. The finite element method using quadrilateral elements allows the simulation of irregular areas with irregular mesh spacing. The model has been applied to real field data and observed to give favourable results (Bobba, A.G., 1993b, 1998(a, b); Bobba, A.G., et al 2000; Piggott, A.R., et al 1994, 1996; Oberdorfer, J.A., et al 1990; Emekli, N., et al 1996; Souza, W.R. and Voss, C.I., 1987; Bush, W.P., 1988; Hutchings, W.C and Tarbox, D.L., 2003).

HST 3D: Kipp, K.L. (1987) developed a 3D finite difference heat and solute transport model (HST3D) for saturated flow conditions with variable density and viscosity. Numerical solutions for pressure, temperature and solute concentrations are solved in sequence. The primary limitation of HST3D emerges from use of finite difference solution techniques for the spatial and temporal derivative approximations and computational burden. The longitudinal and transverse dispersivities limit the cell size to minimize the numerical dispersion or oscillation. The code has been widely used for saltwater intrusion problems. The recent versions of HST3D are supported with pre and post processors.

3D finite element model: Huyakorn, P.S., et al., (1987) developed a 3D finite element model for the simulation of salt water in both unconfined and confined aquifer systems. The model is based on groundwater flow and solute transport equations. The density coupling of these two equations is done using a Picard sequential solution algorithm. A few typical examples are solved under various boundary conditions for model verification. A fairly large number of unknowns (of the order of five to ten thousand) can be handled conveniently on a medium personal computer.

Modflow: McDonald, M.G. and Harbaugh, A.W. (1988); Harbaugh, A.W. and McDonald, M.G. (1996); Harbaugh, A.W., et al., (2000); MODFLOW developed a versatile finite difference groundwater flow model. It is a three dimensional model and the aquifers/layers could be either unconfined or confined systems or a combination of both. The model simulates recharge, evapotranspiration, flow to wells, drains and river beds. The model incorporates a series of separate modules which are independent and one can select modules which are needed for a particular problem. The model is simple to use and is available in CDs. The MODFLOW programmes could also be used for salt water intrusion problems by introducing an additional input file needed to simulate saltwater intrusion. The input file consists of water levels ,density variations and some tip/toe tracking parameters. The details of salt water intrusion package could be found in Bakker, M. (2003).

Seawat model: Guo, W. and Langevin, C.D. (2002) recently developed SEAWAT model by combining the popular MODFLOW (McDonald, M.G. and Harbaugh, W.A., 1988) and MT3D (Zheng, C. and Wang, P.P., 1998) models with modifications to account for variable densities between sea water and fresh water. The pressure heads are converted to equivalent freshwater heads for variable densities in space and time. Initially the flow field is solved by MODFLOW and then is followed by MT3D for solution concentrations. The updated density field is then computed from the new concentration levels and are then incorporated back into MODFLOW as relative density difference terms. The flow and transport equations are solved several times for the same time step until the difference in fluid density between consecutive iterations are less than user specified tolerance. Rao, S.V.N., et al (2009) developed some model scenarios for Mahanadi Delta regions along Orissa Coast (India) using SEAWAT model, the results of which are presented in the later paragraphs of the chapter as a case study.

Feflow: Diersch, H.J.G., 2006 (FEFLOW), developed a 3-D finite element based density dependent flow and transport model. The model allows time dependent and spatially variable recharges/pumping, injection and boundary conditions and therefore is considered appropriate for simulating sea water intrusion problems in coastal systems caused by natural and anthropogenic processes. There have been several recent uses of FEFLOW to model sea water intrusion. Faye, S., et al., (2001); Barazzuoli, P., et al (2008).

Two selected case studies one using method of characteristics (Konikow, L.F. and Bredehoeft , J.D., 1978) in pennar delta region of Andhra Pradesh and the second using SEAWAT (Guo, W. and

Langevin, C.D. (2002)) are presented in the following paragraphs applied to Mahanadi delta region of Orissa along the East coast of India.

Case study 1: Coastal Pennar delta (Andhra Pradesh): The Pennar delta region (Long. 80°00° and 80°12°: Lat. 14°40° and 14°20°) is one of the important delta basins in coastal Andhra Pradesh, with prolific freshwater aquifer systems, supporting vast irrigation potential. The interception of vast amounts of surface water in the upper reaches of the river through downstream anicut (small dam) at Sangham and Nellore and the increased groundwater withdrawal through a number of well fields over the last two decades have caused significant changes in relation to recharge potentials, groundwater flow paths and brackish-water contamination levels. This has been particularly true at the outfall region of the delta where a coastal canal with brackish water runs parallel to the coast, which also acts as a source for chloride contamination. In order to assess the sustainability of this freshwater resource, detailed hydrogeological and hydrogeochemical mapping was carried out to delineate hydrogeological boundaries and assess reservoir characteristics of this freshwater resource (Radhakrishna, I., 1991). The Buckingham coastal canal with brackish water and the estuary region of Pennar river are the source areas for chloride contamination in the outfall area, and could perhaps spread inland in the years to come in view of the increasing groundwater abstraction rates year after year. The newly proposed brackish water aquaculture programme in the outer rim of the delta could trigger this phenomenon further inland. In a bit to understand and quantify the saline water intrusion processes in this part of the coast, a two dimensional code namely Method Of Characteristics (MOC) (Konikow, L.F. and Bredehoeft , J.D., 1978) is applied to the Pennar delta region (Andhra Pradesh), (Radhakrishna, I. and Chowdary, M.V.R., 1998).

Pinder, G.F. and Bredehoeft, J.D. (1968) expressed the following two-dimensional flow equation:

$$\delta / \delta x_i (T_{ij} \delta h / \delta x_j) = S.\delta h / \delta t + W(x, y, t) \qquad .....(6.37)$$

$i, j = 1,2$

where T_{ij} is the transmissivity tensor (L^2/T); h is the hydraulic head(L); S is the storage coefficient (L^0); t is the observation time(T); W is the volume flux per unit area (L/T); (x_i, x_j are the Cartesian coordinates).

$$W(x,y,t) = Q(x,y,t) - K_z/m(H_s - h) \qquad .....(6.38)$$

where Q is the rate of withdrawal (L^3/T); K_z is the vertical hydraulic conductivity of the confining layer (L/T); m is the thickness of the confining layer(L); and H_s is the hydraulic head in source of leakage (L) (L = length in meters; T = time in seconds).

Lohman, S.W. (1972) gave the following expression for average groundwater velocity.

$$V_i = -(K_{ij}/n)(\delta h/\delta x_j) \qquad .....(6.39)$$

where V_i is the velocity in the direction of x_i (L/t) and n is the effective porosity of the aquifer (L^0). Bredehoeft, J.D. and Pinder, G.F. (1973) used the following transport and dispersion equation to account for movement of chemical species.

$$d(C)dt = \delta/\delta x_i(D_{ij}. \delta C/\delta x_j) - \delta/\delta x_i(Cv_i) - C' W/nb \qquad .....(6.40)$$

$i,j = 1,2$

where C is the concentration of the dissolved chemical species (M/L^3); D_{ij} is the dispersion tensor (L^2/T); b is the saturated thickness (l); C′ is the concentration of the dissolved chemical in a source (M/L^3) (M=mass in kilograms).

Scheidegger, A.E. (1961) expressed dispersion coefficient by the following equation:

$$D_{ij} = \alpha_{ijmn}(V_m V_n)/[V] \quad .....(6.41)$$

where α_{ijmn} is the dispersivity of the aquifer (L); V_m and V_n are the components of velocity in the m and n directions, respectively (L/T); and [V] is the magnitude of velocity (L/T).

The dispersivity tensor is further defined in terms of longitudinal (α_1) and transverse (α_2) dispersivities as given below:

$$D_L = \alpha_1[V] \quad .....(6.42)$$

and

$$D_T = \alpha_2[V] \quad .....(6.43)$$

Substituting Eqs. 6.42 and 6.43 in Eq. 6.41, the dispersion coefficient for two-dimensional flow could be stated as follows:

$$D_{xx} = D_L\{(V_x)^2/[v]^2\} + D_T\{(V_y)^2/[V]^2\} \quad .....(6.44$$

$$D_{yy} = D_T\{(V_x)^2/[v]^2\} + D_L\{(V_y)^2/[V]^2\} \quad .....(6.45)$$

$$D_{xx} = D_{yx} = (D_L - D_T) + (V_x V_y)/[V]^2 \quad .....(6.46)$$

Grid design, hydrogeological model boundaries and input parameters

The south Pennar delta is divided into uniformly spaced grids as shown in Fig. 6.36. Each grid block represents 1 km^2, and there are 242 grids in the area meant for model analysis. Each input value to the grid block (transmissivity, leakance, recharge, draft etc.) are considered to be average values of each grid block. “Zero flow” boundaries are assigned on all the outer peripheries of paleo delta grid. Constant water level heads are assigned wherever a high-density canal network exists. Vertical recharge components were introduced wherever high-density canal network and surface tanks are available and also through a leakance factor assigned at different nodal points depending upon specific hydrogeological situations prevailing. The normal recharge parameters and the leakance values in the system are obtained through analysis of subsurface hydrogeological data available at different specific sites and through a process of iteration while calibrating the model to match with the observed flow field. The leakance [(m/s)/m] is defined as ratio of vertical hydraulic conductivity to the thickness of the confining layer. The rate of leakage is a function of the difference between the head of the aquifer and the head in the source bed, stream or lake. The leakance values assigned vary from 0 to 1. Along the outer peripheries of the basin where clay exists, leakance values between 10^{-9} and 10^{-5} are assigned. A maximum leakance value of 1 is given at the constant head cell region (chloride-free) and at the outer boundary along the coastal canal and estuary (with chloride).

The transmissivity values in the system have a range varying from 2500 m^2/day (central region) to 1200 m^2/day in the peripheral region of the delta. The data is used to assign the nodal values in the grid map. A storage coefficient value of 10^{-4}, an effective porosity of 30% and dispersivity value of 30 m are uniformly assigned in the system. An initial background chloride level of 50 mg/l is also specified in all nodal points except along the coastal canal and near the estuary region where chloride values of the order of 1500 mg/l are assigned.

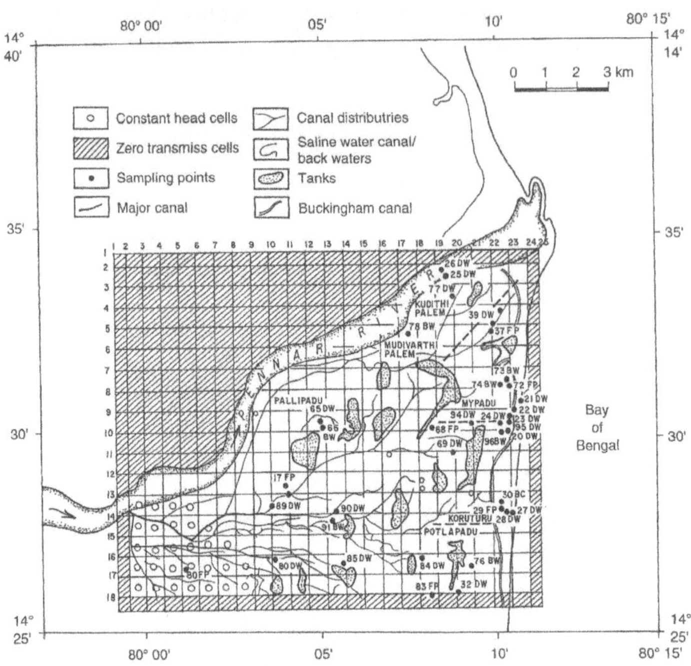

Fig. 6.36 Finite difference scheme

Model calibration and hydraulic stresses

A reliable and accurate groundwater flow model is a prerequisite for obtaining a reliable solute transport model. A good groundwater flow model could be developed provided the input parameters such as recharge potentials (lateral and vertical), transmissivity and storage coefficient variations and hydrogeological boundaries are accurate enough to represent true field situations. A solute transport model is therefore a result of several factors, and consequently demands a high degree of precision with respect to observations and estimations of aquifer characteristics and hydrological boundaries. The initial flow conditions must be known to model the system when it is not subjected to significant hydraulic stresses. Since these conditions are not known, a steady-state initial groundwater flow map was generated taking certain historical hydrological data and morphological situations into consideration and also introducing vertical recharge and leakance at suitable points, which is used for subsequent prognosis. In order to calibrate the model with the observed flow field for the year 1985, draft (negative) and recharge (positive) components at selected nodal points are introduced as shown in Fig. 6.37. The draft calculations were based on the intensity/density of tube well structures in a particular area and the annual pumpage in different well fields of the delta. The recharge components were calculated through hydrograph trends and hydrological test analyses. A total draft of the order of 54 MCM (million cubic metres) per year, introduced into the grid system has brought the simulated and observed flow lines to near match levels. By and large, the estimated draft by the year 1985 is close to the simulated draft in the system. Therefore this map is treated as the calibrated base map for subsequent prognostic studies to assess chloride contamination migration rates under different hydraulic stress conditions. The heads and fluxes computed by the flow model are used as input parameters to the transport model and initial chloride levels of 50 mg/l in the delta area and 1500 mg/l at the estuary and along the canal area are introduced and the chloride particle migration trends are obtained for different hydraulic stress situation.

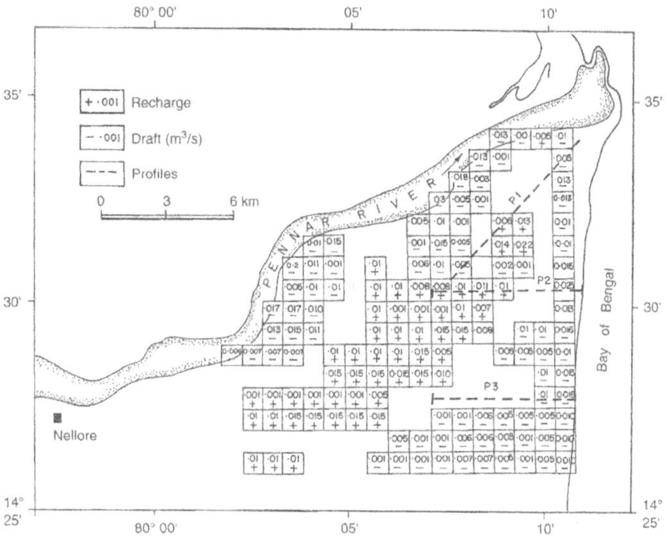

Fig. 6.37 Selected nodal points for draft and recharge

Hydraulic stresses, flow patterns and chloride movement

Groundwater flow patterns in the paleo Pennar delta obtained for two simulated hydraulic stress situations, namely at 54 and 108 MCM per year, and the resultant flow patterns and chloride migration trends are presented and discussed in the following paragraphs.

54-MCM hydraulic stress

The simulated groundwater flow system at 54 MCM hydraulic stress is characterized by a recharge mound in the central part of the delta and radial groundwater flow towards the outfall area as shown in fig.6.38. Both observed and calibrated groundwater levels measured from mean sea level by and large match/agree for the year 1985 (post-monsoon). There are also two elongated subsurface flow channels – one running nearly perpendicular to the river and the other parallel to the river. The flow system is also characterized by localized groundwater depressions, indicating overdraft situations around (1) Pallipadu

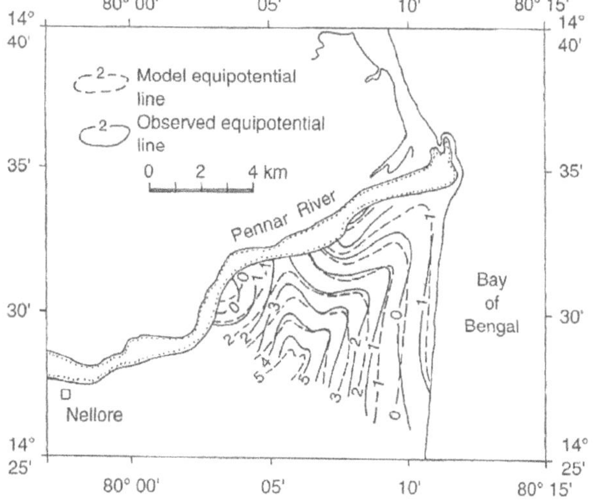

Fig. 6.38 Observed and computed groundwater flows at 54 MCM hydraulic stress

and (2) Mudivarthipalem – Kudithipalem. The groundwater levels vary between 6.00 m above mean sea level and 2.00 m below mean sea level. The maximum level at Pallipadu depression is 0.00 m mean sea level and 2.00 m below mean sea level at Mudivarthipalem - Kudithipalem region. The system also shows that all along the coastal boundary up to 1 km from the coastline, the groundwater levels have fallen below mean sea level.

108-MCM hydraulic stress

The simulated groundwater flow trends at increased hydraulic stress of the order of 108 MCM are shown in Fig. 6.39. By and large the trends remain similar, and the boundaries of recharge mounds and depressions have undergone substantial change. The groundwater flow channel running perpendicular to the river loses its prominence and the flow channel running parallel to the river still remains prominent. The aerial spread of the groundwater depressions at Pallipadu and Mudivarthipalem-Kudithipalem area are further enlarged. The zero mean sea level contour has further transgressed landwards. Groundwater levels vary from 6.00 m above mean sea level to 6.00 m below mean sea level. The water levels in Pallipadu have gone down further up to 4.00 m below mean sea level, and 6.00 m below mean sea level around Mudivarthipalem – Kudithipalem. A new depression south of Mypadu along Koruturu – Potlapudi track appeared.

Chloride migration rates and prognosis

A detailed description of groundwater system characterization and flow configuration for Pennar delta region has been made in the earlier paragraphs. A brackish water canal (Buckingham Canal) runs parallel to the shore boundary which emerges from the estuary region of the delta. The paleo delta region is also characterized by the number of paleo channels which could act as good groundwater repositories and also act as active conduits for fresh groundwater flow from the upstream areas as well as brackish water mixing from the outer periphery of the delta, inland depending upon the groundwater flow gradient and permeability character of the near-surface aquifer system. During the 1970's, when groundwater development was at low key, many of the well structures constructed along the coastal canal region (outfall region of the delta) contained freshwater in both dug wells and shallow tube wells.

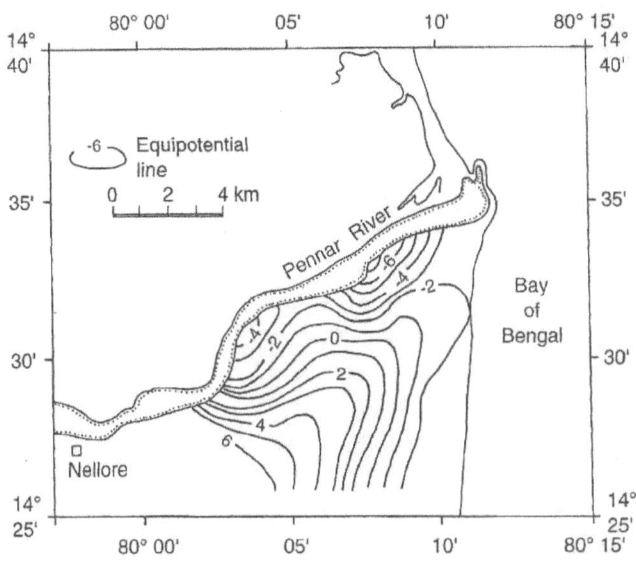

Fig. 6.39 Groundwater flow at 108 MCM hydraulic stress

By the 1980's, when the groundwater development in the outer region of the delta had increased in a big way, the groundwater levels all along the coastal margin had fallen, thereby giving way to brackish-water mixing processes inland. By 1985, several well fields close to the canal boundary and estuary region and become brackish and water levels also fell below mean sea level quite considerably. In order to assess the possible mixing rates in the delta region in the years to come, three profiles have been selected, namely p1, p2, p3. The prognostic levels of chloride ions for different time-periods at different stress situations are presented in Tables 6.13 and 6.14 are also graphically shown in Fig. 6.40 and 6.41. The average migration rate at different hydraulic stresses and for different chloride intensity levels along the three profiles are presented in Table 6.14.

Table 6.13 Chloride movement (in km) at different hydraulic stresses across selected profiles

Hydraulic stress	**54 MCM**				**108 MCM**			
Year/profile	**100 mg/l**	**500 mg/l**	**1000 mg/l**	**1500 mg/l**	**100 mg/l**	**500 mg/l**	**1000 mg/l**	**1500 mg/l**
year								
P1	1.152	0.832	0.640	-	2.048	1.216	0.576	-
P2	0.704	0.576	0.448	-	1.472	0.640	0.448	-
P3	0.960	0768	0.640	-	1.408	0.832	-	-
10 years								
P1	2.304	1.403	0.763	-	3.392	2.368	1.536	-
P2	1.408	0.960	0.512	-	1.600	1.472	1.344	-
P3	1.536	1.344	0.704	-	2.304	1.536	1.280	-
20 years								
P1	2.496	1.920	1.024	-	3.520	2.370	1.768	-
P2	1.472	1.280	0.576	-	1.920	1.344	1.512	-
P3	1.536	1.408	0.768	-	2.688	1.664	1.960	-
30 years								
P1	2.880	2.112	1.472	-	3.840	3.400	2.800	1.288
P2	1.600	1.280	0.640	-	1.920	1.600	1.808	0.640
P3	1.600	1.472	1.152	-	2.880	2.240	1.908	1.024

Table 6.14 Chloride intensity levels and average mixing rates

Draft	**Chloride concentration level**	**Migration rates (km/year)**		
		Profile 1	**Profile 2**	**Profile3**
54 MCM	100 mg/l	0.0576	0.0290	0.0210
	500 mg/l	0.0420	0.0230	0.0230
	1000 mg/l	0.0170	0.0060	0.0290
108 MCM	100 mg/l	0.0590	0.0150	0.0490
	500 mg/l	0.0720	0.0320	0.0460
	1000 mg/l	0.0760	0.0405	-

By and large the simulated chloride migration rates in all the profiles go up from lower stress to higher stress situations. Maximum rates are seen in profile 1 followed by profiles 3 and 2. This may be due to the prevailing relative changes in hydraulic conductivities. The migration rate of the 500 mg/l chloride intensity contour at 54 MCM in the first profile is 0.042 km/year, and 0.023 km/year in the other two profiles. It goes up to 0.072 km/year in profile 1 and 0.032 and 0.046 km/year, respectively, in profiles 2 and 3, when simulated stress is raised to 108 MCM in the system. All the prognostic chloride migration trends obtained for a 30-year period appear reasonable for the type of stress field introduced in the system. The model could serve as a useful tool in assessing or evaluating future management practices, and also provide a basis to understand the migration rates of chloride contamination under specific hydraulic stress situations in a coastal delta.

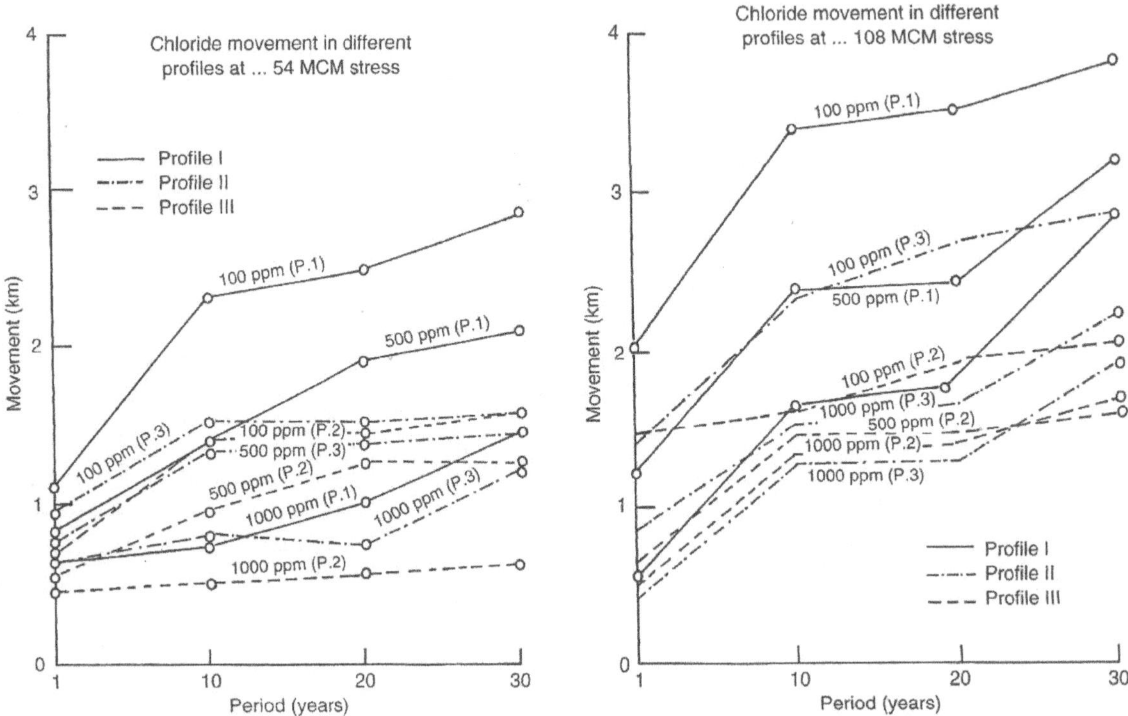

Fig. 6.40 Chloride movement with time at 54 MCM

Fig. 6.41 Chloride movement with time at 108 MCM

Protection and management

In order to protect this paleo Pennar delta free from saline water intrusion hazard and to minimize the chloride migration/mixing phenomenon the following remedial measures are suggested:

- The existing well fields should be redesigned by way of reducing the hydraulic stresses in the outer rim of the delta and increasing withdrawal rates in the central delta.
- The coastal canal should be converted into a freshwater canal by creating the necessary engineering structures at the estuary-canal junction and linking up with a freshwater source through the delta canal system.
- The existing surface tank reservoirs which are located near the coastal boundary should be desilted and kept continuously at full levels to facilitate increased infiltration rates.
- A freshwater canal parallel to the existing brackish water canal could be made so that freshwater infiltration barrier is created to check the moving brackish water front and ultimately dilute it. This should be done in case item no. 2 is not possible.

- The paleo Pennar delta region boundaries identified and demarcated should be designated as a "Groundwater protection zone" and no brackish water aquaculture programs should be allowed with in the boundary. The measures cited here, if adopted, would help raise irrigation potential and agriculture production and keep the soil-land-water system free from threats of brackish water intrusion.

Case study 2: Mahanadi delta region (Orissa)

The Mahanadi delta system is the third largest delta plain along the east coast of India and covers an area of about 0.9×10^4 km^2 which lies between 85°40' : 86°45' E and 19° 40' : 20°35' N. The catchment area of the river covers 1.42×10^5 km^2. The climate setting is tropical with hot and humid monsoon climate. The sediment yield of the catchment is of the order of 200-400 tons/km^2. The region is characterized by tropical monsoon climate. Cyclonic storms, floods and sometimes droughts are rather frequent in the area with an average annual rainfall of the order of 1572 mm. There are two distinct high precipitation peaks-one around Soro area in Balasore district and the other near Pipli in Puri district. The intensity of rainfall is generally high in Balasore and Puri districts and moderate in Cuttack district.

The Mahanadi river after traversing over a distance of about 800 kms spreads itself in the form of delta plain with its apex near Naraj with a network of distributaries and sub distributaries viz: Birupa, Mahanadi/Chitrala, Katjhori/Devi and Koyakhai/Daya/Kusubhadra/Bhargavi. A major portion of the delta region is flat with gentle slope towards Bay of Bengal. A typical characteristic geo-morphological feature is the formation of long narrow islands in the central part of the delta system which could be a reflection of the manifestation of the sub surface structure of the delta system. The entire delta region could be classified into three morphological units namely upper, central and lower delta regions. The upper delta region is characterized by coarse-medium sandy facies with admixtures of clays and sands. The central delta region is characterized by a number of lenticular islands, developed as a result of bifurcation and rejoining of the sub distributaries. The lower delta region adjoining the coastline has a number of low lying swamps, tidal canals and a series of sand ridges and valleys.

The exploratory endeavors so far carried out in this region have brought to light the availability of thick sequences of aquifer systems belonging to Recent, Quaternary and Tertiary stages of geological time scale. The thicknesses of these aquifer systems increase from 100 m to 400 m and beyond towards the out fall region of the delta from the upstream areas with a gentle slope. At the same time the saline contamination hazards reach peak levels in most parts of northern and central Mahanadi delta region up to the depth levels varying between 60 m to 300 m below which fresh water aquifers occuring under deep confined conditions with artesian and sub artesian pressure heads. In contrast the southern parts of the Mahanadi delta region is characterized by fresh water aquifers floating over saline water aquifers with thicknesses varying from 10 m to 120 m with interface depth levels decreasing towards the coast.

The interface structure, its nature and boundary conditions obviously become very important parameters to build up required strategies for sustainable development of fresh water resources. This would also form the basis for an appropriate well field development while estimating the mechanisms involved in understanding saline water intrusion processes. Twelve representative well log sites have been selected where the interface boundaries are clearly identifiable. The nature and characteristics of the interface at each well site are shown in Table 6.15.

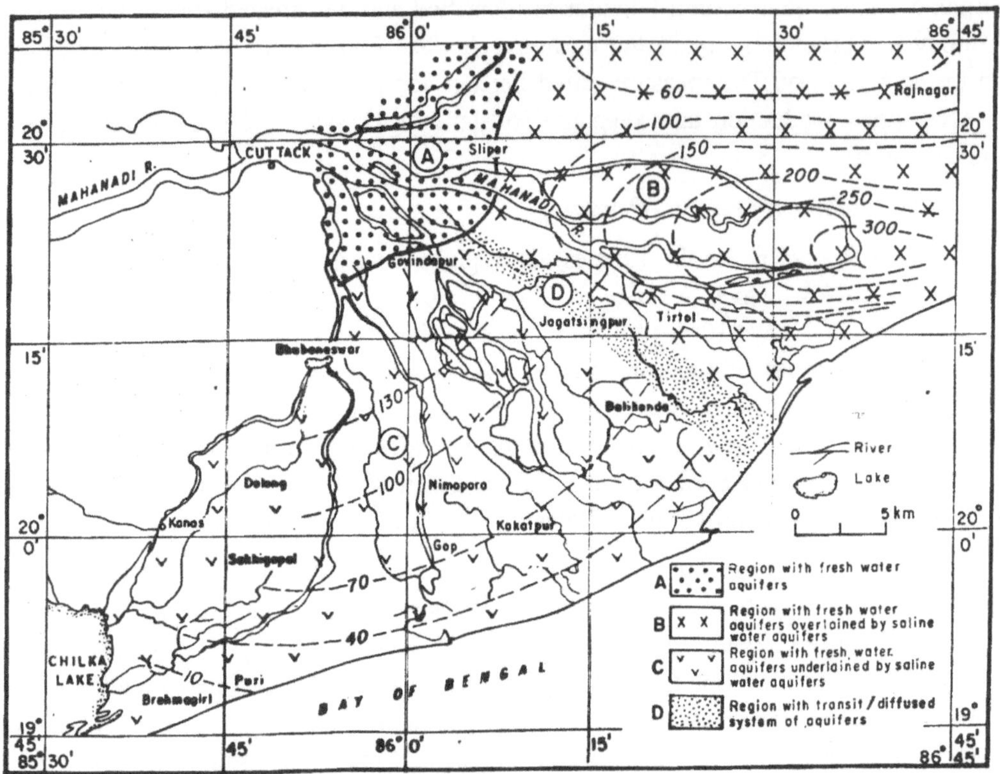

Fig. 6.42 Interface structural map of Mahanadi delta

(Source: Radhakrishna, I., 2001)

Table 6.15 Saline-freshwater interface depths and characteristics

S.no.	Delta region	Well site	Interface depth (m)	Nature of interface disposition
1.	South Mahanadi delta	Puri	28	Diffusion zone with freshwater overlying brackish water
		Konark	30	
		Patalia	100	
		Arilo	100	
2.	North Mahanadi delta	Jajang	120	Aquiclude with brackish water overlying freshwater
		Masakani	290	
		Dubakona	50	
		Madanpur	60	
3.	Baitarni-Brahmani delta	Musanga	60	Aquiclude zone with brackish water overlying freshwater. No interface all through freshwater.
		Kherang	120	
		Dakshinabad	Nil	
		Mandari	Nil	

An interface structure map (Radhakrishna, I., 2001) based on the above data for the entire region is presented in Fig. 6.42. This suggests that the interface could be a hydraulic diffusion boundary with freshwater floating over saline water or a geological aquiclude/aquitard boundary with freshwater underneath the saline water. The nature of the interface and its depths has to be carefully evaluated while designing the well field structures. The south Mahanadi delta region is characterized by a freshwater system sitting over a brackish water system with a diffusion boundary and with depths varying from 10 m to 120 m. The interface depth in general declines towards the coast but becomes greater inland. The north Mahanadi delta is represented by brackish water systems sitting over freshwater systems with aquicludes separating these two fluid systems. These aquicludes have gentle slopes and become thick towards the coast and pinches inland. It is therefore necessary in such systems to adopt cement plugging in the gravel pack area at different depths to avoid vertical migration of saline water while pumping. It is also necessary in such situations to estimate the lateral extension of aquiclude interface boundaries upstream where saline and freshwater systems come into juxtaposition, and to optimize and control the pumping without destabilizing the freshwater regime. The Brahmini-Baitarni region is also characterized by a similar disposition of the interface boundaries with depths varying from 60 m to 150 m. In the present study the southern part of delta system involving rivers Daya, Bhargavi and Kusubhadra is modeled as simplified conceptual unconfined aquifer and in the northern part of the Mahanadi delta system as a confined aquifer system based on borehole and electrical log data. Between these two formations a mixed interface has been observed. The conceptual models representing the confined and unconfined aquifers are shown in Fig 6.43 and 6.44 respectively. At present both these aquifers are prone to unsustainable groundwater development leading to contamination of freshwater with saline water from upper and underneath aquifers in southern (unconfined) and northern (confined) deltaic aquifers respectively.

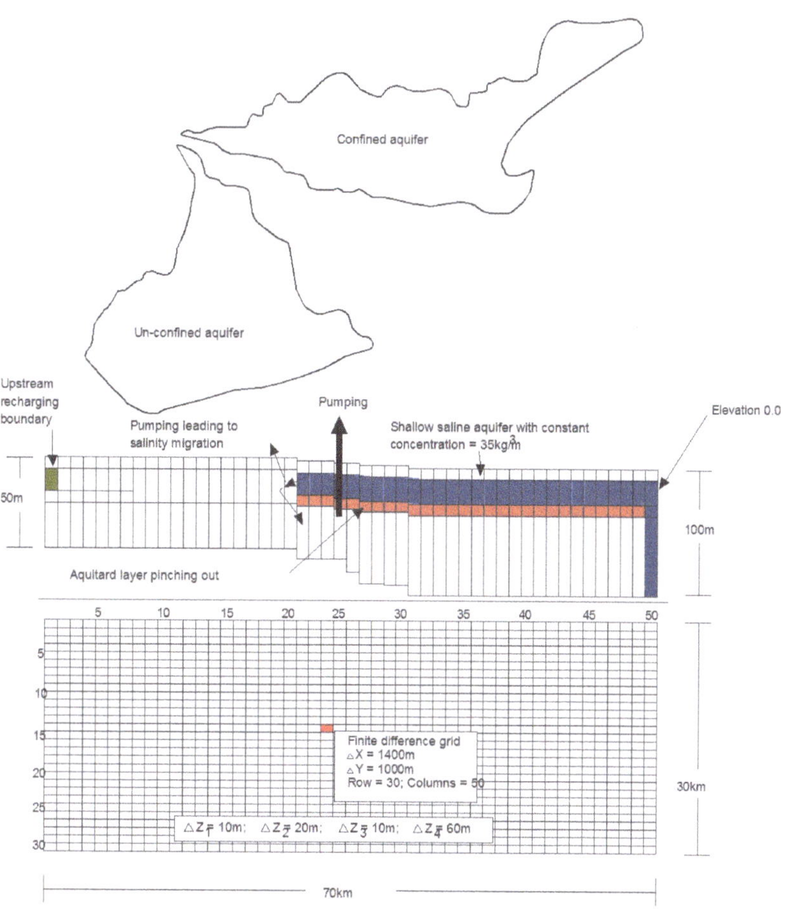

Fig. 6.43 Conceptualized confined aquifer model in section and plan

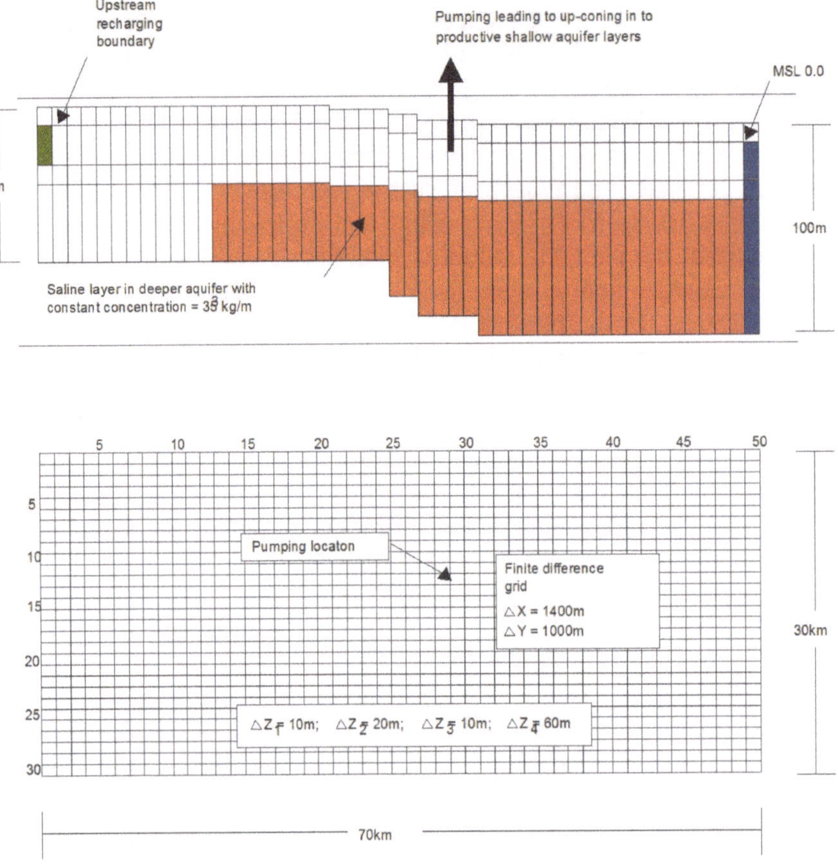

Fig. 6.44 Conceptualized unconfined aquifer model in section and plan

Description of the conceptualized aquifer models

Rao, S.V.N., et al 2009 developed a conceptual model representing the Mahanadi delta system in terms of aquifer geometry, boundary conditions and aquifer properties to obtain insight into the flow dynamics of saline-fresh water using SEAWAT 2000 (Langevin, C.D., et al 2004). The model has been successfully demonstrated for a number of studies both conceptual and application based (Rao, S.V.N., et al., 2003; 2004; 2005).

MODFLOW-SEAWAT 2000

It is a density driven flow and transport model. The SEAWAT 2000 model has been developed by combining the popular MODFLOW and MT3D models. The governing equations for 3-D density-dependent miscible flow and transport model for a coastal aquifer may be written as follows:

Flow equation is represented by,

$$-\nabla.(\rho q)+\bar{\rho}\, q_s=\frac{\partial(\rho\theta)}{\partial t} \qquad .....(6.47)$$

where

Δ is the gradient operator $\frac{\partial}{\partial x}+\frac{\partial}{\partial y}+\frac{\partial}{\partial z}$

q is the specific discharge vector $[LT^{-1}]$, with its components given by

$$q_x = \frac{k_x \partial P}{\mu \partial x} \qquad q_y = \frac{k_y \partial P}{\mu \partial y} \qquad q_z = \frac{k_z}{\mu}[\frac{\partial P}{\partial z} + \rho g]$$

ρ is the variable fluid density $[ML^{-3}]$;

$\bar{\rho}$ is the density of water entering from a source or leaving through a sink $[ML^{-3}]$;

q_s is the volumetric flow per unit volume of aquifer representing sources/sinks $[T^{-1}]$;

θ is porosity [dimensionless];

t is the time [T];

q_x, q_y, q_z are the individual components of specific discharge and k_x, k_y, k_z represent instrinsic permeability's $[L^2]$ in the three coordinate directions;

g is the gravitational constant $[LT^2]$;

μ is the dynamic viscosity $[ML^{-1}T^{-1}]$;

P is the fluid pore pressure $[ML^{-1}T^{-2}]$;

Transport equation is represented by

$$\nabla.D(\nabla c) - \nabla(vc) + q_s = \frac{\partial c}{\partial t} \qquad .....(6.48)$$

where

c is the solute concentration $[ML^{-3}]$

D is the dispersion coefficient $[L^2T^{-1}]$ and

v represents the seepage velocity $[LT^{-1}]$

The empirical equation for density, as a function of concentration is written as,

$$\rho = \rho_f + E_c \qquad .....(6.49)$$

where E is a dimensionless constant having an approximate value of 0.7143 for salt concentrations ranging from zero for freshwater to 35 kg/m^3 for seawater and ρ_f is the fluid density of freshwater.

Both MODFLOW and MT3D use the implicit finite difference approach to solve the flow and transport equations, respectively. The SEAWAT does not solve the flow equation directly. Appropriate modifications are incorporated to account for density variations between seawater and freshwater. The pressure head is converted to equivalent freshwater head for the variable density water in space and time. This approach enables MODFLOW (constant density) to be used with minor changes. During any computational time step, the flow field is first solved by the MODFLOW and this is followed by the solution for concentration variations using MT3D. The updated density field is then determined from the new concentrations and is incorporated back into MODFLOW as relative density difference terms. The flow and transport equations are solved repeatedly through implicit coupling for the same time step until the difference in fluid density between consecutive iterations is less than a user specified tolerance.

Confined aquifers in northern delta system

The borehole and electrical log data indicate confining conditions due to aquitard (clay) layers dipping downwards towards the sea. These clay layers pinch out in the upstream part of the delta system where fresh water and saline water coming into juxtaposition. The farmers have been pumping from deeper productive freshwater aquifers over several years, resulting in contamination of deep fresh water aquifers either through vertical recharge or lateral migration from the shoreline side. The purpose of this simulation is to determine the impact of steady state pumping over a long periods (such as 30 years). Time invariant aquifer properties and boundary conditions were assumed (see table 6.16). The conceptual aquifer model is shown below in Fig. 6.45.

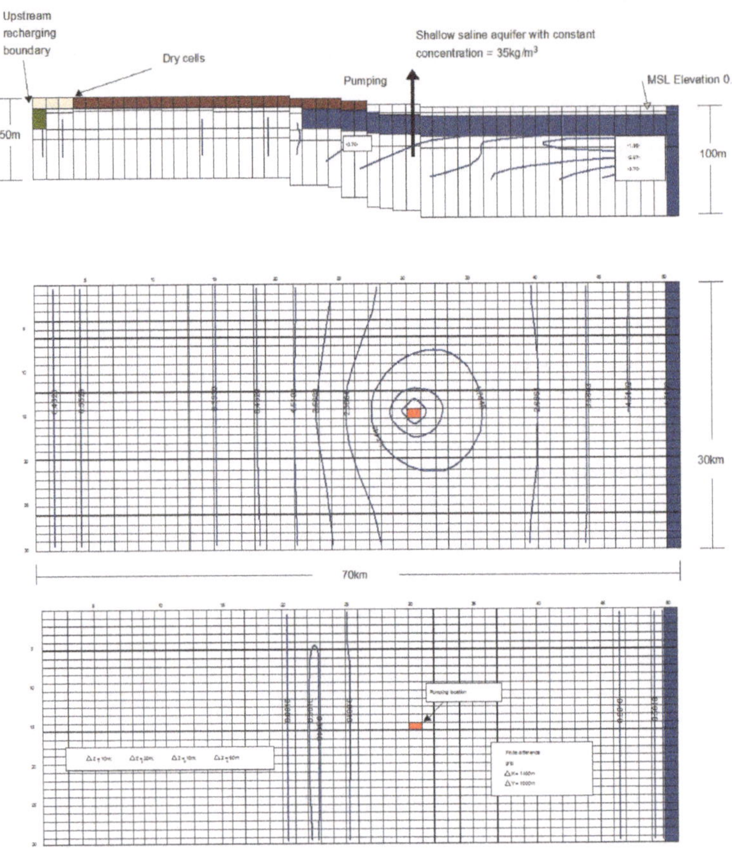

Fig. 6.45 Distribution of heads (msl) and concentration (kg/m^3) in 4th layer due to steady state pumping of 500 m^3/day

Unconfined aquifer in southern delta system

The electrical log profiles in the southern regions of Mahanadi delta indicate presence of potential fresh water in the upper layers which get recharged from deltaic river systems and rainfall and have been successfully tapped by farmers for irrigation since long times. However the deeper aquifer system contains saline water. Therefore excess pumping is resulting in upcoming of saline water besides lateral seawater ingress and hence the need to regulate proper development in a sustainable manner. The unconfined aquifer simulation over a long period of time (30 years) seeks to demonstrate the need for planned and regulated pumping. The conceptual aquifer model is shown below in fig.6.46 and aquifer details in Table 6.16.

Simulation analysis and results

The 4-layer aquifer systems discussed above were simulated for a long duration such as 30 years. The purpose is to evaluate the effect of long term stresses on quality of water (salinity) at well pump screen locations for both confined and unconfined aquifer systems. The aquifer properties and boundary conditions are discussed in table 6.16 and the Fig. 6.43 and 6.44. The initial conditions were simulated by running the model for a long period (3000 days) for average stresses using false transient approach. The numerical model was limited to 50 (row/column) and 4 layers due to authors limited access to evaluation version only of GW Vistas software (Rumbaugh, J.O., et al., 2004). A higher resolution in the horizontal and vertical directions is expected to provide improved results in terms of numerical accuracy.

Table 6.16 Properties of conceptual 4-layer aquifer model

S.no.	Particulars	Values
1	**Confined aquifer in northern part of delta system**	
	(a) Number of rows, columns and layers	30,50,4
	(b) Uniform delta X, Delta Y (in meters)	1000, 1400
	(c) Aquifer thickness (variable as in figure)	-
	(d) Global Hydraulic Conductivity values in X, Y & Z directions except for the aquitard in 3rd layer pinching out midway along X direction (m/day)	10,10, 1
	(e) Hydraulic conductivity of the aquitard in X,Y & Z direction (m/day)	1e-007, 1e-007, 1e-009
	(f) specific yield of upper layer	0.15
	(g) specific storage (/m)	0.0001
	(h) Longitudinal and lateral dispersivity (m)	70, 10
	(i) density of fresh and sea water (kg/m^3)	0, 35
	(j) concentration of fresh and seawater (kg/m^3)	0.71
	(k) slope of concentration –density	
	(l) steady state boundary conditions (i) River recharging head on western end (m) (ii) Sea boundary (m) (iii) Pumping in 4th layer (m^3/day)	 45 4 500-2000
	(m) period of simulation (years)	30
2.	**Unconfined aquifer in the southern part of delta system**	
	(a) Number of rows, columns and layers	30, 50, 4
	(b) Uniform delta X, Delta Y (in meters)	1000, 1400
	(c) Aquifer thickness (variable as in figure)	
	(d) Global Hydraulic Conductivity values in X, Y & Z directions except for the aquitard in 3rd layer pinching out midway along X direction (m/day)	10,10,1
	(e) specific yield of upper layer	0.15
	(f) specific storage (/m)	0.0001
	(g) annual rainfall recharge (m)	0.225
	(h) Longitudinal and lateral dispersivity (m)	70,10
	(i) density of fresh and sea water (kg/m^3)	1000, 1025
	(j) concentration of fresh and seawater (kg/m^3)	0, 35
	(k) slope of concentration –density	0.71
	(l) steady state boundary conditions (i) River recharging head on western end (m) (ii) Sea boundary (m) (iii) Pumping in second layer (m^3/day)	 10 0 500-2000
	(m) period of simulation (years)	30

The confined aquifer was simulated under varying pumping rates under steady state conditions from the productive aquifer each for a period of 30 years for different pumping rates (500-5000 m^3/d). Intuitively the salinity from upper layers migrates down below where the aquitard pinches out. The head and concentration distribution in the productive aquifer (4th layer) for a typical pumping of 500 m^3/day at the end of 30 years is shown in fig.6.45. The concentration values near well pumping screen locations for different rates of pumping indicate the magnitude of contamination coming from upper aquifers rather than from sea side are shown in Fig.6.46 below. At high pumping rates (5000 m^3/d) the saline water contamination is clearly unsustainable as evident from Fig.6.46.

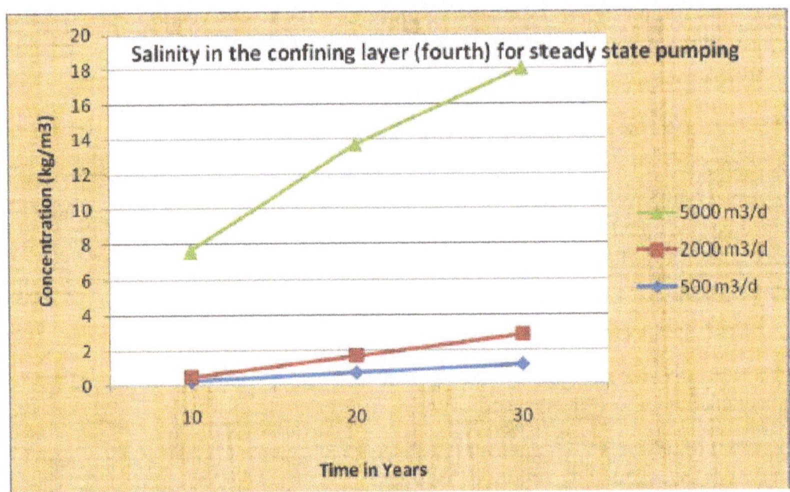

Fig. 6.46 Salinity for varying pumping rates near pump screen location for the confined aquifer

The unconfined aquifer system was simulated similarly for a period of 30 years from the shallow aquifer system. The unconfined aquifers get rainfall recharge but are prone to upcoming phenomena from underneath aquifers. The head and concentration distribution in the productive aquifer (2nd layer) for a typical pumping of 500 m^3/day at the end of 30 years is shown in Fig. 6.47. The concentration values near well pumping screen locations for different rates of pumping indicate the magnitude of contamination arising from bottom most aquifer due to upcoming of saline water for varying steady state pumping rates is shown in Fig.6.48.

Conclusions

The conceptual models presented in the study seek to provide an insight into the sensitive freshwater-saline water dynamics under long-term steady state conditions when subjected to increasing groundwater development. The conceptualization of the aquifer system is based on several hydro-geological and geophysical data bases.

The confined aquifers in the northern part of the delta system are prone to contamination from overlying saline aquifers when the deep fresh water aquifers are subjected to exploitation. The unconfined aquifers get rainfall and river recharge in the southern part of the deltaic system but are prone to upcoming phenomena under similar conditions and hence need regulated development. The model results need to be viewed in a conceptual sense rather than in terms of their absolute values. Detailed model analysis for the multi-layer real-aquifer system is further required to be improved for actual planning and development.

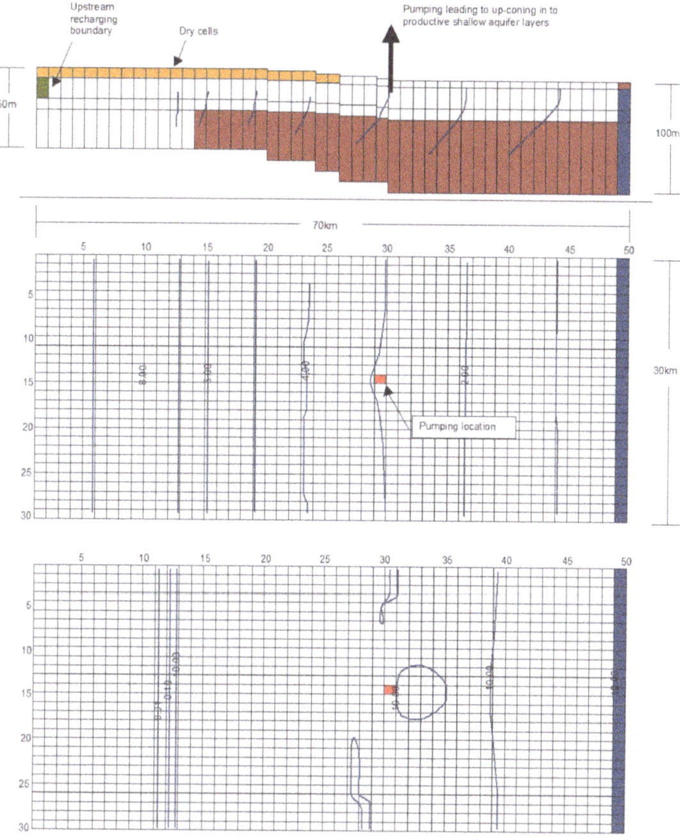

Fig. 6.47 Distribution of heads (msl) and concentration (kg/m^3) in 2nd layer due to steady state pumping of 500 m^3/day.

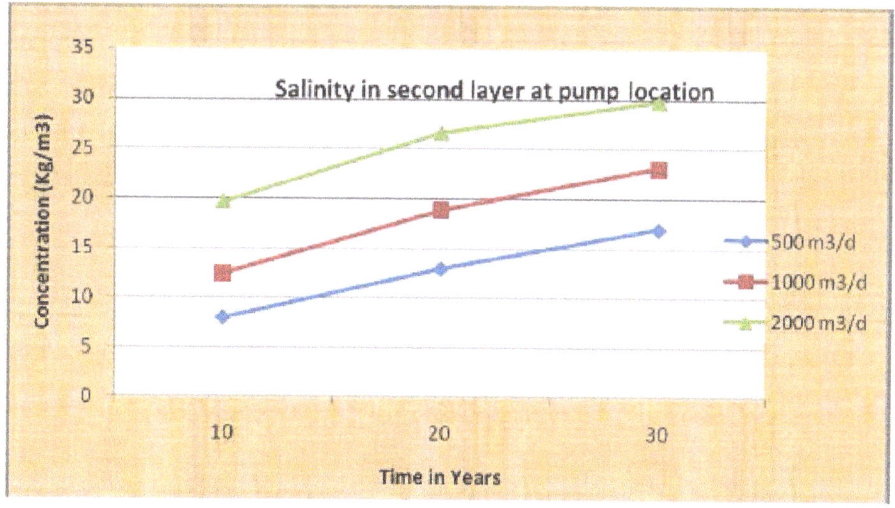

Fig. 6.48 Salinity for varying pumping rates near pump screen location for the unconfined aquifer

6.7.5 Groundwater management models

Groundwater management models take into account both simulation and optimization models and examines the system response as a whole to the external stresses. The simulation models provide the planners with important predictive tools but do not identify the optimal groundwater development strategies where as the optimization models will provide optimal planning, design alternatives etc. Therefore the models which solve flow and transport equations when coupled with the optimization techniques have become powerful tools for management goals of the coastal aquifer systems. Gorelick, S.M. (1983) had given an excellent review of groundwater management models, relevant to salt water intrusion related programmes. A brief description of these models are given below.

Embedded approach: The embedded optimization models use either finite difference or finite element approximations of flow and transport equations as equality constraints within the management model. The hydraulic heads and stresses (recharge/discharge) are decision variables in space and time. The method is simple and straight forward but the constraint matrix becomes large for transient conditions and difficult to handle with normal linear programming packages. The application of embedding technique approach for salt water intrusion problems has several limitations, numerically weak and cannot be used for large scale regional problems (Willis, R. and Finney, B.A., 1988; Das, A. and Datta, B., 1999(a,b)).

Response matrix approach: The response matrix approach is based on principle of superposition and linearity. This approach makes use of an external groundwater flow simulation model to develop unit response of hydraulic head in the system which represents the response of the aquifer system to unit perturbations in the pumpage/recharge at selected points. An assemblage of unit response (response matrix) is included as a constraint in the optimization model and is normally applicable to linear problems only and requires running of the simulation model several times involving large initial computational effort. Maddock, III T. (1972); Morel-Seytoux, H.J. and Daly, C.J., (1975); Illangasekare, T. and Morel-Seytoux, H.J., (1982,1986) developed and applied the unit response approach to simulate flow in a single aquifer system based on linear flow equation and the principle of superposition. Willis, R. and Finney, B.A., (1985, 1988) solved a non linear optimization problem for control of sea water intrusion in a groundwater basin of Taiwan. The aquifer response was simulated by Sharp interface models developed by Mercer, J.W., et al (1980) and used in both response matrix approach and reduced gradient Quasi-Newton method for solving the optimization models.

Linked Simulation-optimization approach: An alternate approach to the above methods is "Linked Simulation-optimization approach" (Gorelick, S.M., 1983; Emch, P.G. and Yeh, W.W.G., 1998) may be useful to solve salt water intrusion management models. The performance of this approach is highly dependent on the simulation model of salt water intrusion because repetitive simulations are involved to achieve an optimal management strategy. Simulation of density dependent salt water intrusion processes involves lot of computational time and memory requirements. Involvement of highly non-linear simulation model in a management approach would take large computational time which could be reduced if one uses parallel computing facilities. A possible alternative is to use a trained ANN (Artificial Neural Network) model as an approximate simulator of the physical processes.

Simulated Annealing(SA) and Genetic Algorithm (GA) methods: More recent trends in non-linear and discrete optimization methods include "Simulated Annealing" (SA) and "Genetic Algorithm" (GA) methods. Simulated Annealing is a simple technique to find optimal solutions for continuous discrete non linear programming problems (Kirkpatrick, S., et al 1983; Johnson, D.S., et al 1989 and Aarts, E. and Korst, J., 1997). The basic concept of this method is to generate a random point and evaluate the objective function. If the trial point is infeasible a new point is generated and if the trial point is feasible, the point is accepted and the best value record is updated. The acceptance/rejection is based on the value of probability density function of Boltzman distribution. Dougherty, D.E. and Marryott, R.A. (1991) and Cunha, M.D.C.A. (1999) have demonstrated the application of SA to hypothetical groundwater management problems. Lee Y-M. and Ellis, J.H. (1994) made a comparative study of eight heuristic algorithms to solve a nonlinear integer optimization problem involving groundwater monitoring network design. Rao, S.V.N., et al., (2004) developed a hypothetical planning model using Simulated Annealing (SA) algorithm and an existing sharp interface model to evolve a conjunctive utilization model for a closely featured delta aquifer system that exists along the east coast of India. In an another scenario Rao, S.V.N., et al., (2004; 2005) used Simulated Annealing (SA) algorithm and the seawater simulation model 'SEAWAT' to evolve a management model in a deltaic environment and estimate the pumping levels in a well by restricting the salinity levels. A trained Artificial Neural Network (ANN) is used to produce the computational problems.

Genetic Algorithms (GA) also could be used as a tool for solving the optimum management models because of its relative efficiency in identifying global optimal solutions for non linear, non convex problems. The GA based optimization approach is suitable for externally linking the numerical simulation model within the optimization model. A trained Artificial Neural Network (ANN) linked to GA based optimization model could be useful to solve some management strategies of coastal systems. The fundamental advantages of SA and GA are that the non convexity can be handled easily for determining a regional solution. GA takes more CPU time than SA for test problems (Huang, M-W. and Arora, J.S., 1997). Parallel processing capabilities can reduce the CPU several times. Mckinney, D.C. and Lin, M-D. (1994) demonstrated the application of GA to a simple hypothetical groundwater remediation problem. Rogers, L.L. and Dowla, F.U. (1994) presented an optimization model for groundwater remediation using Artificial Neural Networks (ANN) and GA. Bhattacharjya, R.K. and Datta, B. (2005) developed a three dimensional transient dependent flow and transport processes approximated using a trained ANN model and linked externally with a GA based optimization model. The objective of the management model is to maximize the withdrawal rate while maintaining the salt concentrations at the same levels. The results obtained are compared with the results obtained through embedded approach and are found to be satisfactory.

6.7.6 Global climate change and sea level rise

- Global warming is now evident as seen through a series of observations of increase in global average air and ocean temperatures, wide spread melting of snow and ice and rise of global mean sea levels. It could be partly natural and partly anthropogenic. A vast majority of the scientists believe that the Global warming and sea level rise is natural and had occurred in the geological time scale several times due to various dynamic forces on the planet. But the present Global warming and sea level rise phenomena to a large extent is related to anthropogenic considerations such as increasing levels of green house gases like Carbon dioxide (CO_2); Methane (CH_4); Nitrous oxide (N_2O); Hydrofluorocarbons (HFCs); Perfluorocarbons (PFCs) and SF_6 (Sulfur hexafluoride). These Green house gases could enhance the properties of earth's atmosphere, allow the solar radiation from the sun to travel through the atmosphere, but prevent the reflected heat to escape back into the space causing the temperature of the earth to rise. These GHG emissions have been rising since 1900's industrialization processes due to increased burning of fossil fuels. The GHG percentages are expected to rise more and more as more number of countries (underdeveloped and developing) are becoming industrialized. The International Energy Agency (IEA) predicts a 53% increase in global energy demand by 2030 and account for further increase in GHGs, temperature and sea level rise events. The GHG emissions are not stable and are increasing due to increased energy consumption levels as a result of population growth and industrialization. The current stock of green house gases in the atmosphere is of the order of 430 ppm as compared to the pre-industrialized concentration levels of the order of 280 ppm which shows an increase of the order of 2.5 ppm per year. Since 1900 Global temperatures have risen by 0.7°C and are continuing to rise at an estimated rate of 0.2 °C per decade and if left unchecked this could increase multi fold in the years to come. The Intergovernmental Panel on

Climate Change (IPCC) concludes that the increase in global temperatures over the next century could accelerate a historical rate of global mean sea level rise from 1 to 2.5 mm/year to about 5 mm/year (IPCC, 1996a). More recent works using new green house gas scenario shows a slightly higher rate of sea level rise (Wigley, T.M.L., 1999). The latest satellite observation suggest an estimated rise of sea level of the order of 3.1 ± 0.7 mm/year, significantly higher than the average rate of sea level rise over the past 40 years. As a result of increasing levels of temperature rise the global sea levels could go upto one meter rise during the 21 century. Three main processes are responsible for the sea level rise. i) Thermal expansion, ii) Melting of glaciers and ice caps and iii) Loss of ice from the Greenland and Antarctica ice sheets. A Global scenario of sea level rise as obtained from wildwildweather.com is shown in Fig. 6.49.

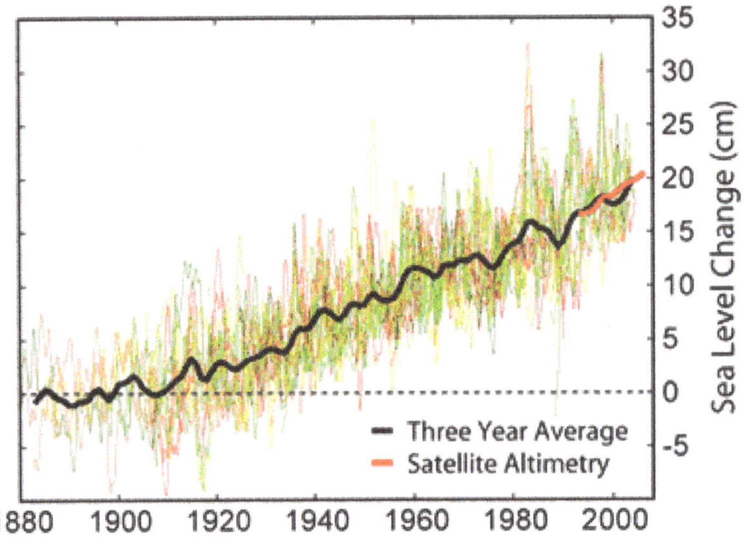

Fig. 6.49 Global scenario of sea level rise based on 23 annual tide guage records

(Source: Wildwildweather.com)

- A small increase in sea level can have a dramatic impact on many coastal environments. Over 600 million people live in coastal areas that are less than 10 m above sea level and two-thirds of the world cities that have population over 5 million are located in these risky areas. A large number of population are required to be relocated considering the accelerated sea level rise in future decades to come. The other serious impacts include coastal erosion, wetland and coastal plain flooding, salinization of aquifers and loss of habitats and plants. The environmental protection agency estimates that 26000 sq.kms of land would be lost if the sea level rises by 0.66 m. The IPCC also estimates that about 33% of coastal land and wetland habitats are likely to be lost in the next hundred years if the level of the ocean continues to rise at its present rates and several oceanic islands are likely to be submerged under the sea. There will also be significant impacts on the availability of fresh water resources. i) Groundwater resources in coastal plains are potentially vulnerable to salinization from rising sea levels, due to their low elevation and hydraulic gradient. ii) Salt water intrusion in the lower reaches of the deltas will be exasperated by predicted rises in sea levels. iii) Reduced protective capacity from extreme storms and floods, as higher sea levels provide a higher base for storm surges. The United Nations Framework Convention on Climate Change (UNFCCC) was convened in 1992 to address the challenges of climate change through the objectives outlined below.

- To stabilize greenhouse gas concentrations to levels that prevent dangerous interference with the global climate system.
- To achieve these reductions within a time frame that allows ecosystems to adapt naturally to climate change, to ensure that food production is not threatened, and to enable economic development to proceed in a sustainable manner.

Kyoto Protocol was developed in 1997 to reinforce the emissions reduction commitments of the UNFCCC. The Protocol came into legal force in 2005 when it was ratified by 30 industrialized nations, creating legally binding targets for a 5 percent reduction in emissions below 1990 levels by 2012.

- Some of the critical areas where the impacts of global climate change are observed are i) *Water*: Rising global temperatures will lead to modifications of the hydrological cycle, resulting in more intense dry seasons and wet rainy seasons, and subsequently heightened risks of more extreme and frequent floods and droughts. Changing climate will also have significant impacts on the availability of water, as well as the quality and quantity of water that is available and accessible. Melting glaciers will increase flood risk during the rainy season, and strongly reduce dry-season water supplies to one-sixth of the World's population. ii) *Agriculture*: Declining crop yields are likely to leave hundreds of millions without the ability to produce or purchase sufficient food supplies, especially in Africa. At mid to high latitudes, crop yields may increase for low levels of change in temperature, but will decline at higher levels of temperature change. iii) *Ecosystems:* changing temperatures will cause ecosystems to shift-forests, land types and plant species will dieback in some areas as temperatures rise, but increase in other areas. However, in many cases, the pace of change in temperature may be too fast for ecosystems to adjust, resulting in the loss of forests and species. iv) *Health:* Higher temperatures expand the range of some dangerous vector-borne diseases, such as malaria, which kills one million people annually, most of whom are children in the developing world. Further, heat waves associated with climate change, and increase in water borne diseases, will result in increased health problems. v) *Coastlines:* Melting ice and thermal expansion of oceans are the key factors driving sea level rise. In addition to exposing coastlines, where the majority of the human population live, to greater erosion and flooding pressures, rising sea levels will also lead to salt water contamination of groundwater supplies, threatening the quality and quantity of freshwater access to large percentages of the population. The vulnerabilities, adaptation processes and mitigation strategies for these sectors in a Global scenario have been extensively described and discussed in the IPCC reports 1990; 1995; 2001 and 2007; which could be referred in the IPCC websites.
- The projected impacts of climate change over India by and large are similar to the contents reflected in the Global scenario but with certain marginal changes which are reflected as follows (Sathaye, J., et al., 2006). (i) *Water resources:* The hydrological cycle is likely to be altered and the severity of droughts and intensity of floods in various parts of India is likely to increase. A general reduction in the quantity of available run-off is predicted. (ii) *Agriculture*: Simulations using dynamic crop models indicate a decrease in yield of crops as temperature increases in different parts of India. However, this could offset by an increase in CO_2 at moderate rise in temperature and negative impact on crop productivity could prevail. (iii) *Forests:* Climate impact assessments using BIOME-3 model and climate projections for the year 2085 show 77% and 68% of the forest grids in India are likely to experience shift in forest types under A2 and B2 scenario, respectively. Indications show a shift towards wetter forest types in the northeastern region and drier forest types in the northwestern region in the absence of human influence. Increasing atmospheric CO_2 concentration and climate warming could also result in a doubling of net primary productivity under the A2 scenario and nearly 70% increase under the B2 scenario. (iv) *Coastal zone:* Simulation models show an increase in frequencies of tropical cyclones in the Bay of Bengal; particularly intense events are projected during the post-monsoon period. Sea

level rise is projected to displace populations in coastal zones, increase flooding in low-lying coastal areas, loss of crop yields, saline water encroachment in both surface and sub surface systems and reduction of fresh water resources. (v) *Human health:* Malaria is likely to persist in many states and new regions may become malaria-prone and the duration of the malaria transmission windows is likely to widen in northern and western states and shorten in southern states. (vi) *Desertification:* Climate change leading to warming and water stress could lead to degradation and desertification processes in the low rainfall areas of the country.

- IPCC, (2007) in its Fourth Assessment report has come out with a very important findings related to coastal systems and low lying areas; some of the salient findings focused in this chapter are highlighted in the following paragraphs.

Global Scenario

- Coasts are very likely to be exposed to increasing risks due to climate change and sea-level rise and the effect will be exacerbated by increasing human-induced pressures on coastal areas.
- It is likely that corals will experience a major decline due to increased bleaching and mortality due to rising sea-water temperatures. Salt marshes and mangroves will be negatively affected by sea-level rise.
- All coastal ecosystems are vulnerable to climate change and sea-level rise, especially corals, salt marshes and mangroves.
- Corals are vulnerable to thermal stress and it is very likely that projected future increases in Sea Surface Temperature (SST) of about 1 to 3°C in the 21st century will result in more frequent bleaching events and widespread mortality, unless there is thermal adaptation or acclimatization by corals.
- Coastal wetlands, including salt marshes and mangroves, are sensitive to sea-level rise, with forecast global losses of 33% given a 36 cm rise in sea level from 2000 to 2080. The largest losses are likely to be on the Atlantic and Gulf of Mexico coasts of the Americas, the Mediterranean, the Baltic, and small-island regions in the Indian ocean areas.
- Ocean acidification is an emerging issue with potential for major impacts in coastal areas, but there is little understanding of the details. It is an urgent topic for further research, especially programmes of observation and measurement.
- Coastal flooding in low-lying areas is very likely to become a greater risk than at present due to sea-level rise and more intense coastal storms, unless there is significant adaptation. Impacts are sensitive to sea-level rise, the socio-economic future, and the degree of adaptation. Without adaptation, more than 100 million people could experience coastal flooding each year by the 2080's due to sea-level rise alone, with the A2 world likely to have the greatest impacts.
- Benefit-cost analysis of responses suggests that it is likely that the potential impacts will be reduced by widespread adaptation. It also suggests that it is likely that impacts and protection costs will fall disproportionately on developing countries.
- Key human vulnerabilities to climate change and sea-level rise exist where the stresses on natural low-lying coastal systems coincide with low human adaptive capacity and/or high exposure and include. (i) Deltas, especially Asian megadeltas (e.g., the Ganges-Brahmaputra in Bangladesh and West Bengal), (ii) Low-lying coastal urban areas, especially areas prone to natural or human-induced subsidence and tropical storm landfall (e.g., New Orleans, Shanghai), (iii) Small islands, especially low-lying atolls (e.g., the Maldives).
- Regionally, the greatest increase in vulnerability is very likely to be in South, South-East and East Asia, and urbanized coastal locations around Africa, and small-island regions. The numbers

affected will be largest in the megadeltas of Asia, but small islands face the highest relative increase in risk.

- Sea-level rise has substantial inertia compared with other climate change factors, and is virtually certain to continue beyond 2100 for many centuries. Stabilisation of climate could reduce, but not stop, sea-level rise. Hence, there is a commitment to adaptation in coastal areas which raises questions about long-term spatial planning and the need to protect versus planned retreat.

National scenario

The Indian coastline stretches 7500 kms including island areas and exhibits several diversified features of coastal zones in terms of morphology, tectonics and water resources. Indian coastline is facing increasing human pressures such as overexploitation of fresh water and marine resources, dumping of industrial and toxic wastes, oil spills and leaks, etc which have resulted in substantial damage to the ecosystems. The impact of global warming - induced sea level rise due to thermal expansion of near surface ocean water has great significance to India due to extensive low-lying densely populated coastal regions. Sea level rise could have a number of physical impacts on coastal areas, including loss of land due to inundation and erosion, increased flooding and salt water intrusion. The climate change and vulnerability assessments for various sectors like greenhouse scenarios, ecosystems, coastal zone, infrastructure and energy systems etc are well documented in the book Shukla, P. R. et al (2003). In another document released "India's Initial National Communication to the United Nations Framework Convention (UNFC)" on Climate Change, (2004). The vulnerability assessment and adaptation methodologies for India have been well documented for different sectors like ecosystems, precipitation run-off levels, monsoon rainfall, water resources, agriculture, forests and health. Mall, R.K., et al (2006) while summarizing water resource potentials in India had analyzed the climate impacts in certain regional scenarios and summarized important vulnerability issues related to regional water resources and groundwater availabilities, changing pattern of rainfall, run-off and recharge situations, sea water intrusion into coastal aquifers and identified risks and responses. Some of the important scenarios related to climate change events in India and the required adaptation methods are outlined in the following paragraphs.

- *Climate related ocean-atmosphere oscillations:* Climate related ocean-atmosphere oscillations can lead to several changes along the coastline. One of the most prominent is the El Nino-Southern Oscillation (ENSO) phenomenon, an interaction between pronounced temperature anomalies and sea level pressure gradients in the equatorial Pacific Ocean, with an average periodicity of 2 to 7 years. Recent research has shown that dominant wind patterns and storminess associated with ENSO may perturb coastal dynamics, influencing (i) Monsoons, (ii) Beach morphodynamics, (iii) Cliff retreats, iv) Groundwater levels, quality changes in estuaries, mangroves and deltas.
- *Climate change and sea level rise:* The Indian coastline stretches about 7500 kms including island areas. The observations made earlier along the Indian coastline indicate a rising trend of sea level of the order of about 1.0 mm/year. However, recent observations suggest a rising trend of 2.5 mm/year. The model simulation studies and GCM outputs indicate that the oceanic region adjoining the Indian subcontinent is likely to warm up by about 1.5 to 2.0 °C by the middle of this century and by about 2.5 to 3.5 °C by the end of the century. The corresponding sea level rise is expected to be between 15 to 38 cm and 46 to 59 cm (Lal, M. and Aggarwal, D., 2000). According to Unnikrishnan, A.S., et al (2006), the estimates of mean sea level rise at Mumbai, Cochin and Vishakhapatnam showed a sea level rise of 0.78, 1.14 and 0.75 mm/year respectively whereas the estimations for Chennai showed a decrease in sea level of the order -0.65 mm/year. The predicted sea level rise of 46 to 59 cm along the Indian coastline is comparable with the projected global mean sea level rise by the end of this century and would have significant impacts on coastal systems of India.

- *Sea surface temperature and tropical cyclones:* Global warming would result in an increase in sea surface temperatures through which changes in frequency and intensity of cyclones in the coastal regions may take place. India is frequently affected by tropical cyclones and storms particularly during monsoon seasons. About 6.5% of the tropical storms in the world are found annually in the Indian ocean (Neumann, C. J., 1993; McBride, J. L., 1995). The frequency of cyclone formation in Bay of Bengal is 5 to 6 times more than the Arabian sea (IMD, 1979). The coastal states Andhra Pradesh, Orissa, Tamil Nadu and West Bengal along the east coast of India are more vulnerable than the west coast in terms of cyclones and storm surges causing huge damage of properties and human losses. Recent study suggest an increase in cyclone intensity of the order of 10 to 20% for a rise in sea surface temperature of 2 to 4 °C. The global warming impacts therefore in future could probably result in increasing frequency levels of cyclones and storm surges with different intensities.
- *Land inundation and population affect:* The potential impact of one meter sea level rise along the Indian coast could probably inundate an area of the order of 5763 km^2 along the coastal states of India affecting 7.1 million population living in the coastal regions (TERI, 1996). The most vulnerable areas along the Indian coastline are Kutch region of Gujarat, Mumbai and South Kerala, Ganges delta (West Bengal), Mahanadi delta (Orissa), Krishna and Godavari deltas (Andhra Pradesh) and Cauvery delta (Tamil Nadu). The vulnerability of a region could be assessed both in terms of physical exposure to the sea level rise and the level of socio-economic development like population density, land use patterns and other investments. A one meter rise of sea level will inundate 1810 sq.km land in Gujarat, 1220 sq.km in West Bengal, 670 sq.km in Tamil Nadu, 550 sq.km in Andhra Pradesh, 480 sq.km in Orissa, 410 sq.km in Maharashtra, 290 sq.km in Karnataka, 160 sq.km in Goa and 120 sq.km in Kerala. The problem therefore appears enormous and the measures to combat with these problems should be initiated early.
- *Salt water intrusion and fresh water resources:* Salt water intrusion is a common problem in most coastal delta regions of India due to excess fresh water withdrawal. The sea level rise events would trigger the salt water inundation in all the low-lying areas, ditches and ponds facilitating infiltration of saline water into the near surface fresh water shallow systems thereby increasing the salinity levels and at the same time rendering the aquifers unsuitable for drinking and irrigation. A one meter rise in sea level could inundate several kilometers upstream and contaminate aquifers in their proximities. These processes would also result in the reduction of recharge potentials of fresh water into the near surface system. Estuary boundaries and the tidal variations will be substantially enhanced into the upstream areas. A one meter rise in sea level could inundate several kilometers upstream in the low-lying estuaries and delta areas and contaminate near surface fresh water aquifers of paleo beach ridges tail end areas of paleo channels etc. There will be new adjustments in the estuary boundaries upstream and the tidal levels as the impact boundaries will further enhance several kilometers upstream and causes reduction in recharge potentials. The deeper fresh water systems which act as potential fresh water resource belts however will not suffer much due to climate change and sea level rise, at least in the present century. However if the present sea level rise scenarios continue beyond the present century, and if the saline water inundation continues to spread beyond the middle reaches of the delta system, the recharge potentials could be drastically changed and could result in contamination of deeper fresh water systems also.
- *Impacts on coastal ecosystems:* Coastal ecosystems contain some of the world's most diverse and productive ecosystems like mangroves, coral reefs and sea grasses. These systems provide natural protection to coasts by dissipating considerable wave energies and are accentuated by anthropogenic activities. Low-lying deltas and coral reefs/atolls are particularly sensitive to changes in frequency and intensity of rainfall and storms. Corals generally grow fast enough to keep pace with sea level rise but may be damaged by warmer sea temperatures. Mangroves such as sunderbans are highly vulnerable to sea level rise since they will change the salinity distribution and productivity levels.

The impact on sea level rise depends upon type of mangrove forest which may keep pace with rising sea level or may be submerged.

- *Adaptation options:* The impacts of sea level rise on the Indian coastline and the ecosystems could be significant which are already under stress due to unsustainable human activities. It is necessary to device appropriate effective response strategies and make vulnerability assessment studies at regional and local scales. The factors related to eustatic sea level rise, hydrological, meteorological, oceanographic changes, and tectonic movements should be considered. There is also a need to make systematic inventory on tidal gauge measurements, tidal flats, changes in estuaries and mangrove systems. It is also necessary to initiate protective actions in respect of beach erosion, salt water intrusion processes, etc and build necessary protective structures for sustainability of ecosystems and fresh water resources. Rules and regulations should be framed to optimize the developmental activities and to keep the systems free from hazards.

6.7.7 Numerical models related to sea level rise

Coastal aquifers within the zone of influence of mean sea level can be threatened by sea level rise which is a world wide phenomena. It is important to estimate the impacts of sea level rise in relation to the coastal aquifer systems/salt water intrusion phenomena. Very few numerical models which are qualitative in nature are available. No serious efforts are made to understand and quantify thc shifts of saline water intrusion in space and time and with reference to global warming and sea level rise phenomena. However a few workers have made some numerical analyses towards this end in a qualitative way. The sea level rise events take place over a period of time of the order of decades. The available reference data indicate that the sea level is rising at a rate of 2.0 mm/year. It should be noted in this context that a small change in the sea level rise could have devastating effects (IPCC 1998). A large number of coastal aquifers are threatened by salt water intrusion and the sea level rise events in future could trigger this phenomena and reduce the fresh groundwater resource potentials. The sea level rise could induce shifting of the saline water interface further inland and reduce the fresh water potentials. As of now only a few quantitative studies on salt water intrusion in coastal aquifer systems induced by sea level rise are available (Meisler, H., et al., 1984; Oude Essink, G.H.P., 1999; Sherif, M.M. and Singh, V.P., 1999; Bobba, A.G., 2002; Ranjan, P., et al., 2005; Ranjan, P., et al 2006). These studies however are site specific and cannot be regionalized and are based on numerical models which are required to be verified through physical parameters. Sherif, M.M. and Singh, V.P. (1999) observed that the 500 mm rise in the Mediterranean sea level will cause an additional saltwater intrusion upto a distance of 9 km in the Nile delta, Egypt, while at the same time a similar rise in sea water level in the Bay of Bengal could trigger the saline water intrusion to an order of 0.4 km only. The differences however are not explained in terms of physical situations. The study also does not indicate any quantitative conclusions about the extent of the sea water intrusion induced by sea level rise. Ranjan, P., et al., (2005) tried to explain the effects of sea level fluctuations on fresh-saline water interface. The simulations were carried out over hundred years period for two case studies (i) West coast of America and (ii) East coast of India using finite difference model. The rise in sea level causes the interface to be shallower and extend further landward. The fresh water loss due to sea level rise over hundred years and for a coastal strip of 2 km width is estimated to be 1-5% of the aquifer volume in the West coast of America and 1-3% in the Bay of Bengal. The results also show that the effect of sea level rise is higher in high latitudes than the middle and equatorial areas. The loss of fresh groundwater resources appears to be more severe in the mid and equatorial latitudes where large diversities of seasonal climates, high density of population and high demands of groundwater resources exist. The results further show that there is an increase in loss of fresh groundwater resource with the reduction of groundwater recharge due to future climate change.

6.7.8 Mitigation and management strategies

Several mitigation methods are in vogue in different parts of the world to control salt water intrusion where withdrawals of fresh water have caused water levels in aquifers to fall significantly below sea levels such overdrafts are normally accompanied by salt water intrusion inland and contaminate fresh water aquifer systems. Some of the mitigation methods are outlined below.

(i) Design and construction of impermeable subsurface barriers in the estuary mouth regions.

(ii) Desiltation of tanks/lakes in the coastal regions.

(iii) Modification of pumping patterns/relocation of wells/redesign of well fields.

(iv) Design and installation of extraction barriers.

(v) Design and installation of injection barriers along the coast.

(vi) Increased Recharge.

(vii) ADR technologies.

(viii) Aquifer Storage and Recovery (ASR) methodologies.

(ix) Artificial Recharge (AR).

(i) ***Design and construction of impermeable subsurface barriers in the estuary mouth regions:*** During the periods of droughts and low river flows the saline water could migrate up along the estuary. Similarly a rise in sea level will also cause saltwater to migrate upstream. In such cases physical barriers could be installed at an appropriate place to stop saline water intrusion upstream. The method consists of constructing a barrier by actually constructing a cut off wall such as a sheet pile wall or a concrete wall or a clay puddle along the coastline to facilitate prevention of saline water movement upstream. The engineering design is illustrated below Fig. 6.50.

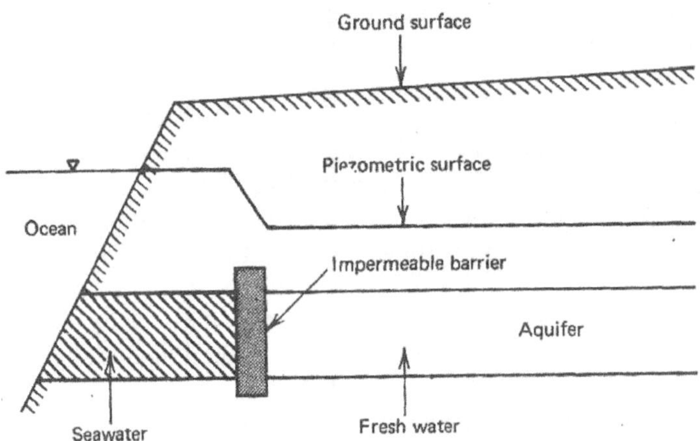

Fig. 6.50 Control of sea water intrusion by an impermeable subsurface barrier parallel to the coast

(Source: Todd, D.K., 1980)

(ii) ***Desiltation of tanks/lakes in the coastal regions:*** In general the coastal delta regions contain huge number of tanks/lakes. Some are natural and some are artificially made and are often filled with silts and clays deposited during different cycles of sedimentation. These silts and clays will reduce the infiltration rates into the underground. Desiltation procedures if adopted will enhance the recharge potentials and help to check the movement of salt water upstream by way of forming through localized pressure ridges. In addition to increase further recharge rates underground, injection wells with appropriate gravel packs to facilitate silt free water recharge the surface aquifer system. This will help to accelerate recharge levels further into the near surface system. If the density of tanks/lakes are high and when the above design is adopted it would facilitate to build up a series of isolated recharge mounds and obstruct the saline water inflow into the system.

(iii) ***Modification of pumping patterns / relocation of wells / redesign of well fields:*** In the event of saline water migration towards upstream it would become necessary to adopt redesigning and relocations of well fields further upstream. This would help to re-establish a seaward hydraulic gradient. Similarly reduction in pumping rates of the existing well fields can also help to check the saline water intrusion inland. In such cases, unconfined or confined aquifers where moderate pumpage already occurs and the effect of a sea level rise is projected to be important, a phased shutdown of wells should be designed as the saltwater intrusion progresses, permitting new wells or new economical surface distribution schemes to be implemented. A modification of pumping patterns will allow water levels to recover in critical areas. This will have an effect of slowing down the saltwater advance. However, the saltwater will intrude until it reaches a new equilibrium, depending upon the recharge rate, the pumping rates, the net overdraft, water levels, aquifer geometry, aquifer characteristics, and the present status of saltwater intrusion. For a saltwater interface currently in equilibrium, an increase in sea level rise may be counteracted by a modification of pumping patterns.

(iv) ***Design and installation of extraction barriers:*** This method involves developing a protective barrier called pumping trough through installation of a line of wells adjacent to sea. The sea water intrusion entering into the land area will be captured by the pumping wells and disallow saltwater intrusion to move upstream. This approach is illustrated in Fig. 6.51. This type of well fields installed while pumping will facilitate brackish water to be discharged back into the sea. The experiences of Kashef, A.I. (1977) and Stone, W.L. (1978) have indicated that the saltwater intrusion caused by pumping overdrafts can be technically controlled but are more expensive than injection barriers. However in certain localized and narrow intrusion belts, the extraction barriers probably could be planned depending upon the socio-economic situation of the area.

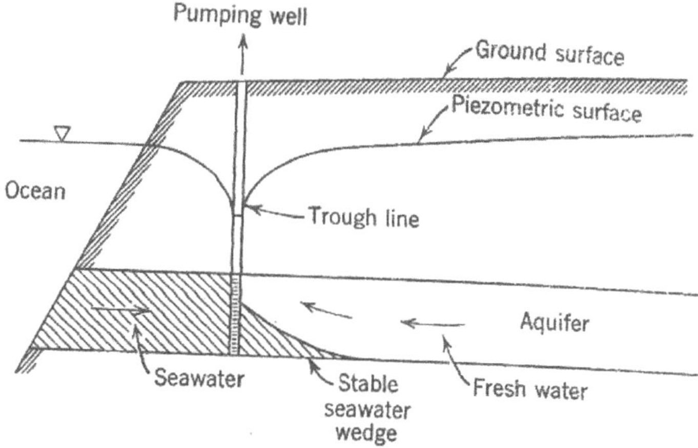

Fig. 6.51 Control of sea water intrusion by extraction barrier forming a pumping trough parallel to the coast

(Source: Todd, D.K., 1980)

(v) ***Design and installation of injection barriers along the coast:*** This method involves building of pressure ridge along the coast by installing series of recharge wells as indicated in Fig. 6.52. It is necessary to recharge the system with high quality water which will move landward to supply water in the pumping fields and at the same time a small amount of the recharge water would also go into the sea. The higher groundwater levels along the injection barrier prevent saltwater intrusion entering inland. A proper design of well spacing and location must be performed to ensure that saltwater does not intrude around the injection barrier, in between individual wells, or move vertically from above or below. The effectiveness of the injection barrier will be maintained if the water levels in the vicinity of the barrier are increased as sea level increases to always maintain them at 1 m above mean sea level. This type of barrier wells to prevent saltwater intrusion are extensively practiced in the coastal regions of U.S.A.

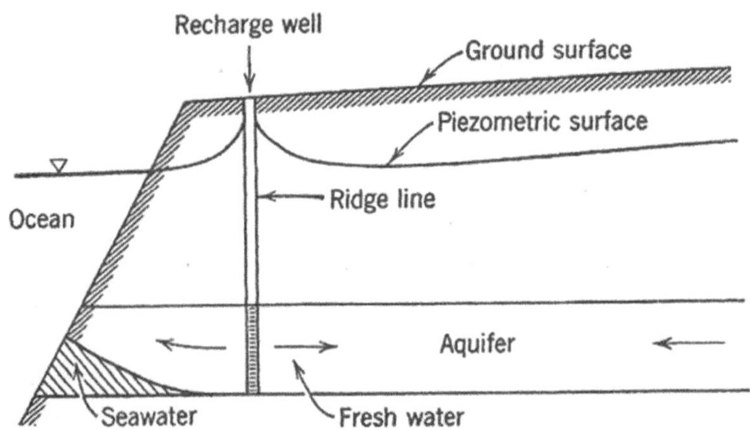

Fig. 6.52 Control of sea water intrusion by an injection barrier forming a pressure ridge parallel to the coast

(Source: Todd, D.K., 1980)

(vi) ***Increased Recharge:*** In several locations, sufficient amounts of fresh water are available for recharge during periods of high precipitation. Although some water is captured during these periods and stored in surface reservoirs, very little water is recharged through groundwater systems for use in drought periods. This extra water which is normally discharged into ocean could be used to replenish the aquifer, build up of groundwater levels, and repel the saltwater intrusion. The additional recharge if exceeds the groundwater withdrawals will keep the fresh saltwater equilibrium in stable position. In several instances, the recharge region of the principal water supply aquifer is far away from the coast and in such cases, it is possible to recharge the confined aquifer far from the shoreline and prevent saltwater intrusion. In the case of unconfined aquifers, the recharge could occur close to the coastline. It is necessary to keep the water levels of both confined and unconfined systems above the mean sea levels and increase the recharge potentials in par with the sea level rise.

(vii) ***ADR technologies:*** Abd-Elhamid, H.F. and Javadi, A.A., (2008) proposed a new methodology to study and control the saltwater intrusion. This methodology aims to overcome the limitations of the previous methods. Abstraction, Desalinization and Recharge (ADR), consists of three steps: a) abstraction of brackish water from the saltwater zone, b) desalination of the abstracted brackish water using RO treatment process, and c) recharge treated water into the aquifer. The main benefits of the (ADR) methodology are to return the mixing zone into the original status and to reach a dynamic balance between fresh and saline groundwater through two processes: (a) Abstraction of brackish groundwater to reduce the volume of saline water, and (b) Recharge treated brackish water to increase the volume of fresh groundwater. The Abstraction-Recharge process helps to move fresh water/salt water interface towards the sea and is considered to be an efficient method to control saltwater intrusion. This process will continue until a state of dynamic equilibrium is reached. Then afterwards a suitable quantity of treated brackish water could be injected into the aquifer to maintain balance between fresh and saline water. The excess treated brackish water if any may be used for different purposes or stored in the aquifer itself. The RO treatment process is considered to be less expensive than other sources of high cost. Desalinization of brackish water is considered to be an efficient alternative to sea water desalinization because the salinity levels of brackish water are less than one-third of that of sea water. It's not known as yet whether this technology has been successfully used either experimentally or as routine anywhere in the world.

(viii) ***Aquifer Storage and Recovery (ASR) methodologies:*** Aquifer Storage and Recovery (ASR) technology refers to a process of recharge and storage of water in an aquifer system as and when plenty of water is available during the wet season and recovery of the stored water at times when it is needed (Pyne, R.D.G., 1995). ASR can function like a traditional surface water reservoir, however the aquifer storage eliminates the evaporative loses. A Schematic diagram describing the processes involved in the ASR technologies is presented below in Fig. 6.53.

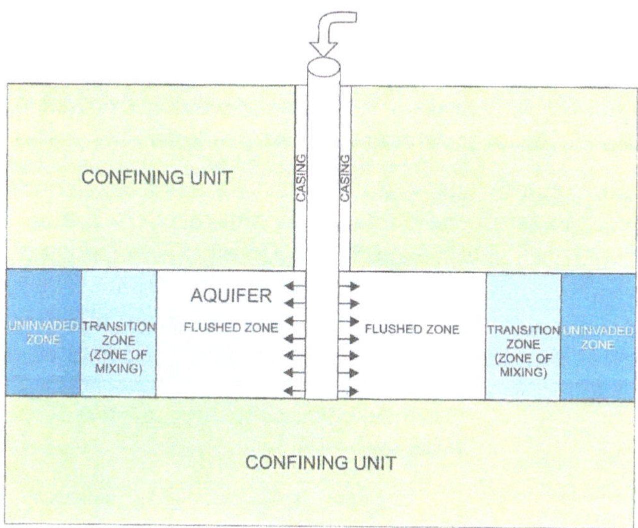

Fig. 6.53 Aquifer storage and recovery well in a confined aquifer depicting idealized flushed and transition zones created by recharge. Flushed zone contains mostly recharged water

(Source: Reese, R.S., 2002, USGS)

ASR technologies have been tested and implemented widely in the coastal regions of U.S.A. Some of the challenging aspects of ASR systems in coastal regions of U.S.A are that the quality of the ambient water in the Upper Floridan aquifer is brackish to saline. The recharge of fresh water into the saline water aquifer system creates a radial zone of diffusion / mixing zone around the well that separates the native saline water from the injected fresh water as shown in the above Fig. 6.53. The shapes of the injected fresh water and mixing zones are dependent on the geology of the injected zone. The mixing between the injected fresh water and the native saline water can reduce the amount of fresh water i.e. recovery during the withdrawal phase and can lower the recovery efficiency of the ASR system. It is required to monitor the salinity levels of the injected and recovered water on a daily basis, during recovery salinity of water increases with time depending upon the volume storage of fresh water and recovery can be terminated when the salinity of the water levels reaches a predetermined level or marginally higher (Reese, R.S., 2002). High salinity levels could also cause buoyant stratification with the injected water rising over the native saline water. Other factors that effect the recovery efficiency of ASR systems include aquifer thickness and permeability, ambient hydraulic gradients, volume of injected water, duration and frequency of injection and withdrawal phases, duration of storage phases, and the total number of injection withdrawal cycles (Reese, R.S., 2002; Pyne, R.D.G., 1995).

Aquifer storage and recovery wells are designed and constructed at 27 sites in southern Florida located in coastal areas. A storage zone of the most of the well sites are located in the Upper Floridan aquifer system. A large scale ASR systems in recent years have been proposed as a part of the Comprehensive Everglades Restoration plan (CERP) and as a part of the cost effective water surplus systems to meet the demands agriculture, recreational purposes and ecological protections. Several technical issues are required to be looked into while implementing the action plan of ASR systems. The issues include (a) a comprehensive hydrogeological information around the areas of proposed ASR well fields, (b) impacts of groundwater level fluctuations on existing wells in relation to surface flow systems, precipitation levels, recharge scenarios etc, (c) water quality concerns regarding the suitability of source water which can be used for recharging into the ASR systems, (d) a number of site specific and regional scale research activities are required to be planned around the ASR pilot programmes to understand the hydrogeological boundaries, optimum well designs and well cluster configurations. Some of the factors which influence the efficiency of ASR system are 1. thicknesses of the aquifer material, 2. transmissivity and storage parameters, 3. leakance parameter from the upper confining zones and chloride concentration levels of the storage system.

The ASR systems are normally operated through several recharge and recovery cycles. There are several ASR well sites under operation with different recovery cycles. The Boynton Beach East Water Treatment Plant well site in Palm Beach County (Florida, U.S.A) appears to have shown the best recovery efficiency which is presented in fig.6.54 below. The recharge volume per cycle ranged from 0.6 to 714 million gallons. The highest recovery efficiency achieved was 84% for cycle 16. The recovery efficiencies at this particular site appear to be the best amongst all ASR sites in the Floridan aquifer systems.

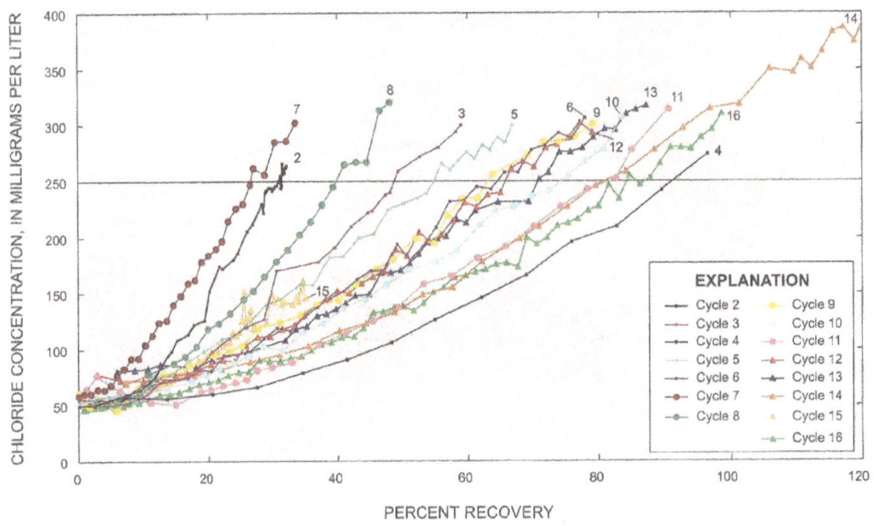

Fig. 6.54 Percent recovery of recharged water during operational cycles in relation to chloride concentration of recovered water at the Boynton Beach East Water Treatment Plant site in Palm Beach County. Recovery for cycle 14 was continued until reaching a chloride concentration of about 1,000 milligrams per liter

(Source: Reese, R.S., 2002, USGS)

Aquifer storage and recovery (ASR) in southern Florida has been proposed as a cost-effective water supply alternative that can help meet the needs of agricultural, municipal, and recreational users and can be used for Everglades ecosystem restoration. Plans have been made to utilize ASR on an unprecedented scale in the Central and Southern Florida Comprehensive Review Study as proposed by the U.S. Army Corps of Engineers and the South Florida Water Management District (1999). This review study is also known as the Comprehensive Everglades Restoration Plan (CERP). About 330 ASR wells have been proposed for southern Florida, each with an assumed capacity of 5Mgal/d during recharge or recovery.

(ix) ***Artificial Recharge (AR):*** Artificial Recharge (AR) methodologies are widely used to improve the quantities and qualities of the subsurface systems. Several direct and indirect methods are in vogue in different basins of the country. Some of the direct methods include surface spreading techniques such as flooding, ditches and furrows, recharge basins, turn off conservations structures, gully plugs and contour barriers, stream channel modification and surface irrigation. The sub surface techniques include injection wells, gravity head recharge wells, collector wells and recharge pits /shafts. Some of the indirect methods include a) induced recharge which could be organized through pumping wells, collector wells and infiltration galleries, b) aquifer modification procedures such as bore blasting, hydrofracturing, etc, c) groundwater conservations structures are organized through groundwater dams/ underground barriers and fracture cementation techniques. The details for implementation of the above structures are available in 1) Manual on Artificial Recharge of Groundwater (CGWB, 1994); 2) Guidelines on Artificial Recharge to Groundwater (CGWB, 1998), 3. Artificial recharge in India (Muralidharan, D. and Athavale, R.N., 1998). All the above techniques in one way or the other could be used effectively based on the hydrological system boundaries and aquifer configuration and saline fresh water interface boundaries. The ASR technologies have not so far been implemented in the Indian context. This technique holds a very good promise if used in most of the saline affected coastal delta regions of India. These techniques coupled with optimization and management models as described in earlier paragraphs would yield significant results for sustainable utilization of fresh groundwater resources in the coastal regions of India and elsewhere in the world.

6.7.9 Future challenges ahead

- A more intensified exploratory efforts are required to characterize the near surface and deep aquifer systems, to assess the fresh water flow patterns and resource potentials and to delineate saline-fresh water interface structural patterns in different delta systems. Some of the latest technologies like electrical resistivity tomography, deep electrical resistivity surveys and electromagnetic methods (Ground and Airborne) coupled with deep drilling, multi geophysical logging approaches and packer yield tests of individual aquifers would help to map the near surface and deep fresh groundwater systems more precisely and facilitate to evolve 3-D hydrogeological maps and optimization schemes.
- A dedicated monitoring systems for both fresh water and saline water aquifers are to be introduced in all the delta systems through specially designed well structures, piezometer nests tapping aquifers at different levels. Several continuous automatic pressure gadgets with computerized storage and retrieval systems are available for this purpose. These gadgets if introduced into the monitoring systems would facilitate to understand and quantify the flow patterns and assess the saline water interface movement in relation to the developmental stresses, anthropogenic and natural impacts.
- A large amount of fresh water from the near surface and deep systems is being discharged into the oceanic basins. The hydrogeological systems analyses presented in this book indicated large

thicknesses of fresh water systems at deeper levels by and large in all the coastal deltas excepting Krishna and Godavari where the deep systems are brackish/saline. A more dedicated work to assess the fresh water discharges into oceanic basins should be undertaken by introducing the latest available techniques such as installation of seepage meters, design and construction of deep piezometers parallel and perpendicular to the coastline and quantify the flow patterns, introduction of tracers such as Radon and Radium isotopes and Methane which would serve a useful purpose to quantify the SGDs both in near surface and deep systems. These studies also should be coupled with appropriate geophysical profiles such as electromagnetic, electrical tomography and deep electrical resistivity surveys etc., along and perpendicular to the coastline and strengthen the database.

- Introduce 2-D and 3-D variable density numerical models/approaches to simulate the flow patterns in both fresh water and saline water aquifer systems and develop predictive analyses related to the movement of fresh and saline water in different aquifer systems and their interfaces in relation to different hydraulic stresses. Several numerical models in recent times have been developed which are illustrated earlier could be tried in all the delta systems, while adopting appropriate calibration strategies with the field situations. It would be also desirable to develop appropriate management schemes by adopting the latest techniques such as embedded, response matrix and linked simulation optimization approaches etc., and also introduce application of genetic algorithm methods. In view of the impending global warming and sea level rise events and their possible impacts on coastal fresh water aquifer systems, appropriate 3-D models should be developed to forecast the possible changing scenarios in relation to the aquifer stresses and sea level rise events.
- Introduce appropriate Artificial Recharge techniques as a part of the management / mitigation strategies as illustrated earlier. Large thicknesses of both fresh water and saline water aquifer systems are abundantly available in different coastal delta systems. In areas where large thicknesses of saline water are available the latest techniques such as Aquifer Storage and Retrieval (ASR) systems could be introduced to store the fresh water inside the saline water reservoir during the periods of monsoon and retrieve the fresh water again as and when required during non monsoon seasons. These practices are being successfully adopted in several coastal regions of U.S.A.
- It is also required to introduce the appropriate legislative methods as a part of the management and mitigation strategies to protect and safeguard the fresh groundwater systems and adopt and establish optimal groundwater consumption policies. It would be necessary in this context to see that the fresh groundwater extraction rates should not exceed the safe yield limits and the available recharge potentials, so as to keep the system sustainable and free from salinity hazards.

References

Aarts, E. and Korst, J., 1997. Simulated Annealing and Boltzman Machines: A stochastic approach to combinatorial optimization and neural computing. John Wiley & Sons, New York.

Abd-Elhamid, H.F. and Javadi, A.A., 2008. An Investigation into control of saltwater intrusion considering the effects of climate change and sea level rise. 20th Salt Water Intrusion Meeting: June 23-27, Naples, Florida, U.S.A., 4 p.

Agarwal, R.P. and Mitra, D.S., 1991. Paleogeographic reconstruction of Bengal delta during quaternary period. Memoirs Geo. Soc. Ind no.22 pp.13-24.

Ahmad Zaharin Aris., et al., 2009. Evolution of groundwater chemistry in the shallow aquifer of a small tropical island in Sabah, Malaysia. Sains Malaysiana v. 38, no. 6, pp. 805-812.

Ahmad, E., 1972. Coastal Geomorphology of India. Orient Longman Ltd., New Delhi, 221 p.

Alder, J., 2003. Distribution of estuaries worldwide. Sea Around Us Project, UBC Vancouver, B.C. (Canada).

Alger, R.P. and Harrison, C.W., 1989. Improved fresh water assessment in sand aquifers utilizing geophysical well logs. The Log Analyst, Jan.-Feb., pp. 31-44.

Alger, R.P., 1966. Interpretation of electric logs in fresh water wells in unconsolidated formations: Soc. Prof. Well Log. Analysts, 7th Annu. Logging Symposium, Tulsa, OK, May, Trans., pp. CC_1-CC_{25}.

Allen, R.G., Pereria, L.S., Dirk Raes and Martin Smith. 1998. Crop evapotranspiration – *guidelines for computing crop water requirements.* FAO Irrigation and Drainage Paper 56.

APHA, 1995. Standard methods for the examination of water and waste water, 17th edn. Washington DC.

Archie, G.E., 1942. The electrical resistivity log as an aid in determining some reservoir characteristics. Jour. Pet. Tech., Vol. 5.

Ashok Kumar, 2011. Hydrogeological conditions and groundwater resources of Surat district, Gujarat. Technical Report: West Central Region C.G.W.B., Ahmedabad.

Athavale, R.N. and Murti, C.S., 1980. Measurement of recharge rates in soils through detection of tritium produced in Chinese thermonuclear tests. In: Proceedings of the Indian Acad. of Sci., vol. 89, no. 2, pp. 145-151.

Athavale, R.N., Murti, C.S. and Chand, R., 1980. Estimation of recharge to the phreatic aquifers of lower Maner basin, India, by using the tritium injection method. Jour. Hydrology, 45, pp.185-202.

Babu, P.V.L.P., 1975. Morphological evolution of the Krishna delta. Ibid., V. 3, no. 1, pp. 21-27.

Babu, P.V.L.P., 1978. Geomorphic evolution of the Mahanadi delta. Proc. Symp. Morphology and evolution of landforms. Dept. of Geology, University of Delhi, Delhi, pp. 217-221.

Babu, P.V.L.P., 1991. Cauvery delta – Its past and present. Memo Geol. Soc. of India. No. 22, pp. 91-101.

Back, W., 1960. Origin of hydrochemical facies of ground water in the Atlantic coastal plain in Internat. Geol. Cong., Geochemical cycles; Internat. Geol. Cong., 21st Copenhagen 1960, proc., pt. 1, pp. 87-95.

Back, W., 1966. Hydrochemical facies and groundwater flow patterns in Northern part of Atlantic coastal plain. Geological Survey Professional Paper 498-A, Washington.

Badhe Shubhada., 1983. An interim report on studies in Ganga Brahmaputra delta-morphology and processes in modern delta. Unpublished Report of ONGC, Dehra Dun.

Bakker, M., 2003. A Dupuit formulation for modeling seawater intrusion in regional aquifer systems. Water resources research.

Balakrishnan, T.S., 1997. Major tectonic elements of the Indian sub continent and contiguous areas: A geophysical view: Geol. Soc. Ind. Mem, 38: 155 p.

Balasubramanyam, M.N., 2006. Geology and tectonics of India: An over view, Mem. No. 9, International Association of Gondwanas Research., 205 p.

Baldwin, A.D., JR., and Miller, J., 1979. Use of Gamma Logger to Delineate Glacial and Bedrock Stratigraphy in Southwestern Ohio. Ground Water v. 17, no.4, pp.385-90.

Barazzuoli, P., Nocchi, M., Rigati, R. and Salleolini, M., 2008. A conceptual and numerical model for groundwater management: a case study on a coastal aquifer in southern Tuscany, Italy. Hydrogeology Journal v. 16, no. 8, pp. 1557-1576.

Baumgartner, F. and Reichel, E., 1975. The world water balance. Vienna and Munich: R. Oldenboury Verlag., 180 p.

Bear, J. and Dagan, G., 1962. The transition zone between fresh and salt waters in coastal aquifers. Technion-Israel Inst. of Technology, Hydraulic Laboratory Progress Rep. No. 1, 124 p.

Bear, J. and Dagan, G., 1963. The transition zone between fresh and salt waters in coastal aquifers. Technion-Israel Inst. of Technology, Hydraulic Laboratory Progress Rep. No. 2, 46 p.

Bear, J. and Dagan, G., 1964a. Moving interface in coastal aquifers. J. Hydraul. Div., Proc. Am. Soc. Civ. Eng., 90 (HY4), pp.193-216.

Bear, J. and Dagan, G., 1964b. Some exact solutions of interface problems by means of the hydrograph method. J. Geophys. Res., 69 (8): pp.1563-1572.

Bear, J. and Kapuler, I., 1981. A numerical solution for the movement of an interface in a layered coastal aquifer. J. Hydrol., 50, pp. 273-298.

Bear, J. and Maulem, Y., 1974. The shape of the interface in steady flow in a stratified aquifer. Water Res. Research, v. 10, no. 6, pp. 1207-1215.

Bear, J. and Todd, D.K., 1960. The transition zone between fresh and salt waters in coastal aquifers. Hydraul. Lab., Univ. of California, Berkeley, Calif., Water Resour. Center Contrib., No. 29, 156 p.

Beck, M.W., Heck, K.L., Able, K.W., Childers, D.L., Eggleston, D.B., 2001. The identification, conservation, and management of estuarine and marine nurseries for fish and invertebrates. *Bioscience,* v. 51, no. 8, pp. 633-641.

Bedi, N. and Vaidyanadhan, R., 1982. Effect of neotectonics on the morphology of the Narmada river in Gujarat, Western India. Zeitschrift fur geomorphologie N.F. 27, pp. 87-102.

Bharali, B., Rath, S. and Sarma, R., 1991. A brief review of Mahanadi delta and the deltaic sediments in Mahanadi basin. Memoirs Geological Society of India. No. 22, pp. 31-49.

Bhatnagar, N.C., et al., 1970. Geology and groundwater resources of parts of Balasore and Cuttack districts, Orissa. Bull. G.S.I. Sr. B, No. 29.

Bhattacharjya, R.K. and Dattta, B., 2005. Optimal management of coastal aquifers using linked simulation optimization approach. Water resources management 19, pp. 295-320.

Bhattacharya, S.K., et al., 1979. Groundwater models for interpretation of silicon-32 and radio-carbon data. Current Trends in Arid Zone Hydrology, Today and Tommorow's Printers and Publishers, New Delhi, pp. 138-147.

Bird, E.C.F., 1993. Submerging coasts: The effects of a rising sea level on coastal environments. John Wiley and Sons, Chichester, United Kingdom, 184 p.

Biswas, S.K, 1987. Regional tectonic framework, structure and evolution of the western marginal basins of India. Tectonophysics, v. 135, pp. 302-317.

Biswas, S.K. and Deshpande, S.V., 1973. Mode of eruption of Deccan trap lavas with special reference ot Kutch. J. Geol. Soc. India, 14, pp. 134-141.

Biswas, S.K., 1982. Rift basins in the western margin of India and their hydrocarbon prospects. Bull. Am. Assoc. Pet. Geol., v. 66, no.10, pp. 1497-1513.

Biswas, S.K., 1988. Structure of the western continental margin of India and their hydrocarbon prospects with special reference to Kutch basin. Amer, Assoc. Petrol. Geol. Bull., v. 66, pp. 1497-1513.

Biswas, S.K., 1992. Tectonic framework and evolution of graben basins of India. Ind. Jour. Petrol Geol., v. 68, no.4.

Biswas, S.K., 1993. Major neotectonic events during the Quaternary in Krishna-Godavari basin. Current science, vol. 64, Nos. 11 & 12, 10 & 25.

Biswas, S.K., 1996. Mesozoic volcanism in the East coast basins of India. Ind. Jour. Geol. V. 68, no. 4.

Biswas, S.K., 1999. Tectonic framework of coastal sedimentary basins and evidences of recent neotectonic movements in Indian peninsula, south of 15° latitude. Indian Journal of Geology, vol. 71, Nos. 1 and 2, pp. 117-132.

Blaney, H.F. and Criddle, W.D., 1950. Determining water requirements in irrigated areas from climatological and irrigation data, U.S. Dept. of Agriculture, Div. Irrigation and Water Conserve., Soil Conservation Service, 96 p.

Bobba, A.G., 1993. Mathematical models for saltwater intrusion in coastal aquifers. Review Article: Water Resources Management 7, pp. 3-37.

Bobba, A.G., 1993b. Field validation of 'SUTRA' groundwater flow model to Lambton County, Ontario, Canada. Water Resources Management, v. 7, pp.289-310.

Bobba, A.G., 1998a. Application of numerical model to predict fresh water depth due to climate change effect in Laccadive islands, India. Supplement to EOS. S75 p.

Bobba, A.G., 1998b. Application of numerical model to predict fresh water depth in islands due to climate change: Agatti island, India, Jour. Env. Hydrology, v. 6, pp.1-13.

Bobba, A.G., 2002. Numerical modeling of salt water intrusion due to human activities and sea level change in the Godavari delta, India. Hydrological Sciences 47(S), S67-S80.

Bobba, A.G., Singh, V.P., Berndtsson, R. and Bengtsson, L., 2000. Numerical simulation of saltwater intrusion into laccadive island aquifers due to climate change. Journal Geological society of India, 55, pp. 589-612.

Boulton, N.S., 1954. The drawdown of the water-table under non-steady conditions near a pumped well in an unconfined formation. Proc. Inst. Civil Engrs., v. 3, pt. 3, pp. 564-579.

Boulton, N.S., 1963. Analysis of data from non-equilibrium pumping tests allowing for delayed yield from storage. Proc. Inst. Civil Engrs., v. 26, pp. 469-482.

Boussinesq, Jr. J., 1887. Essai sur la theorie des Eaux Courantes. Memoirs Presente pair Divers Savants al Acadimic des Sciences de l' Institute National de France, v. 23, no. 1, 680 p.

Bredehoeft, J.D. and Pinder, G.F., 1973. Mass transport in flowing groundwater. Water Resources Research, v. 9, no. 1, pp.194-210.

Breier, J.A., Edmonds, H.N. and Villareal, T.A., 2004. Submarine groundwater discharge and associated nutrient fluxes to the Corpus Christi Bay system: Texas Water Development Board Contract Report 2002-483-416, 50 p.

Bugna, G.C., Chanton, J.P., Cable, J.E., Burnett, W.C. and Cable, P.H., 1996. The importance of groundwater discharge to the methane budgets of nearshore and continental shelf waters of the northeastern Gulf of Mexico: Geochimica et Cosmochimica Acta, v. 60, pp. 4735-4746.

Burke, L., Kura, Y., Kassem, K., Ravenga, C., Spalding, M. and McAllister, D., 2001. Pilot assessment of global ecosystems: coastal ecosystems. World Resources Institute (WRI), Washington, D.C (USA), 94 p.

Burnett, W.C., Bokuniewicz, H., Huettel, M., Moore, W.S and Taniguchi, M., 2003. Groundwater and pore water inputs to the coastal zone. Biogeochemistry 66, pp.3-33.

Bush, W.P., 1988. Simulation of saltwater movement in the Floridian aquifer system Hilton Head Island, South Carolina, USGS Water supply paper 2331.

Casshyap, G.M., 1977. Patterns of sedimentation in gondwana basins (key paper) in. IV. Int. Gond. Symp. (eds. Laskar, B., and Raja Rao, C.S). Calcutta, pp. 1-34.

Central Water Commission, 2006. Integrated hydrological data book. Hydrological Data Directorate Information Systems Organisation Water Planning & Projects Wing, New Delhi.

C.G.W.B., 1982. Hydrogeology and groundwater potential of Howrah district, West Bengal. Technical Report: Series 'D'-No. 31. Eastern Region, Calcutta.

C.G.W.B., 1988. Hydrogeology atlas of West Bengal. Eastern Region, Calcutta.

C.G.W.B., 1991. Groundwater resources and developmental potential of Chengai Anna district, Tamil Nadu. Technical Report: Southern Region, Hyderabad.

C.G.W.B., 1993. Groundwater resources and development prospects in Karaikal region, Union Territory of Pondicherry. Technical Report: Southern Region, Hyderabad.

C.G.W.B., 1993. Groundwater resources and development prospects in Madras district, Tamil Nadu. Technical Report: Southern Region, Hyderabad.

C.G.W.B., 1993. Groundwater resources and development prospects in Pondicherry region, Union Territory of Pondicherry. Technical Report: Southern Region, Hyderabad.

C.G.W.B., 1994. Hydrogeology and groundwater resources of south 24 Paraganas district, West Bengal. Technical Report: Series 'D'. Eastern Region, Calcutta.

C.G.W.B., 1994. Manual on artificial recharge of groundwater. Technical series-M, No. 3. 215 p.

C.G.W.B., 1995. Groundwater resources and development prospects in East Godavari district, Andhra Pradesh. Technical Report: Southern Region, Hyderabad.

C.G.W.B., 1995. Groundwater resources and development prospects in Krishna district, Andhra Pradesh. Southern Region, Hyderabad.

C.G.W.B., 1995. Hydrogeology and groundwater resources of north 24 Paraganas district, West Bengal. Technical Report: Series 'D'-No. 116. Eastern Region, Calcutta.

C.G.W.B., 1997. Groundwater resources and development prospects in Nagappattinam Quaid-E-Milleth district, Tamil Nadu. Technical Report: South Eastern Coastal Region, Chennai.

C.G.W.B., 1998. Guide on artificial recharge to groundwater. Ministry of water resources, New Delhi.

C.G.W.B., 1998. Groundwater exploration in Tamil Nadu and U.T. of Pondicherry. South Eastern Coastal Region, Chennai.

C.G.W.B., 1998. Groundwater year book of West Bengal. Eastern Region, Calcutta.

C.G.W.B., 1999. Groundwater year book (1997-1998). Technical Report: Series 'D'-No. 145. Eastern Region, Calcutta.

C.G.W.B., 1999. Hydrogeological framework and development prospects in West Godavari district, Andhra Pradesh. Technical Report: Southern Region, Hyderabad.

C.G.W.B., 2001. Hydrogeological framework and development prospects in Nellore district, Andhra Pradesh. Southern Region, Hyderabad.

C.G.W.B., 2002. Groundwater resources and development potentials in Ramanathapuram district, Tamil Nadu. Technical Report: Series 'D'. South Eastern Coastal Region, Chennai.

C.G.W.B., 2003. Groundwater resources and development potentials of Kanchipuram district, Tamil Nadu. Technical Report: Series 'D'. South Eastern Coastal Region, Chennai.

C.G.W.B., 2005. Dynamic ground water resources of India. Ministry of Water Resources, Govt. of India, New Delhi.

C.G.W.B., 2005. Hydrogeological framework and groundwater development prospects in Tamil Nadu. South Eastern Coastal Region, Chennai.

C.G.W.B., 2006. Groundwater resources and development potential of Tiruvarur district, Tamil Nadu. Technical Report: Series 'D'. South Eastern Coastal Region, Chennai.

C.G.W.B., 2006. Hydrogeological framework and development prospects of Guntur district, Andhra Pradesh. Southern Region, Hyderabad.

Chadha, D.K., 1999. A proposed new diagram for geochemical classification of natural waters and interpretation of chemical data. Hydrogeol Jour. 7, pp. 431-439.

Chakladar, S. and Mishra, A.K., 1990. Hydrogeology and groundwater resources of Medinipur district, West Bengal. Technical Report of CGWB, Sr. D, No. 77. Eastern Region, Calcutta.

Chakrapani, R., 1989. Groundwater resources and developmental potential of South Arcot district, Tamil Nadu. Technical Report: Southern Region, Hyderabad.

Chandramohan, P., Jena, B.K. and Sanil Kumar, V., 2001. Littoral drift sources and sinks along the Indian coast. Current Science, v. 81, no. 3.

Chandrasekharam, D., 1985. Structure and evolution of the western continental margin of India, deduced from gravity, seismic, geomagnetic, and geochronological studies. Physics of the Earth and Planetary Interiors., v. 41, pp. 186-198.

Charette, M.A., Splivallo, R., Herbold, C., Bollinger, M.S. and Moore, W.S., 2003. Salt marsh submarine groundwater discharge as traced by radium isotopes: Marine Chemistry, v. 84, pp. 113-121.

Chaturvedi, R.S., 1973. A note on the investigation of groundwater resources in Western districts of Uttar Pradesh. Annual Report, U.P. Irr. Res. Inst., pp. 86-122.

Chen, K.P. and Jiao, J.J., 2007. Preliminary study on sea water intrusion and aquifer freshening near reclaimed coastal area of Shenzhen, China. Water Science and Technology: water supply, v. 7, no. 2, pp. 137-145.

Choudhury, K. and Saha, D.K., 2004. Integrated geophysical and chemical study of saline water intrusion. Ground Water v. 42, no. 5, September-October, pp. 671-677.

Chow, V.T., 1952. On the determination of transmissivity and storage coefficients from pumping test data. Am. Geophys. Union Trans., vol. 33, pp. 397-404.

Church, T.M., 1996. An underground route for the water cycle. Nature, 380, 579-80.

Clark, I.D. and Fritz, P., 1997. Environmental isotopes in hydrogeology. Lewis Publishers, New York, 311 p.

Coleman, J.M. and Roberts, H.H., 1989. Deltaic Coastal Wetlands, Coastal Lowlands: Geology and Geotechnology, pp. 1-24.

Coleman, J.M. and Wright, L.D., 1971. Analysis of major river systems and their deltas: procedures and rationale, with two examples – Coastal Studies Institute Louisiana State University, Tech. Report. 95, 125 p.

Coleman, J.M., 1976. Deltas: processes of deposition and models for exploration – continuing Education Publ. (Champaign, Ill.): 102 p.

Collins, W.D., 1923. Graphic representation of analyses. Ind. And Eng. Chem., 15, 394 p.

Cooper, H.H., Jr., and Jacob, C.E., 1946. A generalized graphical method for evaluating formation constants and summarizing well-field history, Trans. Amer. Geophysical Union, v. 27, pp. 526-534.

Cooper, H.H., Kohout, F.A., Henry, H.R. and Glover, R.E., 1964. Sea water in coastal aquifers. U.S. Geol. Surv., Water-supply Pap. 1613-C, 84 p.

Cox, K.G., 1988. Inaugural Address. In: Subbarao, K.V. (Ed) Deccan Flood Basalts. Geol.Soc.India, Mem.10, pp. 15-22.

Cox, K.G., 1989. The role of mantle plumes in the development of continental drainage patterns. Nature, v. 342, pp. 17-33.

CRZ (Coastal Regulation Zone) Notification., 1991. Ministry of Environment and Forests. Gazette of India vide No. S.O.114(E), dated, the 19th February and several amendments subsequently.

CRZ (Coastal Regulation Zone) Notification., 2011. Ministry of Environment and Forests(Govt of India). Gazette of India, Extraordinary, part II, section 3, subsection II 6th January.

Cunha Maria Da Conceio Ao, 1999. On solving aquifer management problems with simulated annealing algorithms. Water res. Mgt., 13. pp. 153-169.

Curray, J.R. and Moore, D.G., 1971. Growth of Bengal deep sea fan and denudation in the Himalayas. Bull. Geol. Soc. Am., 82, pp. 563-572.

Curray, J.R., 1980. The IPOD programme on passive continental margins. Phil. Trans. R.Soc., London, A-294, pp. 17-33.

Custodio, E., et al., 1987. Groundwater problems in Coastal areas. Studies and Reports in hydrology. Series No. 45 (UNESCO).

CWC (Central Water Commission), 2005. Water Sector at a Glance, New Delhi.

Dagan, G. and Bear, J., 1968. Solving the problem of local interface upconing in a coastal aquifer by the method of small perturbations. J. Hydraul. Res., v. 6, no. 1, pp. 15-44.

Das Gupta, A. and Onta, P.R., 1994. Groundwater management models for Asian developing countries. Water Resour. Devl. v. 10, no. 4, pp. 457-474.

Das, A. and Datta, B., 1995. Simulation of density dependent 2-D seawater intrusion in coastal aquifers using nonlinear optimization algorithm. Proc., Am. Water Resour. Assoc. Annual Summer Symp. On Water Resour. and Envir. Hazards: Emphasis on Hydrol. and Cultural Insight in Pacific Rim, Herndon, Va., pp. 227-286.

Das, A. and Datta, B., 1998. Optimization based solution of density dependent seawater intrusion in coastal aquifers. J. Hydrologic Engg., ASCE.

Das, A. and Datta, B., 1999a. Development of multi objective management models for coastal aquifers. J. Water Res. Planning and Mgt. Div. ASCE, v. 125, no. 2, pp. 76-87.

Das, A. and Datta, B., 1999b. Development of management models for sustainable use of coastal aquifers. J. Irrigation and Drainage Engg. Div. ASCE, v. 125, no. 3, pp. 112-121.

Das, A. and Datta, B., 2000. Optimization based solution of density-dependent seawater intrusion in coastal aquifers. J. Hydrol. Eng. v. 5, no. 1, pp. 82-89.

Das, S., 1991. Hydrogeological features of Deltas and Estuarine tracts of India. Memoirs Geological Society of India. No. 22, pp. 183-225.

Datta, A. K., 1974. Neogene and Quaternary paleogeography of the Cambay basin. Abstract Vol. IV, Ind. Colloq. Micropal. And Stratig. Dehra Dun.

Datta, P.S., et al., 1980. Comparative study of groundwater recharge in the Alluvial deposits of Indo-Gangetic and Sabarmati Alluvial plain. Mausam, v. 30, no. 1, pp. 129-133.

Davis, J. L., 1972. Geographical variation in coastal development In: K.M. Clayton (ed) Oliver and Boyd, Edinburgh., 204 p.

Davis, J.C., 1973. Statistics and data analysis in Geology. John Willey and Sons. Inc. New York 550 p.

Davis, S.N. and DeWiest, J.M., 1991. Hydrogeology. Krieger Publishing Company, Malabar, FL, 463 p.

De Glee, G.J., 1930. Over gradvatersteemingen bij vateronttekking door middle van putten, U. Waltman Jr., Delft, 175 p.

DeJosselin DeJong, G., 1959. Vortex theory for multiple phase flow through porous media. Water Resources Center Contribution nr. 23, Univ of California, Berkeley.

Deshmukh, D.S., et al., 1973. Geology and ground water resources of Alluvial areas of West Bengal. Bull. G.S.I. Sr. B, No. 34, Eastern Region, Calcutta.

Diersch, H.J.G., 2006. FEFLOW reference and user manual, Finite element subsurface flow and transport simulation system, WASY Institute for water resource planning and systems research Ltd., Berlin, 278 p.

Dougherty David, E., and Robert A. Maryott, 1991. Optimal groundwater management. 1. simulated annealing. Water resources research vol. 27, no. 10. pp. 2493-2508.

Dowdle, P.R., Laverman, A.M. and Oremland, R.S., 1996. Bacterial dissimilatory reduction of Arsenic(V) to Arsenic(III) in anoxic sediments. Applied Environmental Microbiology, v. 62, no. 5, pp. 1664-1669.

Duncan, R.A. and Pyle, D.G., 1988. Rapid eruption of the Deccan flood basalts, Western India, Geol. Soc. India, Mem. No. 10, pp. 1-10.

Dupuit, J., 1863. Etudes theoriques et pratiques sur le movement des eaux dans les canaux decouverts et a travers les terrains permeables. 2 eme edition; Dunot, Paris, 304 p.

Durov, S.A., 1948. Natural waters and graphical representation of their composition. Dokl Akad Nauk SSSR 59, pp. 87-90.

Dyck, J.H., Keys, W.S. and Menelely, W.A., 1972. Application of geophysical logging to ground-water studies in southeastern Saskatchewan. Can. J. Earth Sci., v. 9, no. 1, pp. 78-94.

Eaton, F.M., 1950. Significance of carbonates in irrigation water. Soil Science, 69, pp. 123-133.

Eldholdm, O. and Todal, A., 1996. The continental margin off western India and the Deccan large igneous province. (Abs). Proc. International Symposium on Geology and Geophysics of the Indian Ocean (GIO-1996), NIO, Goa.

Elliot, T., 1978. Deltas. In: H.G. Reading (ed.): Sedimentary Environments and Facies – Elsevier (New York): pp. 97-142.

Emch, P.G. and Yeh, W.W.G., 1998. Management model for conjunctive use of coastal surface water and groundwater. J. Water Resour. Plan. Manage. ASCE v. 124, no. 3, pp. 129-139.

Emekli, N., Karahanoglu, N., Yazicigil, H. and Doyuran, V., 1996. Numerical simulation of saltwater intrusion in a groundwater basin. Wat. Environ. Res. 68(4), pp.855-866.

Essaid, H.I., 1986. A comparison of the coupled freshwater-saltwater flow and the Ghyben Herzberg sharp interface approaches to modeling of transient behavior in coastal aquifer systems. J.Hydrology, 86, pp. 169-193.

Essaid, H.I., 1990a. A multilayered sharp interface model of coupled freshwater and saltwater flow in coastal systems: Model development and application. Water Resources Res., v. 26, no. 7, pp. 1431-1454.

Estuaries of India, ENVIS publication series-1/2002. Centre of Advanced Study in Marine Biology, Annamalai University, Tamil Nadu.

Ewel, K.C., Twilley, R.R. and Ong, J.E., 1998. Different kinds of mangrove forests provide different goods and services. Global Ecology and Biogeography, v. 7, no. 1, pp. 83-94.

Faye, S., Diop, I.N., Faye, S.C., Evans, G., Pfister, M., Maloszewski, P. and Seiler, K.P., 2001. Seawater intrusion in the dakar (Senegal) confined aquifer: calibration and testing of a 3D finite element model. In New Approaches Characterizing Groundwater Flow, eds. K.P. Seiler and S. Wohnlich, pp. 1183-1186: Swets & Zeitlinger Lisse.

Ferris, J.G., Knowles, D.B., Brown, R.H. and Stallman, R.W., 1962. Theory of aquifer tests U.S. Geol. Survey Water Supply paper 1536-E. pp. 337-353.

Fitterman, D.V. and Stewart, M.T., 1986. Transient electromagnetic sounding for groundwater: Geophysics, v. 51, pp. 995-1005.

Fitterman, D.V., and Deszcz-Pan, M., 1999. Geophysical mapping of saltwater intrusion in everglades national park. 3rd International Symposium on Ecohydraulics, Salt Lake City, Utah, 12-16 July, 18 p. http://SofiaUSGS.gov/publications/papers/geophys.map/

Freeze, R.A. and Cherry, J.A., 1979. Groundwater. Prentice-Hall, Engle-wood Cliffs, 604 p.

Freshwater Year, 2003. Ministry of Water Resources, Govt. of India, New Delhi, 2003, 53p.

Fuloria, R.C., 1993. Geology and hydrocarbon prospects of Mahanadi basin, India. Proc. Second Seminar on Petroliferous Basins of India, Vol. 1. S.K. Biswal et al., (eds), Indian Petroleum Publishers, Dehra Dun, India.

Galeati, G., Gambolati, G. and Neuman, S.P., 1992. Coupled and partially coupled Eulerian-Lagrangian model of freshwater-seawater mixing. Water Resour. Res. v. 28, no. 1, pp. 147-165.

Ganapathi, S. and Pandey, A.N., 1991. Evolution of landforms on Narmada and Tapti estuarine deltas, Gujarat. Mem. Geol. Soc. of India, No. 22, pp. 103-119.

Ganguly, S., 1997. Petroleum geology and exploration history of the Bengal basin in India and Bangladesh. Jour. Geol.Soc. India. V. 69, No. 1, pp. 1-25.

Ganju, J. L. and Mitra, D.S., 1982. Geomorphology of a part of the East Coast between Balasore and Indo-Bangladesh border. Proc. Symp. On Resources Survey for Land Use Planning and Environmental Conservation, Indian Society of Photo interpretation and Remote Sensing, Dehra Dun. 1982.

Garg, S.K., 2007. Hydrology & Water Resources Engineering. Khanna Publishers, pp. 1244.

Garrels, R.M. and MacKenzie, F.T., 1971. Evolution of sedimentary rocks. New York: Norton and Co; 397 p.

Geyh, M., 2000. Groundwater, saturated and unsaturated zone. In: Mook, W.G. (ed.), Environmental Isotopes in the Hydrological Cycle; Principles and Applications. IHP-V, Technical Documents in Hydrology, UNESCO, Paris, Vol. IV, No. 39, 196p.

Ghabayen, S.M.S., et al., 2006. Ionic and isotopic ratios for identification of salinity sources and missing data in the Gaza aquifer. Journal of hydrology 318, pp. 360-373.

Ghosh, D.P., 1971. Inverse filter coefficient for the computation of apparent resistivity: Standard Curves for a Horizontally Stratified Earth, Geophys. Prospecting, v. 19, no. 4, pp. 769-775.

Ghyben, B.W., 1889. Nota in verband met de voorgenomen put boring nabij Amsterdam (Notes on the probable results of the proposed well drilling near Amsterdam). K. Inst. Ing. Tijdschr., The Hague, 21 p.

Glover, R.E., 1959. The pattern of fresh-water flow in a coastal aquifer. J. Geophys. Res., v. 64, no.4, pp. 457-459.

Godio, A., Chiara, P., Gaill, C.C. and Naldi, M., 1998. A Combined geophysical survey for hydrogeological purposes in North-Eastern Italy. Proceedings of IV Meeting Environmental and Engineering Geophysical Society, pp. 209-212.

Goel, A.K., 2008. http://www.manage.gov.in/managelib/faculty/Goel.html.

Goel, P.S. et al., 1975. Tritium tracer studies on groundwater recharge in the alluvial deposits of Indo-Gangetic plains of Western U.P., Punjab and Haryana. Approaches and Methodologies for Development of Groundwater Resoures, proc. Indo-German Workshop, Hyderabad, pp. 319-322.

Gondouin, M., Tixer, M.P. and Simard, G.L., 1957. An experimental study on the influence of the chemical composition of electrolytes on the SP curve. J. Pet. Tech., 9(2), SPE reprint series, No.1, pp. 65-79.

Gorelick, S.M., 1983. A review of distributed parameter groundwater management modeling method. Water Resour. Res. v. 19, no. 2, pp. 78-86.

Guo, W. and Langevin, C.D., 2002. (SEAWAT): A computer program for simulation of three-dimensional variable-density ground-water flow, U.S. Geological Survey Open-File Report 01-434, 79 p.

Hantush, M.S., 1956. Analysis of data from pumping tests in leaky aquifers. Am. Geophys. Union Trans., vol. 37, pp. 702-714.

Hantush, M.S., 1960. Modification of the theory of leaky aquifers. Jour. Geophy, Res. Vol. 65, pp. 3713-3725.

Hantush, M.S., 1962. Aquifer tests on partially penetrating wells. Trans. ASCE. Vol. 127(1), pp. 284-308.

Hantush, M.S., 1964. Hydraulics of wells. Advances in Hydro-science, Academic Press Vol. I, 442 p., New York.

Hantush, M.S. and Jacob, C.E., 1955. Steady three-dimensional flow to a well in a two-layered aquifer Trans. Amer. Geophysical Union, Vol. 36, pp. 286-292.

Hantush, M.S. and Jacob, C.E., 1955. Non-steady radial flow in an infinite leaky aquifer. Am. Geophys. Union Trans., vol. 36: pp. 95-100.

Hantush, M.S., 1968. Unsteady movement of freshwater in thick unconfined saline aquifers. Bull. Int. Assoc. Sci. Hydrol. V. XIII, no. 2, pp. 40-60.

Harbaugh, A.W. and McDonald, M.G., 1996. Users documentation for MODFLOW-96, an update to the U.S.geological survey modular finite-difference groundwater flow model. U.S. Geological Survey Open-File Report 96-0485, 56 p.

Harbaugh, A.W., Banta, E.R., Hill, M.C. and McDonald, M.G., 2000. MODFLOW-2000, the U.S. Geological Survey modular groundwater model user guide to modularization concepts and the groundwater flow process. U.S. Geological Survey Open-File Report 00-92, 121 p.

Hargreaves, H.G., 1968. Consumptive Use Derived from Evaporation Pan Data. Jour. Irr. and Drainage Div., Proc. ASCE, pp. 97-105.

Harrington, J.M., Fendorf, S.B. and Rosenzweig, R.F., 1998. Biotic generation of arsenic(III) in metal(loid)-contaminated sediments. Environmental Science and Technology, v. 32, no. 16, pp. 2425-2430.

Hartwid Kremer and Chris Crossland, 2002. Coastal change and the "Anthropocene". Past and future perspectives of the IGBP-LOICZ project. International Conference on Low-lying Coastal areas and Hydrology and Integrated Coastal Zone Management. Federal Republic of Germany, 9-12 september.

Haubold, R.G., 1975. Approximation for steady interface beneath a well pumping fresh water overlying salt water. Ground Water, v. 13, no. 3, pp. 254-259.

Hayes, H. O., 1975. Morphology of sand accumulation in estuaries. In Crocin L (ed): Estuarine research v. II, New York, Academic Press.

Hearst, J.R. and Nelson, P.H., 1985. Well Logging for Physical Properties. McGraw-Hill, NY.

Hem, J.D., 1989. Study and interpretation of chemical characteristics of natural water. 3rd edn. US Geological Survey of Water Supply, Paper 2254. USGS, Washington DC.

Henry, H.R., 1959. Salt intrusion into fresh-water aquifers. J. Geophys. Res., v. 64, no. 11, pp. 1911-1919.

Henry, H.R., 1962. Transitory movements of the salt-water front in an extensive artesian aquifer. U.S. Geol. Surv., Prof. Pap. 450-B, pp. B87-B88.

Henry, H.R., 1964. Effects of dispersion on salt encroachment in coastal aquifers. In: Sea Water in Coastal Aquifers. U.S. Geol. Surv., Water-supply Pap. 1613-C, pp. 70-84.

Herzberg, A., 1901. Die Wasserversorgung einiger Nordsee Bader. J. Gasbeleucht. Verw. Beleuchtungsarten Wasserversorgung., 44, pp.815-819, 842-844.

Hill, R.A., 1940. Geochemical Patterns in Coachella Valley, California. Amer. Geophys. Union, Trans., 21, pp. 46-49.

Hodlur, G.K., Ratnakar Dhakate and Rolland Andrade, 2006. Correlation of vertical electrical sounding and borehole-log data for delineation of saltwater and freshwater aquifers. Geophysics. vol. 71, no. 1 (January-February):p. g11-g20.6 figs., 7 tables.

Hoekstra, P. and Blohm, M., 1990. Case histories of time domain electromagnetic soundings. In: Ward, S.H. (Ed), Geotechnical and Environmental Geophysics: Environmental and Groundwater, SEG Investigations in Geophysics, vol. 2, no. 5.

Horton, R.E., 1933. The Role of Infiltration in the Hydrologic Cycle. Am. Geophys. Union Trans., 14, pp. 446-460.

Huang Min-wei and Jasbir S Arora., 1997. Optimal design with discrete variables: some numerical experiments. Int. Jr. For numerical methods in engineering, Vol. 40, pp. 165-188.

Hubbert, M.K., 1940. The theory of ground-water motion. J. Geol., v. 48, no. 8, pp. 785-944.

Hutchings, W.C. and Tarbox, D.L., 2003. A model of seawater intrusion in surficial and confined aquifers of northeast Florida. The second International conference on saltwater intrusion and coastal aquifers- Monitoring, Modeling, and Management. Merida, Yucatan, Mexico.

Huyakorn, P.S., Anderson, P.F., Mercer, J.W., White, H.O., 1987. Saltwater intrusion in aquifers: Development and testing of a three dimensional finite element model. Water Resour. Res. v. 23, no. 2, pp. 293-312.

Huyakorn, P.S., Wu, Y.S. and Park, N.S., 1996. Multiphase approach to the numerical solution of a sharp interface saltwater intrusion problem. Water Resources Research, v. 32, no. 1, pp. 93-102.

ICWE, 1992. International Conference on Water and Environment, Dublin (Ireland).

Illangasekare, T. and Morel-Seytoux, H.J., 1982. Stream-aquifer influence coefficients as tools for simulation and management. Water Resour. Res. v. 18, no. 1, pp. 168-176.

Illangasekare, T. and Morel-Seytoux, H.J., 1986. A discrete kernel simulation model for conjunctive management of stream aquifer system. J. Hydrology. v. 85, pp. 319-338.

IMD, 1979. Tracks of storms and depressions in the Bay of Bengal and the Arabian sea, 1877-1970, India Meteorological Department (IMD), New Delhi.

IMD, 2004. www.imd.ernet.in/section/climate/annual-rainfall.html.

IMD, 2011. www.tropmet.res.in/~kolli/mol/Monsoon/Historical/air.html.

India's Initial National Communication to the United Nations Framework Convention on Climate change, 2004. Ministry of Environment and Forests, Govt. of India. Chapter-3: Vulnerability, Assessment and Adaptation.

I.P.C.C., 1996a. Changes in sea level. Climate change 1995: The Science of Climate Change. J.T. Houghton, L.G. Meira Filho, and B.A. Callander, eds. Cambridge University Press, Cambridge, pp. 359-405.

I.P.C.C., 1998. The regional impacts of climate change: An assessment of vulnerability; A special report of IPCC work group, Cambridge University Press, UK. 517 p.

I.P.C.C., 2007. Parry, M.L., Canziani, O.F., Palutikof, J.P. and Co-authors: Climate change 2007: Impacts, adaptation and vulnerability. Contribution of Working Group II to the Fourth Assessment Report. Eds., Cambridge University Press, Cambridge, UK, pp. 23-78.

I.P.C.C., 1990. Working group I: Scientific Assessment of Climate Change (Eds): J.T Houghton, G.J Jenkins, J.J Ephraums, Cambridge University Press, Cambridge (UK), 410 p.

I.P.C.C., 1990. Working group II: Impacts Assessment of Climate Change (Eds): W.J. McG. Tegart, G.W. Sheldon and D.C. Griffiths, Australian Government Publishing Service, Camberra, Australia, 294 p.

I.P.C.C., 2001. Working Group I: The Scientific Basis (Eds): J.T Houghton, Y.Ding, D.J. Griggs, M.Noguer, P.J. van der Linden, X. Dai, K. Maskell, C.A. Johnson (eds.). Cambridge University Press, Cambridge (UK), 892 p.

I.P.C.C., 2001. Working Group II: Impacts, Adaptation and Vulnerability (Eds): McCarthy, J.J.; Canziani, O.F.; Leary, N.A.; Dokken, D.J.; and White, K.S. (eds). Cambridge University Press, Cambridge (UK).

I.P.C.C., 2001. Working Group III: Mitigation (Eds): Metz, B.; Davidson, O.; Swart, R.; and Pan, J., (eds). Cambridge University Press, Cambridge (UK).

Jackson, P., 1993. New developments in airborne electromagnetics: Preview, no. 44, pp. 15-20.

Jacob, C.E., 1940. On the flow of water in an elastic artesian aquifer. Trans. Amer. Geophysical Union, Vol. 21, pp. 574-586.

Jin-Yong Lee and Sung-Ho Song, 2007. Groundwater chemistry and ionic ratios in a western coastal aquifer of Buan, Korea: Implication for sea water intrusion, vol. 11, No. 3, p. 259-270.

Johannes, R.E., 1980. The ecological significance of the submarine discharge of groundwater. Mar Ecol, Prog Ser 3, pp. 365-73.

Johnson, D.S., Aragon, C.R., McGeoch, L.A. and Schevon, C., 1989. Optimization by simulated annealing; Graph partitioning. Operation research, Vol. 37, No. 6. pp. 865-891.

Jones, P.H. and Buford, J.B., 1951. Electric logging applied to Groundwater Exploration. Geophysics, v.16, no. 1, pp. 115-139.

Jupp, D.L.B. and Vozoff, K., 1975. Stable iterative methods for the inversion of geophysical data, Geophys. J.R. Astr. Soc., pp. 957-976.

Kaila, K.L., 1988. Mapping the thickness of Deccan trap flows in India from DSS studies and inferences about a hidden Mesozoic basin in the Narmada-Tapti region. Geol. Soc. India, Mem. No. 10, pp. 91-116.

Kaila, K.L., Reddy, P.R., Dixit, M.M. and Koteswara Rao, P., 1985. Crustal structure across the Narmada-Son lineament, Central India from deep seismic soundings. J. Geol. Soc. India, v. 26, pp. 465-480.

Kaiser, H.F., 1958. The Varimax criteria for analytical rotation in factor analysis, Psychometrika 23, pp. 87-200.

Kameswara Rao, K., 1991. Geomorphic features in the Pennar delta and its evolution. Mem. Geol. Soc. India., 22, pp. 65-74.

Karanth, K.R., 1987. Groundwater assessment development and management. Tata McGraw Hill. 720 p.

Kashef, A.I., 1975a. Comparative study of fresh-salt water interfaces using finite element and simple approaches. Water Resour. Bull., v. 11, no. 4, pp. 651-655.

Kashef, A.I., 1975b. Management of retardation of salt water intrusion in coastal aquifers. Office of Water Research and Technology, Washington, D.C., 281 p.

Kashef, A.I., 1977. Management and control of salt water intrusion in coastal aquifers: Critical reviews in Environmental control. Boca Raton, Fla., CRC Press.

Kauffman, A.A. and Keller, G.V., 1983. Frequency and Transient Soundings. Elsevier Science Publications.

Kauffman, S and Libby, W.S., 1954. The natural distribution of tritium. Physical Review, v. 93, no. 6, pp. 1337-1344.

Kay, R.T., Bayless, E.R. and Solak, R.A., 2002. Use of isotopes to identify sources of ground water, estimate ground-water-flow rates, and assess aquifer vulnerability in the Calumet region of Northwestern Indian and Northeastern Illinois. U.S. Geological Survey Water-Resources Investigation Report 02-4213, Indianapolis, 60 p.

Keeney, D., 1990. Sources of nitrate to ground water. Critical Reviews in Environmental Control, 16, pp. 257-304.

Keller, G.V. and Frischknecht, F.C., 1966. Electrical methods in geophysical prospecting: International series of monographs in electromagnetic waves, Oxford, Pergamon, 519 p.

Kesavarao, et al., 2006. Agroclimatic characterization of APRLP-ICRISAT nucleus watersheds in Nalgonda, Mahabubnagar and Kurnool districts. Global Theme on Agroecosystems Report., no. 30, ICRISAT, 52 p.

Keys, W.S. and Brown, R.F., 1978. The use of temperature logs to trace the movement of injected water. Ground Water, 16, pp. 32-48.

Keys, W.S. and Mac Cary, L.M., 1971. Application of borehole geophysics to water-resources investigations. In Techniques of Water-Resources Investigations, U.S. Geophysical Survey, Book 2, Chap. E1.

Keys, W.S., 1987. Borehole geophysics as applied to groundwater investigations. USGS open file report 87-539., 304 p.

Keys. W.S., 1986. Analysis of geophysical logs of water wells with a microcomputer. Ground water, 24, pp.750-60.

Kipp, K.L., 1987. HST3D: A computer code for simulation of head and solute transport in three-dimensional groundwater flow systems: U.S. Geological Survey Water Resources Investigations Report, 80-26, 517 p.

Kirkpatrick, S., Gelatt, C.D. Jr. and Vecchi, M.P., 1983. Optimization by Simulated annealing. science, Vol. 220, No. 4598, pp. 671-680.

Koefed, O., 1979. Geosounding principles I-Elsevier, Amsterdam, 276 p.

Kohout, F.A., 1964. The flow of fresh water and salt water in the Biscayne aquifer of the Miami area, Florida. In: Sea Water in Coastal Aquifers. U.S. Geol. Surv., Water-Supply Pap. 1613-C, pp.12-32.

Kohout, F.A., 1965. A hypothesis concerning cyclic flow of salt water related to geothermal heating in the Floridan aquifer: New York Academy of Sciences, series II, v. 28, no. 2, pp. 249-271.

Konikow, L.F. and Bredehoeft, J.D., 1978. Computer model of two-dimensional solute transport and dispersion in groundwater: techniques of water resource unvestigations of the United States Geological Survey. Book 7, Chapter 2, U.S. Gov, Washington D.C.

Korzun, V.I., 1978. World water balance and water resources of the earth. UNESCO. Ed, 663 pages.

Kreiger, R.A., Hatchet, J.L. and Poole, J.L., 1957. Preliminary survey of the saline-water resources of the United States: U.S. Geological Survey Water-Supply Paper 1374, 172 p.

Krest, J.M., Moore, W.S. and Rama, 1999. ^{226}Ra and ^{228}Ra in the mixing zones of the Mississippi and Atchafalaya rivers- indicators of groundwater input: Marine Chemistry, v. 64, no. 3, pp. 129-152.

Krishna Rao, B. and Swamy, A.S.R., 1991. Sediment characteristics of environments in the modern Krishna-Godavari deltas. Mem. Geo. Soc of India, No. 22, pp. 121-138.

Krishna Rao, B., Swamy, A.S.R., Murthy, P.B., Srinivasa rao, K. and Rajendra Prasad, P., 1990. Distribution of calcrete in Holocene coastal sediments from Andhra Pradesh, East Coast of India. In: Sea Level Variation and its impact on Coastal Environment (Ed.) G. Victor Rajamanikam, Tamil University, Thanjavur, Pub. No. 131, pp. 97-114.

Krishnan, M.S., 1953. Structural and tectonic history of India. Geol. Surv. India, Mem. No. 81, pp. 103-105.

Krupanidhi, K.V.J.R., 1986. Ground water resources and development potential of Broach (Bharuch) district, Gujarat. Technical Report: Western Region C.G.W.B., Jaipur.

Kruseman, G. P. and De Ridder, N.A., 1970. Analysis and evolution of pumping test data, Intl. Inst. For Land Reclamation and Improvement, Bull, Wageningen, 11, 200 p.

Kwader, T., 1986. The use of geophysical logs for determining formation water quality. Ground Water, 24, pp. 11-15.

Lal, D. and Peters, B., 1962. Cosmic ray produced isotopes and their application to problems in Geophysics. Progress in Elementary Particle and Cosmic ray Physics; North-Holland Publishing Co., Vol. 6, pp. 1-74.

Lal, M. and Aggarwal, D., 2000. Climate change and its impacts in India, Asia-Pacific Jour. Environment & Development (Communicated).

Langevin, C.D., 2001. Simulation of groundwater discharge to Biscayne Bay, Southeastern Florida: Water Resources Investigation Report 00-4251, U.S. Geological Survey, 127 p.

Langevin, C.D., Shoemaker, W.B., Guo, W., and Missimer, C.D.M., 2004. MODFLOW-2000, USGS modular groundwater model-Documentation of the SEAWAT 2000 version with variable density flow process (VDF) and the integrated MT3DMS transport process (IMT). USGS open report pp. 03-426.

Lee Yuh-Ming and Hugh Ellis, J., 1994. Comparison of algorithms for nonlinear integer optimization: Application to monitoring network design. ASCE Jr. of Environment Engineering, vol. 122, no. 6, pp. 524-531.

Lee, D.R., 1977. A device for measuring seepage flux in lakes and estuaries. Limnol Oceanogr; 22, pp. 140-147.

Lohman, S.W., 1972.Groundwater hydraulics. U.S. Geol. Surv., Washington. Professional Paper. 708, 70 p.

Loke, M.H. and Barker, R.D., 1995. Least-squares disconvolution of apparent resistivity pseudo-sections. Geophysics, v. 60, no. 6, pp. 1682-1690.

Loke, M.H. and Barker, R.D., 1996. Rapid least-squares inversion of apparent resistivity pseudosections by a quasi-newton method. Geophysic. Prospect., v. 44, no. 1, pp. 131-152.

Loke, M.H., Wilkinson, P.B. and Chambers, J.E., 2010. Fast computation of optimized electrode arrays for 2D resistivity surveys. J. Comput. Geosci., v. 36, no. 11, pp. 1414-1426.

Loveson, V.J. and Rajamanickam, V., 1987. Coastal geomorphology of the Southern Tamil Nadu, India. Proc. Nat. Symp. On Land Transformation and Management, Hyderabad. (Eds.) S. K. Bahn and V. K. Jha, Indian Society of Remote Sensing, Dehra Dun, pp. 115-129.

Loveson, V.J., Rajamanickam, G.V., and Anbarasu, K., 1990. Remote sensing application in the study of sea-level variation along the Tamilnadu coast, India. In: Sea-level variation and its impact on coastal environment, Eds. G.V. Rajamanickam, Pub. Tamil University, Thanjavur, pp. 179-195.

Lvovitch, M.I., 1974. World water resources in their future. Moscow: Mysl. 263 pages.

Mac Cary, L.M., 1983. Geophysical logging in carbonate aquifers. Ground Water, 21, pp. 334-42.

Maddock, III T., 1972. Algebraic technological function from a simulation model. Water Resour. Res. v. 8, no. 1, pp. 129-134.

Mahadevan, T.M., 1994. Deep continental structure of India: A review. Geol. Soc. India. Mem. 28, 569 p.

Mahalik, N.K., 2000. Mahanadi delta: geology, resources and biodiversity. AIT Alumi. Assoc., New Delhi, 169 p.

Mahender Reddy, K. and Shah, B.M., 1991. Evolution of the Krishna delta, Andhra Pradesh., Mem. Geol. Soc. India., No. 22. pp. 57-64.

Mall, R.K., Gupta, A., Singh, R., Singh, R.S. and Rathore, L.S., 2006. Water resources and climate change: An Indian perspective. Current science, Vol. 90, No.12.

Mandal, B.K., Roy Chowdhury, T., Samanta, G., Basum G.K., Chowdhary, P.P., Chanda, C.R., Lodh F., Karan, N.K., Dhar, R.K., Tamili, D.K., Das, D., Saha, K.C. and Chakraborthi, D., 1996. Arensic in groundwater in seven districts of West Bengal, India – the biggest arsenic calamity in the world. Curr. Sci., v. 70, pp. 976-986.

Mandal, C.K., Mandal, D.K., Srinivas, C.V., Sehgal, J. and Velayutham, M., 1999. Soil-climatic database for crop planning in India. National Bureau of Soil Survey & Land Use Planning. ICAR, Nagpur, India.

Mangroves of India. Centre for advanced study in Marine biology, Annamalai University, Tamilnadu. ENVIS Publication series 2/2002.

Matalas, N.C. and Reicher, B.J., 1967. Some comments on the use of factor analysis. Water Resour Res 3, pp. 213-223.

Mathur, L.P., Rao, K.L.N. and Chaube, A.N., 1968. Tectonic framework of Cambay basin, India: Oil and Nat. Gas Comm. Bull., v. 5, pp. 7-28.

Matthess, G., 1982. The properties of groundwater. Wiley, New York, 498 p.

Maucha, Rezso, 1949. The graphics symbolization of the chemical composition of natural waters. Hidrologie Kozolny, v. 13, p. 117-118.

Maurya, D.M.; Rachna Raj and Chamyal, L.S., 2000. History of tectonic evolution of Gujarat alluvial plains Western India during Quaternary: A review. Jour. Geol. Soc. Of India, vol. 55, pp. 343-366.

McBride, J.L., 1995. Tropical cyclone formation, In Elsberry R.L. (ed) Global perspectives on tropical cyclone, WMO Tech Doc WMO/TD-No. 693. Report no. TCP-38, WMO, Geneva, pp. 65-103.

McDonald, M.G. and Harbaugh, W.A., 1988. A modular 3D finite difference groundwater flow model –techniques of water resources investigations of the USGS, Chapter A1, Book 6.

Mckinney Daene, C., and Min-Der Lin, 1994. Genetic algorithm solution of groundwater management models. Water res. Research, Vol. 30, No. 6, pp. 1897-1906.

McNeill, J.D., 1980a. Electrical conductivity of soils and rocks. Geonics Ltd. Report, TN5, 21 p.

McNeill, J.D., 1980b. Electromagnetic terrain conductivity measurement at low induction numbers. Geonics Ltd. Report, TN.,6, 15 p.

McNeill, J.D., 1986. Electromagnetic geophysical methods applied to ground water exploration and evaluation, Developing World Water, 60 p.

McNeill, J.D., 1990. Use of electromagnetic methods for groundwater studies, in ward, S.H., ed., Geotechnical and environmental geophysics: Tulsa, Soc. Expl. Geophys., pp. 191-218.

Meijerink, A.M.J., 1971. Reconnaissance survey of the Quaternary geology of the Cauvery delta. Jour. Geol. Soc. Ind., v. 12, no. 2, pp. 113-124.

Meisler, H., Leahy, P.P. and Knobel, L.L., 1984. Effect of eustatic sea level changes on saltwater-freshwater in the Northern Atlantic Coastal Plain. USGS Water Supply Paper 2255. Reston, Virginia.

Mercer, J.W., Larson, S.P. and Faust, C.R., 1980. Simulation of saltwater interface motion. Ground water v. 18, no. 4, pp. 374-385.

Meyers, G.D., 1992. A review of nuclear logging. The Log Analyst v.33, no. 3, pp. 228-238.

Ministry of Water Resources, New Delhi, 1999. Integrated water resources development – A plan for action, Report of The National Commission for Integrated Water Resources Development.

Mondal, N.C., Singh, V.P., Singh, V.S. and Saxena, V.K., 2010. Determining the interaction between groundwater and saline water through groundwater major ions chemistry. Journal of Hydrology 388, pp. 100-111.

Mooney, H.M. and Wetzel, W.E., 1956. The potentials about a point electrode and apparent resistivity for a two, three and four layered earth. Univ. of Minnesota Press, Minneapolis, Minn. 146 p.

Moore, W.S., 1996. Large ground water inputs to coastal waters revealed by ^{226}Ra enrichments, Nature, 380, pp. 612-614.

Moore, W.S., 1997. High fluxes of radium and barium from the mouth of the Ganges-Brahmaputra River during low river discharge suggest a large groundwater source. Earth Planet Sci Lett; 150, pp. 141-150.

Morel-Seytoux, H.J. and Daly, C.J., 1975. A discrete kernel generator for stream-aquifer studies. Water Resour. Res. v. 11, no. 2, pp. 253-260.

Morgan, J.P. and McIntire, W.G., 1959. Quaternary geology of the Bengal basin, East Pakistan and India. Bull Geol. Soc. Am., 70, 70, pp. 319-342.

Morgan, W.J., 1981. Hotspot Tracks and the opening of the Atlantic and Indian Oceans. In: C.Emiliani (Ed) The Sea, Wiley, New York, v. 7, pp. 443-487.

Morgan, C.O and Winner, M.D., Jr., 1962. Hydrochemical facies in the "400-foot" and "600-foot" sands of the Baton Rouge area, Louisiana, in short papers in Geology Hydrology and topography; U.S. Geol. Survey Prof. paper 450-B, p. B120-B121.

Mukharjee, M. K., 1985. Morphology of the South Gujarat Coast Its neotectonic significance. In: S.S. Merh and N. M. Vashi (ed) Proc. Sym. Quat. India. M.S.U. Baroda, pp. 26-33.

Munnich, K.O., 1968. Moisture movement measured by isotope tagging; Guidebook on Nuclear Techniques in Hydrology, IAEA, Vienna, pp. 112-117.

Murali Sabnavis, and Patangay, N.S., 2006. Principles and applications of groundwater geophysics. AEG Publ. No. 40, India.

Muralidharan, D. and Athavale, R.N., 1998. Artificial recharge in India. Technical Report, NGRI, Hyderabad.

Murthy, K.S.R., 1997. Some geodynamic complexities related to the evolution of Bengal Fan and the neotectonic activity of the south Indian shield. Curr. Sci., v. 73, pp. 247-251.

Murthy, K.V.S., Singh, R.P. and Nagar, V.K., 1973. Hydrocarbon prospects of Mahanadi basin. Unpublished Report of O.N.G. Commission of India, 30 p.

Muskat, M. and Wyckoff, R.D., 1935. An approximate theory of water-coning in oil production. Trans. Am. Inst. Min. Metall. Pet. Eng., 114, pp.144-163.

Muskat, M., 1937. The flow of homogeneous fluids through porous media. McGraw-Hill, New York, 763 p.

Nabighian, M.N., ed., 1988. Electromagnetic methods in applied geophysics: Tulsa, Soc. Expl. Geophys., v. 1, 503 p.

Nabighian, M.N., ed., 1991. Electromagnetic methods in applied geophysics: Tulsa, Soc. Expl. Geophys., v. 2, 992 p.

Nageswara Rao, J., 1988. Geomorphology, Quaternary geology and Neotectonic structure of Krishna delta, East Coast of Andhra Pradesh. Geol. Surv. Ind., Spl. Pub. Vol II, No. 11, pp. 317-322.

Nageswara Rao, K. and Vaidyanadhan, R., 1978. Geomorphic features in the Krishna delta and its evolution. Symposium on Morphology and Evolution of Landforms. Department of Geology. University of Delhi, Delhi, pp. 120-130.

Nageswara Rao, K. and Vaidyanadhan, R., 1979. Evolution of the coastal landforms in the Krishna delta front, India. Trans. Inst. Indian.

Narasimhan, T.N., 1990. Paleochannels of the Palar river west of Madras city: Possible implications for vertical movement. Jour. Geol. Soc. India, v. 36, pp. 471-474.

Narasimhan, T.N., 2005. Water: science and society; Current Science vol. 89, no. 5, pp. 787-793.

Nayak, S.R., Gupta, M. C., Chauhan, H. B., Shaikh, M. G., Anjali P. and Rao, R. S. 1987. Application of landsat data for coastal zone monitoring: A case study on part of west coast of India IRS-UP/SAC/MCE/SN/07/87 Reports of SAC (ISRO) Ahmedabad, pp. 1-21.

Neumann, C.J., 1993. Global Overview, Chapter: Global Guide to Tropical Cyclone Forecasting, WMO, Geneva.

Niyogi, D., 1975. Quaternary geology of the coastal plain in West Bengal and Orissa. Ind. Jour. Earth. Science (Calcutta), 2, pp. 51-61.

Norris, S.E., 1972. The use of gamma logs in determining the character of unconsolidated sediments and well construction features. Ground Water, v. 10, no. 6, pp. 14-21.

Oberdorfer, J.A., Hogan, P.J. and Buddemeier, R.W., 1990. Atoll island hydrogeology: Flow and freshwater occurrence in a tidally dominated system. J. Hydrol. 120(1-4), pp. 327-340.

Orellana, E. and Mooney, H.M., 1966. Master tables and curves for vertical electrical sounding over layered structures. Madrid Interciccia, pp. 150-166.

Oude Essink, G.H.P., 1999. Impact of sea level rise in the Netherlands. In Seawater Intrusion in Coastal Aquifers-Concepts, Methods, and Practices, ed. J. Bear, A.H.D, Cheng, S. Sorek, D. Ouazar, and I. Herrera, pp. 507-530. Dordrecht, The Netherlands: Kluwer Academic Publishers.

Paine, J.C., Angle, E.S. and Petrossian, R., 2000. Identifying and assessing ground water in the lower Rio grande valley, Texas, using Airborne Electromagnetic induction. Sageep. pp. 83-99.

Palmer, C., 1911. The geochemical interpretation of water analyses. U.S. Geol. Surv. Bull., No. 479, 31 p.

Papadopulos, I.S. and Cooper, H.H. Jr., 1967. Drawdown in a well of large diameter, Water Resources Research, v. 3, pp. 241-244.

Paul M. Barlow, 2003. Groundwater in freshwater-saltwater environments of the Atlantic coast. USGS Circular 1262, Reston, Virginia, 111 p.

Paulsen, R.J., Smith, C.F., O'Rourke, D. and Wong T., 2001. Development and evaluation of an ultrasonic groundwater seepage meter. Ground Water 39, pp. 904-911.

Penman, H.L., 1948. Natural evaporation from open water, bare soil and grass. Proceedings of the Royal Society, Serial A, Vol. 193, pp. 120-145.

Peterson, F.L., 1974. Neutron well logging in Hawaii, Tech. Rept. 75, Water Resources Research Center, Univ. of Hawaii, Honolulu, 42 p.

Piggott, A.R., Bobba, A.G. and Novakowski, K.S., 1996. Regression and Inverse analysis in regional groundwater modeling. J. Wat. Resour. Plan. Manage. ASCE v. 122, no. 1, pp. 1-10.

Piggott, A.R., Bobba, A.G. and Xiang, J., 1994. Inverse analysis of the SUTRA groundwater model. Groundwater v. 32, no. 5, pp. 829-836.

Pinder, G.F. and Cooper, H.H., 1970. A numerical technique for calculating the transient position of the salt water front. Water Resour. Res., v. 6, no. 3, pp. 875-880.

Pinder, G.F. and Page, R.H., 1977. Finite element simulation of salt water intrusion on the South Fork of Long Island. Proceedings of the 1st International Conference on Finite Elements in Water Resources, Pentech Press, London, pp. 251-269.

Piper, A.M., 1944. A graphic procedure in the geochemical interpretation of water analyses. Am. Geophys. Union, Trans., 25, pp. 914-923.

Pofali, R.M., et al., 1991. The soils of the deltas of India. Memo. Geol. Soc. of India, No. 22, pp. 227-241.

Polo, J.F. and Ramis, F.R., 1983. Simulation of saltwater-freshwater interface motion. Water Resources Research, v. 19, no. 1, pp. 61-68.

Prabhakar, K. and Anbarasu, K., 2010. Evolution of Vaigai delta, Tamil Nadu, India (East Coast) during Quaternary, Int. Jour.of Geomatics and Geosciences. Vol. 1, no. 2.

Prabhakar, K.N. and Zutshi, P.L., 1993. Evolution of Southern part of India East Coast basins. Journ. Geol. Soc. India, vol. 41, pp. 215-230.

Putti, M. and Paniconi, C., 1995. Picard and Newton linearization for the coupled model of saltwater intrusion in aquifers. Advances in Water Resour., v. 18, no. 3, pp. 159-170.

Pyne, R.D.G., 1995. Groundwater recharge and wells, a guide to aquifer storage recovery: Boca Raton, Fla., Lewis Publishers, 376 p.

Radhakrishna, I. and Chowdary, M.V.R., 1998. Simulation of chloride migration rates in Paleo Pennar delta region, coastal Andhra Pradesh, India. Environmental Geology, v. 36, no. 1-2.

Radhakrishna, I. and Dutt, D.K., 1976. Groundwater resource evaluation in selected tracts of Coastal Orissa. Technical Report, CGWB, Eastern Region, Calcutta.

Radhakrishna, I. and Gangadhara Rao, T., 1990. Evaluation of hydrogeochemical parameters with spontaneous potential logs. J. Hydrol. 114, pp. 245-257.

Radhakrishna, I. and Krishna, G.S.R., 1990. Groundwater resources and saline water ingression in Pennar delta. Technical Report: NGRI, Hyderabad.

Radhakrishna, I. and Krishna, G.S.R., 1991. Groundwater resources and saline water ingression in Pennar delta (Part I). Technical Report.

Radhakrishna, I. and Venteshwarlu, T., 1980. Aquifer response in the ultramafic field area of sukhinda valley – Orissa (India) Groundwater Vol. 18, No. 5, pp. 479-485.

Radhakrishna, I., 1981. Saline water encroachment in the coastal aquifers of Orissa (India). Saline water intrusion meet (Swim), Hannover. Geol.Jb. C 29, pp. 105-114.

Radhakrishna, I., 1991. Assessment and management of groundwater resources: saline freshwater inter-relationships in coastal delta complexes and saline water movement with time. Groundwater resources and saline water ingression in Pennar delta, Technical Report: Part I. NGRI.

Radhakrishna, I., 1997. Recognition of Paleo Pennar delta: Assessment of Aquifer Parameters and Environment Constraints. Journal Geological Society of India. Vol. 50, Oct, pp. 491-499.

Radhakrishna, I., 2001. Saline fresh water interface structure in Mahanadi delta region, Orissa, India. Environmental Geology v. 40, no. 3, pp. 369-380.

Radhakrishna, I., 2001. Saline/brackish water intrusion processes, identification of fresh groundwater resources: development and optimization techniques for sustained development of fresh groundwater resources in coastal delta complexes. Technical Report: NGRI, Hyderabad.

Radhakrishna, I., 2006. Assessment of saline-freshwater interface structure in the Indian coastal delta complexes sustainability of groundwater resources and its indicators: Proceedings of symposium S3 held during the seventh IAHS scientific assembly at Foz do Iguacu, Brazil, April 2005. IAHS Publ. 302, 2006.

Radhakrishna, I., Rao, T.G., Athavale, R.N., Singh, V.S., Subrahmanyam, K., Anjaneyulu, G.R., Venkateswarlu, T. and Deshmukh, S.D., 1979. Evaluation of shallow aquifers in lower maner basin. Indo-German collaboration project on exploration and management studies of groundwater resources. Technical Report No. GH 10-HG 3, NGRI.

Raiverman, V., Singh, G. and Murti, K.V.S., 1966. Fracture pattern in cauvery basin. Bull. ONGC v. 3, no. 1, pp. 13-20.

Rajmohan, N.J., and Elango, L., 2006. Hydrogeochemistry and its relation to groundwater level fluctuation in the Palar and Cheyyar river basins, southern India. Hydrol. Process, 20, pp. 2415-2427.

Raju, M.B., et al., 1982. Hydrogeological conditions in East Godavari District, Andhra Pradesh. Technical Report, CGWB.

Raju, N.J., 2006. Seasonal evaluation of hydro-geochemical parameters using correlation and regression analysis. Curr. Sci, India v. 91, no. 6, pp. 820-826.

Rallison, R.E., 1980. Origin and evolution of the SCS runoff equation. In 'Symposium on Watershed Management, ASCE ,Vol.II, pp. 912-924, New York.

Rama, and Moore, W.S., 1996. Using the radium quartet for evaluating groundwater input and water exchange in salt marshes: Geochimica at Cosmoschimica Acta, v. 60, pp. 4645-4652.

Ramaswamy S.M., 1991. A remote sensing study of river deltas of Tamil Nadu. Mem. Geol. Soc. Ind. 22, pp. 75-89.

Ramaswamy. S.M., 2006. Holocene tectonics revealed by Tamil Nadu deltas, India. Jour. Geol. Soc. India, v. 67, pp. 637-648.

Ramaswamy. S.M., Saravanavel, J. and Selvakumar, R., 2006. Late Holocene geomorphic evolution of cauvery delta, Tamil Nadu, Jorn. Geol. Soc. India, Vol. 67, pp. 649-657.

Ramaswamy. S.M., Venkatasubramanian, V., Riaz Abdullah, S. and Balaji, S., 1992. The phenomenon of river migration in Northern Tamil Nadu – Evidence from Satellite Data, Archaeology and Tamil Literature. Man and Environment, Pune, v. XVII, no. 2, pp. 13-25.

Ramesh Kumar Gupta, 2008. ccb.colorado.edu/waf/docs/India_presentation.ppts.

Rangarajan, R. and Athavale, R.N., 2000. Annual replenishable ground water potential of India-an estimate based on injected tritium studies. Journal of Hydrology: 234, pp. 38-53.

Ranjan, P., So Kazama and Masaki Sawamoto, 2005. Effect of sea level rise on the loss of fresh groundwater resources: Case studies of Western American Coast and Bay of Bengal. Annual Journal of Hydraulic Engineering, JSCE, Vol. 49, pp. 97-102.

Ranjan, P., So Kazama and Masaki Sawamoto, 2006. Effects of climate change on coastal fresh groundwater resources. Global Environmental Change 16, pp. 388-399.

Rankama, Kalervo, and Sahama, T.G., 1950. Geochemistry: Chicago, University of Chicago Press, 912 p.

Rao, G.N., 1993. Geology and hydrocarbon prospects of East Coast sedimentary basins of India with special reference to Krishna Godavari basin. Geol.Soc.India, vol. 41, May, pp. 444-494.

Rao, K.L., 1975. India's water wealth its assessment, Uses and Projections. Orient longman Ltd. 254 p.

Rao, K.N., George, C.J. and Ramasastri, K.S., 1971. Potential evapotranspiration over India. India Met. Dept. Sci. Rep. No. 136.

Rao, P.S., 1989. Quaternary geology and geomorphology of corommandal coast of Tamil Nadu and Pondicherry. Proc. Vol. Workshop on coastal zone management. Andhra University, Visakhapatnam, pp. 12-14 Oct.

Rao, R.P. and Talukdar, S.N., 1980. Petroleum geology of Bombay High field, India. In: M.T. Halbouty (Editor), Giant Oil and Gas Fields of the Decade 1968-1978 Am. Assoc. Pet. Geol., Mem., 30, pp. 487-506.

Rao, S.V.N., 2003. Planning Models for water resources management in coastal and deltaic systems. A Ph.D thesis, Indian Institute of Technology, Madras.

Rao, S.V.N., Bhallamudi, S.M., Thandaveswara, B.S. and Mishra, G.C., 2004. Conjunctive use of surface and groundwater for coastal and deltaic systems. Journal of water resources planning and management, vol. 130, No. 3, pp. 255-267.

Rao, S.V.N., Bhallamudi, S.M., Thandaveswara, B.S. and Srinivasulu, V., 2005. Planning groundwater development in coastal deltas with paleo channels. Water Resources Management, Kluwer publication 19, pp. 625-639.

Rao, S.V.N., Bhandari, A., Nitin Joshi, Ajay Sharma and Radhakrishna, I., 2009. Conceptual models representing saline-fresh water interface dynamics for Mahanadi delta system: Orissa, India. Unpublished report. pp. 1-23.

Rao, S.V.N., Srinivasulu, V., Bhallamudi, S.M., Thandaveswara, B.S. and Sudheer, K.P., 2004. Planning groundwater development in coastal aquifers. Hydrological Sciences J., v. 49, no. 1, pp. 155-170.

Rao, S.V.N., Thandaveswara, B.S., Bhallamudi, S.M. and Srinivasulu, V., 2003. Optimal groundwater management in deltaic regions using simulated annealing and neural networks. Water resources management, Kluwer Publishers, 17, pp. 409-428.

Ray, D., Basu, Saibal, Suryanarayana, A.V.K. and Samantha, U 1985. Habitat of Oil and Gas in Krishna-Godavari Basin, Unpublished ONGC Report.

Reese, R.S., 2002. Inventory and review of aquifer storage and recovery in southern Florida: U.S. Geological Survey Water-Resources Investigations Report 02-4036, 56 p.

Reese, T.F., et al, 1995. Geohydrology, water quality, and nitrogen geochemistry in the saturated and unsaturated zones beneath various land uses, Riverside and San Bernardino Countries, California, 1991-1993. U.S. Geological Survey Water-Resources Investigations Report 94-4127, Sacramento, CA, 267 p.

Reilly, T.E. and Goodman, A.S., 1985. Quantitative analysis of saltwater-freshwater relationships in groundwater systems- A historical perspective: Journal of Hydrology, v. 80, pp. 125-160.

Rengamannar, V. and Pradhan, P.K., 1991. Geomorphology and evolution of Godavari delta. Memoirs Geological Society of India. No. 22. pp. 51-56.

Reyment, R.A., Joreskog, K.H. Applied factor analysis in the natural sciences. Cambridge, England; New York, NY, USA, Cambridge University Press, 1993. 371 p.

Richard, L.A., 1954. Diagnosis and improvement of saline and alkali soils. Agriculture handbook 60. US Department of Agriculture, Washington DC. 160 p.

Richards, R.T., Troester, J.W. and Martinez, M.I., 1995. A comparision of electromagnetic techniques used in a reconnaissance of the ground-water resources under the coastal phin of isla de mona, puerto rico. Symposium on the Application of Geophysics/Environmental and Engineering Problems, 251 p.

Richter, B.C. and Kreitler, C.W., 1993. Geochemical techniques for identifying source of ground-water salinization. CRC press, 258 p.

Rogers, L.L. and Dowla, F.U., 1994. Optimization of groundwater remediation using artificial neural networks and parallel solute transport modeling. Water Resour. Res. v. 30, no. 2, pp. 457-481.

Ronnback, P., 1999. The Ecological basis for economic value of seafood production supported by mangrove ecosystems. Ecological Economics, v. 29, no. 2, pp. 235-252.

Royer, J.Y., Sclater, J.G. and Sandwell, D.T., 1989. A preliminary tectonic fabric chart of the Indian Ocean. Proc. Indian Acad. Sci. (Earth Planet Sci.), v. 98, pp. 7-24.

Rumbaugh, J.O. and Rumbaugh, D.B., 2004. Guide to using groundwater vistas, Version 4 ESI, Reinhold's, PA, U.S.A.

Sambasiva Rao, M. and Vaidyanadhan, R., 1979. Morphology and evolution of Godavari delta, India. Zeitschrift fur Geomorphologie, N.F.Bd. v. 23, no. 3, pp. 243-255.

Sambasiva Rao, M., 1982. Morphology and evolution of Modern Cauvery delta, Tamil Nadu, India. Trans. Inst. Ind. Geographers (Pune), v. 4, no. 1, pp. 67-78.

Sambasiva Rao, M., Nageswara Rao, K. and Vaidyanadhan, R., 1978. Morphology and evolution of Mahanadi and Brahmani-Baitarani deltas. Symp. Morphology and evolution of landforms. Dept. of Geology, University of Delhi, Delhi, pp. 241-249.

Sanford, W.E. and Konikow, L.F., 1985. A two-constituent solute-transport model for ground water having variable density: U.S. Geological Survey Open-File Report 87-526, 174 p.

Sanil kumar, V., Pathak, K.C., Pednekarm, P., Raju, N.S.N. and Gowthaman, R., 2006. Coastal processes along the Indian Coast line: current science, vol. 91, No.4, pp. 530-536.

Sankaranarayana, P.K. and Ramanujachary, K.R., 1967. Short Note-An Inverse Slope Method of Determining Absolute Resistivity. Geophysics, vol. 33, no. 6, pp. 1036-1040.

Santing, G., 1963. Salt water-fresh water relationships, in "The Development of Ground Water Resources with Special Reference to Deltaic Areas," Water Resources Series, No. 24, UNESCO, pp. 52-62.

Sastri, V.V., Sinha, R.N., Gurcharan Singh and Murty, K.V.S., 1973. Stratigraphy and Tectonics of the sedimentary basins on east coast of peninsular India. Am. Assoc. Pet. Geol. Bull., 574, pp. 655-678.

Sastri, V.V., Venkatachala, B.S. and Narayanan, V., 1981. The evolution of the East Coast of India. Jour. Palaeogeography, Paleoclimatology, Paleoecology, 36, pp. 23-54, Netherlands.

Sathaye, J., Shukla, P.R. and Ravindranath, N.H., 2006. Climate change, sustainable development and India: Global and National concerns. Current Science, Vol. 90, No. 3. pp. 314-368.

Satish, S., Elango, L., Rajesh, R. and Sarma, V.S., 2011. Assessment of seawater mixing in a coastal aquifer by high resolution electrical resistivity tomography. Int. J. Environ. Sci. Tech., v. 8, no. 3, pp. 483-492.

Satpathy, B.N., Athavale, R.N., Krishnamurthy, N.S., Chandra, P.C. and Flathe, H., 1976. Resistivity surveys in some parts of Mahanadi delta for delineation of fresh and saline aquifers. Technical Report No. GH 4 – GP 3, pp. 1-43.

Scheidegger, A.E., 1960. Underground dispersion of miscible liquids. Publication No. 52 of the Int. Ass. of Sc. Hydrology (General Assembly Helsinki 1960), pp. 462-469.

Scheidegger, A.E., 1961. General theory of dispersion in porous media. J. Geophys. Res. 66, pp. 3273-3278.

Schmorak, S. and Mercado, A., 1969. Upconing of fresh water-sea water interface below pumping wells, field study. Water Resour. Res., v. 5, no. 6, pp. 1290-1311.

Schoeller, H., 1935. Utilite de la notion des exchanges de bases pour la comparison des eaux souterraines: France, Societe Geologie Comptes rendus Sommaire et Bulletin Series 5, v. 5, pp. 651-657.

Schoeller, H., 1962. Les eaux souterraines: Paris, France, Massio et Cie, 642 p.

Schoeller, H., 1965. Hydrodynamique lans lekarst (ecoulemented emmagusinement). Actes Colloques Doubronik, I, AIHS et UNESCO, pp. 3-20.

Seaber, P.R., 1962. Cation hydrochemical facies of ground water in the Englishtown Formation, New Jersey, in short papers in geology, hydrology and topography: U.S. Geol. Survey Prof. paper 450-B, pp. B124-B126.

Senthilkumar, G., Ramanathan, A.L., Nainwal, H.C. and Chidarribaram, S., 2008. Evaluation of the hydrogeochemistry of groundwater using factor analysis in the Cuddalore coastal region, Tamil Nadu, India. Indian J. Marine Sci. v. 37, no 2, pp. 181-185.

Shamir, U. and Dagan, G., 1971. Motion of the seawater interface in coastal aquifers: A numerical solution. Water Resour. Res., v. 7, no. 3, pp. 644-657.

Sherif, M.M. and Singh, V.P., 1999. Effect of climate change on sea water intrusion in coastal aquifers. Hydrological processes v. 13, no. 8: pp. 1277-1287.

Shiklomanov, I.A., 1998. Ed: World Water Resources: A New Appraisal and Assessment for the 21st Century. UNESCO.

Shukla, P.R., Sharma, S.K., Ravindranath, N.H., Garg, A. and Sumana Bhattacharya, 2003. Climate change and India: Vulnerability assessment and adaptation. University press (India) private limited, 462 p.

Simmons, C.T., Fenstemaker, T.R. and Sharp, J.M., Jr., 1991. Variable-density groundwater flow and solute transport in heterogeneous porous media-Approaches, Resolutions and Future challenges. Journal of Contaminant Hydrology, v. 52, pp. 245-275.

Singh, A.P. and Gupta, P.K., 2003. Geological characterization for dredging operation of beach placer deposit in india, proceedings of the fourth South Asia geological congress, New Delhi, India (GEOSAS-IV), pp. 343-346.

Singh, I.B and Seeta Rama Swamy, A., 2006. Delta Sedimentation: East Coast of India. Technology Publications Dehradun 392 p.

Sinha Ray, S.P. and Das, C.R., 2002. Role of 'SP' in evaluating the saline/fresh water interface in the sub-surface in saline hazard areas of south 24 parganas district, West Bengal. Bhu-Jal news, Quar. Jour. Volume 17 (1 & 2).

Sivashanmuganathan, K., 2008. Groundwater management studies in Tiruvallur district, Tamil Nadu. Technical Report: South Eastern Coastal Region, Chennai.

Small, C. and Nicholls, R.J., 2003. A Global analysis of human settlement in coastal zones, Journal of Coastal Research, v. 19, no. 3, pp. 584-599.

Souza, W.R. and Voss, C.I., 1987. Analysis of an anisotropic coastal aquifer system using variable density flow and solute transport simulation. Journal of Hydrology, 92, pp. 17-41.

Spalding, M.D., Blasco, F. and Field, C.D., 1997. World mangrove atlas. The International Society for Mangrove Ecosystems, Okinawa (Japan), 178 p.

Spalding, M.D., Ravilious, C. and Green E.P., 2001. World atlas of coral reefs. University of California press, Berkeley, 424 p.

Stewart, M.T. and Gay, M.C., 1986. Evaluation of transient electromagnetic soundings for deep detection of conductive fluids: Ground Water, v. 24, pp. 351-356.

Stewart, M.T., 1982. Evaluation of electromagnetic methods for rapid mapping of salt-water interfaces in coastal aquifers. Ground Water Vol. 20, No.5, September-October.

Stiff, H.A., jr., 1951. The Interpretation of chemical water analysis by means of patterns. Jour. Petroleaum Technology, v. 3, no. 10, pp. 15-17.

Stone, Wayne L., 1978. An assessment of alternate seawater intrusion control strategies for the oxnard plain of Ventura County, California. Report submitted in partial satisfaction of the requirements for the degree of Doctor of Environmental Science and Engineering. Berkeley: University of California.

Storey, M., Mahoney, J.J., Saunders, A.D., Duncan, R.A., Kelley, S.P. and Coffin, M.F., 1995. Timing of hot spot related volcanism and the breakup of Madagascar from India. Sci., v. 267, pp. 852-855.

Strack, O.D.L., 1976. A single-potential solution for regional interface problems in coastal aquifers. Water Resour. v. 12, no. 6, pp.1165-1174.

Subba Rao, N., 2006. Seasonal variation of groundwater quality in a part of Guntur district, Andhra Pradesh, India. Environ Geol 49, pp. 413-429.

Subrahmanya, K.R., 1998. Tectono-Magmatic evolution of the west coast of India. International Association for Gondwana Research, Japan. v. I, Nos. 3-4, pp. 319-327.

Subrahmanya, K.R., 2001. Origin and evolution of the Western Ghats and the west coast of India. In: Sahyadri - The great escarpment of the India subcontinent. Geol. Soc. India, Mem. No. 47, pp. 463-474.

Subrahmanyam, A.V., Anil Kumar, V., Desapti, T., Deshmukh, R.D. and Viswanathan, G., 2005. Discovery of micro-diamonds in beach placers of the east coast, Andhra Pradesh, India, Current Science, Vol. 88, No. 8.

Subramanyam V.P., 1983; "Some aspects of water balance in the tropical monsson climates of India". Hydrology of Humid Tropical Regions with Particular Reference to the hydrological effects of agriculture and forestry practice (Proceedings of the Hamburg Symposium). IAHA Publications no. 140.

Subramanyam, V.P., 1956. Climatic types of India according to the rational classification of Thornthwaite. Ind. Jour. Met. And Geophys., Vol. 7, No. 3, pp. 253-264

Subramanyam, V.P., 1956. Water balance of India according to Thornthwaite's concept of Potential evapotranspiration. Annals Assn. of Amer. Geogrs., Vol. 46, No. 3, pp. 300-311

Subramanyam, V.P., 1982. Water Balance and It's Applications. (With special reference to India). Andhra University, Series no.173. pp. 1-102.

Subramanyam, V.P., Subba Rao, B. and Murty, V.S.N., 1970. Water balance of the Godavari river basin. Hydraulics, Irrigation and Flood Control Group, I.E(I) Journal-CI, Vol. 50.

Sukhija, B.S., Rama, 1973. Evaluation of groundwater recharge in semiarid regions of India using environmental tritium. In: Proceedings of Indian Acad.of Sci., vol. 77, sec. A, no. 6, pp.2 79-292.

Sukhija, B.S., Reddy, D.V., Nagabhushanam, P., Syed Hussain and Giri, V.Y., 1996. The use of environmental isotopes and chloride as natural tracers to investigate the effects of depressurisation of a coastal aquifer for lignite mining, India. Hydrogeology Journal, vol.4, no. 2, pp. 70-88.

Taniguchi, M. and Fukuo, Y., 1993. Continuous measurements of ground-water seepage using an automatic seepage meter. Ground Water 31, pp. 675-679.

Taniguchi, M., Burnett, W.C., Cable, J.E. and Turner, J.V., 2002. Investigation of submarine groundwater discharge. Hydrol. Process. 16, pp. 2115-2129.

T.E.R.I., 1996. The economic impact of one metre sea level rise on Indian coastline-methods and case studies, Report submitted to the Ford Foundation.

The Year book of Agriculture, 1955. Water. The United States Department of Agriculture.

Theis, C.V., 1935. The relation between the lowering of the piezometric surface and the rate and duration of discharge of a well using groundwater storage. Am. Geophys. Union Trans., vol. 16, pp. 519-524.

Thiem, G., 1906. Hydrologische Methoden. Gebhardt, Leipzig, 56 p.

Thornthwaite, C.W. and Mather, J.R., 1955. The water balance. Publications in Climatology, Drexel Institute of Technology (laboratory of Climatology), vol. 8, no. 1.

Thornthwaite, C.W., 1948: An Approach toward a rational classification of climate. Geographical Review, vol. 38, no. 1.

Todd, D.K. and Huisman, L., 1959. Ground water flow in the Netherlands Coastal Dunes. J. Hydraul. Div., Proc. Am. Soc. Civ. Eng., 85 (HY7), pp. 63-81.

Todd, D.K., 1959. Groundwater Hydrology, John Wiley & Sons, Inc., 535 p.

Todd, D.K., 1980. Groundwater hydrology. 2nd edition, University of California, Berkeley. John Wiley & Sons.

U.S. Army Corps of Engineers and South Florida Water Management District, 1999. Central and Southern Florida Project Comprehensive Review Study: Final Integrated Feasibility Report and Programmatic Environment Impact Statement, 27 p.

U.N.C.E.D., 1992. United Nations Conference on Environment and Development, Rio de Janeiro, (Earth Summit).

U.N.D.P., 1974. Groundwater investigation in Tamil Nadu (Area-III).

U.N.D.P., 1974. Groundwater investigation in Tamil Nadu (Area-IV).

U.N.D.P.,1975. Groundwater investigation in Tamil Nadu (Area-I & II).

U.N.E.P., 1992. United Nations Environment Programme, The World Environment 1972-1992: Two decades of Challenge. Chapman & Hall, New York, US, 884 p.

Unnikrishnan, A.S., Rupa kumar, K., Fernandes, S.E., Michael, G.S. and Patwardhan, S.K., 2006. Sea level changes along the Indian coast: Observations and projections. Current science, Vol. 90, No. 3.

Vaidyanadhan, R. and Ghosh, R.N., 1993. Quternary of the east coast of India. Curr. Sci., 64, pp. 804-816.

Vaidyanadhan, R., 1971. Evolution of drainage of Cauvery in south India. Jour. Geol. Soc. India, v. 12, pp. 14-23.

Vaidyanadhan, R., 1987. Coastal geomorphology in India. Jour. Geol. Soc. Ind. v. 29, no. 4, pp. 373-378.

Vaidyanadhan, R., 1989. Need for a new look at our river deltas. Ibid. v. 33, no. 6, pp. 491-495.

Vaidyanadhan, R., 1990. Morphology of east coast modern deltas. In: Field Seminar on Recent Deltas, Geology Department, Andhra University, Visakhapatnam, pp. 31-35.

Vaidyanadhan, R., 1991. Quaternary deltas of India. Memoir 22. Geological Society of India, 291 p.

Valiela, I., Bowen, J.L. and York, J.K., 2001. Mangrove forests: One of the world's threatened major tropical environments. Bioscience, v. 21, no. 10, pp. 807-815.

Valsangkar, A.B., Radhakrishna Murthy, C., Subbarao, K.V. and Beckinsale, R.D., 1981. Paleomagnetism and Potassium-Argon age studies of acid igneous rocks from St.Mary Islands. Geol. Soc. India, Mem.3, pp. 265-276.

Van Der Veer, P., 1977. Analytical solution for steady interface flow in a coastal aquifer involving a phreatic surface with precipitation. Journal of hydrology, vol. 34, pp. 1-11.

Van Lissa, R.V., Van Maanen, H.R.J. and Odera, F.W., 1987. The use of remote sensing and geophysics for groundwater exploration in Nyanza Province, Kenya. Presented at the African Water Technology Conference, Nairobi, Kenya, Feb.

Vappicha, V.N. and Nagaraja, S.H., 1976. An approximate solution for the transient interface in a coastal aquifer. J. Hydrol., 31, pp.161-173.

Victor Rajamanickam, G., 2000. Light heavy minerals on the Indian continental shelf, Including beaches. A Handbook of Marine Mineral deposits, CRC Press, LLC.

Vinod Kumar, K., and Asis Bhattacharya., 2003. Geological evolution of Mahanadi delta, Orissa using high resolution satellite data. Current science, vol. 85, no. 10.

Vogel, J.C., 1967. Investigations of groundwater flow with radiocarbon; Isotopes in Hydrology, IAEA, Vienna, pp. 355-369.

Vonhoff, J.A., 1966. Water quality determinations from spontaneous potential electric log curves. *J. Hydrol.,* 4, pp. 341-47.

Voss, C.I., 1984 (SUTRA); Saturated-unsaturated transport, a finite element simulation model for saturated-unsaturated fluid-density-dependent groundwater flow with energy transport or chemically reactive single species solute transport. U.S. Geol. Surv. Water Resour. Invest. Rep 84-4369, Washington.

Vozoff, K. and Jupp, D.L.B., 1975. Joint inversion of geophysical data, Geophys. J.R. Astr. Soc. pp. 977-991.

Walton, W.C., 1962. Selected analytical methods for well and aquifer evaluation. Illinois State Water Survey Bull., no. 49, 81 p.

Walton, W.C., 1970. Groundwater resources evaluation, McGraw Hill Book Co., New York, 664 p.

Wani, S.P., Balloli, S.S., Kesava Rao, A.V.R. and Sreedevi, T.K., 2004. Combating drought through integrated watershed management for sustainable dryland agriculture. Proc. Of the UNESCAP/NRSA Regional workshop on Agricultural Drought Monitoring & Assessment using Space Technology. May 3-7, Hyderabad.

WCC, 1993. World Coast 2000: Preparing to meet the coastal challenges of the 21st century. Conference statement, World Coast Conference 1993, Noordwijk, The Netherlands, 1-5 November 1993.

Welch, A.H., Westjohn, D.B., Helsel, D.R. and Wanty, R.B., 2000. Arsenic in groundwater of the united states- occurrence and geochemistry. Ground Water, v. 38, no. 4, pp. 589-604.

Wenzel, L.K., 1942. Methods of determining permeability of water bearing materials. U.S. Geol, Surv. Water Supply Paper, 887,192 p.

West, W.D., 1962. The line of the Narmada and Son valleys: Current Sci., v. 31, no. 4, pp. 143-144.

Wicks, C.M., Herman, J.S., Randazzo, A.F and Jee, J.L., 1995. Water-rock interaction in a modern coastal mixing zone. Geol. Soc. Amer. Bull. 107, no. 9, pp.1023-1032.

Wigley, T.M.L., 1999. The science of climate change: global and U.S. Perspectives. Pew Center on Global Climate Change, Arlington, VA.

Wilcox, L.V., 1995. Classification and use of irrigiation waters. USDA Circular No. 696, pp. 19.

Wilkinson, C., 2004. Status of Coral Reefs of the World: 2004. Australian Institute of Marine Science (AIMS), Townsville (Australia).

Willis, R. and Finney, B.A., 1985. Optimal control of nonlinear groundwater hydraulics: Theoretical development and numerical experiments. Water Resour. Res. V. 21, no. 10, pp. 1476-1482.

Willis, R. and Finney, B.A., 1988. Planning model for optimal control of saltwater intrusion. J. Water Resour. Plan. Manage. ASCE v. 114, no. 2, pp. 163-178.

Wilson, J.L. and Sa Da Coata A., 1982. Finite element simulation of a saltwater/freshwater interface with indirect toe tracking. Water Resources Research, v. 8, no. 4, pp. 1069-1080.

Wilson, J.W., 1970. Integration of radar and rain gauge data for improved rainfall measurements. Journal of Applied Meteorology, 9, pp. 489-97.

Wright, L.D., Coleman, J.M. and Ericson, M.W., 1974. Analysis of major river systems and their deltas - coastal studies institute. Tech. Report,156, 114 p.

Zektser, I.S. and Dzhamalov, R.G., 2007. Submarine Groundwater. Taylor & Francis. 466 p.

Zektser, I.S., 2000. Ground water and the environment, Applications for the Global Community. Lewis Publishers, Boca Raton, USA 175 p.

Zektser, I.S., Dzhamalov, R.G. and Safronova, T.I., 1983. Role of submarine groundwater discharge in the water balance of Australia. IAHS-AISH Publication, No. 142. Groundwater in Resources Planning, pp. 209-219.

Zheng, C. and Wang, P.P., 1998. MT3DMS, A modular 3-D multi-species transport model for simulation of advection, dispersion and chemical reactions of contaminants in groundwater systems: Vicksburg, Miss., Waterways Experiment station, Army Corps of Engineers.

Zimmerman, U., Munnich, K.O., Roether, W., 1967. Downward movement of soil moisture traced by means of hydrogen isotopes. Geophys. Momo., Am. Geophys. Union 11, pp. 28-36.

Zohdy, A.A.R., Eaton, G.P. and Mabey, D.R., 1974. Application of surface geophysics to ground-water investigations. In Techniques of Water-Resources Investigations, U.S. Geological Survey, Book 2, Chap. D1.

Index

A

B

C

F

G

O

P

Q

R

S

T

U

V

W

Printed at Repro India Ltd.